Telomeres

SECOND EDITION

Telomeres

SECOND EDITION

EDITED BY

Titia de Lange
The Rockefeller University

Vicki Lundblad
The Salk Institute for Biological Sciences

Elizabeth Blackburn
University of California, San Francisco

COLD SPRING HARBOR LABORATORY PRESS
Cold Spring Harbor, New York

TELOMERES
Second Edition

Monograph 45
©2006 by Cold Spring Harbor Laboratory Press
All rights reserved
Printed in the United States of America

Publisher	John Inglis
Acquisition Editor	Alexander Gann
Project Coordinator	Inez Sialiano
Production Manager	Denise Weiss
Production Editor	Kathleen Bubbeo
Compositor	Compset Inc.
Desktop Editor	Danny deBruin
Interior Book Designer	Denise Weiss and Emily Harste
Cover Designer	Michael Albano

Front Cover Artwork: Chromosome orientation fluorescence in situ hybridization (CO-FISH) of human telomeres. Image provided by Jerry Shay.

Library of Congress Cataloging-in-Publication Data

Telomeres/edited by Titia de Lange, Vicki Lundblad, Elizabeth Blackburn.
   —2nd ed.
      p. cm.—(Cold Spring Harbor monograph series; monograph 45)
   Includes index.
   ISBN 0-87969-734-2 (hardcover: alk. paper) -- ISBN 0-87969-810-1 (pbk.: alk. paper)
      1. Telomere. I. De Lange, Titia. II. Lundblad, Vicki. III. Blackburn, Elizabeth H.
IV. Series: Cold Spring Harbor monograph series; 45.
      [DNLM: 1. Telomere—physiology. 2. Aging—genetics. 3. Neoplasms—genetics.
   QU 470 T277 2006]
   QH600.3T45 2006
   572.8'7—dc22   2005021171

10 9 8 7 6 5 4 3 2

Contents

Preface, vii

1   A History of Telomere Biology, *1*
    E.H. Blackburn

2   The Telomerase Ribonucleoprotein Particle, *21*
    G. Cristofari and J. Lingner

3   Telomerase Biochemistry and Biogenesis, *49*
    J.-L. Chen and C.W. Greider

4   Telomerase and Human Cancer, *81*
    J.W. Shay and W.E. Wright

5   Modeling Cancer and Aging in the
    Telomerase-deficient Mouse, *109*
    K.-K. Wong, S. Chang, and R.A. DePinho

6   Telomerase Deficiency and Human Disease, *139*
    I. Dokal and T. Vulliamy

7   Telomerase-independent Maintenance
    of Mammalian Telomeres, *163*
    A.A. Neumann and R.R. Reddel

8   Telomerase-independent Telomere Maintenance in Yeast, *199*
    M.J. McEachern and J.E. Haber

9   Meiotic Telomeres, *225*
    H. Scherthan

10 Telomere Position Effect: Silencing Near the End, *261*
    M.A. Mondoux and V.A. Zakian

vi   Contents

11   The Structural Biology of Telomeres, *317*
     D. Rhodes

12   Budding Yeast Telomeres, *345*
     V. Lundblad

13   Mammalian Telomeres, *387*
     T. de Lange

14   *Drosophila* Telomeres, *433*
     S. Pimpinelli

15   Ciliate Telomeres, *465*
     C. Price

16   Fission Yeast Telomeres, *495*
     J.P. Cooper and Y. Hiraoka

17   Plant Telomeres, *525*
     D.E. Shippen

Appendix: A Personal Account of the Discovery of Telomerase, *551*
     E.H. Blackburn

Index, *565*

Preface

T EN YEARS AGO, ELIZABETH BLACKBURN and Carol Greider assembled the first monograph on telomeres. In an era focused on gene expression, telomeres were beginning to attract wider attention, largely because of the emerging link between telomerase and cancer. The first edition lent gravitas to this young field, so small at that time that the majority of its members were contributors to the monograph. As telomeres became the focus of interest for an increasing number of investigators, the first edition served as a unique and valuable compendium.

Over the past decade, the field has advanced rapidly. Whereas there were 2,553 publications on telomeres and telomerase by the end of 1995, PubMed informs us that this number is 13,378 today. The telomerase reverse transcriptase, TERT, was cloned and shown to immortalize human cells. The complex role of telomere dynamics in human cancer became clear, in great part because of modeling in the mouse. Mutations leading to diminished telomerase activity were found in patients with progressive bone marrow failure. The t-loop structure of mammalian telomeres was revealed. The mechanism by which telomerase is recruited to chromosome ends emerged in fine detail in budding yeast, and the principles of telomere length homeostasis became known in broad strokes in several settings. New telomeric proteins were identified in yeast, plants, and mammals, and the consequences of telomere dysfunction were established in many systems.

It was obvious to us that simple revision of the first edition was not the right strategy for the current version. The second edition is a completely new compilation of essays with no relationship to the structure of the previous edition other than a focus on telomeres. Thankfully, the fivefold expansion of the literature is not directly reflected in the number of pages of this volume, but the body of knowledge is significantly larger now and so is this book. This edition features a historical treatise, followed by ten chapters on the general attributes of telomeres and telomerase and aspects of telomere biology related to human health. The last six chapters describe telomere biology in the best-studied systems (budding yeast, mammals, *Drosophila*, ciliates, plants, and fission yeast) to highlight both the common themes and the vast differences in telomere biology in different eukaryotes. By necessity, we had to omit several important issues that did not (yet) merit their own chapters but will, we hope, be featured in the third edition.

This collection of essays was brought together through the efforts of the contributing authors, some of whom worked under considerable duress. We thank the authors who delivered on time for doing so and those who were late for working fast after deadlines had passed. The crew at Cold Spring Harbor Laboratory Press, including Kathleen Bubbeo, Inez Sialiano, and Danny deBruin, was instrumental in converting our drafts into coherent texts. This monograph was initiated by Alex Gann, who functioned as navigator and commander, making us stay the course, and moving several (small) mountains.

TITIA DE LANGE
VICKI LUNDBLAD
ELIZABETH BLACKBURN
NEW YORK, JULY 2005

1

A History of Telomere Biology

Elizabeth H. Blackburn

Department of Biochemistry and Biophysics
University of California, San Francisco
San Francisco, California 94143-2200

AN IMAGE FROM EARLY IN THE HISTORY OF MODERN cytogenetics
encapsulates the central elements of telomere biology. It is a meticulously
rendered drawing made of a microscopic image in a 1910 paper by
Theodor Boveri (Fig. 1) (Gall 1996), in which he followed up on his
original observation in the late 1800s that soon after fertilization in the
roundworm *Ascaris*, as cell division commences, the chromosomes
fragment. This of course does not happen in cells destined to become
germ-line cells, but only in a specific subset of the earliest blastomeres.
Cytogenetic studies starting in the 1930s, primarily by McClintock and
Muller, showed that if a chromosome breaks, the broken end usually
becomes unstable. However, McClintock showed that soon after fertilization,
in specific cell types, a broken end can heal in a genetically determined
process. The implication of Boveri's image is that the fragmented *Ascaris*
chromosomes have healed their ends, because they persist throughout the
life of the animal.

The events so vividly captured in Boveri's micrograph now have many
of their molecular underpinnings known, at least in part. The modern
studies of telomere DNA structure that started in the 1970s identified
the DNA that confers stability—by binding proteins—to a newly created
chromosomal end. The discovery of telomerase in the 1980s demonstrated the enzymatic mechanism by which such DNA can be acquired,
and is maintained, at chromosomal termini. This chapter outlines the
history of these advances in telomere and telomerase biology research up
to the early 1990s. By then, many basic concepts of modern telomere
biology in the molecular era had emerged.

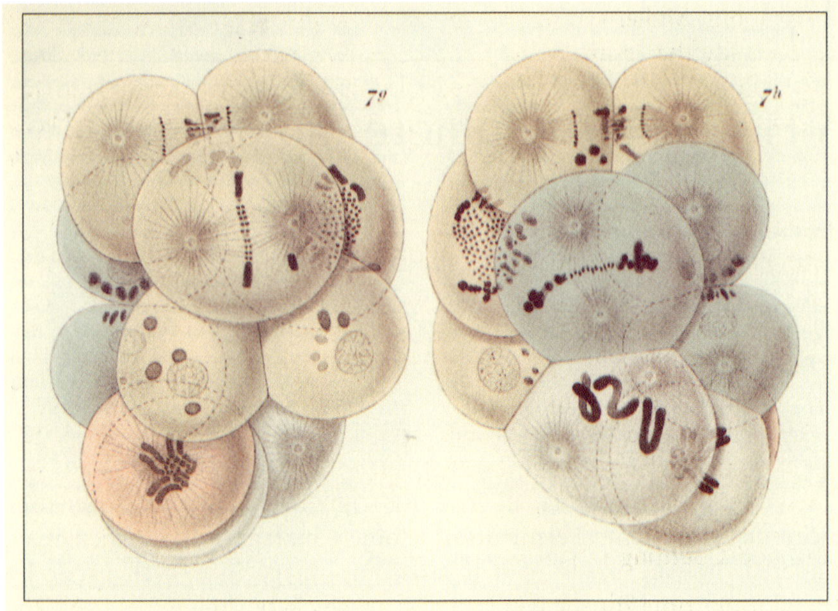

Figure 1. Fragmenting chromosomes are apparent in the round cell at the front of each group of cells. (Reprinted, with permission, from Gall 1996.)

CYTOGENETICS

Fruit Flies and X Rays

The genetic understanding of telomeres was an organic outgrowth of the earliest efforts by geneticists and cytogeneticists to understand the very nature of the gene, the hereditary unit. Chromosomes had become visible to microscopists, and cytogeneticists had realized that they carried Mendel's units of heredity. Hermann Muller's mentor Thomas Hunt Morgan came to envisage genes as a "string of beads" along the length of the chromosomes. Roentgen's discovery of X rays had been applied to medicine, and X rays had been used to guide surgeries to remove shrapnel from the wounded in World War I. It also became clear that X rays could have other uses. In the 1920s, Muller succeeded in mutagenizing fruit flies with X rays, and he used new genetic techniques to recover the products of the broken chromosomes: inversions, translocations, and deficiencies. But conspicuously underrepresented among the deficiencies that were recovered were terminal deficiencies, chromosomes lacking some chunk of the end region all the way out to the terminus (as far as could be determined by microscopy). Drawing together the many observations, Muller synthesized the idea that the recovered chromosomes were usually the result of the rejoining ("reattachment") of two broken ends, which, by becoming joined in some arrangement, created the rearranged chromosome. Reattachments, however, did not occur "between originally free ends, or between a broken and a(n originally) free end, or between any end and a(n originally) free end... ." In his 1938 paper in the scientific journal *The Collecting Net*, Muller and his colleague Darlington and, independently, J.D.S. Haldane (also famous for saying "The Universe is not just stranger than we imagine, it is stranger than we can imagine.") named the "free end" the telomere (Muller 1938). Interestingly, Muller apparently thought of the telomere as a gene (and clearly stated this in his paper in *The Collecting Net*); he thought that each end contained an essential gene which cells could not lose. This topic has been further reviewed by Gall (1995).

Corn and Counting Chromosome Knobs

At the same time during the 1920s as Muller was showing that X rays could cause mutations, Barbara McClintock was turning maize into a powerful system for studies of genes and development. She accomplished this by pioneering innovative methods for microscopic visualization and, most critically, the individual recognition of each of the maize chromosomes in cells. Special "knobs," or locally heterochromatic regions, including some at the chromosome ends, could be seen by her techniques and used as identifiers of individual chromosomes and as landmarks along their arms (Fig. 2). Her work ultimately led to her discovery of mobile genetic elements. Along the way, during the 1930s, it also yielded a rich harvest of original findings about telomeres that form the bedrock of our modern understanding of these special chromosomal parts.

The first clear recognition of the importance of the "natural ends" of the chromosome, as McClintock termed what we now call telomeres, appears in a 1931 report (McClintock 1931). Its purpose was to analyze, cytologically, X-ray-induced chromosome alterations and how they related to genetic and phenotypic changes. Like Muller, McClintock found that translocations, deficiencies, and ring chromosomes were frequent outcomes of X-irradiation. All could be envisaged as resulting from fusions of broken ends with each other. In this report, she wrote: "No case was found of the attachment of a piece of one chromosome to the end of another [intact chromosome]" (McClintock 1931). Thus, by 1931, Barbara McClintock had already recognized that the natural end of an

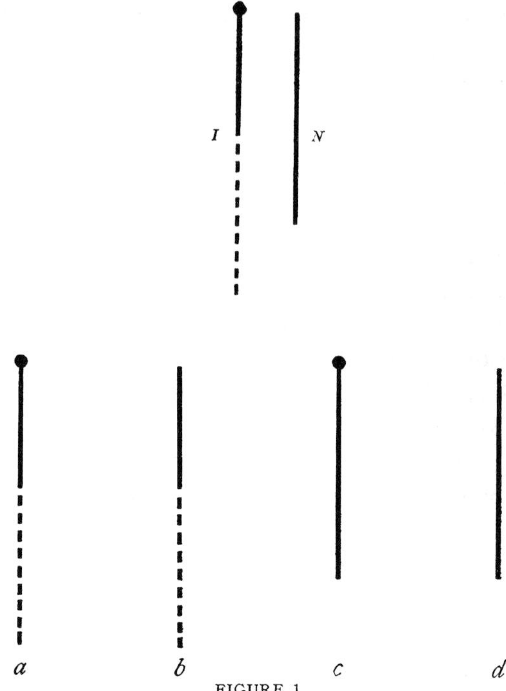

FIGURE 1

Above—Diagram of the chromosomes in which crossing-over was studied.   (Labeled as in figure 1, preceding paper.)

Below—Diagram of chromosome types found in gametes of a plant with the constitution shown above.

> a—Knobbed, interchanged chromosome.
> b—Knobless, interchanged chromosome.
> c—Knobbed, normal chromosome.
> d—Knobless, normal chromosome.
> a and d are non-crossover types.
> b and c are crossover types.

Figure 2. The "knobs" at chromosome ends indicative of heterochromatin at telomeres in maize, as drawn by McClintock (Creighton and McClintock 1931).

intact chromosome has a distinct functional nature such that its properties differ from those of broken ends.

The Rupture of Unbearably Stretched Chromatids: The Breakage–Fusion–Bridge Cycle

If natural chromosome ends appeared to be cytologically inert, what were the properties of broken ends? In a 1932 paper (McClintock 1932),

McClintock was struck by the great efficiency with which broken ends in the same nucleus fuse with each other, producing ring-shaped chromosomes, whose "size and genetic content . . . are not the same in all the nuclei of the plant" (McClintock 1932). This observation further upset the growing notion of genetic constancy in every cell of an organism (although Boveri and Robert Hegner had already documented exceptions: the fragmenting chromosomes in *Ascaris* and the fly *Miastor*, respectively [for review, see Gall 1996]). McClintock reported in this 1932 paper and later in 1938 (McClintock 1938b) that "nothing but rings have been observed to come from rings." She thoroughly analyzed the cytology of the anaphase bridges resulting from the presence of ring chromosomes (McClintock 1938a). In subsequent papers published in 1939 and 1941 (McClintock 1939, 1941a,b), she formulated the breakage–fusion–bridge cycle model (Fig. 3). This model accounted for how the fusion of broken ends generated by ruptures of dicentric chromosomes in anaphase could

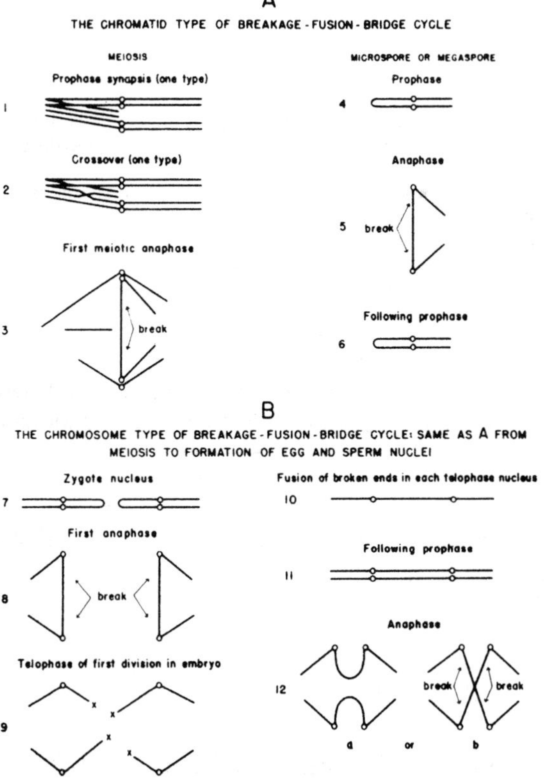

Figure 3. Breakage–fusion–bridge cycle as reviewed by McClintock (1952).

cause the genetic and cytological behavior of these chromosomes. It outlined the mechanism by which ring chromosomes are perpetuated through successive nuclear divisions as follows: If, in a ring chromosome, a crossover within the chromosome occurred after what McClintock called "splitting" of the chromosome (now called chromosomal DNA "replication," but nobody knew yet that the genetic material was DNA), a double-sized ring with two spindle fiber attachment regions (centromeres) would result. In the following nuclear division, as the two centromeres moved to opposite poles of the mitotic spindle, the continuous chromosome arms between them would become stretched, forming "bridges" temporarily connecting the two groups of separating daughter chromosomes on the spindle in anaphase. By either late anaphase or telophase, the stretched bridges would eventually break somewhere along the chromosome arms. Each daughter nucleus would thus receive two broken ends, which would efficiently fuse with each other, regenerating a ring chromosome whose genetic content depended on where the breaks had (randomly) occurred. The next time, following chromosome replication, a crossover occurred between the daughter chromatids of a ring chromosome, the cycle would repeat itself (Fig. 3). The breakage–fusion–bridge cycle was clearly a recipe for continuing genetic instability and is thought to underlie much of the genetic instability generated in cancer cells.

Making New Telomeres

In some of McClintock's experiments producing breakage–fusion–bridge cycles, young embryos were X-irradiated and, importantly, terminal deficiencies were recovered (McClintock 1932). If a deficiency was truly terminal (as opposed to a deletion only extending out to near the chromosome end), this implied formation of a new chromosome end. This finding anticipated McClintock's future work on chromosome end healing.

In a 1939 paper (McClintock 1939), she presented what is arguably the first formulation of the idea that new telomeres can be formed by an active process. She reported that a broken end can, in particular circumstances, lose its tendency to fuse with other broken ends, and she concluded that the broken end had thus permanently "healed," becoming as stable as any normal telomere through all subsequent nuclear divisions. Importantly, the maize tissue type and developmental stage were critical: The broken end seemed to heal only in young plant embryos shortly after the first zygotic division. This tissue specificity and developmental-stage specificity of the capability of a broken end to heal hinted at a specialized process, under developmental control. It echoes Boveri's observation of chromosome fragmentation to produce the thereafter-stable shorter chromosomes, which also occurred in the first cell divisions after fertilization, and only in certain cells. Strikingly, McClintock's statement that in the embryo tissue "broken ends apparently heal and remain permanently healed" implies that healing caused a permanent molecular modification of the broken end. Finally, McClintock (1942–1943) even proposed that a single broken end becomes healed "during the reproductive cycle of the chromosome" and that "experiments should be focused on this period." Fast-forward to our modern understanding that the permanent healing of broken ends often involves the action of telomerase, which normally acts in S phase, and McClintock's insight shows remarkable prescience.

Cytological Associations between Telomeres

Although inert compared with newly broken chromosome ends (McClintock 1931, 1938b), telomeres did show very specific cytological behaviors when examined in interphase, mitosis, or meiosis (for review, see Blackburn and Szostak 1984). Telomeric regions of chromosomes were often heterochromatic in appearance (Bennett 1977) (examples were McClintock's "knobs" in certain maize chromosomes; see her schematic drawing of the maize karyotype in Fig. 2). They were frequently observed to associate with each other at the cytological level. Notably, in meiosis, during leptotene, all of the telomeric regions of the chromosomes characteristically clustered together just under the nuclear envelope and toward the position of the centriole, causing all of the chromosomes to become arranged in the poetically named "bouquet stage." Nor were telomere–telomere associations confined to meiosis—they were seen in interphase and mitotic prophase nuclei (for review, see Ashley and Wagenaar 1974; Dancis and Holmquist 1979), when telomeres were also frequently seen to lie just under the nuclear membrane. In squashed nuclei, cloned telomere-associated DNA sequences of *Drosophila* hybridized to ectopic fibers stretching between the tips of polytene chromosomes, as well as to the tips of the chromosomes themselves (Rubin 1978; Young et al. 1983).

Although telomere associations generally seemed to be transient, in some circumstances, fusions between telomeres appeared to have stabilized, as revealed by karyotype analysis of certain groups of related species. Such "Robertsonian" fusions (for review, see White 1978) were a type of karyotypic change between related species in which the short arms of two acrocentric chromosomes have fused to form a new metacentric chromosome. Any pairwise association of acrocentrics seemed able to occur, to produce a new evolutionarily tolerated metacentric fusion product, paralleling the apparently indiscriminate clustering among telomeres in the bouquet stage. But no one knew what all of these behaviors of telomeres meant in molecular terms.

THE END-REPLICATION PROBLEM AND IDEAS FOR SOLVING IT

The general inertness of telomeres toward broken ends (their "capping" function) and their cytological clustering behaviors awaited molecular explanations. Then, the discovery of the mechanism of DNA replication at the biochemical level raised another problem that telomeres must solve. By then, work on eukaryotic chromosomes in the 1950s and 1960s had converged on the picture that the DNA of every chromosome consisted of a single, long, linear DNA molecule. Watson's well-known articulation of the DNA end-replication problem was laid out in the 1970s, initially in the context of the linear phages that were the subject of much genetic analysis then (Watson 1972). The problem of completing the replication of a linear DNA molecule was as follows: All known DNA polymerases require a polynucleotide primer (either DNA or RNA) bearing a 3′-hydroxyl (3′-OH) group, this primer being removed (if it is RNA) once synthesis has been primed. However, the removal of the primer will result in a 5′-terminal gap being left in one of the daughter strands of a linear DNA molecule. This problem would have to be solved by telomeres. Models for telomere replication started to be put forth.

In the absence of any molecular evidence about telomeres or their DNA, for years it was the cytological observations on telomeres that drove many of the inferences about their replication properties. Through the 1970s and into the 1980s, various and often ingenious solutions to this general problem of complete telomeric DNA replication were proposed, some of which were demonstrated to apply to viral genomes (Blackburn and Szostak 1984). Genomes that are temporarily or permanently circular (viral or bacterial) clearly circumvent this problem. Linear prokaryotic viruses such as T7 and T4 produce concatemers by fusion of the ends of viral genomes so that priming can take place using a neighboring viral DNA strand (Watson 1972; Broker 1973). Priming of replication of the linear genomes of phage ϕ^{29} and the mammalian adenoviruses is accomplished with a protein, covalently attached to the 5′ terminus of each nascent DNA strand, substituting for and bypassing the need for an oligonucleotide bearing a 3′-OH group to prime strand synthesis. An early example of the proposed solutions for telomere replication is that of Cavalier-Smith (1974), who proposed that the sequences at the ends of eukaryotic linear DNA molecules were palindromic. Following chromosome replication, the strand on each daughter chromatid with the 3′-OH terminus was single stranded, because, as discussed above, there would be a gap left at the 5′ end of the complementary strand. This 3′-OH end could then fold back by base-pairing with itself, because its sequence is palindromic, to form a terminal loop. The remaining gap would be closed by the DNA polymerase fill-in and ligation. Then, the opposite strand was nicked by a specific endonuclease to make a new 3′-OH end for priming the last fill-in step, which would occur after the loop had unfolded to become the template for completion of synthesis by DNA polymerase. The feature of this model that has survived is the proposed common telomeric sequence shared by all telomeres in a species, based on the idea that then only one sequence-specific endonuclease would be required to recognize all the telomeres. Bateman (1975) elegantly simplified the model by suggesting that the telomere was normally a self-paired hairpin structure, thus closing off the end by a terminal loop (for review, see Blackburn and Szostak 1984). Although this structure and Cavalier-Smith's and Bateman's models were eventually found to hold for vaccinia virus DNA (Baroudy et al. 1982), it was not true for chromosomal DNAs. However, these models influenced the current thinking about chromosomal DNA termini so that at one point, a reported cross-link at yeast telomeres was taken as evidence for Bateman's model. (In hindsight, modern understanding of the chemical properties of the telomeric DNA repeat tract sequence at yeast telomeres, as described in the next section, suggests the stickiness of the termini may have been a result of in vitro G-quartet DNA formation.)

Dancis and Holmquist combined these general models with the cytological evidence for telomere–telomere interactions (Dancis and Holmquist 1979; Holmquist and Dancis 1979). They suggested that the replication of eukaryotic chromosome ends required pairwise interactions and transient fusions of telomeres. In this model, pairs of telomeres recombine with each other via their common DNA sequences. The fused telomeres, since they contain no ends, could then be replicated completely and resolved into daughter telomeres as in the models of Cavalier-Smith (1974) or Bateman (1975). The evidence for telomeric associations, and the occasional occurrence of telomeric fusions, which could be caused by failure to separate such a pair of replicating telomeres, made this a tempting model for a while.

THE BEGINNING OF THE MOLECULAR ERA OF TELOMERES

Telomeric DNA Sequences

Determining which, if any, of these models might be correct for eukaryotic chromosomes required determining the DNA sequence and structure at chromosomal termini. But most chromosomes are very long, making direct analysis of telomeric DNA technically daunting. The linear DNA minichromosomes that comprise the amplified ribosomal RNA genes (rDNAs) in several simple eukaryotes were the first to yield to analysis of their telomeric sequences and terminal DNA structures, because of their relative shortness (<100 kb) and abundance. These sometimes appeared to have sticky ends (Fig. 4).

The first such sequence defined was that of the amplified rDNA minichromosomes of the somatic nucleus of the ciliated protozoan *Tetrahymena thermophila*. About 10,000 of these 21-kb palindromic linear molecules (Engberg et al. 1976; Karrer and Gall 1976) were present in the somatic nucleus of each *Tetrahymena* cell, which allowed them to be

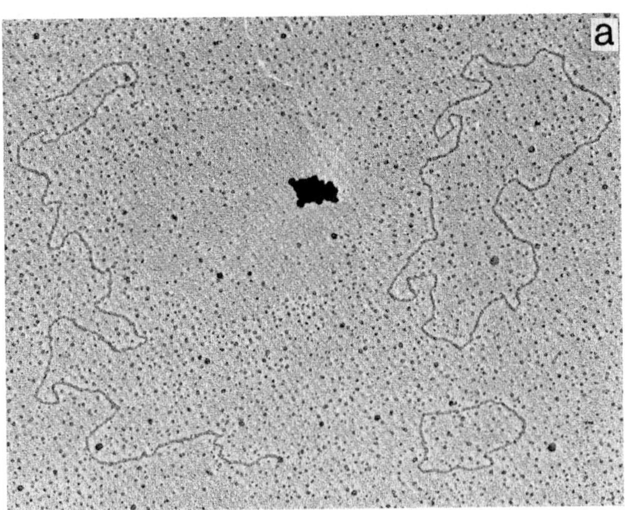

Figure 4. Electron micrograph showing a linear rDNA minichromosomal DNA (*left*), a circularized rDNA (*upper right*), and a circular bacterial plasmid DNA added as a size reference (*lower right*). (Reprinted, with permission, from Gall 1974.)

purified. I determined their end sequences by a combination of in vitro labeling and other analytical techniques. The structure at the termini of these molecules was surprising and complex (Blackburn and Gall 1978; Blackburn et al. 1983). At each end, there were approximately 50 tandem repeats of the hexanucleotide unit CCCCAA/GGGGTT, with the G-rich strand bearing the 3'-OH end of each end of the linear DNA. Oddly, too, the number of tandem repeats per end was heterogeneous among the population of purified molecules, ranging from a minimal estimate of 20 to up to about 70.

Evidence then emerged, in the late 1970s and early 1980s, that similar, extremely simple, tandemly repeated DNA sequences were at the molecular ends of linear DNAs in other eukaryotic nuclei: the linear rDNAs of the slime molds *Physarum* and *Dictyostelium*, the short minichromosomes of the somatic nuclei of a different group of ciliates (the hypotrichous ciliates) such as *Oxytricha*, and the multiple chromosomes in the *Trypanosome* parasite that causes sleeping sickness. All of the telomeric DNAs had a similar type of pattern: tandem repeats of a simple unit containing a cluster of C residues (for review, see Blackburn 1990).

The hypotrichous ciliates were especially useful for determining the details of the very molecular end of the telomeric DNA. David Prescott had shown by the mid 1970s that the DNA of the somatic nuclei of these ciliates was in the form of short, linear molecules. Their shortness and homogeneity allowed direct DNA sequencing by end-labeling methods. Two groups deduced that there was a defined 3' overhang of the GGGGTTTT repeat strand: 14 bases long in one hypotrich species and 16 bases in another species (Oka et al. 1980; Klobutcher et al. 1981; Pluta et al. 1982). In *Tetrahymena* rDNA telomeres, on the other hand, several single-stranded, apparently 1-nucleotide gaps were found in the CCCCAA-repeat strand at specific positions toward the outer end of the cluster of repeats (Blackburn and Gall 1978; Blackburn et al. 1983). Later, in the 1980s, Eric Henderson found that there is a 12-base overhang of the GGGGTT repeat strand in *Tetrahymena* telomeres and a similar length overhang in the telomeres of the linear rDNA of the slime mold *Dydinium*, which has the same repeats as human telomeres: TTAGGG repeats (Henderson et al. 1987).

These findings were not predicted by any of the current telomere replication models (Cavalier-Smith 1974; Bateman 1975; Dancis and Holmquist 1979; Holmquist and Dancis 1979). A biochemical search suggested that there were no tightly or covalently attached proteins at the rDNA ends (Blackburn et al. 1983). In addition, the very tip of the

Tetrahymena rDNA was relatively inaccessible to end-labeling in vitro, and the "stickiness" of these ends had been exhibited under electron microscopy and by the behavior of the terminal restriction fragments on gels. Thus, it seemed reasonable at the time to think that these easily disrupted end-to-end associations were mediated by a terminal hairpin loop, as had been suggested in the chromosome end-replication models.

Conservation of Aspects of Telomere Function

The telomeric restriction fragment of *Tetrahymena* rDNA was able to fulfill certain telomeric functions in a distantly related eukaryote, the budding yeast *Saccharomyces cerevisiae* (baker's yeast). This seemingly rather long-shot possibility was tested in a collaboration between Jack Szostak and myself (Szostak and Blackburn 1982). A linear yeast plasmid was constructed in vitro by ligation of purified terminal regions of *Tetrahymena* rDNA to each end of a linearized yeast plasmid. This plasmid, in its original circular form, had been able to replicate autonomously as a free circle in yeast, but once linearized, it could only be maintained at low frequency, by recombining into the chromosome. However, when the linearized plasmid with its foreign *Tetrahymena* telomeres ligated to both ends was introduced into yeast, it was efficiently recovered and maintained. It could now replicate as a linear molecule. The repeated CCCCAA/GGGGTT sequence was still in the plasmid, although no such sequence was naturally present anywhere in the yeast genome.

This linear plasmid then became the basis for cloning out yeast telomeres. One of its ends was cut off by a restriction enzyme, again preventing the plasmid from being maintained as a free molecule in yeast. However, the cut end could be randomly ligated to a library of yeast restriction fragments, a low fraction of which were predicted to be the natural telomeres of yeast chromosomes. Such fragments were found that functioned as a stable end and hence restored to the linear plasmid the ability to be maintained at high frequency as a free linear plasmid. At the molecular level, they mapped to the telomeres of yeast chromosomes (Szostak and Blackburn 1982). Analysis of these yeast telomeres, both on linear plasmids and in chromosomes, showed that they too had specific single-strand breaks near their molecular ends, and in vitro labeling properties of the yeast ends were very similar to those of the rDNA of *Tetrahymena* (Blackburn and Gall 1978; Szostak and Blackburn 1982). The yeast telomeres were then cloned and sequenced and were shown to consist of about 300 bp of a sequence made up solely of tandem irregular repeats of the sequence formula $C_{1-3}A$ (Shampay et al. 1984).

Unexpectedly, when they were recovered back out of the yeast cells, the *Tetrahymena* rDNA termini themselves had undergone marked structural changes after their sojourn of replication and maintenance in yeast. First, they were longer, having increased by about 100–300 bp (Szostak and Blackburn 1982). Second, they now hybridized with a poly-d(GT) probe, as did the yeast chromosomal telomeres, in contrast to the original rDNA terminal restriction fragments (Walmsley et al. 1983). In addition, after extraction from the yeast cells in which they had been maintained, these end fragments possessed single-stranded discontinuities, but their spacing was now that of the cloned yeast telomere. Direct sequence analysis by Janis Shampay showed that the rDNA telomeres acquired yeast sequences over the course of their stay in yeast; about 200 bp of the $C_{1-3}A$ yeast telomeric repeat was found to be joined directly to the distal end of the *Tetrahymena* CCCCAA repeats (Shampay et al. 1984). Similar experiments were later done using the rDNA termini of *Oxytricha*, which carry a short stretch of terminal CCCCAAAA repeats (Pluta et al. 1984).

It now appeared that telomeric sequences and aspects of their functions would be generally conserved. Unlike centromeres and chromosomal origins of replication, we had shown that telomeres from one species could stabilize linear plasmids or chromosomes in cells of another species, even though the two organisms had different telomeric DNA sequences (for review, see Zakian 1989). Telomeric DNA sequences from more eukaryotes were determined through the 1980s. By the end of the 1980s, it was emerging that telomeric DNA of eukaryotic chromosomes in general consists of simple tandemly repeated sequences characterized by clusters of G residues in one strand, often resulting in G-rich versus C-rich strands. The G-rich strand was oriented 5′ to 3′ toward the chromosome terminus. A given species had a characteristic telomeric repeat sequence common to the ends of all its chromosomes. It became apparent that human telomeres may be like those of *Tetrahymena* at the molecular level. In 1986, Howard Cooke and co-workers reported that restriction fragments of the terminally located pseudoautosomal regions of human X and Y chromosomes also ran as fuzzy bands on gels (Cooke and Smith 1986), and in 1988 the human telomere sequence was shown to consist of tandem repeats of TTAGGG (Moyzis et al. 1988). The same telomeric sequence cropped up in widely divergent species; a tandem TTAGGG repeat tract is the telomeric sequence of humans, other vertebrates, trypanosomes, acellular slime molds, and filamentous fungi. In human germ-line (sperm) nuclei, about 10 kb of this sequence is found at every telomere (Moyzis et al. 1988); thus, the total genome is about 1 part in 3000 telomeric DNA by weight. However, mean telomere length commonly varied (for reviews, see Zakian 1989; de Lange et al. 1990, and references therein).

The similar molecular properties of the telomeric DNAs in all systems so far examined, and the cross-phyla functionality of ciliate telomeres demonstrated by the yeast experiments described above, suggested that the essential functions of telomeres are highly conserved. By 1990, it had been shown that telomeres from organisms as diverse as humans and ciliated protozoa could function to seed telomere addition in *S. cerevisiae*. From such findings arose the idea that features important for telomere function or recognition may be conferred by special structural properties of the simple G-cluster-containing DNA sequences. Z-DNA and later G-quartet DNA were favorite possibilities (for review, see Blackburn 1994), but despite searches through the 1980s and 1990s for hard evidence that either form has a normal biological role in telomere functions, such evidence has remained elusive.

Earliest Glimpses of Telomeric Proteins

Soon after the discovery of the telomeric sequence of *Tetrahymena*, attempts were initiated to see what proteins are associated with this unusual DNA. The somatic nuclear DNA and rDNA telomeres of *Tetrahymena* were found to be protected from nuclease digestion in a manner very different from that of nucleosomally packaged DNA (Blackburn and Chiou 1981). Hence, it was deduced that nonhistone proteins packaged the telomeric DNA tracts to confer this protection. No covalently attached protein could be found on *Tetrahymena* rDNA telomeres, despite a search using sensitive labeling methods that could have detected even submolar amounts of any covalently attached protein at the ends of these minichromosomes (Blackburn et al. 1983). In the ciliate *Oxytricha*, a tightly, but noncovalently, bound protein that was not a histone was then found to protect the short telomeric tracts of hypotrichous ciliate somatic chromosomes (Gottschling and Cech 1984; Gottschling and Zakian 1986). This telomere-specific structural protein recognizes and binds tenaciously to the telomeric G-rich strand overhang in a sequence-specific fashion, also protecting the neighboring duplex telomeric DNA (see Chapter 11). In 1986, Judith Berman (Berman et al. 1986) and colleagues biochemically identified a protein that could bind the yeast DNA repeat duplex, and later this was found by Roger Kornberg and colleagues to be the transcriptional activator repressor protein Rap1p (Buchman et al. 1988). In 1990, Virginia Zakian and co-workers demonstrated that Rap1p was associated with telomeres in vivo and negatively controls telomere length (Conrad et al. 1990). This was the first demonstration of the more general principle that proteins associated with telomeres prevent telomeres from overelongation.

A New Enzyme, Telomerase, That Synthesized Telomeric DNA

In the meantime, the DNA end-replication problem of telomeres was being solved by the discovery of telomerase by Greider and Blackburn, as described in the Appendix. Telomerase could explain how replication of the chromosomal DNA could be completed without the loss of terminal sequences predicted to result from normal semiconservative DNA replication; addition of telomeric DNA to the chromosomal ends by telomerase could counterbalance this terminal DNA attrition. Evidence for such attrition had been vividly seen on certain *Drosophila* chromosomes in which, occasionally, movement of transposons from telomeric regions led to terminal deletions. These chromosomes lost terminal DNA sequences progressively from the affected end at an estimated rate of 50–100 bp/fly generation (Biessmann and Mason 1988; Levis 1989). A model in which telomeres are continually being lengthened and shortened in small increments was proposed to account for the observed dynamic behavior of individual *S. cerevisiae* telomeres in vivo (Shampay and Blackburn 1988). In this model, mean telomere length is determined by the balance between the processes of addition and terminal loss of sequences. Furthermore, variability of telomere lengths could be explained by changes in the relative rates of these processes.

The journey toward identifying telomerase protein components began with the finding of a gene in yeast, *EST1*, identified by Vicki Lundblad and Jack Szostak in 1989. They proposed that *EST1* encodes a protein component of telomerase on the basis of the phenotype of *est1* mutants, which had "ever-shortening" telomeres over several cell generations, with increased rates of chromosome loss preceding senescence of the cell population (Lundblad and Szostak 1989). Additional *est* mutants led to the discovery of other telomerase proteins, including the telomerase core protein, TERT (telomerase reverse transcriptase), as described in Chapter 2.

In 1986, Howard Cooke and co-workers reported that restriction fragments of the terminally located pseudoautosomal regions of human X and Y chromosomes were shorter in adult cells than in germ cells and suggested that telomere shortening might account for this (Cooke and Smith 1986). The tandem repeats of human telomeric DNA, so like the TTGGGG repeats of *Tetrahymena*, suggested that a telomerase-like mechanism might also synthesize human telomeres, a prediction fulfilled by Gregg Morin's 1989 documentation of human telomerase activity (Morin 1989). In 1990, telomere shortening was first implicated in human cellular senescence and cancer (de Lange et al. 1990; Harley et al. 1990; Hastie et al. 1990). Telomerase appeared to be inactive in normal adult human cells, such as human fibroblasts. In contrast, by 1992, telomerase was known to be often activated after immortalization of human cells in culture (Counter et al. 1992), and in 1994, the first report appeared that up-regulated telomerase activation is the norm in human tumors (Kim et al. 1994).

Now, thoughts of manipulating telomerase for medical purposes came to the fore. It was proposed that telomerase inhibition might be useful to block cancer cell proliferation (Kim et al. 1994). Trypanosomes and other human protozoan parasites are effectively immortalized, as is *Tetrahymena*, and hence were thought likely to use telomerase for telomere maintenance; it was thus suggested that inhibiting their telomerase might be one way to attack these pathogens (Blackburn 1991), especially if telomerase was not active in normal adult human cells. Conversely, if telomerase were limiting in normal adult human cells, then perhaps activating telomerase in those cells might prolong their life spans (Levy et al. 1992).

Despite the widespread occurrence of telomerase in eukaryotes, telomeres in certain settings can also be maintained without telomerase, using recombination, as described in Chapters 7 and 8. In a last twist in the story of telomere maintenance, in *Drosophila*, in which Muller's early genetic studies had a crucial role in defining the nature of a telomere, the work of Mary Lou Pardue, Mel Green, James Mason, Harald Biessman, and their colleagues would show that occasional addition of retroelement elements (primarily Het-A elements) accomplishes what telomerase does in the more common eukaryotic situation (for review, see Pardue and DeBaryshe 2003; see Chapter 14).

ACKNOWLEDGMENTS

I am grateful to the many members of my laboratory who have contributed over the years to the work cited here.

REFERENCES

Ashley T. and Wagenaar E.B. 1974. Telomeric associations of gametic and somatic chromosomes in diploid and autotetraploid *Ornithogalum virens*. *Can. J. Genet. Cytol.* **16:** 61–76.

Baroudy B.M., Venkatesan S., and Moss B. 1982. Incompletely base-paired flip-flop terminal loops link the two DNA strands of the vaccinia virus genome into one uninterrupted polynucleotide chain. *Cell* **28:** 315–324.

Bateman A.J. 1975. Letter: Simplification of palindromic telomere theory. *Nature* **253:** 379–380.

Bennett M.D. 1977. Heterochromatin, aberrant endosperm nuclei and grain shrivelling in wheat-rye genotypes. *Heredity* **39:** 411–419.

Berman J., Tachibana C.Y., and Tye B.K. 1986. Identification of a telomere-binding activity from yeast. *Proc. Natl. Acad. Sci.* **83:** 3713–3717.

Biessmann H. and Mason J.M. 1988. Progressive loss of DNA sequences from terminal chromosome deficiencies in *Drosophila melanogaster*. *EMBO J.* **7:** 1081–1086.

Blackburn E.H. 1990. Telomeres and their synthesis. *Science* **249:** 489–490.

———. 1991. Structure and function of telomeres. *Nature* **350:** 569–573.

———. 1994. Telomeres: No end in sight. *Cell* **77:** 621–623.

Blackburn E.H. and Chiou S.S. 1981. Non-nucleosomal packaging of a tandemly repeated DNA sequence at termini of extrachromosomal DNA coding for rRNA in *Tetrahymena*. *Proc. Natl. Acad. Sci.* **78:** 2263–2267.

Blackburn E.H. and Gall J.G. 1978. A tandemly repeated sequence at the termini of the extrachromosomal ribosomal RNA genes in *Tetrahymena*. *J. Mol. Biol.* **120:** 33–53.

Blackburn E.H. and Szostak J.W. 1984. The molecular structure of centromeres and telomeres. *Annu. Rev. Biochem.* **53:** 163–194.

Blackburn E.H., Budarf M.L., Challoner P.B., Cherry J.M., Howard E.A., Katzen A.L., Pan W.C., and Ryan T. 1983. DNA termini in ciliate macronuclei. *Cold Spring Harbor Symp. Quant. Biol.* **47:** 1195–1207.

Broker T.R. 1973. An electron microscopic analysis of pathways for bacteriophage T4 DNA recombination. *J. Mol. Biol.* **81:** 1–16.

Buchman A.R., Kimmerly W.J., Rine J., and Kornberg R.D. 1988. Two DNA-binding factors recognize specific sequences at silencers, upstream activating sequences, autonomously replicating sequences, and telomeres in *Saccharomyces cerevisiae*. *Mol. Cell. Biol.* **8:** 210–225.

Cavalier-Smith T. 1974. Palindromic base sequences and replication of eukaryote chromosome ends. *Nature* **250:** 467–470.

Conrad M.N., Wright J.H., Wolf A.J., and Zakian V.A. 1990. RAP1 protein interacts with yeast telomeres in vivo: Overproduction alters telomere structure and decreases chromosome stability. *Cell* **63:** 739–750.

Cooke H.J. and Smith B.A. 1986. Variability at the telomeres of the human X/Y pseudoautosomal region. *Cold Spring Harbor Symp. Quant. Biol.* **51:** 213–219.

Counter C.M., Avilion A.A., LeFeuvre C.E., Stewart N.G., Greider C.W., Harley C.B., and Bacchetti S. 1992. Telomere shortening associated with chromosome instability is arrested in immortal cells which express telomerase activity. *EMBO J.* **11:** 1921–1929.

Creighton H.B. and McClintock B. 1931. A correlation of cytological and genetical crossing-over in *Zea mays*. *Proc. Natl. Acad. Sci.* **17:** 492–497.

Dancis B.M. and Holmquist G.P. 1979. Telomere replication and fusion in eukaryotes. *J. Theor. Biol.* **78:** 211–224.

de Lange T., Shiue L., Myers R.M., Cox D.R., Naylor S.L., Killery A.M., and Varmus H.E. 1990. Structure and variability of human chromosome ends. *Mol. Cell. Biol.* **10:** 518–527.

Engberg J., Andersson P., Leick V., and Collins J. 1976. Free ribosomal DNA molecules from *Tetrahymena pyriformis* GL are giant palindromes. *J. Mol. Biol.* **104:** 455–470.

Gall J.G. 1974. Free ribosomal RNA genes in the macronucleus of *Tetrahymena*. *Proc. Natl. Acad. Sci.* **71:** 3078–3081.

———. 1995. Beginning of the end: Origins of the telomere concept. In *Telomeres* (ed. E.H. Blackburn and C.W. Greider), pp. 1–10. Cold Spring Harbor Laboratory Press, Cold Spring Harbor, New York.

———. 1996. *Views of the cell: A pictorial history.* The American Society for Cell Biology, Bethesda, Maryland.

Gottschling D.E. and Cech T.R. 1984. Chromatin structure of the molecular ends of *Oxytricha* macronuclear DNA. Phased nucleosomes and a telomeric complex. *Cell* **38:** 501–510.

Gottschling D.E. and Zakian V.A. 1986. Telomere proteins: Specific recognition and protection of the natural termini of *Oxytricha* macronuclear DNA. *Cell* **47:** 195–205.

Harley C.B., Futcher A.B., and Greider C.W. 1990. Telomeres shorten during ageing of human fibroblasts. *Nature* **345:** 458–460.

Hastie N.D., Dempster M., Dunlop M.G., Thompson A.M., Green D.K., and Allshire R.C. 1990. Telomere reduction in human colorectal carcinoma and with ageing. *Nature* **346:** 866–868.

Henderson E., Hardin C.C., Walk S.K., Tinoco I., Jr., and Blackburn E.H. 1987. Telomeric DNA oligonucleotides form novel intramolecular structures containing guanine-guanine base pairs. *Cell* **51:** 899–908.

Holmquist G.P. and Dancis B. 1979. Telomere replication, kinetochore organizers, and satellite DNA evolution. *Proc. Natl. Acad. Sci.* **76:** 4566–4570.

Karrer K.M. and Gall J.G. 1976. The macronuclear ribosomal DNA of *Tetrahymena pyriformis* is a palindrome. *J. Mol. Biol.* **104:** 421–453.

Kim N.W., Piatyszek M.A., Prowse K.R., Harley C.B., West M.D., Ho P.L., Coviello G.M., Wright W.E., Weinrich S.L., and Shay J.W. 1994. Specific association of human telomerase activity with immortal cells and cancer. *Science* **266:** 2011–2015.

Klobutcher L.A., Swanton M.T., Donini P., and Prescott D.M. 1981. All gene-sized DNA molecules in four species of hypotrichs have the same terminal sequence and an unusual 3′ terminus. *Proc. Natl. Acad. Sci.* **78:** 3015–3019.

Levis R.W. 1989. Viable deletions of a telomere from a *Drosophila* chromosome. *Cell* **58:** 791–801.

Levy M.Z., Allsopp R.C., Futcher A.B., Greider C.W., and Harley C.B. 1992. Telomere end-replication problem and cell aging. *J. Mol. Biol.* **225:** 951–960.

Lundblad V. and Szostak J.W. 1989. A mutant with a defect in telomere elongation leads to senescence in yeast. *Cell* **57:** 633–643.

McClintock B. 1931. Cytological observations of deficiencies involving known genes, translocations and an inversion in *Zea mays*. *Mo. Agric. Exp. Res. Stn. Res. Bull.* **163:** 4–30.

———. 1932. A correlation of ring-shaped chromosomes with variegation in *Zea mays*. *Proc. Natl. Acad. Sci.* **18:** 677–681.

———. 1938a. The fusion of broken ends of sister half-chromatids following chromatid breakage at meiotic anaphases. *Mo. Agric. Exp. Res. Stn. Res. Bull.* **290:** 1–48.

———. 1938b. The production of homozygous deficient tissues with mutant characteristics by means of the aberrant mitotic behavior of ring-shaped chromosomes. *Genetics* **23:** 315–376.

———. 1939. The behavior in successive nuclear divisions of a chromosome broken at meiosis. *Genetics* **25:** 405–416.

———. 1941a. Spontaneous alterations in chromosome size and form in *Zea mays. Cold Spring Harbor Symp. Quant. Biol.* **9:** 72–80.

———. 1941b. The stability of broken ends of chromosomes in *Zea mays. Genetics* **26:** 234–282.

———. 1942–1943. Maize genetics. *Carnegie Inst. Wash. Year Book* **42:** 148–152.

———. 1952. Chromosome organization and genic expression. *Cold Spring Harbor Symp. Quant. Biol.* **16:** 13–47.

Morin G.B. 1989. The human telomere terminal transferase enzyme is a ribonucleoprotein that synthesizes TTAGGG repeats. *Cell* **59:** 521–529.

Moyzis R.K., Buckingham J.M., Cram L.S., Dani M., Deaven L.L., Jones M.D., Meyne J., Ratliff R.L., and Wu J.R. 1988. A highly conserved repetitive DNA sequence, (TTAGGG)n, present at the telomeres of human chromosomes. *Proc. Natl. Acad. Sci.* **85:** 6622–6626.

Muller H.J. 1938. The remaking of chromosomes. *The Collecting Net* **8:** 182–195.

Oka Y., Shiota S., Nakai S., Nishida Y., and Okubo S. 1980. Inverted terminal repeat sequence in the macronuclear DNA of *Stylonychia pustulata. Gene* **10:** 301–306.

Pardue M.L. and DeBaryshe P.G. 2003. Retrotransposons provide an evolutionarily robust non-telomerase mechanism to maintain telomeres. *Annu. Rev. Genet.* **37:** 485–511.

Pluta A.F., Kaine B.P., and Spear B.B. 1982. The terminal organization of macronuclear DNA in *Oxytricha fallax. Nucleic Acids Res.* **10:** 8145–8154.

Pluta A.F., Dani G.M., Spear B.B., and Zakian V.A. 1984. Elaboration of telomeres in yeast: Recognition and modification of termini from *Oxytricha* macronuclear DNA. *Proc. Natl. Acad. Sci.* **81:** 1475–1479.

Rubin G.M. 1978. Isolation of a telomeric DNA sequence from *Drosophila melanogaster. Cold Spring Harbor Symp. Quant. Biol.* **42:** 1041–1046.

Shampay J. and Blackburn E.H. 1988. Generation of telomere-length heterogeneity in *Saccharomyces cerevisiae. Proc. Natl. Acad. Sci.* **85:** 534–538.

Shampay J., Szostak J.W., and Blackburn E.H. 1984. DNA sequences of telomeres maintained in yeast. *Nature* **310:** 154–157.

Szostak J.W. and Blackburn E.H. 1982. Cloning yeast telomeres on linear plasmid vectors. *Cell* **29:** 245–255.

Walmsley R.M., Szostak J.W., and Petes T.D. 1983. Is there left-handed DNA at the ends of yeast chromosomes? *Nature* **302:** 84–86.

Watson J.D. 1972. Origin of concatemeric T7 DNA. *Nat. New Biol.* **239:** 197–201.

White M.J.D. 1978. Chain processes in chromosomal speciation. *Syst. Zool.* **27:** 285–298.

Young B.S., Pession A., Traverse K.L., French C., and Pardue M.L. 1983. Telomere regions in *Drosophila* share complex DNA sequences with pericentric heterochromatin. *Cell* **34:** 85–94.

Zakian V.A. 1989. Structure and function of telomeres. *Annu. Rev. Genet.* **23:** 579–604.

2

The Telomerase Ribonucleoprotein Particle

Gaël Cristofari and Joachim Lingner
Swiss Institute for Experimental Cancer Research (ISREC)
École Polytechnique Fédérale de Lausanne (EPFL) and
National Center of Competence in Research "Frontiers in Genetics"
CH-1066 Epalinges s/Lausanne, Switzerland

ALL EUKARYOTES AND A FEW PROKARYOTES KEEP THEIR GENOMES in the form of linear DNA molecules. This requires special mechanisms to fully replicate DNA ends because of the DNA end-replication problem and nucleolytic processing of telomere ends after semiconservative DNA replication (see Chapter 1). The most common solution to the end-replication problem occurs through the ribonucleoprotein (RNP) reverse transcriptase telomerase, discovered in 1985 in the holotrichous ciliate *Tetrahymena thermophila* by Carol Greider and Elizabeth Blackburn (Greider and Blackburn 1985; Chapter 1). The demonstration that the enzyme contains an RNA moiety, which copurifies with enzymatic activity (Greider and Blackburn 1989) and specifies the sequence of telomeric repeats (Yu et al. 1990), unmasked its nature as a cellular reverse transcriptase. The identification of telomerase protein subunits through genetic screens in yeast (Lendvay et al. 1996) and through the biochemical purification of telomerase from the hypotrichous ciliate *Euplotes aediculatus* (Lingner and Cech 1996) uncovered the presence of universally conserved reverse transcriptase sequence motifs in the catalytic subunit of telomerase (Lingner et al. 1997). Other less-well-conserved telomerase-associated proteins have crucial roles in the biogenesis of telomerase and the regulation of the interaction of telomerase with chromosome ends.

In this chapter, we first describe the structural and functional features of telomerase RNA (TR) and the telomerase reverse transcriptase (TERT).

We then give a comprehensive overview of telomerase-associated proteins in ciliates, yeast, and humans. We also summarize biochemical and genetic evidence that telomerase forms higher-order structures in some organisms. For the enzymology and biogenesis of telomerase, see Chapter 3.

TELOMERASE RNAs

TRs have been identified from a large number of holotrichous and hypotrichous ciliated protozoa, several yeast species from the genera *Saccharomyces* and *Kluyveromyces*, and a large number of vertebrates including human (Chen and Greider 2004). Comparison of the TR sequences from different phyla demonstrates that this RNA diverged quickly in evolution. The sizes range from 150–200 nucleotides in ciliates to approximately 450 nucleotides in vertebrates to approximately 1200 nucleotides in budding yeast. TR sequences can be aligned with confidence only among closely related organisms, and secondary structure models were derived for the different groups by phylogenetic comparison (Romero and Blackburn 1991; Lingner et al. 1994; Chen et al. 2000; Dandjinou et al. 2004; Zappulla and Cech 2004). In this most powerful approach, RNA duplexes are considered to exist if compensatory base-pair changes in homologous sequences that maintain pairing potential can be identified at the corresponding positions. In addition, chemical probes and structure-specific RNases have been instrumental in evaluating secondary structure models (Bhattacharyya and Blackburn 1994; Zaug and Cech 1995; Sperger and Cech 2001; Antal et al. 2002; Dandjinou et al. 2004; Forstemann and Lingner 2005). Mutagenic studies have been used to test derived structural models and their importance for function (see below). Recently, nuclear magnetic resonance (NMR) analysis of short TR fragments has been employed to determine the structure of subdomains in the absence of proteins in solution (Comolli et al. 2002; Leeper et al. 2003; Theimer et al. 2003, 2005; Leeper and Varani 2005).

Figure 1 shows the secondary structure models of TRs from the ciliate *T. thermophila* (Romero and Blackburn 1991), *Homo sapiens* (Chen et al. 2000), and the yeast *Saccharomyces cerevisiae* (Dandjinou et al. 2004; Zappulla and Cech 2004). Strikingly, despite the divergence in sequence and size, some common structural features appear to be universally conserved. The template of all telomerases is single stranded, allowing base-pairing with the telomere 3′ end, while residing in the active site of the TERT. The length of the template is approximately 1.5–2 times the telomeric repeat length, thus enabling both annealing of the telomeric 3′ end with the template and addition of one telomeric repeat per elongation

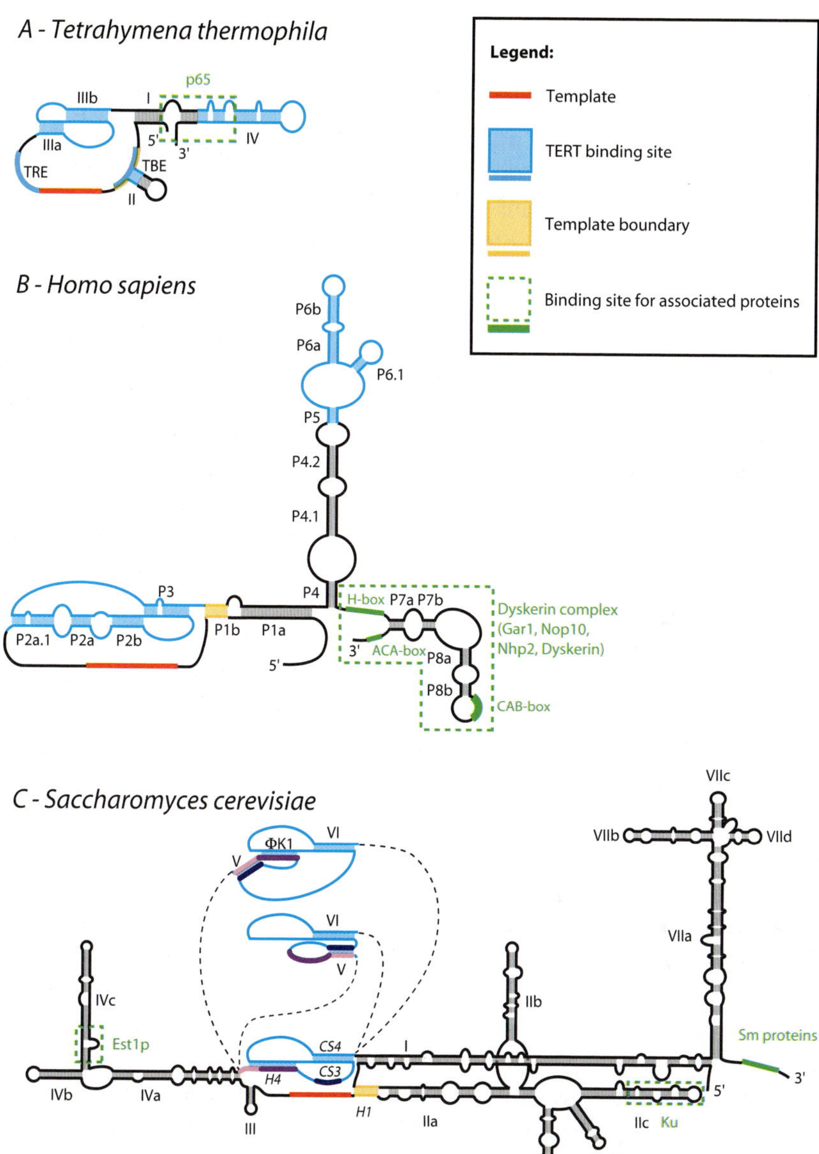

Figure 1. Telomerase RNA secondary structure models for *Tetrahymena thermophila* (*A*), *Homo sapiens* (*B*), and *Saccharomyces cerevisiae* (*C*). The TR lengths and sequences diverged in the different phylogenetic groups, but a single-stranded RNA template (*red*) and the pseudoknot domain are conserved. For *S. cerevisiae*, three alternative secondary structures have been proposed for the putative pseudoknot region. Identical regions in the three models are highlighted with identical colors. Template boundary elements are required to prevent template readthrough at the 5′ end (*yellow*). Stem IV of *T. thermophila* TR and stem loop P6.1 of hTR are required for enzymatic activity and can be provided either in *cis* or in *trans*. Elements bound by TR processing and maturation proteins are required for TR stability (La motif containing proteins [*E. aediculatus* p43 or *T. thermophila* p65] in ciliates, the dyskerin complex in vertebrates, and the Sm proteins in *S. cerevisiae*).

cycle (see Chapter 3). The template is flanked 5' by a long-range base-pairing element and 3' by a pseudoknot domain.

Strikingly, evidence from *Tetrahymena* and vertebrate telomerases indicates that distinct domains that can be separated between two telomerase RNA fragments cooperate for telomere extension. One fragment includes the template region, boundary elements, and pseudoknot, whereas the other (referred to as activator) includes a template distant stem loop (stem loop IV in *Tetrahymena* [Sperger and Cech 2001; Lai et al. 2003] and stem loop P6 in vertebrates [Tesmer et al. 1999; Mitchell and Collins 2000; Chen et al. 2002; Moriarty et al. 2004]). In *Tetrahymena* and humans, telomerase activity can be reconstituted in a cell-free system by combining in *trans* pseudoknot-template and activator stem-loop fragments with TERT, indicating that they contain independent domains (Sperger and Cech 2001; Lai et al. 2003). In addition, the two fragments contact two separate TERT regions (see below). These observations suggest a direct function of the TR moiety in the reaction cycle, apart from its role as a template (Lai et al. 2003).

Pseudoknot

TRs have the potential to form a pseudoknot structure. In ciliates, this is well supported through nucleotide covariation (ten Dam et al. 1991; Lingner et al. 1994). Disruption of pseudoknot base-pairing within the TR from *T. thermophila* prevents in vivo the stable association with TERT, whereas compensatory changes restore telomerase activity (Gilley and Blackburn 1999). In vitro, telomerase activity is affected by mutations and deletions in this domain (Autexier and Greider 1998; Licht and Collins 1999; Lai et al. 2003). The pseudoknot is not required for assembly with TERT in vitro, and other TERT-TR interactions are sufficient for stable association (Licht and Collins 1999; Lai et al. 2001).

Phylogenetic and structural analyses support the notion that vertebrate TRs also form a pseudoknot structure (Chen et al. 2000; Theimer et al. 2005). The solution structure of the human TR (hTR) pseudoknot reveals an extended triple helix surrounding the helical junction, explaining the phylogenetic sequence conservation (Theimer et al. 2005). Chemical probing of hTR in vivo and NMR structural studies indicate that the pseudoknot in hTR is not stably formed; it is possible that it functions as a molecular switch (Antal et al. 2002; Comolli et al. 2002). Indeed, pseudoknots are often thermodynamically metastable, existing in equilibrium between the pseudoknot and a single stem loop. Mutation of the TR pseudoknot impairs assembly of active telomerase in cellular reconstitution assays in mouse and human cells (Martin-Rivera and Blasco 2001; Ly et al. 2003a) and hTERT is interacting with an hTR fragment encompassing the pseudoknot-template region, in addition to binding to the CR4-CR5 domain (Mitchell and Collins 2000; Chen et al. 2002). However, pseudoknot-hTERT interactions have not been clearly mapped (Ly et al. 2003a). The P3 helix of the pseudoknot of hTR can also form in vitro in *trans*, mediating hTR dimerization (Ly et al. 2003b).

Alternative pseudoknot structures have also been proposed for TRs from *Saccharomyces* (three proposed alternative structures are indicated in Fig. 1C) (Chappell and Lundblad 2004; Dandjinou et al. 2004; Lin et al. 2004) and *Kluyveromyces* (Tzfati et al. 2003). Chemical probing of the *S. cerevisiae* Est2p-Tlc1 catalytic telomerase core indicates that stem V (Fig. 1C) is not stably formed and that the H4 helix (lower pseudoknot in Fig. 1C) is shorter than drawn, whereas accessibility to chemical probing is consistent with existence of stem VI (also referred to as CS3/CS4; see Fig. 1C) and stem φK1 (upper pseudoknot) (Forstemann and Lingner 2005). In *S. cerevisiae*, the base-pair potential and the primary sequence of stem VI are important for binding the yeast TERT Est2p, whereas the other stem of the pseudoknot is dispensable (Chappell and Lundblad 2004; Lin et al. 2004). Thus, the pseudoknot structure is not required for Est2p binding and telomerase function. However, the entire Est2p interaction domain can be replaced by the TR pseudoknot from *Oxytricha nova*, which bears no apparent sequence similarity with the replaced region in Tlc1 (Chappell and Lundblad 2004). This indicates that a pseudoknot structure, although not essential for Est2p binding, can be recognized by Est2p or that the *Oxytricha* pseudoknot promotes formation of a stable TR core structure (Zappulla and Cech 2004) helping to present the sequence elements that mediate Est2p binding.

5′ Boundary Elements

To prevent reverse transcription beyond the 5′ boundary of the template, different structures evolved in different phylogenetic groups (Fig. 1, yellow/orange-colored elements). In *Tetrahymena*, an interaction between TERT and a ciliate conserved RNA sequence (5′-UGUCA-3′) just upstream of the template is required to prevent template readthrough (Autexier and Greider 1995; Lai et al. 2002). In the budding yeasts *Kluyveromyces* and *Saccharomyces*, a paired element just upstream of the template specifies the 5′ boundary of the template (Tzfati et al. 2000; Seto et al. 2003). In human telomerase, the more template distant helix P1b is required for template boundary definition, whereas in mouse TR, which lacks helix P1b, the boundary is established through the 5′ end, which abuts the template by only two nucleotides (Chen and Greider 2003). The role of boundary elements is considered in detail in Chapter 3.

Protein-binding Sites

The divergence of TRs outside the template and pseudoknot is remarkable. The divergence of size, sequence, and structure may be due to different requirements for telomere length regulation and function at chromosome ends of highly polyploid ciliates, single-celled yeast, and multicellular organisms. This divergence in function is reflected in part by the nonconserved identity of telomerase-associated proteins, which have different TR-binding sites (see Fig. 1). However, not even the TR-binding sites of the universally conserved TERT subunit were maintained during evolution. In addition, telomerase biogenesis pathways have diverged. TRs are synthesized in ciliates by RNA polymerase III and in yeast and higher eukaryotes by RNA polymerase II (Chapter 3). Telomerase-associated proteins are discussed in detail below.

THE TELOMERASE REVERSE TRANSCRIPTASE

The biochemical purification of telomerase from *E. aediculatus* (Lingner and Cech 1996) and a genetic screen in *S. cerevisiae* (Lendvay et al. 1996) led to the identification of TERT (Lingner et al. 1997). This facilitated the subsequent identification of the TERT orthologs from the fission yeast *Schizosaccharomyces pombe*, human and mouse, *T. thermophila*, and others (Harrington et al. 1997; Kilian et al. 1997; Meyerson et al. 1997; Nakamura et al. 1997; Bryan et al. 1998; Collins and Gandhi 1998). All TERT polypeptides share in their carboxy-terminal half sequence homology with reverse transcriptases (RTs) from retroelements and retroviruses (Fig. 2). All seven defined RT motifs (1, 2, A, B′, C, D, and E) are present in TERT. Phylogenetic analysis suggests that TERT is most related to RTs encoded by non-long terminal repeat (LTR) retrotransposons and that telomerase is among the most ancient eukaryotic RTs (Eickbush 1997; Nakamura and Cech 1998). The evolutionary link between retrotransposon-derived RTs and telomerases is further strengthened by the substitution of a telomerase-based end-replication mechanism through a mechanism involving the retrotransposons HeT-A and TART in *Drosophila* (Biessmann et al. 1992a,b; Levis et al. 1993).

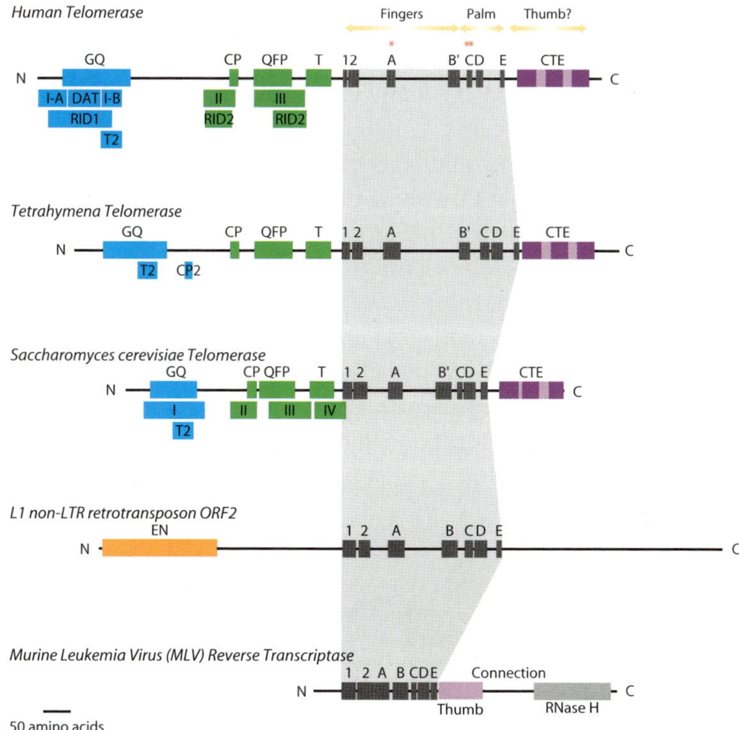

Figure 2. Domain organization of TERTs and comparison to RTs from retroviruses and retrotransposons. The RT domain (*dark gray*) includes conserved RT motifs 1 to E (fingers and palm) and the thumb domain, which is much more divergent among RTs (*pink*). The three aspartates that are part of the active site are indicated by asterisks. The CTE (carboxy-terminal extension) of telomerases contains three blocks of conserved amino acids. It does not show obvious sequence similarity with the RT thumb domain, but functional studies suggest similar roles. The amino-terminal half contains one or two RNA-binding domains. The motif nomenclature originates from different references (Friedman and Cech 1999; Miller et al. 2000; Xia et al. 2000; Armbruster et al. 2001; Moriarty et al. 2002). The region highlighted in *green* corresponds to an RNA-binding domain. The region highlighted in *blue* may bind to another region in TR as shown for the RID1 domain in hTERT and the corresponding region in *Tetrahymena* TERT. Endonuclease (EN, *orange*) and RNase H (*light gray*) domains found in some retrotransposons and retroviruses have not been identified in TERT.

The importance of some of the RT motifs for telomerase activity has been demonstrated previously (Counter et al. 1997; Harrington et al. 1997; Lingner et al. 1997; Weinrich et al. 1997; Nakayama et al. 1998). Thus, the three-dimensional architecture of the domains is predicted to be similar to the well-studied structures from human immunodeficiency virus (HIV) and murine leukemia virus (MLV) RTs, which have been compared to that of a right hand with fingers (motifs 1→B′), palm (motifs C→E), and most probably thumb (CTE) domains, the active site residing in the palm (Figs. 2 and 3) (O'Reilly et al. 1999). Most critical are three aspartic acids of motifs A and C that reside in the active site. They are required by all RTs for enzymatic activity, coordinating the two catalytic metal ions (Mg^{2+} or Mn^{2+}) involved in the polymerization reaction (red asterisks in Fig. 2; red surface in Fig. 3).

The carboxy-terminal extensions (CTEs) of human and *Tetrahymena* TERTs are essential for enzymatic activity, whereas this domain is dispensable for activity in *S. cerevisiae* (Friedman and Cech 1999; Bachand and Autexier 2001; Lai et al. 2001). Nonetheless, biochemical experiments of yeast and human telomerase support a role of this domain in processivity (Huard et al. 2003), and it has been suggested that this domain functions as a thumb

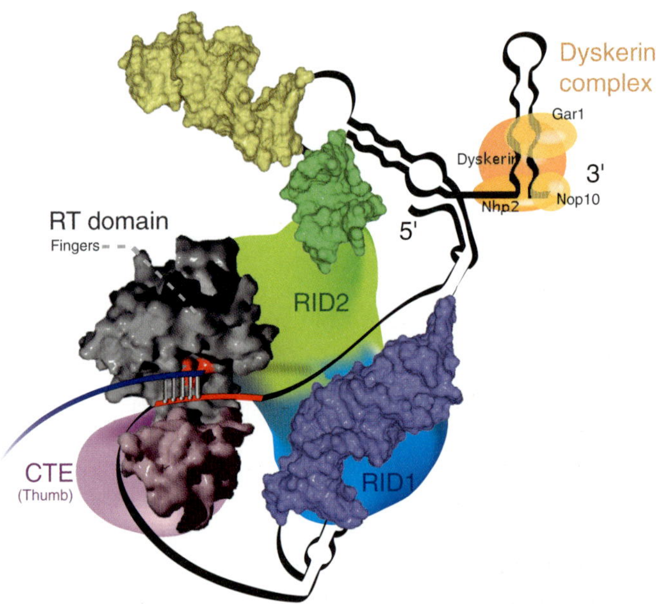

Figure 3. Model of a monomeric human telomerase RNP in the absence of hEST1A/B and other telomerase-associated proteins. The central RT structure (*gray* and *pink*) corresponds to murine leukemia virus RT (Das and Georgiadis 2004) and includes its finger, palm, and thumb domains but not the RNase H domain. The colors are the same as those in Fig. 2. The linker region between RID1 and RID2 is large, and therefore the position of RID1 is highly speculative. The secondary structure of TR is drawn as a ribbon except for fragments for which the three-dimensional structure has been determined by NMR (P6.1 region, *green* [Leeper et al. 2003]; CR4/CR5 region, *yellow* [Leeper and Varani 2005]; part of the pseudoknot, *blue* [Theimer et al. 2005]).

domain for telomerase (Peng et al. 2001). However, no amino acid sequence similarity is evident between the CTE of telomerases and the thumb of RTs.

Analysis of the TERT protein family reveals a conserved structural organization also outside the RT domain. The amino-terminal half of the protein contains several conserved amino acid sequence motifs within TERT that form one or two TR-binding domains (Fig. 2). These interactions may enable the stable association with TR while allowing the template to move through the active site (Bryan et al. 2000).

In *T. thermophila*, mutations of the TERT-specific T motif and CP motifs strongly reduce TR binding, whereas mutations in the RT domain have no apparent effect on TR binding (Bryan et al. 2000). The T and CP motifs are part of a high-affinity RNA-binding domain (RBD) (O'Connor et al. 2005), which interacts with nucleotides upstream of the template near the 5′ end of TR (Lai et al. 2001) (TBE in Fig. 1A) required for definition of the 5′ boundary (Lai et al. 2002). Additional low-affinity TERT-TR interactions involve an amino-terminal RNA interaction domain (Nterm) (O'Connor et al. 2005) and an element 3′ of the template called TRE (for template recognition element) and stem IV (Lai et al. 2003).

Similarly to the results obtained for *Tetrahymena* telomerase, mutations in the corresponding amino-terminal region in the *S. cerevisiae* TERT Est2p interfere with binding to the yeast telomerase RNA, TLC1 (Friedman and Cech 1999). Point mutations in the TERT-conserved motifs CP, QFP, and T abolish or strongly reduce association with TLC1, confirming the importance of the conserved amino acids for RNA binding (Xia et al. 2000; Bosoy et al. 2003). As discussed above, the Est2p-binding site in TLC1 has been proposed to form a pseudoknot structure.

Human TERT binds to the CR4/CR5 domain and to a TR fragment encompassing the pseudoknot and template (Mitchell and Collins 2000) using two separate RBDs, RID1 and RID2 (Moriarty et al. 2004). RID2 interacts with the CR4-CR5 domain, whereas RID1 binds the pseudoknot-template region.

The amino-terminal RBDs of TERT have no recognizable sequence similarity with other proteins. Interestingly, however, the Gag proteins of retroviruses and LTR retrotransposons and ORF1p of non-LTR retrotransposons are encoded upstream of the RT gene within the same RNA molecule. They function as RNA chaperones and recruit the genomic RNA to the reverse transcription complex (Cristofari and Darlix 2002), reminiscent of the putative roles of the RBDs of TERT. Thus, it will be fascinating to see if the three-dimensional structures will uncover structural similarity, which could indicate a common evolutionary origin of TERT RBDs and Gag/ORF1 proteins.

TELOMERASE-ASSOCIATED PROTEINS

Telomerase activity can be reconstituted in vitro in cellular extracts from TERT and TR, suggesting that these two components form the catalytic core of telomerase. However, additional proteins that associate with telomerase have been identified, and they have crucial roles in telomerase assembly, localization, and regulation at chromosome ends. Estimates of the native molecular masses of telomerases based on gel filtration or glycerol and sucrose gradient centrifugation range from 230 kD in *E. aediculatus* to 1.5 MD in *Homo sapiens* to greater than 5 MD in *Euplotes crassus* (Lingner and Cech 1996; Greene and Shippen 1998; Schnapp et al. 1998). Interestingly, the molecular mass of the *E. crassus* telomerase holoenzyme changes from 280 kD in vegetative cells to 1.6 MD during macronuclear development when chromosome healing occurs (see Chapter 15). This change correlates with expression of an alternative TERT variant which might assemble with other telomerase-associated proteins (Karamysheva et al. 2003). Table 1 provides a comprehensive overview of known telomerase-associated and -interacting proteins. The identity and sequence of these proteins have diverged in different organisms. Some of them are found in other RNP complexes, whereas others are telomerase specific.

Ciliate Telomerases

Purification of active endogenous telomerase from *E. aediculatus* identified the telomerase-associated protein p43 (Lingner and Cech 1996; Aigner et al. 2000). In *T. thermophila*, the orthologous p65 was identified through ectopic expression of epitope-tagged TERT and purification (Witkin and Collins 2004). Both proteins are telomerase specific, but they reveal similarity to the La autoantigen, an RNA-binding protein involved in maturation of RNA polymerase III transcripts. p43 (which has a molecular mass of 51 kD) binds with nanomolar affinity to stem I and adjacent nucleotides of *E. aediculatus* TR in vitro (Aigner et al. 2003), whereas p65 appears to contact the 3'-end proximal region of stem IV (Prathapam et al. 2005). The p43/p65 polypeptides may have roles in telomerase maturation and activity. When assembled in vitro with the heterologous *Tetrahymena* TERT-TR complex, p43 stimulates activity and processivity of telomerase (Aigner and Cech 2004). Depletion of *Tetrahymena* p65 leads to reduced levels of TR and TERT (Witkin and Collins 2004) and, in in vitro reconstitution experiments, p65 stimulates, when prebound to TR, the association with TERT (Prathapam et al. 2005).

Table 1. Comprehensive survey of telomerase-associated and -interacting proteins

Organism size of RNP (ref)	Multimeri-zation	Associated/ interacting proteins	Interaction found with (method)	Function	Remarks	Reference
Tetrahymena thermophila 500 kD (Wang and Blackburn 1997)	no	p75	(TAP-TERT purification)	n.d.		Witkin and Collins 2004
		p65	TR (TAP-TERT purification)	TER stability and RNP assembly	ortholog of _E. aediculatus_ p43, contains a La motif	Witkin and Collins 2004; Prathapam et al. 2005
		p45	(TAP-TERT purification)	n.d.		Witkin and Collins 2004
		p20	(TAP-TERT purification)	n.d.		Witkin and Collins 2004
Euplotes crassus vegetative growth: 280 kD macronuclear development: 550 kD, 1.6 MD, >5 MD (Greene and Shippen 1998)	n.d.					
Euplotes aediculatus vegetative growth: 230 kD (Lingner and Cech 1996)	no	p43	TR (copurification with telomerase activity and MS identification)	TER stability? stimulates telomerase activity in vitro	ortholog of _T. thermophila_ p65, contains a La motif	Lingner and Cech 1996; Aigner et al. 2000, 2003; Aigner and Cech 2004

(Continued)

Table 1. (*continued*)

Organism size of RNP (ref)	Multimeri-zation	Associated/ interacting proteins	Interaction found with (method)	Function	Remarks	Reference
Saccharomyces cerevisiae 20S (Lingner et al. 1997)	yes	Est1p	TLC1 (ever shorter telomeres genetic screen, IP)	binding of Cdc13p mediating telomerase recruitment and binding of telomeric DNA (not required for telomerase activity in vitro)		Lundblad and Szostak 1989; Lendvay et al. 1996; Steiner et al. 1996; Hughes et al. 2000; Evans and Lundblad 2002; Seto et al. 2002
		Est3p	Est2p (ever shorter telomeres genetic screen, IP)	telomere elongation in vivo (not required for telomerase activity in vitro)		Lendvay et al. 1996; Morris and Lundblad 1997; Hughes et al. 2000; Friedman et al. 2003
		Ku	TLC1 (allele-specific interactions, in vitro binding)	telomere elongation and chromosome healing		Peterson et al. 2001; Stellwagen et al. 2003; Fisher et al. 2004
		Cdc13p	Est1p (ever shorter telomeres genetic screen, genetic suppression, IP)	binding of Est1p (not required for telomerase activity in vitro)		Lendvay et al. 1996; Nugent et al. 1996; Qi and Zakian 2000; Pennock et al. 2001
		Sm proteins	TLC1 (IP)	TLC1 expression/stability and maturation		Seto et al. 1999
		PinX1/ Gno1p	Est2p (IP)	inhibitor of Est2p-TLC1 assembly	ortholog of human PinX1	Lin and Blackburn 2004

Schizosaccharomyces pombe 35S–1.5 MD (Lue and Peng 1997)	n.d.	SpEst1		telomere elongation telomere capping	ortholog of yeast Est1	Beernink et al. 2003
Homo sapiens 1.5 MD (Schnapp et al. 1998)	yes	hEst1A	telomerase activity (IP)	telomere capping telomere elongation	sequence similarities with budding yeast Est1	Reichenbach et al. 2003; Snow et al. 2003
		hEst1B	telomerase activity (IP)	n.d.	sequence similarities with budding yeast Est1	Reichenbach et al. 2003; Snow et al. 2003
		hGar1p	hTR (bandshift/ UV cross-link)	TR stability RNP assembly		Dragon et al. 2000; Wang and Meier 2004
		hNhp2	hTR (bandshift/ UV cross-link)	TR stability RNP assembly		Pogacic et al. 2000; Wang and Meier 2004
		hNop10	hTR (bandshift/ UV cross-link)	TR stability RNP assembly		Pogacic et al. 2000; Wang and Meier 2004
		dyskerin	hTR (IP, UV cross-link)	TR stability RNP assembly	mutated in X-linked form of dyskeratosis congenita	Mitchell et al. 1999b; Wang and Meier 2004
		SMN	telomerase activity (IP)	RNP assembly? nucleolar localization of hTERT		Bachand et al. 2002
		14-3-3	hTERT (2-hybrid, IP)	nuclear localization of hTERT		Seimiya et al. 2000
		Ku	telomerase activity (IP) hTR (bandshift, IP)	n.d.		Chai et al. 2002; Ting et al. 2005
		La	hTR (UV-cross-link, IP)	n.d.		Ford et al. 2001

(Continued)

Table 1. (*continued*)

Organism size of RNP (ref)	Multimeri-zation	Associated/ interacting proteins	Interaction found with (method)	Function	Remarks	Reference
		Nucleolin	telomerase activity (IP)	nucleolar localization of hTERT		Khurts et al. 2004
		PinX1	TERT (IP)	inhibitor of telomerase activity	found in 2-hybrid screen with Pin2/TRF1	Zhou and Lu 2001; Banik and Counter 2004
		Hsp90	hTERT (IP, 2-hybrid)	RNP assembly in vivo and in vitro		Holt et al. 1999; Forsythe et al. 2001
		Hsp23	hTERT (IP, 2-hybrid)	RNP assembly in vivo and in vitro		Holt et al. 1999; Forsythe et al. 2001
		KIP	hTERT (2-hybrid)	positive regulator of telomere length		Lee et al. 2004
		hnRNP A1	telomerase activity (IP)	telomere elongation		Labranche et al. 1998
		hnRNP C1/C2	hTR (UV cross-link, IP)	n.d.		Ford et al. 2000
		hnRNP D	telomerase activity (IP)	n.d.		Eversole and Maizels 2000
		L22	hTR (3-hybrid)	n.d.		Le et al. 2000
		hStau	hTR (3-hybrid)	n.d.		Le et al. 2000

It is unclear for most proteins whether they are associated stoichiometrically with TR and TERT and whether they have a role in the mature telomerase RNP or during RNP assembly. For some proteins, it cannot be excluded that their association occurs only in vitro, on disruption of cellular compartments. IP, immunoprecipitation; n.d., not determined.

In addition to p65, three other polypeptides—p75, p45, and p20—were identified upon purification of epitope-tagged telomerase in *Tetrahymena* (Witkin and Collins 2004). So far, two of the four identified polypeptides, p65 and p45, are known to be required for telomere maintenance; the functions of p20 and p45 remain to be elucidated. Two proteins previously thought to be associated with *Tetrahymena* telomerase (p80 and p95; Collins et al. 1995) were not observed to cofractionate with the telomerase RNP in the more recent epitope-tag-mediated affinity purification. Although this could reflect the difference in fractionation approaches, in vitro investigation of p80 and p95 suggests that neither is stably associated with the *Tetrahymena* enzyme (Mason et al. 2001). The vertebrate ortholog of p80 named TEP1 is also not required for telomerase activity and telomere maintenance (Liu et al. 2000).

Yeast Telomerase

The ever shorter telomere (EST) screen uncovered in *S. cerevisiae* four genes, *EST1–EST4*, that are essential for telomerase activity in vivo (Lundblad and Szostak 1989; Lendvay et al. 1996; for a detailed discussion, see Chapter 12). Est1p binds directly to a bulged stem in TLC1 (Seto et al. 2002). It recruits telomerase to chromosome ends through an interaction with the single-stranded binding protein Cdc13p (=Est4p) (Evans and Lundblad 1999; Pennock et al. 2001; Bianchi et al. 2004). In addition, Est1p binds specifically to single-stranded G-rich telomeric sequence in vitro (Virta-Pearlman et al. 1996). Est1p orthologs have also been identified in *S. pombe* and *Candida albicans*, and in both of these organisms, not only does Est1p seem to be required for telomerase-mediated telomere extension, but it also has a role in protecting chromosome ends from rapid deletion (Singh et al. 2002; Beernink et al. 2003). Est3p is also essential for telomerase activity in vivo in *S. cerevisiae* and *C. albicans*, but its function is not understood (Hughes et al. 2000; Singh et al. 2002). In *S. cerevisiae*, the association of Est3p with TLC1 depends on the presence of Est2p. Mutations in region I near the amino terminus of Est2p result in loss of Est3p from the telomerase complex, suggesting a direct or indirect physical interaction between region I and Est3p (Friedman et al. 2003).

Two proteins that bind to chromosome ends, Cdc13p and Ku, also interact with telomerase, although they are not stable constituents of the telomerase RNP in cellular extracts (see Chapter 12). The association of Cdc13p with Est1p requires an ionic interaction between Glu-252 of Cdc13p and Lys-444 of Est1p (Pennock et al. 2001). The Ku heterodimer binds to the 48-nucleotide TLC1 stem loop (Stellwagen et al. 2003; Fisher et al. 2004) and through its interaction with telomerase, Ku contributes to telomerase recruitment to telomeres (Fisher et al. 2004) and broken chromosome ends (Stellwagen et al. 2003). However, in the absence of Ku, telomeres are stable in length, albeit very short (Boulton and Jackson 1998), demonstrating that the Ku-mediated recruitment of telomerase is not essential for telomerase activity in vivo as seen for Cdc13p, although it may contribute to telomere extension by increasing the local concentration of telomerase near its chromosome end substrate.

TLC1 contains near its 3′ end a canonical binding site for Sm proteins (Seto et al. 1999), which is a hallmark of small RNPs involved in splicing of mRNA precursors. Immunoprecipitation experiments indicate that most or all yeast telomerase is associated with Sm proteins, which form a heptameric doughnut-shaped structure around the Sm-binding site. The Sm proteins may be required for removal of the poly(A) tail from the TLC1 precursor (Chapon et al. 1997), and they may promote hyper-methylation of the 5′ cap, which is another hallmark shared between TLC1 and small nuclear RNAs (snRNAs) (Seto et al. 1999; see Chapter 3).

Human Telomerase

Human proteins—hEST1A, hEST1B, and hEST1C (also known as SMG6, SMG5, and SMG7)—that contain weak sequence similarity with yeast Est1p were identified by database mining (Fig. 4) (Reichenbach et al. 2003; Snow et al. 2003). The homology domain comprises one and a half tetra-tricopeptide repeats, which are protein–protein interaction modules, and a domain named EST1. The structure of this conserved region in EST1C has been determined by X-ray crystallography, revealing a 14-3-3-like domain that binds phosphoserine-containing polypeptides (Fukuhara et al. 2005). The role of this domain for telomere function is unknown.

Immunoprecipitation experiments in HeLa cell extracts indicate that hEST1A is associated with a large fraction of active telomerase (Reichenbach et al. 2003). Human EST1B also coimmunoprecipitates telomerase activity, and both hEST1A and hEST1B bind hTERT in rabbit reticulocyte lysates independently of hTR (Snow et al. 2003). No association of telomerase with hEST1C has been reported. Overexpression of hEST1A leads to rapid telomere uncapping in HT1080 cells, with cells accumulating chromosomal end-to-end fusions (Reichenbach et al. 2003). Longer periods of hEST1A overexpression in 293T cells affect telomere length (Snow et al. 2003). In addition to an apparent involvement of hEST1A and hEST1B in telomere maintenance, hEST1A, hEST1B, and hEST1C also function in RNA surveillance (Chiu et al. 2003; Gatfield et al. 2003). The RNA surveillance

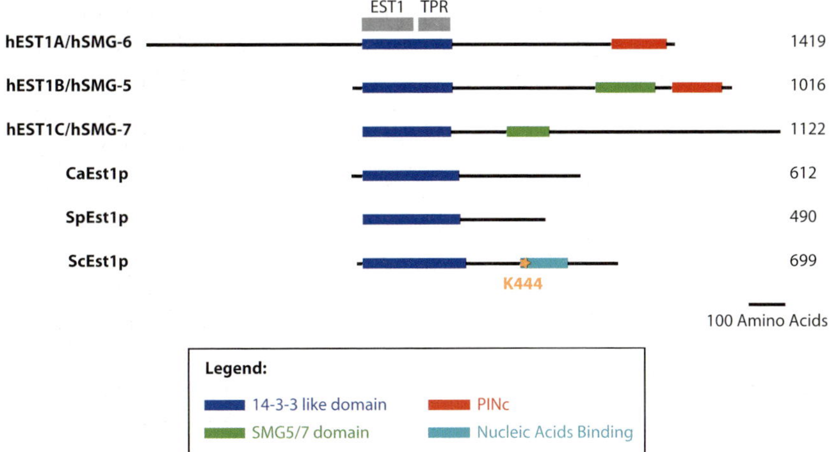

Figure 4. Domain architecture of the EST1 protein family. The conserved EST1 and TPR motifs from EST1C form a 14-3-3-like domain that binds phosphoserine-containing polypeptides (Fukuhara et al. 2005). The PINc domain contains conserved acidic residues and is found in several nucleic-acid-binding proteins (Clissold and Ponting 2000). Lys-444 of *S. cerevisiae* Est1p forms an ionic interaction with Glu-252 of Cdc13p, which is critical for telomerase recruitment (Pennock et al. 2001). The region in *S. cerevisiae* Est1p required for nucleic acid binding is indicated (Virta-Pearlman et al. 1996).

machinery mediates degradation of mRNAs carrying premature stop codons, but it also regulates expression of many physiologic transcripts (Mendell et al. 2004). Thus, perturbations in RNA surveillance could influence telomere metabolism indirectly, regulating the expression levels of telomere factors. Thus, although the physical interaction between telomerase and hEST1A and hEST1B suggests a direct function, unequivocal distinction between direct and indirect effects will require analysis of separation-of-function mutant hEST1 alleles. One could also speculate that a functional link exists between telomere maintenance and RNA surveillance pathways and that hEST1 proteins contribute to telomere maintenance through their associations with both telomerase and the RNA surveillance machinery, which might alter expression of telomere replication factors in response to telomere shortening or damage. Indeed, mutations in genes involved in the nonsense-mediated RNA decay machinery cause altered expression of a subset of telomere factors in *S. cerevisiae* (Lew et al. 1998; Dahlseid et al. 2003).

hTR contains at its 3′ end box H/ACA small nucleolar RNA (snoRNA)-like motifs (Mitchell et al. 1999a). Three nucleotides upstream of the 3′ end resides an ACA trinucleotide and upstream of the terminal stem-loop structure an H box consensus sequence 5′-AGAGGA-3′ (Fig. 2). This region of the RNA is important for hTR stability and 3′-end processing (Mitchell et al. 1999a; Fu and Collins 2003; for a detailed discussion, see Chapter 3). As for other H/ACA-type snoRNPs, coimmunoprecipitation and bandshift experiments have also shown that human telomerase associates with dyskerin, GAR1, NOP10, and NHP2 (Mitchell et al. 1999a; Dragon et al. 2000; Pogacic et al. 2000).

The molecular chaperones p23 and Hsp90 associate with hTERT, promoting the in vitro assembly of hTR and hTERT in rabbit reticulocyte lysates. In addition, both chaperones are associated with mature telomerase in cell extracts, and inhibition of Hsp90 function in vivo with geldanamycin blocks assembly of active telomerase (Holt et al. 1999; Forsythe et al. 2001). As summarized in Table 1, a large number of additional proteins have been identified to interact with human telomerase in extracts or through two- and three-hybrid screens. The exact function for most of them for biogenesis or in the mature RNP remains to be elucidated.

Figure 3 shows a monomeric human TERT-TR complex, assembled with H/ACA proteins, but in the absence of hEST1 and other telomerase-associated proteins for which the molecular interactions are less well established.

MULTIMERIZATION

Studies on *S. cerevisiae* and human telomerases indicate that they can form dimers or higher-order multimers (Prescott and Blackburn 1997; Beattie et al. 2001; Wenz et al. 2001; Arai et al. 2002). In *S. cerevisiae*, an enzymatically inactive template mutant becomes active when assembled with wild-type TLC1 in the same RNP complex (Prescott and Blackburn 1997). Recombinant human telomerase heterodimers that contain one wild-type and one mutant TR template are virtually inactive when compared to the wild-type homodimer (Wenz et al. 2001). These studies indicate an important functional role of telomerase dimerization, with TR templates cooperating during telomere extension. However, the telomerases from *E. aediculatus* and *T. thermophila* are monomeric in solution, whereas the telomerase from *E. crassus* was proposed to multimerize (Wang et al. 2002). It is odd that a fundamentally important property of the enzyme should not be conserved in evolution. However, if the dimerization served to extend the telomeres of sister chromatids in parallel, this function might have been lost in ciliates, because they contain highly polyploid minichromosomes in their macronuclei and distinct telomere length regulation pathways and chromosome segregation mechanisms (see Chapter 15).

The hTERT polypeptide has been implicated in multimerization. Full-length hTERT can coimmunoprecipitate labeled full-length hTERT and hTERT fragments. Studies with hTERT fragments suggest interactions between the amino and carboxyl termini of hTERT (Beattie et al. 2001; Arai et al. 2002). In addition, it has been proposed that human telomerase dimerization is mediated by RNA–RNA interactions involving pseudoknot formation in *trans* (Ly et al. 2003b). hTR can dimerize in vitro, forming a *trans*-pseudoknot involving P3 (Fig. 2). Two different P3 mutants that disrupt intramolecular pairing but allowed base-pairing in *trans* were inactive on their own but active when combined, forming heterodimeric telomerases with a *trans*-pseudoknot. Whether in wild-type telomerase the pseudoknot also forms in *trans* is unknown. It has been observed in vitro that hTR dimerization can also occur through the internal loop J7b/8a, which contains a self-complementary sequence (Ren et al. 2003). However, self-complementarity of loop J7b/8a is not conserved in mice and other vertebrate TR sequences, suggesting that this interaction may not be physiologically relevant.

PROSPECTIVE

During the past several years, much progress has been made through the identification of numerous telomerase proteins and RNA moieties from different phylogenetic groups. The TR secondary structure models, developed mostly through the phylogenetic comparison of RNA sequences, have begun to provide a framework to test the function of TR structure and sequence, for protein binding and during the reaction cycle. The secondary structure models provide a static view of the RNA, and it will be interesting to determine whether the pseudoknot and other parts of the RNA change their conformation during the reaction cycle or in response to protein assembly during biogenesis. Another big challenge will be the elucidation of the three-dimensional structure of the telomerase core and the holoenzyme. Two obstacles must be overcome. First, recombinant systems must be identified to express large amounts of full-length TERT to make structural analysis feasible. Second, the in vitro reconstitution efficiency of enzymatically active hTERT-hTR complexes must be improved. A better understanding of the telomerase biogenesis pathway may help to reproduce and improve in vitro assembly. The structural elucidation of TERT and TR fragments has become a feasible approach and promises to provide additional important insight soon.

A large number of functional studies will be required to understand the role of telomerase multimerization and the role of telomerase-associated proteins. It is unclear which proteins are part of the holoenzyme. The generation of cellular extracts may disrupt unstable but physiologically relevant interactions within telomerase, whereas new unphysiological interactions may form upon disruption of cellular compartments. Mutational analysis of telomerase-associated proteins in the different experimental systems combined with cell biological methods and determination of telomerase association and activity should help to elucidate their function during biogenesis and regulation. In addition, the comparison of different experimental systems should provide further insight into the evolution of telomere maintenance systems in distant organisms.

ACKNOWLEDGMENTS

G.C. was supported by an EMBO postdoctoral fellowship. The research conducted in the authors' laboratory was supported by grants from the Swiss National Science Foundation, the Swiss Cancer League, and the Sixth Framework Programme of the European Commission. The authors declare that they have no competing financial interests.

REFERENCES

Aigner S. and Cech T.R. 2004. The *Euplotes* telomerase subunit p43 stimulates enzymatic activity and processivity *in vitro*. *RNA* **10**: 1108–1118.

Aigner S., Postberg J., Lipps H.J., and Cech T.R. 2003. The *Euplotes* La motif protein p43 has properties of a telomerase-specific subunit. *Biochemistry* **42**: 5736–5747.

Aigner S., Lingner J., Goodrich K.J., Grosshans C.A., Shevchenko A., Mann M., and Cech T.R. 2000. *Euplotes* telomerase contains an La motif protein produced by apparent translational frameshifting. *EMBO J.* **19**: 6230–6239.

Antal M., Boros E., Solymosy F., and Kiss T. 2002. Analysis of the structure of human telomerase RNA in vivo. *Nucleic Acids Res.* **30**: 912–920.

Arai K., Masutomi K., Khurts S., Kaneko S., Kobayashi K., and Murakami S. 2002. Two independent regions of human telomerase reverse transcriptase are important for its oligomerization and telomerase activity. *J. Biol. Chem.* **277**: 8538–8544.

Armbruster B.N., Banik S.S., Guo C., Smith A.C., and Counter C.M. 2001. N-terminal domains of the human telomerase catalytic subunit required for enzyme activity in vivo. *Mol. Cell. Biol.* **21**: 7775–7786.

Autexier C. and Greider C.W. 1995. Boundary elements of the *Tetrahymena* telomerase RNA template and alignment domains. *Genes Dev.* **9**: 2227–2239.

———. 1998. Mutational analysis of the *Tetrahymena* telomerase RNA: Identification of residues affecting telomerase activity *in vitro*. *Nucleic Acids Res.* **26**: 787–795.

Bachand F. and Autexier C. 2001. Functional regions of human telomerase reverse transcriptase and human telomerase RNA required for telomerase activity and RNA-protein interactions. *Mol. Cell. Biol.* **21**: 1888–1897.

Bachand F., Boisvert F.M., Cote J., Richard S., and Autexier C. 2002. The product of the survival of motor neuron (SMN) gene is a human telomerase-associated protein. *Mol. Biol. Cell* **13:** 3192–3202.

Banik S.S. and Counter C.M. 2004. Characterization of interactions between PinX1 and human telomerase subunits hTERT and hTR. *J. Biol. Chem.* **279:** 51745–51748.

Beattie T.L., Zhou W., Robinson M.O., and Harrington L. 2001. Functional multimerization of the human telomerase reverse transcriptase. *Mol. Cell. Biol.* **21:** 6151–6160.

Beernink H.T., Miller K., Deshpande A., Bucher P., and Cooper J.P. 2003. Telomere maintenance in fission yeast requires an Est1 ortholog. *Curr. Biol.* **13:** 575–580.

Bhattacharyya A. and Blackburn E.H. 1994. Architecture of telomerase RNA. *EMBO J.* **13:** 5721–5723.

Bianchi A., Negrini S., and Shore D. 2004. Delivery of yeast telomerase to a DNA break depends on the recruitment functions of Cdc13 and Est1. *Mol. Cell* **16:** 139–146.

Biessmann H., Champion L.E., O'Hair M., Ikenaga K., Kasravi B., and Mason J.M. 1992a. Frequent transpositions of *Drosophila melanogaster* HeT-A transposable elements to receding chromosome ends. *EMBO J.* **11:** 4459–4469.

Biessmann H., Valgeirsdottir K., Lofsky A., Chin C., Ginther B., Levis R.W., and Pardue M.L. 1992b. HeT-A, a transposable element specifically involved in "healing" broken chromosome ends in *Drosophila melanogaster. Mol. Cell. Biol.* **12:** 3910–3918.

Bosoy D., Peng Y., Mian I.S., and Lue N.F. 2003. Conserved N-terminal motifs of telomerase reverse transcriptase required for ribonucleoprotein assembly in vivo. *J. Biol. Chem.* **278:** 3882–3890.

Boulton S.J. and Jackson S.P. 1998. Components of the Ku-dependent non-homologous end-joining pathway are involved in telomeric length maintenance and telomeric silencing. *EMBO J.* **17:** 1819–1828.

Bryan T.M., Goodrich K.J., and Cech T.R. 2000. Telomerase RNA bound by protein motifs specific to telomerase reverse transcriptase. *Mol. Cell* **6:** 493–499.

Bryan T.M., Sperger J.M., Chapman K.B., and Cech T.R. 1998. Telomerase reverse transcriptase genes identified in *Tetrahymena thermophila* and *Oxythricha trifallax. Proc. Natl. Acad. Sci.* **95:** 8479–8484.

Chai W., Ford L.P., Lenertz L., Wright W.E., and Shay J.W. 2002. Human Ku70/80 associates physically with telomerase through interaction with hTERT. *J. Biol. Chem.* **277:** 47242–47247.

Chapon C., Cech T.R., and Zaug A.J. 1997. Polyadenylation of telomerase RNA in budding yeast. *RNA* **3:** 1337–1351.

Chappell A.S. and Lundblad V. 2004. Structural elements required for association of the *Saccharomyces cerevisiae* telomerase RNA with the Est2 reverse transcriptase. *Mol. Cell. Biol.* **24:** 7720–7736.

Chen J.L. and Greider C.W. 2003. Template boundary definition in mammalian telomerase. *Genes Dev.* **17:** 2747–2752.

———. 2004. An emerging consensus for telomerase RNA structure. *Proc. Natl. Acad. Sci.* **101:** 14683–14684.

Chen J.L., Blasco M.A., and Greider C.W. 2000. Secondary structure of vertebrate telomerase RNA. *Cell* **100:** 503–514.

Chen J.L., Opperman K.K., and Greider C.W. 2002. A critical stem-loop structure in the CR4-CR5 domain of mammalian telomerase RNA. *Nucleic Acids Res.* **30:** 592–597.

Chiu S.Y., Serin G., Ohara O., and Maquat L.E. 2003. Characterization of human Smg5/7a: A protein with similarities to *Caenorhabditis elegans* SMG5 and SMG7 that functions in the dephosphorylation of Upf1. *RNA* **9:** 77–87.

Clissold P.M. and Ponting C.P. 2000. PIN domains in nonsense-mediated mRNA decay and RNAi. *Curr. Biol.* **10:** R888–890.

Collins K. and Gandhi L. 1998. The reverse transcriptase component of the *Tetrahymena* telomerase ribonucleoprotein complex. *Proc. Natl. Acad. Sci.* **95:** 8485–8490.

Collins K., Kobayashi R., and Greider C.W. 1995. Purification of *Tetrahymena* telomerase and cloning of genes encoding the two protein components of the enzyme. *Cell* **81:** 677–686.

Comolli L.R., Smirnov I., Xu L., Blackburn E.H., and James T.L. 2002. A molecular switch underlies a human telomerase disease. *Proc. Natl. Acad. Sci.* **99:** 16998–17003.

Counter C.M., Meyerson M., Eaton E.N., and Weinberg R.A. 1997. The catalytic subunit of yeast telomerase. *Proc. Natl. Acad. Sci.* **94:** 9202–9207.

Cristofari G. and Darlix J.L. 2002. The ubiquitous nature of RNA chaperone proteins. *Prog. Nucleic Acid Res. Mol. Biol.* **72:** 223–268.

Dahlseid J.N., Lew-Smith J., Lelivelt M.J., Enomoto S., Ford A., Desruisseaux M., McClellan M., Lue N., Culbertson M.R., and Berman J. 2003. mRNAs encoding telomerase components and regulators are controlled by UPF genes in *Saccharomyces cerevisiae*. *Eukaryot. Cell* **2:** 134–142.

Dandjinou A.T., Levesque N., Larose S., Lucier J.F., Elela S.A., and Wellinger R.J. 2004. A phylogenetically based secondary structure for the yeast telomerase RNA. *Curr. Biol.* **14:** 1148–1158.

Das D. and Georgiadis M.M. 2004. The crystal structure of the monomeric reverse transcriptase from Moloney murine leukemia virus. *Structure* **12:** 819–829.

Dragon F., Pogacic V., and Filipowicz W. 2000. In vitro assembly of human H/ACA small nucleolar RNPs reveals unique features of U17 and telomerase RNAs. *Mol. Cell. Biol.* **20:** 3037–3048.

Eickbush T.H. 1997. Telomerase and retrotransposons: Which came first? *Science* **277:** 911–912.

Evans S.K. and Lundblad V. 1999. Est1 and Cdc13 as comediators of telomerase access. *Science* **286:** 117–120.

———. 2002. The Est1 subunit of *Saccharomyces cerevisiae* telomerase makes multiple contributions to telomere length maintenance. *Genetics* **162:** 1101–1115.

Eversole A. and Maizels N. 2000. In vitro properties of the conserved mammalian protein hnRNP D suggest a role in telomere maintenance. *Mol. Cell. Biol.* **20:** 5425–5432.

Fisher T.S., Taggart A.K., and Zakian V.A. 2004. Cell cycle-dependent regulation of yeast telomerase by Ku. *Nat. Struct. Mol. Biol.* **11:** 1198–1205.

Ford L.P., Shay J.W., and Wright W.E. 2001. The La antigen associates with the human telomerase ribonucleoprotein and influences telomere length in vivo. *RNA* **7:** 1068–1075.

Ford L.P., Suh J.M., Wright W.E., and Shay J.W. 2000. Heterogeneous nuclear ribonucleoproteins C1 and C2 associate with the RNA component of human telomerase. *Mol. Cell. Biol.* **20:** 9084–9091.

Forstemann K. and Lingner J. 2005. Telomerase limits the extent of base pairing between template RNA and telomeric DNA. *EMBO Rep.* **6:** 361–366.

Forsythe H.L., Jarvis J.L., Turner J.W., Elmore L.W., and Holt S.E. 2001. Stable association of hsp90 and p23, but Not hsp70, with active human telomerase. *J. Biol. Chem.* **276:** 15571–15574.

Friedman K.L. and Cech T.R. 1999. Essential functions of amino-terminal domains in the yeast telomerase catalytic subunit revealed by selection for viable mutants. *Genes Dev.* **13:** 2863–2874.

Friedman K.L., Heit J.J., Long D.M., and Cech T.R. 2003. N-terminal domain of yeast telomerase reverse transcriptase: Recruitment of Est3p to the telomerase complex. *Mol. Biol. Cell* **14:** 1–13.

Fu D. and Collins K. 2003. Distinct biogenesis pathways for human telomerase RNA and H/ACA small nucleolar RNAs. *Mol. Cell* **11:** 1361–1372.

Fukuhara N., Ebert J., Unterholzner L., Lindner D., Izaurralde E., and Conti E. 2005. SMG7 is a 14-3-3-like adaptor in the nonsense-mediated mRNA decay pathway. *Mol. Cell* **17:** 537–547.

Gatfield D., Unterholzner L., Ciccarelli F.D., Bork P., and Izaurralde E. 2003. Nonsense-mediated mRNA decay in *Drosophila:* At the intersection of the yeast and mammalian pathways. *EMBO J.* **22:** 3960–3970.

Gilley D. and Blackburn E.H. 1999. The telomerase RNA pseudoknot is critical for the stable assembly of a catalytically active ribonucleoprotein. *Proc. Natl. Acad. Sci.* **96:** 6621–6625.

Greene E.C. and Shippen D.E. 1998. Developmentally programmed assembly of higher order telomerase complexes with distinct biochemical and structural properties. *Genes Dev.* **12:** 2921–2931.

Greider C.W. and Blackburn E.H. 1985. Identification of a specific telomere terminal transferase activity in *Tetrahymena* extracts. *Cell* **43:** 405–413.

———. 1989. A telomeric sequence in the RNA of *Tetrahymena* telomerase required for telomere repeat synthesis. *Nature* **337:** 331–337.

Harrington L., Zhou W., McPhail T., Oulton R., Yeung D.S., Mar V., Bass M.B., and Robinson M.O. 1997. Human telomerase contains evolutionarily conserved catalytic and structural subunits. *Genes Dev.* **11:** 3109–3115.

Holt S.E., Aisner D.L., Baur J., Tesmer V.M., Dy M., Ouellette M., Trager J.B., Morin G.B., Toft D.O., Shay J.W., Wright W.E., and White M.A. 1999. Functional requirement of p23 and Hsp90 in telomerase complexes. *Genes Dev.* **13:** 817–826.

Huard S., Moriarty T.J., and Autexier C. 2003. The C terminus of the human telomerase reverse transcriptase is a determinant of enzyme processivity. *Nucleic Acids Res.* **31:** 4059–4070.

Hughes T.R., Evans S.K., Weilbaecher R.G., and Lundblad V. 2000. The Est3 protein is a subunit of yeast telomerase. *Curr. Biol.* **10:** 809–812.

Karamysheva Z., Wang L., Shrode T., Bednenko J., Hurley L.A., and Shippen D.E. 2003. Developmentally programmed gene elimination in *Euplotes crassus* facilitates a switch in the telomerase catalytic subunit. *Cell* **113:** 565–576.

Khurts S., Masutomi K., Delgermaa L., Arai K., Oishi N., Mizuno H., Hayashi N., Hahn W.C., and Murakami S. 2004. Nucleolin interacts with telomerase. *J. Biol. Chem.* **279:** 51508–51515.

Kilian A., Bowtell D.D., Abud H.E., Hime G.R., Venter D.J., Keese P.K., Duncan E.L., Reddel R.R., and Jefferson R.A. 1997. Isolation of a candidate human telomerase catalytic subunit gene, which reveals complex splicing patterns in different cell types. *Hum. Mol. Genet.* **6:** 2011–2019.

Labranche H., Dupuis S., Bendavid Y., Bani M.R., Wellinger R.J., and Chabot B. 1998. Telomere elongation by hnRNP A1 and a derivative that interacts with telomeric repeats and telomerase. *Nat. Genet.* **19:** 199–202.

Lai C.K., Miller M.C., and Collins K. 2002. Template boundary definition in *Tetrahymena* telomerase. *Genes Dev.* **16:** 415–420.

———. 2003. Roles for RNA in telomerase nucleotide and repeat addition processivity. *Mol. Cell* **11:** 1673–1683.

Lai C.K., Mitchell J.R., and Collins K. 2001. RNA binding domain of telomerase reverse transcriptase. *Mol. Cell. Biol.* **21:** 990–1000.

Le S., Sternglanz R., and Greider C.W. 2000. Identification of two RNA-binding proteins associated with human telomerase RNA. *Mol. Biol. Cell* **11:** 999–1010.

Lee G.E., Yu E.Y., Cho C.H., Lee J., Muller M.T., and Chung I.K. 2004. DNA-protein kinase catalytic subunit-interacting protein KIP binds telomerase by interacting with human telomerase reverse transcriptase. *J. Biol. Chem.* **279:** 34750–34755.

Leeper T.C. and Varani G. 2005. The structure of an enzyme-activating fragment of human telomerase RNA. *RNA* **11:** 394–403.

Leeper T., Leulliot N., and Varani G. 2003. The solution structure of an essential stem-loop of human telomerase RNA. *Nucleic Acids Res.* **31:** 2614–2621.

Lendvay T.S., Morris D.K., Sah J., Balasubramanian B., and Lundblad V. 1996. Senescence mutants of *Saccharomyces cerevisiae* with a defect in telomere replication identify three additional EST genes. *Genetics* **144:** 1399–1412.

Levis R.W., Ganesan R., Houtchens K., Tolar L.A., and Sheen F.M. 1993. Transposons in place of telomeric repeats at a *Drosophila* telomere. *Cell* **75:** 1083–1093.

Lew J.E., Enomoto S., and Berman J. 1998. Telomere length regulation and telomeric chromatin require the nonsense-mediated mRNA decay pathway. *Mol. Cell. Biol.* **18:** 6121–6130.

Licht J.D. and Collins K. 1999. Telomerase RNA function in recombinant *Tetrahymena* telomerase. *Genes Dev.* **13:** 1116–1125.

Lin J. and Blackburn E.H. 2004. Nucleolar protein PinX1p regulates telomerase by sequestering its protein catalytic subunit in an inactive complex lacking telomerase RNA. *Genes Dev.* **18:** 387–396.

Lin J., Ly H., Hussain A., Abraham M., Pearl S., Tzfati Y., Parslow T.G., and Blackburn E.H. 2004. A universal telomerase RNA core structure includes structured motifs required for binding the telomerase reverse transcriptase protein. *Proc. Natl. Acad. Sci.* **101:** 14713–14718.

Lingner J. and Cech T.R. 1996. Purification of telomerase from *Euplotes aediculatus*: Requirement of a primer 3′ overhang. *Proc. Natl. Acad. Sci.* **93:** 10712–10717.

Lingner J., Hendrick L.L., and Cech T.R. 1994. Telomerase RNAs of different ciliates have a common secondary structure and a permuted template. *Genes Dev.* **8:** 1984–1998.

Lingner J., Hughes T.R., Shevchenko A., Mann M., Lundblad V., and Cech T.R. 1997. Reverse transcriptase motifs in the catalytic subunit of telomerase. *Science* **276:** 561–567.

Liu Y., Snow B.E., Hande M.P., Baerlocher G., Kickhoefer V.A., Yeung D., Wakeham A., Itie A., Siderovski D.P., Lansdorp P.M., Robinson M.O., and Harrington L. 2000. Telomerase-associated protein TEP1 is not essential for telomerase activity or telomere length maintenance in vivo. *Mol. Cell. Biol.* **20:** 8178–8184.

Lue N.F. and Peng Y. 1997. Identification and characterization of a telomerase activity from *Schizosaccharomyces pombe*. *Nucleic Acids Res.* **25:** 4331–4337.

Lundblad V. and Szostak J.W. 1989. A mutant with a defect in telomere elongation leads to senescence in yeast. *Cell* **57:** 633–643.

Ly H., Blackburn E.H., and Parslow T.G. 2003a. Comprehensive structure-function analysis of the core domain of human telomerase RNA. *Mol. Cell. Biol.* **23:** 6849–6856.

Ly H., Xu L., Rivera M.A., Parslow T.G., and Blackburn E.H. 2003b. A role for a novel "trans-pseudoknot" RNA-RNA interaction in the functional dimerization of human telomerase. *Genes Dev.* **17:** 1078–1083.

Martin-Rivera L. and Blasco M.A. 2001. Identification of functional domains and dominant negative mutations in vertebrate telomerase RNA using an in vivo reconstitution system. *J. Biol. Chem.* **276:** 5856–5865.

Mason D.X., Autexier C., and Greider C.W. 2001. *Tetrahymena* proteins p80 and p95 are not core telomerase components. *Proc. Natl. Acad. Sci.* **98:** 12368–12373.

Mendell J.T., Sharifi N.A., Meyers J.L., Martinez-Murillo F., and Dietz H.C. 2004. Nonsense surveillance regulates expression of diverse classes of mammalian transcripts and mutes genomic noise. *Nat. Genet.* **36:** 1073–1078.

Meyerson M., Counter C.M., Eaton E.N., Ellisen L.W., Steiner P., Dickinson Caddle S., Ziaugra L., Beijershergen R.L., Davidoff M.J., Liu Q., Bacchetti S., Haber D.A., and Weinberg R.A. 1997. *hEST2*, the putative human telomerase catalytic subunit gene, is up-regulated in tumor cells and during immortalization. *Cell* **90:** 785–795.

Miller M.C., Liu J.K., and Collins K. 2000. Template definition by *Tetrahymena* telomerase reverse transcriptase. *EMBO J.* **19:** 4412–4422.

Mitchell J.R. and Collins K. 2000. Human telomerase activation requires two independent interactions between telomerase RNA and telomerase reverse transcriptase. *Mol. Cell* **6:** 361–371.

Mitchell J.R., Cheng J., and Collins K. 1999a. A box H/ACA small nucleolar RNA-like domain at the human telomerase RNA 3′ end. *Mol. Cell. Biol.* **19:** 567–576.

Mitchell J.R., Wood E., and Collins K. 1999b. A telomerase component is defective in the human disease dyskeratosis congenita. *Nature* **402:** 551–555.

Moriarty T.J., Marie-Egyptienne D.T., and Autexier C. 2004. Functional organization of repeat addition processivity and DNA synthesis determinants in the human telomerase multimer. *Mol. Cell. Biol.* **24:** 3720–3733.

Moriarty T.J., Huard S., Dupuis S., and Autexier C. 2002. Functional multimerization of human telomerase requires an RNA interaction domain in the N terminus of the catalytic subunit. *Mol. Cell. Biol.* **22:** 1253–1265.

Morris D.K. and Lundblad V. 1997. Programmed translational frameshifting in a gene required for yeast telomere replication. *Curr. Biol.* **7:** 969–976.

Nakamura T.M. and Cech T.R. 1998. Reversing time—Origin of telomerase. *Cell* **92:** 587–590.

Nakamura T.M., Morin G.B., Chapman K.B., Weinrich S.L., Andrews W.H., Lingner J., Harley C.B., and Cech T.R. 1997. Telomerase catalytic subunit homologs from fission yeast and human. *Science* **277:** 955–959.

Nakayama J.I., Tahara H., Tahara E., Saito M., Ito K., Nakamura H., Nakanishi T., Tahara E., Ide T., and Ishikawa F. 1998. Telomerase activation by hTRT in human normal fibroblasts and hepatocellular carcinomas. *Nat. Genet.* **18:** 65–68.

Nugent C.I., Hughes T.R., Lue N.F., and Lundblad V. 1996. Cdc13p: A single-strand telomeric DNA-binding protein with a dual role in yeast telomere maintenance. *Science* **274:** 249–252.

O'Connor C.M., Lai C.K., and Collins K. 2005. Two purified domains of telomerase reverse transcriptase reconstitute sequence-specific interactions with RNA. *J. Biol. Chem.* **280:** 17533–17539.

O'Reilly M., Teichmann S.A., and Rhodes D. 1999. Telomerases. *Curr. Opin. Struct. Biol.* **9:** 56–65.

Peng Y., Mian I.S., and Lue N.F. 2001. Analysis of telomerase processivity: Mechanistic similarity to HIV-1 reverse transcriptase and role in telomere maintenance. *Mol. Cell* **7:** 1201–1211.

Pennock E., Buckley K., and Lundblad V. 2001. Cdc13 delivers separate complexes to the telomere for end protection and replication. *Cell* **104:** 387–396.

Peterson S.E., Stellwagen A.E., Diede S.J., Singer M.S., Haimberger Z.W., Johnson C.O., Tzoneva M., and Gottschling D.E. 2001. The function of a stem-loop in telomerase RNA is linked to the DNA repair protein Ku. *Nat. Genet.* **27:** 64–67.

Pogacic V., Dragon F., and Filipowicz W. 2000. Human H/ACA small nucleolar RNPs and telomerase share evolutionarily conserved proteins NHP2 and NOP10. *Mol. Cell. Biol.* **20:** 9028–9040.

Prathapam R., Witkin K.L., O'Connor C.M., and Collins K. 2005. A telomerase holoenzyme protein enhances telomerase RNA assembly with telomerase reverse transcriptase. *Nat. Struct. Mol. Biol.* **12:** 252–257.

Prescott J. and Blackburn E.H. 1997. Functionally interacting telomerase RNAs in the yeast telomerase complex. *Genes Dev.* **11:** 2790–2800.

Qi H. and Zakian V.A. 2000. The *Saccharomyces* telomere-binding protein Cdc13p interacts with both the catalytic subunit of DNA polymerase alpha and the telomerase-associated Est1 protein. *Genes Dev.* **14:** 1777–1788.

Reichenbach P., Hoss M., Azzalin C.M., Nabholz M., Bucher P., and Lingner J. 2003. A human homolog of yeast Est1 associates with telomerase and uncaps chromosome ends when overexpressed. *Curr. Biol.* **13:** 568–574.

Ren X., Gavory G., Li H., Ying L., Klenerman D., and Balasubramanian S. 2003. Identification of a new RNA.RNA interaction site for human telomerase RNA (hTR): Structural implications for hTR accumulation and a dyskeratosis congenita point mutation. *Nucleic Acids Res.* **31:** 6509–6515.

Romero D.P. and Blackburn E.H. 1991. A conserved secondary structure for telomerase RNA. *Cell* **67:** 343–353.

Schnapp G., Rodi H.P., Rettig W.J., Schnapp A., and Damm K. 1998. One-step affinity purification protocol for human telomerase. *Nucleic Acids Res.* **26:** 3311–3313.

Seimiya H., Sawada H., Muramatsu Y., Shimizu M., Ohko K., Yamane K., and Tsuruo T. 2000. Involvement of 14-3-3 proteins in nuclear localization of telomerase. *EMBO J.* **19:** 2652–2661.

Seto A.G., Livengood A.J., Tzfati Y., Blackburn E.H., and Cech T.R. 2002. A bulged stem tethers Est1p to telomerase RNA in budding yeast. *Genes Dev.* **16:** 2800–2812.

Seto A.G., Zaug A.J., Sobel S.G., Wolin S.L., and Cech T.R. 1999. *Saccharomyces cerevisiae* telomerase is an Sm small nuclear ribonucleoprotein particle. *Nature* **401:** 177–180.

Seto A.G., Umansky K., Tzfati Y., Zaug A.J., Blackburn E.H., and Cech T.R. 2003. A template-proximal RNA paired element contributes to *Saccharomyces cerevisiae* telomerase activity. *RNA* **9:** 1323–1332.

Singh S.M., Steinberg-Neifach O., Mian I.S., and Lue N.F. 2002. Analysis of telomerase in *Candida albicans*: Potential role in telomere end protection. *Eukaryot. Cell* **1:** 967–977.

Snow B.E., Erdmann N., Cruickshank J., Goldman H., Gill R.M., Robinson M.O., and Harrington L. 2003. Functional conservation of the telomerase protein Est1p in humans. *Curr. Biol.* **13:** 698–704.

Sperger J.M. and Cech T.R. 2001. A stem-loop of *Tetrahymena* telomerase RNA distant from the template potentiates RNA folding and telomerase activity. *Biochemistry* **40:** 7005–7016.

Steiner B.R., Hidaka K., and Futcher B. 1996. Association of the Est1 protein with telomerase activity in yeast. *Proc. Natl. Acad. Sci.* **93:** 2817–2821.

Stellwagen A.E., Haimberger Z.W., Veatch J.R., and Gottschling D.E. 2003. Ku interacts with telomerase RNA to promote telomere addition at native and broken chromosome ends. *Genes Dev.* **17:** 2384–2395.

ten Dam E., van Belkum A., and Pleij K. 1991. A conserved pseudoknot in telomerase RNA. *Nucleic Acids Res.* **19:** 6951.

Tesmer V.M., Ford L.P., Holt S.E., Frank B.C., Yi X., Aisner D.L., Ouellette M., Shay J.W., and Wright W.E. 1999. Two inactive fragments of the integral RNA cooperate to assemble active telomerase with the human protein catalytic subunit (hTERT) in vitro. *Mol. Cell. Biol.* **19**: 6207–6216.

Theimer C.A., Blois C.A., and Feigon J. 2005. Structure of the human telomerase RNA pseudoknot reveals conserved tertiary interactions essential for function. *Mol. Cell* **17**: 671–682.

Theimer C.A., Finger L.D., Trantirek L., and Feigon J. 2003. Mutations linked to dyskeratosis congenita cause changes in the structural equilibrium in telomerase RNA. *Proc. Natl. Acad. Sci.* **100**: 449–454.

Ting N.S., Yu Y., Pohorelic B., Lees-Miller S.P., and Beattie T.L. 2005. Human Ku70/80 interacts directly with hTR, the RNA component of human telomerase. *Nucleic Acids Res.* **33**: 2090–2098.

Tzfati Y., Fulton T.B., Roy J., and Blackburn E.H. 2000. Template boundary in a yeast telomerase specified by RNA structure. *Science* **288**: 863–867.

Tzfati Y., Knight Z., Roy J., and Blackburn E.H. 2003. A novel pseudoknot element is essential for the action of a yeast telomerase. *Genes Dev.* **17**: 1779–1788.

Virta-Pearlman V., Morris D.K., and Lundblad V. 1996. Est1 has the properties of a single-stranded telomere end-binding protein. *Genes Dev.* **10**: 3094–3104.

Wang C. and Meier U.T. 2004. Architecture and assembly of mammalian H/ACA small nucleolar and telomerase ribonucleoproteins. *EMBO J.* **23**: 1857–1867.

Wang H. and Blackburn E.H. 1997. De novo telomere addition by *Tetrahymena* telomerase in vitro. *EMBO J.* **16**: 866–879.

Wang L., Dean S.R., and Shippen D.E. 2002. Oligomerization of the telomerase reverse transcriptase from *Euplotes crassus. Nucleic Acids Res.* **30**: 4032–4039.

Weinrich S.L., Pruzan R., Ma L., Ouellette M., Tesmer V.M., Holt S.E., Bodnar A.G., Lichtsteiner S., Kim N.W., Trager J.B., Taylor R.D., Carlos R., Andrews W.H., Wright W.E., Shay J.W., Harley C.B., and Morin G.B. 1997. Reconstitution of human telomerase with the template RNA component hTR and the catalytic protein subunit hTRT. *Nat. Genet.* **17**: 498–502.

Wenz C., Enenkel B., Amacker M., Kelleher C., Damm K., and Lingner J. 2001. Human telomerase contains two cooperating telomerase RNA molecules. *EMBO J.* **20**: 3526–3534.

Witkin K.L. and Collins K. 2004. Holoenzyme proteins required for the physiological assembly and activity of telomerase. *Genes Dev.* **18**: 1107–1118.

Xia J., Peng Y., Mian I.S., and Lue N.F. 2000. Identification of functionally important domains in the N-terminal region of telomerase reverse transcriptase. *Mol. Cell. Biol.* **20**: 5196–5207.

Yu G.L., Bradley J.D., Attardi L.D., and Blackburn E.H. 1990. In vivo alteration of telomere sequences and senescence caused by mutated *Tetrahymena* telomerase RNAs. *Nature* **344**: 126–132.

Zappulla D.C. and Cech T.R. 2004. Yeast telomerase RNA: A flexible scaffold for protein subunits. *Proc. Natl. Acad. Sci.* **101**: 10024–10029.

Zaug A.J. and Cech T.R. 1995. Analysis of the structure of *Tetrahymena* nuclear RNAs in vivo: Telomerase RNA, the self-splicing rRNA intron, and U2 snRNA. *RNA* **1**: 363–374.

Zhou X.Z. and Lu K.P. 2001. The Pin2/TRF1-interacting protein PinX1 is a potent telomerase inhibitor. *Cell* **107**: 347–359.

3

Telomerase Biochemistry and Biogenesis

Jiunn-Liang Chen

Department of Chemistry & Biochemistry, and
School of Life Sciences
Arizona State University
Tempe, Arizona 85287-1604

Carol W. Greider

Department of Molecular Biology and Genetics
Johns Hopkins University School of Medicine
Baltimore, Maryland 21205

THE TELOMERASE ENZYME

Telomerase is a specialized DNA polymerase that adds telomere repeats onto chromosome ends. The maintenance of telomeres by telomerase is conserved in most eukaryotes, and the enzymatic reaction catalyzed by telomerase is conserved. Thus, the essential function of telomerase in maintenance of telomere length likely arose early in evolution and had a critical role in allowing the establishment of linear chromosomes in eukaryotes.

The telomerase enzyme is a ribonucleoprotein (RNP) that consists of two essential core components: a catalytic protein component, telomerase reverse transcriptase (TERT), and an essential RNA component, telomerase RNA (TR). As the name implies, the TERT protein has regions of homology with the catalytic motifs of reverse transcriptase (Lingner et al. 1997b; Nakamura et al. 1997). This makes sense since telomerase copies a small template region of the RNA component to synthesize telomere DNA repeats. In addition to the reverse transcriptase motifs, as described in detail below, additional conserved motifs in TERT have other specific roles. The RNA component contains a short region that is complementary to the telomeric repeat sequence (Greider and Blackburn 1989; Shippen-Lentz and Blackburn 1990). This template region is copied using conventional Watson–Crick base-pairing to specify the telomeric sequence. Altering the sequence of the template region alters the telomere repeats synthesized by telomerase (Shippen-Lentz and Blackburn 1990; Yu et al. 1990). Unlike the TERT component, the sequence of the TR is not conserved across species. As described below, most TRs have a core structural domain containing the template region and a pseudoknot domain. However, in different groups of species, different additional domains have become part of the TR. The sequence, number, and size of these additional domains vary significantly between groups of species. For example, the human TR (hTR) contains a BoxH/ACA short nucleolar RNA (snoRNA) domain, whereas ciliate and yeast RNAs do not have this domain (Mitchell et al. 1999; Chen et al. 2000). The structural conservation of the core RNA with the addition of different RNA domains in different lineages is again consistent with the very early evolutionary origin of telomerase.

In addition to the core telomerase enzyme components, there are a number of additional proteins in the telomerase holoenzyme complex. These additional components vary from species to species, and some are specialized to bind the species-specific RNA domains in TR. These additional proteins have specific roles in the localization and stability of telomerase, whereas the core components specify telomerase polymerase activity.

THE TELOMERASE REACTION

The telomerase reaction proceeds by the sequential addition of deoxynucleotide triphosphates onto the 3′ end of a telomeric DNA sequence. In vivo, this 3′ end is the GT-rich single-stranded overhang present at the extreme terminus of the telomere. In this chromosomal setting, the access of telomerase to the 3′ end is regulated by telomere-binding proteins as described in Chapter 12. The in vitro telomerase reaction utilizes single-stranded DNA oligonucleotides that mimic the telomere GT-rich overhang (Greider and Blackburn 1985). In the standard assay, telomerase is incubated with a telomeric DNA primer, radiolabeled deoxynucleotide triphosphates, buffer, and magnesium. The products are then resolved on a polyacrylamide gel, and a ladder of bands corresponding to the telomere repeat is typically seen (Fig. 1A). For the *Tetrahymena* telomerase, an 18-mer oligonucleotide containing three repeats of the telomeric (TTGGGG)$_3$ sequence is an efficient primer. The primer is bound by telomerase, and each nucleotide in the repeat is added one base at a time

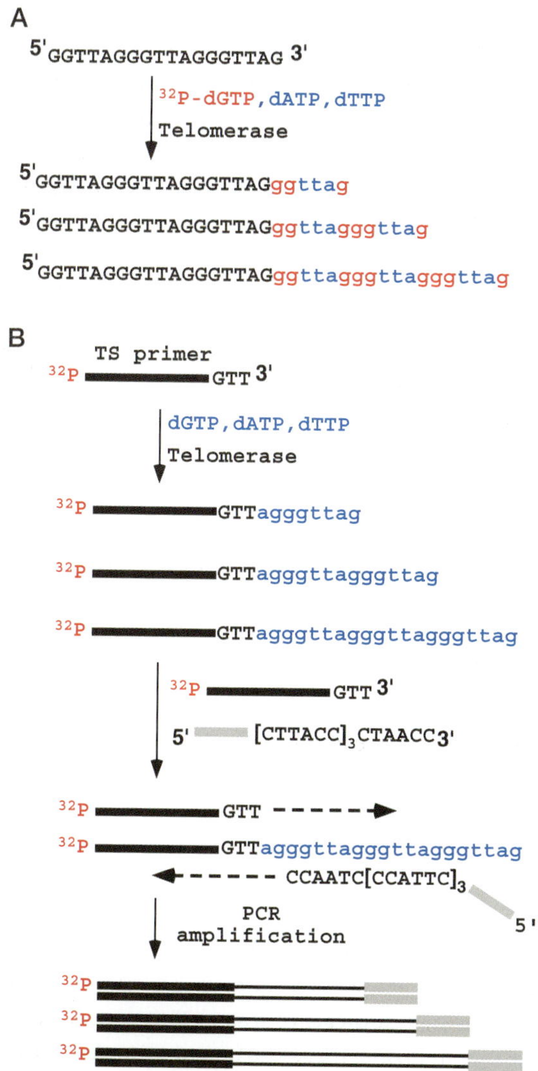

Figure 1. Telomerase activity assays. (*A*) The conventional telomerase activity assay for human telomerase. A telomeric sequence (TTAGGG) primer oligonucleotide is incubated with buffer, Mg^{2+}, dATP, dTTP, and ^{32}P-labeled dGTP and telomerase. Telomerase adds the appropriate nucleotide, one base at a time, onto the primer 3' end. The products are visualized on a high-resolution polyacrylamide gel. (*B*) The TRAP assay. The first step of a TRAP assay is like the conventional assay. A telomeric (TS) primer is incubated with buffer, Mg^{2+}, and unlabeled nucleotides. The products from the reaction are then amplified in a second step by the addition of a primer complementary to the TTAGGG sequence added, and several rounds of PCR are carried out. The complementary primer has a short sequence of nontelomeric DNA on the 5' end to avoid misalignment of the repetitive sequences (Kim et al. 1994; Kim and Wu 1997).

by copying the template region of the RNA. When the end of the template is reached, there is a translocation step in which the 3′ end of the product DNA is repositioned. This elongation generates a characteristic ladder of repeated sequences because of the pausing and dissociation of some of the enzyme-primer complexes at specific steps during elongation. For example, the *Tetrahymena* enzyme has a higher affinity for dGTP than it does for dTTP; a pattern of pauses is seen in the addition reaction at each position where a dT residue is incorporated (Greider 1991). In addition, during the translocation step, some fraction of the elongating telomere product may dissociate from the enzyme. This dissociation generates a strong band in the elongation products that corresponds to the end of the template (Greider 1991).

Reconstitution of Telomerase

Telomerase activity can be reconstituted in an in vitro assay using only TERT and TR. The TERT protein is synthesized by in vitro transcription and translation in rabbit reticulocyte lysate. The in-vitro-transcribed TR is then added, and an active telomerase complex is assembled. The assembled telomerase is then immunoprecipitated and assayed to characterize the properties of the purified core enzyme.

Many of the functional studies of the protein and RNA domains described below were carried out using this in vitro reconstitution system. For many assays, it would be advantageous to have purified protein and RNA to assay their interaction without having to synthesize and assemble them in rabbit reticulocyte lysate. However, this has been problematic as the TERT protein is not well-behaved during purification; it precipitates upon purification from many different purification protocols. In addition, some assisted assembly may be required for the RNA and protein to form an active complex. This assembly appears to be facilitated by the chaperone proteins present in the rabbit reticulocyte lysate (Holt et al. 1999). In one study, human TERT (hTERT) was assembled in vitro with T7-transcribed hTR in the presence of insect lysate, although the components required for this assembly are not yet clear (Wenz et al. 2001).

Telomerase activity can also be reconstituted in vivo in cells that lack endogenous TERT and TR. The telomerase components are transfected, expressed, and assembled in vivo with other telomerase components to form a functional holoenzyme (Counter et al. 1998a). Mouse cells that are deleted for the RNA component of telomerase can be grown in culture and have mutant RNAs added back. The mutant RNAs then associate with the endogenous TERT, and enzyme activity can be assayed (Blasco et al. 1995). A number of functional regions of the TR described below have been identified using the in vivo reconstitution system (Martin-Rivera and Blasco 2001; Chen et al. 2002). For probing functional regions of the human RNA, the human cell line VA13 is often used. This cell line lacks expression of both hTERT and hTR and maintains its telomeres via the alternative lengthening of telomeres (ALT) pathway (see Chapters 7 and 8). Transfection of the gene encoding hTR into this cell line restores telomerase activity (Mitchell et al. 1999). Transfection of mutant gene constructs allows testing of the functional region of the RNA for RNA–protein interaction and telomerase activity (Mitchell and Collins 2000; Chen and Greider 2003a,b).

Telomerase Activity Assays

The basis for all telomerase activity assays is the addition of telomere repeats to the 3′ end of DNA primer by telomerase. In practice, two types of assays are most frequently used: the direct assay and the telomeric repeat amplification protocol (TRAP) assay. In the direct assay, radiolabeled nucleotides are incorporated into a telomeric primer, and the products of the telomerase reaction are visualized on a denaturing polyacrylamide gel. This assay allows the pattern of elongation products to be examined. If mutations are made in the template that alter the nucleotide incorporation or the position of translocation, the pattern of elongation products will be altered. In addition, the pattern generated by primers with different 3′ ends will differ in this assay. Altering the 3′-end sequence can allow experimental analysis of the interaction between telomeric DNA primer and RNA template throughout the template (Fig. 1A) (Greider 1991). Thus, the direct assay is the ideal tool for analyzing the details of the telomerase reaction mechanism.

In addition to the direct assay, telomerase activity can be visualized by the TRAP assay. This assay uses polymerase chain reaction (PCR) to amplify the products of the telomerase reaction (Kim et al. 1994). TRAP is most frequently used to assay human and mouse telomerase activity, although it has been adapted to other organisms as well (Fitzgerald et al. 1996; Magnenat et al. 1999). A TRAP assay has, in practice, two steps: First, a primer is elongated by telomerase and, second, additional complementary primers and DNA polymerase are used to amplify the telomeric products generated by telomerase (Fig. 1B). The amplification products are then resolved on a gel. This reaction also generates a banding pattern with a 6-base periodicity; however, that periodicity is generated by the PCR and does not reflect the mechanism of repeat addition by telomerase. Because of the exponential power of PCR amplification, TRAP is very sensitive and useful for the identification of low levels of telomerase activity such as in human tissue or tumor samples (Kim et al. 1994).

Telomerase Processivity and Translocation

The telomerase catalytic mechanism, like other reverse transcriptases, adds one deoxynucleotide at a time to the 3′ end of the DNA primer. For telomerase to synthesize multiple telomere repeats without dissociating, it must proceed in a processive fashion. Polymerase processivity refers to the probability that the enzyme will remain bound to the substrate after each nucleotide addition. For telomerase, the reiterative copying of the short template region involves two distinct types of processivities: nucleotide addition processivity and repeat addition processivity. After the addition of each nucleotide, there is a probability that the substrate will dissociate from the enzyme. When dissociation happens at a high rate, the enzyme is said to have low processivity. In contrast to nucleotide processivity, the repeat addition processivity describes the probability of dissociation of the growing primer during the translocation step after the end of the template is reached. The total number of repeats added before the enzyme dissociates will increase as repeat addition processivity increases (Greider 1991). These two kinds of processivities can be thought of as involving two different types of translocations: type I and type II (Fig. 2). Type I translocation involves the movement between the DNA–RNA hybrid and the catalytic site of the TERT protein, whereas type II translocation refers to the repositioning of the primer 3′ end after the entire RNA template has been copied. The telomeric DNA primer interacts with a site separate from the template on the TERT protein termed the anchor site (Harrington and Greider 1991; Morin 1991). The anchor site allows the primer to remain associated with telomerase during type II translocation. Very short primers, 6–8 nucleotides long, are not bound by both the anchor site and catalytic site and thus dissociate from the template after elongation (Collins and Greider 1993). The anchor site has binding specificity for telomeric sequences (Harrington and Greider 1991; Morin 1991; Hammond et al. 1997; Lue and Peng 1998). Although primers of nontelomeric sequence can be bound and elongated, telomeric sequences are more efficiently bound. In addition, the base-pairing interaction of the 3′ end of the primer with the template favors complementary telomeric primers (Harrington and Greider 1991; Morin 1991).

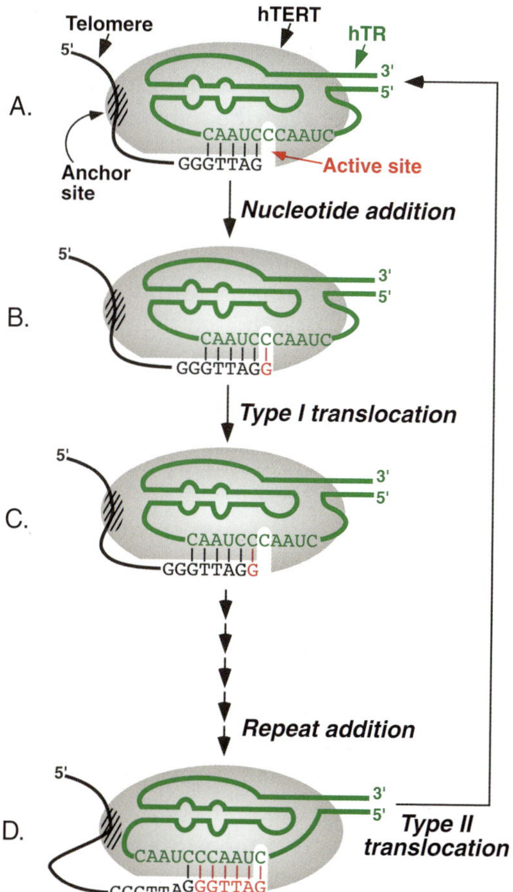

Figure 2. Processive elongation by telomerase. (*A*) The telomerase enzyme is shown at the top. The telomeric primer is bound on the 5′ end at the anchor site, and the 3′ end is positioned in the active site of telomerase and is base-paired to the alignment region of the RNA template. (*B*) Nucleotide addition occurs at the catalytic site; one nucleotide is added at a time. (*C*) The addition of each nucleotide must be accompanied by a small translocation of the active site along the template RNA that we refer to as type I translocation. The probability that telomerase remains bound during this nucleotide addition is referred to as nucleotide addition processivity. (*D*) Once the end of the template is reached, the 3′ end must be repositioned on the RNA. This involves transient dissociation and realignment of the primer 3′ end. We refer to this as a type II translocation event. For most telomerases, the primer frequently remains bound during this translocation and the enzyme continues to elongate it. Repeat addition processivity refers to the probability that the primer will dissociate during type II translocation.

The telomerase elongation reaction requires only a primer, deoxynucleotide triphosphates, Mg^{2+}, and telomerase. No high-energy cofactor such as rATP or rGTP is required. Yet, although there is a potential for a long stretch of base-pairing between the template and elongated primer, this long region of pairing, which would require energy to melt, does not form during the elongation cycle. Nucleotide addition by telomerase is likely coupled to melting of the base-pairing between the template and growing primer (Collins and Greider 1993; Hammond and Cech 1998; Benjamin et al. 2000). Chemical probing of the accessibility of the TR template region in yeast indicates that although the template is 16 nucleotides long, the length of the DNA–RNA hybrid is kept constant at 7–8 bp after primer binding and elongation (Forstemann and Lingner 2005). Therefore, when the 5′ end of the template is reached, no substantial energy is required to disrupt the base-pairing interaction and realign the primer 3′ end for a second round of addition.

The telomerase processivity measured in vitro can be affected by reaction conditions such as temperature, ionic strength, and dGTP concentration. The repeat addition processivity can be affected by the base-pairing interactions between the telomere product and the RNA template (Chen and Greider 2003b). The processivity increases as temperature and salt concentration decrease, which likely promotes realignment of the 3′ end of an extended primer to the RNA template and therefore results in efficient translocation and a processive reaction (Sun et al. 1999). A separate role in promoting translocation has been proposed for dGTP, but the actual mechanism remains unclear (Hammond and Cech 1997, 1998; Maine et al. 1999; Hardy et al. 2001). In addition, some intrinsic properties of the telomerase enzyme also contribute to the translocation efficiency and processivity. Furthermore, in *Euplotes*, an accessory telomerase protein, p43, can also enhance repeat addition processivity of telomerase (Aigner and Cech 2004).

The translocation step usually occurs after the elongation reaction reaches the extreme 5′ end of RNA template. In some organisms, such as *Euplotes* and yeast, which have very long templates ranging from 15 to 30 nucleotides, telomere synthesis can be discontinuous along the template and stall at different sites within the template under different dGTP concentrations (Shippen-Lentz and Blackburn 1990; Hammond and Cech 1997; Fulton and Blackburn 1998; Underwood et al. 2004). When translocation occurs at sites within the RNA template, a heterogeneous sequence is generated in the telomere product. For example, the *Saccharomyces cerevisiae* telomere repeat is an irregular arrangement of TG, TGG, and TGGG sequences.

Telomerase Cleavage Activity

In addition to nucleotide incorporation, telomerase also possesses a cleavage activity (Collins and Greider 1993). In *Tetrahymena*, a primer with the sequence 5′-(TTGGGG)$_n$TTG-3′ that anneals to the 5′ end of the template will have the terminal dG residue removed and a radiolabeled dG residue will be added. This cleavage activity is not simply the reverse of the polymerization step as removal of several nucleotides at once can also occur. When a primer with a nontelomeric 3′ end anneals to the RNA template, the unpaired nucleotide can be cleaved off, and the new 3′ end of the paired nucleotide can be used as substrate for telomere elongation. Cleavage activity is present in human, yeast, and ciliate telomerases (Collins and Greider 1993; Melek et al. 1996; Wang and Blackburn 1997; Niu et al. 2000; Huard and Autexier 2004; Oulton and Harrington 2004) and may process the 3′ end of nontelomeric DNA primer to generate a suitable primer for telomerase. The cleavage activity is important for chromosome healing in vivo (Bednenko et al. 1997) as the broken ends of chromosome can be processed and ready for de novo telomere addition by telomerase.

TELOMERASE RNA

TRs are surprisingly divergent among eukaryotes. The size of the RNA can vary from ~150 nucleotides (in ciliates) to ~500 nucleotides (in vertebrates) to ~1300 nucleotides (in yeast). In addition to large size variation, there is also dramatic sequence variation. It is not possible to align the primary sequences of the RNAs between these groups of species. More surprisingly, although ciliate TR is transcribed by RNA polymerase III, the yeast and vertebrate TRs are transcribed by RNA polymerase II. With such size and sequence divergence, it was not possible to identify the structural and functional core for TR common to all organisms through sequence comparisons.

The dramatic variation in size among TRs was easier to understand after the establishment of the secondary structure models for ciliate, vertebrate, and yeast TRs. These secondary structures were independently established using phylogenetic comparative analysis within each group of organisms (Romero and Blackburn 1991; Chen et al. 2000; Chappell and Lundblad 2004; Dandjinou et al. 2004). The ciliate RNA contains a pseudoknot structure (helix IIIa and IIIb), two conserved base-paired helices (helix I and IV), and helix II (Fig. 3A). This helix II, however, may not be essential as it is absent in *Tetrahymena paravorax*

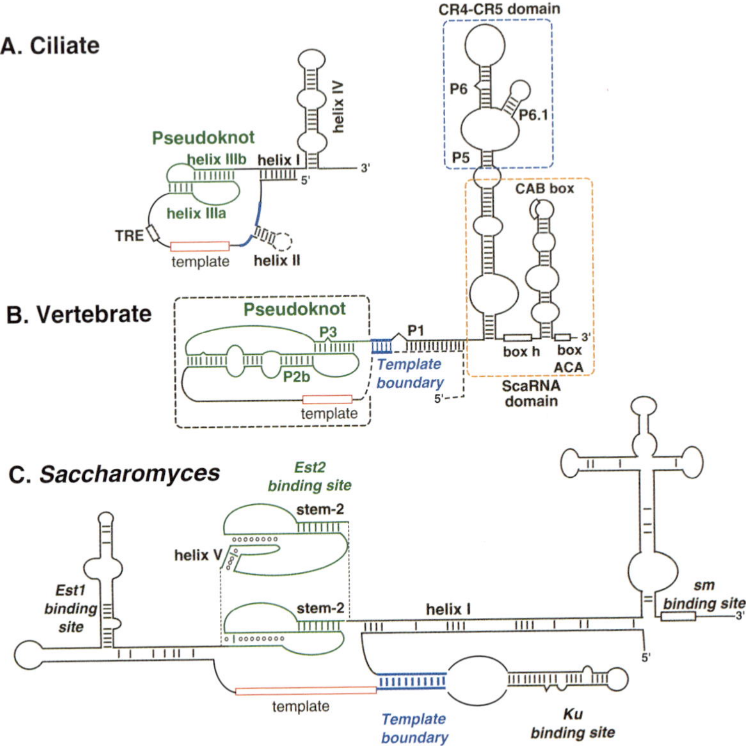

Figure 3. Secondary structure and functional regions of telomerase RNAs. (*A*) The ciliate TR has a pseudoknot domain that contains the pseudoknot (*green*) and the template sequence (*red*). The template boundary (*blue*) is determined by a sequence flanking helix II. TERT protein recognizes the RNA template via the 5′ template boundary element and 3′ template recognition element (TRE). (*B*) The vertebrate TR also has a pseudoknot domain (*black dashed box*) that contains the pseudoknot (*green*) and the template sequence (*red*). The template boundary (*blue*) is determined by the length of the P1 stem. The CR4-CR5 domain essential for enzyme activity and the scaRNA domain that binds the dyskerin complex, which is present in vertebrate RNAs but not other TRs, are highlighted with *blue* and *orange* dashed boxes, respectively. (*C*) The yeast TR contains a pseudoknot, although the exact structure is not established, and thus two alternative structures are shown (*green*). The template sequence is shown in *red* and the template boundary (*blue*) is immediately adjacent to the template. This large RNA contains addition domains for binding of the Est1 protein, Ku protein, and Sm protein complex.

(McCormick-Graham and Romero 1995). The vertebrate TR consists of three conserved structural elements: the pseudoknot, the CR4-CR5, and the scaRNA domains, a conserved structure shared with small Cajal body RNA (scaRNA) and snoRNA (Fig. 3B) (Mitchell et al. 1999; Chen et al. 2000; Jady et al. 2004). For yeast TR, several local stem-loop structures were initially identified in *Saccharomyces* and *Kluyveromyces* RNAs as binding sites for various telomerase-associated proteins (Fig. 3C) (Tzfati et al. 2000, 2003; Peterson et al. 2001; Seto et al. 2002). Pseudoknot structures were proposed in *Kluyveromyces* RNA (Tzfati et al. 2003) and later in *Saccharomyces* RNA (Chappell and Lundblad 2004; Dandjinou et al. 2004; Lin et al. 2004). On the basis of the RNA sequence alignment from several *Saccharomyces* species, the secondary structure model of yeast TR, however, includes two possible alternative pseudoknot structures (Chappell and Lundblad 2004). The details of the yeast pseudoknot structure remains to be clarified. Nevertheless, the presence of a pseudoknot in the yeast RNA structure would unify the overall structure of TRs as it is present in the ciliates and in vertebrate telomerases as well (Lin et al. 2004).

The proposed secondary structures of TRs from most species share a similar structural configuration near the template region. Flanking the template, there is an upstream long-range base-pairing (called helix I or P1) and a downstream pseudoknot structure (Fig. 3; the pseudoknot structure is highlighted in green). However, the larger vertebrate and yeast RNAs apparently have acquired additional RNA structural domains during evolution that serve as binding sites for species-specific telomerase proteins (Zappulla and Cech 2004), such as p43 protein (ciliate), Sm protein (yeast), Est1 protein (yeast), Ku protein (yeast), and dyskerin complex (vertebrate) (see Chapter 2). Each of these proteins and their corresponding RNA-binding elements likely have distinct roles in different species.

Template Boundary Definition

One of the most important functions of TR is to provide a template for telomere DNA synthesis. The sequence of the RNA template is complementary to the sequence of telomere DNA synthesized. Telomerase uses only a small portion of its intrinsic RNA as a template. The number of nucleotides in the template region is usually 1.5 copies of the telomere repeat but ranges from 8 nucleotides in mouse to 30 nucleotides in *Kluyveromyces lactis*.

For the template to accurately specify telomere DNA synthesis, the RNA template must reside in the catalytic site of TERT, base-pair with the telomere primer, and have a defined boundary where polymerization stops and/or translocation occurs. These functions are specified in the pseudoknot domain of the RNA. This template domain is composed of the pseudoknot structure, a primer alignment region, and a region of the RNA that specifies the template's 5′ boundary. Although all of these features are present in TRs from various species, the detailed mechanisms used differ among species.

Within the pseudoknot domain, the RNA template is adjacent to and at the 5′ end of the pseudoknot structure that is a binding site for the TERT protein in vertebrate and yeast RNA (Fig. 3). This specific RNA–protein interaction allows correct positioning of the RNA template in the catalytic pocket of the TERT protein. Interestingly, in ciliates, the TERT protein recognizes the RNA template via the 5′ motif (i.e., template boundary element) and the 3′ motif (i.e., template recognition element [TRE]) adjacent to the template (Lai et al. 2002; Miller and Collins 2002). The ciliate pseudoknot structure, unlike the vertebrate RNAs, seems to be required for telomerase assembly in vivo (Gilley and Blackburn 1999), but not TERT protein binding.

The primer alignment region of the template domain has an important role in accurately positioning the primer 3′ end on the template for telomere repeat synthesis (Fig. 4). Telomere DNA primers interact with the 3′ region of the RNA template via Watson–Crick base-pairing. Although the sequence of the template region is complementary to telomeric substrates, it is not used as a template for synthesis of telomeric repeats. Mutations in the alignment positions do not become incorporated into growing telomere repeats (Autexier and Greider 1995). Permutations or mutations in the alignment domain in *Tetrahymena* TR disturb the ability of the base-pairing of the 3′ end of the primer with the RNA and impair repeat synthesis (Autexier and Greider 1995). In human and mouse RNAs, the number of base-pairings between primer and RNA template affects translocation efficiency and thus repeat addition processivity (Chen and Greider 2003b).

The 3′ boundary of the template is defined by its sequence complementarity to the telomere, whereas the 5′ boundary is actively specified by unique RNA elements. The template boundary region of the RNA helps specify the site at which translocation occurs. Surprisingly, the structural elements that define the 5′ boundary of the RNA template are quite divergent in different species (Autexier and Greider 1995; Tzfati et al. 2000; Lai et al. 2002; Chen and Greider 2003a; Seto et al. 2003). The elements responsible for template boundary definition in ciliate, yeast, and vertebrate TRs are all located close to and upstream of the template sequence (Fig. 4). In ciliate TR, the boundary definition is mediated by a conserved sequence motif upstream of the template (Autexier and Greider 1995) that is bound by the TERT protein (Miller et al. 2000; Lai et al. 2002). For the human TR, the structure and position of the helix P1b are important for boundary definition (Fig. 4A). Interestingly, mouse TR lacks helix P1b, and the boundary is established by its 5′ end located only 2 nucleotides upstream of the boundary (Chen and Greider 2003a). In yeast, a base-paired structure upstream of the RNA template defines

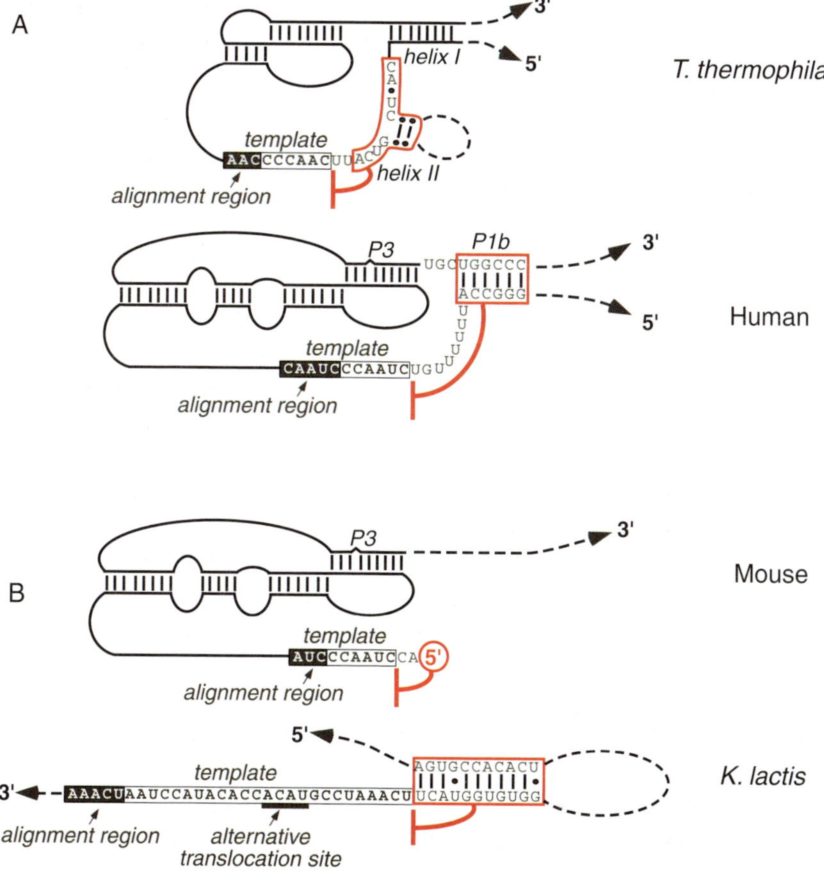

Figure 4. Mechanisms for template boundary definition. TRs from different organisms use different methods for defining the template boundary. For all of the RNAs, the template sequence contains an alignment region near the 5′ end (*black box with white letters*) that is not copied during elongation. (*A*) The *Tetrahymena* RNAs and hTRs restrict the template movement. In ciliates, this is due to TERT binding to the sequence at the base of helix II. In humans, it is due to the distance of helix P1b from the end of the template. (*B*) The mouse and *K. lactis* RNAs limit the availability of the upstream flanking unpaired sequence. In the mouse, there are only 2 nucleotides upstream of the template, and in *K. lactis* there is a base-paired helix immediately adjacent to the template that blocks further copying of the RNA 5′ of the template sequence.

the 5′ boundary in both *Kluyveromyces* and *Saccharomyces* (Fig. 4B) (Tzfati et al. 2000; Seto et al. 2003).

These divergent boundary definition elements suggest diverse mechanisms for template boundary definition. For example, the ciliate and human elements may constrain the movement of the RNA template through RNA–protein interaction or RNA–RNA base-pairing. In contrast, the mouse and *Kluyveromyces* budding yeast elements may define the template boundary by limiting the availability of the single-stranded region adjacent to the template.

Species-specific RNA Domains

In addition to the conserved core structure of the TR, there are species-specific domains that account for the extreme size differences in TRs from different species. Many of these RNA domains have been shown to be specific binding sites for different proteins in different species and are important for functional RNP architecture.

Ciliates

In addition to the template domain, ciliate TR contains a unique helix IV (Fig. 3A). The distal stem loop of helix IV is a potential binding site for TERT protein and appears to have a role in both nucleotide and repeat addition processivity (Lai et al. 2003; Mason et al. 2003). TERT also exhibits high affinity at a conserved sequence motif (i.e., template boundary element) that flanks the helix II (Miller et al. 2000; Lai et al. 2001) and a short sequence motif (called TRE) downstream from the template (Miller and Collins 2002). Helix I, and perhaps the proximal helix IV, appears to be the binding site for the p43 protein in *Euplotes* (Aigner et al. 2000, 2003). Interestingly, the p65 protein, a *Tetrahymena* homolog of p43, appears to recognize the sequence of the proximal region of stem IV (Witkin and Collins 2004; Prathapam et al. 2005). Although both proteins bind to a similar region in the ciliate RNA, it would be interesting to know if they share the same function for telomerase.

Yeast

In addition to the potential pseudoknot structure required for binding the yeast TERT protein Est2 (Livengood et al. 2002), two essential stem-loop structures in *Saccharomyces* TR (i.e., TLC1) were identified as binding sites for the telomerase-associated proteins Est1 and Ku (Fig. 3C) (Seto et al. 2002; Stellwagen et al. 2003). This Est1–TLC1 interaction might allow Est1 protein to recruit or activate telomerase at the telomere (Seto et al. 2002). Similarly, the interaction between Ku and TLC1 RNA enables telomerase to act at both broken and normal chromosome ends (Stellwagen et al. 2003). Moreover, a binding site for the Sm protein was identified near the 3′ end of the RNA that is important for TR biogenesis (Seto et al. 1999). The large size of the yeast RNA seems to provide a flexible scaffold for various telomerase protein subunits to form a functional complex in vivo (Zappulla and Cech 2004).

Vertebrate

The 5′ half of vertebrate TR contains the central template/pseudoknot domain, whereas the 3′ half contains the CR4-CR5 and scaRNA domains (Fig. 3B). Both the CR4-CR5 and the pseudoknot domains are essential for telomerase activity in vitro (Tesmer et al. 1999). More specifically, a highly conserved stem loop, p6.1, in the CR4-CR5 domain is critical for telomerase activity (Chen et al. 2002) and might interact with the template region (Ueda and Roberts 2004). However, the actual function of this structural element is still unknown. The 3′-distal scaRNA domain contains two conserved motifs, Box H and ACA, which are binding sites for the dyskerin protein complex and are essential for RNA stability in vivo (Mitchell et al. 1999). An extra structural element in the scaRNA domain, initially termed CR7, shares sequence similarity with the CAB box motif identified in scaRNAs that are responsible for posttranscriptional modification of snRNA in the Cajal body (Jady et al. 2004). This motif may be a binding site for a specific protein involved in TR localization to the Cajal body.

CATALYTIC TELOMERASE REVERSE TRANSCRIPTASE

The catalytic TERT component of telomerase RNP was identified initially in the ciliate *Euplotes aediculatus* and was shown to have homology with the yeast Est2 protein (Lingner et al. 1997b). Homologs of TERT have since been found in a variety of eukaryotic organisms, including vertebrates—human, mouse, rat, hamster, dog, chicken, and *Xenopus* (Meyerson et al. 1997; Nakamura et al. 1997; Greenberg et al. 1998; Guo et al. 2001; Kuramoto et al. 2001; Wong et al. 2003; Delany and Daniels 2004; Nasir et al. 2004); plants—rice and *Arabidopsis* (Fitzgerald et al. 1999; Oguchi et al. 1999; Heller-Uszynska et al. 2002); yeast—*Schizosaccharomyces pombe, Candida albicans,* and *S. cerevisiae* (Lendvay et al. 1996; Lingner et al. 1997b; Nakamura et al. 1997; Metz et al. 2001); and ciliates— *Oxytricha, Tetrahymena,* and *Paramecium* (Bryan et al. 1998; Collins and Gandhi 1998; Takenaka et al. 2001). In contrast to the TR, the TERT protein sequence is conserved. Sequence analysis of TERT proteins from distantly related organisms reveals several regions of conserved sequence motifs (Fig. 5). These motifs can be grouped into three distinct regions:

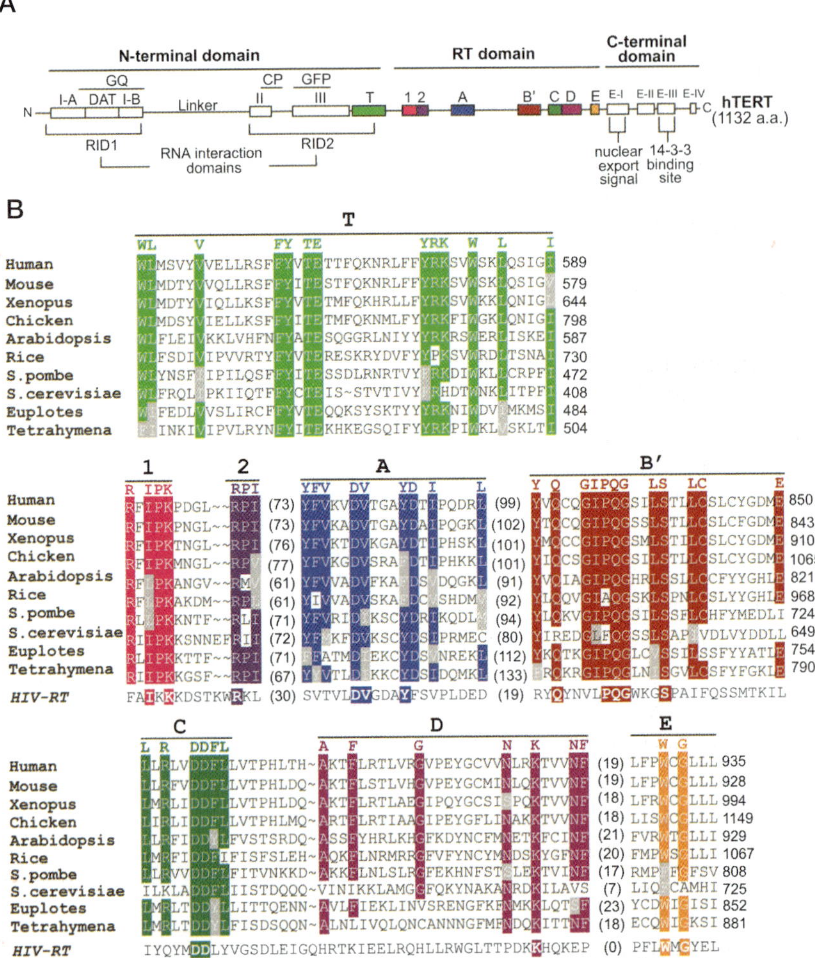

Figure 5. TERT functional domains. (*A*) The conserved domains in the TERT protein are shown for the amino-terminal region, the reverse transcriptase domain (RT domain), and the carboxy-terminal domain. These functional regions are described in detail in the text. (*B*) An alignment of the conserved sequence of the T domain and the seven conserved RT domains is shown. The conserved residues of the RT domains are shown aligned with the HIV reverse transcriptase to highlight the conserved catalysis residues.

the amino-terminal domain, reverse transcriptase (RT) domain, and carboxy-terminal domain. The motifs (1, 2, A, B′, C, D, E) in the RT domain are universally conserved among all RTs. Motifs in the amino-terminal and carboxy-terminal domains are unique to telomerase and not as conserved as the RT domain. Each domain of telomerase has distinct functional roles, including catalysis, nucleolar localization, RNA binding, dimerization, and recruitment to the telomere.

Amino-Terminal Domain

A major feature that is different in TERT compared to other reverse transcriptases is the basic amino-terminal domain. This amino-terminal domain consists of the conserved motif I, followed by a large, hypervariable linker region and the motifs II, III, and T (Fig. 5). In hTERT, motif I can be further divided into motif I-A, motif DAT, and motif I-B. Mutations in the motif DAT (dissociates activities of telomerase) disrupt the in vivo function but not the in vitro activity of telomerase (Beattie et al. 2000; Armbruster et al. 2001, 2003, 2004; Banik et al. 2002; Bosoy et al. 2003). Recent evidence suggests that this DAT region has a role in primer-binding specificity and may be an anchor site for binding of telomerase to telomeric substrates (Moriarty et al. 2005). In yeast, the TERT (Est2) has an amino-terminal motif I that is required for telomere maintenance and viability in vivo but is dispensable for telomerase activity in vitro (Friedman and Cech 1999). This may be due to the higher sensitivity of the in vivo assay for telomerase activity than the in vitro assay. A mutation in the motif I in the yeast TERT results in loss of Est3 protein association with the telomerase complex (Friedman et al. 2003).

The amino-terminal domain is involved in localization of TERT to the nucleolus. When fluorescently labeled hTERT proteins (GFP-hTERT or YFP-hTERT) are overexpressed in human cells, they localize to nucleoli (Etheridge et al. 2002; Yang et al. 2002). This nucleolar localization of the GFP fusion proteins occurs in the absence of hTR, indicating an intrinsic localization signal in the TERT protein. Interestingly, one group mapped the nucleolar localization domain (NoLD) to motifs II, III, and T in the amino-terminal domain (Etheridge et al. 2002), whereas another group mapped this nucleolar localization domain to the extreme amino terminus (Yang et al. 2002). Thus, the mechanism of nucleolar localization is not yet clear.

The TERT amino-terminal domain is also involved in RNA binding. The assembly of a functional telomerase RNP complex requires specific, stable interactions between the TERT protein and its integral RNA component. A minimal region from residue 326 to 613 containing motifs II, III, and T is required for hTR-specific binding by hTERT (Beattie et al. 2000; Bryan et al. 2000b; Armbruster et al. 2001; Bachand and Autexier 2001; Lai et al. 2001). A similar RNA-binding region has been identified in the *Tetrahymena thermophila* TERT (Lai et al. 2001) and *S. cerevisiae* TERT (Est2 protein) (Bryan et al. 2000b). However, it is not known if the nature of the RNA–protein interactions is the same among these distantly related organisms. For example, in yeast TERT, mutations in motif I, not just motifs II, III, and T, also affect binding to yeast TR (*TLC1* RNA) (Friedman and Cech 1999).

The Reverse Transcriptase Domain Is the Catalytic Center of Telomerase

Motifs (1, 2, A, B′, C, D, E) in the telomerase RT domain are conserved in all RTs and are essential for catalytic activity (Nakamura and Cech 1998). The conserved aspartic acid residues (D) in motifs A and C together coordinate the two active-site metal ions found in all polymerases. These two aspartic acid residues are essential for catalytic activity for all RTs and also for telomerase (Counter et al. 1997; Lingner et al. 1997a,b). This suggests that telomerase and RT might share a similar three-dimensional structure in the catalytic core. Residues important for nucleotide binding (lysine residue in motif 1 and arginine residue in motif 2) and nucleotide-addition processivity (conserved residues in motifs C, E, and the carboxy-terminal extension region) also appear to be conserved between viral RTs and telomerase (Fig. 5B) (Miller et al. 2000; Bosoy and Lue 2001; Peng et al. 2001). Together, these conserved residues form the catalytic site with a nucleotide-binding pocket that binds the correct nucleotides with high affinity, but discriminates against noncognate nucleotides. In yeast TERT, specific residues within motifs C and E in the RT domain and the carboxy-terminal domain are critical for this nucleotide addition processivity (Miller et al. 2000; Bosoy and Lue 2001; Peng et al. 2001).

Some conserved residues in the RT domain of the TERT also have a role in repeat addition processivity. For example, in the *Tetrahymena* TERT, a residue substitution at position 813 in motif C from leucine to tyrosine was shown to increase repeat addition processivity (Bryan et al. 2000a). In yeast TERT, a mutation between motifs A and B′ in the telomerase-specific region, named IFD (insertion in finger domain), impairs the ability of telomerase to add multiple repeats (Lue et al. 2003). These residues in the TERT RT domain might facilitate DNA dissociation from the RNA template after repeat addition increasing the dissociation and thus decreasing repeat addition processivity.

Carboxy-Terminal Domain

The conservation of the motifs in the carboxy-terminal domain is relatively weak compared to the conservation in the RT and the amino-terminal domains. This domain is thought to be equivalent to the thumb domain of HIV RT, but it likely has a functional role specific to telomerase (Peng et al. 2001). Mutations in the carboxyl terminus of the hTERT affect telomerase processivity (Huard et al. 2003). The carboxy-terminal domains of human and *Tetrahymena* TERT are essential for enzymatic activity, but dispensable for telomerase activity in yeast (Friedman and Cech 1999; Beattie et al. 2000, 2001; Bachand and Autexier 2001; Lai et al. 2001; Peng et al. 2001; Banik et al. 2002). A DAT region was also mapped to the carboxyl terminus of the TERT protein. The addition of a carboxy-terminal hemagglutinin (HA) tag or mutations near the carboxyl terminus impair the in vivo telomere elongation, but not the in vitro catalytic activity of telomerase (Counter et al. 1998b; Banik et al. 2002). This suggests that this region may also be involved with the recruitment of telomerase to the telomere. In addition, the carboxyl terminus of hTERT contains binding sites for the 14-3-3 and CRM1 proteins, which regulate the intracellular localization of TERT protein (Seimiya et al. 2000).

Dimerization Domain

Although the yeast and human telomerases appear to function as higher-order multimers, the telomerases from *Euplotes aediculatus* and *Tetrahymena* act as monomers (Lingner and Cech 1996; Bryan et al. 2003). Functional studies indicate that the yeast telomerase complex contains two active sites (Prescott and Blackburn 1997). Biochemical analysis of the recombinant human telomerase suggests that the human enzyme complex contains two hTERT proteins and two TR molecules (Wenz et al. 2001). The full-length hTERT can coimmunoprecipitate labeled full-length hTERT or hTERT fragments, providing evidence for multimeric hTERT interactions (Beattie et al. 2001). This multimerization seems to be mediated through interaction between the amino- and carboxy-terminal domains (Beattie et al. 2001; Arai et al. 2002).

SPECIES-SPECIFIC PROTEINS AND BIOGENESIS OF TELOMERASE RNP

In addition to the core components, TERT and TR, telomerase consists of a variety of species-specific proteins (see Chapter 2). Many of these proteins have roles in biogenesis, assembly, and regulation of telomerase. Some of them are stably associated components of the holoenzyme, whereas others might only transiently interact with telomerase during the biogenesis of telomerase. Similar to the RNA component, these proteins vary dramatically from species to species; they are discussed below according to their functions in different groups of species.

Telomerase RNA Biogenesis

The ciliate TR is transcribed by RNA polymerase III, unlike yeast and vertebrate TRs which are transcribed by RNA polymerase II. Moreover, yeast TR contains an Sm-protein-binding site (Seto et al. 1999), a characteristic of small nuclear RNA (snRNA), and might share a similar biogenesis pathway with snRNA. In contrast, the 3′-half of vertebrate RNA contains a structure and motifs similar to Box H/ACA, snoRNA, and scaRNA and might share a biogenesis pathway with snoRNA and scaRNA. These differences in transcription machinery and structural features suggest diverse mechanisms for TR biogenesis between species.

Ciliates

In the ciliate *E. aediculatus*, telomerase affinity purification identified a telomerase-associated protein, p43, related to the La protein family that contains a La motif with RNA-binding activity (Lingner and Cech 1996; Aigner et al. 2000). The p43 protein binds specifically to the central region (stem I and adjacent nucleotides) of ciliate TR (Aigner et al. 2003) and might function in RNP assembly (Aigner et al. 2000) and telomerase nuclear retention (Mollenbeck et al. 2003). In *Tetrahymena,* an affinity purification using tagged TERT protein identified four telomerase-associated proteins: p20, p45, p65, and p75 (Witkin and Collins 2004). The p65 subunit contains a La motif and appears to be a potential homolog to the *Euplotes* p43. The *Tetrahymena* p65 protein binds similarly to the central stem I region of the RNA (Witkin and Collins 2004). Genetic depletion of p65 reduces TR accumulation in vivo and might also be involved in RNP biogenesis and function. The binding between p65 protein and TR facilitates telomerase complex assembly in vitro (Prathapam et al. 2005) and might have a critical role in telomerase RNP biogenesis. The detailed functions of p20 and p75 remain to be revealed.

Yeast

Yeast TR is synthesized as a poly(A)-tailed precursor and later processed to a mature form (Chapon et al. 1997). Yeast TR contains an Sm-binding site and its binding to the Sm protein is important for its in vivo maturation and accumulation (Seto et al. 1999). Yeast TR and snRNAs, both RNA polymerase II transcripts, have a 5′-2,2,7-trimethylguanosine (TMG) cap and might share a similar assembly pathway. Most snRNAs are exported into the cytoplasm for processing and maturation. Since yeast TR can be transiently detected in the cytoplasm, RNA maturation might occur in the cytoplasm. The importin protein Mtr10, which is involved in nuclear transport, has a role in the yeast TR nuclear localization and maturation (Ferrezuelo et al. 2002). In addition, several yeast telomerase-associated proteins, including Est1 and Est3, have been identified using genetic screening and shown to be essential for telomerase function and telomere maintenance (Lundblad and Szostak 1989; Lendvay et al. 1996). Since Est1, Est2, and yeast TR are localized independently to the nucleus, the assembly of the telomerase complex likely occurs in the nucleus, not the cytoplasm (Teixeira et al. 2002).

Human

A conserved secondary structure and sequence motifs are shared among vertebrate TRs, snoRNAs, and scaRNAs (Mitchell et al. 1999; Jady et al. 2004). This Box H/ACA motif is a binding site for dyskerin protein complex and is essential for precursor RNA 3′-end processing and mature RNA accumulation (Mitchell et al. 1999). An additional motif, the CAB box, exists both in the CR7 domain of vertebrate TR and in scaRNA (Jady et al. 2004). scaRNA localizes specifically to the Cajal body, an intranuclear structure where processing and assembly of a number of small functional RNPs occur. Consistent with the presence of the CAB box, the endogenous hTR also accumulates in Cajal bodies (Jady et al. 2004; Zhu et al. 2004). Accumulation of properly processed TR requires an intact CR7 domain (Fu and Collins 2003). It would be interesting to know if the Cajal body is the primary assembly site for telomerase RNP. Three human homologs of Est1 protein have been identified via sequence searching and cloning (Reichenbach et al. 2003; Snow et al. 2003). However, human Est1 protein does not seem to interact directly with hTR (Snow et al. 2003), but instead interacts with TERT protein for recruiting telomerase to the telomere.

TERT Protein Biogenesis

Although expression of hTERT protein is the limiting factor for enzymatic activity of telomerase in a variety of cell types, intracellular trafficking of telomerase may also have a regulatory role for enzyme activity. hTERT localizes to nucleoli (Etheridge et al. 2002; Yang et al. 2002) and is thought to shuttle between nucleoli and nucleoplasm upon transformation and DNA damage (Wong et al. 2002). The nucleoli therefore might be a reservoir of telomerase ready for delivery of the telomere upon regulatory signals.

There is some evidence that TERT is present in both nucleoli and Cajal bodies. GFP-hTERT or yellow fluorescent protein (YFP)-hTERT fusion proteins expressed in cells accumulates in both nucleoli (Etheridge et al. 2002; Wong et al. 2002; Yang et al. 2002) and Cajal bodies (Zhu et al. 2004). However, the assembly site of functional telomerase remains unclear. It is not yet known what factors may facilitate the export of assembled telomerase RNP from the Cajal body or nucleolus.

The 14-3-3 protein, which is involved in nuclear shuttling of a number of proteins, binds to hTERT through the carboxyl terminus (Seimiya et al. 2000). An hTERT mutant that cannot bind to 14-3-3 fails to localize into the nucleus. It was therefore suggested that 14-3-3 enhances the nuclear localization of hTERT by interfering with the access of CRM1/exportin 1 protein, a receptor for the nuclear export machinery, to the hTERT nuclear export signal (NES)-like motif (residues 970–981) near the 14-3-3 binding site (residues 1030–1047). This suggests that 14-3-3 might control the intracellular localization of hTERT. The mutations in either hTERT or 14-3-3 that disrupt interaction between these two proteins did not eliminate or reduce telomerase activity, suggesting that 14-3-3 might control the shuttling of the assembled telomerase, but not disrupt the activity of the telomerase complex.

Another protein involved in nucleolar localization, nucleolin, interacts with telomerase and affects its subnuclear localization via both protein–protein and RNA–protein interactions (Khurts et al. 2004). Although nucleolin might be responsible for the nucleolar shuttling of the telomerase, its role in telomerase RNP assembly remains to be explored.

Another nucleolar protein, PinX, regulates telomerase in yeast by sequestering TERT in the nucleolus in an inactive complex lacking TR (Lin and Blackburn 2004). In humans, PinX1 represses telomerase activity in vivo by binding to both hTERT and hTR (Banik and Counter 2004).

RNP Assembly

The chaperone proteins p23 and Hsp90 have critical roles in vertebrate telomerase assembly in vitro and in vivo (Holt et al. 1999). Inhibition of Hsp90 activity leads to reduction of both telomerase activity and steady level of TERT protein. The functional chaperone-mediated assembly of telomerase is ATP dependent and involves at least five chaperone components—hsp90, p23, hsp70, p60, and hsp40 (Forsythe et al. 2001). Hsp90 and p23 stably associate with the active telomerase complex, whereas other chaperone proteins only transiently interact with the telomerase complex during assembly (Fig. 6). The unassembled TERT protein seems to be ubiquitinated and degraded through the proteasome-mediated degradation pathway (Kim et al. 2005).

The assembly of functional telomerase might involve the survival of motor neuron (SMN) protein complex that functions in the assembly of

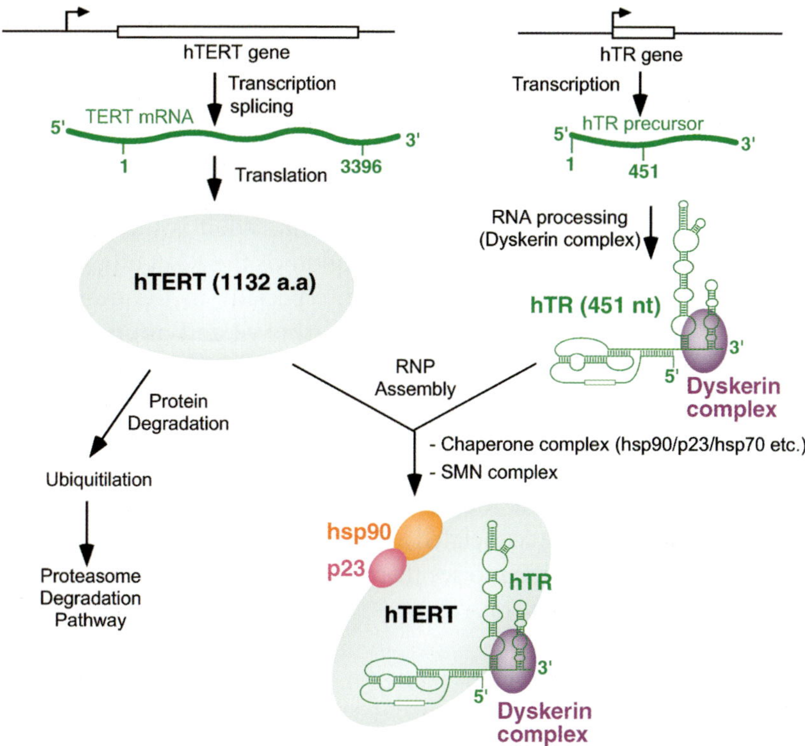

Figure 6. Schematic representation of telomerase biogenesis and assembly. The core telomerase components TERT and TR are transcribed and translated separately. The hTR is processed at the 3′ end and bound by the proteins in the dyskerin complex. TERT and hTR assemble with the assistance of chaperone proteins Hsp90, p23, Hsp70, and others. The details of the subcellular sites of assembly are described in the text.

various RNPs, including spliceosomes, snoRNP, and transcriptosomes (for review, see Paushkin et al. 2002). In nuclei, the SMN complex is localized in spherical nuclear particles, called gems, that often associate with Cajal bodies. Consistent with this, the SMN protein colocalizes with TR (Jady et al. 2004) and associates with the telomerase complex (Bachand et al. 2002), possibly via interaction with GAR1, a member of the dyskerin complex (Pellizzoni et al. 2001).

CONCLUDING REMARKS

Telomerase is a unique and fascinating enzyme. Although the broad outlines of how telomerase synthesizes telomere repeats onto chromosome ends is described here in this chapter, many important details are still unknown. A number of very interesting questions about telomerase biochemistry still remain to be answered. Telomerase is unique among reverse transcriptase enzymes in copying only a small region of a tightly bound RNA template. It will be very interesting to understand how the RNA moves past the catalytic site or how the catalytic site moves within the enzyme to move along the RNA. Next, there are regions on telomerase RNAs, such as Stem p6.1 in hTR, that are required for telomerase activity but are not located in the pseudoknot domain. What is the mechanism by which the conserved residues in this region of the RNA affect enzyme activity? The TERT protein is conserved, especially in the catalytic RT domain. What are the functions of the carboxy- and amino-terminal domains? Why can so few residues be deleted from either end without a reduction in activity? The telomerase-associated proteins vary greatly between different groups of species. What does this tell us about the role of telomerase in these different organisms? The association of hTR with dyskerin and the fact that mutations in both result in human disease indicate that telomerase-associated proteins have a profound effect on telomerase. It will be exciting to see the answers to many of these questions revealed over the next several years.

REFERENCES

Aigner S. and Cech T.R. 2004. The *Euplotes* telomerase subunit p43 stimulates enzymatic activity and processivity in vitro. *RNA* **10:** 1108–1118.

Aigner S., Postberg J., Lipps H.J., and Cech T.R. 2003. The *Euplotes* La motif protein p43 has properties of a telomerase-specific subunit. *Biochemistry* **42:** 5736–5747.

Aigner S., Lingner J., Goodrich K.J., Grosshans C.A., Shevchenko A., Mann M., and Cech T.R. 2000. *Euplotes* telomerase contains an La motif protein produced by apparent translational frameshifting. *EMBO J.* **19:** 6230–6239.

Arai K., Masutomi K., Khurts S., Kaneko S., Kobayashi K., and Murakami S. 2002. Two independent regions of human telomerase reverse transcriptase are important for its oligomerization and telomerase activity. *J. Biol. Chem.* **277:** 8538–8544.

Armbruster B.N., Etheridge K.T., Broccoli D., and Counter C.M. 2003. Putative telomere-recruiting domain in the catalytic subunit of human telomerase. *Mol. Cell. Biol.* **23:** 3237–3246.

Armbruster B.N., Banik S. S., Guo C., Smith A.C., and Counter C.M. 2001. N-terminal domains of the human telomerase catalytic subunit required for enzyme activity in vivo. *Mol. Cell. Biol.* **21:** 7775–7786.

Armbruster B.N., Linardic C.M., Veldman T., Bansal N.P., Downie D.L., and Counter C.M. 2004. Rescue of an hTERT mutant defective in telomere elongation by fusion with hPot1. *Mol. Cell. Biol.* **24:** 3552–3561.

Autexier C. and Greider C.W. 1995. Boundary elements of the *Tetrahymena* telomerase RNA template and alignment domains. *Genes Dev.* **9:** 2227–2239.

Bachand F. and Autexier C. 2001. Functional regions of human telomerase reverse transcriptase and human telomerase RNA required for telomerase activity and RNA-protein interactions. *Mol. Cell. Biol.* **21:** 1888–1897.

Bachand F., Boisvert F.M., Cote J., Richard S., and Autexier C. 2002. The product of the survival of motor neuron (SMN) gene is a human telomerase-associated protein. *Mol. Biol. Cell* **13:** 3192–3202.

Banik S.S. and Counter C.M. 2004. Characterization of interactions between PinX1 and human telomerase subunits hTERT and hTR. *J. Biol. Chem.* **279:** 51745–51748.

Banik S.S., Guo C., Smith A.C., Margolis S.S., Richardson D.A., Tirado C.A., and Counter C.M. 2002. C-terminal regions of the human telomerase catalytic subunit essential for in vivo enzyme activity. *Mol. Cell. Biol.* **22:** 6234–6246.

Beattie T.L., Zhou W., Robinson M.O., and Harrington L. 2000. Polymerization defects within human telomerase are distinct from telomerase RNA and TEP1 binding. *Mol. Biol. Cell* **11:** 3329–3340.

———. 2001. Functional multimerization of the human telomerase reverse transcriptase. *Mol. Cell. Biol.* **21:** 6151–6160.

Bednenko J., Melek M., Greene E.C., and Shippen D.E. 1997. Developmentally regulated initiation of DNA synthesis by telomerase: Evidence for factor-assisted de novo telomere formation. *EMBO J.* **16:** 2507–2518.

Benjamin S., Baran N., and Manor H. 2000. Interference footprinting analysis of telomerase elongation complexes. *Mol. Cell. Biol.* **20:** 4224–4237.

Blasco M.A., Funk W., Villeponteau B., and Greider C.W. 1995. Functional characterization and developmental regulation of mouse telomerase RNA. *Science* **269:** 1267–1270.

Bosoy D. and Lue N.F. 2001. Functional analysis of conserved residues in the putative "finger" domain of telomerase reverse transcriptase. *J. Biol. Chem.* **276:** 46305–46312.

Bosoy D., Peng Y., Mian I.S., and Lue N.F. 2003. Conserved N-terminal motifs of telomerase reverse transcriptase required for ribonucleoprotein assembly in vivo. *J. Biol. Chem.* **278:** 3882–3890.

Bryan T.M., Goodrich K.J., and Cech T.R. 2000a. A mutant of *Tetrahymena* telomerase reverse transcriptase with increased processivity. *J. Biol. Chem.* **275:** 24199–24207.

———. 2000b. Telomerase RNA bound by protein motifs specific to telomerase reverse transcriptase. *Mol. Cell* **6:** 493–499.

———. 2003. *Tetrahymena* telomerase is active as a monomer. *Mol. Biol. Cell* **14:** 4794–4804.

Bryan T.M., Sperger J.M., Chapman K.B., and Cech T.R. 1998. Telomerase reverse transcriptase genes identified in *Tetrahymena thermophila* and *Oxytricha trifallax*. *Proc. Natl. Acad. Sci.* **95:** 8479–8484.

Chapon C., Cech T.R., and Zaug A.J. 1997. Polyadenylation of telomerase RNA in budding yeast. *RNA* **3:** 1337–1351.

Chappell A.S. and Lundblad V. 2004. Structural elements required for association of the *Saccharomyces cerevisiae* telomerase RNA with the Est2 reverse transcriptase. *Mol. Cell. Biol.* **24:** 7720–7736.

Chen J.-L. and Greider C. W. 2003a. Template boundary definition in mammalian telomerase. *Genes Dev.* **17:** 2747–2752.

———. 2003b. Determinants in mammalian telomerase RNA that mediate enzyme processivity and cross-species incompatibility. *EMBO J.* **22:** 304–314.

Chen J.-L., Blasco M.A., and Greider C.W. 2000. Secondary structure of vertebrate telomerase RNA. *Cell* **100:** 503–514.

Chen J.-L., Opperman K.K., and Greider C.W. 2002. A critical stem-loop structure in the CR4-CR5 domain of mammalian telomerase RNA. *Nucleic Acids Res.* **30:** 592–597.

Collins K. and Gandhi L. 1998. The reverse transcriptase component of the *Tetrahymena* telomerase ribonucleoprotein complex. *Proc. Natl. Acad. Sci.* **95:** 8485–8490.

Collins K. and Greider C.W. 1993. *Tetrahymena* telomerase catalyzes nucleolytic cleavage and nonprocessive elongation. *Genes Dev.* **7:** 1364–1376.

Counter C.M., Meyerson M., Eaton E.N., and Weinberg R.A. 1997. The catalytic subunit of yeast telomerase. *Proc. Natl. Acad. Sci.* **94:** 9202–9207.

Counter C.M., Meyerson M., Eaton E.N., Ellisen L.W., Caddle S.D., Haber D.A., and Weinberg R.A. 1998a. Telomerase activity is restored in human cells by ectopic expression of hTERT (hEST2), the catalytic subunit of telomerase. *Oncogene* **16:** 1217–1222.

Counter C.M., Hahn W.C., Wei W., Caddle S.D., Beijersbergen R.L., Lansdorp P.M., Sedivy J.M., and Weinberg R.A. 1998b. Dissociation among in vitro telomerase activity, telomere maintenance, and cellular immortalization. *Proc. Natl. Acad. Sci.* **95:** 14723–14728.

Dandjinou A.T., Levesque N., Larose S., Lucier J.F., Abou Elela S., and Wellinger R.J. 2004. A phylogenetically based secondary structure for the yeast telomerase RNA. *Curr. Biol.* **14:** 1148–1158.

Delany M.E. and Daniels L.M. 2004. The chicken telomerase reverse transcriptase (chTERT): Molecular and cytogenetic characterization with a comparative analysis. *Gene* **339:** 61–69.

Etheridge K.T., Banik S.S., Armbruster B.N., Zhu Y., Terns R.M., Terns M.P., and Counter C.M. 2002. The nucleolar localization domain of the catalytic subunit of human telomerase. *J. Biol. Chem.* **277:** 24764–24770.

Ferrezuelo F., Steiner B., Aldea M., and Futcher B. 2002. Biogenesis of yeast telomerase depends on the importin mtr10. *Mol. Cell. Biol.* **22:** 6046–6055.

Fitzgerald M.S., McKnight T.D., and Shippen D.E. 1996. Characterization and developmental patterns of telomerase expression in plants. *Proc. Natl. Acad. Sci.* **93:** 14422–14427.

Fitzgerald M.S., Riha K., Gao F., Ren S., McKnight T.D., and Shippen D.E. 1999. Disruption of the telomerase catalytic subunit gene from *Arabidopsis* inactivates telomerase and leads to a slow loss of telomeric DNA. *Proc. Natl. Acad. Sci.* **96:** 14813–14818.

Forstemann K. and Lingner J. 2005. Telomerase limits the extent of base pairing between template RNA and telomeric DNA. *EMBO Rep.* **6:** 361–366.

Forsythe H.L., Jarvis J.L., Turner J.W., Elmore L.W., and Holt S.E. 2001. Stable association of hsp90 and p23, but Not hsp70, with active human telomerase. *J. Biol. Chem.* **276:** 15571–15574.

Friedman K.L. and Cech T.R. 1999. Essential functions of amino-terminal domains in the yeast telomerase catalytic subunit revealed by selection for viable mutants. *Genes Dev.* **13:** 2863–2874.

Friedman K.L., Heit J.J., Long D.M., and Cech T.R. 2003. N-terminal domain of yeast telomerase reverse transcriptase: Recruitment of Est3p to the telomerase complex. *Mol. Biol. Cell* **14:** 1–13.

Fu D. and Collins K. 2003. Distinct biogenesis pathways for human telomerase RNA and H/ACA small nucleolar RNAs. *Mol. Cell* **11:** 1361–1372.

Fulton T.B. and Blackburn E.H. 1998. Identification of *Kluyveromyces lactis* telomerase: Discontinuous synthesis along the 30-nucleotide-long templating domain. *Mol. Cell. Biol.* **18:** 4961–4970.

Gilley D. and Blackburn E.H. 1999. The telomerase RNA pseudoknot is critical for the stable assembly of a catalytically active ribonucleoprotein. *Proc. Natl. Acad. Sci.* **96:** 6621–6625.

Greenberg R.A., Allsopp R.C., Chin L., Morin G.B., and DePinho R.A. 1998. Expression of mouse telomerase reverse transcriptase during development, differentiation and proliferation. *Oncogene* **16:** 1723–1730.

Greider C.W. 1991. Telomerase is processive. *Mol. Cell. Biol.* **11:** 4572–4580.

Greider C.W. and Blackburn E.H. 1985. Identification of a specific telomere terminal transferase activity in *Tetrahymena* extracts. *Cell* **43:** 405–413.

———. 1989. A telomeric sequence in the RNA of *Tetrahymena* telomerase required for telomere repeat synthesis. *Nature* **337:** 331–337.

Guo W., Okamoto M., Lee Y.M., Baluda M.A., and Park N.H. 2001. Enhanced activity of cloned hamster TERT gene promoter in transformed cells. *Biochim. Biophys. Acta* **1517:** 398–409.

Hammond P.W. and Cech T.R. 1997. dGTP-dependent processivity and possible template switching of *Euplotes* telomerase. *Nucleic Acids Res.* **25:** 3698–3704.

———. 1998. *Euplotes* telomerase: Evidence for limited base-pairing during primer elongation and dGTP as an effector of translocation. *Biochemistry* **37:** 5162–5172.

Hammond P.W., Lively T.N., and Cech T.R. 1997. The anchor site of telomerase from *Euplotes aediculatus* revealed by photo-cross-linking to single- and double-stranded DNA primers. *Mol. Cell. Biol.* **17:** 296–308.

Hardy C.D., Schultz C.S., and Collins K. 2001. Requirements for the dGTP-dependent repeat addition processivity of recombinant *Tetrahymena* telomerase. *J. Biol. Chem.* **276:** 4863–4871.

Harrington L.A. and Greider C.W. 1991. Telomerase primer specificity and chromosome healing. *Nature* **353:** 451–454.

Heller-Uszynska K., Schnippenkoetter W., and Kilian A. 2002. Cloning and characterization of rice (*Oryza sativa* L) telomerase reverse transcriptase, which reveals complex splicing patterns. *Plant J.* **31:** 75–86.

Holt S.E., Aisner D.L., Baur J., Tesmer V.M., Dy M., Ouellette M., Trager J.B., Morin G.B., Toft D.O., Shay J.W., et al. 1999. Functional requirement of p23 and Hsp90 in telomerase complexes. *Genes Dev.* **13:** 817–826.

Huard S. and Autexier C. 2004. Human telomerase catalyzes nucleolytic primer cleavage. *Nucleic Acids Res.* **32:** 2171–2180.

Huard S., Moriarty T.J., and Autexier C. 2003. The C terminus of the human telomerase reverse transcriptase is a determinant of enzyme processivity. *Nucleic Acids Res.* **31:** 4059–4070.

Jady B.E., Bertrand E., and Kiss T. 2004. Human telomerase RNA and box H/ACA scaRNAs share a common Cajal body-specific localization signal. *J. Cell Biol.* **164:** 647–652.

Khurts S., Masutomi K., Delgermaa L., Arai K., Oishi N., Mizuno H., Hayashi N., Hahn W.C., and Murakami S. 2004. Nucleolin interacts with telomerase. *J. Biol. Chem.* **279:** 51508–51515.

Kim J.H., Park S.M., Kang M.R., Oh S.Y., Lee T.H., Muller M.T., and Chung I.K. 2005. Ubiquitin ligase MKRN1 modulates telomere length homeostasis through a proteolysis of hTERT. *Genes Dev.* **19:** 776–781.

Kim N.W. and Wu F. 1997. Advances in quantification and characterization of telomerase activity by the telomeric repeat amplification protocol (TRAP). *Nucleic Acids Res.* **25:** 2595–2597.

Kim N.W., Piatyszek M.A., Prowse K.R., Harley C.B., West M.D., Ho P.L., Coviello G.M., Wright W.E., Weinrich S.L., and Shay J.W. 1994. Specific association of human telomerase activity with immortal cells and cancer. *Science* **266:** 2011–2015.

Kuramoto M., Ohsumi K., Kishimoto T., and Ishikawa F. 2001. Identification and analyses of the *Xenopus* TERT gene that encodes the catalytic subunit of telomerase. *Gene* **277:** 101–110.

Lai C.K., Miller M.C., and Collins K. 2002. Template boundary definition in *Tetrahymena* telomerase. *Genes Dev.* **16:** 415–420.

———. 2003. Roles for RNA in telomerase nucleotide and repeat addition processivity. *Mol. Cell* **11:** 1673–1683.

Lai C.K., Mitchell J.R., and Collins K. 2001. RNA binding domain of telomerase reverse transcriptase. *Mol. Cell. Biol.* **21:** 990–1000.

Lendvay T.S., Morris D.K., Sah J., Balasubramanian B., and Lundblad V. 1996. Senescence mutants of *Saccharomyces cerevisiae* with a defect in telomere replication identify three additional EST genes. *Genetics* **144:** 1399–1412.

Lin J. and Blackburn E.H. 2004. Nucleolar protein PinX1p regulates telomerase by sequestering its protein catalytic subunit in an inactive complex lacking telomerase RNA. *Genes Dev.* **18:** 387–396.

Lin J., Ly H., Hussain A., Abraham M., Pearl S., Tzfati Y., Parslow T.G., and Blackburn E.H. 2004. A universal telomerase RNA core structure includes structure motifs required for binding the telomerase reverse transcriptase protein. *Proc. Natl. Acad. Sci.* **101:** 14713–14718.

Lingner J. and Cech T.R. 1996. Purification of telomerase from *Euplotes aediculatus*: Requirement of a primer 3′ overhang. *Proc. Natl. Acad. Sci.* **93:** 10712–10717.

Lingner J., Cech T.R., Hughes T.R., and Lundblad V. 1997a. Three Ever Shorter Telomere (*EST*) genes are dispensable for *in vitro* yeast telomerase activity. *Proc. Natl. Acad. Sci.* **94:** 11190–11195.

Lingner J., Hughes T.R., Shevchenko A., Mann M., Lundblad V., and Cech T.R. 1997b. Reverse transcriptase motifs in the catalytic subunit of telomerase. *Science* **276:** 561–567.

Livengood A.J., Zaug A.J., and Cech T.R. 2002. Essential regions of *Saccharomyces cerevisiae* telomerase RNA: Separate elements for Est1p and Est2p interaction. *Mol. Cell. Biol.* **22:** 2366–2374.

Lue N.F. and Peng Y. 1998. Negative regulation of yeast telomerase activity through an interaction with an upstream region of the DNA primer. *Nucleic Acids Res.* **26:** 1487–1494.

Lue N.F., Lin Y.C., and Mian I.S. 2003. A conserved telomerase motif within the catalytic domain of telomerase reverse transcriptase is specifically required for repeat addition processivity. *Mol. Cell. Biol.* **23:** 8440–8449.

Lundblad V. and Szostak J.W. 1989. A mutant with a defect in telomere elongation leads to senescence in yeast. *Cell* **57:** 633–643.

Magnenat L., Tobler H., and Muller F. 1999. Developmentally regulated telomerase activity is correlated with chromosomal healing during chromatin diminution in *Ascaris suum*. *Mol. Cell. Biol.* **19:** 3457–3465.

Maine I.P., Chen S.F., and Windle B. 1999. Effect of dGTP concentration on human and CHO telomerase. *Biochemistry* **38:** 15325–15332.

Martin-Rivera L. and Blasco M.A. 2001. Identification of functional domains and dominant negative mutations in vertebrate telomerase RNA using an in vivo reconstitution system. *J. Biol. Chem.* **276:** 5856–5865.

Mason D.X., Goneska E., and Greider C.W. 2003. Stem-loop IV of *Tetrahymena* telomerase RNA stimulates processivity in *trans*. *Mol. Cell. Biol.* **23:** 5606–5613.

McCormick-Graham M. and Romero D.P. 1995. Ciliate telomerase RNA structural features. *Nucleic Acids Res.* **23:** 1091–1097.

Melek M., Greene E.C., and Shippen D.E. 1996. Processing of nontelomeric 3′ ends by telomerase: Default template alignment and endonucleolytic cleavage. *Mol. Cell. Biol.* **16:** 3437–3445.

Metz A.M., Love R.A., Strobel G.A., and Long D.M. 2001. Two telomerase reverse transcriptases (TERTs) expressed in *Candida albicans*. *Biotechnol. Appl. Biochem.* **34:** 47–54.

Meyerson M., Counter C.M., Eaton E.N., Ellisen L.W., Steiner P., Caddle S.D., Ziaugra L., Beijersbergen R.L., Davidoff M.J., Liu Q., et al. 1997. hEST2, the putative human telomerase catalytic subunit gene, is up-regulated in tumor cells and during immortalization. *Cell* **90:** 785–795.

Miller M.C. and Collins K. 2002. Telomerase recognizes its template by using an adjacent RNA motif. *Proc. Natl. Acad. Sci.* **99:** 6585–6590.

Miller M.C., Liu J.K., and Collins K. 2000. Template definition by *Tetrahymena* telomerase reverse transcriptase. *EMBO J.* **19:** 4412–4422.

Mitchell J.R. and Collins K. 2000. Human telomerase activation requires two independent interactions between telomerase RNA and telomerase reverse transcriptase. *Mol. Cell* **6:** 361–371.

Mitchell J.R., Cheng J., and Collins K. 1999. A box H/ACA small nucleolar RNA-like domain at the human telomerase RNA 3′ end. *Mol. Cell. Biol.* **19:** 567–576.

Mollenbeck M., Postberg J., Paeschke K., Rossbach M., Jonsson F., and Lipps H.J. 2003. The telomerase-associated protein p43 is involved in anchoring telomerase in the nucleus. *J. Cell Sci.* **116:** 1757–1761.

Moriarty T.J., Ward R.J., Taboski M.A., and Autexier C. 2005. An anchor site-type defect in human telomerase that disrupts telomere length maintenance and cellular immortalization. *Mol. Biol. Cell* **16:** 3152–3161.

Morin G.B. 1991. Recognition of a chromosome truncation site associated with α-thalassaemia by human telomerase. *Nature* **353:** 454–456.

Nakamura T.M. and Cech T.R. 1998. Reversing time: Origin of telomerase. *Cell* **92:** 587–590.

Nakamura T.M., Morin G.B., Chapman K.B., Weinrich S.L., Andrews W.H., Lingner J., Harley C.B., and Cech T.R. 1997. Telomerase catalytic subunit homologs from fission yeast and human. *Science* **277:** 955–959.

Nasir L., Gault E., Campbell S., Veeramalai M., Gilbert D., McFarlane R., Munro A., and Argyle D.J. 2004. Isolation and expression of the reverse transcriptase component of the *Canis familiaris* telomerase ribonucleoprotein (dogTERT). *Gene* **336:** 105–113.

Niu H., Xia J., and Lue N.F. 2000. Characterization of the interaction between the nuclease and reverse transcriptase activity of the yeast telomerase complex. *Mol. Cell. Biol.* **20:** 6806–6815.

Oguchi K., Liu H., Tamura K., and Takahashi H. 1999. Molecular cloning and characterization of AtTERT, a telomerase reverse transcriptase homolog in *Arabidopsis thaliana*. *FEBS Lett.* **457:** 465–469.

Oulton R. and Harrington L. 2004. A human telomerase-associated nuclease. *Mol. Biol. Cell* **15:** 3244–3256.

Paushkin S., Gubitz A.K., Massenet S., and Dreyfuss G. 2002. The SMN complex, an assemblyosome of ribonucleoproteins. *Curr. Opin. Cell Biol.* **14:** 305–312.

Pellizzoni L., Baccon J., Charroux B., and Dreyfuss G. 2001. The survival of motor neurons (SMN) protein interacts with the snoRNP proteins fibrillarin and GAR1. *Curr. Biol.* **11:** 1079–1088.

Peng Y., Mian I.S., and Lue N.F. 2001. Analysis of telomerase processivity: Mechanistic similarity to HIV-1 reverse transcriptase and role in telomere maintenance. *Mol. Cell* **7:** 1201–1211.

Peterson S.E., Stellwagen A.E., Diede S.J., Singer M.S., Haimberger Z.W., Johnson C.O., Tzoneva M., and Gottschling D.E. 2001. The function of a stem-loop in telomerase RNA is linked to the DNA repair protein Ku. *Nat. Genet.* **27:** 64–67.

Prathapam R., Witkin K.L., O'Connor C.M., and Collins K. 2005. A telomerase holoenzyme protein enhances telomerase RNA assembly with telomerase reverse transcriptase. *Nat. Struct. Mol. Biol.* **12:** 252–257.

Prescott J. and Blackburn E.H. 1997. Functionally interacting telomerase RNAs in the yeast telomerase complex. *Genes Dev.* **11:** 2790–2800.

Reichenbach P., Hoss M., Azzalin C.M., Nabholz M., Bucher P., and Lingner J. 2003. A human homolog of yeast Est1 associates with telomerase and uncaps chromosome ends when overexpressed. *Curr. Biol.* **13:** 568–574.

Romero D.P. and Blackburn E.H. 1991. A conserved secondary structure for telomerase RNA. *Cell* **67:** 343–353.

Seimiya H., Sawada H., Muramatsu Y., Shimizu M., Ohko K., Yamane K., and Tsuruo T. 2000. Involvement of 14-3-3 proteins in nuclear localization of telomerase. *EMBO J.* **19:** 2652–2661.

Seto A.G., Livengood A.J., Tzfati Y., Blackburn E.H., and Cech T.R. 2002. A bulged stem tethers Est1p to telomerase RNA in budding yeast. *Genes Dev.* **16:** 2800–2812.

Seto A.G., Zaug A.J., Sobel S.G., Wolin S.L., and Cech T.R. 1999. *Saccharomyces cerevisiae* telomerase is an Sm small nuclear ribonucleoprotein particle. *Nature* **401:** 177–180.

Seto A.G., Umansky K., Tzfati Y., Zaug A.J., Blackburn E.H., and Cech T.R. 2003. A template-proximal RNA paired element contributes to *Saccharomyces cerevisiae* telomerase activity. *RNA* **9:** 1323–1332.

Shippen-Lentz D. and Blackburn E.H. 1990. Functional evidence for an RNA template in telomerase. *Science* **247:** 546–552.

Snow B.E., Erdmann N., Cruickshank J., Goldman H., Gill R.M., Robinson M.O., and Harrington L. 2003. Functional conservation of the telomerase protein Est1p in humans. *Curr. Biol.* **13:** 698–704.

Stellwagen A.E., Haimberger Z.W., Veatch J.R., and Gottschling D.E. 2003. Ku interacts with telomerase RNA to promote telomere addition at native and broken chromosome ends. *Genes Dev.* **17:** 2384–2395.

Sun D., Lopez-Guajardo C.C., Quada J., Hurley L.H., and Von Hoff D.D. 1999. Regulation of catalytic activity and processivity of human telomerase. *Biochemistry* **38:** 4037–4044.

Takenaka Y., Matsuura T., Haga N., and Mitsui Y. 2001. Expression of telomerase reverse transcriptase and telomere elongation during sexual maturation in *Paramecium caudatum*. *Gene* **264:** 153–161.

Teixeira M.T., Forstemann K., Gasser S.M., and Lingner J. 2002. Intracellular trafficking of yeast telomerase components. *EMBO Rep.* **3:** 652–659.

Tesmer V.M., Ford L.P., Holt S.E., Frank B.C., Yi X., Aisner D.L., Ouellette M., Shay J.W., and Wright W.E. 1999. Two inactive fragments of the integral RNA cooperate to assemble active telomerase with the human protein catalytic subunit (hTERT) in vitro. *Mol. Cell. Biol.* **19:** 6207–6216.

Tzfati Y., Fulton T.B., Roy J., and Blackburn E.H. 2000. Template boundary in a yeast telomerase specified by RNA structure. *Science* **288:** 863–867.

Tzfati Y., Knight Z., Roy J., and Blackburn E.H. 2003. A novel pseudoknot element is essential for the action of a yeast telomerase. *Genes Dev.* **17:** 1779–1788.

Ueda C.T. and Roberts R.W. 2004. Analysis of a long-range interaction between conserved domains of human telomerase RNA. *RNA* **10:** 139–147.

Underwood D.H., Zinzen R.P., and McEachern M.J. 2004. Template requirements for telomerase translocation in *Kluyveromyces lactis. Mol. Cell. Biol.* **24:** 912–923.

Wang H. and Blackburn E.H. 1997. De novo telomere addition by *Tetrahymena* telomerase in vitro. *EMBO J.* **16:** 866–879.

Wenz C., Enenkel B., Amacker M., Kelleher C., Damm K., and Lingner J. 2001. Human telomerase contains two cooperating telomerase RNA molecules. *EMBO J.* **20:** 3526–3534.

Witkin K.L. and Collins K. 2004. Holoenzyme proteins required for the physiological assembly and activity of telomerase. *Genes Dev.* **18:** 1107–1118.

Wong J.M., Kusdra L., and Collins K. 2002. Subnuclear shuttling of human telomerase induced by transformation and DNA damage. *Nat. Cell Biol.* **4:** 731–736.

Wong S.C., Ong L.L., Er C.P., Gao S., Yu H., and So J.B. 2003. Cloning of rat telomerase catalytic subunit functional domains, reconstitution of telomerase activity and enzymatic profile of pig and chicken tissues. *Life Sci.* **73:** 2749–2760.

Yang Y., Chen Y., Zhang C., Huang H., and Weissman S.M. 2002. Nucleolar localization of hTERT protein is associated with telomerase function. *Exp. Cell Res.* **277:** 201–209.

Yu G.L., Bradley J.D., Attardi L.D., and Blackburn E.H. 1990. *In vivo* alteration of telomere sequences and senescence caused by mutated *Tetrahymena* telomerase RNAs. *Nature* **344:** 126–132.

Zappulla D.C. and Cech T.R. 2004. Yeast telomerase RNA: A flexible scaffold for protein subunits. *Proc. Natl. Acad. Sci.* **101:** 10024–10029.

Zhu Y., Tomlinson R.L., Lukowiak A.A., Terns R.M., and Terns M.P. 2004. Telomerase RNA accumulates in Cajal bodies in human cancer cells. *Mol. Biol. Cell* **15:** 81–90.

4

Telomerase and Human Cancer

Jerry W. Shay and Woodring E. Wright
Department of Cell Biology
The University of Texas Southwestern Medical Center at Dallas
Dallas, Texas 75309-9039

THE CONTINUOUS GROWTH OF ADVANCED HUMAN CARCINOMAS almost universally correlates with the up-regulation of telomerase. As such, telomerase is a very attractive target for cancer therapeutics. Although the role of telomerase in cell immortalization is well established, it is less clear whether the expression of telomerase in cancer is simply a passive component of the neoplastic process, a critical requirement for stabilizing the genome to prevent catastrophic genomic instability, or a fundamental carcinogenic mechanism enhancing genomic instability. These concepts and several other key issues still remain to be resolved and are discussed in this chapter. In addition, the pros and cons of the most promising antitelomerase approaches currently being investigated are reviewed.

BACKGROUND

The ends of human chromosomes (telomeres) are composed of thousands of TTAGGG DNA repetitive sequences. Telomerase is a ribonucleoprotein enzyme complex (a cellular reverse transcriptase) that maintains telomere length in cancer cells by adding TTAGGG repeats onto the telomeric ends, thus compensating for the normal erosion of telomeres that occurs in all dividing cells. Telomerase is expressed during early development and remains fully active in specific germ-line cells, but it is undetectable in most normal somatic cells except for proliferative cells of renewal tissues (e.g., bone marrow cells, basal cells of the epidermis, proliferative endometrium, and intestinal crypt cells). In all nonreproductive proliferative cells, progressive telomere shortening is observed, and when telomeres become sufficiently short, further cell division is blocked, a process often referred to as replicative senescence (Harley et al. 1990). Telomere shortening is the molecular means to count the number of times a cell has divided and appears to function to initially protect human cells against the development of cancer by limiting the maximal number of permitted cell divisions. Thus, telomere shortening may be a potent tumor suppressor pathway. The telomere-telomerase hypothesis of aging and cancer is based on the findings that nearly all cancer cells have engaged a mechanism to maintain stable telomere lengths, almost universally by reactivating or up-regulating telomerase activity (Table 1).

Table 1. Telomerase activity in samples obtained from malignant tissues

Type of malignancy	No. samples positive/no. tested	% positive
Acute myeloid leukemia	47/64	73
Basal cell carcinoma	73/77	95
Bladder carcinoma	172/185	92
Breast carcinoma (ductal and lobular)	300/339	88
Cervical carcinoma	16/16	100
Chronic myeloid leukemia		
chronic	30/42	71
blast	21/21	100
Colorectal carcinoma	123/138	89
Gastric carcinoma	72/85	85
Head and neck squamous cell carcinoma	112/120	86
Hepatocellular carcinoma	149/173	86
Lymphoma		
low grade	12/14	86
high grade	16/16	100
Melanoma	6/7	85
Neuroblastoma	99/105	94
Non-small-cell lung carcinoma	98/125	78
Ovarian carcinoma	21/23	91
Pancreatic carcinoma	41/43	95
Prostate carcinoma	52/58	90
Renal carcinoma	95/115	83
Retinoblastoma	17/34	50
Small-cell lung carcinoma	15/15	100
Squamous cell carcinoma	15/18	83

Data are summarized from Shay and Bacchetti (1997). For primary study citations and details, please see this reference.

PREVENTING THE SHORTENING OF
TELOMERES PREVENTS CELLULAR AGING

Hermann Muller first described telomeres in a lecture in 1938 (Muller 1962), and Muller's work was the basis of much of the effort of Barbara McClintock (1941). However, it was not known at that time if telomeres had a role in cell replication. It was generally agreed that a cellular mechanism was needed to maintain telomeres in immortal single-cell organisms (Blackburn and Gall 1978) and that a similar mechanism was needed to preserve the germ line in multicellular organisms. The answer emerged in studies with *Tetrahymena thermophila* by Greider and Blackburn (1985; also see Chapter 3), who discovered the ribonucleoprotein enzyme activity that they referred to as terminal telomere transferase, now called telomerase. Greider and Blackburn found that telomeres are synthesized de novo by telomerase, by extending the 3′ end of telomeres and thus elongating them. Telomerase was later found to occur in extracts of immortal human cell lines (Morin 1989) and in about 90% of all human primary tumors (Kim et al. 1994).

Shortly after the discovery of telomerase, Howard Cooke discussed the variability of human telomeres (Cooke et al. 1985) and their implications at a Cold Spring Harbor Symposium (Cooke and Smith 1986). These provocative discussions probably increased interest in investigating telomeres, which in turn led to reports that normal human cells progressively lose telomeres and might lead to replicative senescence (Harley et al. 1990). In addition, it was reported that cancer cells have shorter telomeres than do adjacent normal cells (de Lange et al. 1990; Hastie et al. 1990), thus providing the first link for the role of telomeres in cancer biology.

Many studies have indicated a correlation between telomere shortening and proliferative failure of human cells, but the evidence for a causal relationship was not demonstrated until 1998. The formal proof that telomeres were rate-limiting for indefinite cell proliferation was shown by the introduction of the telomerase catalytic protein component—human telomerase reverse transcriptase (hTERT)—into normal telomerase-silent human cells. This resulted in restoration of telomerase activity (Bodnar et al. 1998). Normal human cells stably expressing transfected telomerase maintained or elongated telomeres, demonstrated an indefinite life span, maintained a normal chromosome complement, and continued to grow in a normal manner (Jiang et al. 1999; Morales et al. 1999). These observations provided the first direct evidence for the hypothesis that telomere length determines the proliferative capacity of human cells.

TELOMERASE REGULATION IN NORMAL AND CANCER CELLS

Telomerase is a multimeric ribonucleoprotein complex (Greider and Blackburn 1985; Morin 1989; Nugent and Lundblad 1998; Collins and Mitchell 2002), in which an essential protein component (hTERT) (Nakamura et al. 1997; Lingner and Cech 1998) uses an internal RNA— human telomerase RNA (hTR) or human telomerase RNA component (hTERC)—as a template (Feng et al. 1995) to catalyze the addition of telomeric sequences to the ends of chromosomes (Cech 2000). The enzyme activity was originally described by Greider and Blackburn in 1985, but it was almost a decade until hTERC was cloned (Feng et al. 1995), and this was subsequently followed by the cloning of the catalytic portion of the enzyme (Nakamura et al. 1997).

Telomerase activity is expressed in embryonic stem cells (Wright et al. 1996) and some proliferative reproductive cells, and these cells fully maintain their telomere lengths (Fig. 1). Although telomerase activity is detected in fetal, newborn, and adult testes and in fetal ovary, it is not detected in mature spermatozoa and oocytes (Wright et al. 1996). This is consistent with the general observation of the lack of telomerase expression in quiescent cells regardless of their competency to express telomerase.

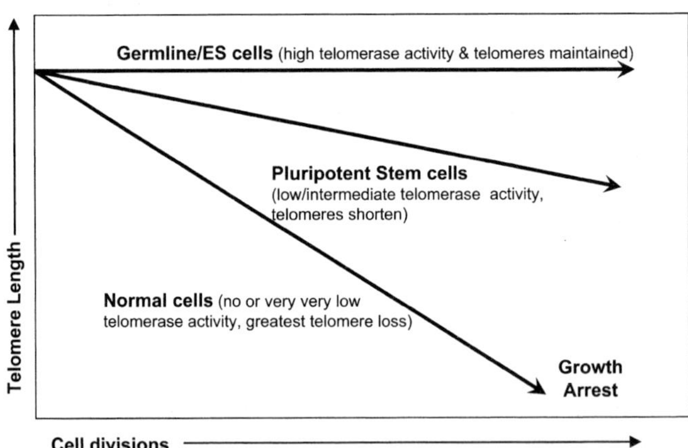

Figure 1. Regulation of telomerase. Only in specific germ-line cells and totipotent embryonic stem (ES) cells does telomerase fully maintain telomere length (Wright et al. 1996). In pluripotent stem cells of renewal tissues (e.g., blood, skin, and intestine), there is regulated telomerase activity that may help extend the proliferative capacity of these cells. However, these proliferative renewal cells show progressive telomere shortening throughout life. Many tissues and cells have either extremely low levels of telomerase or no detectable telomerase activity, and these cell types may show the greatest rate of telomere loss (Aisner et al. 2002; Cong et al. 2002; Forsyth et al. 2002).

Telomerase expression is detected at high levels by the blastocyst stage and is also present in essentially all human tissues tested during the first trimester. As gestation proceeds, telomerase activity continues to be expressed in a subset of fetal somatic tissues (liver, intestine, lung, skin, muscle, adrenal glands, and kidney), but it is not detected in brain and bone extracts after 16 weeks of gestation, despite ongoing cell division in these tissues. Telomerase activity is switched off in additional tissues as fetal progression continues (adrenal gland, muscle, lung, skin, and liver). Fetal tissues that have detectable telomerase activity display a rapid loss of detectable enzyme levels on explant into cell culture (Wright et al. 1996). This suggests that culture conditions might initiate a response leading to differentiation-like quenching of telomerase activity.

Several types of rapidly dividing adult pluripotent stem cells also express some telomerase (Forsyth et al. 2002), but these tissues exhibit progressive telomere shortening throughout life (Fig. 1). Finally, telomerase activity is greatly reduced or transcriptionally silenced in most adult tissues (Aisner et al. 2002; Cong et al. 2002), and as a consequence, telomeres shorten with each cell division with increased age, perhaps at a faster rate compared to stem cells of renewal tissues such as in the skin, intestine, and blood (Fig. 1).

To explore telomerase regulation, Lin and Elledge (2003) employed a general genetic screen to identify negative regulators of *hTERT* (Fig. 2). They discovered a variety of tumor suppressor/oncogenes involved in *hTERT* repression. One, *Mad1/c-Myc*, had been previously implicated in *hTERT* regulation. A second, *SIP1* (a transcriptional target of the *TGF-β* pathway), mediates *TGF-β*-regulated repression of *hTERT*. A third, the tumor suppressor Menin, is a direct repressor of *hTERT*. Depleting Menin-immortalized primary human fibroblasts caused a transformation phenotype when coupled with expression of SV40 large and small T antigen and oncogenic *ras*. In addition, there was a novel protein with two BRCT domains identified (BRIT1) and a protein kinase that interacts with Rb (Rak). These studies suggest that multiple tumor suppressor/oncogene pathways coordinately repress *hTERT* expression and imply that telomerase is reactivated in human tumors through oncogenic mutations.

ARE SHORTENED TELOMERES AN ONCOGENIC OR TUMOR SUPPRESSOR MECHANISM?

In normal cells, once a shortened telomere length is attained, a DNA damage signal cascade correlating with DNA repair foci leads to cellular senescence (Fig. 3) (d'Adda di Fagagna 2003; Takai et al. 2003; Zou

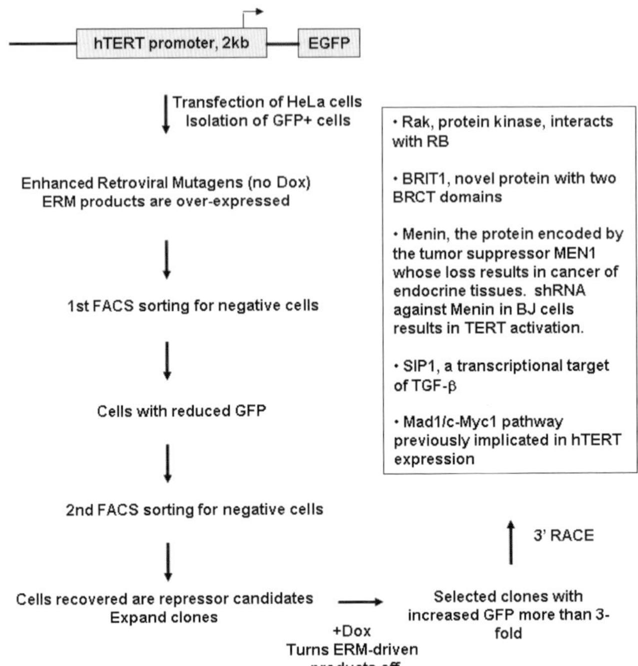

Figure 2. Enhanced retroviral mutagenesis to identify multiple tumor suppressor pathways negatively regulating telomerase. Reporter HeLa cells were infected with enhanced retroviral mutagenesis (ERM) retroviral vectors containing an epitope tag and splice donor under the control of a tet-responsive element. Cells with reduced green fluorescent protein (GFP) intensity due to the expression of the ERM-driven insertion were FACS (fluorescence-activated cell sorting) sorted and expanded. As a secondary screen, doxycycline was added to repress the expression directed by the ERM retroviral insertions in these cells. Cells that restored GFP expression in the presence of doxycycline were sorted, and candidate telomerase repressors were characterized (Lin and Elledge 2003).

et al. 2004). Although it was formerly thought to be possible that a single sentinel telomere was responsible for the initiation of senescence, recent work has shown that a group of approximately 10 of the 92 human telomere ends form associations with each other (Zou et al. 2004). In any particular senescent cell, it is only one or a few short "uncapped" telomeres that lead to the DNA-damage signal-induced growth arrest that is responsible for M1 (mortality stage 1) replicative senescence (Zou et al. 2004).

A central question is whether the short telomeres that induce replicative senescence act as an initial antitumor protection mechanism

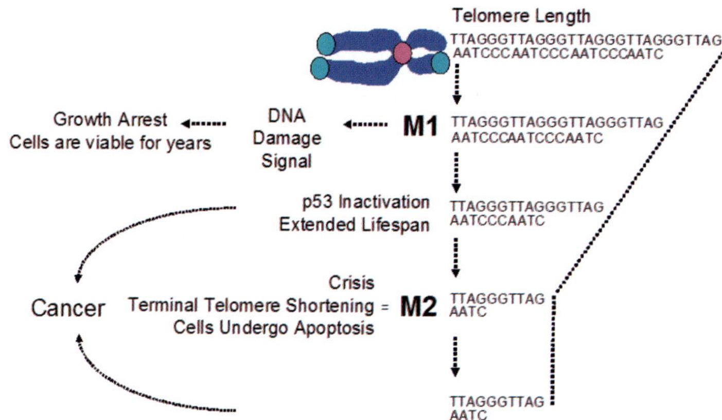

Figure 3. Senescence, crisis, and immortalization. When a few telomeres are short, end associations trigger a DNA-damage signaling pathway leading to a long-term growth arrest (often referred to as replicative senescence, the Hayflick limit, or M1 for mortality stage 1). In cells that have alterations in the p53 signaling pathway, there is a bypass of senescence leading to an extended life span (Wright et al. 1989). There are some instances where the abrogation of M1 senescence can lead directly to the cancer state, but in most instances, there is eventually a second growth-arrest state referred to as crisis or mortality stage 2 (M2). The vast majority of cells at M2 (crisis) have undergone terminal telomere shortening, frequently resulting in the induction of apoptosis. In a rare case, however, there is escape from M2, cellular immortality, and the expression of telomerase activity. Because most cancer cells are immortal and express telomerase activity, this is the common pathway for oncogenic transformation (Kim et al. 1994; Shay and Bacchetti 1997).

in the absence of other genetic and epigenetic alterations or actually drive genomic instability leading to cancer. It is known that human cells can bypass M1 senescence when cell cycle checkpoint (e.g., pRb and p53) pathways are disabled (Wright et al. 1989; Shay et al. 1991a,b; Counter et al. 1992), but ultimately the continuing cell division in the absence of telomerase leads to terminal telomere shortening (crisis, M2 [mortality stage 2]), chromosome end fusions, chromosome breakage-fusion-bridge cycles, and mitotic catastrophe resulting in loss of cell viability (Fig. 3).

The ability to circumvent telomere-based growth limitations is thought to be one critical and rate-limiting step in the evolution of most malignancies (Harley 1991; Shay et al. 1991b). Clinically significant tumors must have a mechanism for telomere maintenance to have the unlimited proliferative capacity that appears to be required for most advanced cancers. The almost universal path to overcoming the telomere mitotic clock is the reactivation or up-regulation of telomerase (Fig. 3) (Kim et al. 1994; Shay and Bacchetti 1997). In addition to demonstrating that ectopic expression of hTERT is sufficient to immortalize normal human cells (Bodnar et al. 1998), hTERT expression results in telomere maintenance without oncogenic transformation (Jiang et al. 1999; Morales et al. 1999; Harley 2002) as long as cells are cultured under adequate conditions (Ramirez et al. 2001). Introduction of hTERT either before or after M1 senescence results in direct immortalization, suggesting that the initiation of both M1 (senescence) and M2 (crisis) is mechanistically based on telomere dysfunction (Shay and Wright 2004). In summary, although a few short telomeres lead to M1 senescence and may be thought of as a potent anticancer protection mechanism, in cells that have bypassed critical cell cycle checkpoint pathways, telomeres not only continue to shorten, but may also lead to genomic instability and cancer progression, including telomerase activation.

WHY DO CELLS AGE?

To understand the role of telomeres in protecting against the development of cancer and the role of telomerase in cancer, we first must consider the more fundamental question of why human cells age. In 1825, Benjamin Gopertz described that after the age of 30 the likelihood of dying doubles every 7 years (Gompertz 1825). As we age, there is a gradual decline in performance of organ systems, resulting in the loss of reserve capacity, leading to an increased chance of death. In 1891, August Weisman wrote "death takes place because a worn-out tissue cannot forever renew itself, and because a capacity for increase by means of cell division is not everlasting but finite" (Weisman 1891). Formal evidence that normal cells could not divide forever and thus could not infinitely renew tissues came in 1961, when Hayflick demonstrated that normal human cells had a limited replicative life span and became senescent after approximately 50–70 doublings (Hayflick and Moorhead 1961). One rationale for the evolution of this limited life span is that it serves as an antitumor mechanism. Most tumor cells require at least four to six mutations to become malignant. Each mutation probably uses up at least 20–30 doublings (both to attain an adequate population size for the next mutation and for the usual necessity of eliminating wild-type alleles for recessive mutations). Replicative senescence would thus inhibit the progression of premalignant cells after they had accumulated a few mutations. Indeed, many correlative reports show that precancerous lesions have critically short telomeres (Kitada et al. 1995; O'Sullivan et al. 2002; Wiemann 2002; Wu et al. 2003;

Meeker and De Marzo 2004; Meeker et al. 2004). However, the benefit of limiting the number of cell doublings must be balanced against the need of the organism for cell turnover to maintain and repair its tissues. It is likely that the number of permitted doublings has evolved to provide sufficient divisions for reasonably adequate reserves during our "historical" expected life span of 30–40 years, but that increases beyond that would come at the expense of increased cancer. As sanitation and medical advances have increased our expected life span, decreased proliferative reserves may begin to affect tissue maintenance and contribute to the physiology of aging.

OF MICE AND MEN

Although most fundamental pathways are highly conserved between different species, it is important to recognize that there are likely to be differences between short-lived species such as the laboratory mouse and long-lived species such as higher primates, including humans. If replicative aging evolved as a mechanism to limit the number of available cell divisions, and thus acts as a brake against the accumulation of the multiple mutations needed for a cell to become malignant, then a 70-kg man who lives for 80 years must be approximately 140,000 times more resistant to cancer than a 0.02-kg mouse that lives for 2 years ([70 kg ÷ 0.02 kg] × [80 years ÷ 2 years] = 140,000). Although mice and humans have about the same lifetime frequency of cancer (~30%), there is also compelling evidence that rodent cells do not use the same mechanism as human cells to regulate the maximal number of cell divisions (Wright and Shay 2000; Shay and Wright 2004). Not only are mouse and rat telomeres five to ten times longer than telomeres in human cells, but many rodent tissues also have detectable telomerase (Shay and Wright 2001). We believe that this represents a fundamental biological difference between normal human cells (which count cell divisions) and rodent cells (which do not). Appreciating this difference is essential for designing and interpreting experiments concerning the role of replicative aging, telomeres, and telomerase in aging and cancer.

CANCER STEM CELLS AND TELOMERASE

It has recently been reported that most cancers are derived from a small subset of rare "stem cells" that have special biological properties and are responsible for tumor maintenance (Reya et al. 2001; Beachy et al. 2004). The hallmark of these stem cells is their ability to self-renew, not their ability to differentiate into more mature cells. More mature or differentiated cells form the bulk of tumor masses and display the phenotypic characteristics that are frequently obtained by microarray and proteomics analyses. However, these more mature cells may not be responsible for resistance to treatment. Because these cancer stem cells are quite rare (estimated to be perhaps 1/1500 cells in primary human breast carcinomas [Al-Hajj et al. 2003] and 10^{-4} or 10^{-5} in hematological malignancies [R. Jones and M. Clarke, pers. comm.]), it has recently been proposed that standard cancer therapies may not be targeting cancer stem cells. There are techniques to isolate these rare breast and hematological cancer stem cells (Al-Hajj et al. 2003), and it has been demonstrated that they are telomerase-positive (M. Clarke, pers. comm.). Thus, telomerase inhibitors being developed and discussed later in this chapter are likely to target both cancer stem cells and the more mature cancer cells. At present, it is not known if the regulation of telomerase in cancer stem cells is the same or different from that of the more differentiated cancer cells.

TELOMERASE DIAGNOSIS IN CANCER

The development of a highly sensitive telomere repeat amplification protocol (referred to as the TRAP assay) allows screening of a wide variety of human tumors, utilizing very small amounts of tissue (Kim et al. 1994). Telomerase activity has been detected in approximately 85% of the malignant human tumors tested (see Table 1). Some highly proliferative normal cells have detectable telomerase activity, but most exhibit levels of activity significantly lower than that detected in high-grade malignant tumors. At the present time, it is not known whether all human tumors arise from preexisting stem-like cells that are telomerase competent, leading to the up-regulation or dysregulation of telomerase (Greaves 1996), or whether tumors are derived from telomerase-silent more differentiated cells with a concomitant reactivation of telomerase activity (Shay and Wright 1996a). Of potential interest is the use of telomerase as a screen for the early detection of human cancer, prior to progression to a high-grade malignant status (Shay 1998). For example, some precancerous lesions have detectable telomerase activity (Shay and Wright 1996b), yet are classified as premalignant by standard cytopathology and histopathology. Detection of telomerase activity in these samples may indicate that these seemingly benign lesions contain a small subset of malignant cells and may assist in patient risk stratification. Thus, the hope is that telomerase measurements may have a utility in early cancer diagnostics and prognostics. However, additional studies on the level and distribution of telomerase activity in the early stages of cancer are needed to determine the utility of telomerase detection in early diagnosis, especially as it relates to clinical outcome.

Several studies have proposed detection of telomerase activity as a predictor of poor clinical outcome (E. Hiyama et al. 1995a,b, 1996; K. Hiyama et al. 1995; Clark et al. 1997; Carey et al. 1999). One example of this is the childhood cancer, neuroblastoma, where certain subclasses (stage III and IV) have high levels of detectable telomerase activity and poor prognosis (E. Hiyama et al. 1995a). Interestingly, one very malignant subclass (stage IV-S) has been shown to spontaneously regress and is actually telomerase-negative (Fig. 4). Without telomerase, the tumor undergoes telomere shortening, the cells presumably enter crisis, and the cancer is eventually cleared from the body. This study also demonstrates that although telomerase or another mechanism is likely to be required for the sustained growth of advanced tumors, it is not required for the initiation of cancer (Fig. 4). In another example, telomerase activity is only rarely detected at low levels in normal lung tissue, but it is detected at high levels in almost all cancer specimens (Fig. 5; Table 1). Small-cell lung carcinoma and malignant mesothelioma specimens almost universally have detectable telomerase activity, but there are a significant number (~15%) of non-small-cell lung carcinomas that are telomerase negative (K. Hiyama et al. 1995). In some instances, primary lung carcinomas are negative or weakly positive for telomerase activity whereas metastasis from the primary lesions are strongly telomerase-positive. This suggests that telomerase is not necessarily required for the initiation of cancer but is likely to be required for the sustained proliferation that characterizes most advanced malignancies. These and most other published studies report that telomerase shows high sensitivity and specificity for cancer detection.

Another example of the importance of developing additional diagnostic screens is the use of telomerase in detecting bladder cancer as an adjuvant to normal cytology. Almost 50% of early-stage bladder tumor cases are initially missed because of the lack of effective cytological markers for detection. Telomerase can be detected in the vast majority (approaching 90%) of bladder cancers using less invasive biopsy techniques (obtaining cells from bladder washings or voided urine) (Lin et al. 1996; Dalbagni et al. 1997; Kinoshita et al. 1997; Yoshida et al. 1997). Thus, detection of telomerase activity relative to human cancer development is potentially an important and novel method, in combination with traditional cytology, for early cancer diagnosis.

Another area of interest is monitoring patients for residual or recurrence of disease using telomerase as a marker. Telomerase activity

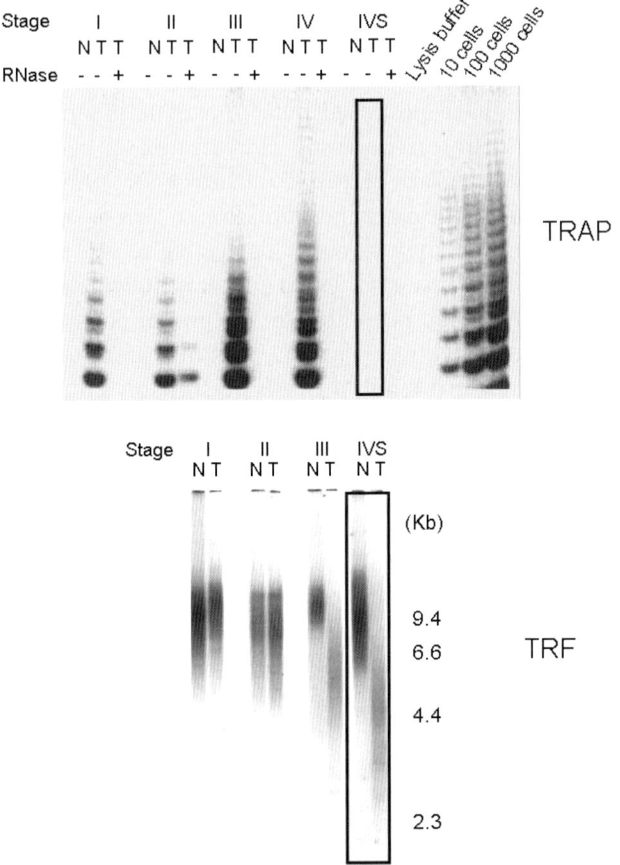

Figure 4. Pediatric neuroblastoma (K. Hiyama et al. 1995). Telomerase polymerase chain reaction (PCR)-based activity assay (TRAP) indicates that some metastatic tumors are negative (*top*), but these have very short telomeres as measured by the TRF assay (*bottom*). Patients without detectable telomerase activity and short telomeres have better outcomes. This suggests that although telomerase is not required for the initiation of cancer, a mechanism to maintain telomere stability is required for the long-term growth of the advanced tumor cell. (N) Normal tissue; (T) tumor.

is detected in adjacent, presumably cleared, regions in more than 10% of biopsied tumor specimens. In these instances, occult micrometastases of the primary tumor can invade the surrounding normal tissue, and small amounts of tumor may remain undetected by pathologists. Recurrence of cancer could be monitored by fine-needle aspiration of cancer patient tissues to assess the progression of the tumor due to this micrometastatic event using telomerase detection.

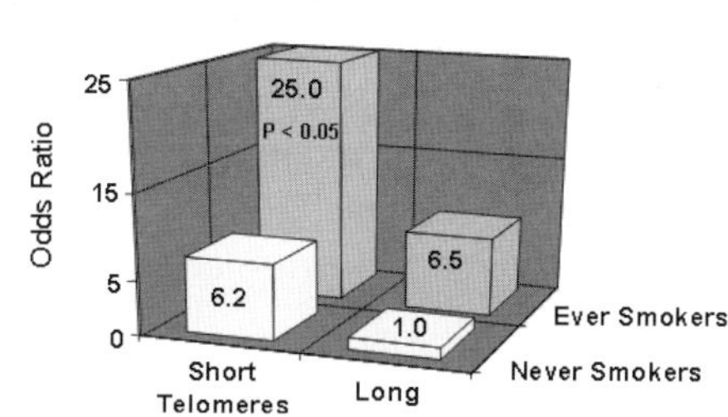

Figure 5. (*A*) Lung cancer and telomerase (K. Hiyama et al. 1995). Although patients with advanced pleural effusions and small-cell lung cancer (SCLC) almost universally have telomerase activity, non-small-cell lung cancers (NSCLC) are telomerase positive in only 85% of the cases. This could be due to experimental procedures or that a substantial fraction of lung tumors are telomerase negative perhaps engaging the ALT pathway or, similarly to the pediatric neuroblastoma (E. Hiyama et al. 1995a) studies, are initially telomerase silent. (*B*) Wu et al. (2003) describe telomere length and risk of developing smoke-related cancers. Odds ratios indicate that current and former smokers have a large increased risk for developing smoke-related cancer if their telomeres are short. However, those who have never smoked also have an increased risk for the development of lung cancer if their telomeres are short.

By comparing variance of telomere length in peripheral blood lymphocytes (PBLs) of monozygotic ($n = 52$) versus dizygotic ($n = 26$) twins, it was determined that the estimated heritability was 0.735 for the monozygotic twins compared to 0.425 for the dizygotic twins, with an overall heritability of 0.624 (Wu et al. 2003). These results indicated that telomere length appears to be largely a heritable trait. Using a molecular epidemiological approach and quantitative fluorescent in situ hybridization (FISH) laser-scanning cytometry assay, it was then demonstrated that individuals who were never smokers with short telomere lengths had a sixfold significantly elevated risk for bladder cancer (Wu et al. 2003). For current and former smokers with short telomeres, the odds ratios for elevated risk for cancer increased to 25-fold (Fig. 5). There was a dose–response relationship between bladder cancer risk and the degree of telomere shortening (Wu et al. 2003). Similar findings were also observed in lung, renal cell, and head and neck cancers (Wu et al. 2003). These findings illustrate the genetic and epigenetic contributions of telomere variations and provide suggestive evidence that telomere dysfunction may be a predisposing factor for cancer.

TELOMERASE INHIBITOR APPROACHES

Several telomerase targeting approaches are currently in preclinical and clinical trials (Fig. 6), including using the hTERT or hTR promoter to increase the cancer specificity of oncolytic virus gene therapy (Fig. 6). There are direct telomerase enzyme inhibitors, including gene therapy, oligonucleotide approaches, and small-molecule inhibitors. Finally, several ongoing clinical trials are testing a cancer immunotherapy using hTERT peptides, the whole hTERT gene priming dendritic cells, direct vaccines by injecting in the skin hTERT peptides (R. Vonderheide and G. Gaudernack, pers. comm.) and by mixing a plasmid containing hTERT directly into lymphocytes (M. Zanetti, pers. comm.). There are several potential outcomes of these therapies, such as a period of telomere attrition before cell growth arrest or death occurs (Shay 2003). The possibility also exists that there could be catastrophic or rapid telomere deletion events in some cancer cell types that normally would not have an adverse effect on cancer cells expressing telomerase, but which could have immediate effects in the presence of a telomerase inhibitor (Fig. 7). Finally, if telomerase is important in stabilizing the genome, then inhibiting telomerase can in theory result in increased genomic instability and possible engagement of a mechanism that is an alternative to telomerase for telomere length maintenance called ALT (Fig. 6).

Using mutated (dominant/negative) versions of hTERT and oligonucleotides directed at the hTR template region, it is predicted that there would be a lag phase until the telomeres of the cancer cells shortened sufficiently to result in critical telomere-shortening effects. In these instances, telomerase inhibitors may be most effective in combinations with other conventional cancer treatments or in a setting of minimal residual disease (White et al. 2001; Chen et al. 2003). However, in some primary human tumors, there may be some very short telomeres, and the lag phase before the telomerase inhibitor produces proliferative deficiencies in vivo may be more rapid (Fig. 7). This is based on some unexpected results from early preclinical research. There is significant

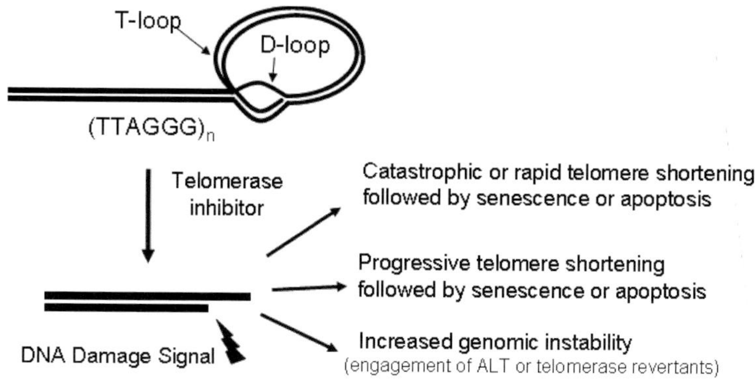

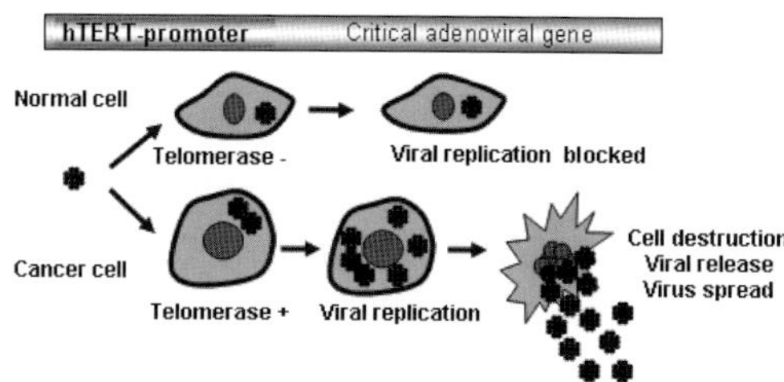

Direct Enzyme Inhibitor
- Gene therapy
- Oligonucleotide
- Small molecule inhibitor

Cancer Immunotherapy (in clinical trials)
- hTERT gene (ex vivo or in vivo)
- hTERT peptide

Figure 6. Telomerase inhibitor approaches. (*Top*) There are several potential outcomes when human tumors are treated with telomerase inhibitors (Corey 2000, 2002; White et al. 2001; Granger et al. 2002; Shay and Wright 2002; Shay 2003; Keith et al. 2004). There is the possibility for very rapid induction of cell growth arrest or death, a time delay depending on initial tumor telomere length until cells growth-arrest or undergo apoptosis, as well as the possibility that more genomic instability could occur leading perhaps to a drug resistance DNA recombination pathway called ALT (alternative lengthening of telomeres). (*Middle*) Oncolytic viruses are used to selectively kill telomerase-expressing cancer cells. Using the hTERT promoter driving critical adenoviral genes could result in highly selective replication in only tumor cells, sparing telomerase-negative normal cells. To avoid affecting stem-like cells, an oncolytic virus that incorporates two cancer-specific pathways (e.g., alteration in Rb/E2F or p53 along with the hTERT promoter) could provide added safety against killing of normal telomerase-expressing proliferative stem cells. (*Bottom*) Other telomere approaches include enzymatic inhibitors and immunotherapy.

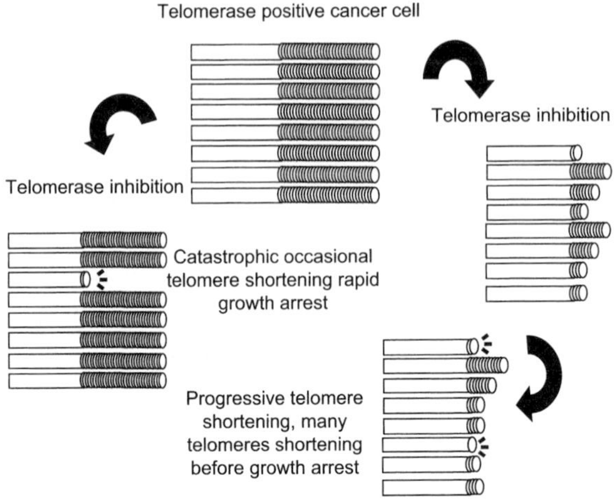

Figure 7. Different effects of telomerase inhibition. Small-molecule telomerase inhibitors would be predicted to take time (perhaps 1–3 months of treatment) before cell death occurred. This is based on most cell culture experiments where up to 3 months of continuous treatment is required to affect cell proliferation. However, there is also evidence of rapid telomere deletion events that could occur after telomerase inhibition. It is possible that some tumor types could respond rapidly to telomerase inhibition by "uncapping" an already very short telomere or by some catastrophic telomere loss event.

telomere length heterogeneity in cancer cells (Holt and Shay 1999), a few dysfunctional telomeres can trigger growth arrest or apoptosis (Kim et al. 2001; d'Adde di Fagagna et al. 2003; Zou et al 2004), and telomerase works preferentially on short telomeres (Ouellette et al 2000; Hemann et al. 2001). These reports suggest that telomerase inhibitors could have rapid, telomere-dependent effects in some tumors without necessarily changing overall bulk telomere length (E. Blackburn, pers. comm.).

Although telomerase is greatly reduced or not detected in most normal tissues (Wright et al. 1996; Forsyth et al. 2002), inhibitors of telomerase could have detrimental effects on those human cells that do express telomerase, such as hematopoietic progenitor cells, germ-line cells, and other cells of the renewal tissues such as the epidermis and intestinal crypts (White et al. 2001; Forsyth et al. 2002; Granger et al. 2002).

Although the potential complications are the same as those of other cancer therapeutic agents currently in use, clinical trials with telomerase immunotherapy approaches have not found stem cell effects to be a serious problem (Minev et al. 2000; Nair et al. 2000; Vonderheide et al. 2001a,b; Vonderheide 2002; M. Zanetti and G. Gaudernack, pers. comm.).

These studies have examined patients for more than 2 years without any evidence of serious adverse effects. Patients have not shown evidence of effects on bone marrow stem cells, and there has not been evidence of autoimmune disease in long-term survivors who continue to receive monthly booster vaccines (G. Gaudernack, pers. comm.).

Using progressive telomere attrition approaches, telomerase-specific inhibitors would shorten the telomeres in the proliferating transient amplifying stem cells; this would not cause their immediate death and would not affect quiescent stem cells. Thus, there could be a window of opportunity to eliminate cancer cells with short telomeres while having only minimal effects on stem cells with longer telomeres. At the present time, it is not known if cancer stem cells have long or short telomeres.

Another issue that has been raised about telomerase inhibitors is that alternative mechanisms for telomere maintenance (ALT) have been reported in other organisms, in experimentally derived human immortalized cell lines, and in some rare human cancers (Lundblad and Blackburn 1993; Bryan et al. 1995, 1997; Bryan and Reddel 1997; Henson et al. 2002; Lundblad 2002; Chapter 7). Telomerase inhibitors might thus result in the emergence of drug resistance telomerase-independent cancer cells. Although this is certainly a possibility, until recently there were no published reports of telomerase-positive human tumor cells being experimentally converted to a telomerase-independent pathway using telomerase inhibitors. Even in this recent report (Bechter et al. 2004), the engagement of an ALT-like phenotype was a rare event and occurred in cells with multiple other alterations including defects in mismatch repair pathways (Bechter et al. 2004). This and other reports have shown that ALT cancer lines are much less oncogenic compared to the same ALT cells expressing an ectopically introduced hTERT (Stewart et al. 2002; Bechter et al. 2004). Thus, although ALT is a potential resistance pathway to telomerase inhibitors, it is unlikely, at present, to be a major factor limiting the usefulness of telomerase inhibitors in a clinical setting.

It has also been suggested that telomerase inhibitors could lead to increased genomic instability in surviving cells (Hackett and Greider 2002; Harrington and Robinson 2002), leading to more advanced or aggressive tumors. Although some experimental animal model evidence supports this, especially in p53-deficient tumors (Chin et al. 1999; Gonzalez-Suarez et al. 2001), most of the experiments treating human cancer cell lines with telomerase inhibitors and experimental models producing progressively shortened telomeres indicate that these approaches do not lead to a more aggressive tumor (deLange and Jacks 1999; Greenberg et al. 1999;

Hahn et al. 1999; Herbert et al. 1999; Zhang et al. 1999; Gonzalez-Suarez et al. 2000; Kim et al. 2001; Boklan et al. 2002; Franco et al. 2002; Goytisolo and Blasco 2002; Shay and Wright 2002).

There are many creative approaches being tested for telomerase inhibition. Some of these are still very early in the preclinical stages, including inhibition of telomerase assembly (e.g., interfering with p23 Hsp90 [Holt et al 1999]), hammerhead ribozymes directed against hTR (Yokoyama et al. 1998), mutant template RNA gene therapy (Kim et al. 2001), and reverse transcriptase inhibitor approaches (Strahl and Blackburn 1996; Gomez et al. 1998; Murakami et al. 1998). For many of these novel approaches, there is a crucial need for better animal-based methods to screen potential telomerase therapeutic modalities prior to initiating clinical trials. Our lack of detailed knowledge of cancer drug uptake in vivo is another primary obstacle for understanding the physiology of normal and cancerous tissue. Bioluminescence assays have recently been developed for studying cancer telomerase therapeutic modalities in xenograft systems. For example, we have introduced a luciferase reporter gene into human breast, prostate, pancreatic, and lung cancer cell lines and have established (our unpublished data) that once introduced into nude mice, in both heterotopic and orthotopic locations, human tumor cells progressively increase in volume and that the tumor burden is directly proportional to the light emission after a single injection of the luciferase substrate, D-luciferin (bioluminescence imaging). We have also established, using standard chemotherapy protocols, that we can produce a setting of minimal residual tumor disease in mice, and we are testing long-term effects of novel antitelomerase cancer therapeutic protocols combined with conventional therapies (our unpublished data). This will permit the development of better protocols for cancer drug delivery by repetitively imaging human tumors in living mice using bioluminescence imaging. These studies will also allow us to monitor spatiotemporal distribution of bioluminescence in human tumor cells longitudinally during the disease course from minimal residual disease to late stages of disease and to obtain quantitative information and not just qualitative information. Combining imaging with animal modeling will allow the animals to be used as their own controls and permit acquisition of molecular data from tumors that are interacting on a molecular level with their microenvironment. This will permit a cohort of animals to be imaged for the entire course of an experiment, which includes tumor initiation, growth, treatment, and regrowth. Animal models should provide useful information for the design of preclinical and clinical telomerase cancer trials.

Other telomerase therapeutics areas that appear to be close to clinical trials and that have already undergone animal preclinical testing include blocking telomere accessibility (G-quadraplex stabilizers [Sun et al. 1997; Read 1999; Riou et al. 2002; Gowan et al. 2002]); combining the transcriptional regulatory sequences from the hTERT or hTR genes to regulate the expression of a cytotoxic drug (Abdul-Ghani et al. 2000; Majumdar et al. 2001; Plumb et al. 2001); using a tumor-specific oncolytic virus approach (Koga et al 2000, 2001; Komata et al. 2001; Gu et al. 2001, 2002); and oligonucleotide approaches (Norton et al. 1996; Pitts and Corey 1998; Matthes and Lehman 1999; Herbert et al. 1999, 2001, 2002; Gryaznov et al. 2001).

Finally, phase I clinical trials using telomerase immunotherapy/vaccine approaches have been very encouraging and are moving forward with additional studies (Minev et al. 2000; Nair et al. 2000; Vonderheide et al. 2001a,b; Vonderheide 2002; M. Zanetti and G. Gaudernack, pers. comm.). Some of these trials have followed patients for more than 2 years without any evidence of serious adverse effects and with some partial responses in patients with advanced malignancies (G. Gaudernack, pers. comm.). Randomized phase II clinical trials are being planned and hopefully will show objective responses in patients and move this approach in the future to vaccinating patients with less advanced malignancies.

Details about telomerase therapy approaches can be found in a series of other reviews (McKenzie et al. 1999; Corey 2000, 2002; Folini et al. 2000; Hodes 2001; Kelland 2001; Granger et al. 2002; Helder et al. 2002; Neidles and Parkinson 2002; Pathak et al. 2002; Shay and Wright 2002; Keith et al. 2004).

SUMMARY

Telomerase is a critical, rate-limiting enzyme that overcomes the growth limitation due to telomere loss that occurs in most cancers. Many precancerous tissues have critically shortened telomeres prior to telomerase detection (Kitada et al. 1995; Wiemann et al. 2002; O'Sullivan et al. 2002; Wu et al. 2003; Meeker and De Marzo 2004; Meeker et al. 2004). This suggests that short telomeres may limit the growth of precancerous cells and that only when telomerase is up-regulated or reactivated do additional cancerous changes occur. However, by itself, telomerase does not cause growth deregulation or promote a more aggressive phenotype in the absence of other alterations (Harley 2002). Many cell types have been immortalized by introduction of hTERT, and these do not have changes that are considered cancerous (Jiang et al. 1999; Morales et al. 1999).

Cancer remains the major cause of death in the United States for those under the age of 85 (98% of the population) despite substantial progress in understanding its molecular mechanisms and the development of an array of powerful treatments. The discovery of validated targets and new drugs is therefore a high priority. Telomerase-based drugs have the potential to act by novel mechanisms that will provide new options for cancer therapy and allow for unprecedented therapeutic specificity and efficacy. Telomerase inhibitors, for example, might not only directly limit or stop the growth of human tumors, but also act in a synergistic fashion with existing therapeutic modalities and amplify their effectiveness (Chen et al. 2003). After initial chemotherapy or surgery, telomerase inhibitors might be used to inhibit the recovery of residual cancer cells, making them more susceptible to attack by the immune system or killing by existing chemotherapeutic agents or other novel therapies (Chen et al. 2003). Similarly, the action of angiogenic inhibitors, which permit ongoing cell division/apoptosis while preventing tumors from outgrowing their blood supply, might be useful when combined with telomerase inhibitors (J. Shay, unpubl.). The consequent telomere shortening might convert the "tumoristatic" effect of angiogenesis inhibitors into a "tumoricidal" effect of the combination therapy.

Immunotherapy directed against telomerase-positive cells, as well as strategies using oncolytic viruses containing a gene promoter targeted to hTERT expressing cells, is currently under active investigation and in some instances in early stage clinical trials. These modalities have the advantage of avoiding the lag phase required with what may be considered the classic mode of telomerase inhibition. However, these treatments may also prove to be more toxic to normal somatic cells expressing telomerase. The use of telomerase template antagonists would usually require a lag phase prior to any detrimental effects on the cells and may not be a stand-alone strategy in patients with a large tumor burden. Telomerase template antagonists and other small-molecule approaches to inhibit telomerase would be most effective in a situation of minimal residual disease or as an adjuvant treatment in combination with conventional therapies.

Telomerase inhibitors will probably be used following standard therapies in which there is no clinical evidence of disease in order to treat possible micrometastases or in patients at high risk to develop cancer in a chemoprevention approach. In these situations, which would likely require prolonged treatment, it will be imperative that the drugs have a low toxicity profile and are easily administered by perhaps an oral route. At the present time, there is still a large body of basic research to pursue, but it is encouraging that formal preclinical investigations of telomerase inhibitors and hTERT-based oncolytic viruses have begun, and there are already clinical studies under way with hTERT vaccines and immunotherapy. Although the vast majority of studies are encouraging, the ultimate utility of these and future drugs based on telomere biology will be determined only after completion of clinical trials and use in a broad range of cancer patients. However, it is reasonable to anticipate that some surprises may arise as important lessons from cell culture and animal models are validated in clinical trials.

ACKNOWLEDGMENTS

This work was supported by the Ellison Medical Foundation, Geron Corporation, NASA NSCOR grant to UT Southwestern Medical Center, and National Cancer Institute grants CA74908 and CA70907.

REFERENCES

Abdul-Ghani R., Ohana P., Matouk I., Ayesh S., Ayesh B., Laster M., Bibi O., Giladi H., Molnar-Kimber K., Sughayer M.A., de Groot N., and Hochberg A. 2000. Use of transcriptional regulatory sequences of telomerase (hTER and hTERT) for selective killing of cancer cells. *Mol. Ther.* **2:** 539–544.

Aisner D.L., Wright W.E., and Shay J.W. 2002. Telomerase regulation: Not just flipping the switch. *Curr. Opin. Genet. Dev.* **12:** 80–85.

Al-Hajj M., Wichas M.S., Benito-Hernandez A., Morrison S.J., and Clarke M.F. 2003. Prospective identification of tumorigenic breast cancer cells. *Proc. Natl. Acad. Sci.* **100:** 3983–3988.

Beachy P.A., Karhadkar S.S., and Berman D.M. 2004. Tissue repair and stem cell renewal in carcinogenesis. *Nature* **432:** 324–331.

Bechter O., Zou Y., Walker W., Wright W.E., and Shay J.W. 2004. Telomeric recombination in MSH6 deficient human colon cancer cells following telomerase inhibition. *Cancer Res.* **64:** 3444–3451.

Blackburn E.H. and Gall J.G. 1978. A tandemly repeated sequence at the termini of the extrachromosomal ribosomal RNA genes in *Tetrahymena*. *J. Mol. Biol.* **120:** 33–53.

Bodnar A.G., Chiu C., Frolkis M., Harley C.B., Holt S.E., Lichtsteiner S., Morin G.B., Ouellette M., Shay J.W., and Wright W.E. 1998. Extension of life span by introduction of telomerase into normal human cells. *Science* **279:** 349–352.

Boklan J., Nanjangud G., MacKenzie K.L., May C., Sadelain M., and Moore M.A. 2002. Limited proliferation and telomere dysfunction following telomerase inhibition in immortal murine fibroblasts. *Cancer Res.* **62:** 2104–2114.

Bryan T.M. and Reddel R.R. 1997. Telomere dynamics and telomerase activity in in vitro immortalised human cells. *Eur. J. Cancer* **33:** 767–773.

Bryan T.M., Engelzou A., Gupta J., Bacchetti S., and Reddel R.R. 1995. Telomere elongation in immortal human cells without detectable telomerase activity. *EMBO J.* **14:** 4240–4248.

Bryan T.M., Marusic L., Bacchetti S., Namba M., and Reddel R.R. 1997. The telomere lengthening mechanism in telomerase-negative immortal human cells does not involve the telomerase RNA subunit. *Hum. Mol. Genet.* **6:** 921–926.

Carey L.A., Kim N.W., Goodman S., Marks J., Henderson G., Umbricht C.B., Dome J.S., Dooley W., Amshey S.R., and Sukumar S. 1999. Telomerase activity and prognosis in primary breast cancer. *J. Clin. Oncol.* **17:** 3075–3081.

Cech T.R. 2000. Life at the end of the chromosome: Telomeres and telomerase. *Angew. Chem. Int. Ed.* **39:** 34–43.

Chen Z., Koeneman K.S., and Corey D.R. 2003. Consequences of telomerase inhibition and combination treatments for the proliferation of cancer cells. *Cancer Res.* **63:** 5917–5925.

Chin L., Artandi S.E., Shen Q., Tam A., Lee S.L., Gottlieb G.J., Greider C.W., and DePinho R.A. 1999. p53 deficiency rescues the adverse effects of telomere loss and cooperates with telomere dysfunction to accelerate carcinogenesis. *Cell* **97:** 527–538.

Clark G.M., Osborne C.K., Levitt D., Wu F., and Kim N.W. 1997. Telomerase activity and survival of patients with node-positive breast cancer. *J. Natl. Cancer Inst.* **89:** 1874–1881.

Collins K. and Mitchell J.R. 2002. Telomerase in the human organism. *Oncogene* **21:** 564–579.

Cong Y., Wright W.E., and Shay J.W. 2002. Human telomerase and its regulation. *Microbiol. Mol. Biol. Rev.* **66:** 407–425.

Cooke H.J. and Smith B.A. 1986. Variability at the telomeres of the human X/Y pseudoautosomal region. *Cold Spring Harbor Symp. Quant. Biol.* **51:** 213–219.

Cooke H.J., Brown W.R., and Rappold G.A. 1985. Hypervariable telomeric sequences from the human sex chromosomes are pseudoautosomal. *Nature* **317:** 687–692.

Corey D.R. 2000. Telomerase: An unusual target for cytotoxic agents. *Chem. Res. Toxicol.* **13:** 957–960.

———. 2002. Telomerase inhibition, oligonucleotides, and clinical trials. *Oncogene* **21:** 631–637.

Counter C.M., Avilion A.A., LeFeuvre C.E., Stewart N.G., Greider C.W., Harley C.B., and Bacchetti S. 1992. Telomere shortening associated with chromosome instability is arrested in immortal cells which express telomerase activity. *EMBO J.* **11:** 1921–1919.

d'Adda di Fagagna F., Reaper P.M., Clay-Farrace L., Fiegler H., Carr P., Von Zglinicki T., Saretzki G., Carter N.P., and Jackson S.P. 2003. A DNA damage checkpoint response in telomere-initiated senescence. *Nature* **426:** 194–198.

Dalbagni G., Han W., Zhang Z.-F., Cordon-Cardo C., Saigo P., Fair W.R., Herr H., Kim N., and Moore M.A.S. 1997. Evaluation of the telomeric repeat amplification protocol (TRAP) assay for telomerase as a diagnostic modality in recurrent bladder cancer. *Clin. Cancer Res.* **3:** 1593–1598.

de Lange T. and Jacks T. 1999. For better or worse? Telomerase inhibition and cancer. *Cell* **98:** 273–275.

de Lange T., Shiue L., Myers R.M., Cox D.R., Naylor S.L., Killery A.M., and Varmus H.E. 1990. Structure and variability of human chromosome ends. *Mol. Cell Biol.* **10:** 518–527.

Feng J., Funk W.D., Wang S.S., Weinrich S.L., Avilion A.A., Chiu C.-P., Adams R.R., Chang E., Allsopp R.C., Yu J.H., et al. 1995. The RNA component of human telomerase. *Science* **269:** 1236–1241.

Folini M., Colella G., Villa R., Lualdi S., Daidone M.G., and Zaffaroni N. 2000. Inhibition of telomerase activity by a hammerhead ribozyme targeting the RNA component of telomerase in human melanoma cells. *J. Invest. Dermatol.* **114:** 259–267.

Forsyth N.R., Wright W.E., and Shay J.W. 2002. Telomerase and differentiation in multicellular organisms: Turn it off, turn it on, and turn it off again. *Differentiation* **69:** 188–197.

Franco S., Segura I., Riese H.H., and Blasco M.A. 2002. Decreased B16F10 melanoma growth and impaired vascularization in telomerase-deficient mice with critically short telomeres. *Cancer Res.* **6:** 552–559.

Gomez D.E., Tejera A.M., and Olivera O.A. 1998. Irreversible telomere shortening by 3′azido-2′,3′dideoxythymidine (AZT) treatment. *Biochem. Biophys. Res. Commun.* **246:** 107–110.

Gompertz B. 1825. On the nature and function expressivity of the law of human mortality and on a new mode of determining life contingencies. *Philos. Trans. R. Soc. Lond.* **115:** 513–585.

Gonzalez-Suarez E., Samper E., Flores J.M., and Blasco M.A. 2000. Telomerase-deficient mice with short telomeres are resistant to skin tumorigenesis. *Nat. Genet.* **26:** 114–117.

Gonzalez-Suarez E., Samper E., Ramirez A., Flores J.M., Martin-Caballero J., Jorcano J.L., and Blasco M.A. 2001. Increased epidermal tumors and increased skin wound healing in transgenic mice overexpressing the catalytic subunit of telomerase, mTERT, in basal keratinocytes. *EMBO J.* **20:** 2619–2630.

Gowan S.M., Harrison J.R., Patterson L., Valenti M., Read M.A., Neidle S., and Kelland L.R. 2002. A G-quadruplex-interactive potent small-molecule inhibitor of telomerase exhibiting in vitro and in vivo antitumor activity. *Mol. Pharmacol.* **61:** 1154–1162.

Goytisolo F.A. and Blasco M.A. 2002. Many ways to telomere dysfunction: In vivo studies using mouse models. *Oncogene* **21:** 584–591.

Granger M.P., Wright W.E., and Shay J.W. 2002. Telomerase in cancer and aging. *Crit. Rev. Oncol. Hematol.* **4:** 29–40.

Greaves M. 1996. Is telomerase activity in cancer due to selection of stem cells and differentiation arrest? *Trends Genet.* **12:** 127–128.

Greenberg R.A., Chin L., Femino A., Lee K.H., Gottlieb G.J., Singer R.H., Greider C.W., and DePinho R.A. 1999. Short dysfunctional telomeres impair tumorigenesis in the INK4a cancer-prone mouse. *Cell* **97:** 515–525.

Greider C.W. and Blackburn E.H. 1985. Identification of a specific telomere terminal transferase activity in *Tetrahymena* extracts. *Cell* **43:** 405–413.

Gryaznov S., Pongracz K., Matray T., Schultz R., Pruzan R., Aimi J., Chin A., Harley C., Shea-Herbert B., Shay J., et al. 2001. Telomerase inhibitors—Oligonucleotide phosphoramidates as potential therapeutic agents. *Nucleosides Nucleotides Nucleic Acids* **20:** 401–410.

Gu J., Andreff M., Roth J.A., and Fang B. 2002. hTERT promoter induces tumor-specific Bax gene expression and cell killing in syngenic mouse tumor model and prevents systemic toxicity. *Gene Ther.* **9:** 30–37.

Gu J., Kagawa S., Takakura M., Kyo S., Inoue M., Roth J.A., and Fang B. 2001. Tumor-specific transgene expression from the human telomerase reverse transcriptase promoter enables targeting of the therapeutic effects of the Bax gene to cancers. *Cancer Res.* **60:** 5359–5364.

Hackett J.A. and Greider C.W. 2002. Balancing instability: Dual roles for telomerase and telomere dysfunction in tumorigenesis. *Oncogene* **21:** 619–626.

Hahn W.C., Stewart S.A., Brooks M.W., York S.G., Eaton E., Kurachi A., Beijersbergen R.L., Knoll J.H., Meyerson M., and Weinberg R.A. 1999. Inhibition of telomerase limits the growth of human cancer cells. *Nat. Med.* **5:** 1164–1170.

Harley C.B. 1991. Telomere loss: Mitotic clock or genetic time bomb? *Mutat. Res.* **256:** 271–282.

———. 2002. Telomerase is not an oncogene. *Oncogene* **21:** 494–502.

Harley C.B., Fletcher A.B., and Greider C.W. 1990. Telomeres shorten during aging. *Nature* **345:** 458–460.

Harrington L. and Robinson M.O. 2002. Telomere dysfunction: Multiple paths to the same end. *Oncogene* **21:** 592–597.

Hastie N.D., Dempster M., Dunlop M.G., Thompson A.M., Green D.K., and Allshire R.C. 1990. Telomere reduction in human colorectal carcinoma and with ageing. *Nature* **346:** 866–868.

Hayflick L. and Moorhead P.S. 1961. The limited in vitro lifetime of human diploid cell strains. *Exp. Cell Res.* **25:** 585–621.

Helder M.N., Wisman G.B.A., and van der Zee A.G.J. 2002. Telomerase and telomeres: From basic biology to cancer treatment. *Cancer Investig.* **20:** 82–101.

Hemann M.T., Strong M.A., Hao L.Y., and Greider C.W. 2001. The shortest telomere, not average telomere length, is critical for cell viability and chromosome stability. *Cell* **107:** 67–77.

Henson J.D., Neumann A.A., Yeager T.R., and Reddel R.R. 2002. Alternative lengthening of telomeres in mammalian cells. *Oncogene* **21:** 598–610.

Herbert B.-S., Pongracz K., Shay J.W., and Gryaznov S.M. 2002. Oligonucleotide N3′-P5′ phosphoramidates as efficient telomerase inhibitors. *Oncogene* **21:** 638–642.

Herbert B.-S., Pitts A.E., Baker S.I., Hamilton S.E., Wright W.E., Shay J.W., and Corey D.R. 1999. Inhibition of human telomerase in immortal human cells leads to progressive telomere shortening and cell death. *Proc. Natl. Acad. Sci.* **96:** 14276–14281.

Herbert B.-S., Wright A.C., Passons C.M., Ali I., Wright W.E., Kopelovich L., and Shay J.W. 2001. Inhibition of the spontaneous immortalization of breast epithelial cells from individuals predisposed to breast cancer: Effects of chemopreventive and anti-telomerase agents. *J. Natl. Cancer Inst.* **93:** 39–45.

Hiyama E., Hiyama K., Yokoyama T., Matsuura Y., Piatyszek M.A., and Shay J.W. 1995a. Correlating telomerase activity level with human neuroblastoma outcomes. *Nat. Med.* **1:** 249–255.

Hiyama E., Gollahon L., Kataoka T., Kuroi K., Yokoyama T., Gazdar A.F., Hiyama K., Piatyszek M.A., and Shay J.W. 1996. Telomerase activity in human breast tumors. *J. Natl. Cancer Inst.* **88:** 116–122.

Hiyama E., Yokoyama T., Tatsumoto N., Hiyama K., Imamura Y., Murakami Y., Kodama T., Piatyszek M., Shay J.W., and Matsuura Y. 1995b. Telomerase activity in gastric cancer. *Cancer Res.* **55:** 3258–3262.

Hiyama K., Hiyama E., Ishioka S., Yamakido M., Inai K., Gazdar A.F., Piatyszek M.A., and Shay J.W. 1995. Telomerase activity in small-cell and non-small-cell lung cancers. *J. Natl. Cancer Inst.* **87:** 895–902.

Hodes R. 2001. Molecular targeting of cancer: Telomeres as targets. *Proc. Natl. Acad. Sci.* **98:** 7649–7651.

Holt S.E. and Shay J.W. 1999. Role of telomerase in cellular proliferation and cancer. *J. Cell. Physiol.* **180:** 10–18.

Holt S.E., Aisner D.L., Baur J., Tesmer V.M., Dy M., Ouellette M., Trager J.B., Morin G.B., Toft D.O., Shay J.W., Wright W.E., and White M.A. 1999. Functional requirement of p23 and Hsp90 in telomerase complexes. *Genes Dev.* **13:** 817–826.

Jiang X.R., Jimenez G., Chang E., Frolkis M., Kusler B., Sage M., Beeche M., Bodnar A.G., Wahl G.M., Tlsty T.D., and Chiu C.P. 1999. Telomerase expression in human somatic cells does not induce changes associated with a transformed phenotype. *Nat. Genet.* **21:** 111–114.

Keith W.N, Bilsland A., Hardie M., and Jeffry Evans T.R. 2004. Drug insight: Cancer cell immortality—Telomerase as a target for novel cancer gene therapies. *Nat. Clin. Pract. Oncol.* **1:** 1–9.

Kelland L.R. 2001. Telomerase: Biology and phase I trials. *Lancet Oncol.* **2:** 95–102.

Kim M.M., Rivera M.A., Botchkina I.L., Shalaby R., Thor A.D., and Blackburn E.H. 2001. A low threshold level of expression of mutant-template telomerase RNA inhibits human tumor cell proliferation. *Proc. Natl. Acad. Sci.* **98:** 7982–7987.

Kim N.W., Pietyszek M.A., Prowse K.R., Harley C.B., West M.D., Ho P.L.C., Coviello G.M., Wright W.E., Weinrich S.L., and Shay J.W. 1994. Specific association of human telomerase activity with immortal cells and cancer. *Science* **266:** 2011–2015.

Kinoshita H., Ogawa O., Kakehi Y., Mishina M., Mitsumori K., Itoh N., Yamada H., Terachi T., and Yoshida O. 1997. Detection of telomerase activity in exfoliated cells in urine from patients with bladder cancer. *J. Natl. Cancer Inst.* **89:** 724–730.

Kitada T., Seki S., Kawakita N., Kuroki T., and Monna T. 1995. Telomere shortening in chronic liver diseases. *Biochem. Biophys. Res. Commun.* **211:** 33–39.

Koga S., Hirohata S., Kondo Y., Komata T., Takakura M., Inoue M., Kyo S., and Kondo S. 2000. A novel telomerase-specific gene therapy: Gene transfer of caspase-8 utilizing the human telomerase catalytic subunit gene promoter. *Hum. Gene Ther.* **11:** 1397–1406.

———. 2001. FADD gene therapy using the human telomerase catalytic subunit (hTERT) gene promoter to restrict induction of apoptosis to tumors in vitro and in vivo. *Anticancer Res.* **21:** 1937–1943.

Komata T., Kondo Y., Kanzawa T., Hirohata S., Koga S., Sumiyoshi H., Srinivasula S.M., Barna B.P., Germano I.M., Takakura M., Inoue M., Alnemri E.S., Shay J.W., Kyo S., and Kondo S. 2001. Treatment of malignant glioma cells with the transfer of constitutively active caspase-6 using the human telomerase catalytic subunit (human telomerase reverse transcriptase) gene promoter. *Cancer Res.* **61:** 5796–5802.

Lin S.-Y. and Elledge S.J. 2003. Multiple tumor suppressor pathways negatively regulate telomerase. *Cell* **113:** 881–889.

Lin Y., Miyamoto H., Fujinami K., Uemura H., Hosaka M., Iwasaki Y., and Kubota Y. 1996. Telomerase activity in human bladder cancer. *Clin. Cancer Res.* **2:** 929–932.

Lingner J. and Cech T.R. 1998. Telomerase and chromosome end maintenance. *Curr. Opin. Genet. Dev.* **8:** 226–232.

Lundblad V. 2002. Telomere maintenance without telomerase. *Oncogene* **21:** 522–531.

Lundblad V. and Blackburn E.H. 1993. An alternative pathway for yeast telomere maintenance rescues est1⁻ senescense. *Cell* **73:** 347–360.

Majumdar A.S., Hughes D.E., Lichtsteiner S.P., Wang Z., Lebkowski J.S., and Vasserot A.P. 2001. The telomerase reverse transcriptase promoter drives efficacious tumor suicide gene therapy while preventing hepatotoxicity encountered with constitutive promoters. *Gene Ther.* **8:** 568–578.

Matthes E. and Lehmann C. 1999. Telomerase protein rather than its RNA is the target of phosphorothioate-modified oligonucleotides. *Nucleic Acids Res.* **27:** 1152–1158.

McClintock B. 1941. The stability of broken ends of chromosomes in *Zea mays. Genetics* **26:** 234–282.

McKenzie K.E., Umbricht C.B., and Sukumar S. 1999. Applications of telomerase research in the fight against cancer. *Mol. Med. Today* **5:** 114–122.

Meeker A.K. and De Marzo A.M. 2004. Recent advances in telomere biology: Implications for human cancer. *Curr. Opin. Oncol.* **16:** 32–38.

Meeker A.K., Hicks J.L., Iacobuzio-Donahue C.A., Montgomery E.A., Westra W.H., Chan T.Y., Ronnett B.M., and De Marzo A.M. 2004. Telomere length abnormalities occur early in the initiation of epithelial carcinogenesis. *Clin. Cancer Res.* **10:** 317–326.

Minev B., Hipp J., Firat H., Schmidt J.D., Langlade-Demoyen P., and Zanetti M. 2000. Cytotoxic T cell immunity against telomerase reverse transcriptase in humans. *Proc. Natl. Acad. Sci.* **97:** 4796–4801.

Morales C.P., Holt S.E., Ouellette M., Kaur K.J., Ying Y., Wilson K.S., White M.A., Wright W.E., and Shay J.W. 1999. Lack of cancer-associated changes in human fibroblasts after immortalization with telomerase. *Nat. Genet.* **21:** 115–118.

Morin G.B. 1989. The human telomerase terminal transferase enzyme is a ribonucleo-protein that synthesizes TTAGGG repeats. *Cell* **59:** 521–529.

Muller H.J. 1962. The remaking of chromosomes. In *Studies of genetics: The selected papers of H.J. Muller*, pp. 384–408. Indiana University Press, Bloomington.

Murakami J., Nagai N., Shigemasa K., and Ohama K. 1998. Inhibition of telomerase activity and cell proliferation by a reverse transcriptase inhibitor in gynaecological cancer cell lines. *Eur. J. Cancer* **35:** 1027–1034.

Nair S.K., Heiser A., Boczkowski D., Majumdar A., Naoe M., Lebkowski J.S., Vieweg J., and Gilboa E. 2000. Induction of cytotoxic T lymphocyte responses and tumor immunity against unrelated tumors using telomerase reverse transcriptase RNA transfected dendritic cells. *Nat. Med.* **6:** 1011–1017.

Nakamura T.M., Morin G.B., Chapman K.B., Weinrich S.L., Andrews W.H., Lingner J., Harley C.B., and Cech T.R. 1997. Telomerase catalytic subunit homologs from fission yeast and humans. *Science* **277:** 955–959.

Neidle S. and Parkinson G. 2002. Telomere maintenance as a target for anticancer drug discovery. *Nat. Rev. Drug Discov.* **1:** 383–393.

Norton J.C., Piatyszek M.A., Wright W.E., Shay J.W., and Corey D.R 1996. Inhibition of human telomerase activity by peptide nucleic acids. *Nat. Biotechnol.* **14:** 615–619.

Nugent C.I. and Lundblad V. 1998. The telomerase reverse transcriptase: Components and regulation. *Genes Dev.* **12:** 1073–1085.

O'Sullivan J.N., Bronner M.P., Brentnall T.A., Finley J.C., Shen W.-T., Emerson S., Emond M.J., Gollahon K.A., Moskovitz A.H., Crispin D.A., Potter J.D., and Rabinovitch P.S. 2002. Chromosomal instability in ulcerative colitis is related to telomere shortening. *Nat. Genet.* **32:** 280–284.

Ouellette M.M., Liao M., Herbert B.-S., Johnson M., Holt S.E., Liss H.S., Shay J.W., and Wright W.E. 2000. Sub-senescent telomere lengths in fibroblasts immortalized by limiting amounts of telomerase. *J. Biol. Chem.* **275:** 10072–10076.

Pathak S., Multani A.S., Furlong C.L., and Sohn S.H. 2002. Telomere dynamics, aneuploidy, stem cells, and cancer (review). *Int. J. Oncol.* **20:** 637–641.

Pitts A.E. and Corey D.R. 1998. Inhibition of human telomerase by 2′-O-methyl RNA oligonucleotides. *Proc. Natl. Acad. Sci.* **95:** 11549–11554.

Plumb J.A., Bilsland A., Kakani R., Zhao J., Glasspool R.M., Knox R.J., Evens T.R.J., and Keith W.N. 2001. Telomerase-specific suicide gene therapy vectors expressing bacterial nitroreductase sensitize human cancer cells to the pro-drug CB1954. *Oncogene* **20:** 7797–7803.

Ramirez R.D., Morales C.P., Herbert B.S., Rohde J.M., Passons C., Shay J.W., and Wright W.E. 2001. Putative telomere-independent mechanisms of replicative aging reflect inadequate growth conditions. *Genes Dev.* **1:** 398–403.

Read M.A., Wood A.A., Harrison J.R., Gowan S., Kelland S.R., Doganjh H.S., and Neidle S. 1999. Molecular modeling studies on G-quadruplex complexes of telomerase inhibitors: Structure-activity relationships. *J. Med. Chem.* **42:** 4538–4546.

Reya T., Morrison S.J., Clarke M.F., and Weissman I.L. 2001. Stem cells, cancer, and cancer stem cells. *Nature* **414:** 105–111.

Riou J.F., Guittat L., Mailliet P., Laoui A., Renou E., Petitgenet O., Megnin-Chanet F., Helene C., and Mergny J.L. 2002. Cell senescence and telomere shortening induced by a new series of specific G-quadruplex DNA ligands. *Proc. Natl. Acad. Sci.* **99:** 2672–2677.

Shay J.W. 1998. Telomerase in cancer: Diagnostic, prognostic and therapeutic implications. *Cancer J. Sci. Am.* (suppl. 1) **4:** S26–S34.

———. 2003. Telomerase therapeutics: Telomeres recognized as a DNA damage signal. *Clinical Can. Res.* **9:** 3521–3525.

Shay J.W. and Bacchetti S. 1997. A survey of telomerase activity in human cancer. *Eur. J. Cancer* **5:** 787–791.

Shay J.W. and Wright W.E. 1996a. The reactivation of telomerase activity in cancer progression. *Trends Genet.* **12:** 129–131.

———. 1996b. Telomerase activity in human cancer. *Curr. Opin. Oncol.* **8:** 66–71.

———. 2001. When do telomeres matter? *Science* **291:** 839–840.

———. 2002. Telomerase: A target for cancer therapy. *Cancer Cell* **2:** 257–265.

———. 2004. Senescence and immortalization: Role of telomeres and telomerase. *Carcinogenesis* **25:** 1–8.

Shay J.W., Pereira-Smith O.M., and Wright W.E. 1991a. A role for both Rb and p53 in the regulation of human cellular senescence. *Exp. Cell Res.* **196:** 33–39.

Shay J.W., Wright W.E., and Werbin H. 1991b. Defining the molecular mechanisms of human cell immortalization. *Biochim. Biophys. Acta* **1072:** 1–7.

Stewart S.A., Hahn W.C., O'Connor B.F., Banner E.N., Lundberg A.S., Modha P., Mizuno H., Brooks M.W., Fleming M., Zimonjic, D.B., et al. 2002. Telomerase contributes to tumorigenesis by a telomere-independent mechanism. *Proc. Natl. Acad. Sci.* **99:** 12606–12611.

Strahl C. and Blackburn E.H. 1996. Effects of reverse transcriptase inhibitors on telomere length and telomerase activity in two immortal human cell lines. *Mol. Cell. Biol.* **16:** 53–56.

Sun D., Thompson B., Cathers B.E., Salazar M., Kerwin S.M., Trent J.O., Jenkins T.C., Neidle S., and Hurley L.M. 1997. Inhibition of human telomerase by a G-quadruplex-interactive compound. *J. Med. Chem.* **40:** 2113–2116.

Takai H., Smogprzewska A., and de Lange T. 2003. DNA damage foci at dysfunctional telomeres. *Curr. Biol.* **13:** 1549–1556.

Vonderheide R.H. 2002. Telomerase as a universal tumor-associated antigen for cancer immunotherapy. *Oncogene* **21:** 674–679.

Vonderheide R.H., Anderson K.S., Hahn W.C., Butler M.O., Schultze J.L., and Nadler, L.M. 2001a. Characterization of HLA-A3–restricted cytotoxic T lymphocytes reactive against the widely expressed tumor antigen telomerase. *Clinical Cancer Res.* **7:** 3343–3348.

Vonderheide R.H., Schultze J.L., Anderson K.S., Maecker B., Butler M.O., Xia Z., Kuroda M.J., von Bergwelt-Baildon M.S., Bedor M.M., Hoar K.M., Schnipper D.R., Brooks

M.W., Letvin N.L., Stephans K.F., Wucherpfennig K.W., Hahn W.C., and Nadler L.M. 2001b. Equivalent induction of telomerase-specific cytotoxic T lymphocytes from tumor-bearing patients and healthy individuals. *Cancer Res.* **61:** 8366–8370.

Weisman A. 1891. *Essays upon heredity and kindred biological problems*, 2nd edition. Clarendon Press, Oxford.

White L., Wright W.E., and Shay J.W. 2001. Telomerase inhibitors. *Trends Biotechnol.* **3:** 146–149.

Wiemann S.U., Satyanayayana A., Tsahuridu M., Tillmann H.L., Zender L., Klempnauer J., Flemming P., Franco S., Blasco M.A., Manns M.P., and Rudolph K.L. 2002. Hepatocyte telomere shortening and senescence are general markers of human liver cirrhosis. *FASEB J.* **16:** 935–942.

Wright W.E. and Shay J.W. 2000. Telomere dynamics in cancer progression and prevention: Fundamental differences in human and mouse telomere biology. *Nature Med.* **6:** 849–851.

Wright W.E., Pereira-Smith O.M, and Shay J.W. 1989. Reversible cellular senescence: A two-stage model for the immortalization of normal human diploid fibroblasts. *Mol. Cell. Biol.* **9:** 3088–3092.

Wright W.E., Piatyszek M.A., Rainey W.E., Byrd W., and Shay J.W. 1996. Telomerase activity in human germline and embryonic tissues. *Dev. Genet.* **18:** 173–179.

Wu X., Amos C.I., Zhu Y., Zhao H., Grossman B.H., Shay J.W., Swan G.E., Benowitz N.L., Luo S., and Spitz M.R. 2003. Telomere dysfunction: A potential cancer predisposition factor. *J. Natl. Cancer Inst.* **95:** 1211–1218.

Yoshida K., Sugino T., Tahara H., Woodman A., Bolodeoku J., Nargund V., Fellows G., Goodison S., Tahara E., and Tarin D. 1997. Telomerase activity in bladder carcinoma and its implications for noninvasive diagnosis by detection of exfoliated cancer cells in urine. *Cancer* **79:** 362–369.

Yokoyama Y., Takahashi Y., Shinohara A., Lian Z., Wan X., Niwa K., and Tamaya T. 1998. Attenuation of telomerase activity by a hammerhead ribozyme targeting the template region of telomerase RNA in endometrial carcinoma cells. *Cancer Res.* **58:** 5406–5410.

Zhang X., Mar V., Zhou W., Harrington L., and Robinson M.O. 1999. Telomere shortening and apoptosis in telomerase-inhibited human tumor cells. *Genes Dev.* **13:** 2388–2399.

Zou Y., Sfeir A., Shay J.W., and Wright W.E. 2004. Does a sentinel or groups of short telomeres determine replicative senescence? *Mol. Biol. Cell* **15:** 3709–3718.

5

Modeling Cancer and Aging in the Telomerase-deficient Mouse

Kwok-Kin Wong

Department of Medical Oncology, Dana Farber Cancer Institute
Harvard Medical School
Boston, Massachusetts 02115

Sandy Chang

Department of Molecular Genetics
MD Anderson Cancer Center
Houston, Texas 77030

Ronald A. DePinho

Department of Medical Oncology, Dana Farber Cancer Institute
and
Department of Genetics and Medicine, Harvard Medical School
Boston, Massachusetts 02115

THE YEAR 2007 WILL MARK THE CENTENNIAL ANNIVERSARY of mouse genetics. In these 100 years, repeated waves of technological innovation and genetic insight have positioned the mouse as a major experimental system for the study of complex human diseases and biological processes such as cancer, aging, and chronic degenerative disorders. The capacity to utilize the mouse to maximum experimental advantage stems from our capacity to precisely manipulate the mouse germ line, control and homogenize genetic variations, and assess genotype–phenotype associations in the context of the whole organism. Indeed, the mouse has contributed greatly to our understanding of the basis for virtually all major human diseases, shedding first light on central biological processes and often providing more surprises than fulfilled prophecies. At the same time, as with all model organisms, full utilization of the mouse necessitates an appreciation for cross-species similarities and differences in biology, genetics, and genomics and, where possible, how such differences can be exploited experimentally to dissect fundamental issues relating to key biological processes. In the study of telomeres and their roles in governing normal physiology and disease, it has been possible to take advantage of species-specific differences in mouse and human telomere dynamics to explore the relevance and specific contributions of telomeres and telomerase in biology and pathophysiology. As a result of efforts from many laboratories, the study of telomeres in the mouse has resulted in a steady stream of insights that are now being confirmed in human tissues with increasing regularity. It is reasonable to anticipate that mouse models will shape future clinical trials designed to reduce these basic biological insights into practical application.

TELOMERES AND TELOMERASE IN HUMANS AND MICE

As a prelude to presentation of the lessons learned from the mouse, it is instructive to begin with a summary of cross-species similarities and differences in telomere structure and telomerase regulation. Despite the great similarities between human and mouse in the telomeric DNA sequence, telomeric structure, and telomeric proteins, there are some differences between these two species (Kipling and Cooke 1990; Griffith et al. 1999). The first notable difference is in the average length of telomeres. Human telomere repeat lengths range from 10 to 15 kb, while those in laboratory mouse, *Mus musculus*, measure from 40 to 80 kb with some chromosomes harboring shorter telomeres, particularly in certain inbred strains (Hastie et al. 1990; Blasco et al. 1997; Zijlmans et al. 1997). In humans, telomerase activity is expressed at robust levels in germ cells, but at low levels in most somatic cells, with notable exceptions being activated leukocytes and tissue stem cells (Harley et al. 1994; Wright et al. 1996; Newbold 1997; Weng et al. 1997; Masutomi et al. 2003). In human tissues and derived cells, telomerase activity has been shown to correlate primarily with the transcriptional state of the hTert gene (Meyerson et al. 1997). In the mouse, less stringent regulation of mTert gene expression contributes to the more readily detectable levels of telomerase activity in various tissues and cultured cells (Kipling and Cooke 1990; Prowse et al. 1993; Prowse and Greider 1995; Greenberg et al. 1998; Martin-Rivera et al. 1998). Although murine telomerase expression is more evident in somatic tissues, telomere shortening has been documented in aging tissues, although the ample telomere reserves appear to avoid critical telomere shortening and cellular compromise (Prowse and Greider 1995; Rudolph et al. 1999). In this context, the aging telomerase-deficient mouse—engineered with more limited telomere reserves—has proven useful for the in vivo analysis of critical telomere shortening, providing an opportunity to examine the role of telomeres and telomerase in normal development, aging and age-related disorders, chronic degenerative conditions, and cancer. And, as is often the case with in vivo genetic studies in model organisms, there have emerged unanticipated observations, and these experimental observations have in turn influenced our views of human biology and disease.

Telomeres and Genome Stability

The cloning of *mTerc* enabled the construction of mice null for this gene and hence telomerase deficient (Blasco et al. 1995, 1997). On phenotypic characterization of *mTerc*$^{-/-}$ mice arising from heterozygous intercrosses, the first notable observation was the lack of cytogenetic, morphological, and physiological abnormalities in tissues and in derivative cells in culture (Blasco et al. 1997). Telomere length determinations in *mTerc*$^{-/-}$ mice confirmed a steady decline in telomere length at a rate of ~120 bp per cell division, a rate analogous to the attrition rate of primary human cells lacking detectable telomerase. Suspecting that the lack of phenotypes stemmed from generous telomere reserves, a mating scheme of successive *mTerc*$^{-/-}$ intercrosses was instituted to drive telomeres to shorter and potentially dysfunctional lengths in later generations (Fig. 1). Cytogenetic analysis of primary bone marrow cells, peripheral blood lymphocytes, and embryonic fibroblasts derived from fourth-generation (G$_4$) *mTerc*$^{-/-}$ mice revealed chromosomal end-to-end fusion with loss of telomere repeats at the junction, in accord with the significant reductions in telomere length in this model (Blasco et al. 1997; Lee et al. 1998; Hande et al. 1999; Rudolph et al. 1999). Chromosomal end-to-end fusions with the loss became evident and the frequency of these fusions increased in the subsequent *mTerc*$^{-/-}$ generations, particularly in the fifth- and sixth-generation (G$_{5-6}$) *mTerc*$^{-/-}$ mice. Subsequent spectral karyotype analyses established a link between telomere dysfunction and complex cytogenetic rearrangements, including nonreciprocal translocations—a common feature of human cancer chromosomes (Hande et al. 1999; Artandi et al. 2000; Chang et al. 2003). Cytogenetic analysis by McClintock on breakage and fusion of maize chromosomes provided the first evidence that telomeres are important in the maintenance of chromosomal integrity (McClintock 1941). Dysfunctional telomeric

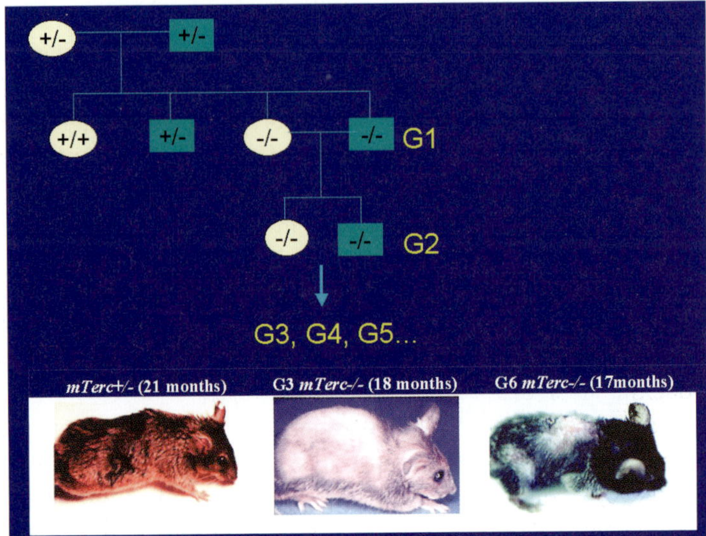

Figure 1. (*Top*) A mating schematic for the breeding of the telomerase mutant mouse through successive generations. (*Bottom*) Representative telomerase mutant mice from the different generations.

ends are highly recombinogenic, leading to the improper chromosomal fusions and generation of anaphase bridges. Subsequent random breakage will generate broken chromosomes in the daughter cells capable of fusing to other free chromosomal ends, thereby establishing the breakage–fusion–bridge (BFB) cycle. Because the break presumably occurs anywhere along the anaphase bridge, this type of genomic instability will result in rapid imbalances in gene dosage in the daughter cells. The continued formation of dicentric chromosomes, fueled by ongoing telomere dysfunction and associated BFB cycles, provides an efficient means of generating novel chromosomal variants with each cell division throughout the cell population. It appears that decline in telomere length, rather than the absence of telomerase activity per se, is the most important parameter dictating chromosomal integrity, because early-generation $mTerc^{-/-}$ mice (which still possess long functional telomeres) are cytogenetically intact. These results further highlight the fact that telomerase preserves genome stability primarily by maintaining telomere structure and averting BFB cycles and their associated adverse cellular consequences. However, emerging evidence has implicated additional nontelomeric functions for telomerase reverse transcriptase (TERT) that may be relevant to stem cell biology and neoplasia (see below).

Coincident with these telomere-linked cytogenetic anomalies, many tissues experienced impaired proliferation and/or increased apoptosis, particularly those tissues with high renewal kinetics. In the hematopoietic system, telomere dysfunction was associated with a marked decline in hematopoietic precursor cell numbers, reduced mitogen-induced proliferation of T and B lymphocytes, and splenic atrophy with a reduction in germinal center function (Lee et al. 1998; Herrera et al. 2000). In the reproductive system, decreased fecundity, first evident in G_4 $mTerc^{-/-}$ mice, progressed to sterility in the majority of G_6 $mTerc^{-/-}$ animals. The late-generation $mTerc^{-/-}$ males experienced progressive marked testicular atrophy stemming from apoptotic germ cell depletion and the resultant severe decline in spermatogenesis. Analogously, G_6 $mTerc^{-/-}$ females showed a decrease in the number of oocytes on ovulation and impaired capacity of most fertilized embryos to progress beyond early development (Lee et al. 1998; Hemann et al. 2001). Reconstitution of telomerase activity in late-generation $mTerc^{-/-}$ mice increases telomere lengths preferentially at the shortest telomeres and is associated with the striking rescue of pathologies associated with telomere dysfunction (Hemann et al. 2001; Samper et al. 2001). Nearly all of the lessons learned in the mTerc mutant model have emerged in analogous studies in mice null for the mTert gene, although possible existence of mTert-specific phenotypes may be uncovered by in-depth analyses (Liu et al. 2000; Chiang et al. 2004; Yuan et al. 2005). Taken together, these results strongly support the view that the long-term homeostasis of renewing organ systems over a lifetime depends on the telomerase-mediated maintenance of telomere length and function.

TELOMERES AND AGING

Cells derived from most multicellular organisms have finite replicative potential in cell culture, and considerable debate has arisen over whether this in vitro phenomenon is relevant to the biology of human aging. This potential link between replicative senescence and human aging has been stimulated by observations such as the accumulation of senescent cells with advancing age (Dimri et al. 1995), correlations between in vitro replicative potential and donor age (Martin et al. 1970), and the diminished in vitro replicative potential of cells derived from individuals with premature aging syndromes (Faragher et al. 1993). Although a definitive connection between telomere dynamics and normal aging in humans has yet to be established, accumulating evidence has strengthened the view that accelerated telomere attrition contributes directly to acquired and inherited degenerative conditions and several premature aging syndromes.

Table 1. Comparison of symptoms observed with human dyskeratosis congenita (DC) and late-generation $mTerc^{-/-}$ mouse

Phenotype	Human DC	G_6 $mTERC^{-/-}$ mouse
Abnormal skin pigmentation	yes	no
Bone marrow failure	yes	yes[a]
Small stature	yes	yes
Alopecia	yes	yes
Hypogonadism	yes	yes
Pulmonary fibrosis	yes	n.d.
Gastrointestinal disorders	yes	yes
Genomic instability	yes	yes
Increased malignancy	yes	yes

[a]When challenged with 5-FU.
n.d., not determined.

To date, the most direct evidence linking telomere dynamics and age-related degenerative conditions in humans comes from the investigations of the rare human disease, dyskeratosis congenita (DC) (Dokal 2001; see Chapter 6). DC is a multisystem disorder characterized by the triad of cutaneous abnormalities: abnormal skin pigmentation, nail dystrophy, and mucosal leukoplakia. DC patients also experience increased cancer risk and most often succumb to bone marrow failure by the fourth decade of life. These phenotypes bear a striking resemblance to those observed in the *mTerc*-null mouse (Table 1) and, in a satisfying confirmation of the relevance of mice in modeling human disease, the autosomal dominant form of DC was found to be linked to mutations in genes governing *Terc* stability (Mitchell et al. 1999) or in the human *Terc* gene itself (Vulliamy et al. 2001). These mutations result in diminished telomerase activity in critical stem cell compartments, leading to telomere dysfunction, premature bone marrow depletion, and ultimately bone marrow failure. Similar to the generational effects seen in the *mTerc*-null mouse, patients with DC are more severely affected in later generations, most likely because of the inheritance of progressively shorter telomeres (Vulliamy et al. 2004). This disease anticipation, in which the onset of disease occurs at progressively younger ages in subsequent generations, is observed only in siblings inheriting a defective copy of *hTERC*. These genetic observations strongly implicate critical telomere shortening in successive generations as a mechanism for disease anticipation and lend support to the notion that a steady increase in the level of telomere dysfunction could contribute to age-related disease processes in normal elderly individuals. Finally, the concordant biological features between the telomerase knockout mouse and DC patients provided a measure of validation that the mouse can serve as a relevant model organism to dissect the complex roles of telomeres in human pathobiology, including normal aging, age-related disorders, inherited premature aging syndromes, and chronic-disease-induced degenerative conditions such as liver cirrhosis.

The Telomerase-Null Mouse and Premature Aging/Degenerative Syndromes

Mice carrying one of several mutant genes linked to human premature aging disorders have failed to recapitulate the spectrum and severity of the corresponding human disease. A prominent example of this cross-species disconnect was observed in efforts to model Werner's syndrome in mice, a well-known human segmental progeric syndrome resulting from deficiency in the Wrn helicase (Lombard et al. 2000). The complete absence of premature aging phenotypes in the Werner's-deficient mouse (Lombard et al. 2000) raised the obvious possibilities that mice either possess functionally redundant helicase activities or that the ample mouse telomere reserves could impede the onset of age-related degenerative phenotypes. The latter possibility gained added support with the observations that cells derived from patients afflicted with premature aging diseases often show increased telomere erosion/dysfunction and premature senescence on passage in culture (Tahara et al. 1997) and that these cellular phenotypes can be reversed solely by the enforced TERT expression and telomere maintenance (Wyllie et al. 2000).

As a first step in dissecting the relevance of telomeres to organismal aging and age-related disorders, extensive physiological analyses were first conducted in successive generations of telomerase knockout mice as well as in aging mice of an intermediate $mTerc^{-/-}$ generation (Rudolph et al. 1999). Late-generation $mTerc^{-/-}$ animals exhibited a shortened life span, a range of aging phenotypes including alopecia, hair graying, a greatly reduced capacity to cope with acute and chronic stress, and a modest increased incidence of cancer (Rudolph et al. 1999). Age-matched animals at earlier generations did not manifest these phenotypes, suggesting that telomere dysfunction contributes to the emergence of these organismal aging phenotypes. The link between telomeres and the stress response is intriguing in light of the fact that aged individuals exhibit markedly reduced capacity to cope with acute stresses such as those incurred by chemotherapeutic treatment or surgical procedures. Although a link between diminished telomeres and stress responses (and aging sequelae in general) has yet to be forged in humans, the $mTerc^{-/-}$ model has provided a system in which to explore age-related biological phenomena and how such phenomena are modulated by telomere status (Cawthon et al. 2003).

Along these lines, while the utility of this mouse model in normal aging continues to be evaluated, the $mTerc^{-/-}$ mouse has contributed directly to the study of inherited premature aging diseases and age-related degenerative conditions. A case in point is Werner's syndrome (WS). In humans, mutations in the Wrn RecQ helicase gene cause WS, a disease characterized by impaired wound healing, osteoporosis, hypogonadism, cataract formation, type II diabetes, increased cancer incidence, and premature death (Epstein et al. 1966; Goto et al. 1996). These phenotypes are consistent with biochemical studies establishing roles for Wrn in many aspects of DNA metabolism, including replication, repair, and recombination (Ogburn et al. 1997; Yamagata et al. 1998; Yan et al. 1998). The pervasive roles of Wrn in genomic integrity are consistent with the diverse spectrum of premature aging phenotypes observed in WS patients. Curiously, despite the wide-ranging activities of the Wrn helicase in human cells, $Wrn^{-/-}$ mice and cells maintain normal karyotypes and do not exhibit any WS cellular or clinical phenotypes. In contrast, placement of the Wrn mutation through successive generations of $mTerc^{-/-}$ mice resulted in a dramatic emergence of many complex in vivo and cellular WS phenotypes as telomeres became limiting, including early onset of many age-related disorders characteristic of human WS (Table 2) (Chang et al. 2004; Du et al. 2004). On the

Table 2. Comparison of symptoms observed with human Werner's syndrome (WS) and various mouse models

Phenotype	Human WS	$WRN^{-/-}$ mouse	$G_{4-6}\ WRN^{+/+}$ mouse	$G_{4-6}\ WRN^{-/-}$ mouse[a]
Osteoporosis	yes	no	no	yes
Cataracts	yes	no	no	yes
Type II diabetes	yes	no	no	yes
Skin defects	yes	no	yes[b]	yes
Hypogonadism	yes	no	yes[b]	yes
Atherosclerosis	yes	no	no	no
Arteriosclerosis	yes	no	no	no
Genomic instability	yes	no	yes[b]	yes
Mesenchymal tumors	yes	no	no	yes[c]

[a]Phenotypes detected by 8 months of age.
[b]Majority of these phenotypes were observed in mice older than 8 months of age.
[c]Tumors observed in aged $G_{1-3}\ Wrn^{-/-}$ mice.

cytogenetic level, Wrn deficiency and telomere dysfunction cooperate to accelerate telomere shortening, resulting in increased chromosomal fusions, production of nonreciprocal translocations (NRTs), and elevated mesenchymal cancers such as osteosarcomas. Similar to the reduced replicative life span observed in human WS fibroblasts, late-generation $mTerc^{-/-}$ $Wrn^{-/-}$ mouse embryonic fibroblasts also exhibit premature replicative senescence.

It is important to emphasize that the $mTerc^{-/-}$ $Wrn^{-/-}$ compound mutant mouse is not simply a worsening of the telomerase-null phenotype, but a recapitulation of the many phenotypes encountered in WS patients. Specifically, late-generation $mTerc^{-/-}$ $Wrn^{-/-}$ mice develop cataracts, hypogonadism, type II diabetes, and a preponderance of osteosarcomas that were not observed in similarly aged $mTerc^{-/-}$ animals (Table 2). In addition, it is notable that mesechymal-derived tissues are primarily affected in $mTerc^{-/-}$ $Wrn^{-/-}$ animals. This observation is interesting in light of the broad expression of the Wrn helicase in mouse and human tissues, raising the possibility of functional redundancy by other helicases in nonmesenchymal compartments (Du et al. 2004). In this regard, the Bloom's syndrome (BLM) helicase exhibits robust expression in highly proliferative compartments (Kaneko 1999) and may therefore attenuate the impact of Wrn deficiency in those tissues. Consistent with this notion of tissue-preferential activities of these helicases is the observation that mice and patients heterozygous for BLM experience phenotypes in epithelial tissues, including a propensity to develop gastrointestinal malignancies (Goss et al. 2002).

Another interesting observation regarding the Wrn-telomerase knockout mouse is the fact that, despite the many nontelomeric DNA functions of Wrn such as those linked to general aspects of genome maintenance, there is a striking absence of genomic instability in the setting of intact telomere function. However, in late-generation $mTerc^{-/-}$ $Wrn^{-/-}$ cells with telomere dysfunction, cytogenetic features are consistent with defects in the DNA repair process, such as a preponderance of many interstitial double-strand DNA breaks (well above those seen in late-generation $mTerc^{-/-}$ cells with comparable telomere dysfunction) as well as accumulation of γH2AX-positive DNA damage foci (Chang et al. 2004). On the basis of these observations, we speculate that a key factor underlying WS may be an impaired DNA repair capacity in the setting of telomere dysfunction (Goytisolo et al. 2000; Wong et al. 2000; Chang et al. 2004). In the setting of this telomere-induced DNA repair defect, these other DNA maintenance functions of Wrn would become manifest, leading to many classical cytogenetic abnormalities such as interstitial breaks. Thus, a further analysis of the cytogenetic and biological phenomena in the Wrn-telomerase compound mouse model may continue to inform us about the molecular elements necessary to trigger clinical aspects of WS. It may be that the critical transition from normalcy at birth to the onset of premature aging phenotypes observed at the second decade of life in WS patients relates to the erosion of telomeres to a critical threshold that precipitates impaired DNA repair, thereby revealing the essential need for Wrn function in many aspects of DNA metabolism.

This view is in line with the emerging consensus that telomere maintenance, DNA repair, and DNA metabolism pathways are inter-twined and function to either suppress or precipitate aging and cancer phenotypes depending on their status and physiological context. Thus, in addition to the generation of a useful model of WS, these studies established that most, if not all, of the pleiotropic effects of Wrn deficiency are induced by critical telomere shortening. By extension, these findings in the mouse suggest that the progressive WS pathology in humans relates significantly to the onset of telomere dysfunction in various cellular compartments. If this is indeed the case, then precise measurements of telomere reserve in vivo may provide an important biomarker for disease onset and progression and for assessment of preventive and interventional therapies.

The Telomere Checkpoint Response and the Initiation of Aging Phenotypes in Mice

A large body of genetic evidence in the mouse has provided increasing support for the concept that aberrant DNA damage signaling and assoc-iated cellular checkpoints can promote the development of degenerative phenotypes similar to those observed in aging. Given these genetic interactions, significant effort has been directed toward delineating how dysfunctional telomeres signal to cellular checkpoint pathways and whether such signaling is involved in the development of age-related phe-notypes. Whereas the cellular and organismal analyses of telomerase-deficient mice as well as cultured human cells have revealed that telomere dysfunction provokes robust DNA damage-like checkpoint responses, it is important to appreciate that the precise nature of the response is highly dependent on cell type and physiological context. Moreover, the type of response appears to be dependent on parameters such as cell type, state of cellular differentiation, tissue context, and associated genetic mutations, among others. Given the complex and context-dependent nature of the telomere checkpoint response, the use of mouse genetics has provided an important in vivo framework for the study of the molecular wiring of this "DNA damage activation response" in relation to aging and how such information may inform the aging process in humans. Characterization of the type of cellular response and dissection of the molecular wiring of the linked "telomere checkpoint pathway" represent intensive areas of investigation in the telomere field.

Recent studies in human cell culture have shown that dysfunctional telomeres can engage canonical DNA damage response pathways (Maser and DePinho 2004). Senescent human fibroblasts display molecular markers characteristic of cells bearing DNA double-strand breaks (Jackson 2002), including phosphorylated γH2AX, 53BP1, NBS1, and CHK2 (d'Adda di Fagagna et al. 2003; Takai et al. 2003). Many of these proteins colocalize with dysfunctional telomeres, and their inactivation enables senescent cells to reenter the cell cycle (d'Adda di Fagagna et al. 2003; Takai et al. 2003). Indeed, there is now strong evidence that global activation of the DNA damage machinery throughout proliferating cellular compartments (by dysfunctional telomeres or other DNA structural defects) initiates cellular checkpoint responses, such as replicative senescence or apoptosis, that in turn lead to organ degeneration and aging phenotypes. The relevance of the telomere checkpoint and DNA damage signaling to organ degeneration was revealed in the late-generation mice doubly null for *mTerc* and *p53*. The p53 tumor suppressor was selected in part on the basis of its role as an essential element in sensing DNA damage and executing appropriate cellular checkpoint responses. In line with this role, the elimination of p53 function in the late-generation *mTerc*$^{-/-}$ mouse resulted in a near-complete eradication of the cellular growth arrest and/or apoptosis responses, attenuation of the progressive organ degeneration and multisystem failure, and amelioration of virtually all of the organismal aging phenotypes associated with telomere dysfunction (Chin et al. 1999). Equally notable was the observation that this phenotypic rescue came at the expense of increased cancer incidence, a finding that ran counter to the prevailing view at the time that telomere-based crisis functioned solely to block the development of cancer and that telomerase served simply to fuel malignant transformation (see below). This study, and complementary data from human cell culture (Karlseder et al. 1999), established that *p53* is a critical genetic element in the telomere checkpoint response.

Because p53 stands at the intersection of many known DNA damage signaling pathways, the findings of the mTerc-p53 cross stimulated efforts to dissect further the wiring of the p53-dependent telomere response in vivo. Previous work from the cancer field has established the existence of two major signaling pathways feeding into p53: (1) the ARF-MDM2 circuit, functioning to monitor and ensure appropriate cell cycle entry; and (2) the ATM-ATR (ataxia telangiectasia mutated-ataxia telangiectasia and Rad3 related) sensors, involved in surveying the integrity of DNA (Shiloh and Kastan 2001; Iwakuma and Lozano 2003; Lowe and Sherr 2003). To assess more directly the role of these genetic elements in the p53-dependent telomere checkpoint response, aging phenotypes were audited in late-generation $mTerc^{-/-}$ mice that were deficient for either INK4a/ARF or ATM. Consistent with the known minor role of ARF in the classical DNA damage response (Kamijo et al. 1997), late-generation $mTerc^{-/-}$ $Ink4a/Arf^{-/-}$ mice showed no amelioration in cellular arrest and apoptosis or in organ compromise (Greenberg et al. 1999; C.M. Khoo and R.A. DePinho, unpubl.). The impact of ATM deficiency on the telomere checkpoint response proved to be much more complex than the picture observed for *Ink4a/Arf* or *p53* mutant crosses. In intermediate-generation $mTerc^{-/-}$ $Atm^{-/-}$ mice (possessing moderate levels of telomere dysfunction), there was a partial rescue of cellular arrest and apoptosis and preservation of organ function—an observation consistent with findings in human cell cultures expressing dominant-negative TRF2 (Karlseder et al. 1999; Wong et al. 2003). With increasing telomere dysfunction and genomic instability, late-generation $mTerc^{-/-}$ $Atm^{-/-}$ mice sustained generalized proliferation defects across all cellular compartments, onset of premature aging phenotypes, and early death (Wong et al. 2003). Importantly, an examination of hematopoietic, gastrointestinal, and neural stem cell compartments revealed a depletion of reserves and high rates of apoptosis and/or decreased proliferative potential of these cells. It is noteworthy that late-generation $mTerc^{-/-}$ $Atm^{-/-}$ neuronal stem cells showed impaired renewal potential and an impaired ability to differentiate into viable neurons (as opposed to astrocytes)—an observation of likely relevance to the neurodegenerative phenotypes observed in ataxia-telangiectasia (A-T) patients.

In stark contrast to the increased cancer in the telomerase-p53 model, there was a near total suppression of lymphomas in late-generation $mTerc^{-/-}$ $Atm^{-/-}$ mice with limiting telomeres (Qi et al. 2003; Wong et al. 2003). These contrasting phenotypes can be rationalized by the fact that dysfunctional telomeres can signal to, and activate, p53 via Atm-independent mechanisms, leading to apoptosis and tumor suppression. It is worth noting that, whereas the Atm-independent p53 activation likely contributes significantly to these degenerative phenotypes and cancer suppression, the extent to which the rampant genomic instability contributes to diminished cellular viability is not known, and this generic instability could be an important contributor to the diminished cell viability and progressive degenerative phenotypes in this model and in A-T patients. Although the relative contribution of this genomic instability will require further study, several lines of evidence support the view that p53 activation contributes directly to the above degenerative phenotypes. First, in late-generation $mTerc^{-/-}$ mice, p53 is robustly activated by dysfunctional telomeres in a number of cellular compartments (Chin et al. 1999; Rudolph et al. 1999; Gonzalez-Suarez et al. 2000). Second, constitutive activation of a hyperfunctional mutant p53 in the mouse germ line has been shown to suppress tumorigenesis while promoting the onset of premature aging phenotypes in mice (Tyner et al. 2002). Thus, on the basis of these observations, it appears reasonable to propose that inappropriate activation of p53 by dysfunctional telomeres contributes significantly to telomere-linked aging phenotypes in vivo (Sharpless and DePinho 2002). Nevertheless, the production and detailed analysis of $mTerc^{-/-}$ $Atm^{-/-}$ $p53^{-/-}$ mice will be required to dissect the relative contributions of the p53-dependent telomere checkpoint and generic chromosomal imbalances on tissue homeostasis in the aging organism.

In summary, these various models and compound mutant crosses have placed p53 at the center of the telomere checkpoint response and have emphasized the complexity of the interrelationship of telomeres and the p53 node. These studies also revealed the importance of the genetic and cell physiological context in dictating the cellular response to telomere dysfunction—either entry into replicative senescence or apoptosis with progression to premature aging or into increased cell survival and genomic instability with progression toward malignancy (Fig. 2).

TELOMERES, TELOMERASE, AND TUMORIGENESIS

The robust and frequent activation of telomerase in advanced human cancers (Shay and Bacchetti 1997), coupled with the requirement of telomerase to sustain the replicative potential of primary cells and tumor cell lines (Bodnar et al. 1998; Counter et al. 1998; Liu et al. 2000), provided the foundation for the widely held view that telomerase activation plays a critical role in the development of cancer. Correspondingly, enforced telomerase activity enabled the efficient malignant transformation of primary cultured cells by activated RAS and SV40 small and large T antigen (Hahn et al. 1999; Elenbaas et al. 2001). An early counterpoint to this procarcinogenic role of telomerase

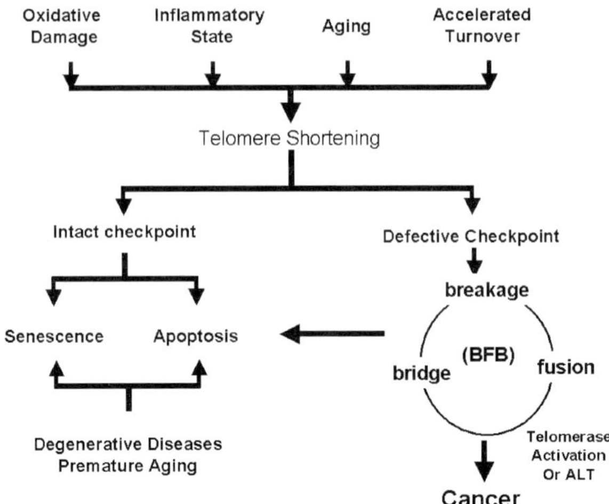

Figure 2. A schematic of the roles of telomere maintenance in cancer, chronic diseases, and aging.

was raised by investigators who speculated that the absence of telomerase and resultant telomere-mediated crisis could provide a cancer-promoting genomic instability (de Lange et al. 1990). It is now recognized that both of these opposing views have merit, as shown by genetic evidence in the telomerase knockout mouse and emerging correlative data from staged human cancers.

The complex roles of telomeres and telomerase in the development of cancer were experimentally elucidated in the aging telomerase-deficient mouse. In this model, rather than the anticipated resistance to age-related spontaneous tumors, there was an unexpected increase in age-dependent cancers, providing genetic evidence that the lack of telomerase and resultant chromosomal instability functioned to promote, albeit modestly, the development of cancer (Rudolph et al. 1999). In subsequent studies designed to understand the telomere–cancer connection, it became clear that the complex interrelationship was highly context-dependent, namely, telomere-based crisis can serve either a cancer-promoting or -suppressing role depending on several key factors, including the integrity of the p53-dependent DNA damage checkpoint and the cell type.

In the setting of long functional telomeres, mice deficient for *Ink4a/Arf* or *p53* are highly cancer-prone and similarly develop a tumor spectrum of lymphomas and sarcomas (Serrano et al. 1997). In *mTerc*$^{-/-}$ *Ink4a/Arf*$^{-/-}$ mice with short dysfunctional telomeres, there was a marked reduction in spontaneous and carcinogen-induced cancer incidence and no alteration in the tumor spectrum (Greenberg et al. 1999; C.M. Khoo and R.A. DePinho, unpubl.). Similarly, primary cells derived from these late-generation *mTerc*$^{-/-}$ *Ink4a/Arf*$^{-/-}$ mice exhibited resistance to transformation by cellular oncogenes, indicating that the inhibitory impact of telomere dysfunction was in part related to a cell-autonomous process. These observations were in line with the prevailing view that telomere-based crisis impedes the development of cancer. At the same time, examination of the *mTerc*$^{-/-}$ *p53*$^{-/-}$ mice revealed that telomere dysfunction resulted in the opposite outcome—a dramatic increase in cancer and a marked shift in tumor spectrum toward epithelial cancers (Chin et al. 1999; Artandi et al. 2000). Importantly, primary cells derived from these mice showed a corresponding increase in oncogene-induced focus formation; however, these transformed foci possessed diminished subcloning efficiency compared with telomerase-reconstituted controls. Taken together, the comparative analyses of the Ink4a/Arf and p53 crosses indicated that the status of tumor suppressor pathways dictates whether telomere dysfunction suppresses (*Ink4a/Arf* null) or enhances (*p53* null) tumorigenesis in vivo and in vitro.

The p53 studies also provided initial indications that, despite enhanced tumor initiation, intact telomeres may be critical for full malignant progression. The opposing role of telomeres in cancer initiation and progression was evidenced further by skin carcinogenesis studies (Gonzalez-Suarez et al. 2000) and by the Apc^{Min} model of intestinal carcinogenesis (Rudolph et al. 2001). In the highly quantitative *mTerc*$^{-/-}$ Apc^{Min} model, telomere dysfunction was associated with an increase in the number of early benign neoplasms, yet a significant decline in the multiplicity and size of more advanced macroscopic adenomas. Consistent with a constrained progression phenotype, these adenomas exhibited a large number of anaphase bridges and an associated intratumoral proliferative arrest and increased apoptosis; accordingly, the p53 gene was retained in these neoplasms and presumably functioned to inhibit tumor progression—an assumption that remains to be directly tested in this model. In more recent cell culture studies designed to explore in detail the impact of telomere-based crisis in tumor progression, it was demonstrated that extensively passaged Myc/RAS-transformed *mTerc*$^{-/-}$ *Ink4a/Arf*$^{-/-}$ fibroblasts will metastasize to the lung following tail vein injection only upon mTerc-reconstituted telomerase activity and restoration of telomere function (Chang et al. 2003).

The above studies led to a model in which the lack of telomerase and associated telomere attrition provide a mutator mechanism enabling would-be cancer cells to achieve a threshold of cancer-promoting changes required to traverse the benign-to-malignant transition. It is also apparent from these studies that the impact of telomere dysfunction is highly dependent on the status of p53 as well as the cell type. With regard to cell type, most notable is the observation that the $mTerc^{-/-}$ $p53^{+/-}$ mice experienced dramatic increase in epithelial cancers relative to lymphomas—a differential impact that appears to relate to the relative sensitivities of various cell types to telomere dysfunction. Along these lines, several early reports established that, at equivalent levels of telomere dysfunction, hematopoietic cell growth and survival are more adversely affected than epithelial cells in the telomerase knockout mouse, even in the setting of p53 deficiency (Lee et al. 1998; Rudolph et al. 1999; R.A. DePinho, unpubl. observations).

Are these findings in the mouse relevant to processes driving the initiation and progression of human cancers, particularly epithelial cancers? First, telomeres of human cancer cells are often significantly shorter than their normal tissue counterparts (Hastie et al. 1990), suggesting that telomere attrition has occurred at some time during the life history of these cancers, presumably during early phases when telomerase activity is low. The subsequent reactivation of telomerase appears to restore telomere function, albeit at a shorter set length. Thus, although reactivation of telomerase is critical to the emergence of immortal human cells, a preceding and transient period of telomere shortening and dysfunction appears to contribute to carcinogenesis by leading to the formation of chromosomal rearrangements through BFB cycles. Serial BFB cycles in turn beget rapid and wholesale genetic changes in the population, which presumably enables rare cells to incur a threshold number of procarcinogenic changes needed to initiate the transformation process. Although, at first glance, the cancer-promoting effects of telomere-based crisis seems contradictory to the role of telomerase in cancer progression, this mechanism is less paradoxical if one considers that many early-stage epithelial cancers deactivate the p53 pathway, which would increase the survival and proliferation of cells at a time of increasing genomic instability (Maser and DePinho 2002; Feldser et al. 2003). In humans, the accumulation of oncogenic lesions during normal aging or accelerated accumulation of DNA damage (e.g., environmental carcinogen exposure or oxidative damage) might deactivate the telomere checkpoint response, allow continued proliferation and telomere attrition, and drive the premalignant cell population into crisis. A rare cell might then acquire the right constellation of procarcinogenic changes and emerge fully transformed. Thus, in human cells, it appears that telomeric shortening can be viewed both as a barrier to cancer development in the presence of intact checkpoint responses and as a facilitator for numerous genetic changes necessary for the emergence of nascent cancer cells in the absence of the checkpoint response pathways.

Modeling Age-associated Cancers in the Mouse

The study of telomeres in the mouse has provided some clues concerning the strong association of increased cancer risk and advancing age in humans. Humans experience a dramatic escalation in cancer risk between the ages 40 and 80, an increase comprising largely epithelial cancers— primarily breast, lung, colon, and prostate cancers. A conventional view has been that the cancer-prone phenotype of older humans reflects the combined effects of cumulative mutational load, decreased DNA repair capabilities, increased epigenetic gene silencing, and altered hormonal and stromal milieus. Although these factors are likely important contributors to increasing cancer incidence with advancing age, it is less evident how such processes would spur the preferential development of epithelial cancers in older humans and, by extension, why aging mice would sustain a remarkably low number of epithelial cancers.

Known tissue- and species-specific differences in the rates and types of somatic mutation appear insufficient to account for these observed cancer patterns. On a similar note, it is not clear whether differences in the relative prominence of key tumor suppressor and repair pathways in humans versus mice would engender these age and/or species variances. Moreover, these mechanisms do not readily explain one of the most distinctive features of epithelial carcinomas of aged humans, as opposed to pediatric cancers and most murine cancers—the presence of a radically altered genome typified by marked aneuploidy and complex nonreciprocal translocations in human epithelial cancers. As mentioned above, the *mTerc p53* mutant mouse has provided a potential explanation for the observed tumor spectrum and cytogenetic profiles in aged humans. In these *Terc p53* mutant mice, the presence of telomere dysfunction results in a dramatic shift in the tumor spectrum toward epithelial cancers including those of the lung, colon, and skin. Moreover, in contrast to the somewhat normal cytogenetic profiles of cancers arising in mice with intact telomeres, the cancers generated in the Terc p53 mutant mice possess cytogenetic profiles with striking resemblance to human epithelial cancer genomes (Fig. 3) (Artandi et al. 2000).

In attempting to assign relevance of these murine studies to humans, it is worth considering that the typical adult cancer (an epithelial carcinoma) derives from a compartment that has undergone continued renewal throughout the human life span. Against this backdrop of physiological cell

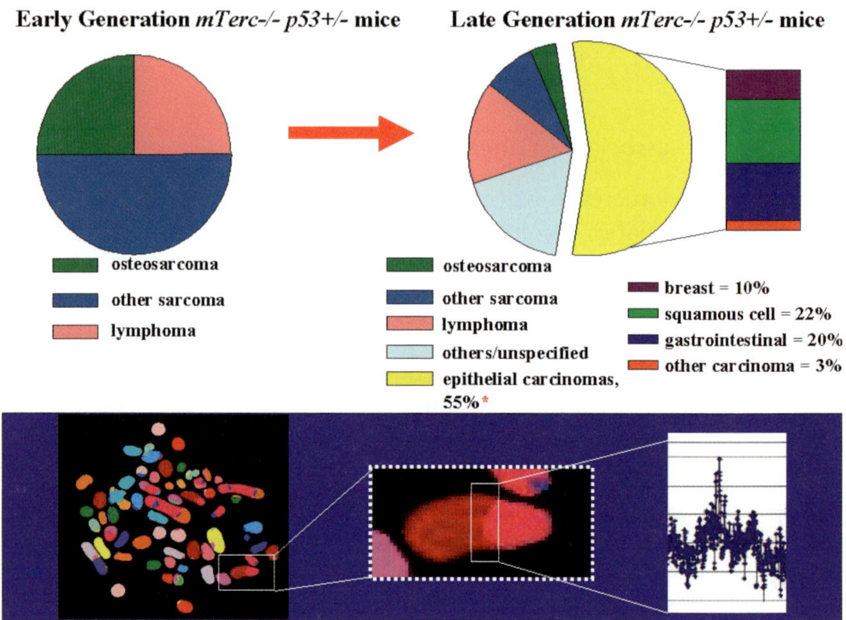

Figure 3. Clinically apparent tumor spectra in early-generation and late-generation $mTerc^{-/-}$ $p53^{+/-}$ mice. *Red asterisk*, random autopsies showed that the late-generation $mTerc^{-/-}$ $p53^{+/-}$ mice have significantly much higher incidences and rates of occult epithelial malignancies, particularly gastrointestinal carcinomas. (*Lower panel*) The spectral karyotype (SKY) profile of a late-generation $mTerc^{-/-}$ $p53^{+/-}$ mouse tumor cell with dysfunctional telomeres and harboring numerous nonreciprocal transloca-tions (NRTs). One of the NRTs (2:6) is magnified with the corresponding amplifi-cation locus shown by an array comparative hybridization profile of the same mouse tumor cell.

turnover and the occasional oncogenic mutation that further drives cellular proliferation, telomere lengths would shorten in self-renewing progenitor cells of these epithelial tissues. If somatic mutations also neutralize RB/INK4a/p53-dependent senescence checkpoints, then con-tinued growth will further drive telomere erosion and loss of the capping function, culminating in cellular crisis with attendant genomic instability. In this manner, telomere-based crisis provides the means to generate additional requisite mutations necessary to initiate carcinogenesis. The subsequent reactivation of telomerase in transformed clones would serve to stabilize the genome to a level compatible with cell viability, allowing these initiated neoplasms to mature further. It is unclear whether additional somatic mutations, beyond telomerase activation, would be needed to produce a fully malignant phenotype with invasive

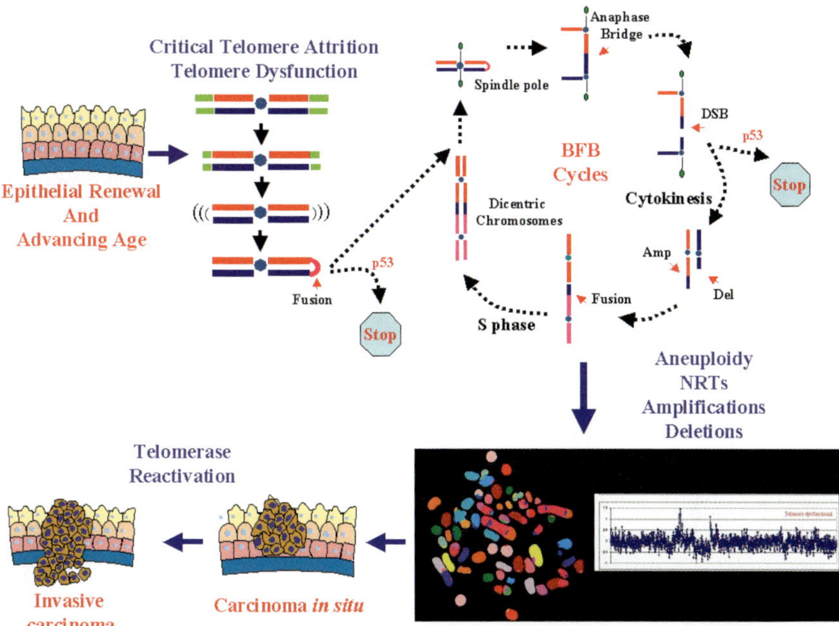

Figure 4. Dysfunctional telomere-induced genomic instability model of epithelial carcinogenesis. Continuous epithelial turnover during aging coupled with somatic mutations inactivating checkpoint responses are thought to lead to critical telomere (*green*) erosion, chromatid (*red* and *blue*) fusions, and the initiation of the breakage-fusion-bridge (BFB) cycles. Here a fusion between two sister chromatids forms a dicentric chromosome, which results in anaphase bridging during segregation in mitosis. The dicentric chromosome is broken when pulled to opposite spindle poles, creating a double-strand break (DSB) and a nidus for amplifications (Amp) and deletions (Del) for the resulting daughter cells. The broken chromosome might become fused to another chromosome (*pink*), generating a second dicentric chromosome and perpetuating the BFB cycle. This facilitation of the accumulation of genetic changes (via aneuploidy, nonreciprocal translocations [NRTs], amplifications, and deletions) by the BFB cycles and the reactivation of telomerase enable cells to emerge from crisis and proceed to malignancy. (Inset, *lower right*) A spectral karyotype (SKY) profile of a mouse tumor cell with dysfunctional telomeres and harboring numerous NRTs and an array comparative hybridization profile of the same mouse tumor cell.

and metastatic potential. Thus, a transient period of explosive chromosomal instability before telomerase reactivation may be required for the stochastic acquisition of the relatively high number of mutations thought to be required for adult epithelial carcinogenesis (Fig. 4).

Our "episodic instability" model of epithelial carcinogenesis fits well with current knowledge regarding the timing of telomerase activation and evolving genomic changes during various stages of human carcinoma development, particularly those of the breast, esophagus, pancreas, and colon. In vitro studies with human mammary epithelial cells recapitulated the stepwise model of progressive telomere shortening with progressive passage, entering into senescence and emerging from this checkpoint with chromosomal aberrations consistent with crisis (Romanov et al. 2001). Significantly, detailed analyses of the various stages of breast cancer progression revealed a phase of anaphase bridging and instability at the transition from ductal hyperplasia to ductal carcinoma in situ (Chin et al. 2004). Most notably, and consistent with the episodic instability model, there was no increase in the number of chromosomal alterations in more advanced disease tissues. The documentation of telomere erosion and anaphase bridging at this transition, coupled with the known activation of telomerase activity in ductal carcinoma in situ (DCIS) and invasive cancers, bolsters the idea that telomere-based crisis is a crucial event in breast tumorigenesis and plays a major role in driving genomic instability (Gray 2003). Similar studies in the analyses of telomere maintenance with human pancreatic tissues spanning the spectrum of early pancreatic intraepithelial neoplasis also showed that telomere shortening is the earliest event for these noninvasive ductal lesions (van Heek et al. 2002).

In addition, comparative genome hybridization (CGH) has also confirmed that dysplastic human breast, esophageal, and colon lesions have already sustained widespread gains and losses of regions of chromosomes early in their development, often well before these tissues demonstrate invasive growth or tumor formation (Al-Mulla et al. 1999; Buerger et al. 1999; Yen et al. 2001). The ploidy changes detected by CGH appear to correlate tightly with the presence of complex chromosomal rearrangements, and these markers of genomic instability are evident in the stages of advanced dysplasia of these tissues (e.g., DCIS, Barrett esophagus, etc.). As these cancers progress through invasive and metastatic stages, genomic instability continues, apparently at a moderate rate, but further mutations would be predicted to derive from non-telomere-based mechanisms. Correspondingly, the measurement of telomerase activity in adenomatous polyps and colorectal cancers has established that telomerase activity is low or undetectable in small- and intermediate-sized polyps, reflecting less intact telomere function; telomerase increases markedly in large adenomas and colorectal carcinomas, reflecting stabilization of telomere function (Tang et al. 1998). Therefore, it appears that there is widespread and severe chromosomal instability early on during human tumorigenesis at a time when telomerase activity is low. Additional support for this model derives from the documentation of anaphase

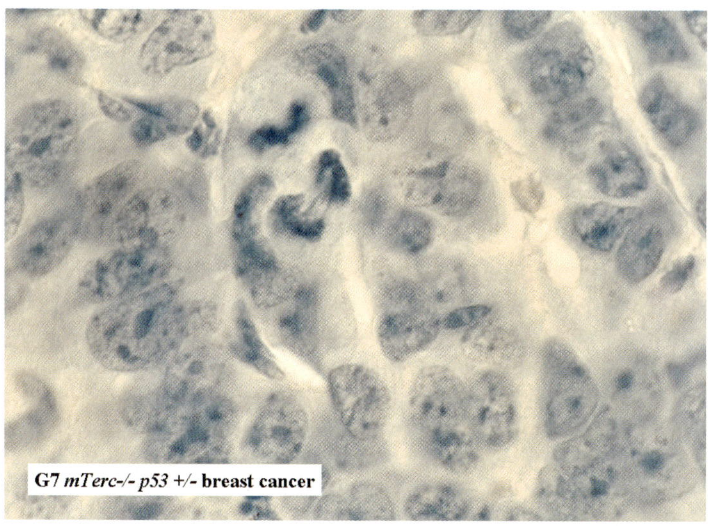

Figure 5. A representative anaphase bridge from a G_7 *mTerc*$^{-/-}$ *p53*$^{+/-}$ breast cancer cell undergoing mitosis.

bridging (a reasonable correlate of telomere-based crisis) in evolving human colorectal cancers (Rudolph et al. 2001) and in genomically unstable pancreatic cancers (Gisselsson et al. 2001; O'Sullivan et al. 2002). This suggests that the double-strand-break (DSB)-induced conditions (including but not limited to telomere dysfunction), coupled with mutations that allow survival in the face of a DSB, could provide an amplification/deletion mechanism across the genome (O'Hagan et al. 2002). Biological forces would in turn lead to the selection of clones with the amplifications and deletions that target cancer-relevant loci. Studies in the telomerase mutant mouse have begun to provide mechanistic insight into how BFB produces cancer-relevant changes as they have demonstrated that telomerase p53 compound mutant mice with telomere dysfunction have increased end-to-end fusions and that the ensuing BFB process is associated with regional amplifications and deletions that appear linked to nonreciprocal translocations (Artandi et al. 2000; O'Hagan et al. 2002) (Figs. 4 and 5).

Telomeres in Cancer-prone Degenerative Conditions

The role of telomeres in driving genomic instability in spontaneous cancers may also be relevant to the high cancer incidence in diseases characterized by chronic cell destruction and renewal. One notable example is the high incidence of hepatocellular carcinoma in the setting of late-stage liver cirrhosis. Cirrhosis is the phenotypic end point of prolonged cycles of hepatocyte destruction and regeneration; cirrhotic livers show a documented reduction in telomere length over time (Kitada et al. 1995; Miura et al. 1997; Wiemann et al. 2002). In addition to extensive cell turnover, accelerated telomere attrition may also be driven by increased oxidative stress and from the altered inflammatory microenvironment milieu. That telomere attrition is causal to the development of cirrhosis was established in the telomerase knockout mouse wherein critical reductions in telomere length and function were shown to accelerate the development of cirrhosis in the setting of chronic liver injury (Rudolph and DePinho 2001). With regard to cancer, chronic liver injury in mice with telomere dysfunction promoted increased cancer initiation, yet impaired malignant progression—observations in line with the data from human studies (Farazi et al. 2003).

Another intriguing case in point is ulcerative colitis, an inflammatory condition characterized by rapid cell turnover and oxidative injury to the intestines, and a high incidence of intestinal dysplasia or cancer (O'Sullivan et al. 2002). Cytogenetic studies and telomere analyses with quantitative fluorescence in situ hybridization on nondysplastic intestinal mucosa from patients with ulcerative colitis showed that tissue samples from patients with known dysplasia or cancer have significantly shorter telomeres and greater chromosomal losses. Thus, chronic inflammation and the resultant increased proliferation appear to promote accelerated telomere shortening and entry into a period of episodic genomic instability.

These observations raise the intriguing possibility that pharmacologic agents that diminish the rate of telomere erosion by reducing oxidative stress or activating telomerase may function as potent cancer preventives. At a minimum and more practically, serial telomere length analyses in diseased tissues may provide biomarkers of cancer risk. Along these lines, several studies have explored the utility of telomere length determination in accessible peripheral blood lymphocytes (PBLs) rather than the diseased tissue per se. Such studies in nonneoplastic diseases have suggested that PBL telomere lengths can provide predictive information on the risk of developing atherosclerosis, premature myocardial infarctions, coronary artery disease, Alzheimer's disease status, and overall mortality (Samani et al. 2001; Brouilette et al. 2003; Cawthon et al. 2003; Obana et al. 2003; Panossian et al. 2003). Recently, it has also been shown that the presence of short telomeres in PBLs is associated with increased risk for the development of carcinomas of the head and neck, kidney, bladder, and lung (Wu et al. 2003). Further substantiation of the utility of PBL telomere lengths as a reliable and noninvasive surrogate marker that predicts risk of epithelial cancers could prove invaluable in identifying populations at significant risk and requiring closer monitoring.

Guiding Translational Opportunities in the Telomere Field

The collective information derived from the mouse will serve as a useful guide in the design and analyses of the many anticipated clinical trials in cancer. With regard to clinical trial design, increased response rates might be predicted if patient populations are screened for p53 status, making eligible those with retention of a p53-dependent DNA damage response. Conversely, tumors deficient for p53 would generate the concern for high levels of genomic instability and acquisition of resistance to other agents in the regimen and the emergence of ALT, the telomerase-independent telomere maintenance pathway. With regard to safety, the organ degeneration and increased rate of spontaneous malignancy in mice with telomere dysfunction encourage close monitoring of highly proliferative tissues and secondary malignancies. The telomerase knockout mouse model has also demonstrated that telomerase deficiency and associated telomere dysfunction enhance sensitivity to ionizing radiation and chemotherapeutic agents that generate double-stranded DNA breaks (Goytisolo et al. 2000; Wong et al. 2000). On the basis of this collective body of information, one particularly attractive clinical trial design would be the potentially synergistic combination of telomerase inhibitors and radiation or break-inducing chemotherapeutic agents. Again, however, particular care is warranted here as the combination of increased DNA damage with reduced capacity for normal repair may produce marked increases in the toxicity of chemoradiotherapy (Goytisolo et al. 2000; Wong et al. 2000).

Based on collective knowledge from the mouse and human cell culture models, we suggest that the ideal clinical trial design may be structured as follows: (1) lymphoid malignancies, as this organ compartment has shown the greatest sensitivity to telomere dysfunction in telomerase-deficient mouse models; (2) tumors that retain a competent p53-response to DNA damage (e.g., ARF-deficient tumors wherein a robust p53-dependent telomere checkpoint response is retained); and (3) tumors with uniform robust telomerase activity yet short telomeres. It remains to be seen whether such trials should couple the telomerase inhibitor with a DSB-inducing chemotherapeutic agent such as doxorubicin to achieve maximal synergistic therapeutic effect. Additionally, clinical assays with appropriate biomarkers for monitoring the effectiveness of the tested agents need to be developed to help guide and facilitate these clinical trials.

SUMMARY

In recent years, we have witnessed significant progress in the telomere biology field that is now maturing into new opportunities for improved diagnostics and novel therapeutic applications in human diseases. Discoveries in mouse telomere biology have already provided new mechanistic insights into the pathogenesis of human cancer and of inherited and acquired degenerative disorders. The role of telomere dysfunction driving episodic genomic instability in epithelial cancers, first seen in the telomerase-null mouse, has now been substantiated in the study of several human cancer types but still requires additional study and confirmation in other cancer types. The telomerase-deficient mouse cancer models also provide a strong platform to discover new cancer-relevant genes using high-resolution genomic technologies capable of cataloging and defining the recurrent amplifications and deletions seen in the genome of the epithelial cancers derived from these mice and their human counterparts. The phenotypes of the telomerase mutant mouse with telomere dysfunction will also be instructive in anticipating and monitoring potential untoward effects from telomerase inhibitors in the cancer clinic. The pivotal role of telomere attrition in the pathogenesis of organ degeneration and aging seen in the mouse may now provide avenues for the development of cancer risk biomarkers and preventive strategies, cancer diagnostics, and cancer therapies.

Although many fundamental questions remain concerning how telomeres and linked pathways influence the development of degenerative diseases, aging, and cancer, there is now clear evidence that telomeres are a focal point in the pathogenesis of these diseases. Further advances in the basic understanding of this rapidly expanding arena will no doubt fuel new opportunities for clinical applications in these human diseases. An integration of laboratory and clinical approaches, combined with rational clinical trial designs incorporating mechanistically based biological end points, will allow these new agents and applications to be tested in a more rapid, cost-effective, and safe manner.

ACKNOWLEDGMENTS

We are grateful to Steven Artandi, William C. Hahn, and Richard Maser for their helpful discussions and manuscript revisions. K.-K.W. is supported by National Institutes of Health (NIH) grant K08AG 2400401 and is a Sidney Kimmel Foundation for Cancer Research Scholar. S.C. is a 2005 Sidney Kimmel Foundation for Cancer Research Scholar and is supported by an Ellison New Scholars award, a Beeson Career Development award from the NIA/NIH, and a grant from the Abraham and Phyllis Katz Foundation. R.A.D is an American Cancer Society Research Professor and an Ellison Medical Foundation Senior Scholar and is supported by NIH grants R01CA84628, P01CA95616, and U01 CA84313.

REFERENCES

Al-Mulla F., Keith W.N., Pickford I.R., Going J.J., and Birnie G.D. 1999. Comparative genomic hybridization analysis of primary colorectal carcinomas and their synchronous metastases. *Genes Chromosomes Cancer* **24:** 306–314.

Artandi S.E., Chang S., Lee S.L., Alson S., Gottlieb G.J., Chin L., and DePinho R.A. 2000. Telomere dysfunction promotes non-reciprocal translocations and epithelial cancers in mice. *Nature* **406:** 641–645.

Blasco M.A., Funk W., Villeponteau B., and Greider C.W. 1995. Functional characterization and developmental regulation of mouse telomerase RNA. *Science* **269:** 1267–1270.

Blasco M.A., Lee H.W., Hande M.P., Samper E., Lansdor P.M., DePinho R.A., and Greider C.W. 1997. Telomere shortening and tumor formation by mouse cells lacking telomerase RNA. *Cell* **91:** 25–34.

Bodnar A.G., Ouellette M., Frolkis M., Holt S.E., Chiu C.P., Morin G.B., Harley C.B., Shay J.W., Lichtsteiner S., and Wright W.E. 1998. Extension of life-span by introduction of telomerase into normal human cells. *Science* **279:** 349–352.

Brouilette S., Singh R.K., Thompson J.R., Goodall A.H., and Samani N.J. 2003. White cell telomere length and risk of premature myocardial infarction. *Arterioscler. Thromb. Vasc. Biol.* **23:** 842–846.

Buerger H., Otterbach F., Simon R., Poremba C., Diallo R., Decker T., Riethdorf L., Brinkschmidt C., Dockhorn-Dworniczak B., and Boecker W. 1999. Comparative genomic hybridization of ductal carcinoma in situ of the breast-evidence of multiple genetic pathways. *J. Pathol.* **187:** 396–402.

Cawthon R.M., Smith K.R., O'Brien E., Sivatchenko A., and Kerber R.A. 2003. Association between telomere length in blood and mortality in people aged 60 years or older. *Lancet* **361:** 393–395.

Chang S., Khoo C.M., Naylor M.L., Maser R.S., and DePinho R.A. 2003. Telomere-based crisis: Functional differences between telomerase activation and ALT in tumor progression. *Genes Dev.* **17:** 88–100.

Chang S., Multani A.S., Cabrera N.G., Naylor M.L., Laud P., Lombard D., Pathak S., Guarente L., and DePinho R.A. 2004. Essential role of limiting telomeres in the pathogenesis of Werner syndrome. *Nat. Genet.* **36:** 877–882.

Chiang Y.J., Hemann M.T., Hathcock K.S., Tessarollo L., Feigenbaum L., Hahn W.C., and Hodes R.J. 2004. Expression of telomerase RNA template, but not telomerase reverse transcriptase, is limiting for telomere length maintenance in vivo. *Mol. Cell. Biol.* **24:** 7024–7031.

Chin K., de Solorzano C.O., Knowles D., Jones A., Chou W., Rodriguez E.G., Kuo W.L., Ljung, B.M., Chew K., Myambo K., et al. 2004. In situ analyses of genome instability in breast cancer. *Nat. Genet.* **36:** 984–988.

Chin L., Artandi S.E., Shen Q., Tam A., Lee S.L., Gottlieb G.J., Greider C.W., and DePinho R.A. 1999. p53 deficiency rescues the adverse effects of telomere loss and cooperates with telomere dysfunction to accelerate carcinogenesis. *Cell* **97:** 527–538.

Counter C.M., Meyerson M., Eaton E.N., Ellisen L.W., Caddle S.D., Haber D.A., and Weinberg R.A. 1998. Telomerase activity is restored in human cells by ectopic expression of hTERT (hEST2), the catalytic subunit of telomerase. *Oncogene* **16:** 1217–1222.

d'Adda di Fagagna F., Reaper P.M., Clay-Farrace L., Fiegler H., Carr P., Von Zglinicki T., Saretzki G., Carter N.P., and Jackson S.P. 2003. A DNA damage checkpoint response in telomere-initiated senescence. *Nature* **426:** 194–198.

de Lange T., Shiue L., Myers R.M., Cox D.R., Naylor S.L., Killery A.M., and Varmus H.E. 1990. Structure and variability of human chromosome ends. *Mol. Cell. Biol.* **10:** 518–527.

Dimri G.P., Lee X., Basile G., Acosta M., Scott G., Roskelley C., Medrano E.E., Linskens M., Rubelj I., Pereira-Smith O., et al. 1995. A biomarker that identifies senescent human cells in culture and in aging skin in vivo. *Proc. Natl. Acad. Sci.* **92:** 9363–9367.

Dokal I. 2001. Dyskeratosis congenita. A disease of premature ageing. *Lancet* (suppl.) **358:** S27.

Du X., Shen J., Kugan N., Furth E.E., Lombard D.B., Cheung C., Pak S., Luo G., Pignolo R.J., DePinho R.A., et al. 2004. Telomere shortening exposes functions for the mouse Werner and Bloom syndrome genes. *Mol. Cell. Biol.* **24:** 8437–8446.

Elenbaas B., Spirio L., Koerner F., Fleming M.D., Zimonjic D.B., Donaher J.L., Popescu N.C., Hahn W.C., and Weinberg R.A. 2001. Human breast cancer cells generated by oncogenic transformation of primary mammary epithelial cells. *Genes Dev.* **15:** 50–65.

Epstein C.J., Martin G.M., Schultz A.L., and Motulsky A.G. 1966. Werner's syndrome a review of its symptomatology, natural history, pathologic features, genetics and relationship to the natural aging process. *Medicine* **45:** 177–221.

Faragher R.G., Kill I.R., Hunter J.A., Pope F.M., Tannock C., and Shall S. 1993. The gene responsible for Werner syndrome may be a cell division "counting" gene. *Proc. Natl. Acad. Sci.* **90:** 12030–12034.

Farazi P.A., Glickman J., Jiang S., Yu A., Rudolph K.L., and DePinho R.A. 2003. Differential impact of telomere dysfunction on initiation and progression of hepatocellular carcinoma. *Cancer Res.* **63:** 5021–5027.

Feldser D.M., Hackett J.A., and Greider C.W. 2003. Telomere dysfunction and the initiation of genome instability. *Nat. Rev. Cancer* **3:** 623–627.

Gisselsson D., Jonson T., Petersen A., Strombeck B., Dal Cin P., Hoglund M., Mitelman F., Mertens F., and Mandahl N. 2001. Telomere dysfunction triggers extensive DNA fragmentation and evolution of complex chromosome abnormalities in human malignant tumors. *Proc. Natl. Acad. Sci.* **98:** 12683–12688.

Gonzalez-Suarez E., Samper E., Flores J.M., and Blasco M.A. 2000. Telomerase-deficient mice with short telomeres are resistant to skin tumorigenesis. *Nat. Genet.* **26:** 114–117.

Goss K.H., Risinger M.A., Kordich J.J., Sanz M.M., Straughen J.E., Slovek L.E., Capobianco A.J., German J., Boivin G.P., and Groden J. 2002. Enhanced tumor formation in mice heterozygous for Blm mutation. *Science* **297:** 2051–2053.

Goto M., Miller R.W., Ishikawa Y., and Sugano H. 1996. Excess of rare cancers in Werner syndrome (adult progeria). *Cancer Epidemiol. Biomark. Prev.* **5:** 239–246.

Goytisolo F.A., Samper E., Martin-Caballero J., Finnon P., Herrera E., Flores J.M., Bouffler S.D., and Blasco M.A. 2000. Short telomeres result in organismal hypersensitivity to ionizing radiation in mammals. *J. Exp. Med.* **192:** 1625–1636.

Gray D.A. 2003. Strategies for engineered negligible senescence. *Sci. Aging Knowledge Environ.* **2003:** pe30.

Greenberg R.A., Allsopp R.C., Chin L., Morin G.B., and DePinho R.A. 1998. Expression of mouse telomerase reverse transcriptase during development, differentiation and proliferation. *Oncogene* **16:** 1723–1730.

Greenberg R.A., Chin L., Femino A., Lee K.H., Gottlieb G.J., Singer R.H., Greider C.W., and DePinho R.A. 1999. Short dysfunctional telomeres impair tumorigenesis in the INK4a(delta2/3) cancer-prone mouse. *Cell* **97:** 515–525.

Griffith J.D., Comeau L., Rosenfield S., Stansel R.M., Bianchi A., Moss H., and de Lange T. 1999. Mammalian telomeres end in a large duplex loop. *Cell* **97:** 503–514.

Hahn W.C., Counter C.M., Lundberg A.S., Beijersbergen R.L., Brooks M.W., and Weinberg R.A. 1999. Creation of human tumour cells with defined genetic elements. *Nature* **400:** 464–468.

Hande M.P., Samper E., Lansdorp P., and Blasco M.A. 1999. Telomere length dynamics and chromosomal instability in cells derived from telomerase null mice. *J. Cell Biol.* **144:** 589–601.

Harley C.B., Kim N.W., Prowse K.R., Weinrich S.L., Hirsch K.S., West M.D., Bacchetti S., Hirte H.W., Counter C.M., Greider C.W., et al. 1994. Telomerase, cell immortality, and cancer. *Cold Spring Harbor Symp. Quant. Biol.* **59:** 307–315.

Hastie N.D., Dempster M., Dunlop M.G., Thompson A.M., Green D.K., and Allshire R.C. 1990. Telomere reduction in human colorectal carcinoma and with ageing. *Nature* **346:** 866–868.

Hemann M.T., Rudolph K.L., Strong M.A., DePinho R.A., Chin L., and Greider C.W. 2001. Telomere dysfunction triggers developmentally regulated germ cell apoptosis. *Mol. Biol. Cell* **12:** 2023–2030.

Herrera E., Martinez A.C., and Blasco M.A. 2000. Impaired germinal center reaction in mice with short telomeres. *EMBO J.* **19:** 472–481.

Iwakuma T. and Lozano G. 2003. MDM2, an introduction. *Mol. Cancer Res.* **1:** 993–1000.

Jackson S.P. 2002. Sensing and repairing DNA double-strand breaks. *Carcinogenesis* **23:** 687–696.

Kamijo T., Zindy F., Roussel M.F., Quelle D.E., Downing J.R., Ashmun R.A., Grosveld G., and Sherr C.J. 1997. Tumor suppression at the mouse INK4a locus mediated by the alternative reading frame product p19ARF. *Cell* **91:** 649–659.

Kaneko H., Matsui E., Fukao T., Kasahara K., Morimoto W., and Kondo N. 1999. Expression of the BLM gene in human haematopoietic cells. *Clin. Exp. Immunol.* **118:** 285–289.

Karlseder J., Broccoli D., Dai Y., Hardy S., and de Lange T. 1999. p53- and ATM-dependent apoptosis induced by telomeres lacking TRF2. *Science* **283:** 1321–1325.

Kipling D. and Cooke H.J. 1990. Hypervariable ultra-long telomeres in mice. *Nature* **347:** 400–402.

Kitada T., Seki S., Kawakita N., Kuroki T., and Monna T. 1995. Telomere shortening in chronic liver diseases. *Biochem. Biophys. Res. Commun.* **211:** 33–39.

Lee H.W., Blasco M.A., Gottlieb G.J., Horner J.W., II, Greider C.W., and DePinho R.A. 1998. Essential role of mouse telomerase in highly proliferative organs. *Nature* **392:** 569–574.

Liu Y., Snow B.E., Hande M.P., Yeung D., Erdmann N.J., Wakeham A., Itie A., Siderovski D.P., Lansdorp P.M., Robinson M.O., and Harrington L. 2000. The telomerase reverse transcriptase is limiting and necessary for telomerase function in vivo. *Curr. Biol.* **10:** 1459–1462.

Lombard D.B., Beard C., Johnson B., Marciniak R.A., Dausman J., Bronson R., Buhlmann J.E., Lipman R., Curry R., Sharpe A., et al. 2000. Mutations in the WRN gene in mice accelerate mortality in a p53-null background. *Mol. Cell. Biol.* **20:** 3286–3291.

Lowe S.W. and Sherr C.J. 2003. Tumor suppression by Ink4a-Arf: Progress and puzzles. *Curr. Opin. Genet. Dev.* **13:** 77–83.

Martin G.M., Sprague C.A., and Epstein C.J. 1970. Replicative life-span of cultivated human cells. Effects of donor's age, tissue, and genotype. *Lab. Invest.* **23:** 86–92.

Martin-Rivera L., Herrera E., Albar J.P., and Blasco M.A. 1998. Expression of mouse telomerase catalytic subunit in embryos and adult tissues. *Proc. Natl. Acad. Sci.* **95:** 10471–10476.

Maser R.S. and DePinho R.A. 2002. Connecting chromosomes, crisis, and cancer. *Science* **297:** 565–569.

———. 2004. Telomeres and the DNA damage response: Why the fox is guarding the henhouse. *DNA Repair* **3:** 979–988.

Masutomi K., Yu E.Y., Khurts S., Ben-Porath I., Currier J.L., Metz G.B., Brooks M.W., Kaneko S., Murakami S., DeCaprio J.A., et al. 2003. Telomerase maintains telomere structure in normal human cells. *Cell* **114:** 241–253.

McClintock B. 1941. The stability of broken ends of chromosomes in *Zea mays. Genetics* **26:** 234–282.

Meyerson M., Counter C.M., Eaton E.N., Ellisen L.W., Steiner P., Caddle S.D., Ziaugra L., Beijersbergen R.L., Davidoff M.J., Liu Q., et al. 1997. hEST2, the putative human telomerase catalytic subunit gene, is up-regulated in tumor cells and during immortalization. *Cell* **90:** 785–795.

Mitchell J.R., Cheng J., and Collins K. 1999. A box H/ACA small nucleolar RNA-like domain at the human telomerase RNA 3′ end. *Mol. Cell. Biol.* **19:** 567–576.

Miura N., Horikawa I., Nishimoto A., Ohmura H., Ito H., Hirohashi S., Shay J.W., and Oshimura M. 1997. Progressive telomere shortening and telomerase reactivation during hepatocellular carcinogenesis. *Cancer Genet. Cytogenet.* **93:** 56–62.

Newbold R.F. 1997. Genetic control of telomerase and replicative senescence in human and rodent cells. *Ciba Found. Symp.* **211:** 177–189; discussion 189–197.

Obana N., Takagi S., Kinouchi Y., Tokita Y., Sekikawa A., Takahashi S., Hiwatashi N., Oikawa S., and Shimosegawa T. 2003. Telomere shortening of peripheral blood mononuclear cells in coronary disease patients with metabolic disorders. *Intern. Med.* **42:** 150–153.

Ogburn C.E., Oshima J., Poot M., Chen R., Hunt K.E., Gollahon K.A., Rabinovitch P.S., and Martin G.M. 1997. An apoptosis-inducing genotoxin differentiates heterozygotic carriers for Werner helicase mutations from wild-type and homozygous mutants. *Hum. Genet.* **101:** 121–125.

O'Hagan R.C., Chang S., Maser R.S., Mohan R., Artandi S.E., Chin L., and DePinho R.A. 2002. Telomere dysfunction provokes regional amplification and deletion in cancer genomes. *Cancer Cell* **2:** 149–155.

O'Sullivan J.N., Bronner M.P., Brentnall T.A., Finley J.C., Shen W.T., Emerson S., Emond M.J., Gollahon K.A., Moskovitz A.H., Crispin D.A., et al. 2002. Chromosomal instability in ulcerative colitis is related to telomere shortening. *Nat. Genet.* **32:** 280–284.

Panossian L.A., Porter V.R., Valenzuela H.F., Zhu X., Reback E., Masterman D., Cummings J.L., and Effros R.B. 2003. Telomere shortening in T cells correlates with Alzheimer's disease status. *Neurobiol. Aging* **24:** 77–84.

Prowse K.R., and Greider C.W. 1995. Developmental and tissue-specific regulation of mouse telomerase and telomere length. *Proc. Natl. Acad. Sci.* **92:** 4818–4822.

Prowse K.R., Avilion A.A., and Greider C.W. 1993. Identification of a nonprocessive telomerase activity from mouse cells. *Proc. Natl. Acad. Sci.* **90:** 1493–1497.

Qi L., Strong M.A., Karim B.O., Armanios M., Huso D.L., and Greider C.W. 2003. Short telomeres and ataxia-telangiectasia mutated deficiency cooperatively increase telomere dysfunction and suppress tumorigenesis. *Cancer Res.* **63:** 8188–8196.

Romanov S.R., Kozakiewicz B.K., Holst C.R., Stampfer M.R., Haupt L.M., and Tlsty T.D. 2001. Normal human mammary epithelial cells spontaneously escape senescence and acquire genomic changes. *Nature* **409:** 633–637.

Rudolph K.L. and DePinho R.A. 2001. Telomeres and telomerase in experimental liver cirrhosis. In *Liver: Biology and pathobiology* (ed. I.M. Arias et al.), pp. 999–1009. Lippincott Williams & Wilkins, Philadelphia.

Rudolph K.L., Millard M., Bosenberg M.W., and DePinho R.A. 2001. Telomere dysfunction and evolution of intestinal carcinoma in mice and humans. *Nat. Genet.* **28:** 155–159.

Rudolph K.L., Chang S., Lee H.W., Blasco M., Gottlieb G.J., Greider C., and DePinho R.A. 1999. Longevity, stress response, and cancer in aging telomerase-deficient mice. *Cell* **96:** 701–712.

Samani N.J., Boultby R., Butler R., Thompson J.R., and Goodall A.H. 2001. Telomere shortening in atherosclerosis. *Lancet* **358:** 472–473.

Samper E., Flores J.M., and Blasco M.A. 2001. Restoration of telomerase activity rescues chromosomal instability and premature aging in Terc$^{-/-}$ mice with short telomeres. *EMBO Rep.* **2:** 800–807.

Serrano M., Lin A.W., McCurrach M.E., Beach D., and Lowe S.W. 1997. Oncogenic ras provokes premature cell senescence associated with accumulation of p53 and p16INK4a. *Cell* **88:** 593–602.

Sharpless N.E. and DePinho R.A. 2002. p53: Good cop/bad cop. *Cell* **110:** 9–12.

Shay J.W. and Bacchetti S. 1997. A survey of telomerase activity in human cancer. *Eur. J. Cancer* **33:** 787–791.

Shiloh Y. and Kastan M.B. 2001. ATM: Genome stability, neuronal development, and cancer cross paths. *Adv. Cancer Res.* **83:** 209–254.

Tahara H., Tokutake Y., Maeda S., Kataoka H., Watanabe T., Satoh M., Matsumoto T., Sugawara M., Ide T., Goto M., et al. 1997. Abnormal telomere dynamics of B-lymphoblastoid cell strains from Werner's syndrome patients transformed by Epstein-Barr virus. *Oncogene* **15:** 1911–1920.

Takai H., Smogorzewska A., and de Lange T. 2003. DNA damage foci at dysfunctional telomeres. *Curr. Biol.* **13:** 1549–1556.

Tang R., Cheng A.-J., Wang J.-Y., and Wang T.-C.V. 1998. Close correlation between telomerase expression and adenomatous polyp progression in multistep colorectal carcinogenesis. *Cancer Res.* **58:** 4052–4054.

Tyner S.D., Venkatachalam S., Choi J., Jones S., Ghebranious N., Igelmann H., Lu X., Soron G., Cooper B., Brayton C., et al. 2002. p53 mutant mice that display early ageing-associated phenotypes. *Nature* **415:** 45–53.

van Heek N.T., Meeker A.K., Kern S.E., Yeo C.J., Lillemoe K.D., Cameron J.L., Offerhaus G.J., Hicks J.L., Wilentz R.E., Goggins M.G., et al. 2002. Telomere shortening is nearly universal in pancreatic intraepithelial neoplasia. *Am. J. Pathol.* **161:** 1541–1547.

Vulliamy T., Marrone A., Szydlo R., Waln A., Mason P.J., and Dokal I. 2004. Disease anticipation is associated with progressive telomere shortening in families with dyskeratosis congenita due to mutations in TERC. *Nat. Genet.* **36:** 447–449.

Vulliamy T., Marrone A., Goldman F., Dearlove A., Bessler M., Mason P.J., and Dokal I. 2001. The RNA component of telomerase is mutated in autosomal dominant dyskeratosis congenita. *Nature* **413:** 432–435.

Weng N.P., Granger L., and Hodes R.J. 1997. Telomere lengthening and telomerase activation during human B cell differentiation. *Proc. Natl. Acad. Sci.* **94:** 10827–10832.

Wiemann S.U., Satyanarayana A., Tsahuridu M., Tillmann H.L., Zender L., Klempnauer J., Flemming P., Franco S., Blasco M.A., Manns M.P., and Rudolph K.L. 2002. Hepatocyte telomere shortening and senescence are general markers of human liver cirrhosis. *FASEB J.* **16:** 935–942.

Wong K.K., Maser R.S., Bachoo R.M., Menon J., Carrasco D.R., Gu Y., Alt F.W., and DePinho R.A. 2003. Telomere dysfunction and Atm deficiency compromises organ homeostasis and accelerates ageing. *Nature* **421:** 643–648.

Wong K.K., Chang S., Weiler S.R., Ganesan S., Chaudhuri J., Zhu C., Artandi S.E., Rudolph K.L., Gottlieb G.J., Chin L., et al. 2000. Telomere dysfunction impairs DNA repair and enhances sensitivity to ionizing radiation. *Nat. Genet.* **26:** 85–88.

Wright W.E., Piatyszek M.A., Rainey W.E., Byrd W., and Shay J.W. 1996. Telomerase activity in human germ line and embryonic tissues and cells. *Dev. Genet.* **18:** 173–179.

Wu X., Amos C.I., Zhu Y., Zhao H., Grossman B.H., Shay J.W., Luo S., Hong W.K., and Spitz M.R. 2003. Telomere dysfunction: A potential cancer predisposition factor. *J. Natl. Cancer Inst.* **95:** 1211–1218.

Wyllie F.S., Jones C.J., Skinner J.W., Haughton M.F., Wallis C., Wynford-Thomas D., Faragher R.G., and Kipling D. 2000. Telomerase prevents the accelerated cell ageing of Werner syndrome fibroblasts. *Nat. Genet.* **24:** 16–17.

Yamagata K., Kato J., Shimamoto A., Goto M., Furuichi Y., and Ikeda H. 1998. Bloom's and Werner's syndrome genes suppress hyperrecombination in yeast sgs1 mutant: Implication for genomic instability in human diseases. *Proc. Natl. Acad. Sci.* **95:** 8733–8738.

Yan H., Chen C.Y., Kobayashi R., and Newport J. 1998. Replication focus-forming activity 1 and the Werner syndrome gene product. *Nat. Genet.* **19:** 375–378.

Yen C.C., Chen Y.J., Chen J.T., Hsia J.Y., Chen P.M., Liu J.H., Fan F.S., Chiou T.J., Wang W.S., and Lin C.H. 2001. Comparative genomic hybridization of esophageal squamous cell carcinoma: Correlations between chromosomal aberrations and disease progression/ prognosis. *Cancer* **92:** 2769–2777.

Yuan R., Astle C.M., Chen J., and Harrison D.E. 2005. Genetic regulation of hematopoietic stem cell exhaustion during development and growth. *Exp. Hematol.* **33:** 243–250.

Zijlmans J.M., Martens U.M., Poon S.S., Raap A.K., Tanke H.J., Ward R.K., and Lansdorp P.M. 1997. Telomeres in the mouse have large inter-chromosomal variations in the number of T2AG3 repeats. *Proc. Natl. Acad. Sci.* **94:** 7423–7428.

6

Telomerase Deficiency and Human Disease

Inderjeet Dokal and Tom Vulliamy

Department of Haematology–Division of Investigative Science
Imperial College London
Hammersmith Hospital
London, United Kingdom

THE CRITICAL IMPORTANCE OF TELOMERES AND TELOMERASE in humans has been recently illustrated by studies on the rare multisystem disorder dyskeratosis congenita (DC). In this disease, two genetic subtypes have been found to be due to mutations in genes that encode components of the telomerase complex. This results in telomerase deficiency, accelerated telomere shortening, and the development of multisystem abnormalities including bone marrow failure and many features of premature aging.

DC and related disorders can now be regarded as disorders principally of telomerase deficiency. In this chapter, clinical and genetic features are described of classical DC as well as other diseases (the "occult/cryptic" and atypical forms of DC) that have been recently linked to DC. The genetic defects found in these patients and their effect on telomerase and telomeres are reviewed here. These developments have implications for the management of patients with DC and related disorders. They also provide new opportunities for exploring the role of telomeres and telomerase in more common processes such as aging and cancer. Some of the topics in this chapter have also been reviewed by Marciniak and Guarente (2001).

CLASSICAL DYSKERATOSIS CONGENITA

Clinical Aspects

Classical DC (also known as Zinsser–Engman–Cole syndrome) is an inherited bone-marrow-failure syndrome characterized by the mucocutaneous

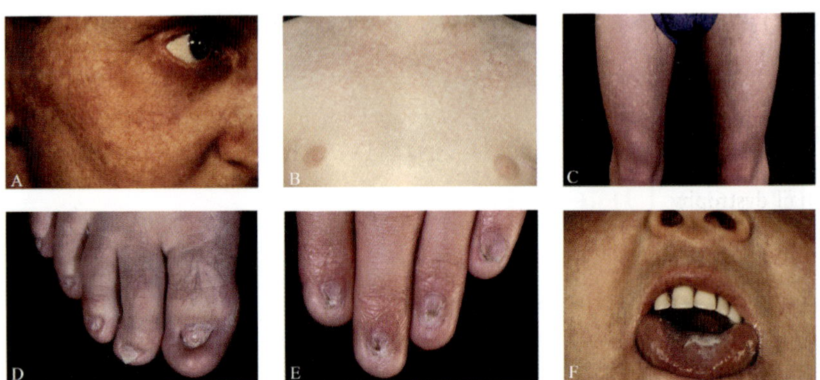

Figure 1. Photographs of DC patients showing abnormal skin pigmentation (*A,B,C*), nail dystrophy (*D,E*), and leukoplakia of tongue (*E*).

triad of abnormal skin pigmentation, nail dystrophy, and mucosal leukoplakia (Fig. 1) (Zinsser 1906; Engman 1926; Cole et al. 1930). A variety (dental, gastrointestinal, genitourinary, hair graying/loss, immunological, neurological, ophthalmic, pulmonary, and skeletal) of other somatic abnormalities have also been reported (Table 1) (Sirinavin and Trowbridge 1975; Drachtman and Alter 1995; Knight et al. 1998a; Solder et al. 1998; Dokal 2000). Bone marrow failure is the principal cause of early mortality, with an additional predisposition to malignancy and fatal pulmonary complications (Dokal 2000). The malignancies usually develop in the third decade and include carcinomas (bronchus, colon, larynx, esophagus, pancreas, skin, tongue), leukemias, and lymphomas. Malignancies in DC patients are difficult to treat, as conventional chemotherapeutic drugs are associated with considerable drug-related toxicity.

During the last 10 years, there have been significant advances in our understanding of DC. Many of these have been facilitated by the DC registry established at the Hammersmith Hospital (London) in 1995. As of September 2004, 226 DC families comprising 352 patients (267 male, 85 female) from 40 countries have been recruited. In approximately 72% of the families, there were only affected males, with 18% of the families having two or more affected males and 54% affected sporadic male cases. As would be predicted, the majority of the families with two or more affected males have been found to represent X-linked recessive DC (because of mutations in *DKC1*, see below), as do some of the families with affected sporadic male cases. In the remaining 28% of the families, there is one or more affected female. In some of these, the inheritance pattern is compatible with autosomal dominant transmission, and this has been substantiated at the genetic level (see below). In others,

Table 1. Somatic abnormalities/complications in patients with classical dyskeratosis congenita (DC)

Abnormality	% of patients
Classical presentation	
abnormal skin pigmentation	89
nail dystrophy	88
bone marrow failure	85.5
leukoplakia	78
Other somatic features	
epiphora	30.5
learning difficulties/developmental delay/mental retardation	25.4
pulmonary disease	20.3
short stature	19.5
extensive dental caries/loss	16.9
esophageal stricture	16.9
premature hair loss/graying/sparse eyelashes	16.1
hyperhiderosis	15.3
malignancy	9.8
intrauterine growth retardation	7.6
liver disease/peptic ulceration/enteropathy	7.3
ataxia/cerebellar hypoplasia	6.8
hypogonadism/undescended testes	5.9
microcephaly	5.9
urethral stricture/phimosis	5.1
osteoporosis/aseptic necrosis/scoliosis	5.1
deafness	0.8

In the future, as patients identified to have DC on the basis of mutation analysis are included, these percentages are likely to change.

the transmission indicates autosomal recessive inheritance: In some of these families with affected females, there is parental consanguinity and/or two or more affected cases in the same generation. These families therefore provide further evidence for autosomal recessive form(s) of the disease, in addition to those published previously (Sorrow and Hitch 1963; Sirinavin and Trowbridge 1975; Ling et al. 1985; Juneja et al. 1987; Pai et al. 1989; Drachtman and Alter 1992; Joshi et al. 1994; Elliot et al. 1999).

Clinical manifestations in DC often appear during childhood, although there is a wide age range. The skin pigmentation and nail changes typically appear first, usually by the age of 10 years. Bone marrow failure usually develops before the age of 20 years; 80–90% of patients will have developed bone marrow abnormalities by the age of 30 years (Knight et al.

1998a; Dokal 2000). In some patients, the bone marrow abnormalities may appear before the mucocutaneous manifestations (occasionally in the first year of life) and can lead to an initial diagnosis of "idiopathic aplastic anemia" (De Boeck et al. 1981; Philips et al. 1992; Forni et al. 1993; Dokal 2000; Fogarty et al. 2003). Considerable clinical variability exists between patients, sometimes even within the same family. A given patient may have diverse combinations of the features listed in Table 1. However, many patients will have one or more abnormalities related to tissues that have a high turnover (e.g., bone marrow, skin, and gastrointestinal tract). Although it is difficult to make generalizations, the X-linked recessive form appears to be associated with a more severe phenotype (more abnormalities and a younger age of onset) than the autosomal dominant form; patients with autosomal dominant DC tend to have fewer abnormalities and a later age of onset. The mucocutaneous abnormalities in autosomal dominant DC can also be "mild" and can make diagnosis difficult. The autosomal recessive families also show considerable heterogeneity, with some patients having severe bone marrow failure by the age of 10 years, yet others having no hematological abnormalities even by the age of 40 years. The main causes of death are bone marrow failure/ immunodeficiency (~60–70%), pulmonary complications (~10–15%), and malignancy (~5–10%) (Knight et al. 1998a; Dokal 2000).

Bone marrow failure/immunodeficiency is the principal cause of premature mortality in DC patients. Oxymetholone (an anabolic steroid) can produce an improvement in hematopoietic function in many patients (~60%) for a variable period of time (Smith et al. 1979; I. Dokal, unpubl.). The main current treatment for severe bone marrow failure is allogeneic hematopoietic stem cell transplantation, and there is some experience using both sibling and alternative stem cell donors (Berthou et al. 1991; Dokal et al. 1992; Langston et al. 1996; Yabe et al. 1997; Rocha et al. 1998; Ghavamzadeh et al. 1999; Lau et al. 1999). Unfortunately, because of early and late fatal pulmonary/vascular complications, the results of allogeneic stem cell transplantation have been less successful. The presence of pulmonary disease in a significant proportion of DC patients (Table 1) (Paul et al. 1992; Verra et al. 1992; Knight et al. 1998a; Safa et al. 2001; Kilic et al. 2003) explains the high incidence of fatal pulmonary complications in the setting of stem cell transplantation. These observations of high transplant-related toxicity in DC patients highlight the fact that many tissues (such as the lungs, liver, skin, and gastrointestinal tract) have difficulties in repair following chemical or radiation stress. This again emphasizes that DC cells from all tissues have generalized defects in their

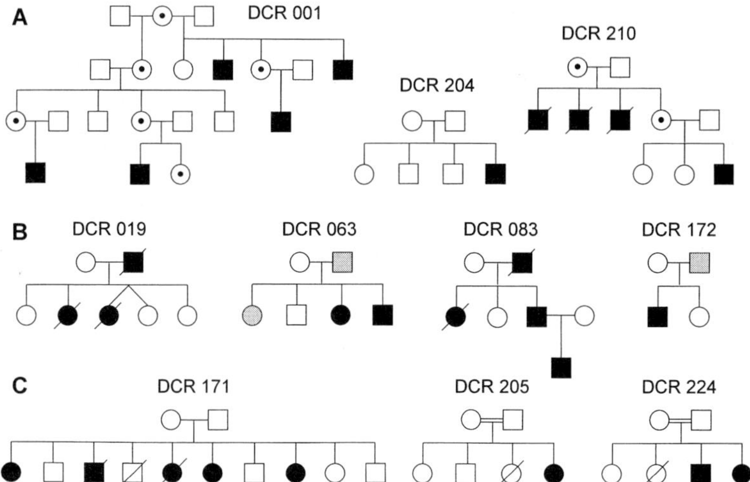

Figure 2. DC registry (DCR) families showing different forms of inheritance. (*A*) X-linked DC due to *DKC1* mutations. DCR 001 shows clear X-linked inheritance with multiple female carriers (shown as targets). The sporadic affected male (*black square*) in DCR 204 arises by de novo mutation; his mother is not a carrier (*open circle*). In family DCR210, the index case has HH syndrome, and his three maternal uncles died before the age of 7 months. (*B*) Autosomal dominant inheritance. In DCR 019, we have been unable to find a TERC mutation. In the other families, with TERC mutations, there is clear male-to-female transmission in family DCR 083, but, in both DCR 063 and DCR 172, the affected father was asymptomatic (*gray squares*). (*C*) Autosomal recessive inheritance is seen in the large family DCR 171 and in family DCR 224 with consanguineous marriage. Recessive inheritance is assumed in DCR 205 in which the affected female (*black circle*) is a sporadic case, but the marriage is consanguineous.

ability to regenerate and repair, which is particularly marked in tissues that have a high turnover.

Genetics and Link to Telomerase

DC is a genetically heterogeneous disorder. X-linked recessive (MIM [Mendelian Inheritance in Man] 305000), autosomal dominant (MIM 127550), and autosomal recessive (MIM 224230) genetic subtypes are recognized. Figure 2 shows DC registry families demonstrating X-linked recessive, autosomal dominant, and autosomal recessive inheritance. Analysis of the DC registry suggests that each of the autosomal forms of DC may be genetically heterogeneous: For example, preliminary homozygosity mapping studies in affected individuals from consanguineous marriages suggest that there is more than one genetic locus for the autosomal recessive category.

X-linked DC

DKC1 *and dyskerin.* Linkage analysis and positional cloning resulted in identification of the *DKC1* gene in Xq28 that is mutated in X-linked DC (Connor et al. 1986; Knight et al. 1996, 1998b; Heiss et al. 1998; Hassock et al. 1999). In X-linked DC patients, mutations are spread throughout the *DKC1* gene, which encodes a 514-amino-acid protein, referred to as dyskerin. Dyskerin is a nucleolar protein (Heiss et al. 1999; Youssoufian et al. 1999) and is highly conserved, with many homologs including yeast (Cbf5p), rat (NAP57), *Drosophila* (mfl), and mouse (Dkc1) (Jiang et al. 1993; Meier and Blobel 1994; Cadwell et al. 1997; Philips et al. 1998; Giordano et al. 1999; Ruggero et al. 2003). Dyskerin has a number of domains including a TruB domain at amino acids 107–247 (named after an *Escherichia coli* pseudouridine synthase), a PUA domain (pseudouridine synthase and archaeosine-specific transglycosylases) at amino acids 296–371, and two nuclear localization signals at amino acids 11–20 and 446–458 (Fig. 3). The TruB domain is the catalytic domain of pseudouridine synthases, which is shared with bacterial TruB proteins, yeast Pus4p, and

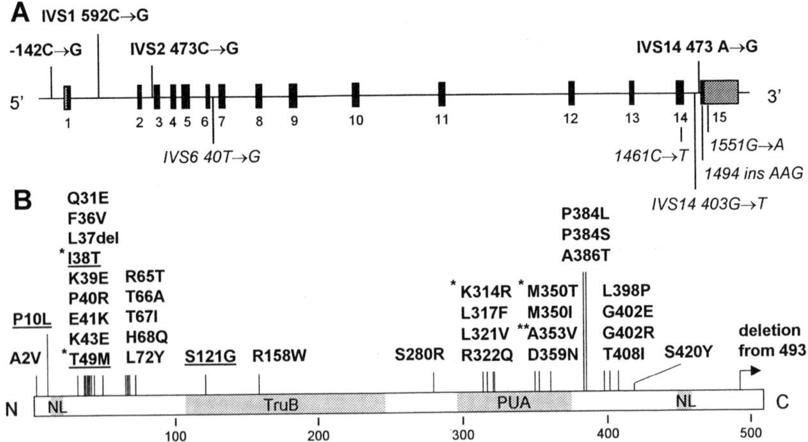

Figure 3. Mutations in the *DKC1* gene. (*A*) A scale drawing of the *DKC1* gene, which is composed of 15 exons (*black squares*) and spans 15 kb within Xq28. Polymorphisms are shown in italics below; a promoter mutation and three splicing mutations that have been found only in DC patients are shown above. (*B*) The locations of the missense mutations and a 3′ deletion are shown above a linear representation of the dyskerin protein. (*Gray boxes*) Location of inferred domains of the protein (see text). Substitutions marked with an asterisk have been identified in more than one family. The A353V mutation is a common recurrent de novo mutation, which has been observed in ~40% of X-linked DC families. The substitutions underlined were detected in patients with the HH syndrome.

the Cb5fp family (Koonin 1996). The PUA domain is a putative RNA-binding domain (Arvind and Koonin 1999). Dyskerin interacts with three proteins (GAR1, NHP2, and NOP10) to form the core of H/ACA ribonucleoprotein particles. In yeast, the dyskerin homolog Cbf5p interacts with similar nucleolar proteins (Gar1, Nhp2, Nop10), and these associate with the H/ACA small nucleolar RNAs (snoRNAs) to form small nucleolar ribonucleoprotein (snoRNP) particles. H/ACA snoRNAs are characterized by having a "hairpin–Hinge–hairpin–ACA" (H/ACA) motif. As a part of these H/ACA ribonucleoprotein particles, yeast Cbf5p has been shown to be a pseudouridine synthase guiding the conversion of uridine to pseudouridine at specific sites of rRNA (pseudouridylation). Yeast *cbf5* mutants demonstrate a pronounced defect in rRNA biosynthesis and are deficient in cytoplasmic ribosomal subunits (Cadwell et al 1997). As dyskerin also has a TruB domain, like its homologs (Tollervey and Kiss 1997; Henras et al. 1998; Lafontaine et al. 1998; Watkins et al. 1998; Zebrarjadian et al 1999; Filipowicz and Pogacic 2002; Wang and Meier 2004), dyskerin has been predicted to have pseudouridylation activity. This is an essential step for ribosome biogenesis, and it has therefore been suggested that part of the pathology in X-linked DC probably relates to defective ribosome biogenesis (Luzzatto and Karadimitris 1998). Recent studies on the mouse dyskerin gene (*Dkc1*) (Ruggero et al. 2003; Mochizuki et al. 2004) provide additional support for this (Meier 2003). Further studies are necessary in human cells to clarify this issue.

Several *DKC1* mutations have been identified (Knight et al. 1999a, 2001; Vulliamy et al. 1999; Heiss et al. 2001; Cossu et al. 2002; Hiramatsu et al. 2002; Lin et al. 2002; Salowsky et al. 2002; Wong et al. 2004) and these are shown in Figure 3. The majority of mutations in *DKC1* cause single-acid substitutions, one of which (A353V) accounts for approximately 40% of X-linked DC cases. The phenotype in these patients with the same dyskerin mutation can vary considerably (e.g., some patients present with severe bone marrow failure by the age of 5 years, yet others have a normal blood count up to the fourth decade) and suggests that other genetic/environmental factors influence the DC phenotype. Although the mutations are spread throughout the *DKC1* gene, two prominent clusters involve amino acids 31–72 and 314–420, encoded in exons 3–6 and exons 9–12. It is noteworthy that these two clusters of mutations lie outside the predicted TruB domain. In future studies, it will be interesting to determine whether dyskerin mutations have different molecular consequences (e.g., to what extent each mutation disrupts the interaction with GAR1, NOP10, NHP2), depending on their precise physical location within the dyskerin molecule. Presently, there are few data regarding dyskerin and other protein changes in X-linked DC. The studies to date have shown that mutant dyskerin protein is detectable at relatively normal levels in the case of the few *DKC1* mutations investigated (Mitchell et al. 1999b; Wong et al. 2004).

Link to telomerase. It has been shown that dyskerin and the other three proteins (GAR1, NHP2, and NPO10) that form the core of RNP particles also associate with the telomerase RNA component (TERC) (Pogacic et al. 2000; Dez et al. 2001), which also contains an H/ACA motif (Mitchell et al. 1999a,b). The structure and function of telomerase are the focus of other chapters in this volume. Together with telomerase reverse transcriptase (TERT), TERC forms the core of the active telomerase complex. Mitchell et al. (1999b) found that in one fibroblast and four lymphoblast cell lines from patients with X-linked DC, the level of TERC was reduced, although no significant defect in rRNA processing or site-specific pseudouridylation was detected. It now appears that telomerase RNA processing and stable RNA accumulation require dyskerin binding to the H/ACA motif of TERC (Mitchell et al 1999b; Mitchell and Collins 2000; Fu and Collins 2003), and dyskerin mutations therefore lead to reductions in TERC. Furthermore, telomere lengths in the lymphoblast lines were shorter than expected for age-matched normal individuals. Similar results (i.e., reduced levels of TERC) have recently been obtained for the peripheral blood mononuclear cells (PBMCs) of one X-linked DC patient (Wong et al. 2004). Telomerase activity, induced by overexpression of TERT in the X-linked DC fibroblast cell line, was shown to be reduced compared to similarly treated lines from DC carriers (Mitchell et al. 1999b). However, in the PBMCs from five X-linked DC patients, telomerase activity seems to vary over a range similar to that seen in five normal controls (Vulliamy et al. 2001a). It has also been shown that in blood cells from patients with autosomal forms of DC, telomeres are shorter than in controls (Vulliamy et al. 2001a). The combination of these findings suggested that DC might be principally a disease of telomere maintenance. Further clarification has come from the elucidation of the genetic basis of autosomal dominant DC (see below).

Autosomal Dominant DC: TERC and DC Pathology

Linkage analysis in one large DC family showed that the gene for autosomal dominant DC is on chromosome 3q, in the same area where the gene for TERC had been previously mapped. This led to TERC mutation analysis in this and other DC families and the demonstration that autosomal

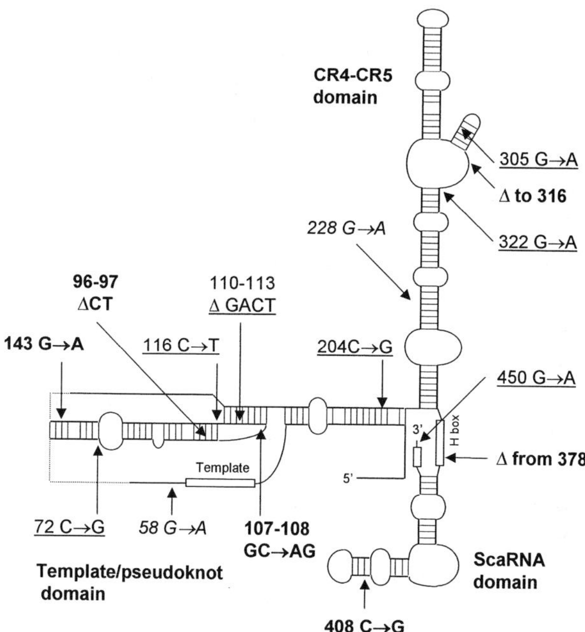

Figure 4. Model of TERC showing position of mutations. The predicted structure of the RNA molecule and the location of the different domains are taken from Chen and Grieder (2004). Mutations shown in boldface were observed in autosomal dominant DC families; those underlined were identified in aplastic anemia patients; two polymorphisms observed in subjects of African origin are shown in italics.

dominant DC is due to mutations in the TERC gene (Vulliamy et al. 2001b). TERC is a 451-nucleotide RNA. Figure 4 shows the position of these mutations on a schematic representation of TERC. TERC consists of four structural domains: the pseudoknot domain (this domain is made up of four short helices: P2a.1, P2a, P2b, and P3), the CR4-CR5 domain, the H/ACA domain, and the CR7 domain. Structure-function analysis reveals that the pseudoknot and CR4-CR5 domains (together with TERT) are required for the catalytic function, whereas the H/ACA and CR7 domains are required for TERC RNA accumulation (as described in Chapters 2 and 3).

Figure 4 shows that in the various autosomal dominant DC families, in addition to two large deletions that remove the 5' and 3' ends of TERC, point mutations and small deletions have been found in several TERC domains. To determine the functional consequences of these mutations, their effect on telomerase activity (either in a cell-free system or after introduction into W138VA13 cells that do not express telomerase) and/or on the structure of telomerase RNA (e.g., using nuclear magnetic resonance [NMR] spectroscopy) has been determined. Collectively, these functional studies (Comolli et al. 2002; Fu and Collins 2003; Ly et al. 2003; Ren et al. 2003; Theimer et al. 2003a,b; Marrone at al. 2004) have demonstrated that TERC mutations seen in autosomal dominant DC patients result in reduced telomerase activity through either impaired RNA accumulation/stability (found to be the case for the following mutations: deletion from nucleotide 378, 408 C → G) or a catalytic defect (found to be the case for the following mutations: 72C → G, 96–97 del CT, 107/108 GC → AG, 110–113 del GACT). Northern blot analysis has demonstrated reduced TERC accumulation only for two mutations (deletion from nucleotide 378 and 408 C → G) (Vulliamy et al. 2001b; Fu and Collins 2003) and suggests that these two mutations affect RNA accumulation/stability. Experiments reconstituting telomerase with both normal and mutant TERC molecules showed no evidence for a dominant-negative effect. Instead, the data suggest that the TERC mutations act via haploinsufficiency (Fu and Collins 2003; Marrone et al. 2004). Additionally, it has been established for some of these mutations (e.g., 72C → G; 107–108 GC → AG, 408 C → G), which are located within the conserved stem cell structure, that it is their effect on the secondary structure as well as on the primary sequence that determines the functional consequences (Comolli et al. 2002; Ly et al. 2003; Marrone et al. 2004). Analysis of these disease-causing TERC mutations has thus also provided important confirmatory data on the different functional TERC domains.

Because the *DKC1*-encoded protein dyskerin and TERC are both components of the telomerase complex and all DC patients have very short telomeres, it is currently believed that DC arises principally from an abnormality in telomerase activity (Marciniak and Guarente 2001). This telomerase deficiency results in accelerated telomere shortening in DC cells (Mitchell et al. 1999b; Vulliamy et al. 2001a) and is associated with increased loss of cells (Montanaro et al. 2002), particularly from tissues that need constant renewal, such as the hematopoietic and dermatological systems that bear the brunt of DC. Evidence for a hematopoietic stem/progenitor cell defect (Fig. 5) in DC has been established in several studies (Colvin et al. 1984; Friedland et al. 1985; Alter et al. 1992; Marsh et al. 1992; Marley et al. 1999) and is therefore consistent with a basic defect in telomerase.

The effect of mutation of one allele of human telomerase RNA in families with autosomal dominant DC can be compared with the effect on laboratory mice of knocking out both alleles of the gene coding for telomerase RNA (*Terc*). Such mice showed no significant abnormalities in early generations, but later generations (beyond fourth generation)

A   **B**

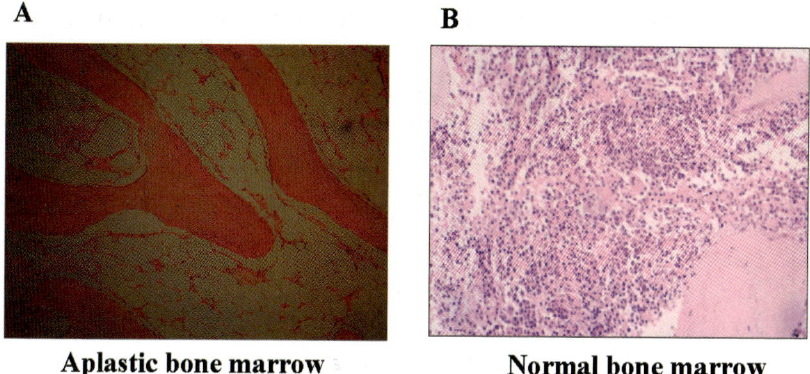

Aplastic bone marrow   **Normal bone marrow**

Figure 5. Bone marrow trephine from a patient with DC (*A*) showing marked reduction (hypocellularity) of hematopoietic cells compared to that from a normal individual (*B*).

showed progressive telomere shortening and a variety of defects including reduced proliferative capacity of hematopoietic cells (Blasco et al. 1997; Lee et al. 1998; Rudolph et al. 1999; Artandi et al. 2000). Heterozygous mice (*Terc*$^{+/-}$) appear asymptomatic but they do have difficulty in elongating their telomeres in interspecies crosses (Hathcock et al. 2002). It is noteworthy that targeted disruption of *Dkc1* causes embryonic lethality in mice (He et al. 2002), and patients with X-linked DC generally tend to be more severe compared to patients with autosomal dominant DC, particularly with regard to the mucocutaneous features. This suggests that dyskerin mutations in humans, as well as producing a major defect in telomerase, might also have an effect (perhaps to a smaller degree) on rRNA biosynthesis. Further work on this is necessary on human cells.

It is interesting to note that the autosomal dominant DC families show an earlier age of onset and a greater number of disease features in succeeding generations. This increase in severity of the disease (anticipation) in successive generations has been associated with progressive telomere shortening (Vulliamy et al. 2004), as is the case in the *Terc*$^{-/-}$ mice. Both DC patients and *Terc*$^{-/-}$ mice also have an increased incidence of cancer. It is likely that in both cases, the increased incidence of malignancy is caused by chromosome instability, which in turn is a result of critically shortened telomeres. Indeed, end-to-end chromosome fusions have been observed both in *Terc*$^{-/-}$ mice (Artandi et al. 2000) and in DC patients (Scappaticci et al. 1989; Dokal et al. 1992; Kehrer et al. 1992; Demiroglu et al. 1997).

Table 2. Dyskeratosis congenita (DC) genetic subtypes: Data based on the DC registry

DC subtype	Approximate % of DC patients	Chromosome location	RNA/protein product	Mutations identified
X-linked recessive	35	Xq28	dyskerin	34
Autosomal dominant	5	3q26	TERC	10
Autosomal recessive[a]	60	?	?	?

[a] The genetic basis of this subtype is unknown and may represent more than one genetic locus.

Autosomal Recessive DC

The genetic basis of the autosomal recessive form(s) of DC is presently unknown. Since dyskerin and TERC have been shown to be key components of the telomerase complex, other molecules associated with this complex or involved in rRNA biosynthesis would seem to be obvious candidates. Further studies, including linkage analysis on consanguineous autosomal recessive families, are ongoing to identify the genes mutated in autosomal recessive DC.

Implications for DC Patients

The demonstration of *DKC1* and *TERC* mutations in DC families provides an accurate diagnostic test, including antenatal diagnosis, in approximately 40% of cases (Table 2). It also provides the basis for designing the much needed new treatments. Since in any given patient, DC is a single gene disorder and the cells that need to be targeted (hematopoietic stem cells) are accessible, DC is a good candidate for hematopoietic gene therapy. Furthermore, there is evidence from fibroblast culture studies and from the skewed patterns of X-chromosome inactivation seen in carriers of X-linked DC (Devriendt et al. 1997; Vulliamy et al. 1997) that cells transfected with the normal gene would have growth/survival advantage compared to the uncorrected cells. Such an advantage would also be predicted from the role of dyskerin and TERC in telomere maintenance.

OTHER HUMAN DISEASES RECENTLY LINKED TO DC AND TELOMERASE DEFICIENCY

Hoyeraal-Hreidarsson Syndrome

Analyses of patients on the DC registry have identified considerable clinical heterogeneity. They have also revealed that some DC patients have features that overlap with those observed in patients with the Hoyeraal–Hreidarsson (HH) syndrome (MIM 600545). This is a severe multisystem disorder that can present in the neonatal period and infancy. It is characterized by severe growth retardation, bone marrow failure, immunodeficiency, and neurological abnormalities (Hoyeraal et al. 1970; Hreidarsson et al. 1988; Berthet et al. 1994; Aalfs et al. 1995; Ohga et al. 1995; Nespoli et al. 1997). The overlap of these HH features with some DC patients led to analysis of the *DKC1* gene in HH patients. These studies demonstrated that some male HH cases are a severe variant of DC where death from bone marrow failure/immunodeficiency occurs before the appearance of the diagnostic features of DC (Knight et al. 1999b). It also highlighted the immunological defects that can be seen in DC, ranging from the severe "T+B-NK-severe immunodeficiency" in HH patients (Knight et al. 1999b; Cossu et al. 2002) to the more variable immunological abnormalities observed in other DC patients (Wiedemann et al. 1984; Rose and Kern 1992; Solder et al. 1998; Dokal 2000; Knudson et al. 2005). Several mutations in dyskerin have now been identified in HH patients (see Fig. 3) (Knight et al. 1999b; Yagahmai et al. 2000; Cossu et al. 2002; Sznajer et al. 2003).

Female cases of HH syndrome are also recognized (Mahmood et al. 1998; Akaboshi et al. 2000; DC registry families), and it is likely that they represent a severe variant of the autosomal recessive form(s) of DC, the genetic basis of which presently remains unknown.

Aplastic Anemia and Myelodysplasia

In some of the DC registry families, affected members have died of severe aplastic anemia before the age of 10 years, and a diagnosis of DC was made subsequently, only when other members of the family survived long enough to develop the classical mucocutaneous features. If it were not for the presence of subsequent members, these would have been characterized as idiopathic aplastic anemia. In patients with idiopathic aplastic anemia, the bone marrow is usually hypocellular and is unable to produce adequate numbers of mature blood cells. The primary defect in idiopathic aplastic anemia is believed to be at the stem cell level, but its precise cause is not known. Like patients with DC, patients with idiopathic aplastic anemia also have short telomeres compared to age-matched controls (Ball et al. 1998; Brummendorf et al. 2001). These observations led us to analyze the *DKC1* and *TERC* genes in idiopathic aplastic anemia patients. Although the *DKC1* gene screen was found to be normal, mutations in *TERC* have been found in some cases of aplastic anemia (including paroxysmal nocturnal hemoglobinuria, which is a clonal hematopoietic disorder that usually evolves on a background of aplastic anemia) and

Table 3. Characteristics of DC, HH, idiopathic AA, and MDS

	DC	HH	AA	MDS
Inheritance pattern	XLR, AR, AD	XLR, AR	?	?
Somatic abnormalities	yes	yes	?no	?no
Bone marrow dysfunction	yes	yes	yes	yes
Short telomeres	yes	yes	yes	yes
Malignancy	yes	yes	yes	yes
Chromosome instability	yes	yes	yes	yes
Number of genes	at least 3	?	?	?
Genes cloned	2	1	?	?

Abbreviations: AA, aplastic anemia; DC, dyskeratosis congenita; AD, autosomal dominant; AR, autosomal recessive; HH, Hoyeraal–Hreidarsson syndrome; MDS, myelodysplasia; XLR, X-linked recessive.

myelodysplasia (Vulliamy et al. 2002, 2004; Yamaguchi et al. 2003; Keith et al. 2004). It is noteworthy that patients with myelodysplasia also have short telomeres (Boultwood et al. 1997), as well as an inability to produce adequate numbers of mature blood cells, which leads to increased mortality from infections and bleeding. The marrow cellularity in patients with myelodysplasia is usually increased, and there are marked abnormalities in morphology ("dysplasia") of the different cell types. Patients with myelodysplasia, like patients with aplastic anemia, are believed to have a defect at the level of the stem cell, but the primary pathology again remains unknown in the majority of cases. The characteristics of patients with DC, HH syndrome, aplastic anemia, and myelodysplasia are compared in Table 3. Patients with *TERC* mutations who present predominantly with features of aplastic anemia or myelodysplasia can be regarded as having "mild/cryptic" DC. It is possible that *TERC* mutations in this group of patients are "less severe" and have their main clinical impact only on the most proliferative tissue in the body, the hematopoietic tissue. An alternative possibility is that in these patients, there may be an interaction with environmental factors. Such factors could include an insult such as infection, which results in maximal cumulative damage to the hematopoietic tissue, or factors such as chronic stress; one study has linked duration and severity of long-term psychological stress with both decreased telomerase levels and shorter telomeres in peripheral blood mononucleocytes (Epel et al. 2004).

In summary, these findings show that in some cases of aplastic anemia/myelodysplasia, the primary defect is in the maintenance of telomeres, and this has implications for the management of patients failing conventional therapies such as immunosuppressive therapy. They also highlight the clinical and genetic heterogeneity of DC and a possible rationale for screening the DC genes in uncharacterized patients (e.g., unexplained pulmonary or liver disease) who have clinical features that overlap with DC.

CONCLUSIONS AND FUTURE PROSPECTIVES

The study of DC has demonstrated that mutations in genes (*DKC1, TERC*) encoding two telomerase holoenzyme components reduce the maximal level telomerase activity and dramatically compromise the proliferative renewal of hematopoietic and epithelial tissue resulting in multisystem abnormalities. It has thus highlighted the critical role of telomerase in human growth and development. From the perspective of bone-marrow-failure syndromes (including cases of idiopathic aplastic anemia), these studies have shown that in some cases of aplastic anemia, the primary defect relates to telomere maintenance. This has implications for the diagnosis of patients with atypical features and in the development of therapies for those failing conventional treatments.

The precise contribution of defects in pseudouridylation in the pathogenesis of X-linked DC remains unclear. Recent studies using mouse embryonic stem cells suggest that mouse dyskerin mutations affect both telomerase and pseudouridylation pathways, and the extent to which these functions were altered varied for the two mutations (A353V and G402E) investigated. Further studies are necessary to clarify this question for human disease.

Many classical patients with DC remain uncharacterized. The elucidation of the genetic basis of these patients may help to clarify the relative importance of telomere maintenance versus ribosomal biogenesis in the pathogenesis of DC. In turn, these studies may identify novel molecules that have a direct or indirect role in the maintenance of telomeres and/or ribosomal biogenesis.

Because the phenotype of DC includes premature aging as well as increased occurrence of malignancy, further studies that result in improving our understanding of DC might have important ramifications for our understanding of these more common processes (Cawthon et al. 2003).

Correction of the molecular defect in DC (e.g., by a hematopoietic gene therapy approach) could pave the way for developing the much-needed new therapies for this group of patients. From the perspective of telomere biology, such an approach represents on exciting opportunity to determine whether telomere rejuvenation is possible in human cells. If successful, this could have implications for the treatment of pathologies outside the field of DC.

SUMMARY

DC is an inherited bone-marrow-failure syndrome exhibiting considerable clinical and genetic heterogeneity. X-linked recessive, autosomal dominant, and autosomal recessive forms are recognized. The gene mutated in X-linked DC (*DKC1*) encodes a highly conserved nucleolar protein called dyskerin. Dyskerin associates with the H/ACA class of snoRNAs in snoRNP particles that are important in guiding the conversion of uracil to pseudouracil during the maturation of rRNA. Dyskerin also associates with TERC, which is important in the maintenance of telomeres. Mutations in TERC were recently demonstrated in patients with autosomal dominant DC and in a subset of patients with aplastic anemia and myelodysplasia. Additionally, some patients with HH syndrome, a severe multisystem disorder, have been found to have *DKC1* mutations. These findings have demonstrated that classical DC, the HH syndrome, and a subset of aplastic anemia are due to a primary defect in telomerase. The multisystem abnormalities seen in these patients, including the increased incidence of malignancy, have highlighted the critical role of telomeres and telomerase in humans. From a clinical perspective, the link between DC and aplastic anemia, and in turn to defective telomerase, suggests that treatments directed at correction of telomerase activity might benefit DC/aplastic anemia patients who do not respond to conventional therapy. Studies on the uncharacterized cases of DC can be expected to improve our understanding of the bone-marrow-failure syndromes further, as well as provide new insights on telomere biology.

ACKNOWLEDGMENTS

We thank Philip Mason, Stuart Knight, Anna Marrone, Amanda Walne, and David Stevens, whose contribution over the past 10 years has been critical to DC research. We are also grateful to the DC families and all our colleagues (doctors and nurses) for their support in establishing the Dyskeratosis Congenita Registry and to the Wellcome Trust for financial support.

REFERENCES

Aalfs C.M., van den Berg H., Barth P.G., and Hennekam R.C.M. 1995. The Hoyeraal-Hreidarsson syndrome: The fourth case of a separate entity with prenatal growth retardation, progressive pancytopenia and cerebellar hypoplasia. *Eur. J. Pediatr.* **154:** 304–308.

Akaboshi S., Yoshimura M, Hara T., Kageyama H., Nishikwa K., Kawakami T, Leshima A., and Takeshita K. 2000. A case of Hoyeraal-Hreidarsson syndrome: Delayed myelination and hypoplasia of corpus callosum are other important signs. *Neuropediatrics* **31:** 141–144.

Alter B.P., Knobloch M.E., He L., Gillio A.P., O'Reilly R.J., Reilly L.K., and Weinberg R.S. 1992. Effect of stem cell factor on in vitro erythropoiesis in patients with bone marrow failure syndromes. *Blood* **80:** 3000–3008.

Aravind L. and Koonin E.V. 1999. Novel predicted RNA-binding domains associated with translation machinery. *J. Mol. Evol.* **48:** 291–302.

Artandi S.E., Chang S., Lee S.L., Alson S., Gottlieb G.J., Chin L., and DePinho R.A. 2000. Telomere dysfunction promotes non-reciprocal translocations and epithelial cancers in mice. *Nature* **406:** 641–645.

Ball S.E., Gibson F.M., Rizzo S., Tooze J.A., Marsh J.C., and Gordon-Smith E.C. 1998. Progressive telomere shortening in aplastic anemia. *Blood* **91:** 3582–3592.

Berthet F., Caduff R., Schaad U.B., Roten H., Tuchscmid P., Boltshauser E., and Seger R.A. 1994. A syndrome of primary combined immunodeficiency with microcephaly, cerebellar hypoplasia, growth failure and progressive pancytopenia. *Eur. J. Pediatr.* **153:** 333–338.

Berthou C., Devergie A., D'Agay M.F., Sonsino E., Scrobohaci M.L., Loirat C., and Gluckman E. 1991. Late vascular complications after bone marrow transplantation for dyskeratosis congenita. *Br. J. Haematol.* **79:** 335–336.

Blasco M.A., Lee H.W., Hande M.P., Samper E., Lansdorp P.M., DePinho R.A., and Greider C.W. 1997. Telomere shortening and tumor formation by mouse cells lacking telomerase RNA. *Cell* **91:** 25–34.

Boultwood J., Fidler C., Kusec R., Rack K., Elliott P.J., Atoyebi O., Chapman R., Oscier D.G., and Wainscoat J.S. 1997. Telomere length in myelodysplastic syndromes. *Am. J. Hematol.* **56:** 266–271.

Brummendorf T.H., Maciejewski J.P., Mak J., Young N.S., and Lansdorp P.M. 2001. Telomere length in leukocyte subpopulations of patients with aplastic anemia. *Blood* **97:** 895–900.

Cadwell C., Yoon H.J., Zebarjadian Y., and Carbon J. 1997. The yeast nucleolar protein Cbf5p is involved in rRNA biosynthesis and interacts genetically with the RNA polymerase I transcription factor RRN3. *Mol. Cell. Biol.* **17:** 6175–6183.

Cawthon R.M., Smith K.R., O'Brien E., Sivatchenko A., and Kerber R.A. 2003. Association between telomere length in blood and mortality in people aged 60 yrs or older. *Lancet* **361:** 393–395.

Chen J.L. and Greider C.W. 2004. Telomerase RNA structure and function: Implications for dyskeratosis congenita. *Trends Biochem. Sci.* **29:** 183–192.

Cole H.N., Rauschkolb J.C., and Toomey J. 1930. Dyskeratosis congenita with pigmentation, dystrophiaunguis and leukokeratosis oris. *Arch. Dermatol. Syph.* **21:** 71–95.

Colvin B.T., Baker H., Hibbin, J.A., Gordon-Smith E.C., and Gordon M.Y. 1984. Haemopoietic progenitor cells in dyskeratosis congenita. *Br. J. Haematol.* **56:** 513–515.

Comolli L.R., Smirnov I., Xu L., Blackburn E.H., and James T.L. 2002. A molecular switch underlies a human telomerase disease. *Proc. Natl. Acad. Sci.* **99:** 16998–17003.

Connor J.M., Gatherer D., Gray F.C., Pirrit L.A., and Affara N.A. 1986. Assignment of the gene for dyskeratosis congenita to Xq28. *Hum. Genet.* **72:** 348–351.

Cossu F., Vulliamy T.J., Marrone A., Badiali M., Cao A., and Dokal I. 2002. A novel DKC1 mutation, severe combined immunodeficiency (T+B-NK- SCID) and bone marrow transplantation in an infant with Hoyeraal-Hreidarsson syndrome. *Br. J. Haematol.* **119**: 765–768.

De Boeck K., Degreef H., Verwilghen R., Corbeel L., and Casteels-Van Daele M. 1981. Thrombocytopenia: First symptom in a patient with dyskeratosis congenita. *Pediatrics* **67**: 898–903.

Demiroglu H., Alikasifoglu M., and Dundar S. 1997. Dyskeratosis congenita with an unusual chromosomal abnormality. *Br. J. Haematol.* **97**: 243–244.

Devriendt K., Matthijs G., Legius E., Schollen E., Blockmans D., van Geet C., Degreef H., Cassiman J.-J., and Fryns J.P. 1997. Skewed X-chromosome inactivation in female carriers of dyskeratosis congenita. *Am. J. Hum. Genet.* **60**: 581–587.

Dez C., Henras A., Faucon B., Lafontaine D.L.J., Caizergues-Ferrer M., and Henry Y. 2001. Stable expression in yeast of the mature form of human telomerase RNA depends on its association with the box H/ACA small nucleolar RNP proteins Cbf5p, Nhp2p and Nop10p. *Nucleic Acids Res.* **29**: 598–603.

Dokal I. 2000. Dyskeratosis congenita in all its forms. *Br. J. Haematol.* **110**: 768–779.

Dokal I., Bungey J., Williamson P., Oscier D., Hows J., and Luzzatto L. 1992. Dyskeratosis congenita fibroblasts are abnormal and have unbalanced chromosomal rearrangements. *Blood* **80**: 3090–3096.

Drachtman R.A. and Alter B.P. 1992. Dyskeratosis congenita: Clinical and genetic heterogeneity. Report of a new case and review of the literature. *Am. J. Pediatr. Hematol. Oncol.* **14**: 297–304.

———.1995. Dyskeratosis congenita. *Dermatol. Clin.* **13**: 33–39.

Elliott A.M., Graham G.E, Bernstein M., Mazer B., and Teebi A.S. 1999. Dyskeratosis congenita: An autosomal recessive variant. *Am. J. Med. Genet.* **83**: 178–182.

Engman M.F. 1926. A unique case of reticular pigmentation of the skin with atrophy. *Arch. Dermatol. Syph.* **13**: 685–687.

Epel E.S., Blackburn E.H., Lin J., Dhabar F., Adler N., Morrow J., and Cawthon R. 2004. Accelerated telomere shortening in response to life stress. *Proc. Natl. Acad. Sci.* **101**: 17312–17315.

Filipowicz W. and Pogacic V. 2002. Biogenesis of small nucleolar ribonucleoproteins. *Curr. Opin. Cell Biol.* **14**: 319–327.

Fogarty P.F., Yamaguchi H., Wiestner A., Baerlocher G.B., Sloand E., Zeng W.S., Read E.J., Lansdorp P.M., and Young N.S. 2003. Late presentation of dyskeratosis congenita as apparently acquired aplastic anaemia due to mutations in telomerase RNA. *Lancet* **362**: 1628–1630.

Forni G.L., Melevendi C., Jappelli S., and Rasore-Quartino A. 1993. Dyskeratosis congenita: Unusual presenting features within a kindred. *Pediatr. Hematol. Oncol.* **10**: 145–149.

Friedland M., Lutton J.D., Spitzer R., and Levere R.D. 1985. Dyskeratosis congenita with hypoplastic anemia: A stem cell defect. *Am. J. Hematol.* **20**: 85–87.

Fu D. and Collins K. 2003. Distinct biogenesis pathways for human telomerase RNA and H/ACA small nucleolar RNAs. *Mol. Cell* **11**: 1361–1372.

Ghavamzadeh A., Alimoghadam K., Nasseri P., Jahani M., Khodabandeh A., and Ghahremani G. 1999. Correction of bone marrow failure in dyskeratosis congenita by bone marrow transplantation. *Bone Marrow Transplant.* **23**: 299–301.

Giordano E., Peluso I., Senger S., and Furia M. 1999. minifly, a *Drosophila* gene required for ribosome biogenesis. *J. Cell Biol.* **144**: 1123–1133.

Hassock S., Vetrie D., and Giannelli, F. 1999. Mapping and characterization of X-linked dyskeratosis congenita (DKC) gene. *Genomics* **55**: 21–27.

Hathcock K.S., Hemann M.T., Opperman K.K., Strong M.A., Greider C.W., and Hodes R.J. 2002. Haploinsufficiency of mTR results in defects in telomere elongation. *Proc. Natl. Acad. Sci.* **99:** 3591–3596.

He J., Navarrete S., Jasinski M., Vulliamy T., Dokal I., Bessler M., and Mason P.J. 2002. Targeted disruption of *Dkc1*, the gene mutated in X-linked dyskeratosis congenita, causes embryonic lethality in mice. *Oncogene* **21:** 7740–7744.

Heiss N.S., Girod A., Saowsky R., Wiemann S., Pepperkok R., and Poustka A. 1999. Dyskerin localises to the nucleolus and its mislocalisation is unlikely to play a role in the pathogenesis of dyskeratosis congenita. *Hum. Mol. Genet.* **8:** 2515–2524.

Heiss N.S., Megarbane A., Klauck S.M., Kreuz F.R., Makhoul E., Majewski F., and Poustka A. 2001. One novel and two recurrent missense DKC1 mutations in patients with dyskeratosis congenita (DKC). *Genet. Couns.* **12:** 129–136.

Heiss N.S., Knight S.W., Vulliamy T.J., Klauck S.M., Wiemann S., Mason P.J., Poustka A., and Dokal I. 1998. X-linked dyskeratosis congenita is caused by mutations in a highly conserved gene with putative nucleolar functions. *Nat. Genet.* **19:** 32–38.

Henras A., Henry Y., Bousquet-Antonelli C., Noaillac-Depeyre J., Gelugne J.P., and Caizergues-Ferrer. 1998. Nhp2p and Nop10p are essential for the function of H/ACA snoRNPs. *EMBO J.* **17:** 7078–7090.

Hiramatsu H., Fujii T., Kitoh T., Sawada M., Osaka M., Koami K., Irino T., Miyajima T., Ito M., Sugiyama T., and Okuno T. 2002. A novel missense mutation in the DKC1 gene in a Japanese family with X-linked dyskeratosis congenita. *Pediatr. Hematol. Oncol.* **19:** 413–419.

Hoyeraal H.M., Lamvik J., and Moe P.J. 1970. Congenital hypoplastic thrombocytopenia and cerebral malformations in two brothers. *Acta Paediatr. Scand.* **59:** 185–191.

Hreidarsson S., Kristjansson K., Johannesson G., and Johannsson J.H. 1988. A syndrome of progressive pancytopenia with microcephaly, cerebellar hypoplasia and growth failure. *Acta Paediatr. Scand.* **77:** 773–775.

Jiang W., Middleton K., Yoon H.J., Fouquet C., and Carbon J. 1993. An essential yeast protein, CBF5p, binds in vitro to centromeres and microtubules. *Mol. Cell. Biol.* **13:** 4884–4893.

Joshi R.K., Atukorala D.N., Abanmi A., and Kudwah A. 1994. Dyskeratosis congenita in a female. *Br. J. Dermatol.* **130:** 520–522.

Juneja H.S., Elder F.F.B., and Gardner F.H. 1987. Abnormality of platelet size and T-lymphocyte proliferation in an autosomal recessive form of dyskeratosis congenita. *Eur. J. Haematol.* **39:** 306–310.

Kehrer H., Krone W., Schindler D., Kaufmann R., and Schrezenmeier H. 1992. Cytogenetic studies of skin fibroblast cultures from a karyotypically normal female with dyskeratosis congenita. *Clin. Genet.* **41:** 129–134.

Keith W.N., Vulliamy T., Zhao J., Ar C., Erzik C., Bisland A., Ulku B., Marrone A., Mason P.J., Bessler M., Serakinci N., and Dokal I. 2004. A mutation in a functional Sp1 binding site of the telomerase RNA gene (hTERC) promoter in a patient with paroxysmal nocturnal haemoglobinuria. *BMC Blood Disord.* **4:** 3–11.

Kilic S., Kose H., and Ozturk H. 2003. Pulmonary involvement in patient dyskeratosis congenita. *Pediatr. Int.* **45:** 740–742.

Knight S., Vulliamy T., Copplestone A., Gluckman E., Mason P., and Dokal I. 1998a. Dyskeratosis Congenita (DC) Registry: Identification of new features of DC. *Br. J. Haematol.* **103:** 990–996.

Knight S.W., Vulliamy T., Forni G.L., Oscier D., Mason P.J., and Dokal I. 1996. Fine mapping of the dyskeratosis congenita locus in Xq28. *J. Med. Genet.* **33:** 993–995.

Knight S.W., Vulliamy T.J., Morgan B., Devriendt K., Mason P.J., and Dokal I. 2001. Identification of novel DKC1 mutations in patients with dyskeratosis congenita: Implications for pathophysiology and diagnosis. *Hum. Genet.* **108:** 299–303.

Knight S.W., Vulliamy T.J., Heiss N.S., Matthijs G., Devriendt K., Connor J.M., D'Urso M., Poustka A., Mason P.J., and Dokal, I. 1998b. 1.4 Mb candidate gene region for X-linked dyskeratosis congenita defined by combined haplotype and X-chromosome inactivation analysis. *J. Med. Genet.* **35:** 993–996.

Knight S.W., Heiss N.S., Vulliamy T.J., Aalfs C.M., McMahon C., Richmond P., Jones A., Hennekam R.C., Poustka A., Mason P.J., and Dokal I. 1999a. Unexplained aplastic anaemia, immunodeficiency, and cerebellar hypoplasia (Hoyeraal-Hreidarsson syndrome) due to mutations in the dyskeratosis congenita gene, DKC1. *Br. J. Haematol.* **107:** 335–339.

Knight S.W., Heiss N.S., Vulliamy T.J., Greschner S., Stavrides G., Pai G.S., Lestringant G., Varma N., Mason P.J., Dokal I., and Poustka A. 1999b. X-linked dyskeratosis congenita is predominantly caused by missense mutations in the DKC1 gene. *Am. J. Hum. Genet.* **65:** 50–58.

Knudson M., Kulkarni S., Ballas Z., Bessler M., and Goldman F. 2005. Association of immune abnormalities with telomere shortening in autosomal dominant dyskeratosis congenita. *Blood* **105:** 682–688.

Koonin E.V. 1996. Pseudouridine synthases: Four families of enzymes containing a putative uridine-binding motif also conserved in dUTPases and dCTP deaminases. *Nucleic Acids Res.* **24:** 2411–2415.

Lafontaine D.L., Bousquet-Antonelli C., and Henry Y. 1998. The box H+ACA snoRNAs carry Cbf5p, the putative pseudouridine synthase. *Genes Dev.* **12:** 527–537.

Langston A.A., Sanders J.E., Deeg H.J., Crawford S.W., Anasetti C., Sullivan K.M., Flowers M.E.D., and Storb R. 1996. Allogeneic marrow transplantation for aplastic anaemia associated with dyskeratosis congenita. *Br. J. Haematol.* **92:** 758–765.

Lau Y.L., Ha S.Y., Chan C.F., Lee A.C.W., Liang R.H.S., and Yuen H.L. 1999. Bone marrow transplant for dyskeratosis congenita. *Br. J. Haematol.* **105:** 571.

Lee H.W., Blasco M.A., Gottlieb G.J., Horner J.W., II, Greider C.W., and DePinho R.A. 1998. Essential role of mouse telomerase in highly proliferative organs. *Nature* **392:** 569–574.

Lin J.H., Lee J.Y., Tsao C.J., and Chao S.C. 2002. DKC1 gene mutation in a Taiwanese kindred with X-linked dyskeratosis congenita. *Kaohsiung J. Med. Sci.* **18:** 573–577.

Ling N.S., Fenske N.A., Julius R.L., Espinoza C.G., and Drake L.A. 1985. Dyskeratosis congenita in a girl simulating chronic graft-vs-host disease. *Arch. Dermatol.* **21:** 1424–1428.

Luzzatto L. and Karadimitris A. 1998. Dyskeratosis and ribosomal rebellion. *Nat. Genet.* **19:** 6–7.

Ly H., Blackburn E.H., and Parslow T.G. 2003. Comprehensive structure-function analysis of the core domain of human telomerase RNA. *Mol. Cell. Biol.* **23:** 6849–6856.

Mahmood F., King M.D., Smyth O.P., and Farrell M.A. 1998. Familial cerebellar hypoplasia and pancytopenia without chromosomal breakages. *Neuropediatrics* **29:** 302–306.

Marciniak R. and Guarente L. 2001. Testing telomerase. *Nature* **413:** 370–373.

Marley S.B., Lewis J.L., Davidson R.J., Roberts I.A.G., Dokal I., Goldman J.M., and Gordon M.Y. 1999. Evidence of a continuous decline in haemopoietic function from birth: Application to evaluating bone marrow failure in children. *Br. J. Haematol.* **106:** 162–166.

Marrone A., Stevens D., Vulliamy T., Dokal I., and Mason P.J. 2004. Heterozygous telomerase RNA mutations found in dyskeratosis congenita and aplastic anaemia reduce telomerase activity via haploinsufficiency. *Blood* **104:** 3936–3942.

Marsh C.W., Will A.J., Hows J.H., Sartori P., Derbyshire P., Williamson P.J., Oscier D.G., Dexter T.M., and Testa N.G. 1992. "Stem cell" origin of the hematopoietic defect in dyskeratosis congenita. *Blood* **79:** 3138–3144.

Meier U.T. 2003. Dissecting dyskeratosis. *Nat. Genet.* **33:** 116–117.

Meier U.T. and Blobel G. 1994. NAP57, a mammalian nucleolar protein with a putative homolog in yeast and bacteria. *J. Cell Biol.* **127:** 1505–1514.

Mitchell J.R. and Collins K. 2000. Human telomerase activation requires two independent interactions between telomerase RNA and telomerase reverse transcriptase in vivo and in vitro. *Mol. Cell* **6:** 361–371.

Mitchell J.R., Cheng J., and Collins K. 1999a. A box H/ACA small nucleolar RNA-like domain at the human telomerase RNA 3′ end. *Mol. Cell. Biol.* **19:** 567–576.

Mitchell J.R., Wood E., and Collins K. 1999b. A telomerase component is defective in the human disease dyskeratosis congenita. *Nature* **402:** 551–555.

Mochizuki Y., He J., Kulkarni S., Bessler M., and Mason P.J. 2004. Mouse dyskerin mutations affect accumulation of telomerase RNA and small nucleolar RNA, telomerase activity, and ribosomal RNA processing. *Proc. Natl. Acad. Sci.* **101:** 10756–10761.

Montanaro L., Chillo A., Trere D., Pession A., Gouoni M., Tazzari P.L., and Derenzini M. 2002. Increased mortality rate and not impaired ribosomal biogenesis is responsible for proliferation defect in dyskeratosis congenita cell lines. *J. Invest. Dermatol.* **118:** 193–198.

Nespoli L., Lascari C., Maccario R., Nosetti L., Broggi U., Locatelli F., Binda S., Gaudio F., Casalone R., and Bosi F. 1997. The Hoyeraal-Hreidarsson syndrome: The presentation of the seventh case. *Eur. J. Pediatr.* **156:** 818–820.

Ohga S., Kai T., Honda K., Nakayama H., Inamitsu T., and Ueda K. 1997. What are the essential symptoms in the Hoyeraal-Hreidarsson syndrome? *Eur. J. Pediatr.* **156:** 80–81.

Pai G.S., Morgan S., and Whetsell C. 1989. Etiologic heterogeneity in dyskeratosis congenita. *Am. J. Med. Genet.* **32:** 63–66.

Paul S.R., Perez-Atayde A., and Williams D.A. 1992. Interstitial pulmonary disease associated with dyskeratosis congenita. *Am. J. Hematol. Oncol.* **14:** 89–92.

Philips B., Billin A.N., Cadwell C., Buchholz R., Erickson C., Merriam J.R., Carbon J., and Poole S.J. 1998. The Nop60B gene of *Drosophila* encodes an essential nucleolar protein that functions in yeast. *Mol. Gen. Genet.* **260:** 20–29.

Philips R.J., Judge M., Webb D., and Harper J.I. 1992. Dyskeratosis congenita: Delay in diagnosis and successful treatment of pancytopenia by bone marrow transplantation. *Br. J. Dermatol.* **127:** 278–280.

Pogacic V., Dragon F., and Filipowicz W. 2000. Human H/ACA small nucleolar RNPs and telomerase share evolutionarily conserved proteins NHP2 and NOP10. *Mol. Cell. Biol.* **20:** 9028–9040.

Ren X., Gavory G., Li H., Ying L., Klenerman D., and Balasubramanian S. 2003. Identification of a new RNA.RNA interaction site for human telomerase RNA (hTR): Structural implications for hTR accumulation and a dyskeratosis congenita point mutation. *Nucleic Acids Res.* **31:** 6509–6515.

Rocha V., Devergie A., Socie G., Ribaud P., Esperou H., Parquet N., and Gluckman E. 1998. Unusual complications after bone marrow transplantation for dyskeratosis congenita. *Br. J. Haematol.* **103:** 243–248.

Rose C. and Kern W.V. 1992. Another case of *Pneumocystis carinii* pneumonia in a patient with dyskeratosis congenita (Zinsser-Cole-Engman syndrome)[letter]. *Clin. Infect. Dis.* **15:** 1056–1057.

Rudolph K.L., Chang S., Lee H.W., Blasco M., Gottlieb G.J., Greider C., and DePinho R.A. 1999. Longevity, stress response, and cancer in aging telomerase-deficient mice. *Cell* **96:** 701–712.

Ruggero D., Grisendi S., Piazza F., Rego E., Mari F., Rao P.H., Cordon-Cardo C., and Pandolfi P.P. 2003. Dyskeratosis congenita and cancer in mice deficient in ribosomal RNA modification. *Science* **299:** 259–262.

Safa W.F., Lestringant G.G., and Frossard P.M. 2001. X-linked dyskeratosis congenita: Restrictive pulmonary disease and a novel mutation. *Thorax* **56:** 891–894.

Salowsky R., Heiss N.S., Benner A., Wittig R., and Poustka A. 2002. Basal transcription activity of the dyskeratosis congenita gene is mediated by Sp1 and Sp3 and a patient mutation in a Sp1 binding site is associated with decreased promoter activity. *Gene* **293:** 9–19.

Scappaticci S., Fraccaro M., and Cerimele D. 1989. Chromosomal abnormalities in dyskeratosis congenita. *Am. J. Med. Genet.* **34:** 609–610.

Sirinavin C. and Trowbridge A. 1975. Dyskeratosis congenita: Clinical features and genetic aspects. Report of a family and review of the literature. *J. Med. Genet.* **12:** 339–354.

Smith C.M., Ramsay N.K.C., Branda R., Nesbit M.E., and Krivit W. 1979. Response to androgens in the constitutional aplastic anemia of dyskeratosis congenita. *Pediatr. Res.* **13:** 441.

Solder B., Weiss M., Jager A., and Belohradsky B.H. 1998. Dyskeratosis congenita: Multisystem disorder with special consideration of immunologic aspects. *Clin. Pediatr.* **19:** 32–38.

Sorrow J.M. and Hitch J.M. 1963. Dyskeratosis congenita: First report of its occurrence in a female and a review of the literature. *Arch. Dermatol.* **88:** 340–347.

Sznajer Y., Baumann C., David A., Journel H., Lacombe D., Perel Y., Segura J.F., Cezard J.P., Peuchmaur M., Vulliamy T., Dokal I., and Verloes A. 2003. Further delineation of the congenital form of X-linked dyskeratosis congenita (Hoyeraal-Hreidarsson syndrome). *Eur. J. Pediatr.* **162:** 863–867.

Theimer C.A., Finger L.D., and Feigon J. 2003a. YNMG tetraloop formation by a dyskeratosis congenita mutation in human telomerase RNA. *RNA* **9:** 1446–1455.

Theimer C.A., Finger L.D., Trantirek L., and Feigon J. 2003b. Mutations linked to dyskeratosis congenita cause changes in the structural equilibrium in telomerase RNA. *Proc. Natl. Acad. Sci.* **100:** 449–454.

Tollervey D. and Kiss T. 1997. Function and synthesis of small nucleolar RNAs. *Curr. Opin. Cell Biol.* **9:** 337–342.

Verra F., Kouzan S., Saiag B., Bignon J., and de-Cremoux H. 1992. Bronchoalveolar disease in dyskeratosis congenita. *Eur. Respir. J.* **5:** 497–499.

Vulliamy T.J., Knight S.W., Dokal I., and Mason P.J. 1997. Skewed X-inactivation in carriers of X-linked dyskeratosis congenita. *Blood* **90:** 2213–2216.

Vulliamy T.J., Knight S.W., Mason P.J., and Dokal I. 2001a. Very short telomeres in the peripheral blood of patients with X-linked and autosomal dyskeratosis congenita. *Blood Cells Mol. Dis.* **27:** 353–357.

Vulliamy T., Marrone A., Dokal I., and Mason P.J. 2002. Association between aplastic anaemia and mutations in telomerase RNA. *Lancet* **359:** 2168–2170.

Vulliamy T., Marrone A., Szydlo R., Walne A., Mason P.J., and Dokal I. 2004. Disease anticipation is associated with progressive telomere shortening in families with dyskeratosis congenita due to mutations in *TERC*. *Nat. Genet.* **36:** 447–449.

Vulliamy T.J., Knight S.W., Heiss N.S., Smith O.P., Poustka A., Dokal I., and Mason P.J. 1999. Dyskeratosis congenita caused by a 3′ deletion: Germline and somatic mosaicism in a female carrier. *Blood* **94:** 1254–1260.

Vulliamy T., Marrone A., Goldman F., Dearlove A., Bessler M., Mason P.J., and Dokal I. 2001b. The RNA component of telomerase is mutated in autosomal dominant dyskeratosis congenita. *Nature* **413:** 432–435.

Wang C. and Meier U.T. 2004. Archticture and assembly of mammalian H/ACA small nucleolar and telomerase ribonucleoproteins. *EMBO J.* **23:** 1857–1867.

Watkins N.J.A., Gottschalk G., Neubauer B., Kastner P., Fabrizio M., Mann M., and Luhrmann R. 1998. Cbf5p, a potential pseudouridine synthase, and Nhp2p, a putative RNA-binding protein, are present together with Gar1p in all H BOX/ACA-multi snoRNPs and constitute a common bipartite structure. *RNA* **4:** 1549–1568.

Wiedemann H.P., McGuire J., Dwyer J.M., Sabetta J., Gee J.B., Smith G.J., and Loke J. 1984. Progressive immune failure in dyskeratosis congenita. Report of an adult in whom *Pneumocystis carinii* and fatal disseminated candidiasis developed. *Arch. Int. Med.* **144:** 397–399.

Wong J.M., Kyasa M.J., Hutchins L., and Collins K. 2004 Telomerase RNA deficiency in peripheral blood mononuclear cells in X-linked dyskeratosis congenita. *Hum. Genet.* **115:** 448–455.

Yabe M., Yabe H., Hattori K., Morimoto T., Hinoara T., Takakura I., Shimamura K., Tang X., and Kato S. 1997. Fatal interstitial pulmonary disease in a patient with DC after allogeneic bone marrow transplantation. *Bone Marrow Transplant.* **19:** 389–392.

Yaghmai R., Kimyai-Asadi A., Rostamiani K., Heiss N.S., Poustka A., Eyaid W., Bodurtha J., Nousari H.C., Hamosh A., and Metzenberg A. 2000. Overlap of dyskeratosis congenita with the Hoyeraal-Hreidarsson syndrome. *J. Pediatr.* **136:** 390–393.

Yamaguchi H., Baerlocher G.M., Lansdorp P.M., Chanock S.J., Nunez O., Sloand E., and Young N.S. 2003. Mutations of the human telomerase RNA gene (*TERC*) in aplastic anemia and myelodysplastic syndrome. *Blood* **102:** 916–918.

Youssoufian H., Gharibyan V., and Qatanani M. 1999. Analysis of epitope-tagged forms of the dyskeratosis congenita protein (dyskerin): Identification of a nuclear localization signal. *Blood Cells Mol. Dis.* **25:** 305–309.

Zebarjadian Y., King T., Fournier M.J., Clarke L., and Carbon J. 1999. Point mutations in yeast CBF5 can abolish in vivo pseudouridylation of rRNA. *Mol. Cell. Biol.* **19:** 7461–7472.

Zinsser F. 1906. Atropha cutis reticularis cum pigmentatione, dystrophia ungiumet leukoplakia oris. *Ikonogr. Dermatol.* (Hyoto) **5:** 219–223.

7

Telomerase-independent Maintenance of Mammalian Telomeres

Axel A. Neumann and Roger R. Reddel
Cancer Research Unit
Children's Medical Research Institute
Sydney, New South Wales 2145, Australia

IN MOST NORMAL SOMATIC CELLS, TELOMERES SHORTEN with every cell division because of the end replication problem (Harley et al. 1990). The cumulative loss of telomeric DNA sequence is thought to act as a "mitotic clock" that limits the replicative potential of cells. Prevention of continued telomere loss by activation of a telomere maintenance mechanism appears to be essential for the development of most human cancers. The majority of cancers and cancer cell lines maintain their telomeres via activity of the telomerase ribonucleoprotein holoenzyme complex (Kim et al. 1994). Some immortalized mammalian cell lines and tumors, however, are able to maintain their telomere lengths over many population doublings in the absence of telomerase activity, indicating the existence of one or more non-telomerase-based mechanisms for telomere maintenance that have been termed alternative lengthening of telomeres (ALT) (Bryan and Reddel 1997). So far, ALT has only been detected in anomalous situations, such as human cancers and immortalized cell lines, telomerase-null mouse cell lines and tumors, and an immortalized Indian muntjac cell line (Bryan et al. 1995, 1997a; Hande et al. 1999; Niida et al. 2000; Zou et al. 2002; Chang et al. 2003). However, indirect evidence was obtained for in vivo ALT activity in splenocytes from late-generation telomerase-null mice, which showed reduced germinal center formation after immunization, but, surprisingly, had elongated telomeres (Herrera et al. 2000). Detailed analyses of telomere length fluctuations in telomerase-null embryonic fibroblasts were also consistent with an

ALT-like recombinational telomere lengthening mechanism (Hande et al. 1999). Here we summarize what is known about non-telomerase-based mechanisms of telomere length maintenance in mammalian cells.

The Telomere Length Phenotype of ALT Cells

In addition to being immortalized without detectable telomerase activity, ALT cells are characterized by an extremely heterogeneous telomere length phenotype. Telomeres of normal human somatic cells cultured in vitro usually shorten at an average rate of 40–200 bp per cell division until the terminal restriction fragments are ~5–8 kb long at senescence (Harley et al. 1990; Harley 1997; Wright et al. 1997; Martens et al. 2000). Human germ line cells generally maintain their telomere lengths via telomerase at ~15 kb (Allshire et al. 1989; de Lange et al. 1990). Telomeres of telomerase-positive human cancers and immortal cell lines are relatively homogeneous in length with the mean length usually <10 kb (de Lange et al. 1990; Hastie et al. 1990; Counter et al. 1992). In contrast, all of the human ALT cell lines and tumors analyzed to date have a very broad telomere length distribution ranging from <3 kb to >50 kb, with an average length of ~20 kb (Murnane et al. 1994; Bryan et al. 1995, 1997a; Grobelny et al. 2000; Opitz et al. 2001). When the ALT mechanism is activated during in vitro immortalization of cells in culture, there is a close temporal correlation between the immortalization event and the occurrence of the characteristic ALT telomere length phenotype (Fig. 1). Fluorescence in situ hybridization (FISH) with a telomere-repeat-specific probe (Fig. 2) shows that the heterogeneous telomere length phenotype characteristic of ALT cell populations is also seen within individual cells: Some chromosome ends have undetectable amounts of telomeric sequence, i.e., very short telomeres, whereas others within the same cell have very strong telomeric signals, i.e., very long telomeres (Lansdorp et al. 1997; Perrem et al. 2001). Heterogeneity within individual ALT cells is also observed when telomeres are visualized by immunofluorescence with antibodies against the telomere-specific binding proteins, TRF1 and TRF2.

The telomere length distribution in ALT cells is dynamic, with lengths fluctuating during cellular proliferation. Murnane and colleagues examined in detail the telomere length dynamics of a human ALT cell line (Murnane et al. 1994) and observed that the frequency of telomere length fluctuations varied greatly among different subclones of this line. The rate of chromosome fusion events was proportional to the frequency of rapid telomere length changes. They concluded that these observations are consistent with a recombinational mechanism. It was subsequently shown

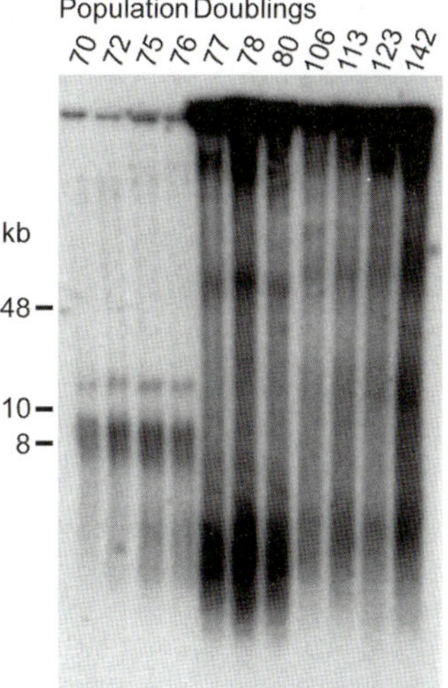

Figure 1. Telomere length heterogeneity in ALT cells. Terminal restriction fragment (TRF) length analysis of an ALT cell line that entered crisis at population doubling 76 and emerged from crisis at population doubling 77, demonstrating the temporal correlation between immortalization and occurrence of the telomere length pattern that is characteristic of ALT. (Reprinted, with permission, from Yeager et al. 1999 [© AACR].)

that the short arm telomere to long arm telomere length ratio for an individual chromosome varied more than 100-fold within an ALT cell population, whereas the ratio varied less than twofold in a comparable telomerase-positive cell line (Fig. 3). However, recent Flow-FISH analysis of telomere length fluctuations in cell subpopulations of mouse and human cell lines also revealed significant gains and losses of telomeric

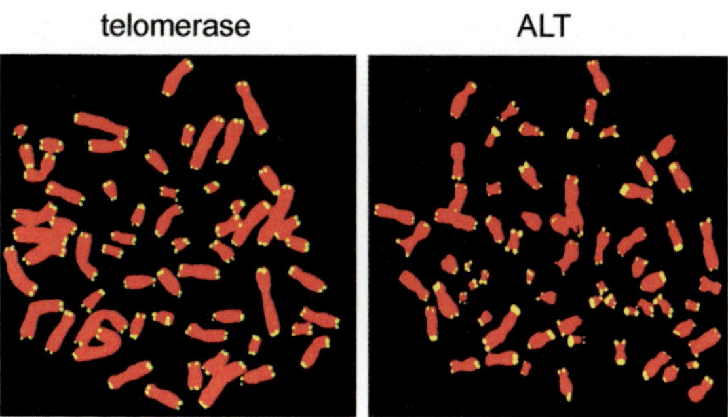

Figure 2. Telomere length phenotype of ALT cells. Fluorescence in situ hybridization (FISH) with a telomere-specific probe (*yellow*) on metaphase chromosomes (*red*) of telomerase-positive and ALT cells, illustrating the highly heterogeneous telomere length phenotype in ALT cells.

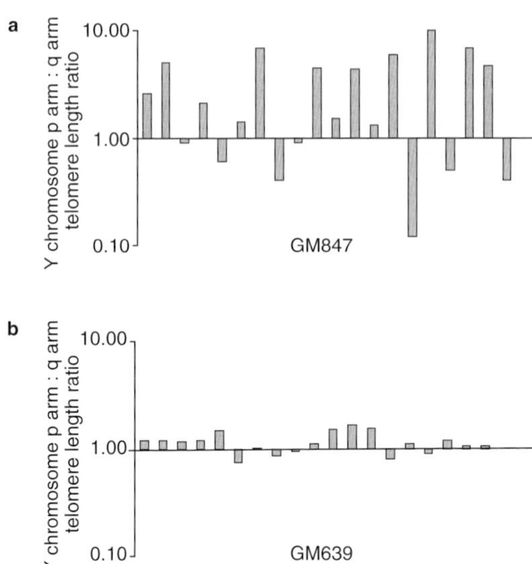

Figure 3. Telomere length fluctuation in ALT cells. Telomere length ratios of the short and the long arm of the Y chromosome from 20 metaphase spreads in an ALT cell line GM847 (*a*) and a comparable telomerase-positive cell line GM639 (*b*) are shown. Each bar represents the ratio for an individual metaphase. The ratio was calculated from the Y-chromosome telomere fluorescence signal intensities on the p and q arms. (Reprinted, with permission, from Perrem et al. 2001 [© American Society for Microbiology].)

DNA in 1 out of 19 non-ALT cell lines (Cabuy et al. 2004); the mechanism of these changes has not been elucidated.

ALT-associated PML Bodies

A second hallmark of human ALT cell lines is the occurrence of specific nuclear structures, referred to as ALT-associated PML bodies (APBs), i.e., PML nuclear bodies with ALT-specific contents (Fig. 4) (Yeager et al. 1999). PML nuclear bodies are dynamic structures present in many, but not all, tissues. PML bodies seem to play a multifaceted role in various cellular processes, including cell cycle regulation, senescence, apoptosis, tumor suppression, immune and inflammatory responses, protein refolding, and degradation, as well as differentiation. In addition, PML bodies are involved in transcriptional regulation and chromatin modification (Hodges et al. 1998; Maul et al. 2000; Ruggero et al. 2000; Zhong et al. 2000; Borden 2002; Dellaire and Bazett-Jones 2004). To facilitate this vast array of cellular functions, PML bodies may sequester and

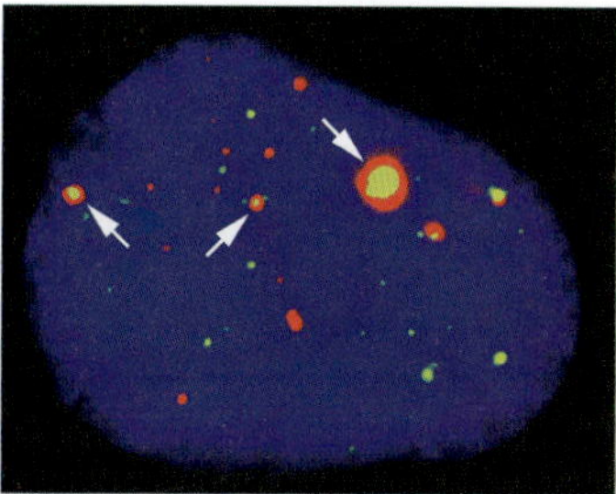

Figure 4. ALT-associated PML bodies (APBs). Distinctive colocalization (white arrows) of TRF2 (*green*) and PML (*red*) proteins detected by immunostaining of an interphase nucleus (*blue*) of an ALT cell line.

release proteins, transporting them to sites of action and assisting protein–protein interactions.

APBs differ from other PML bodies in that they contain specific components, including telomeric DNA and the telomere repeat binding factors TRF1 and TRF2 (Yeager et al. 1999). APBs were detected in 17/17 ALT cell lines, 0/20 telomerase-positive cell lines, and 0/5 mortal cell strains (Yeager et al. 1999; Henson et al. 2002). It should be noted that within a given ALT cell line population, APBs were detected in only ~5–10% of the cells (Yeager et al. 1999), most likely because of their formation during specific phases of the cell cycle (Grobelny et al. 2000; Wu et al. 2000). APBs became evident in an in vitro immortalized cell line at the same time that the characteristic heterogeneous telomere length phenotype of ALT occurred (Yeager et al. 1999). They are there-fore a very useful marker for the presence of ALT activity. APBs were observed in human tumor cell lines and in nude mouse tumors formed by a human ALT cell line (Yeager et al. 1999). Moreover, with a combi-nation of immunohistochemistry and telomere FISH, APBs were detected in human tumor specimens, including fine needle aspirates, frozen sections, and paraffin-embedded blocks, even after storage for many years (Yeager et al. 1999; Henson et al. 2002, 2004). There was an excellent correlation between tumor telomere length heterogeneity and the pres-ence of APBs: 62/62 tumors that were positive for ALT according to telomere length analysis had detectable APBs (Henson et al. 2004).

At least some of the telomeric DNA found in APBs is extrachromo-somal (Yeager et al. 1999), and some of this has free ends and is there-fore linear (T. Yeager et al., unpubl.). It is not yet known whether all of the extrachromosomal telomeric repeat (ECTR) DNA found in ALT cells (Ogino et al. 1998; Tokutake et al. 1998; Cesare and Griffith 2004) is sequestered within APBs. It is possible that there is dynamic interchange of telomeric DNA between various nuclear subcompartments. Live imaging of ALT cells showed that telomeric DNA undergoes dynamic association and dissociation with other telomeric DNA inside and outside of PML bodies (Molenaar et al. 2003). ECTR DNA is not usually found in telomerase-positive immortalized cells or normal human cells (Ogino et al. 1998), but it has been detected in fibroblasts from patients with ataxia telangiectasia and in ATM-null mouse cells for reasons that are not clear (Hande et al. 2001).

In addition to the telomere-binding proteins, TRF1 and TRF2, APBs contain many other proteins involved in telomere binding and in DNA replication, repair, and recombination processes. These include RAD51, RAD52, and RPA (Yeager et al. 1999); RAD51D (Tarsounas et al. 2004); BLM (Yankiwski et al. 2000; Stavropoulos et al. 2002); WRN (Johnson et al. 2001); hRap1 and BRCA1 (Wu et al. 2003); MRE11, RAD50, and NBS1 (MRN complex) (Wu et al. 2000; Zhu et al. 2000); PARP2 (Dantzer et al. 2004); ERCC1 and XPF (Zhu et al. 2003); RAD1, RAD9, RAD17, and HUS1 (Nabetani et al. 2004); and RIF1 (Silverman et al. 2004). Phosphorylated histone H2AX (H2AXγ), a molecular marker of double-strand breaks (DSBs), also colocalizes with a subset of APBs (Nabetani et al. 2004), which suggests that some telomeric DNA in APBs is recognized as a DSB and is consistent with the observation that some of this DNA is linear. Formation of APBs has been shown to require NBS1, which then sequentially recruits MRE11, RAD50, and BRCA1 into these structures (Wu et al. 2003). Association of the MRN complex with telomeres and ALT-associated PML bodies (APBs) may be due to its association with telomere-binding proteins (Lombard and Guarente 2000; Wu et al. 2000; Zhu et al. 2000). In view of the suggested functions of PML bodies in general, and the nature of the specific contents of APBs, it seems feasible that the latter may concentrate and/or modify DNA and proteins required for the ALT mechanism and possibly provide a macromolecular platform that facilitates ALT-mediated telomere maintenance, or alternatively that they are involved in the removal of extrachromosomal telomeric chromatin generated as a consequence of the ALT mechanism and play no further part in telomere maintenance.

Increased Telomeric Exchanges in ALT Cells

Another recently described characteristic of ALT cells is the frequent occurrence of postreplicative exchanges specifically between telomeric repeats (Bechter et al. 2004b; Londoño-Vallejo et al. 2004). These exchanges

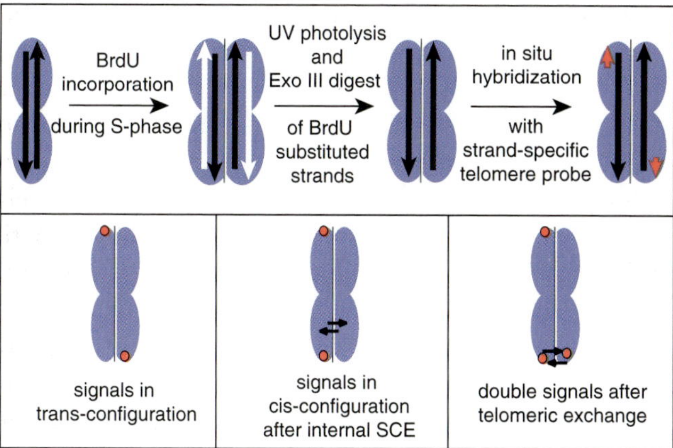

Figure 5. Detection of postreplicative telomeric exchange by chromosome orientation-FISH (CO-FISH). Metaphase chromosomes are harvested after one round of DNA replication in the presence of BrdU. Fixed chromosomes are stained with Hoechst 33258 and photolyzed under UV light to nick the BrdU substituted strand, followed by digestion with exonuclease III. This procedure effectively removes the newly synthesized DNA strand, but leaves the two parental strands intact. A telomere-specific single-stranded probe hybridizes to complementary telomeric DNA on the parental strand of one chromatid of each chromosome arm. This results in a two-signal-hybridization pattern, either a *trans*-configuration, where signals are on opposite sides of each sister chromatid, or a *cis*-configuration, where both signals are on opposite sides of the same sister chromatid as the result of an internal sister chromatid exchange (SCE). Never, or rarely, does an exchange occur within the telomeric repeats of the sister chromatids, resulting in a double signal on both chromatids in the same chromosome arm. In ALT cells, however, this phenotype can be seen very often, which indicates frequent telomeric exchanges.

are undetectable by conventional sister chromatid exchange (SCE) analysis, but can be visualized by a strand-specific in situ hybridization technique, termed chromosome orientation (CO)-FISH (Fig. 5) (Meyne et al. 1994; Bailey et al. 2001, 2004b). This technique cannot distinguish, however, whether the telomeric DNA exchanges occur between sister chromatids, between different chromosomes, or between a chromosome and extrachromosomal telomeric DNA. Detailed CO-FISH analysis of a panel of mortal cell strains, in vitro immortalized cell lines, and cancer-derived cell lines showed that postreplicative exchanges involving a telomere and another TTAGGG-repeat tract occurred frequently in ALT cells but only rarely or never in non-ALT cells (Table 1). CO-FISH analyses of mouse cells with various DNA repair gene deficiencies revealed low levels of telomeric exchange events, presumably representing

Table 1. Telomere CO-FISH and SCE analyses in different cell types

Phenotype cells (total number of metaphases)	SCE mean/ metaphase	No. double CO-FISH signals/ 100 metaphases
Mortal and TERT-immortalized		
diploid cell lines		
PBL[a] (50)	10.0	0
BJ (30)	5.3	2
BJ+ TERT (40)	6.2	3
WI38+ TERT (35)	4.9	2
GM16859[b] (30)	93.2	0
Transformed (telomerase[+])		
LBL[c] (25)	5.1	0
LBL + mitomycin (25)	30.8	0
293T (30)	19.0	0
293T+ mitomycin (30)	61.7	0
NIH3T3 (55)	N.D.	5–9
SKOV3 (35)	23.4	0
R970.5 (30)	23.3	0
HT1080 (40)	13.8	0
HT1080 + hPOT1 (25)	10.2	2
JFCF-6T/2H 2FL-6 (20)	22.9	0
Transformed (ALT[+])		
VA13[d] (80)	25.3	45–150
VA13 (+ telom.)[d] (55)	23.2	170–280
GM847 (70)	20.9	28–65
GM847-C3 (+ telom.)[d] (40)	21.2	56–90
GM847-C6 (+ telom.)[d] (45)	19.1	85–125
U-2 OS (33)	12.0	85–105
SAOS-2 (35)	10.1	167–240

[a]Phytohemagglutinin-stimulated peripheral blood lymphocytes from a healthy donor.
[b]Primary fibroblast cell line from a Bloom's syndrome patient.
[c]Lymphoblastoid cell line "B" (healthy donor).
[d]Clonal origin.
Modified table reprinted, with permission, from Londoño-Vallejo et al. 2004 (© AACR).

values typical of non-ALT cells (Bailey et al. 2004a). Even in telomerase-positive cells with telomeres that were lengthened, for example, by expression of exogenous POT1 or TERT, only occasional telomeric exchanges were seen with CO-FISH, indicating that, although long telomeres may be a substrate for telomeric exchange, additional factors are required to generate the high rate of exchanges observed in ALT cells (Londoño-Vallejo et al. 2004). The ALT cells had no increase in SCEs at interstitial genomic sites, in agreement with a study showing that ALT cells do not have higher rates of homologous recombination overall than comparable telomerase-positive cells (Bechter et al. 2003, 2004a).

Although frequent telomeric exchanges occurred in all of the ALT cell lines examined (Londoño-Vallejo et al. 2004), the relationship of these exchanges to ALT is not yet clear. Following repression of telomerase activity in human colon cancer cells, a survivor population had periodic increases in telomere length and exhibited increased telomeric exchanges, but did not have other hallmarks of ALT such as APBs (Bechter et al. 2004b).

Hypothetically, the high frequency of telomeric exchanges in ALT cells could be explained by the recruitment to the telomere of proteins involved in mitotic homologous recombination for ALT-mediated telomere lengthening, thereby resulting in the accumulation of proteins that are also capable of generating telomeric SCEs. To generate exchanges that are interchromosomal or involve extrachromosomal telomeric repeats, however, and to increase telomere length by recombination-mediated DNA replication may require additional processes that are specific to ALT (Londoño-Vallejo et al. 2004). In this view, the increased exchanges may be a side effect of the necessary accumulation of recombination proteins and may not be sufficient for ALT.

It is interesting to consider the possibility, however, that telomeric SCEs may actually be the basis of the ALT mechanism. In this view, unequal telomeric exchanges between sister chromatids, nonsister chromatids, or extrachromosomal telomeric repeats could elongate the telomeres of some separating chromatids before mitosis (and reciprocally shorten others), without any synthesis of additional telomeric DNA. This could allow a subpopulation of cells to continue proliferating at the expense of other daughter cells that have shortened telomeres and undergo senescence or crisis (Bailey et al. 2004a).

Telomeric Recombination in ALT Cells

The telomere length dynamics of human ALT cells (Murnane et al. 1994; Perrem et al. 2001) suggested that telomeres may be maintained by a process involving recombination. Strong evidence for intertelomeric recombination in such cells was acquired by experiments where a plasmid tag was targeted into telomeres (Dunham et al. 2000). FISH analysis demonstrated that a progressive increase in the number of chromosomes in clonally derived cells containing tagged telomeres occurred with increasing population doublings. These experiments showed that DNA sequences can be copied from telomere to telomere in human ALT cells

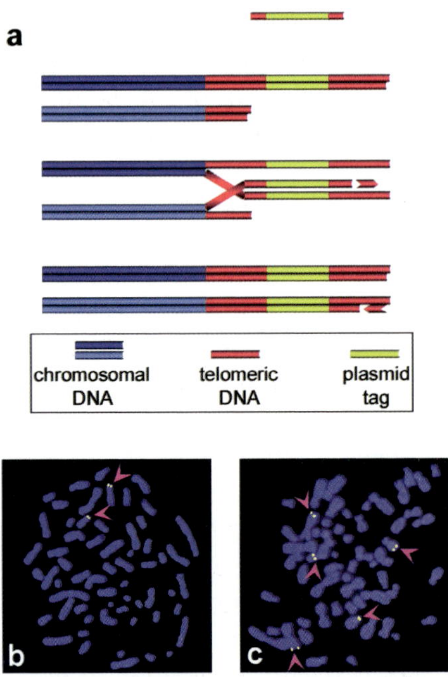

Figure 6. Detection of intertelomeric copying of DNA. (*a*) Detection strategy. If a DNA tag is inserted into a telomere by targeting, intertelomeric recombination proximal (i.e., centromeric) to the tag will result in copying of the tag (*yellow*) to the telomere (*red*) of another chromosome. FISH detection of an intratelomeric plasmid DNA tag in a clonal ALT cell line, at (*b*) early passage, and (*c*) later passage. (Panels *b* and *c* reprinted, with permission, from Dunham et al. 2000 [© Nature Publishing Group].)

(Fig. 6). In contrast, such intertelomeric recombination events were not observed in telomerase-positive cells or in ALT cells in which a chromosome truncation/telomere seeding strategy was used to insert a plasmid tag centromeric to telomeric DNA repeats (Dunham et al. 2000). It has been shown in yeast that subtelomeric regions may be involved in intertelomeric recombination events (Louis and Haber 1990; Lundblad and Blackburn 1993; McEachern and Iyer 2001). FISH analysis with chromosome-specific subtelomeric probes that do not hybridize to the region of degenerate repeats immediately centromeric to the telomere did not detect involvement of subtelomeric DNA (Dunham et al. 2000). However, PCR-mediated sequence analyses of the subtelomere–telomere junction region in cells before and after ALT-mediated immortalization indicated that TTAGGG sequences within the region of degenerate repeats were involved occasionally in recombination-mediated telomere elongation (Varley et al. 2002). Interestingly, although ALT cells exhibited instability at some minisatellite loci that are regarded as recombinogenic (Tsutsui et al. 2003; Jeyapalan et al. 2005), telomerase-positive and ALT cells were found to have the same overall frequency of homologous recombination elsewhere in the genome (Bechter et al. 2003). Thus, the increased recombination seen in ALT cells is restricted to loci including the telomere.

The results of the experiments described above are consistent with ALT being a recombination-mediated DNA replication mechanism in which the 3′ single-stranded DNA end of one telomere invades the double-stranded DNA of another telomere using it as a copy template, which results in a net increase in telomeric DNA (Fig. 7a). These data do not exclude the possibility that the copy template may be linear or circular

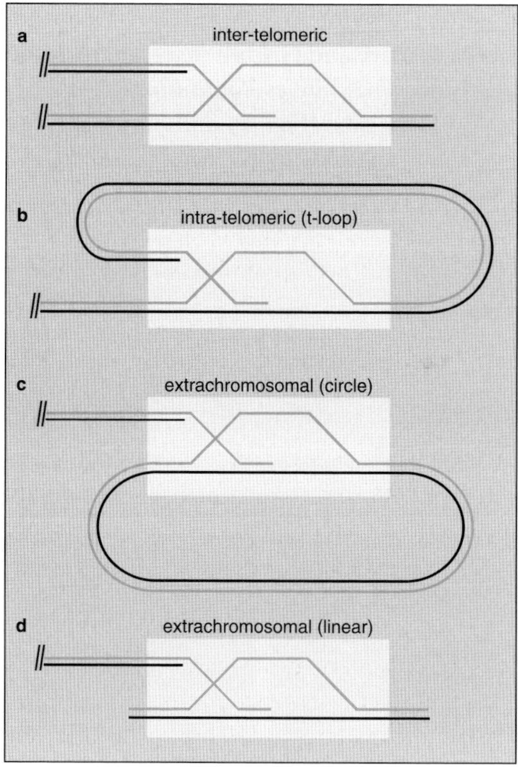

Figure 7. Templates for recombination-mediated telomere lengthening in ALT cells. Four potential templates, described in more detail in the text, include another telomere (*a*), the proximal, i.e., centromeric, region of the same telomere via t-looping (*b*), and extrachromosomal telomeric DNA that is either circular (*c*) or linear (*d*). (Modified figure reprinted, with permission, from Henson et al. 2002 [© Nature Publishing Group].)

extrachromosomal telomeric DNA sequences or even the same telomere via t-loop formation (Fig. 7b–d).

t-loops

Human and mouse telomeres can form lariat structures, called t-loops (Griffith et al. 1999; see also Chapter 13). It is possible that t-loops facilitate ALT activity: Extension of the invading 3′overhang strand together with lagging strand synthesis could potentially allow extensive telomere lengthening (Fig. 7b).

Paradoxically, t-loops may also contribute indirectly to ALT-mediated telomere lengthening by facilitating rapid telomere *shortening* events that generate circular or linear ECTR DNA, which may then act as a copy template for telomere lengthening (Fig. 7c,d). Electron microscopic analyses of ALT cells (Cesare and Griffith 2004) detected both t-loops and abundant extrachromosomal circles containing telomeric DNA ranging from <1 kb to >50 kb in size (Fig. 8), and abundant circular telomeric

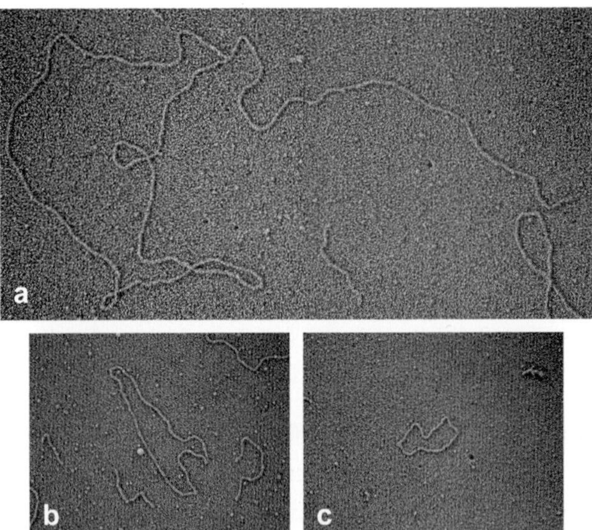

Figure 8. t-loops and t-circles in ALT cells. Electron microscope (EM) images of (*a*) a t-loop molecule and (*b,c*) extrachromosomal DNA circles from the telomere-enriched fractions of ALT cells (GM847). DNA was prepared for EM by surface spreading with cytochrome *c* and rotary shadow casting; images are shown in negative contrast. Loop and tail sizes are 19.0 and 8.4 kb in *a*, and circle lengths are 8.9 and 3.9 kb in *b* and *c*, respectively. (Courtesy of Anthony Cesare and Jack Griffith; reprinted, with permission, from Cesare and Griffith 2004 [© American Society for Microbiology].)

DNA has also been detected in ALT cells by 2D gel electrophoresis (Cesare and Griffith 2004; Wang et al. 2004). Rapid reductions of telomere length have been observed in ALT cells (Murnane et al. 1994), as well as in hybrid cells in which ALT is repressed (Perrem et al. 1999, 2001). Telomeric rapid deletion (TRD), a process whereby elongated telomeres are shortened to the average length of telomeres on other chromosomes by intrachromatid telomere repeat excision in yeast (Li and Lustig 1996), has been postulated to make use of a t-loop structure (Rubelj and Vondracek 1999; Bucholc et al. 2001) and involves RAD52, RAD50, and MRE11, genes that are required for telomerase-deficient yeast type II survivors. In human cells, rapid telomere deletion events are accompanied by generation of t-loop-sized circles, and it is very likely that these events are inhibited by TRF2 and that they involve the RAD51 paralog XRCC3 as well as NBS1 and/or its partners in the MRN complex (Wang et al. 2004). The abundant telomeric circles found in ALT cells suggest that these cells undergo frequent t-loop homologous recombination events.

Rolling Circle

Circular ECTR generated, for example, by excision of t-loops could potentially be used for recombinational telomere lengthening by a rolling circle of replication, where the single-stranded telomeric 3′ overhang invades a circle of ECTR DNA (Fig. 7c). Branch migration of the invading 3′ overhang could subsequently allow rolling of the circle and result in extensive telomere elongation. Telomere lengthening by rolling-circle gene conversion employing a small circle of telomeric DNA as a template has been demonstrated in telomerase-deficient *Kluyveromyces lactis* cells (Natarajan and McEachern 2002). In these experiments, a circular template containing two types of telomeric repeats resulted in an elongated telomere showing a repeating pattern based on the unit structure of the circle being copied. In a second, more frequent type of event, other intertelomeric gene conversions using the elongated telomere as a copy template spread its sequence to other telomeres; this is referred to as the "roll and spread" mechanism (McEachern 2001; Natarajan and McEachern 2002; Tomaska et al. 2004). These results suggested that very small DNA circles can be used as a template for rolling-circle replication, and although they have not been identified to this point in telomerase-deficient *K. lactis* cells, they have been detected in certain mutants with highly elongated telomeres (Underwood et al. 2004) and, as noted above, abundant ECTR DNA circles have been found in human ALT cells (Cesare and Griffith 2004; Wang et al. 2004).

Linear ECTR DNA

Small dispersed, low molecular weight telomeric DNA that was shown to be double-stranded and linear has been isolated from ALT cells (Ogino et al. 1998). As mentioned above, it is possible that t-loops facilitate the production of both circular and linear ECTR DNA. The latter could conceivably participate in telomere elongation by end-joining or homologous recombination and copy templating (Fig. 7d), although the small size of most of the linear ECTR DNA means that it is unlikely to account for the large, rapid increases in telomere length seen in ALT cells. ECTR DNA could be responsible for sequestering telomere-binding proteins in the cell. Moreover, its colocalization with proteins involved in recombination within APBs suggests that ECTRs are either directly involved in the ALT mechanism or are generated as its by-products. Interestingly, APBs were reported to become prevalent in the late S/G$_2$/M phase of the cell cycle (Grobelny et al. 2000; Wu et al. 2000), i.e., around the same time that recombinational repair preferentially occurs in the cell (Takata et al. 1998).

Is There More Than One ALT Mechanism?

Some telomerase-negative human and mouse cells appear to have a telomere maintenance mechanism but lack at least some of the hallmarks described for human ALT cell lines and tumors (Argilla et al. 2004; Bechter et al. 2004b; Fasching et al. 2005; Marciniak et al. 2005). This raises the question whether there may be more than one ALT mechanism in mammalian cells.

Telomerase deletion mutants in the yeast species *Saccharomyces cerevisiae*, *K. lactis*, and *Schizosaccharomyces pombe* undergo gradual telomere shortening followed by an eventual loss of cell viability. Rare telomerase-null *S. pombe* survivors solve the dilemma of shortened telomeres either by generating circular chromosomes through fusing subtelomeric regions that have lost all telomeric repeats (Kimmel et al. 1992; Nakamura et al. 1998) or by recombination resulting in linear chromosomes with long heterogeneous amplified telomeres. So far, there is no evidence for the phenomenon of chromosome circularization in telomerase deletion mutants of higher eukaryotes.

Recombination-mediated telomere lengthening is also a survivor mechanism in *S. cerevisiae* and *K. lactis* (Lundblad and Blackburn 1993; McEachern and Blackburn 1996). In all cases, survivors are dependent on Rad52p function, which has been shown to be essential for recombination

(Lundblad and Blackburn 1993). In *S. cerevisiae*, there are at least two classes of survivors.

Type I telomerase-null yeast survivors show tandem amplification of a subtelomeric Y′ repeat element with adjacent blocks of intrachromosomal telomeric repeats and a short terminal telomeric repeat (Lundblad and Blackburn 1993; Le et al. 1999; Teng and Zakian 1999; Teng et al. 2000). Mouse embryonic stem cells made deficient for mTERC by sequential gene targeting have been shown to activate a telomerase-independent mechanism for telomere length maintenance (Niida et al. 2000), and analysis of the telomeric structures in one of the survivor mTERC$^{-/-}$ cell lines showed amplification of telomeric and nontelomeric tandem array sequences on most chromosome ends, a phenotype reminiscent of telomerase-negative yeast type I survivors. Interestingly, the cells of one telomerase-negative simian virus 40 (SV40) immortalized WRN-deficient human cell line contain substantial amounts of nontelomeric DNA (SV40 sequences) within their telomeres (Fasching et al. 2005; Marciniak et al. 2005). Like other ALT cells, they show evidence of intertelomeric recombination but lack APBs. They do, however, contain nuclear aggregates of telomeric DNA and many other components found in APBs, including MRE11, NBS1, RAD50, RAD51, RAD52, and TRF2, but not PML or SP100 (Fasching et al. 2005). It will be of interest to determine the extent to which the telomere maintenance mechanism in these cells resembles that of type I yeast survivors.

Type II *S. cerevisiae* and *K. lactis* telomerase-null survivors show elongation of terminal telomeric repeat tracts with no detectable amplification of any other sequences (Lundblad and Blackburn 1993; Le et al. 1999; Teng and Zakian 1999; Teng et al. 2000). Type II survivors are proposed to arise through intra- or interchromatid exchanges involving the shortening telomeric repeats themselves (Chen et al. 2001), through interactions with extrachromosomal circles of telomeric DNA (Teng et al. 2000) or through a combination of both (Natarajan and McEachern 2002; Tomaska et al. 2004). Interestingly, the heterogeneous telomere length phenotype of telomerase-deficient type II survivors is similar to that of human ALT cells (McEachern and Blackburn 1996; Teng and Zakian 1999).

Exonuclease1 (Exo1), a member of the RAD2 nuclease family, contributes to telomere maintenance in type I and II telomerase-negative *S. cerevisiae* survivors (Bertuch and Lundblad 2004; Maringele and Lydall 2004a) and plays a role in preventing replication of senescent yeast cells (Maringele and Lydall 2004a). Loss of Exo1p may result in another telomerase-null survivor pathway in yeast, characterized by loss of telomeric and subtelomeric repeats and formation of large DNA palindromes (Maringele and Lydall 2004b) (see Chapter 8). It is not known at present whether a similar mechanism plays a role in any telomerase-negative mammalian cells.

Evidence for the Existence of ALT Repressor Genes

The observation that ALT activity can be detected in a subset of immortalized human cell lines and tumors but not in normal cells implies that the normal cells contain ALT repressor genes and/or that there are genes that need to be activated for ALT to occur. Fusion of immortal ALT cells with mortal cells resulted in senescent hybrids that had lost the characteristic ALT telomere phenotype, demonstrating that activation of ALT requires one or more recessive mutations that presumably result in loss of one or more ALT repressor genes (Perrem et al. 1999). Some telomerase-positive immortalized cell lines also appear to have ALT repressors, because ALT was repressed in some immortal hybrids produced by fusing ALT and telomerase-positive cells (Perrem et al. 1999, 2001). Fusion of some pairs of ALT and telomerase-positive cells, however, produced hybrids in which telomerase and not ALT was repressed (Katoh et al. 1998), indicating that some ALT cells (like normal cells) contain a repressor of telomerase. Microcell-mediated chromosome transfer (MMCT) of chromosome 7 into ALT cells suppressed both the immortal and the ALT phenotype (Ogata et al. 1993; Nakabayashi et al. 1997). However, the putative gene on chromosome 7 may not specifically repress ALT, because chromosome 7 also induced senescence of a telomerase-positive cell line (Ogata et al. 1995). MMCT of chromosome 6, but not 7, caused senescence of a different ALT cell line (Kumata et al. 2002). A loss of heterozygosity (LOH) study on all 22 autosomes and the X chromosome was performed by polymerase chain reaction (PCR) analysis with polymorphic microsatellite markers in a panel of in vitro immortalized human cell lines with the aim of identifying candidate chromosomal regions containing ALT repressor genes (Shigeeda et al. 2003). LOH on chromosome 8 was observed at 40% of informative loci in 61% of the ALT cell lines examined, and the percentage was higher at a locus adjacent to the Werner's syndrome gene on chromosome 8p12. Understanding the mechanisms whereby ALT is repressed in normal cells may make it possible to design anticancer strategies for restoring this repression.

Telomerase Activity Usually Does Not Repress ALT

The absence of telomerase activity in ALT cells is associated with lack of expression of TERT (telomerase reverse transcriptase, the catalytic subunit of telomerase) and sometimes also TR (telomerase RNA; encoded by the telomerase RNA component gene [TERC]). ALT cells have undetectable levels of full-length TERT transcript (Kilian et al. 1997) as a result of methylation of the TERT CpG island, whereas in telomerase-negative normal cells the CpG island was found to be unmethylated (Dessain et al. 2000). Direct evidence that telomere length maintenance in ALT cells is telomerase independent came from the observation that some human ALT cell lines have undetectable TR expression levels (Bryan et al. 1997b), which is associated with methylation of the TERC promoter region (Hoare et al. 2001). In ALT cells that do express the TERC gene, its sequence was shown to be wild type (Bryan et al. 1997b).

Exogenous TERT expression in ALT cells that express TR induces telomerase activity in vitro (Wen et al. 1998), indicating that adequate levels of other essential telomerase components are expressed in these cells, whereas in TR-negative cells exogenous expression of both TR and TERT was necessary and sufficient to induce telomerase activity (Wen et al. 1998). These findings initiated studies of the effects of exogenous telomerase expression (Cerone et al. 2001; Ford et al. 2001; Grobelny et al. 2001; Perrem et al. 2001), as well as of mutant TR template sequence (Guiducci et al. 2001) in ALT-proficient cells. Expression of exogenous telomerase in ALT cells was usually compatible with continued ALT activity, even though the shortest telomeres underwent lengthening by telomerase (Cerone et al. 2001; Ford et al. 2001; Grobelny et al. 2001; Perrem et al. 2001; Stewart et al. 2002). The very long telomeres characteristic of ALT cells persisted, however, as did APBs and ECTR circles (Cerone et al. 2001; Grobelny et al. 2001; Perrem et al. 2001; Cesare and Griffith 2004). Although there is no clear example of spontaneous activation of both mechanisms within the same cell, these findings suggest that ALT and telomerase activity are able to coexist in human cells, and that ALT can still act on telomeres that are not critically short. In contrast, the telomere phenotype of telomerase-deficient yeast survivors reverted to wild type many generations after reexpression of telomerase activity (Teng and Zakian 1999). One study with human cells, however, observed reduced ALT activity after exogenous telomerase expression in two out of nine clones (Ford et al. 2001), which could be explained by ALT being switched off as a stochastic event in some cells (as has been shown for telomerase in telomerase-positive cells [Bryan et al. 1998]), or that ALT could be repressed in the situation where particularly high TERT expression levels might sequester molecules common to both pathways of telomere maintenance. A further explanation could be that in some ALT cells telomeric recombination can occur only at telomeres that are shorter than the new telomere length set point established by exogenous telomerase activity.

Genes Potentially Involved in Repression of ALT

No ALT repressor genes have yet been identified, but genetic changes known to permit the activation of ALT (or telomerase) include expression of SV40 or human papillomavirus (HPV) oncogenes (Bryan et al. 1995). Among other effects, the SV40 and HPV oncogenes disrupt the function of p53 and retinoblastoma (Rb) protein family members (Bryan and Reddel 1994). This results in an extended proliferative life span, but additional genetic changes (shown by somatic cell hybridization studies to involve loss of one or more repressors) are usually required for activation of a telomere maintenance mechanism. Disruption of p53 and Rb function by other means, e.g., spontaneous loss of p53 and p16 [INK4A] function in Li-Fraumeni syndrome cells (Rogan et al. 1995; Vogt et al. 1998) or the combination of p53 inactivation and cyclin D1 overexpression (Opitz et al. 2001), produces a similar outcome. These changes foster a milieu of genetic instability and therefore facilitate immortalization by increasing the probability of the additional genetic changes required for activation of ALT (or telomerase) (Reddel 2001). This may also apply to normal diploid fibroblasts that became immortalized using the ALT mechanism after transduction of cyclin A2 or cdk1 (Luo et al. 2004). Notably, cytogenetic analysis of these cells revealed loss of chromosome 6, to which a putative ALT repressor has been localized (Kumata et al. 2002).

It is also possible, however, that normal function of the p53 and Rb family proteins may have a more direct role in repressing ALT activity. A connection between telomere length regulation and the Rb protein family members—namely, Rb1, Rbl1 (Rb-like 1; p107), and Rbl2 (Rb-like 2; p130)—was demonstrated in mouse embryonic fibroblasts (MEFs) that were either doubly deficient in Rbl1 and Rbl2 or triply deficient in Rbl1, Rbl2, and Rb1 (Garcia-Cao et al. 2002). Cells with either genotype had noticeably elongated telomeres in comparison to wild-type or Rb1-deficient cells. The lengthening of telomeres was not associated with increased levels of telomerase activity, nor did the telomeres appear to have lost their normal end-capping function. Possible explanations for the observed phenotype could be an altered accessibility of the telomeres to telomerase or partial activation of an ALT-like phenotype, which could coexist in these cells with telomerase activity, as has been shown to be possible for human cells (Cerone et al. 2001; Grobelny et al. 2001; Perrem et al. 2001). Strikingly, late-passage triple-knockout MEFs had long and heterogeneous telomeres, a typical feature of ALT cells. Further studies are needed to determine how the Rb family proteins are involved in telomere length regulation and whether they directly repress ALT activity in normal cells.

The possibility that p53 is a direct repressor of ALT is also an attractive concept, especially because of p53's ability to inhibit inappropriate homologous recombination (Sengupta and Harris 2005). Using inducible alleles of human p53, Razak et al. (2004) showed that expression of trans-activation-incompetent p53 inhibited DNA synthesis in ALT cell lines, but did not affect telomerase-positive cell lines. The inhibition of DNA synthesis in ALT cells by p53 required intact specific DNA binding in addition to suppression of recombination functions. It will be of great interest to determine whether p53 specifically represses the ALT mechanism in these cells.

It is possible that alterations in telomeric chromatin may facilitate the activation of ALT by making telomeres more accessible to proteins involved in the recombination machinery. In *S. cerevisiae*, inactivation of the HHO1 gene, which encodes the linker histone Hho1p, resulted in a less pronounced period of proliferative arrest and an increased survival of telomerase-deficient yeast cells (Downs et al. 2003). Furthermore, primary cells derived from mice doubly deficient for both the histone methyltransferases Suv39h1 and Suv39h2, which govern methylation of histone H3 Lys9 primarily in centromeric heterochromatin regions, have abnormally long telomeres compared to wild-type controls, presumably because of loss of heterochromatic features, a more "open" telomere chromatin configuration, and, thus, altered accessibility of telomerase (Blasco 2004; Garcia-Cao et al. 2004). It is interesting to speculate whether such inactivation or deregulation of linker histones, or other genetic and epigenetic chromatin modifications, will turn out to be similarly implicated in the telomerase-independent maintenance of telomeres in mammalian cells (d'Adda di Fagagna et al. 2004).

Mismatch repair (MMR) proteins have been considered as potential ALT repressors. They are evolutionarily highly conserved and have important functions in maintaining eukaryotic genome stability. The MMR machinery not only is necessary to reduce mutation frequencies by correcting replication errors, but also has an important role in inhibiting recombination, especially between homeologous DNA sequences, because of its ability to recognize mismatches in recombination intermediates. Defects in the MMR machinery enhanced proliferation in the period before activation of a telomerase-independent survivor pathway in telomerase-null yeast cells (Rizki and Lundblad 2001). It has not yet been reported whether MMR-deficient human cells undergo prolonged proliferation prior to crisis, but presumably such cells would be competent to activate either telomerase or ALT. It was perhaps not surprising, therefore, that when tumors with and without MMR deficiency (as manifested by microsatellite instability) were analyzed, telomerase activity was found to be equally prevalent in each group (De Caceres et al. 2004). Furthermore, five ALT cell lines had no evidence of microsatellite instability (Tsutsui et al. 2003). There is therefore no clear relationship between MMR deficiency and ALT in human cells.

Genes Potentially Involved in the ALT Mechanism

It is conceivable, but not yet proven, that many of the proteins found in APBs could participate in the ALT mechanism. Most of them, such as RAD51, RAD52, RPA, the MRN complex, and the RecQ-type helicases WRN and BLM, are functionally compatible with homologous recombination and recombination-dependent replication. Overexpression of SP100, a constituent of PML nuclear bodies, has recently been demonstrated to result in sequestration of the MRE11, RAD50, and NBS1 (MRN) recombination proteins (Jiang et al. 2005). This was associated with repression of the ALT mechanism, as evidenced by progressive telomere shortening at ~120 bp per population doubling, suppression of rapid changes in telomere length, and inhibition of APB formation in growth-arrested cells. Carboxyl-terminal-truncated SP100 proteins that did not sequester the MRN complex failed to inhibit ALT. These findings strongly support the notion that the MRN complex is involved in the ALT mechanism.

For a number of other proteins it is not clear whether they have a specific role in ALT or a more general contribution to telomere protection. For example, one of the constituents of PML bodies is the SUMO-1 (small ubiquitin-related modifier) protein that can covalently bind to other proteins including PML, RAD51, RAD52, and PCNA (Shen et al. 1996; Tanaka et al. 1999; Yeh et al. 2000; Lallemand-Breitenbach et al. 2001). Mutation of the *S. pombe* homolog of SUMO-1 results in an elongated telomere phenotype (Tanaka et al. 1999).

Another APB component, the RAD51 recombination paralog, RAD51D, has a role in telomere length regulation, but this was shown to be independent of the telomerase status of cells. Depletion of RAD51D by short interfering RNA caused significant telomere shortening in ALT cells, and telomerase-positive MEFs that were doubly deficient for RAD51D and p53 had even more pronounced telomere shortening (Tarsounas et al. 2004). Telomere shortening has also been observed in

MEFs from RAD54 null mice (Jaco et al. 2003). It seems likely that RAD51D and RAD54 both participate in telomere end protection rather than having a direct role in recombination-mediated telomere lengthening.

Sgs1 is involved in the type II survivor pathway in budding yeast. In this organism, Sgs1 is the sole member of the RecQ family of DNA helicases, which include five human orthologs, three of which—WRN, BLM, and RECQL4—are known to be mutated in familial syndromes featuring premature aging and increased cancer susceptibility, i.e., Werner's, Bloom's, and Rothmund-Thomson syndromes, respectively (Johnson et al. 2001). WRN and BLM are both present in APBs (Yankiwski et al. 2000; Johnson et al. 2001; Stavropoulos et al. 2002). The BLM RecQ helicase might function at telomeres to regulate D-loop formation following strand invasion and/or resolve recombined or entangled telomeres in ALT cells, but it might also play a role in resolving recombination events in mortal and telomerase-positive cells. TRF2 stimulated BLM unwinding of two telomere substrates in vitro, a 3′ overhang and a telomere D-loop structure; TRF1, in contrast, inhibited unwinding (Opresko et al. 2002; Stavropoulos et al. 2002; Lillard-Wetherell et al. 2004). WRN is unique among human RecQ helicases in that it also contains 3′ to 5′ exonuclease activity (Mohaghegh et al. 2001), although it is not known whether helicase and exonuclease cooperate to resolve DNA structures in vivo. WRN may be involved both in recombinational repair at telomeres and in the dissociation of a 3′ telomeric tail from inappropriate recombination intermediates by resolving telomeric D-loops (Opresko et al. 2002, 2004). TRF1 and TRF2 cooperate to promote WRN-mediated D-loop unwinding but limit processing of the 3′ end by the WRN exonuclease activity. WRN physically interacts with human Exo1 protein and dramatically stimulates both its 5′ to 3′ exonuclease and flap endonuclease activities (Sharma et al. 2003). Both BLM and WRN are able to unwind various forms of G-quadruplex DNA, a configuration that can form at G-rich sequences such as telomeres (Mohaghegh et al. 2001). Transient overexpression of a green fluorescent protein (GFP)-BLM fusion protein resulted in ALT cell-specific increases in telomeric DNA, and it was therefore concluded that BLM is directly involved in ALT (Stavropoulos et al. 2002).

Intriguingly, cells lacking the WRN RecQ helicase have recently been shown to display telomeric deletions from single sister chromatids. Only telomeres replicated by lagging strand synthesis were affected, and prevention of loss of individual telomeres was dependent on the helicase activity of WRN (Crabbe et al. 2004). Telomerase activity was able to counteract telomere loss. Collectively, these results suggested that WRN is necessary for efficient and complete lagging strand replication of the G-rich telomeric strand, thus preventing telomere dysfunction and consequent genomic instability. It is very likely that there are redundant helicases performing similar activities in resolution of G-quadruplex DNA, such as the BRCA1-binding helicase-like protein BACH1 (Cantor et al. 2001) or the human novel helicase-like (NHL) gene, which belongs to the RAD3/ERCC2 subfamily (Bai et al. 2000), the mouse homolog of which, Rtel1, has been shown to be essential in murine telomere length regulation (Ding et al. 2004). The role, if any, of such helicases in ALT remains to be determined.

ALT in Human Tumors

ALT has been detected in a range of human tumors and tumor cell lines from bone, soft tissue, brain, lung, kidney, adrenal cortex, breast, and ovary (Mehle et al. 1996; Bryan et al. 1997a; Hakin-Smith et al. 2003; Ulaner et al. 2003; Henson et al. 2004). All immortalized cell lines analyzed so far utilize a single telomere maintenance mechanism, i.e., they either have telomerase activity or they have an ALT mechanism, with the possible exception of some lymphocytic cell lines that may have both (Strahl and Blackburn 1996). In tumors, however, the situation is more complex: Approximately 85% of all human tumors have telomerase activity (Shay and Bacchetti 1997), but few surveys have been carried out to estimate the prevalence of ALT in human tumors (Bryan et al. 1997a; Hakin-Smith et al. 2003; Ulaner et al. 2003; Henson et al. 2004). It cannot be concluded that the 15% of tumors that are telomerase negative must use ALT, because it is likely that some cancers do not require activation of a telomere maintenance mechanism and immortalization (Reddel 2000; Seger et al. 2002). In addition, some tumors have both ALT and telomerase activity (Bryan et al. 1997a; Hakin-Smith et al. 2003; Ulaner et al. 2003; Henson et al. 2004); it is unclear whether this means that both telomere maintenance mechanisms are active in the same tumor cells or only in different subpopulations within the same tumor.

Approximately 50% of sarcomas overall are telomerase negative (Henson et al. 2002, and references therein), especially those that have complex karyotypes (Scheel et al. 2001; Montgomery et al. 2004; Ulaner et al. 2004). Thus, ALT appears to be more common in tumors derived from tissues of mesenchymal origin; this is mirrored in the higher incidence of ALT in in vitro immortalized cell lines, many of which are fibroblastic in origin, than in tumor cell lines, which are mostly derived from epithelial cancers. Mesenchymal cell compartments generally have a slower cell turnover and therefore less telomere shortening than most epithelial tissues, and may consequently repress telomerase more tightly, whereas normal epithelial cells show low levels of telomerase activity (Yasumoto et al. 1996). So, even if the probability of ALT being activated is the same during the development of carcinomas and sarcomas, tighter repression of telomerase in mesenchymal cells may result in ALT being more common in sarcomas.

ALT and Prognosis of Cancer Patients

Two studies in cancer models have suggested that ALT cells may be less malignant than telomerase-positive cells (Stewart et al. 2002; Chang et al. 2003), and there is accumulating evidence that telomerase might provide advantages to tumor cells in addition to telomere length maintenance (Broccoli et al. 1996; Gonzalez-Suarez et al. 2001; Blasco 2002). However, human cancers that use ALT may be just as aggressive as their telomerase-positive counterparts. Of 115 osteosarcomas analyzed in two studies (Ulaner et al. 2003; Sanders et al. 2004), 17% had telomerase only, 50% had ALT only, 22% had both ALT and telomerase, and 11% had neither (Table 2). Neither study found that ALT was associated with a more favorable prognosis, and both of the studies identified metastatic tumors that had ALT as the sole telomere maintenance mechanism. The absence of any telomere maintenance mechanism correlated more strongly with survival than tumor stage or response to chemotherapy. Thus, it appears that ALT-positive and telomerase-positive osteosarcomas are similarly aggressive in their clinical behavior.

A counterexample is the high-grade brain tumor, glioblastoma multiforme (GBM), where the presence of ALT is associated with a more favorable outcome. In one report, 19 of 77 (25%) GBMs were ALT positive and four of these were also telomerase positive (Table 2). Although almost all of these tumors were rapidly lethal, there was a highly significant correlation between the presence of ALT and survival (Hakin-Smith et al. 2003). There is evidence, however, that GBMs are genetically heterogeneous, and it needs to be determined whether ALT is directly correlated with less aggressive behavior of this tumor type or whether it is instead a marker for a subset of GBMs that have accumulated a distinct set of genetic alterations.

ALT and Cancer Treatment

Telomerase inhibitors are being developed as potential anticancer treatments for telomerase-positive cancers, but there is concern that activation of ALT in the treated tumor cells may result in resistance to

Table 2. Prevalence of telomerase and ALT in human tumors

Reference	Tumor type	No. of tumors	Telo$^+$ (%)	Telo$^-$ (%)	ALT$^+$ (%)	Telo$^+$/ALT$^-$ (%)	Telo$^-$/ALT$^+$ (%)	Telo$^+$/ALT$^+$ (%)	Telo$^-$/ALT$^-$ (%)
Ulaner et al. (2003)	osteosarcoma	71	44	56	66	17	39	27	17
Sanders et al. (2004)	osteosarcoma	44	32	68	80	18	66	14	2
Henson et al. (2004)	osteosarcoma	58			47				
Ulaner et al. (2004)	Ewing's Sarcoma	30	70	30	0	70	0	0	30
Henson et al. (2004)	soft tissue sarcoma	101			35				
	Subtypes:								
	mixed fibrous histiocytoma	22			77				
	leiomyosarcoma	13			62				
	liposarcoma	9			33				
	synovial sarcoma	11			9				
	rhabdomyosarcoma	35			6				
Hakin-Smith et al. (2003)	glioblastoma multiforme	77	34	66	24	29	19	5	47
Henson et al. (2004)	astrocytoma	50			34				
Henson et al. (2004)	papillary carcinoma of the thyroid	17			0				

these drugs. This concern has been addressed in several studies. Resistance to adriamycin, which resulted in decreased TERT mRNA expression and a tenfold reduction in telomerase activity in a human gastric adenocarcinoma cell line, was associated with telomeres lengthening up to ~50 kb (Kim et al. 2002). Human ovarian SKOV-3 cells in which telomerase was repressed either by the reverse transcriptase inhibitor AZT (~80% inhibition) or by expression of a TR antisense construct (complete inhibition) were able to maintain the length of their telomeres, but in the absence of any other characteristics of the ALT phenotype (Gan et al. 2002). Inhibition of telomerase by expression of a dominant negative TERT allele in HCT15 colon cancer cells resulted in survivors with a phenotype that included episodic moderate telomere elongation, as well as increased telomeric exchanges as visualized by telomere-specific CO-FISH (Bechter et al. 2004b). The resistant cells did not have APBs, nor did they show the telomere length heterogeneity characteristic of ALT. Although these experiments have not provided clear evidence for the activation of ALT in response to inhibition of telomerase, the evidence they provide for telomere maintenance and even telomere lengthening and prolonged survival in this context does not allay the concern.

It is expected that most telomerase inhibitors will be ineffective against ALT or mixed ALT/telomerase tumors (see Chapter 4). Gene therapy approaches in which promoter fragments from either the RNA component (TR) or the catalytic subunit (TERT) of telomerase are used to drive the expression of therapeutic transgenes, such as proapoptotic caspase-8 (Komata et al. 2002), caspase-6 (Komata et al. 2001), TRAIL (Katz et al. 2003) and Bax (Gu et al. 2000), pro-drug-activating enzymes (Majumdar et al. 2001; Bilsland et al. 2003; Takeda et al. 2003), or the lysis-activating E1A gene in adenoviral vectors (Huang et al. 2004), are promising cancer-cell-specific treatments that will also be ineffective against ALT tumors. Repression of ALT in immortalized cell lines results in senescence and cell death (Nakabayashi et al. 1997; Perrem et al. 1999), and therefore ALT, like telomerase, may be an attractive target for telomere-directed cancer therapy. Combined treatment with ALT and telomerase inhibitors may also help prevent the emergence of drug resistance.

To predict the potential toxicity of ALT inhibitory drugs, it will be important to determine whether there is a normal counterpart of ALT. It seems possible that ALT tumors are using a dysregulated form of a normal mechanism, analogous to the dysregulated telomerase activity in telomerase-positive tumors. For instance, ALT could be based on a putative normal recombinational repair mechanism that processes broken or damaged telomeres, and thus prevents irregular and untimely telomere uncapping, and might be especially important in telomerase-negative somatic cells. It has been proposed that t-loops enabled the evolutionary transition from circular to linear chromosomes in early eukaryotes, both by providing a mechanism for chromosome end protection and by facilitating ALT-like recombination-mediated telomere lengthening (as illustrated in Fig. 7b) via a preexisting prokaryotic pathway, and that the ALT-like activity was repressed when it was superseded by the telomerase reverse transcriptase mechanism in later eukaryotes (de Lange 2004). t-loops have persisted in eukaryotes, providing telomere protection and potentially regulating telomerase access, but are generally not thought to be used for replicative telomere extension. It is interesting to speculate, however, that tightly regulated ALT-like activity may persist in normal eukaryotic cells and that dysregulation of this process results in the abnormal telomere phenotype seen in ALT tumors.

ACKNOWLEDGMENTS

Work in the authors' laboratory is funded by project grants and a fellowship from the National Health and Medical Research Council of Australia, and by the Nippon Boehringer Ingelheim Virtual Research Institute of Aging. We thank Nicola Royle, José-Arturo Londoño-Vallejo, members of the Reddel laboratory, and the editors of this book for their helpful comments on the manuscript and Elizabeth Collins for help with the bibliography. We have no competing financial interests.

REFERENCES

Allshire R.C., Dempster M., and Hastie N.D. 1989. Human telomeres contain at least three types of G-rich repeat distributed non-randomly. *Nucleic Acids Res.* **17:** 4611–4627.

Argilla D., Chin K., Singh M., Hodgson J.G., Bosenberg M., De Solorzano C.O., Lockett S., DePinho R.A., Gray J., and Hanahan D. 2004. Absence of telomerase and shortened telomeres have minimal effects on skin and pancreatic carcinogenesis elicited by viral oncogenes. *Cancer Cell* **6:** 373–385.

Bai C., Connolly B., Metzker M.L., Hilliard C.A., Liu X., Sandig V., Soderman A., Galloway S.M., Liu Q., Austin C.P., and Caskey C.T. 2000. Overexpression of M68/DcR3 in human gastrointestinal tract tumors independent of gene amplification and its location in a four-gene cluster. *Proc. Natl. Acad. Sci.* **97:** 1230–1235.

Bailey S.M., Brenneman M.A., and Goodwin E.H. 2004a. Frequent recombination in telomeric DNA may extend the proliferative life of telomerase-negative cells. *Nucleic Acids Res.* **32:** 3743–3751.

Bailey S.M., Goodwin E.H., and Cornforth M.N. 2004b. Strand-specific fluorescence in situ hybridization: The CO-FISH family. *Cytogenet. Genome Res.* **107:** 14–17.

Bailey S.M., Cornforth M.N., Kurimasa A., Chen D.J., and Goodwin E.H. 2001. Strand-specific postreplicative processing of mammalian telomeres. *Science* **293:** 2462–2465.

Bechter O.E., Shay J.W., and Wright W.E. 2004a. The frequency of homologous recombination in human ALT cells. *Cell Cycle* **3:** 547–549.

Bechter O.E., Zou Y., Shay J.W., and Wright W.E. 2003. Homologous recombination in human telomerase-positive and ALT cells occurs with the same frequency. *EMBO Rep.* **4:** 1138–1143.

Bechter O.E., Zou Y., Walker W., Wright W.E., and Shay J.W. 2004b. Telomeric recombination in mismatch repair deficient human colon cancer cells after telomerase inhibition. *Cancer Res.* **64:** 3444–3451.

Bertuch A.A. and Lundblad V. 2004. EXO1 contributes to telomere maintenance in both telomerase-proficient and telomerase-deficient *Saccharomyces cerevisiae. Genetics* **166:** 1651–1659.

Bilsland A.E., Anderson C.J., Fletcher-Monaghan A.J., McGregor F., Evans T.R.J., Ganly I., Knox R.J., Plumb J.A., and Keith W.N. 2003. Selective ablation of human cancer cells by telomerase-specific adenoviral suicide gene therapy vectors expressing bacterial nitroreductase. *Oncogene* **22:** 370–380.

Blasco M.A. 2002. Telomerase beyond telomeres. *Nat. Rev. Cancer* **2:** 627–633.

———. 2004. Telomere epigenetics: A higher-order control of telomere length in mammalian cells. *Carcinogenesis* **25:** 1083–1087.

Borden K.L. 2002. Pondering the promyelocytic leukemia protein (PML) puzzle: Possible functions for PML nuclear bodies. *Mol. Cell. Biol.* **22:** 5259–5269.

Broccoli D., Godley I.A., Donehower L.A., Varmus H.E., and de Lange T. 1996. Telomerase activation in mouse mammary tumors: Lack of detectable telomere shortening and evidence for regulation of telomerase RNA with cell proliferation. *Mol. Cell. Biol.* **16:** 3765–3772.

Bryan T.M. and Reddel R.R. 1994. SV40-induced immortalization of human cells. *Crit. Rev. Oncogenesis* **5:** 331–357.

———. 1997. Telomere dynamics and telomerase activity in *in vitro* immortalised human cells. *Eur. J. Cancer* **33:** 767–773.

Bryan T.M., Englezou A., Dunham M.A., and Reddel R.R. 1998. Telomere length dynamics in telomerase-positive immortal human cell populations. *Exp. Cell Res.* **239:** 370–378.

Bryan T.M., Englezou A., Dalla-Pozza L., Dunham M.A., and Reddel R.R. 1997a. Evidence for an alternative mechanism for maintaining telomere length in human tumors and tumor-derived cell lines. *Nat. Med.* **3:** 1271–1274.

Bryan T.M., Englezou A., Gupta J., Bacchetti S., and Reddel R.R. 1995. Telomere elongation in immortal human cells without detectable telomerase activity. *EMBO J.* **14:** 4240–4248.

Bryan T.M., Marusic L., Bacchetti S., Namba M., and Reddel R.R. 1997b. The telomere lengthening mechanism in telomerase-negative immortal human cells does not involve the telomerase RNA subunit. *Hum. Mol. Genet.* **6:** 921–926.

Bucholc M., Park Y., and Lustig A.J. 2001. Intrachromatid excision of telomeric DNA as a mechanism for telomere size control in *Saccharomyces cerevisiae. Mol. Cell. Biol.* **21:** 6559–6573.

Cabuy E., Newton C., Roberts T., Newbold R., and Slijepcevic P. 2004. Identification of subpopulations of cells with differing telomere lengths in mouse and human cell lines by flow FISH. *Cytometry* **62:** 150–161.

Cantor S.B., Bell D.W., Ganesan S., Kass E.M., Drapkin R., Grossman S., Wahrer D.C.R., Sgroi D.C., Lane W.S., Haber D.A., and Livingston D.M. 2001. BACH1, a novel helicase-like protein, interacts directly with BRCA1 and contributes to its DNA repair function. *Cell* **105:** 149–160.

Cerone M.A., Londono-Vallejo J.A., and Bacchetti S. 2001. Telomere maintenance by telomerase and by recombination can coexist in human cells. *Hum. Mol. Genet.* **10:** 1945–1952.

Cesare A.J. and Griffith J.D. 2004. Telomeric DNA in ALT cells is characterized by free telomeric circles and heterogeneous T-loops. *Mol. Cell. Biol.* **24:** 9948–9957.

Chang S., Khoo C.M., Naylor M.L., Maser R.S., and DePinho R.A. 2003. Telomere-based crisis: Functional differences between telomerase activation and ALT in tumor progression. *Genes Dev.* **17:** 88–100.

Chen Q., Ijpma A., and Greider C.W. 2001. Two survivor pathways that allow growth in the absence of telomerase are generated by distinct telomere recombination events. *Mol. Cell. Biol.* **21:** 1819–1827.

Counter C.M., Avilion A.A., LeFeuvre C.E., Stewart N.G., Greider C.W., Harley C.B., and Bacchetti S. 1992. Telomere shortening associated with chromosome instability is arrested in immortal cells which express telomerase activity. *EMBO J.* **11:** 1921–1929.

Crabbe L., Verdun R.E., Haggblom C.I., and Karlseder J. 2004. Defective telomere lagging strand synthesis in cells lacking WRN helicase activity. *Science* **306:** 1951–1953.

d'Adda di Fagagna F., Teo S.H., and Jackson S.P. 2004. Functional links between telomeres and proteins of the DNA-damage response. *Genes Dev.* **18:** 1781–1799.

Dantzer F., Giraud-Panis M.J., Jaco I., Ame J.C., Schultz I., Blasco M., Koering C.E., Gilson E., Menissier-De Murcia J., de Murcia G., and Schreiber V. 2004. Functional interaction between Poly(ADP-Ribose) Polymerase 2 (PARP-2) and TRF2: PARP activity negatively regulates TRF2. *Mol. Cell. Biol.* **24:** 1595–1607.

De Caceres I.I., Frolova N., Varkonyi R.J., Dulaimi E., Meropol N.J., Broccoli D., and Cairns P. 2004. Telomerase is frequently activated in tumors with microsatellite instability. *Cancer Biol. Ther.* **3:** 289–292.

de Lange T. 2004. Opinion: T-loops and the origin of telomeres. *Nat. Rev. Mol. Cell Biol.* **5:** 323–329.

de Lange T., Shiue L., Myers R.M., Cox D.R., Naylor S.L., Killery A.M., and Varmus H.E. 1990. Structure and variability of human chromosome ends. *Mol. Cell. Biol.* **10:** 518–527.

Dellaire G. and Bazett-Jones D.P. 2004. PML nuclear bodies: Dynamic sensors of DNA damage and cellular stress. *BioEssays* **26:** 963–977.

Dessain S.K., Yu H., Reddel R.R., Beijersbergen R.L., and Weinberg R.A. 2000. Methylation of the human telomerase gene CpG island. *Cancer Res.* **60:** 537–541.

Ding H., Schertzer M., Wu X., Gertsenstein M., Selig S., Kammori M., Pourvali R., Poon S., Vulto I., Chavez E., Tam P.P., Nagy A., and Lansdorp P.M. 2004. Regulation of murine telomere length by Rtel: An essential gene encoding a helicase-like protein. *Cell* **117:** 873–886.

Downs J.A., Kosmidou E., Morgan A., and Jackson S.P. 2003. Suppression of homologous recombination by the *Saccharomyces cerevisiae* linker histone. *Mol. Cell* **11:** 1685–1692.

Dunham M.A., Neumann A.A., Fasching C.L., and Reddel R.R. 2000. Telomere maintenance by recombination in human cells. *Nat. Genet.* **26:** 447–450.

Fasching C.L., Bower K., and Reddel R.R. 2005. Telomerase-independent telomere length maintenance in the absence of alternative lengthening of telomeres–associated promyelocytic leukemia bodies. *Cancer Res.* **65:** 2722–2729.

Ford L.P., Zou Y., Pongracz K., Gryaznov S.M., Shay J.W., and Wright W.E. 2001. Telomerase can inhibit the recombination-based pathway of telomere maintenance in human cells. *J. Biol. Chem.* **276:** 32198–32203.

Gan Y., Mo Y., Johnston J., Lu J., Wientjes M., and Au J. 2002. Telomere maintenance in telomerase-positive human ovarian SKOV-3 cells cannot be retarded by complete inhibition of telomerase. *FEBS Lett.* **527:** 10.

Garcia-Cao M., Gonzalo S., Dean D., and Blasco M.A. 2002. A role for the Rb family of proteins in controlling telomere length. *Nat. Genet.* **32:** 415–419.

Garcia-Cao M., O'Sullivan R., Peters A.H., Jenuwein T., and Blasco M.A. 2004. Epigenetic regulation of telomere length in mammalian cells by the Suv39h1 and Suv39h2 histone methyltransferases. *Nat. Genet.* **36:** 94–99.

Gonzalez-Suarez E., Samper E., Ramirez A., Flores J.M., Martin-Caballero J., Jorcano J.L., and Blasco M.A. 2001. Increased epidermal tumors and increased skin wound healing in transgenic mice overexpressing the catalytic subunit of telomerase, mTERT, in basal keratinocytes. *EMBO J.* **20:** 2619–2630.

Griffith J.D., Comeau L., Rosenfield S., Stansel R.M., Bianchi A., Moss H., and de Lange T. 1999. Mammalian telomeres end in a large duplex loop. *Cell* **97:** 503–514.

Grobelny J.V., Godwin A.K., and Broccoli D. 2000. ALT-associated PML bodies are present in viable cells and are enriched in cells in the G_2/M phase of the cell cycle. *J. Cell Sci.* **113:** 4577–4585.

Grobelny J.V., Kulp-McEliece M., and Broccoli D. 2001. Effects of reconstitution of telomerase activity on telomere maintenance by the alternative lengthening of telomeres (ALT) pathway. *Hum. Mol. Genet.* **10:** 1953–1961.

Gu J., Kagawa S., Takakura M., Kyo S., Inoue M., Roth J.A., and Fang B. 2000. Tumor-specific transgene expression from the human telomerase reverse transcriptase promoter enables targeting of the therapeutic effects of the *Bax* gene to cancers. *Cancer Res.* **60:** 5359–5364.

Guiducci C., Cerone M.A., and Bacchetti S. 2001. Expression of mutant telomerase in immortal telomerase-negative human cells results in cell cycle deregulation, nuclear and chromosomal abnormalities and rapid loss of viability. *Oncogene* **20:** 714–725.

Hakin-Smith V., Jellinek D.A., Levy D., Carroll T., Teo M., Timperley W.R., McKay M.J., Reddel R.R., and Royds J.A. 2003. Alternative lengthening of telomeres and survival in patients with glioblastoma multiforme. *Lancet* **361:** 836–838.

Hande M.P., Samper E., Lansdorp P., and Blasco M.A. 1999. Telomere length dynamics and chromosomal instability in cells derived from telomerase null mice. *J. Cell Biol.* **144:** 589–601.

Hande M.P., Balajee A.S., Tchirkov A., Wynshaw-Boris A., and Lansdorp P.M. 2001. Extra-chromosomal telomeric DNA in cells from $Atm^{-/-}$ mice and patients with ataxia-telangiectasia. *Hum. Mol. Genet.* **10:** 519–528.

Harley C.B. 1997. Human ageing and telomeres. *Ciba Found. Symp.* **211:** 129–144.

Harley C.B., Futcher A.B., and Greider C.W. 1990. Telomeres shorten during ageing of human fibroblasts. *Nature* **345:** 458–460.

Hastie N.D., Dempster M., Dunlop M.G., Thompson A.M., Green D.K., and Allshire R.C. 1990. Telomere reduction in human colorectal carcinoma and with ageing. *Nature* **346:** 866–868.

Henson J.D., Neumann A.A., Yeager T.R., and Reddel R.R. 2002. Alternative lengthening of telomeres in mammalian cells. *Oncogene* **21:** 598–610.

Henson J.D., Hannay J.A., McCarthy S.W., Royds J.A., Yeager T.R., Robinson R.A., Wharton S.B., Jellinek D.A., Arbuckle S.M., Yoo J., Robinson B.G., Learoyd D.L., Stalley P.D., Bonar S.F., Yu D., Pollock R.E., and Reddel R.R. 2004. A robust assay for alternative lengthening of telomeres (ALT) in tumors demonstrates the significance of ALT in sarcomas and astrocytomas. *Clin. Cancer Res.* **11:** 217–225.

Herrera E., Martinez C., and Blasco M.A. 2000. Impaired germinal center reaction in mice with short telomeres. *EMBO J.* **19:** 472–481.

Hoare S.F., Bryce L.A., Wisman G.B.A., Burns S., Going J.J., Van der Zee A.G.J., and Keith W.N. 2001. Lack of telomerase RNA gene *hTERC* expression in alternative lengthening of telomeres cells is associated with methylation of the *hTERC* promoter. *Cancer Res.* **61:** 27–32.

Hodges M., Tissot C., Howe K., Grimwade D., and Freemont P.S. 1998. Structure, organization, and dynamics of promyelocytic leukemia protein nuclear bodies. *Am. J. Hum. Genet.* **63:** 297–304.

Huang Q., Zhang X., Wang H., Yan B., Kirkpatrick J., Dewhrist M.W., and Li C.Y. 2004. A novel conditionally replicative adenovirus vector targeting telomerase-positive tumor cells. *Clin. Cancer Res.* **10:** 1439–1445.

Jaco I., Munoz P., Goytisolo F., Wesoly J., Bailey S., Taccioli G., and Blasco M.A. 2003. Role of mammalian Rad54 in telomere length maintenance. *Mol. Cell. Biol.* **23:** 5572–5580.

Jeyapalan J.N., Varley H., Foxon J.L., Pollock R.E., Jeffreys A.J., Henson J.D., Reddel R.R., and Royle N.J. 2005. Activation of ALT pathway for telomere maintenance can affect other sequences in the human genome. *Hum. Mol. Genet.* **14:** 1785–1794.

Jiang W.Q., Zhong Z.H., Henson J.D., Neumann A.A., Chang A.C., and Reddel R.R. 2005. Suppression of alternative lengthening of telomeres by Sp100-mediated sequestration of MRE11/RAD50/NBS1 complex. *Mol. Cell. Biol.* **25:** 2708–2721.

Johnson F.B., Marciniak R.A., McVey M., Stewart S.A., Hahn W.C., and Guarente L. 2001. The *Saccharomyces cerevisiae* WRN homolog Sgs1p participates in telomere maintenance in cells lacking telomerase. *EMBO J.* **20:** 905–913.

Katoh M., Katoh M., Kameyama M., Kugoh H., Shimizu M., and Oshimura M. 1998. A repressor function for telomerase activity in telomerase-negative immortal cells. *Mol. Carcinog.* **21:** 17–25.

Katz M.H., Spivack D.E., Takimoto S., Fang B., Burton D.W., Moossa A.R., Hoffman R.M., and Bouvet M. 2003. Gene therapy of pancreatic cancer with green fluorescent protein and tumor necrosis factor-related apoptosis-inducing ligand fusion gene expression driven by a human telomerase reverse transcriptase promoter. *Ann. Surg. Oncol.* **10:** 762–772.

Kilian A., Bowtell D.D.L., Abud H.E., Hime G.R., Venter D.J., Keese P.K., Duncan E.L., Reddel R.R., and Jefferson R.A. 1997. Isolation of a candidate human telomerase catalytic subunit gene, which reveals complex splicing patterns in different cell types. *Hum. Mol. Genet.* **6:** 2011–2019.

Kim J.H., Lee G.E., Kim J.C., Lee J.H., and Chung I.K. 2002. A novel telomere elongation in an adriamycin-resistant stomach cancer cell line with decreased telomerase activity. *Mol. Cells* **13:** 228–236.

Kim N.W., Piatyszek M.A., Prowse K.R., Harley C.B., West M.D., Ho P.L.C., Coviello G.M., Wright W.E., Weinrich S.L., and Shay J.W. 1994. Specific association of human telomerase activity with immortal cells and cancer. *Science* **266:** 2011–2015.

Kimmel M., Axelrod D.E., and Wahl G.M. 1992. A branching process model of gene amplification following chromosome breakage. *Mutat. Res.* **276:** 225–239.

Komata T., Kondo Y., Kanzawa T., Ito H., Hirohata S., Koga S., Sumiyoshi H., Takakura M., Inoue M., Barna B.P., Germano I.M., Kyo S., and Kondo S. 2002. Caspase-8 gene therapy using the human telomerase reverse transcriptase promoter for malignant glioma cells. *Hum. Gene Ther.* **13:** 1015–1025.

Komata T., Kondo Y., Kanzawa T., Hirohata S., Koga S., Sumiyoshi H., Srinivasula S.M., Barna B.P., Germano I.M., Takakura M., Inoue M., Alnemri E.S., Shay J.W., Kyo S., and Kondo S. 2001. Treatment of malignant glioma cells with the transfer of constitutively active caspase-6 using the human telomerase catalytic subunit (human telomerase reverse transcriptase) gene promoter. *Cancer Res.* **61:** 5796–5802.

Kumata M., Shimizu M., Oshimura M., Uchida M., and Tsutsui T. 2002. Induction of cellular senescence in a telomerase negative human immortal fibroblast cell line, LCS-AF.1-3, by human chromosome 6. *Int. J. Oncol.* **21:** 851–856.

Lallemand-Breitenbach V., Zhu J., Puvion F., Koken M., Honore N., Doubeikovsky A., Duprez E., Pandolfi P.P., Puvion E., Freemont P., and de The H. 2001. Role of promyelocytic leukemia (PML) sumolation in nuclear body formation, 11S proteasome recruitment, and As$_2$O$_3$-induced PML or PML/retinoic acid receptor α degradation. *J. Exp. Med.* **193:** 1361–1372.

Lansdorp P.M., Poon S., Chavez E., Dragowska V., Zijlmans M., Bryan T., Reddel R., Egholm M., Bacchetti S., and Martens U. 1997. Telomeres in the hematopoietic system. In *Telomeres and telomerase* (ed. D.J. Chadwick and G. Cardew), pp. 209–218. John Wiley & Sons, West Sussex, United Kingdom.

Le S., Moore J.K., Haber J.E., and Greider C.W. 1999. *RAD50* and *RAD51* define two pathways that collaborate to maintain telomeres in the absence of telomerase. *Genetics* **152:** 143–152.

Li B. and Lustig A.J. 1996. A novel mechanism for telomere size control in *Saccharomyces cerevisiae*. *Genes Dev.* **10:** 1310–1326.

Lillard-Wetherell K., Machwe A., Langland G.T., Combs K.A., Behbehani G.K., Schonberg S.A., German J., Turchi J.J., Orren D.K., and Groden J. 2004. Association and regulation of the BLM helicase by the telomere proteins TRF1 and TRF2. *Hum. Mol. Genet.* **13:** 1919–1932.

Lombard D.B. and Guarente L. 2000. Nijmegen breakage syndrome disease protein and MRE11 at PML nuclear bodies and meiotic telomeres. *Cancer Res.* **60:** 2331–2334.

Londoño-Vallejo J.A., Der-Sarkissian H., Cazes L., Bacchetti S., and Reddel R.R. 2004. Alternative lengthening of telomeres is characterized by high rates of telomeric exchange. *Cancer Res.* **64:** 2324–2327.

Louis E.J. and Haber J.E. 1990. Mitotic recombination among subtelomeric Y′ repeats in *Saccharomyces cerevisiae*. *Genetics* **124:** 547–559.

Lundblad V. and Blackburn E.H. 1993. An alternative pathway for yeast telomere maintenance rescues *est1⁻* senescence. *Cell* **73:** 347–360.

Luo P., Tresini M., Cristofalo V., Chen X., Saulewicz A., Gray M.D., Banker D.E., Klingelhutz A.L., Ohtsubo M., Takihara Y., and Norwood T.H. 2004. Immortalization in a normal foreskin fibroblast culture following transduction of cyclin A2 or cdk1 genes in retroviral vectors. *Exp. Cell Res.* **294:** 406–419.

Majumdar A.S., Hughes D.E., Lichtsteiner S.P., Wang Z., Lebkowski J.S., and Vasserot A.P. 2001. The telomerase reverse transcriptase promoter drives efficacious tumor suicide gene therapy while preventing hepatotoxicity encountered with constitutive promoters. *Gene Ther.* **8:** 568–578.

Marciniak R.A., Cavazos D., Montellano R., Chen Q., Guarente L., and Johnson F.B. 2005. A novel telomere structure in a human alternative lengthening of telomeres cell line. *Cancer Res.* **65:** 2730–2737.

Maringele L. and Lydall D. 2004a. EXO1 plays a role in generating type I and type II survivors in budding yeast. *Genetics* **166:** 1641–1649.

————2004b. Telomerase- and recombination-independent immortalization of budding yeast. *Genes Dev.* **18:** 2663–2675.

Martens U.M., Chavez E.A., Poon S.S., Schmoor C., and Lansdorp P.M. 2000. Accumulation of short telomeres in human fibroblasts prior to replicative senescence. *Exp. Cell Res.* **256:** 291–299.

Maul G.G., Negorev D., Bell P., and Ishov A.M. 2000. Review: Properties and assembly mechanisms of ND10, PML bodies, or PODs. *J. Struct. Biol.* **129:** 278–287.

McEachern M.J. 2001. Recombinational telomere elongation in the yeast *K. lactis*. In *Telomeres and telomerases: Cancer and biology* (ed. G. Krupp), pp. 1–12. Landes Bioscience, Georgetown, Texas.

McEachern M.J. and Blackburn E.H. 1996. Cap-prevented recombination between terminal telomeric repeat arrays (telomere CPR) maintains telomeres in *Kluyveromyces lactis* lacking telomerase. *Genes Dev.* **10:** 1822–1834.

McEachern M.J. and Iyer S. 2001. Short telomeres in yeast are highly recombinogenic. *Mol. Cell* **7:** 695–704.

Mehle C., Piatyszek M.A., Ljungberg B., Shay J.W., and Roos G. 1996. Telomerase activity in human renal cell carcinoma. *Oncogene* **13:** 161–166.

Meyne J., Goodwin E.H., and Moyzis R.K. 1994. Chromosome localization and orientation of the simple sequence repeat of human satellite I DNA. *Chromosoma* **103:** 99–103.

Mohaghegh P., Karow J.K., Brosh R.M., Jr., Bohr V.A., and Hickson I.D. 2001. The Bloom's and Werner's syndrome proteins are DNA structure-specific helicases. *Nucleic Acids Res.* **29:** 2843–2849.

Molenaar C., Wiesmeijer K., Verwoerd N.P., Khazen S., Eils R., Tanke H.J., and Dirks R.W. 2003. Visualizing telomere dynamics in living mammalian cells using PNA probes. *EMBO J.* **22:** 6631–6641.

Montgomery E., Argani P., Hicks J.L., DeMarzo A.M., and Meeker A.K. 2004. Telomere lengths of translocation-associated and nontranslocation-associated sarcomas differ dramatically. *Am. J. Pathol.* **164:** 1523–1529.

Murnane J.P., Sabatier L., Marder B.A., and Morgan W.F. 1994. Telomere dynamics in an immortal human cell line. *EMBO J.* **13:** 4953–4962.

Nabetani A., Yokoyama O., and Ishikawa F. 2004. Localization of hRad9, hHus1, hRad1 and hRad17, and caffeine-sensitive DNA replication at ALT (alternative lengthening of telomeres)-associated promyelocytic leukemia body. *J. Biol. Chem.* **279:** 25849–25857.

Nakabayashi K., Ogata T., Fujii M., Tahara H., Ide T., Wadhwa R., Kaul S.C., Mitsui Y., and Ayusawa D. 1997. Decrease in amplified telomeric sequences and induction of senescence markers by introduction of human chromosome 7 or its segments in SUSM-1. *Exp. Cell Res.* **235:** 345–353.

Nakamura T.M., Cooper J.P., and Cech T.R. 1998. Two modes of survival of fission yeast without telomerase. *Science* **282:** 493–496.

Natarajan S. and McEachern M.J. 2002. Recombinational telomere elongation promoted by DNA circles. *Mol. Cell. Biol.* **22:** 4512–4521.

Niida H., Shinkai Y., Hande M.P., Matsumoto T., Takehara S., Tachibana M., Oshimura M., Lansdorp P.M., and Furuichi Y. 2000. Telomere maintenance in telomerase-deficient mouse embryonic stem cells: Characterization of an amplified telomeric DNA. *Mol. Cell. Biol.* **20:** 4115–4127.

Ogata T., Ayusawa D., Namba M., Takahashi E., Oshimura M., and Oishi M. 1993. Chromosome 7 suppresses indefinite division of nontumorigenic immortalized human fibroblast cell lines KMST-6 and SUSM-1. *Mol. Cell. Biol.* **13:** 6036–6043.

Ogata T., Oshimura M., Namba M., Fujii M., Oishi M., and Ayusawa D. 1995. Genetic complementation of the immortal phenotype in group D cell lines by introduction of chromosome 7. *Jpn. J. Cancer Res.* **86:** 35–40.

Ogino H., Nakabayashi K., Suzuki M., Takahashi E.I., Fujii M., Suzuki T., and Ayusawa D. 1998. Release of telomeric DNA from chromosomes in immortal human cells lacking telomerase activity. *Biochem. Biophys. Res. Commun.* **248:** 223–227.

Opitz O.G., Suliman Y., Hahn W.C., Harada H., Blum H.E., and Rustgi A.K. 2001. Cyclin D1 overexpression and p53 inactivation immortalize primary oral keratinocytes by a telomerase-independent mechanism. *J. Clin. Invest.* **108:** 725–732.

Opresko P.L., Von Kobbe C., Laine J.P., Harrigan J., Hickson I.D., and Bohr V.A. 2002. Telomere binding protein TRF2 binds to and stimulates the Werner and Bloom syndrome helicases. *J. Biol. Chem.* **277:** 41110–41119.

Opresko P.L., Otterlei M., Graakjaer J., Bruheim P., Dawut L., Kolvraa S., May A., Seidman M.M., and Bohr V.A. 2004. The Werner syndrome helicase and exonuclease cooperate to resolve telomeric D loops in a manner regulated by TRF1 and TRF2. *Mol. Cell* **14:** 763–774.

Perrem K., Colgin L.M., Neumann A.A., Yeager T.R., and Reddel R.R. 2001. Coexistence of alternative lengthening of telomeres and telomerase in hTERT-transfected GM847 cells. *Mol. Cell. Biol.* **21:** 3862–3875.

Perrem K., Bryan T.M., Englezou A., Hackl T., Moy E.L., and Reddel R.R. 1999. Repression of an alternative mechanism for lengthening of telomeres in somatic cell hybrids. *Oncogene* **18:** 3383–3390.

Razak Z.R., Varkonyi R.J., Kulp-McEliece M., Caslini C., Testa J.R., Murphy M.E., and Broccoli D. 2004. p53 differentially inhibits cell growth depending on the mechanism of telomere maintenance. *Mol. Cell. Biol.* **24:** 5967–5977.

Reddel R.R. 2000. The role of senescence and immortalization in carcinogenesis. *Carcinogenesis* **21:** 477–484.

———. 2001. An alternative lifestyle for immortalized oral keratinocytes. *J. Clin. Invest.* **108:** 665–667.

Rizki A. and Lundblad V. 2001. Defects in mismatch repair promote telomerase-independent proliferation. *Nature* **411:** 713–716.

Rogan E.M., Bryan T.M., Hukku B., Maclean K., Chang A.C.M., Moy E.L., Englezou A., Warneford S.G., Dalla-Pozza L., and Reddel R.R. 1995. Alterations in p53 and p16^{INK4} expression and telomere length during spontaneous immortalization of Li-Fraumeni syndrome fibroblasts. *Mol. Cell. Biol.* **15:** 4745–4753.

Rubelj I. and Vondracek Z. 1999. Stochastic mechanism of cellular aging—Abrupt telomere shortening as a model for stochastic nature of cellular aging. *J. Theor. Biol.* **197:** 425–438.

Ruggero D., Wang Z.G., and Pandolfi P.P. 2000. The puzzling multiple lives of PML and its role in the genesis of cancer. *BioEssays* **22:** 827–835.

Sanders R.P., Drissi R., Billups C.A., Daw N.C., Valentine M.B., and Dome J.S. 2004. Telomerase expression predicts unfavorable outcome in osteosarcoma. *J. Clin. Oncol.* **22:** 3790–3797.

Scheel C., Schaefer K.-L., Jauch A., Keller M., Wai D., Brinkschmidt C., van Valen F., Boecker W., Dockhorn-Dworniczak B., and Poremba C. 2001. Alternative lengthening of telomeres is associated with chromosomal instability in osteosarcomas. *Oncogene* **20:** 3835–3844.

Seger Y.R., Garcia-Cao M., Piccinin S., Cunsolo C.L., Doglioni C., Blasco M.A., Hannon G.J., and Maestro R. 2002. Transformation of normal human cells in the absence of telomerase activation. *Cancer Cell* **2:** 401–413.

Sengupta S. and Harris C.C. 2005. p53: Traffic cop at the crossroads of DNA repair and recombination. *Nat. Rev. Mol. Cell Biol.* **6:** 44–55.

Sharma S., Sommers J.A., Driscoll H.C., Uzdilla L., Wilson T.M., and Brosh R.M., Jr. 2003. The exonucleolytic and endonucleolytic cleavage activities of human exonuclease 1 are stimulated by an interaction with the carboxyl-terminal region of the Werner syndrome protein. *J. Biol. Chem.* **278:** 23487–23496.

Shay J.W. and Bacchetti S. 1997. A survey of telomerase activity in human cancer. *Eur. J. Cancer* **33:** 787–791.

Shen Z., Pardington-Purtymun P.E., Comeaux J.C., Moyzis R.K., and Chen D.J. 1996. UBL1, a human ubiquitin-like protein associating with human RAD51/RAD52 proteins. *Genomics* **36:** 271–279.

Shigeeda N., Uchida M., Barrett J.C., and Tsutsui T. 2003. Candidate chromosomal regions for genes involved in activation of alternative lengthening of telomeres in human immortal cell lines. *Exp. Gerontol.* **38:** 641–651.

Silverman J., Takai H., Buonomo S.B., Eisenhaber F., and de Lange T. 2004. Human Rif1, ortholog of a yeast telomeric protein, is regulated by ATM and 53BP1 and functions in the S-phase checkpoint. *Genes Dev.* **18:** 2108–2119.

Stavropoulos D.J., Bradshaw P.S., Li X., Pasic I., Truong K., Ikura M., Ungrin M., and Meyn M.S. 2002. The Bloom syndrome helicase BLM interacts with TRF2 in ALT cells and promotes telomeric DNA synthesis. *Hum. Mol. Genet.* **11:** 3135–3144.

Stewart S.A., Hahn W.C., O'Connor B.F., Banner E.N., Lundberg A.S., Modha P., Mizuno H., Brooks M.W., Fleming M., Zimonjic D.B., Popescu N.C., and Weinberg R.A. 2002. Telomerase contributes to tumorigenesis by a telomere length-independent mechanism. *Proc. Natl. Acad. Sci.* **99:** 12606–12611.

Strahl C. and Blackburn E.H. 1996. Effects of reverse transcriptase inhibitors on telomere length and telomerase activity in two immortalized human cell lines. *Mol. Cell. Biol.* **16:** 53–65.

Takata M., Sasaki M.S., Sonoda E., Morrison C., Hashimoto M., Utsumi H., Yamaguchi-Iwai Y., Shinohara A., and Takeda S. 1998. Homologous recombination and non-homologous end-joining pathways of DNA double-strand break repair have overlapping roles in the maintenance of chromosomal integrity in vertebrate cells. *EMBO J.* **17:** 5497–5508.

Takeda T., Inaba H., Yamazaki M., Kyo S., Miyamoto T., Suzuki S., Ehara T., Kakizawa T., Hara M., DeGroot L.J., and Hashizume K. 2003. Tumor-specific gene therapy for undifferentiated thyroid carcinoma utilizing the telomerase reverse transcriptase promoter. *J. Clin. Endocrinol. Metab.* **88:** 3531–3538.

Tanaka K., Nishide J., Okazaki K., Kato H., Niwa O., Nakagawa T., Matsuda H., Kawamukai M., and Murakami Y. 1999. Characterization of a fission yeast SUMO-1 homologue, Pmt3p, required for multiple nuclear events, including the control of telomere length and chromosome segregation. *Mol. Cell. Biol.* **19:** 8660–8672.

Tarsounas M., Munoz P., Claas A., Smiraldo P.G., Pittman D.L., Blasco M.A., and West S.C. 2004. Telomere maintenance requires the RAD51D recombination/repair protein. *Cell* **117:** 337–347.

Teng S.C. and Zakian V.A. 1999. Telomere-telomere recombination is an efficient bypass pathway for telomere maintenance in *Saccharomyces cerevisiae*. *Mol. Cell. Biol.* **19:** 8083–8093.

Teng S.C., Chang J., McCowan B., and Zakian V.A. 2000. Telomerase-independent lengthening of yeast telomeres occurs by an abrupt Rad50p-dependent, Rif-inhibited recombinational process. *Mol. Cell* **6:** 947–952.

Tokutake Y., Matsumoto T., Watanabe T., Maeda S., Tahara H., Sakamoto S., Niida H., Sugimoto M., Ide T., and Furuichi Y. 1998. Extra-chromosome telomere repeat DNA in telomerase-negative immortalized cell lines. *Biochem. Biophys. Res. Commun.* **247:** 765–772.

Tomaska L., McEachern M.J., and Nosek J. 2004. Alternatives to telomerase: Keeping linear chromosomes via telomeric circles. *FEBS Lett.* **567:** 142–146.

Tsutsui T., Kumakura S., Tamura Y., Tsutsui T.W., Sekiguchi M., Higuchi T., and Barrett J.C. 2003. Immortal, telomerase-negative cell lines derived from a Li-Fraumeni syndrome patient exhibit telomere length variability and chromosomal and minisatellite instabilities. *Carcinogenesis* **24:** 953–965.

Ulaner G.A., Hoffman A.R., Otero J., Huang H.Y., Zhao Z., Mazumdar M., Gorlick R., Meyers P., Healey J.H., and Ladanyi M. 2004. Divergent patterns of telomere maintenance mechanisms among human sarcomas: Sharply contrasting prevalence of the alternative lengthening of telomeres mechanism in Ewing's sarcomas and osteosarcomas. *Genes Chromosomes Cancer* **41:** 155–162.

Ulaner G.A., Huang H.Y., Otero J., Zhao Z., Ben-Porat L., Satagopan J.M., Gorlick R., Meyers P., Healey J.H., Huvos A.G., Hoffman A.R., and Ladanyi M. 2003. Absence of a telomere maintenance mechanism as a favorable prognostic factor in patients with osteosarcoma. *Cancer Res.* **63:** 1759–1763.

Underwood D.H., Carroll C., and McEachern M.J. 2004. Genetic dissection of the *Kluyveromyces lactis* telomere and evidence for telomere capping defects in TER1 mutants with long telomeres. *Eukaryot. Cell* **3:** 369–384.

Varley H., Pickett H.A., Foxon J.L., Reddel R.R., and Royle N.J. 2002. Molecular characterization of inter-telomere and intra-telomere mutations in human ALT cells. *Nat. Genet.* **30:** 301–305.

Vogt M., Haggblom C., Yeargin J., Christiansen-Weber T., and Haas M. 1998. Independent induction of senescence by $p16^{INK4a}$ and $p21^{CIP1}$ in spontaneously immortalized human fibroblasts. *Cell Growth Differ.* **9:** 139–146.

Wang R.C., Smogorzewska A., and de Lange T. 2004. Homologous recombination generates T-loop-sized deletions at human telomeres. *Cell* **119:** 355–368.

Wen J., Cong Y.S., and Bacchetti S. 1998. Reconstitution of wild-type or mutant telomerase activity in telomerase negative immortal human cells. *Hum. Mol. Genet.* **7:** 1137–1141.

Wright W.E., Tesmer V.M., Huffman K.E., Levene S.D., and Shay J.W. 1997. Normal human chromosomes have long G-rich telomeric overhangs at one end. *Genes Dev.* **11:** 2801–2809.

Wu G., Lee W.H., and Chen P.L. 2000. NBS1 and TRF1 colocalize at promyelocytic leukemia bodies during late S/G_2 phases in immortalized telomerase-negative cells: Implication of NBS1 in alternative lengthening of telomeres. *J. Biol. Chem.* **275:** 30618–30622.

Wu G., Jiang X., Lee W.H., and Chen P.L. 2003. Assembly of functional ALT-associated promyelocytic leukemia bodies requires Nijmegen breakage syndrome 1. *Cancer Res.* **63:** 2589–2595.

Yankiwski V., Marciniak R.A., Guarente L., and Neff N.F. 2000. Nuclear structure in normal and Bloom syndrome cells. *Proc. Natl. Acad. Sci.* **97:** 5214–5219.

Yasumoto S., Kunimura C., Kikuchi K., Tahara H., Ohji H., Yamamoto H., Ide T., and Utakoji T. 1996. Telomerase activity in normal human epithelial cells. *Oncogene* **13:** 433–439.

Yeager T.R., Neumann A.A., Englezou A., Huschtscha L.I., Noble J.R., and Reddel R.R. 1999. Telomerase-negative immortalized human cells contain a novel type of promyelocytic leukemia (PML) body. *Cancer Res.* **59:** 4175–4179.

Yeh E.T.H., Gong L., and Kamitani T. 2000. Ubiquitin-like proteins: New wines in new bottles. *Gene* **248:** 1–14.

Zhong S., Salomoni P., and Pandolfi P.P. 2000. The transcriptional role of PML and the nuclear body. *Nat. Cell Biol.* **2:** E85–E90.

Zhu X.D., Kuster B., Mann M., Petrini J.H.J., and de Lange T. 2000. Cell-cycle-regulated association of RAD50/MRE11/NBS1 with TRF2 and human telomeres. *Nat. Genet.* **25:** 347–352.

Zhu X.D., Niedernhofer L., Kuster B., Mann M., Hoeijmakers J.H., and de Lange T. 2003. ERCC1/XPF removes the 3′ overhang from uncapped telomeres and represses formation of telomeric DNA-containing double minute chromosomes. *Mol. Cell* **12:** 1489–1498.

Zou Y., Yi X., Wright W.E., and Shay J.W. 2002. Human telomerase can immortalize Indian muntjac cells. *Exp. Cell Res.* **281:** 63–76.

8

Telomerase-independent Telomere Maintenance in Yeast

Michael J. McEachern
Department of Genetics
Fred Davidson Life Sciences Complex
University of Georgia
Athens, Georgia 30602

James E. Haber
Department of Biology and Rosenstiel Center
Brandeis University
Waltham, Massachusetts 02454–9110

Prior to identification of telomerase, recombination was considered a possible mechanism for telomere maintenance (see Chapter 1). Although normal telomeres are generally thought to be refractory to recombination, homologous recombination can provide a mechanism for telomere maintenance when telomerase is absent and telomeres become progressively shorter. Here, we review the mechanisms of telomerase-independent telomere maintenance in budding yeasts with particular emphasis on *Saccharomyces cerevisiae* and *Kluyveromyces lactis*. Other chapters in this book discuss telomerase-independent telomere maintenance in fission yeast, mammals, and *Drosophila*. Work on *S. cerevisiae* has revealed two major recombination-based pathways that produce rare survivors among senescent cells lacking telomerase. These two *RAD52*-dependent pathways differ in the sequences they add to chromosome ends and in the proteins that participate in the elongation process. Both pathways appear to be closely related to break-induced replication (BIR), a recombination pathway that can repair chromosomal double-strand breaks (DSBs). In addition, an alternative, apparently recombination-independent pathway has been described in which chromosome ends are maintained without any canonical telomeric sequences. Work on *K. lactis* has provided additional insights concerning recombinational telomere elongation (RTE), suggesting a major role for rolling-circle replication between shortened telomere ends and extrachromosomal telomeric circles, followed by recombination-mediated spreading of sequences from one telomere to all others.

RECOMBINATION-MEDIATED TELOMERE MAINTENANCE IN *S. CEREVISIAE*

Telomeres in *S. cerevisiae* and *K. lactis* are normally a few hundred base pairs in length and shorten at rates of 3–5 bp per cell division in the absence of telomerase. Mutants missing one or more of the telomerase components display a characteristic slow decline in the ability to grow over 20–50 generations that eventually results in the failure of most cells to divide when telomeres are less than approximately 100 bp, but from these senescing cells, rare survivors arise (Lundblad and Szostak 1989; Singer and Gottschling 1994; Lendvay et al. 1996). The formation of survivors almost always depends on homologous recombination, as all or virtually all survivors are eliminated when the central recombination gene *RAD52* is deleted. The frequency of formation of survivors has not been possible to determine accurately. This stems from a number of factors including the variability in the timing of survivor formation, the variable and often continuously changing growth rates of both senescent cells and survivor cells, and the multiple recombination events needed to produce a survivor. Few lineages of senescing cells passaged either by serial dilution in liquid or by serial restreaking on plates will completely die out. Yet some senescent yeast colonies of approximately 1 mm in size (equivalent to 10^5–10^6 cells) produce no survivors. This suggests that the rate of survivor formation may at times be very low.

There are two prominent *RAD52*-dependent pathways that yield survivors in *S. cerevisiae*; they are distinguished both by their genetic requirements and by the structures of the telomeres themselves. In one pathway originally described by Lundblad and Blackburn (1993)—now called Type I events—essentially all telomeric ends are extended by the acquisition or amplification of subtelomeric Y′ sequences that preexist at some chromosome ends and neighboring short tracts of telomeric repeat DNA (Fig. 1A,B). A second pathway, also initially described by Lundblad and Blackburn (1993) and subsequently characterized in detail by Teng and Zakian (1999), results in the extensive elongation of telomere sequences themselves, with few alterations of subtelomeric sequences

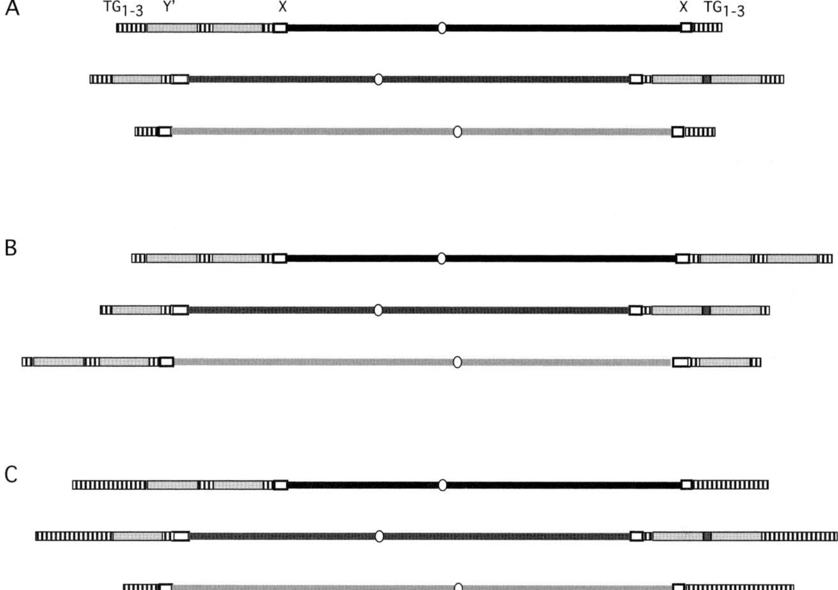

Figure 1. Two types of survivors in budding yeast lacking telomerase. (*A*) Structure of *S. cerevisiae* telomeres. TG_{1-3} telomere sequences are appended either to subtelomeric X or Y′ sequences at each end. Some ends have multiple Y′ sequences, many of which are separated by short TG_{1-3} tracts. Intrachromosomal recombination between Y′ elements can produce extrachromosomal Y′ circles, which are autonomously replicating. (*B*) Type I survivors lacking telomerase arise in a Rad52- and Rad51-dependent process in which Y′ elements, often multiple copies, are proliferated to most chromosome ends. (*C*) Type II survivors arise from a Rad52-dependent, Rad51-independent process in which telomere sequences themselves are elongated.

(Fig. 1C). These two classes of survivors are each considered in detail below.

Both types of survivors appear to make use of two alternative homologous recombination pathways known as Rad51-dependent and Rad51-independent BIR (Dunn et al. 1984; Malkova et al. 1996, 2005; Bosco and Haber 1998; Davis and Symington 2004). BIR appears to be fundamentally important not only in the restarting of stalled and broken replication forks (Haber 1999; Michel 2000; Michel et al. 2004), but also in cases where only one of two ends of the DSB succeeds in strand invasion of a homologous sequence. In BIR, recombination-mediated strand invasion of the broken end into a template sequence can set up a repair replication fork that can proceed all the way to a chromosome end (Fig. 2).

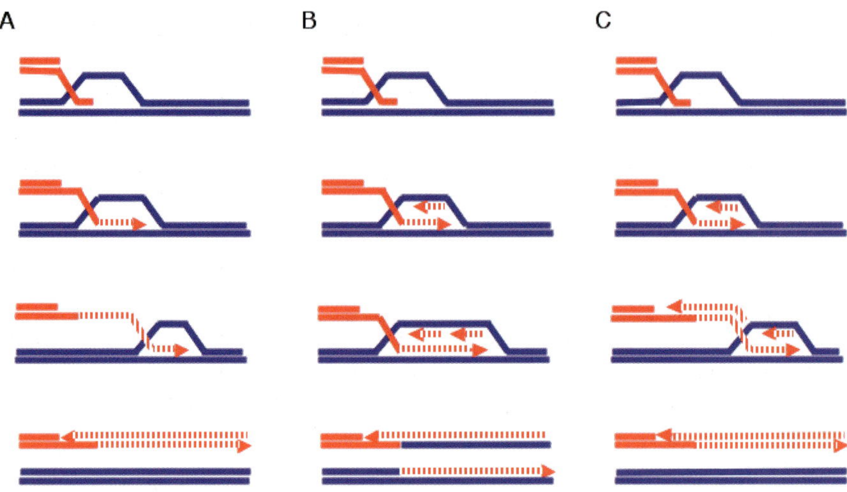

Figure 2. Possible mechanisms of BIR. (*A*) Strand invasion and primer extension lead to the synthesis of a single new strand that is later converted into a double-strand product. (*B*) Strand invasion sets up a unidirectional replication fork that proceeds to the end to the template chromosome. Resolution of a Holliday junction leaves two semiconservative replication products. Note that this process would be accompanied by an apparent sister-chromatid exchange. (*C*) Branch migration of the Holliday junction will displace two newly synthesized strands.

THE TYPE I PATHWAY FOR RECOMBINATION-MEDIATED TELOMERE MAINTENANCE

In the Type I survivors pathway, two families of Y′ elements (Y′ long and Y′ short) can each become amplified, although typically to different extents, within a given survivor. The extent of Y′ amplification is highly variable but can be greater than 100-fold (Lundblad and Blackburn 1993). The apparent movements of Y′ elements, which harbor some features of a degenerate transposable element, are best explained as a recombination event between a deprotected telomere end and similar TG_{1-3} sequences that can occur between Y′ elements (Fig. 1B) (Louis et al. 1994). In other cases, recombination may be between Y′ sequences themselves. As discussed further below, the template for the proliferation of Y′ sequences could be a circular, autonomously replicating form of Y′ that also carries a telomeric repeat tract sequence.

The appearance of Type I survivors depends on the "canonical" homologous recombination proteins, Rad51, Rad54, Rad55, and Rad57 (Le et al. 1999; Teng and Zakian 1999). Rad51 is budding yeast's homolog of the bacterial strand exchange protein, RecA, and is presumed to be the principal agent in promoting the strand invasion of an eroding telomere into donor sites. Rad51-mediated recombination is facilitated by Rad55 and Rad57, Rad51 paralogs that form a heterodimer, and by Rad54, a member of the family of Swi2/Snf2 chromatin remodeling factors (Symington 2002). A notable feature of Type I survivors is that a characteristic terminal restriction fragment (a 1.3-kb *Xho*I fragment) is still present and carries TG_{1-3} sequences. These telomeric repeat arrays are maintained at sizes substantially below those of wild-type cells. This short telomere size may contribute to the fact that Type I survivors are slower growing than wild-type cells or Type II survivors.

How the terminal TG_{1-3} sequences of Type I survivors are maintained remains unanswered. One possibility is that strand invasion of a short telomere into a longer telomere followed by DNA synthesis could lead to an extended telomere longer than either parental molecule. It is likely that restoration and maintenance of telomeric tracts at lengths above a critically short size are more important to the emergence of the improved growth of Type I survivors than is the amplification of Y' sequences. The Y' amplification may simply be a by-product of BIR events whose primary value to the cell is achieving modest elongation of the terminal telomeric repeat tracts. Alternatively, proliferation of Y' elements at chromosome ends might provide some selective advantage. By homogenizing subtelomeric and (possibly) telomeric sequences at chromosome ends, recombination at telomeres may become enhanced. Another possibility is that the helicase protein encoded by Y' elements and strongly induced in Type I survivors might in some way promote survivor formation (Yamada et al. 1998). Y' elements might even contribute directly to end protection, much as the HeT-A and TART transposons protect *Drosophila* telomeres without any telomerase (Biessmann and Mason 1997; Chapter 14). Perhaps such a selective advantage contributes to the improved growth of Type I survivors relative to senescent cells with similarly short terminal telomeric tracts.

Chromosomes of Type I survivors examined on pulsed-field gels are longer than wild-type chromosomes and exhibit a smeared appearance (Liti and Louis 2003). Some of the increased size presumably stems from the sizable amount of extra DNA, but it would appear that most chromosomes do not enter the pulsed-field gel at all. Although Y' elements could add as much as approximately 100 kb per chromosome if randomly distributed among 16 chromosomes (Lundblad and Blackburn 1993), this addition cannot account for the failure of Type I chromosomes to enter the gel. The smeared appearance of Type I chromosomes may reflect ongoing instability in the Y' arrays in these survivors, perhaps brought on by the chronic presence of short dysfunctional terminal telomeric tracts.

Alternatively, the end structures in Type I chromosomes may be more complex than illustrated in Figure 1B. Analysis of *Not*I digests of Type I survivor chromosomes on pulsed-field gels indicates that their terminal fragments have an altered structure, as they do not enter the gels as efficiently as the internal fragments (E.J. Louis, pers. comm.).

THE TYPE II PATHWAY FOR RECOMBINATION-MEDIATED TELOMERE MAINTENANCE

A second Rad52-dependent pathway for telomere maintenance, called the Type II pathway, leads to the substantial elongation of the telomeric sequences themselves (Fig. 1C) (Lundblad and Blackburn 1993; McEachern and Blackburn 1996; Teng and Zakian 1999; Teng et al. 2000). Although highly variable, this telomere elongation can result in telomeres being extended by 10 kb or more of G-rich repeat sequences in *S. cerevisiae*. Type II survivors arise less frequently than Type I survivors; however, their faster rate of growth leads them to overgrow Type I survivors in liquid cultures.

Direct evidence that recombination events that form Type II survivors involve telomeres that still retain some telomeric repeats came from work of Lingner and co-workers (Teixeira et al. 2004). Cloned telomeres from senescing *S. cerevisiae* cells were examined for changes from the starting sequence. The irregularity of *S. cerevisiae* telomeric sequences allows any new additions by telomerase or recombination to be readily identified. These authors found telomeres from 43 to 226 bp that had acquired up to 179 bp of new telomeric sequence at their termini in the absence of telomerase. These data could suggest that BIR events that copy sequence from one telomere to another occur frequently in senescing cells and typically generate only modest elongation of telomeres. Interestingly, not all arrays of Rap1-binding sites are capable of serving as a telomeric terminus despite being able to negatively regulate telomere length (Grossi et al. 2001). Such synthetic Rap1 arrays always became "capped" with about 100 bp of genuine telomeric repeats, presumably the minimal size for a functional protective cap.

In *S. cerevisiae*, the Type II pathway depends on another set of "Rad" proteins: the Mre11-Rad50-Xrs2 (MRX) complex and Rad59 (Le et al. 1999; Teng and Zakian 1999; Teng et al. 2000; Chen et al. 2001; Tsukamoto et al. 2001). As discussed below, these genetic requirements are generally reflected in BIR repair induced at other chromosome locations. As predicted from the individual analyses, nearly all survivors are eliminated in a *rad51Δ rad50Δ* double mutant (Le et al. 1999; Teng and Zakian 1999;

Teng et al. 2000; Chen et al. 2001; Tsukamoto et al. 2001), although *rad52Δ* was more defective. The MRX complex and Rad59 both have strand-annealing activity in vitro, as does Rad52 (for review, see Symington 2002), but it remains unclear how strand invasion is accomplished without Rad51. A recent paper from Shinohara's lab (Tsukamoto et al. 2003) showed that a carboxy-terminal truncation of Rad52, lacking its Rad51-interaction domain, was still competent to promote Type II survivors. One possibility is that the telomere ends are opened up by a helicase; indeed, the Sgs1 helicase, related to human BLM and WRN helicases, is required for Type II events (Watt et al. 1996; Cohen and Sinclair 2001; Huang et al. 2001; Johnson et al. 2001). An alternative possibility for the role of Sgs1 is that it is involved in removing G quadruplexes or other structures from single stranded 3′ telomeric tails formed after exonucleolytic processing of telomeric ends.

The two PI-3-kinase-related protein kinases (PIKKs), Tel1 and Mec1, are also important to the formation of Type II survivors. Single mutants in genes for either of these proteins have reduced ability to form Type II survivors, and *tel1 mec1* double mutants can only form Type I survivors (Tsai et al. 2002). How these PIKKs regulate Type II repair events is not known, but it is possible that DNA-damage-induced cell cycle arrest and other checkpoint responses are more critical without the proliferation of long, nonessential Y′ sequences at chromosome ends. Interestingly, a *tel1* mutation partially suppresses the senescence of a telomerase deletion mutant (Ritchie et al. 1999). Tel1 has been suggested to act at least in part via MRX both in telomere maintenance and in the DNA damage response, because Xrs2 protein is required to recruit Tel1 to sites of both DNA damage and telomeres (Ritchie and Petes 2000; Usui et al. 2001). MRX recruitment to telomeres and DSB ends is Tel1 independent (Shima et al. 2005; Takata et al. 2005; Tsukamoto et al. 2005).

An important point that should be kept in mind is that although investigators routinely label *S. cerevisiae* survivor clones as being either Type I or Type II, this division is not always clear-cut. Survivors labeled as Type I based on their lack of elongated telomeric repeat arrays do not always show extensive Y′ amplification. In addition, survivors labeled as Type II may have some telomeres that are still short. Furthermore, the long telomeric repeat arrays of Type II survivors may form in Type I survivors that still retain amplified Y′ elements (Teng and Zakian 1999). Thus, *S. cerevisiae* survivors retaining both *RAD50* and *RAD51* pathways available to them may have histories of using both pathways.

It should be stressed that postsenescence survivors in both *S. cerevisiae* and *K. lactis* are unstable. Over time, their telomeres continue to shorten and may initiate further rounds of recombination-dependent repair and recombinational telomere elongation. This is particularly obvious with Type II survivors, the long telomeres of which routinely shorten gradually without becoming lengthened until they first again become very short (McEachern and Blackburn 1996; Teng et al. 2000). If all telomeres in a survivor again become critically short at the same time, the cells may undergo a secondary round of severe growth senescence and survivor formation. However, most survivors do not exhibit secondary rounds of senescence as severe as the initial senescence. In some cases, this could reflect the presence of suppressor mutations. In most cases, it likely reflects the heterogeneity of telomere size. *K. lactis* telomerase deletion mutants that begin senescence with telomeres both shorter and more uniform in size than those of wild-type cells undergo senescence not only more rapidly, but also more uniformly and more severely (McEachern 2002). Additionally, as discussed more fully below, the presence of even a single long telomere in a telomerase deletion mutant can substantially reduce the severity of senescence (Topcu et al. 2005).

RELATIONSHIP OF Rad51-DEPENDENT AND Rad51-INDEPENDENT TELOMERE MAINTENANCE TO BIR INVOLVING NONTELOMERIC SEQUENCES

The genetic requirements of BIR have been determined by examining diploids in which there is a single HO-endonuclease-induced DSB in the middle of the right arm of chromosome III. Normally, such a DSB would be repaired by "short-patch" gene conversion. A *rad52Δ* diploid shows almost no repair of the broken chromosome; it is simply lost, creating a strain derivative lacking that homolog of the chromosome. But surprisingly, a *rad51Δ* strain eliminates gene conversions but still allows BIR to proceed (Malkova et al. 1996). Signon et al. (2001) showed that a similar phenotype is found in *rad54Δ*, *rad55Δ*, and *rad57Δ* mutants, all of which eliminate gene conversions but allow BIR. Further genetic analysis of this *RAD51*-independent BIR pathway revealed that it is largely dependent on another set of recombination genes: *RAD50*, *MRE11*, *XRS2*, *RAD59*, and *TID1* (*RDH54*) (Signon et al. 2001). To examine *RAD51*-dependent BIR, Malkova et al. (2005) created a modified diploid in which there is little homology distal to the DSB on the template chromosome. *RAD51*-mediated BIR is significantly more efficient than what is seen in the absence of *RAD51*; a large proportion of the colonies appear to reflect repair events that occurred within the first cell cycle. A version of BIR has also been studied on a plasmid that contains inverted repeated sequences, one of which is interrupted by an HO endonuclease cleavage site (Ira and Haber 2002). When the homology on both sides of the cleavage site is substantial (several hundred base pairs), recombination proceeds predominantly by Rad51-mediated gene conversion. However, when homology is reduced to 70–100 bp on either side, Rad51-mediated gene conversions are severely impaired; it appears that to initiate gene conversion, Rad51 requires about 100 bp of homology. With very short homology, there is instead a Rad51-independent repair pathway that appears to involve first BIR and then single-strand annealing. The ability of a *RAD51*-independent, *RAD52*- and *RAD50*-dependent pathway to initiate BIR at very short regions of homology is consistent with the origin of Type II recombinants. Because *S. cerevisiae* telomere sequences are degenerate, one would not expect there to be long perfectly matched sequence between a single-stranded resected telomere end and another telomere sequence. This may also explain how mutations in mismatch repair genes facilitate survivor formation (Rizki and Lundblad 2001). By this same argument, however, one would expect a deletion of *SGS1* also to promote Type II survivors, as this helicase has been shown to prevent homologous recombination (Myung et al. 2001; Sugawara et al. 2004); but in fact, *sgs1*Δ strains are unable to carry out Type II repair. This indicates that Sgs1 has a more important role in facilitating Type II recombination.

TELOMERE MAINTENANCE IN *S. CEREVISIAE* BY ROLL AND SPREAD?

As detailed below, work on telomerase mutants in *K. lactis* has provided strong evidence for a roll-and-spread model for recombinational telomere maintenance. In this model, a tiny telomeric circle (t-circle) is used as a template to first generate the first elongated telomere and later BIR events copying this sequence then elongate all other telomeres. The extent to which rolling-circle synthesis or spreading of sequence from a single telomere source occurs during formation of Type I or Type II survivors of *S. cerevisiae* remains unclear at this point. The abrupt emergence of Type II survivors with highly elongated telomeric repeat tracts without obvious intermediate forms led to the suggestion that a rolling-circle mechanism might produce *S. cerevisiae* Type II survivors (Teng et al. 2000). Certainly, the spread of the first elongated telomeric sequence to many or even all other telomeres could potentially explain why Type I and Type II survivors tend to have similarly structured telomeres within a given survivor clone, yet are so different from one another. If formation of Type I and Type II arrays was independent at each telomere, almost all survivors would probably be expected to contain both kinds of arrays among the 32 telomeres in a haploid *S. cerevisiae* cell. Large circles of DNA containing subtelomeric Y′ elements have long been recognized in wild-type *S. cerevisiae* cells (Horowitz and Haber 1985). Such circles conceivably could be copied to generate Type I-type arrays. However, *S. cerevisiae* cells typically contain two size classes of subtelomeric Y′ elements and both of these classes become amplified, to differing extents, during formation of Type I survivors (Lundblad and Blackburn 1993). This indicates that there is not a single source of amplified Y′ elements in most Type I survivors. It would be useful to determine if amplified copies of the two types of Y′ elements are interspersed together or if they exist primarily in uniform tandem arrays. This could provide clues to the mechanism of amplification.

Very recent work by Teng and colleagues has led them to suggest that a roll-and-spread mechanism is responsible for at least the Type II survivors of *S. cerevisiae* telomerase deletion mutants. These authors showed that circles of DNA containing a telomeric repeat tract and a Kan^r determinant that were produced by Cre-Lox excision will promote formation of tandem arrays of the telomere-Kan^r sequence at telomeres (Lin et al. 2005). In the transformation assay used, *rad50* and *rad52* mutations greatly reduced array formation, but a *rad51* mutation did not. Wild-type cells also readily formed arrays at a rate only two- to threefold less than a telomerase mutant. It was also shown that excision of Y′ circles became greatly elevated at the point when survivor formation was occurring (Lin et al. 2005). This elevated rate of circle formation was *RAD50* dependent. However, the survivors in these experiments were Type II and thus could not be directly derived from the Y′ circles. These results are consistent with the possibility that the requirement for Rad50 in Type II survivor formation stems from the need for the MRX complex to form t-circles and/or utilize them as templates for rolling-circle BIR events. Strikingly, a recent paper on the formation of t-circles in mammalian cells by telomere rapid deletion (Wang et al. 2004) found that circle formation by recombination was prevented in cells lacking Nbs1 (the Xrs2 substitute in mammalian cells, which complexes with Mre11 and Rad50).

RECOMBINATIONAL TELOMERE ELONGATION IN *K. LACTIS*

In *K. lactis*, survivor formation is simpler in nature than in *S. cerevisiae* as only Type II survivors are observed (McEachern and Blackburn 1996). The lack of Type-I-like survivors stems from the lack of subtelomeric blocks of homologous telomere sequence such as those that are present in *S. cerevisiae*. Artificial generation of *K. lactis* cells that have a single chromosome end with an extra block of telomeric repeats inserted in the subtelomeric sequence can lead to formation of telomeric tandem arrays of the sequence unit between the original two blocks of telomeric repeats (Natarajan et al. 2003). This arrangement, which rapidly spreads to multiple telomeres, is highly reminiscent of Type I survivors. The Type II mechanism of *K. lactis* (referred to as RTE, for recombinational telomere elongation), although generally similar to that of *S. cerevisiae*, does appear to differ in some ways. The extent of telomere elongation is generally less, seldom more than 2 kb and often much less. The shorter telomeres of *K. lactis* survivors probably contribute to their being less stable in their growth rate than seems to often be the case with *S. cerevisiae* Type II survivors. The genetic requirements for RTE in *K. lactis*, although minimally characterized to date, appear not to be identical to those of *S. cerevisiae*. *RAD59*, the only gene other than *RAD52* studied thus far, affects, but is not required for, *K. lactis* survivor formation (S. Iyer and M. McEachern, unpubl.). Another interesting difference from *S. cerevisiae* is that *RAD52* itself is not absolutely required for RTE. Mutants lacking telomerase and *RAD52* are greatly reduced in their ability to generate survivors; however, a few survivors with modestly elongated telomeres can be found (McEachern and Blackburn 1996). This could stem from the simple fact that *K. lactis* cells have fewer telomeres than *S. cerevisiae* (12 vs. 32) cells, or it may represent yet another pathway for survivor formation through RTE. It will be interesting to learn whether any of the differences from *S. cerevisiae* stem from the different types of telomeric repeats in the two yeasts (highly heterogeneous in *S. cerevisiae* vs. highly uniform 25-bp repeats in *K. lactis*).

Important insights into the recombination events that generate RTE have come from experiments using *K. lactis* cells with mutationally tagged telomeres. The long (25-bp) perfect telomeric repeats of *K. lactis* have permitted the identification of phenotypically silent mutations that create restriction sites within each telomeric repeat. The mutation known as "Bcl," a single-base change that creates a *Bcl*I restriction site, has been particularly useful (Roy et al. 1998; McEachern et al. 2002). Replacement of the wild-type telomerase RNA gene (*TER1*) with the *TER1–7C(Bcl)* allele leads to the production of a telomerase with a mutant template that synthesizes only Bcl repeats. *TER1–7C(Bcl)* cells therefore have telomeres that terminate with Bcl repeats but retain wild-type repeats in the more internal region of the telomeres.

Deletion of *TER1–7C(Bcl)* from cells with telomeres that retained only a small number of basal wild-type repeats in each telomere led to the production of senescent cells with very short telomeres that sometimes retained Bcl repeats at their termini (Natarajan and McEachern 2002). These cells produced postsenescent survivors in a manner indistinguishable from *ter1Δ* cells containing only wild-type repeats. A crucial result was that survivors that retained any Bcl repeats had telomeres with repeating patterns of Bcl and wild-type repeats. Moreover, the same pattern appeared to be present in most or all telomeres of a particular survivor clone but varied between different independent survivor clones. This startling observation led to what is now called the roll-and-spread model (Fig. 3). According to this model, the rampant recombination at the critically short telomeres of senescing cells occasionally produces a tiny circle of telomeric DNA. Somehow, this circle is used as a template for rolling-circle DNA synthesis that generates a long tract of telomeric DNA in essentially a single step. Once one chromosome end acquired the elongated telomeric tract, all other telomeres in the cell could acquire copies of the elongated sequence through intertelomeric BIR events. Rolling-circle replication around a telomeric circle composed of both wild-type and Bcl repeats would produce a repeating pattern matching the sequence of the circle. As the sequence patterns in telomeres of post-senescence survivors could have a periodicity at least as small as 100 bp (Natarajan and McEachern 2002), this predicted that circles of the same tiny size could be produced and copied in *ter1Δ* cells.

A number of predictions of the roll-and-spread model have now been confirmed. Circles of DNA containing telomeric repeats have been shown to promote telomere elongation; 1.6-kb circles composed of a cloned *K. lactis* telomere and an *S. cerevisiae URA3* gene were found to lead to long tandem arrays of the *URA3*-telomere unit being added onto chromosome ends after transformation into *K. lactis*. In transformants of cells with short recombination-prone telomeres (including *ter1Δ*), the *URA3*-telomere arrays were invariably present at multiple telomeres as soon as cells could be examined (~25–30 cell divisions after transformation) (Natarajan and McEachern 2002). In contrast, transformants of wild-type *K. lactis* typically had a *URA3*-telomere array at only a single telomere. These results suggested that even wild-type cells could undergo a rolling-circle event but that the short telomere mutants underwent more frequent recombination events that copied the *URA3*-telomere arrays onto many other telomeres. Consistent with this interpretation, transformations with mixtures of circles differing by only a single restriction site showed that all copies of the *URA3*-telomere unit in a cell, even when present at multiple telomeres, were derived from a single transforming molecule.

Other experiments have shown that much smaller circles can also promote telomere elongation in *K. lactis* (Natarajan et al. 2003). Circles

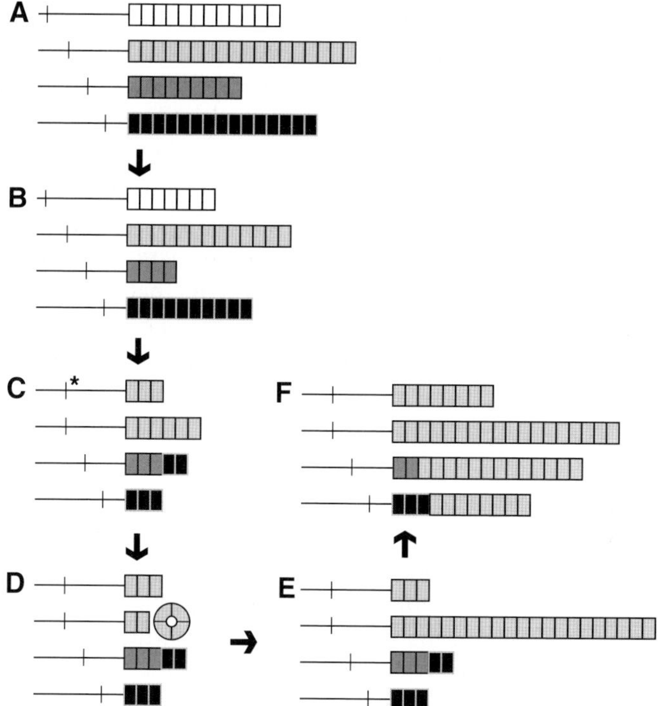

Figure 3. The roll-and-spread model. (*A*) Four telomeres are depicted at the point when telomerase is deleted; boxes represent telomeric repeats. Different shades are shown to permit visualization of events where sequence from one telomere is copied onto another telomere. Short vertical lines indicate polymorphisms within homologous subtelomeric sequence (*horizontal line*). Size differences between telomeres reflect upper and lower size limits in the presence of telomerase. (*B*) Telomeres gradually shorten in the absence of telomerase. Size differences between telomeres continue to reflect those present at the point when telomerase was deleted. (*C*) As particular telomeres shorten to critically short sizes, they become prone to initiating recombination. In this way, a telomere still retaining a few telomeric repeats may acquire additional repeats by a BIR-like event. Alternatively, the recombination event may result in replacement of both subtelomeric and telomeric DNA with sequence from another telomere (resulting in subtelomeric gene conversion; *asterisk*). This type of telomere–telomere recombination can probably generate slightly longer telomeres and may be responsible for maintaining the short terminal telomeric repeat arrays in *S. cerevisiae* Type I survivors. (*D*) Formation of a tiny telomeric circle, possibly through an intratelomeric deletion (see Fig. 4). (*E*) Rolling-circle replication using the tiny telomeric circle as a template results in the creation of at least one appreciably elongated telomere. Such events may or may not be localized to the same telomere from which the circle was derived (see Fig. 4). (*F*) Sequence of the first elongated telomere is spread to most or all other telomeres through additional BIR-like events. The resulting survivor has all elongated telomeres containing terminal repeats derived from the sequence of the original tiny telomeric circle.

composed of a 75-nucleotide single-stranded region with three *K. lactis* telomeric repeats and a nontelomeric double-stranded region of 25 bp were constructed and cotransformed with an ARS (autonomous replication sequence) plasmid into *K. lactis* cells with short recombinogenic telomeres. About 1% of transformants had acquired short tandem arrays of sequence at their telomeres that were derived from the cotransformed minicircle. This demonstrated that telomeric circles at least as small as 100 nucleotides could be used to elongate telomeres. Additionally, the telomere elongation observed in the transformants that acquired sequence from the 100-nucleotide circle was just hundreds of base pairs, similar to the extent of elongation that occurs in *K. lactis ter1Δ* survivors.

Whether tiny t-circles are produced in senescing cells remains to be determined. No small extrachromosomal telomeric DNA is obvious in Southern blots of DNA from *ter1Δ* cells. As the formation of elongated telomeres (and survivors) is the exception rather than the rule for senescing cells, it is expected that t-circle formation will be rare in senescing cells. However, tiny t-circles prove to be very abundant in some *K. lactis* mutants with long dysfunctional telomeres (Underwood et al. 2004). A *ter1–16T* mutant (altering the Rap1-binding site of the telomeric repeat) contains small circles of extrachromosomal telomeric DNA that were either double stranded or single stranded and ranged down to about 100 bp or nucleotides in size (Groff-Vindman et al. 2005). The single-stranded t-circles were specifically of the G-rich telomeric strand. Models for t-circle formation and rolling-circle synthesis are shown in Figure 4. The extent to which recombination or t-circles contribute to the telomere elongation in the *ter1–16T* mutant is not known. Telomere elongation in *ter1–16T* can clearly occur independently of Rad52 function (Underwood et al. 2004); however, this does not preclude the possibility that recombination might also be involved in the lengthening.

The roll-and-spread model also predicted that most or all elongated telomeres in a postsenescence survivor arise from copying a single elongated telomere. Recent work has confirmed that such extensive sequence spreading can readily occur during survivor formation (McEachern and Iyer 2001; Topcu et al. 2005). *K. lactis* cells were created that contained a single telomere composed of the phenotypically silent Bcl repeats. Upon deletion of the wild-type *TER1* gene, about 10% of survivors had acquired Bcl repeats on all of their telomeres. The other 90% had not spread Bcl repeats to other telomeres. These results showed unequivocally that spreading of sequence from one to all telomeres during survivor formation can occur and that it is essentially an all-or-none phenomenon. The frequency of sequence spreading from the Bcl telomere was consistent

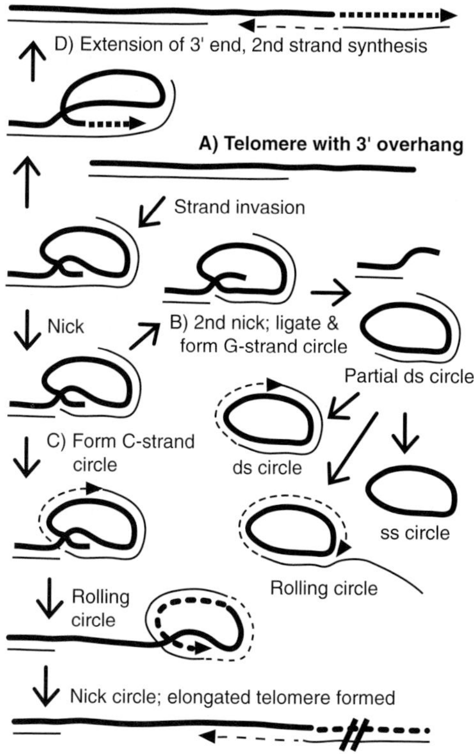

Figure 4. Two models showing formation of a small telomeric circle from a t-loop-like intermediate. (*A*) A telomere processed to have a 3′ overhang can undergo intramolecular strand invasion into more internal telomeric repeats sequence. G-rich and C-rich strands (G-strand and C-strand) are shown as *bold* and *thin* lines, respectively. (*B*) A nick in the C-strand followed by ligation of the newly cut 5′ end to the strand-invaded 3′ end leads to detachment of a G-strand circle with a partial C-strand. This could potentially be processed to be fully single-stranded, fully double-stranded, or become engaged in rolling-circle DNA synthesis. (*C*) In an alternative model, a nick in the C-strand is extended and ligated to generate a C-strand circle still base-paired to the telomere. The invaded 3′ end could directly engage in rolling-circle DNA synthesis that could generate substantial telomere elongation. After synthesis ends, the C-strand circle might detach from the telomere or, if nicked (shown), the circle could become part of the extended telomere. Unlike the first model where circles might be free to diffuse to be used by any telomere, this second model predicts that the first extended telomere will be the same telomere that produced the small telomeric circle. (*D*) Extension of the invaded 3′ end to generate a modestly elongated telomere without rolling-circle synthesis.

with 1 of the 12 *K. lactis* telomeres being randomly "chosen" as the source of sequence to be spread.

A hypothetical intermediate proposed by the roll-and-spread model is a cell with one relatively long telomere but otherwise having very short telomeres. Such a state is likely to be very transient and very difficult to directly identify during the normal course of senescence and survivor formation. However, it has been possible to artificially construct *K. lactis* cells with this telomere structure. A telomeric fragment composed of approximately 40 Bcl repeats was transformed into *K. lactis* cells where it replaced a single native telomere of shorter size (Topcu et al. 2005). In *ter1Δ* cells with one long Bcl telomere, spreading of the Bcl repeats to all other telomeres occurred at 95% frequency. Thus, the sequence from one long telomere is highly preferentially spread during survivor formation. Moreover, growth senescence was suppressed. At a time point when growth was normally very poor in ordinary *ter1Δ* cells, cells with the one long telomere grew appreciably better. This result suggests that the preexisting presence of a single long telomere bypassed the need for rolling-circle synthesis and led to the earlier formation of cells with all elongated telomeres.

A major question regarding RTE concerns: What governs the transition between a capped telomeric state resistant to initiating recombination and an uncapped (or partially uncapped) state capable of initiating recombination? It is clear that in cells with normally functioning telomeric proteins, this transition can occur when some telomeric repeats still remain at a chromosome end. By determining the number of wild-type repeats remaining in telomeres that had acquired Bcl repeats in the spreading experiments described above, it was concluded that once *K. lactis* telomeres dropped below approximately 100 bp in size, they became highly prone to initiating recombination with other telomeres (Topcu et al. 2005). The relatively sharp transition point between recombination-resistant and recombination-prone states that was inferred by these experiments suggests that the transition involves a particular structural change in the telomere such as loss of a t-loop or undoing of a folded structure.

ATYPICAL RECOMBINATION-MEDIATED TELOMERE MAINTENANCE

Telomere maintenance by recombination in yeast is not limited to mutants lacking genes for telomerase components. Certain other mutations also produce recombination-mediated survivors, and some of these have telomere lengths and growth properties more closely resembling those of human ALT cells than typical yeast telomerase deletion mutants. In an

S. cerevisiae tel1 mec1 double mutant, telomerase is detectable in vitro but very poorly able to add repeats onto chromosome ends in vivo. Consequently, *tel1 mec1* mutants undergo senescence and survivor formation that is similar to that seen in telomerase deletion mutants (Ritchie et al. 1999). However, the senescence of *tel1 mec1* mutants is slightly slower and can be suppressed by mutations increasing telomerase's ability to add repeats such as deletion of *RIF1* and *RIF2* (Chan et al. 2001). Similarly, *Schizosaccharomyces pombe* cells deleted for *tel1* and *mec1* equivalents also display senescence (Naito et al. 1998).

In *K. lactis*, a number of *ter1* mutations that cripple, but do not eliminate, in vivo telomerase activity also can cause RTE (M. McEachern, unpubl.). These mutants typically linger in a partially senescent state without producing the high rate of cell death seen in a telomerase deletion mutant. However, they sometimes form derivatives with moderately elongated telomeres. This elongation is clearly due to RTE as the elongated telomeres have interspersed wild-type and mutant repeats despite the absence of a wild-type telomerase, similar to that seen in telomerase deletion mutants (Natarajan and McEachern 2002) and supporting the idea that RTE in these cases occurs through the roll-and-spread mechanism.

Combining the *cdc13–1* allele with a *yku70Δ* mutation at the semi-permissive temperature of 25°C led immediately to a large cell and slow growth phenotype characteristic of late senescence of telomerase deletion mutants (Grandin and Charbonneau 2003). These effects were suppressed by a *mec3* mutation, consistent with the formation of single-stranded DNA and checkpoint-mediated arrest in the double mutant. The poor growth of *cdc13–1 yku70Δ* mutants persisted for many cell divisions but eventually gave way to faster-growing smaller cells. Telomere lengths in the poorly growing double mutants were no shorter than in *yku70Δ* single mutants, arguing that extremely short telomere length was not the source of the senescent-like growth. The fast-growing *cdc13–1 yku70Δ* survivors were found to have elongated telomeric repeat tracts similar in length but often more heterogeneous in size than those of a Type II survivor. The increased heterogeneity was suggested to be due to increased instability of the telomeres. *tlc1 yku70Δ* double mutants exhibited accelerated senescence (Nugent et al. 1998) that had Type II telomere structure as soon as cells from germinating spores could be examined (Grandin and Charbonneau 2003). These results suggested that the *yku70Δ* mutation specifically promoted the Type II survivor pathway. Both *cdc13–1* and *ykuΔ* mutations produce increased levels of single-stranded telomeric tails (Garvik et al. 1995; Gravel et al. 1998;

Polotnianka et al. 1998; Maringele and Lydall 2002), and the increased size of such tails might be expected to promote telomeric recombination. Surprisingly, the Type II pathway of *cdc13–1 yku70Δ* and *tlc1 yku70Δ* mutants was found to be more dependent on *RAD51* than on *RAD50/RAD59* (Grandin and Charbonneau 2003). These data clearly indicate that there is more flexibility in the pathways for generating Type II survivors than was previously appreciated.

What causes these variations in the Type II survivor pathway? Disrupting the function of Cdc13 and Ku, proteins known to protect telomeres from DNA damage responses, could be expected to lead to telomeres being able to initiate recombination at longer lengths than would otherwise be the case. Additionally, the longer the length of the telomeric repeat tracts that are initiating strand invasion, the more likely any resulting recombination event could involve Rad51. Rad51-dependent events require longer stretches of homology than Rad50-dependent events (Ira and Haber 2002). Longer telomeric tracts initiating recombination might also act to delay the formation of single-stranded DNA in subtelomeric Y′ elements. That potentially could reduce the likelihood of Type I survivors. Finally, the short terminal repeat tracts of Type I survivors might be expected to always be below the length threshold that achieves adequate telomere protection in cells with compromised telomere proteins. Therefore, even if Type I survivors did occur in such cells, they might not be capable of producing better growing survivors.

Another unusual case of RTE has been found to occur in *K. lactis* cells carrying a mutation in the *STN1* gene (Iyer et al. 2005). This mutation leads to a chronic growth condition that resembles a moderately senescent state as well as to extreme telomere elongation that occurs independently of telomerase. The telomeres in *stn1-M1* cells are very similar to those of human ALT cells, extending as a constant smear from very short sizes up to limit mobility in gels. The telomere elongation in this mutant was labeled Type IIR for the runaway elongation that distinguishes it from *K. lactis* telomerase deletion mutants with their much more modest elongation. It was suggested that the ability of telomeres in *stn1-M1* cells to initiate recombination is largely or entirely independent of telomere length. Thus, the regulation that effectively occurs in telomerase deletion mutants, whereby RTE is blocked once a telomere is elongated to a size above approximately 100 bp, appears to be lost in *stn1-M1* cells. Consistent with this, *stn1-M1* telomeres display many symptoms of severe capping defects, including long single-stranded tails, very high rates of subtelomeric gene conversion, and extreme levels of a telomeric rapid deletion (TRD)-like phenomenon upon reintroduction of a wild-type

STN1 gene (Iyer et al. 2005). It will be very interesting to determine whether the Type IIR RTE of *stn1-M1* cells involves either rolling-circle DNA synthesis or spreading of sequence from a single telomere source.

Another example of what could be called type IIR RTE has recently been described in *K. lactis*. Telomerase deletion mutants constructed to have telomeres composed largely of "Kpn" mutant repeats were found to produce highly elongated telomeres (Topcu et al. 2005). The two base changes that produce this mutation within each 25-bp telomeric repeat have previously been shown to disrupt the ability of telomeric repeats to negatively regulate telomerase-mediated telomere maintenance (McEachern and Blackburn 1995; Underwood et al. 2004). This suggests that some of the same telomeric features that regulate telomerase's access to the telomeric end are also crucial to block telomeres from initiating recombination.

SURVIVAL BY TERMINAL PALINDROME FORMATION

Recent work by Maringele and Lydall (2004) has demonstrated that *S. cerevisiae* cells under certain conditions can maintain chromosome ends in the absence of both telomerase and recombination. Ordinarily, *tlc1Δ rad52Δ* cells produce no detectable survivors; however, *tlc1Δ rad52Δ exo1Δ* cells exhibited a slow emergence of survivors. The *exo1Δ* mutation also permitted proliferation of *mre11Δ yku70Δ* mutants, cells that normally senesce rapidly because of defective telomere function. Interestingly, deletion of *MRE11*, whose nuclease contributes to the formation of 3' telomeric overhangs, greatly increased the number of survivors in a *tlc1Δ rad52Δ exo1Δ* mutant. In the absence of the *exo1Δ* mutation, however, *tlc1Δ rad52Δ mre11Δ* mutants produced no survivors. These results indicated that reducing the nucleolytic processing that occurs at telomeres with capping defects permitted proliferation of survivors that were independent of both telomerase and recombination.

Analysis of these survivors showed that they maintained linear but abnormally sized chromosomes that had generally lost telomeric and subtelomeric DNA as well as up to tens of kilobases of unique DNA from chromosomal termini, depending on the position of the first telomere-adjacent essential gene. These survivors often developed large inverted duplications (palindromes) at some chromosome ends. The palindromes appear to have developed from small naturally occurring inverted repeats. It was postulated that once a truncated chromosome end was processed to generate a single-stranded inverted repeat, that repeat could fold back on itself to form a hairpin and the 3' end could act as a primer to initiate

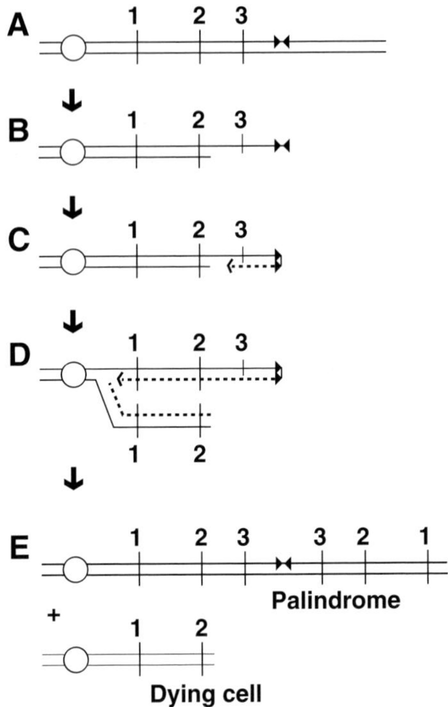

Figure 5. Suggested mechanism for generation of terminal palindromic sequences in PAL survivors (based on Maringele and Lydall 2004). (*A*) Both strands of one arm of a chromosome are shown with centromere (*circle*). Numbers indicate genetic markers and arrowheads indicate a short inverted repeat (shown on only one strand for clarity). (*B*) In the absence of telomerase, Rad52, and *Exo*I, loss of sequence from the end occurs. Degradation of the 5′ end produces 3′ single-stranded tails. (*C,D*) Single-stranded inverted repeat folds into hairpin with the 3′ end serving as a primer for DNA synthesis that copies much of the chromosome arm. (*E*) Replication of the hairpin end results in the formation of a large palindrome at the chromosome end in one daughter cell.

extensive DNA synthesis (Fig. 5). Upon normal DNA replication, the large hairpin structure would be converted into a large terminal palindrome. Terminal palindromes could be unstable and prone to further sequence amplification. By extending a chromosome end, the formation of a terminal palindrome would counteract the gradual loss of sequence that would otherwise eventually lead to the deletion of essential genes.

THE FLIP SIDE OF RECOMBINATIONAL TELOMERE MAINTENANCE: TELOMERE RAPID DELETIONS

Work in recent years has shown that recombination can occur in apparently normal yeast telomeres, sometimes at surprisingly high frequencies.

Lustig and colleagues have demonstrated that unusually elongated telomeres created in wild-type *S. cerevisiae* cells are subject to high rates of recombination events that tend to truncate the telomeres down to near the size of normal-length telomeres (Li and Lustig 1996). This phenomenon was named telomere rapid deletion (TRD). Through the use of a sequence tag embedded within telomeric sequences, TRD was shown to be due to intratelomeric recombination (Bucholc et al. 2001). TRD events are largely dependent on both Rad52 and MRX (the complex of Mre11, Rad50, and Xrs2) and have been hypothesized to be initiated from an invasion of the 3' end of the telomere into more internal regions of the same telomere (Bucholc et al. 2001; Lustig 2003). The resulting recombination event is proposed to result in a telomere shortened by a terminal deletion and possibly also the production of a telomeric circle derived from the deleted sequence. Interestingly, the rate of TRD is elevated 30–70-fold in meiosis, reaching levels that can exceed 20% per meiosis (Joseph et al. 2005). Deletion of Ndj1, a protein known to localize to yeast telomeres in meiosis and function in meiotic telomere bouquet formation, greatly reduces meiotic TRD and mildly reduces mitotic TRD.

Although TRD has thus far been studied only with telomeres that are longer than normal, it conceivably could also be relevant to cells with only normal-length telomeres and senescing cells lacking telomerase. The slow turnover of repeats at positions deep within functionally wild-type *K. lactis* telomeres was suggested to be due to occasional truncation of telomeres to very short sizes following addition of new sequences from telomerase or intertelomeric recombination (McEachern et al. 2002). In addition, as discussed above, telomeric circles produced by a TRD-like mechanism may be crucial to at least one recombination-dependent mechanism that lengthens telomeres in the absence of telomerase.

TELOMERASE-INDEPENDENT TELOMERE MAINTENANCE AND THE EVOLUTIONARY ORIGINS OF TELOMERES

The occurrence of alternative telomere maintenance pathways in yeasts and some other organisms has helped lead to suggestions that an alternative mechanism may have predated telomerase. de Lange (2004) has suggested that the t-loop structure of many modern telomeres may represent the ancestral telomeric form. This might have originated in the absence of telomerase and specific telomere-binding proteins but would have required telomeres composed of directly repeated sequences to permit t-loop formation. Telomere maintenance in this scenario would presumably come from some form of RTE. The repeated emergence of linear mitochondrial

DNAs in yeasts argues that RTE can still arise and be successful as a telomere maintenance mechanism (Nosek and Tomaska 2003).

Maringele and Lydall (2004) suggested that the PAL mechanism might be an ancestral pathway for telomere maintenance. Some early theories of telomere maintenance suggested roles for palindromes similar to what seems to occur in PAL survivors (Cavalier-Smith 1974; Bateman 1975). In addition, PAL-like mechanisms may replicate the ends of some bacterial chromosomes, organellar DNAs, and viruses (Ellis and Day 1986; Nosek et al. 1998; Qin and Cohen 2000; Cotmore and Tattersall 2003). It will be interesting to determine whether something resembling the PAL mechanism might also sometimes occur in human cancer cells.

CONCLUDING REMARKS

The study of yeast cells lacking telomerase has provided startling examples of the extremes that cells can go to when faced with critically shortened telomeres. Recent years have seen considerable progress in both determining the conditions where recombinational telomere maintenance arises and elucidating the mechanisms by which it occurs. This work will provide important directions for helping to understand not only ALT in mammalian cells, but also the general nature of telomeres and how they are capped to protect chromosome ends from repair processes such as homologous recombination.

ACKNOWLEDGMENTS

We thank David Lydall and Laura Maringele for comments on the manuscript.

REFERENCES

Bateman A.J. 1975. Letter: Simplification of palindromic telomere theory. *Nature* **253:** 379–380.

Biessmann H. and Mason J.M. 1997. Telomere maintenance without telomerase. *Chromosoma* **106:** 63–69.

Bosco G. and Haber J.E. 1998. Chromosome break-induced DNA replication leads to non-reciprocal translocations and telomere capture. *Genetics* **150:** 1037–1047.

Bucholc M., Park Y., and Lustig A.J. 2001. Intrachromatid excision of telomeric DNA as a mechanism for telomere size control in *Saccharomyces cerevisiae*. *Mol. Cell. Biol.* **21:** 6559–6573.

Cavalier-Smith T. 1974. Palindromic base sequences and replication of eukaryotic chromosome ends. *Nature* **250:** 267–470.

Chan S.W., Chang J., Prescott J., and Blackburn E.H. 2001. Altering telomere structure allows telomerase to act in yeast lacking ATM kinases. *Curr. Biol.* **11:** 1240–1250.

Chen Q., Ijpma A., and Greider C.W. 2001. Two survivor pathways that allow growth in the absence of telomerase are generated by distinct telomere recombination events. *Mol. Cell. Biol.* **21:** 1819–1827.

Cohen H. and Sinclair D.A. 2001. Recombination-mediated lengthening of terminal telomeric repeats requires the Sgs1 DNA helicase. *Proc. Natl. Acad. Sci.* **98:** 3174–3179.

Cotmore S.F. and Tattersall P. 2003. Resolution of parvovirus dimer junctions proceeds through a novel heterocruciform intermediate. *J. Virol.* **77:** 6245–6254.

Davis A.P. and Symington L.S. 2004. RAD51-dependent break-induced replication in yeast. *Mol. Cell. Biol.* **24:** 2344–2351.

de Lange T. 2004. T-loops and the origin of telomeres. *Nat. Rev. Mol. Cell Biol.* **5:** 323–329.

Dunn B., Szauter P., Pardue M.L., and Szostak J.W. 1984. Transfer of yeast telomeres to linear plasmids by recombination. *Cell* **39:** 191–201.

Ellis T.H.N. and Day A. 1986. A hairpin plastid genome in barley. *EMBO J.* **5:** 2769–2774.

Garvik B., Carson M., and Hartwell L. 1995. Single-stranded DNA arising at telomeres in *cdc13* mutants may constitute a specific signal for the *RAD9* checkpoint. *Mol. Cell. Biol.* **15:** 6128–6138.

Grandin N. and Charbonneau M. 2003. The Rad51 pathway of telomerase-independent maintenance of telomeres can amplify TG1–3 sequences in *yku* and *cdc13* mutants of *Saccharomyces cerevisiae. Mol. Cell. Biol.* **23:** 3721–3734.

Gravel S., Larrivee M., Labrecque P., and Wellinger R.J. 1998. Yeast Ku as a regulator of chromosomal DNA end structure. *Science* **280:** 741–744.

Groff-Vindman C., Natarajan S., Cesare A., Griffith J.D., and McEachern M.J. 2005. Recombination at dysfunctional long telomeres forms tiny double and single stranded t-circles. *Mol. Biol. Cell* **25:** 4406–4412.

Grossi S., Bianchi A., Damay P., and Shore D. 2001. Telomere formation by Rap1p binding site arrays reveals end-specific length regulation requirements and active telomeric recombination. *Mol. Cell. Biol.* **21:** 8117–8128.

Haber J.E. 1999. DNA recombination: The replication connection. *Trends Biochem. Sci.* **24:** 271–275.

Horowitz H. and Haber J.E. 1985. Identification of autonomously replicating circular subtelomeric Y′ elements in *Saccharomyces cerevisiae. Mol. Cell. Biol.* **5:** 2369–2380.

Huang P., Pryde F.E., Lester D., Maddison R.L., Borts R.H., Hickson I.D., and Louis E.J. 2001. SGS1 is required for telomere elongation in the absence of telomerase. *Curr. Biol.* **11:** 125–129.

Ira G. and Haber J.E. 2002. Characterization of RAD51-independent break-induced replication that acts preferentially with short homologous sequences. *Mol. Cell. Biol.* **22:** 6384–6392.

Iyer S., Chadha A.D., and McEachern M.J. 2005. A mutation in the *STN1* gene triggers an alternative lengthening of telomere-like runaway recombinatorial telomere elongation and rapid deletion in yeast. *Mol. Cell. Biol.* **25:** 8064–8073.

Johnson F.B., Marciniak R.A., McVey M., Stewart S.A., Hahn W.C., and Guarente L. 2001. The *Saccharomyces cerevisiae* WRN homolog Sgs1p participates in telomere maintenance in cells lacking telomerase. *EMBO J.* **20:** 905–913.

Joseph I., Jia D., and Lustig A.J. 2005. Ndj1p-dependent epigenetic resetting of telomere size in yeast meiosis. *Curr. Biol.* **15:** 231–237.

Le S., Moore J.K., Haber J.E., and Greider C.W. 1999. *RAD50* and *RAD51* define two pathways that collaborate to maintain telomeres in the absence of telomerase. *Genetics* **152:** 143–152.

Lendvay T.S., Morris D.K., Sah J., Balasubramanian B., and Lundblad V. 1996. Senescence mutants of *Saccharomyces cerevisiae* with a defect in telomere replication identify three additional EST genes. *Genetics* **144:** 1399–1412.

Li B. and Lustig A.J. 1996. A novel mechanism for telomere size control in *Saccharomyces cerevisiae*. *Genes Dev.* **10:** 1310–1326.

Lin C.Y., Chang H.H., Wu K.J., Tseng S.F., Lin C.C., Lin C.P., and Teng S.C. 2005. Extrachromosomal telomeric circles contribute to Rad52-, Rad50-, and polymerase δ-mediated telomere-telomere recombination in *Saccharomyces cerevisiae*. *Eukaryot. Cell* **4:** 327–336.

Liti G. and Louis E.J. 2003. *NEJ1* prevents NHEJ-dependent telomere fusions in yeast without telomerase. *Mol. Cell* **11:** 1373–1378.

Louis E.J., Naumova E.S., Lee A., Naumov G., and Haber J.E. 1994. The chromosome end in yeast: Its mosaic nature and influence on recombinational dynamics. *Genetics* **136:** 789–802.

Lundblad V. and Blackburn E.H. 1993. An alternative pathway for yeast telomere maintenance rescues *est1⁻* senescence. *Cell* **73:** 347–360.

Lundblad V. and Szostak J.W. 1989. A mutant with a defect in telomere elongation leads to senescence in yeast. *Cell* **57:** 633–643.

Lustig A.J. 2003. Clues to catastrophic telomere loss in mammals from yeast telomere rapid deletion. *Nat. Rev. Genet.* **4:** 916–923.

Malkova A., Ivanov E.L., and Haber J.E. 1996. Double-strand break repair in the absence of *RAD51* in yeast: A possible role for break-induced DNA replication. *Proc. Natl. Acad. Sci.* **93:** 7131–7136.

Malkova A., Naylor M., Yamaguchi M., Ira G., and Haber J.E. 2005. *RAD51*-dependent break-induced replication differs in kinetics and checkpoint responses from *RAD51*-mediated gene conversion. *Mol. Cell. Biol.* **25:** 933–944.

Maringele L. and Lydall D. 2002. *EXO1*-dependent single-stranded DNA at telomeres activates subsets of DNA damage and spindle checkpoint pathways in budding yeast *yku70Δ* mutants. *Genes Dev.* **16:** 1919–1933.

———. 2004. Telomerase- and recombination-independent immortalization of budding yeast. *Genes Dev.* **18:** 2663–2675.

McEachern M.J. 2002. Recombinational telomere elongation in the yeast *Kluyveromyces lactis*. In *Telomerases, telomeres, and cancer* (ed. G. Krupp and R. Parwaresch), pp. 347–358. Landes Bioscience/Kluwer/Eurekah.com, New York.

McEachern M.J. and Blackburn E.H. 1995. Runaway telomere elongation caused by telomerase RNA gene mutations. *Nature* **376:** 403–409.

———. 1996. Cap-prevented recombination between terminal telomeric repeat arrays (telomere CPR) maintains telomeres in *Kluyveromyces lactis* lacking telomerase. *Genes Dev.* **10:** 1822–1834.

McEachern M.J. and Iyer S. 2001. Short telomeres in yeast are highly recombinogenic. *Mol. Cell* **7:** 695–704.

McEachern M.J., Underwood D.H., and Blackburn E.H. 2002. Dynamics of telomeric DNA turnover in yeast. *Genetics* **160:** 63–73.

Michel B. 2000. Replication fork arrest and DNA recombination. *Trends Biochem. Sci.* **25:** 173–178.

Michel B., Grompone G., Flores M.J., and Bidnenko V. 2004. Multiple pathways process stalled replication forks. *Proc. Natl. Acad. Sci.* **101:** 12783–12788.

Myung K., Datta A., Chen C., and Kolodner R.D. 2001. SGS1, the *Saccharomyces cerevisiae* homologue of BLM and WRN, suppresses genome instability and homologous recombination. *Nat. Genet.* **27:** 113–116.

Naito T., Matsuura A., and Ishikawa F. 1998. Circular chromosome formation in a fission yeast mutant defective in two ATM homologues. *Nat. Genet.* **20:** 203–206.

Natarajan S. and McEachern M.J. 2002. Recombinational telomere elongation promoted by DNA circles. *Mol. Cell. Biol.* **22:** 4512–4521.

Natarajan S., Groff-Vindman C., and McEachern M.J. 2003. Factors influencing the recombinational expansion and spread of telomeric tandem arrays in *Kluyveromyces lactis. Eukaryot. Cell* **2:** 1115–1127.

Nosek J. and Tomaska L. 2003. Mitochondrial genome diversity: Evolution of the molecular architecture and replication strategy. *Curr. Genet.* **44:** 73–84.

Nosek J., Tomaska L., Fukuhara H., Suyama Y., and Kovac L. 1998. Linear mitochondrial genomes: 30 years down the line. *Trends Genet.* **14:** 184–188.

Nugent C.I., Bosco G., Ross L.O., Evans S.K., Salinger A.P., Moore J.K., Haber J.E., and Lundblad V. 1998. Telomere maintenance is dependent on activities required for end repair of double-strand breaks. *Curr. Biol.* **8:** 657–660.

Polotnianka R.M., Li J., and Lustig A.J. 1998. The yeast Ku heterodimer is essential for protection of the telomere against nucleolytic and recombinational activities. *Curr. Biol.* **8:** 831–834.

Qin Z. and Cohen S.N. 2000. Long palindromes formed in *Streptomyces* by nonrecombinational intra-strand annealing. *Genes Dev.* **14:** 1789–1796.

Ritchie K.B. and Petes T.D. 2000. The Mre11p/Rad50p/Xrs2p complex and the Tel1p function in a single pathway for telomere maintenance in yeast. *Genetics* **155:** 475–479.

Ritchie K.B., Mallory J.C., and Petes T.D. 1999. Interactions of *TLC1* (which encodes the RNA subunit of telomerase), *TEL1*, and *MEC1* in regulating telomere length in the yeast *Saccharomyces cerevisiae. Mol. Cell. Biol.* **19:** 6065–6075.

Rizki A. and Lundblad V. 2001. Defects in mismatch repair promote telomerase-independent proliferation. *Nature* **411:** 713–716.

Roy J., Fulton T.B., and Blackburn E.H. 1998. Specific telomerase RNA residues distant from the template are essential for telomerase function. *Genes Dev.* **12:** 3286–3300.

Shima H., Suzuki M., and Shinohara M. 2005. Isolation and characterization of novel *xrs2* mutations in *Saccharomyces cerevisiae. Genetics* **170:** 71–85.

Signon L., Malkova A., Naylor M., and Haber J.E. 2001. Genetic requirements for *RAD51*- and *RAD54*-independent break-induced replication repair of a chromosomal double-strand break. *Mol. Cell. Biol.* **21:** 2048–2056.

Singer M.S. and Gottschling D.E. 1994. *TLC1*: Template RNA component of *Saccharomyces cerevisiae* telomerase. *Science* **266:** 404–409.

Sugawara N., Goldfarb T., Studamire B., Alani E., and Haber J.E. 2004. Heteroduplex rejection during single-strand annealing requires Sgs1 helicase and mismatch repair proteins Msh2 and Msh6 but not Pms1. *Proc. Natl. Acad. Sci.* **101:** 9315–9320.

Symington L.S. 2002. Role of RAD52 epistasis group genes in homologous recombination and double-strand break repair. *Microbiol. Mol. Biol. Rev.* **66:** 630–670.

Takata H., Tanaka Y., and Matsuura A. 2005. Late S phase-specific recruitment of Mre11 complex triggers hierarchical assembly of telomere replication proteins in *Saccharomyces cerevisiae. Mol. Cell* **17:** 573–583.

Teixeira M.T., Arneric M., Sperisen P., and Lingner J. 2004. Telomere length homeostasis is achieved via a switch between telomerase-extendible and -nonextendible states. *Cell* **117:** 323–335.

Teng S.C. and Zakian V.A. 1999. Telomere-telomere recombination is an efficient bypass pathway for telomere maintenance in *Saccharomyces cerevisiae. Mol. Cell. Biol.* **19:** 8083–8093.

Teng S., Chang J., McCowan B., and Zakian V.A. 2000. Telomerase-independent lengthening of yeast telomeres occurs by an abrupt Rad50p-dependent, Rif-inhibited recombinational process. *Mol. Cell* **6:** 947–952.

Topcu Z., Nickles K., Davis C., and McEachern M.J. 2005. Abrupt disruption of capping and a single source for recombinationally elongated telomeres in *Kluyveromyces lactis. Proc. Natl. Acad. Sci.* **102:** 3348–3353.

Tsai Y.L., Tseng S.F., Chang S.H., Lin C.C., and Teng S.C. 2002. Involvement of replicative polymerases, Tel1p, Mec1p, Cdc13p, and the Ku complex in telomere-telomere recombination. *Mol. Cell. Biol.* **22:** 5679–5687.

Tsukamoto M., Yamashita K., Miyazaki T., Shinohara M., and Shinohara A. 2003. The N-terminal DNA-binding domain of Rad52 promotes *RAD51*-independent recombination in *Saccharomyces cerevisiae. Genetics* **165:** 1703–1715.

Tsukamoto Y., Taggart A.K., and Zakian V.A. 2001. The role of the Mre11-Rad50-Xrs2 complex in telomerase-mediated lengthening of *Saccharomyces cerevisiae* telomeres. *Curr. Biol.* **11:** 1328–1335.

Tsukamoto Y., Mitsuoka C., Terasawa M., Ogawa H., and Ogawa T. 2005. Xrs2p regulates Mre11p translocation to the nucleus and plays a role in telomere elongation and meiotic recombination. *Mol. Biol. Cell* **16:** 597–608.

Underwood D.H., Zinzen R.P., and McEachern M.J. 2004. Template requirements for telomerase translocation in *Kluyveromyces lactis. Mol. Cell. Biol.* **24:** 912–923.

Usui T., Ogawa H., and Petrini J.H. 2001. A DNA damage response pathway controlled by Tel1 and the Mre11 complex. *Mol. Cell* **7:** 1255–1266.

Wang R.C., Smogorzewska A., and de Lange T. 2004. Homologous recombination generates T-loop-sized deletions at human telomeres. *Cell* **119:** 355–368.

Watt P.M., Hickson I.D., Borts R.H., and Louis E.J. 1996. *SGS1*, a homologue of the Bloom's and Werner's syndrome genes, is required for maintenance of genome stability in *Saccharomyces cerevisiae. Genetics* **144:** 935–945.

Yamada M., Hayatsu N., Matsuura A., and Ishikawa F. 1998. Y'-Help1, a DNA helicase encoded by the yeast subtelomeric Y' element, is induced in survivors defective for telomerase. *J. Biol. Chem.* **273:** 33360–33366.

9

Meiotic Telomeres

Harry Scherthan

Institute for Radiation Biology Bw
80937 Munich, Germany
and
Max-Planck-Institute for Molecular Genetics
D-14195 Berlin, Germany

THE PECULIAR BEHAVIOR OF CHROMOSOMES and their ends during the prophase of the first meiotic division fascinated cytologists of the late 19th century, long before the capping function of telomeres was realized a half century later (for a historical overview, see Chapter 1). At that time, investigations centered around reproductive tissues, because a high rate of cell division and the condensation of meiotic chromosomes within the intact nuclear envelope (NE) facilitated chromosome analysis. Moreover, germ cells were a favorite subject of cytology in the search for a process that would compensate for the genome doubling that occurs at fertilization. The acidic heavy metal staining techniques of the early days (Flemming 1895) left somatic nuclei with a more-or-less reticulate staining, but revealed threadlike chromosomes that had their ends bundled together at a limited region of the nuclear periphery in differentiating germ cells (e.g., Platner 1885; Schreiner and Schreiner 1905; Gelei 1921; for an overview see Wilson 1925). This remarkable clustering of meiotic telomeres was baptized "chromosomal bouquet" because of its resemblance to bundled flower stems (Fig. 1A) (Eisen 1900). Telomere clustering occurs near the centrosome in species with a locally defined microtubule-organizing center (MTOC), as is the case in animals, algae, and fungi (Fig. 1A,B). It was soon reasoned that the polarization of meiotic chromosomes could represent a stage where homologous chromosomes undergo homology search (Boveri 1904), which ascribed a

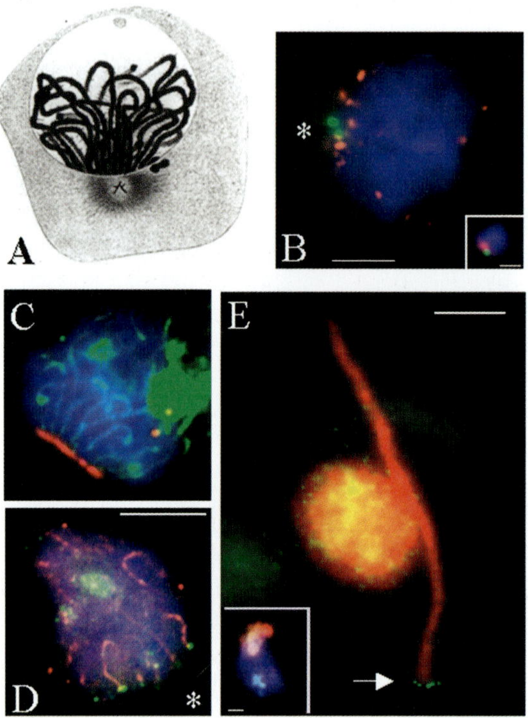

Figure 1. (*A*) Pachytene bouquet of *Batrachoseps* with telomeres clustered near the centrosome (reproduced from Eisen 1900). (*B*) Telomere (*red*, TRF2 immunofluorescence [IF]) clustering near the centrosome (*green*; γ-tubulin IF, *asterisk*) in a human spermatocyte. Bar, 5 μm. *Inset*: Telomeres form a tight cluster (*red*, X′Y telomere repeat fluorescence in situ hybridization) at the SPB (*green*, Tub4 IF) in a 3D-preserved *S. cerevisiae* meiocyte. Bar, 1 μm. (*C*) Tight telomere (*red*) clustering in a late zygotene oocyte of cattle (*green*, SCP3). (*D*) Human zygotene spermatocyte with Mre11 IF signals (*green*) at ends of axial elements (*red*, SCP3 IF). *Asterisk* denotes the bouquet base. Bar, 5 μm; it also applies to *C*. (*E*) Extended meiotic stage 3 micronucleus (crescent) of *Tetrahymena thermophila* with telomeres (*green, arrow*) clustered at the head region (see Loidl and Scherthan 2004 for details). Bar, 5 μm. *Inset*: Spread horsetail nucleus of *S. pombe* with telomeres (*red*) and the telomere-associated nucleolus (*orange*) at the tip of a horsetail nucleus. Clustered centromeres (*green*) locate in the distal part of the nucleus. Bar, 1 μm.

special role to meiotic telomeres. Despite more than a century of research, however, the mechanisms underlying meiotic telomere behavior remain largely speculative. This chapter summarizes what has recently been learned about telomere biology in meiosis. For other aspects of meiosis, like homolog search and pairing, recombination, segregation, and developmental signals, the reader will be referred to excellent reviews elsewhere in the literature.

MEIOSIS

Meiosis has evolved to compensate for the genome doubling that occurs at fertilization and to instigate genetic diversity in sexually reproducing eukaryotes. Two successive meiotic divisions without an intervening DNA replication accomplish reduction of genome size and create the haploid chromosome number common to gametes or spores.

Meiosis differs from mitosis in that mitosis, after one round of DNA replication, segregates one copy of each parental chromosome to opposite poles, thereby creating genetically identical daughter cells. Meiosis I, on the other hand, segregates replicated parental chromosomes from each other, which leads to a reshuffled chromosomal constitution in the resulting daughter cells. The latter skip DNA replication and segregate sister chromatids in a mitosis-like second meiotic division, which creates genetically diverse haploid gametes or spores (Fig. 2). Instrumental for the reductional meiosis I division is the side-by-side pairing and reciprocal recombination between homologous chromosomes (homologs) during a prolonged first meiotic prophase. Reciprocal recombination between homologs generates physical connections (chiasmata) that ensure their correct segregation (Roeder 1997; Smith and Nicolas 1998; Page and Hawley 2003).

Entry into meiosis is controlled by nutrient conditions in yeasts (see Honigberg and Purnapatre 2003), whereas in multicellular organisms associated cells regulate differentiation of meiotic stem cells and entry into the meiotic program, which, however, is poorly understood (see Mackay 2000; Wolgemuth 2003; West and Daley 2004). In nonhaplontic organisms, meiosis initiates when the premeiotic cell has passed a protracted phase of DNA synthesis and enters the special G_2 phase that precedes the first meiotic division. During this prophase I, chromosomes become visible inside the nucleus as threadlike structures which align side by side according to homology. Because, in most species, homologs are spatially separated at the onset of prophase I, they undergo homology search and prealignment leading to the high levels of intimate pairing specific to meiosis (for review, see Zickler and Kleckner 1998; McKee 2004). In the majority of sexually reproducing organisms, homologs become stably connected by a proteinaceous zipperlike structure termed the synaptonemal complex (SC; Fawcett 1956; Moses 1956), that runs along paired homologs, thereby forming so-called bivalents (for review, see von Wettstein 1984; Page and Hawley 2004). In some species meiosis may proceed in the

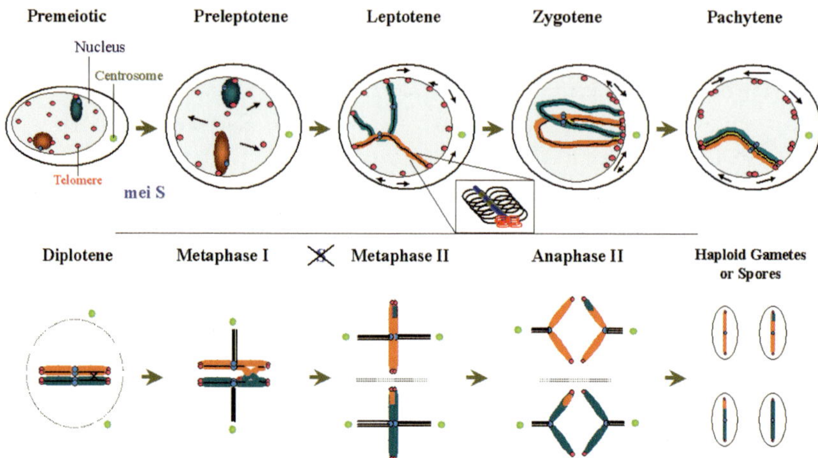

Figure 2. A generalized scheme of chromosome and telomere dynamics during meiosis. Telomeres are shown in *red*, centromeres in *blue*, and one pair (for simplicity) of homologous chromosomes in *orange* and *blue*. *Premeiotic cells*: Chromosomes occupy compact territories and are usually spatially separated. Telomeres are scattered throughout the nuclear volume, as is typical for most mammalian and some plant cells. At variance, yeast telomeres form a few peripheral clusters in premeiotic cells (see text). *Preleptotene*: After premeiotic DNA replication (meiS), chromosome territories start to elongate and telomeres transit to the nuclear envelope (NE) to which they attach, while axial elements (AEs) have not yet developed. *Leptotene*: Chromosomes assemble AEs (*black lines*) along their length with DNA looping out from the AEs that consist of special cohesin complexes and AE-specific proteins. Telomere loops are usually shorter than more internal DNA loops (*magnifying box*). In late leptotene phase, elongated chromosomes have attached to the NE with their telomeres that are distributed over the NE. Telomere movements along the NE lead to congregation of telomeres near the centrosome. During these movements chromosomes engage in homology testing and recognition, which is DNA double-strand-break-dependent in most species. *Zygotene*: Telomeres congregate opposite to the centrosome (animals) or SPB (yeast) during the onset of zygotene forming a chromosomal bouquet. The duration of the bouquet stage may vary between species or gender (see text). *Pachytene*: All homologs are paired lengthwise (bivalents) and are stably connected by the synaptonemal complex (*yellow dotted lines*) while telomeres are again scattered over the NE. In some conditions/species, clustering is released only during pachytene (see text). *Diplotene*: Telomere attachment is released as the NE dissolves and chromosomes first decondense and then condense in reparation for the MI division. *Metaphase I*: Sites of reciprocal recombination become visible as chiasmata that provide physical connections between homologs when these undergo monopolar attachment with both sister centromeres so that centromeres of homologous chromosomes face opposite spindle poles. After segregation in anaphase I (not shown), cells skip an S phase (S) and attach sister kinetochores to opposite spindle poles in the MII division. *Anaphase II* (which is similar to anaphase of mitotic cells) separates sister chromatids, thereby creating haploid gametes or spores.

absence of the SC, which is, e.g., the case in *Schizosaccharomyces pombe*, *Aspergillus nidulans*, and *Tetrahymena thermophila* (Wolfe et al. 1976; Olson et al. 1978; Egel-Mitani et al. 1982), or in the absence of recombination, as is the case in *Drosophila* males (Meyer 1960; Rasmussen 1973) and silkworm females (Sturtevant 1915; Maeda 1939).

Classically, the first meiotic prophase is subdivided into successive substages according to consecutive changes in chromosome morphology seen in the light microscope (John 1990). After premeiotic S phase, cells pass a preleptotene stage where chromosomes are indistinguishable by conventional staining methods (Fig. 2). At leptotene stage, chromosomes become visible as thin threads along which protein cores (axial elements; also known as axial cores) assemble. Axial elements are made of cohesin complexes that consist of cohesin SMC1β (instead of mitotic SMC1), SMC3, REC8 (which replaces mitotic Rad21/Scc1/Mcd1), and STAG3 (instead of Scc3/SA1 or SA2) (Klein et al. 1999; Watanabe and Nurse 1999; Pezzi et al. 2000; Pelttari et al. 2001, Prieto et al. 2001, 2002; Siomos et al. 2001; James et al. 2002; Eijpe et al. 2003; Manheim and McKim 2003; Molnar et al. 2003; Pasierbek et al. 2003; Couteau et al. 2004), condensin (Yu and Koshland 2003), and specific axial element (AE) proteins like, e.g., SCP3 and SCP2 of the mouse (Lammers et al. 1994; Offenberg et al. 1998; Yuan et al. 2000), *C. elegans* Him3 (Zetka et al. 1999), yeast Red1 (Smith and Roeder 1997), or *Arabidopsis* and *Brassica* Asy1 (Armstrong et al. 2002).

When AEs have approached each other to ~300 nm they become interconnected by transverse filament proteins, such as mammalian SCP1 or budding yeast Zip1 (Meuwissen et al. 1992; Sym et al. 1993; Dobson et al. 1994), giving rise to tripartite SC (see Heyting 1996; Page and Hawley 2004), which is growing during the zygotene stage. In the context of the SC, axial elements are called lateral elements. Once synapsis is complete between all homologs, the cell has reached the pachytene stage. The synaptonemal complex is disassembled during the following diplotene stage in which chromosomes decondense. In some species this leads to an ill-defined appearance of chromosomes, known as the "diffuse stage" (Zickler and Kleckner 1999). Prior to the meiosis I (MI) division chromosomes condense and sites of crossing-over become visible as chiasmata that represent physical connections between chromatids of homologs (Fig. 2). Maintenance of sister chromatid cohesion allows the attachment of parental centromeres (which consist of tightly associated sister kinetochores) to opposite spindle poles (monopolar attachment; Fig. 2). Arm, but not centromere, cohesion is lost during anaphase I, which leads to the segregation of homologs. The subsequent meiosis

II division resembles a mitotic division and segregates sister chromatids (for review, see Petronczki et al. 2003; Marston and Amon 2004).

PASSAGE THROUGH MEIOSIS REQUIRES INTACT TELOMERES

Telomere repeats and associated proteins together form a protective cap at the ends of linear DNA strands, which prevents chromosome ends from fusion and degradation (see Chapters 1 and 13). Telomere repeats are lost during replication but repeats can be replenished by the specialized reverse transcriptase telomerase (see Chapters 2 and 3). Manipulation of telomerase activity in budding yeast and mouse models has revealed that loss of telomeric repeats compromises passage through meiosis (Hemann et al. 2001; Maddar et al. 2001; Liu et al. 2004). The latter has also been observed in *Caenorhabditis elegans Mrt-2* mutants and the *tel1 rad3* double mutant of *S. pombe* (Naito et al. 1998; Ahmed and Hodgkin 2000).

In the mouse, telomerase is highly expressed in premeiotic germ cells of the testes (spermatogonia) where it adds terminal T_2AG_3 repeats to chromosome ends to replenish telomere repeat tracts before cells enter the first meiotic prophase (Ravindranath et al. 1997; Achi et al. 2000). Mice deficient for the RNA subunit of telomerase, *Terc* (see Chapter 5), display organ failure when telomeres are shortened to critical size (Blasco et al. 1997; Lee et al. 1998). Late generation *Terc*$^{-/-}$ mice also exhibit impaired spermatogenesis, because spermatogonia with critically shortened telomeres are eliminated by programmed cell death (Hemann et al. 2001). Meiocytes of generation 4 (G4) *Terc*$^{-/-}$ mice with less severely reduced telomere repeat tracks manage to enter meiosis, but, under some conditions, experience defects in AE formation and synapsis (as determined by SCP3 staining). This is particularly evident in females with less stringent prophase I control (Liu et al. 2004) and indicates a role for telomeres in mediating AE formation and synaptic pairing (see below).

It is possible that the telomere/NE association requires only a short stretch of telomeric DNA. In the budding yeast *Kluyveromyces lactis*, it has been found that a minimal number of telomere repeats is sufficient to pass through meiosis (Maddar et al. 2001). This finding aligns with the faithful telomere/NE attachment in those G6 *Terc*$^{-/-}$ spermatocytes that manage to enter meiosis despite a significantly reduced number of telomere repeats (Franco et al. 2002). It will be interesting to learn about the telomere repeat length of individual chromosomes of late *Terc*$^{-/-}$ meiocytes and their respective capacity to attach to the NE. Because telomereless chromosomes elicit defects in AE formation and ring chromosomes compromise meiotic chromosome segregation (Ishikawa and Naito 1999), it appears that linear chromosomes with telomeres are instrumental for passage through meiosis.

MEIOTIC TELOMERES ATTACH TO THE NUCLEAR ENVELOPE

In mammalian cells telomeres associate with the filamentous nuclear matrix (de Lange 1992) and are scattered throughout the nuclear volume (Vourc'h et al. 1993; Luderus et al. 1996). In some cell types, telomeres locate within the nuclear hemisphere opposite to the centromeres that face the MTOC as a consequence of anaphase polarization from the previous cell division, a nuclear topology known as the Rabl configuration (Rabl 1885). The Rabl configuration is particularly evident in rapidly dividing yeast cells (Funabiki et al. 1993; Jin et al. 1998, 2000) and has been observed in some cell types of animals (Sperling and Ludtke 1981; Hochstrasser et al. 1986; Haaf and Ward 1995b) and several plant species (Stack and Clark 1974; Fussell 1987; Martinez-Perez et al. 2000). However, the Rabl configuration is not a universal motif of 3D nuclear organization in animals and plants (Vourc'h et al. 1993; Dong and Jiang 1998; Bass et al. 2000; Weierich et al. 2003).

When cells enter meiosis, premeiotic nuclear architecture is dissolved and telomeres reposition to the NE to which they attach during lep-

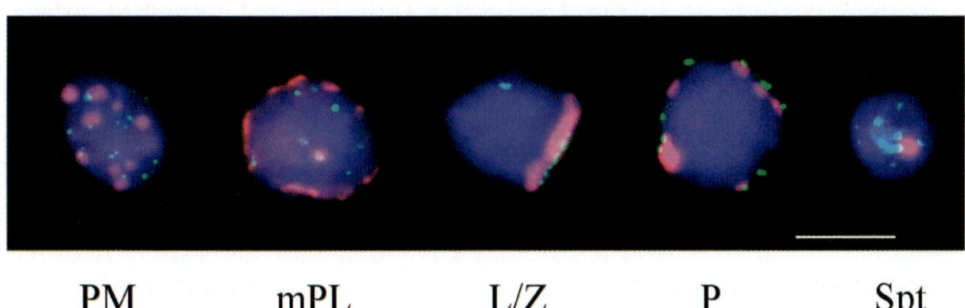

Figure 3. Redistribution of telomeres (*green*, FITC) and pericentric major satellite DNA (*red*, Cy3) in premeiotic (PM) and prophase I nuclei of the male mouse as revealed by fluorescence in situ hybridization. Nuclear equator is shown for all nuclei. (PM) Nucleus of a spermatogonium (premeiotic cell) displays scattered telomeres and pericentric satellite clusters. (mPL) Mid-preleptotene nucleus with a perinuclear layer of major satellite DNA and interior telomeres. (L/Z) Leptotene/early zygotene nucleus with tight telomere and heterochromatin clustering. Telomere signals at the bouquet base are small, whereas the one off the base is larger because of homologous pairing. Telomere clustering is rapidly dissolved in male mouse meiosis leading to peripherally dispersed telomeres and satellite DNA clusters in pachytene nuclei (P), as is also the case in budding yeast and maize (see text). (Spt) Haploid spermatid with a central pericentromeric heterochromatin cluster and few telomere signals. Bar, 10 μm.

totene stage (Figs. 2 and 3) (see von Wettstein 1984). Three-dimensional reconstruction of electron microscope serial sections has shown that telomeres attach with the wide end of the conical thickening of the AE end to the nucleoplasmic face of the inner nuclear membrane, which is the case in insects (Moens 1969b; Rasmussen 1977b; Wandall and Svendsen 1985), mammals (Woollam et al. 1966; Esponda and Gimenez-Martin 1972), plants (Moens 1969a; Gillies 1975; Thomas and Kaltsikes 1976), and fungi (Byers and Goetsch 1975; Zickler and Olson 1975; Zickler 1977).

Meiotic telomeres seem to be firmly anchored in the nuclear periphery through filament bundles that link telomeres at the inner nuclear membrane to dense spherical structures at the cytoplasmic face of the NE. Such filaments have been observed in mammalian and *Neurospora* meiocytes (Bojko 1983, 1990; Liebe et al. 2004; and references therein). In fission yeast meiosis, such linkages seem to tether telomeres to the spindle pole body (the fungal MTOC) (see below; Shimanuki et al. 1997; Flory et al. 2004). A tight telomere/cytoskeleton interaction may explain why telomere attachment to the meiotic NE resists pulling forces applied by micromanipulation or centrifugation (Gelei 1921; Hiraoka 1952).

The telomere of meiotic chromosomes is organized such that telomeric DNA repeats form numerous short double-stranded DNA loops that have their bases fixed to the tips of the AEs (Fig. 2, Leptotene), as was shown by fluorescence in situ hybridization (FISH) to spread meiotic nuclei (Moens and Pearlman 1990; Heng et al. 1996). Telomeres in spread pachytene meiocytes of mammals appear as round FISH or immunofluorescence (IF) signals (Fig. 4A), which are revealed by scanning near-field optical microscopy (SNOM), a variant of atomic force microscopy (see Hausmann et al. 2003), as a globular structure with the T_2AG_3 telomere repeat signals buried in the center (Fig. 4B). However, telomere repeats in 3D-preserved meiocytes locate within an ~200-nm-wide attachment plate at the interface of AE end and inner nuclear membrane (Fig. 4C,D) (Liebe et al. 2004).

A role of telomere repeats in NE attachment is also suggested by genetic evidence. For instance, perturbation of telomere repeat sequence or complete loss of terminal repeats in the mouse leads to elimination of spermatogonia prior to entry into meiosis (Hemann et al. 2001) or to synaptic errors in oocytes (Liu et al. 2004), and yeast cells without functional terminal repeats fail in meiosis (Naito et al. 1998; Maddar et al. 2001; Alexander and Zakian 2003). Meiosis in haploid budding yeast is delayed if an extra chromosome (disome) is present, perhaps to allow time for interhomolog interactions to be established. However, if the disomic chromosome is a ring lacking telomere repeats, no homology-dependent

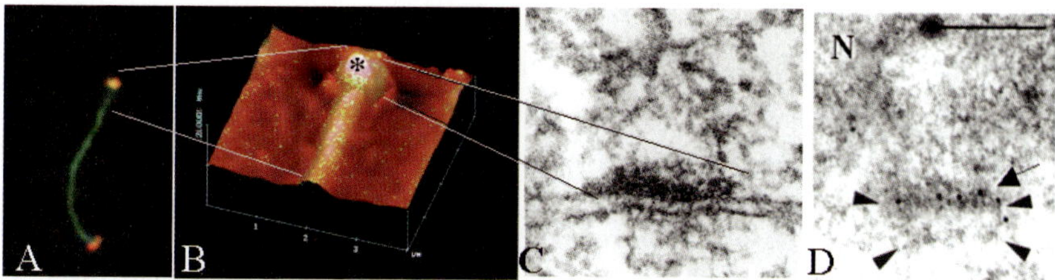

Figure 4. (*A*) A SCP3-stained synaptonemal complex (SC) (*green*) from a human spread spermatocyte carries two round telomere signals (*red*, TRF1 immunofluorescence) at its ends. (*B*) Scanning near-field optical microscopy (SNOM) image of the telomere region of an SC end. Topological image (*orange red*) and fluorescence channel (TRF1 fluorescence; Cy3, *yellow dots*) are superimposed. The telomere at the SC end is present as a globular structure with the telomere fluorescence peaking in the center (asterisked peak for intensity; for details see Hausmann et al. 2003). (*C*) EM analysis of 3D-preserved meiocytes shows the telomere in flat contact with the nuclear membrane. The telomere attachment in a $Sycp3^{-/-}$ spermatocyte is present as a dark electron-dense plate at the nucleoplasmic face of the inner nuclear membrane. Fibrillar material spans the nuclear envelope toward the cytoplasm in the lower detail. (*D*) A telomere attachment plate after EM-in situ hybridization with a T_2AG_3 repeat probe carries telomere-repeat signals (*black gold grains; arrow*) at the attachment plate at the inner NM (*upper arrowheads*). N, nucleus. (*C,D* reprinted with permission, from Liebe et al. 2004.) Bar for *C* & *D*, 0.2 μm.

delay is elicited (Rockmill and Roeder 1998), underlining the requirement for functional telomeres in homology searching.

3D-FISH analysis of mammalian mini-ring chromosomes, which lack physical ends and telomeric repeats, has shown that these fail to colocalize with the perinuclear telomere cluster of mouse spermatocytes (Voet et al. 2003). On the other hand, ring chromosomes of maize and fission yeast that carry telomere repeats or form a telomere protein complex at the fusion site still participate in meiotic telomere attachment and clustering (Carlton and Cande 2002; Sadaie et al. 2003), whereas ring chromosomes of *S. pombe* that lack telomere repeats fail to pass meiosis (Naito et al. 1998; Maddar et al. 2001). Altogether, this indicates that telomere repeats are required for telomere attachment and function during meiosis, which raises the question about the protein components involved in telomere/NE attachment.

TELOMERE PROTEINS AND NUCLEAR ENVELOPE ATTACHMENT

Immunofluorescence has shown that meiotic telomeres contain proteins that are also present at their somatic counterparts. This is the case for the mammalian duplex $(T_2AG_3)_n$ repeat binding proteins TRF1 and

TRF2 (Scherthan et al. 2000b), their fission yeast homolog Taz1 (Cooper et al. 1998; Nimmo et al. 1998), and budding yeast Rap1 (Klein et al. 1992). Meiotic telomeres furthermore contain telomere-interacting proteins such as Tin2, mRap1, tankyrase1, and components of the homologous recombination machinery like Nbs1 and Mre11 (Fig.1D) (Lombard and Guarente 2000; Scherthan et al. 2000b; Liebe et al. 2004) as well as the tumor suppressor menin (Suphapeetiporn et al. 2002) and the cyclin-dependent kinase Cdk2 (Ashley et al. 2001). However, little is known about the role of these proteins at meiotic telomeres.

In fission yeast, genetic manipulation has revealed that telomere attachment and clustering depend on the presence of telomere repeat binding protein Taz1 (Cooper et al. 1998; Nimmo et al. 1998) and its interacting partner Rap1 (Chikashige and Hiraoka 2001; Kanoh and Ishikawa 2001). Recent analysis suggests that Taz1 is attached to the spindle pole body (SPB) via its interaction with the SMC protein Ccp1, which itself is linked to the SPB via the pericentrin Pcp1p (Flory et al. 2004). These findings and the presence of TRF2 in isolated frog oocyte NEs (Podgornaya et al. 2000) agree with the localization of T_2AG_3 repeats in telomere attachment plates of mammalian spermatocytes (Liebe et al. 2004) and suggest that telomere-binding proteins likely mediate NE attachment in higher animals too.

Saccharomyces cerevisiae Rap1 directly binds telomere repeats (Longtine et al. 1989) and is associated with meiotic telomeres (Klein et al. 1992). Rap1 binding is crucial for progression through meiosis, as strains with altered telomere sequence and Rap1-binding sites have defective meiosis (Alexander and Zakian 2003). Induction of meiosis, furthermore, induces significant expression of the telomere protein Ndj1/Tam1 that locates to meiotic telomeres and is required for wild-type levels of crossing-over and homolog disjunction, telomere peripheralization, and clustering, as well as sister chromatid cohesion (Chua and Roeder 1997; Conrad et al. 1997; Trelles-Sticken et al. 2000). Ndj1 has been found to interact with the spindle pole body component Msp3, which itself interacts with the telomere-associated protein Est1 as revealed by a high-throughput two-hybrid screen (Uetz et al. 2000). Ndj1/Msp3 interactions may play a role in telomere clustering at the SPB, which occurs during the bouquet stage of *S. cerevisiae* meiosis (Trelles-Sticken et al. 1999).

A defect of sister chromatid cohesion in *ndj1Δ* meiosis (Chua and Roeder 1997; Conrad et al. 1997) raises the question whether telomere adhesion to the NE involves cohesin function or promotes cohesion.

Cohesin is a protein complex that is required for morphogenesis of meiotic chromosome cores and for homolog segregation (for review, see Jessberger 2002; Nasmyth 2002). Cohesin has been localized by IF to telomere attachments of $Sycp3^{-/-}$ mice (Liebe et al. 2004) that lack the AE protein SCP3 (Yuan et al. 2000). Furthermore, aberrant intranuclear localization of a subset of zygotene/pachytene telomeres in mouse spermatocytes deficient for meiosis-specific Smc1-β cohesin suggests a role for cohesin in mammalian meiotic telomere/NE attachment (Revenkova et al. 2004). However, it is currently not known whether the $Smc1\beta^{-/-}$ telomere defect results from initial attachment failure during leptotene or whether some telomeres lose attachment when AEs and SCs hypercondense (Revenkova et al. 2004). Budding yeast $rec8\Delta$ meiocytes, on the other hand, maintain telomere clustering throughout prophase I whereas $rec8\Delta$ meiocytes expressing mitotic Scc1 relax telomere clustering, suggesting that cohesin is required for exit from the bouquet stage (Trelles-Sticken et al. 2005).

Factors that were assumed to play a role in telomere attachment, like the yeast nuclear pore and telomere chromatin-associated proteins Mlp1 and Mlp2 (Strambio-de-Castillia et al. 1999; Galy et al. 2000), have nevertheless been found to undergo fine telomere clustering (Trelles-Sticken E., U. Nehrbass, and H. Scherthan, unpubl.). Normal telomere attachment and clustering were also observed in mouse spermatocytes lacking proteins of the inner nuclear membrane (Table 1), which is the case for the mouse homolog of the *Drosophila* germ-cell-less protein Mgcl-1 (Kimura et al. 2003) that associates with the NE through LAP2-β (Nili et al. 2001) and the spermatocyte-specific A-type lamin isoform C2 (Alsheimer et al. 2004) that accumulates at spermatocyte telomeres at the inner nuclear membrane (Alsheimer et al. 1999). These observations suggest that the proteins of the meiocyte NE so far tested are dispensable for telomere attachment.

In fission yeast, meiotic telomere clustering requires Rik1-dependent recruitment of the chromo- and SET-domain Clr4 histone methyltransferase to meiotic telomeres—in the absence of the two proteins, telomere attachment fails and chromosomes missegregate (Tuzon et al. 2004). A milder telomere-clustering defect was noted in mutants compromised in heterochromatin function because of the absence of the RNAi machinery (Hall et al. 2003), suggesting a link between histone methylation, centromeric RNAi processing, and meiotic telomere clustering. Functional heterochromatin seems to be a conserved requirement for meiotic telomere clustering, because we observed that the Set1 histone methyltransferase of budding yeast (Briggs et al. 2001; Boa et al. 2003), which contributes to telomeric silencing, recombination, and meiotic gene expression (Sollier et al. 2004), is also required for telomere clustering (Trelles-Sticken et al. 2005b).

However, meiotic telomere clustering occurs in the absence of the Sir3 protein (Trelles-Sticken et al. 2003), which is a component of repressive telomeric heterochromatin in *S. cerevisiae* (Hecht et al. 1995). Likewise, telomere attachment and clustering were observed to occur in mice double null for the *Suv39h1* and *Suv39h2* histone H3 methyltransferases (H. Scherthan and T. Jenuwein, unpubl.), which are required for heterochromatin formation (for a recent review, see Lachner and Jenuwein 2002). However, *Suv39h* double-null spermatocytes displayed illegitimate AE-like connections between SC ends (Peters et al. 2001), suggesting that meiotic telomeres are deprotected in the absence of histone methylation. This indicates that telomere attachment and capping function can be separated in meiocytes.

Perinuclear telomere redistribution has so far not been observed in early prophase I of *Drosophila*. This species displays premeiotic homolog association (see Metz 1916; Hiraoka et al. 1993) and lacks canonical telomere repeats (Biessmann et al. 1990; see Chapter 14). In recombinogenic meiosis of *Drosophila* females, SC ends are nonrandomly distributed near the NE to which they occasionally attach (Carpenter 1975). In *Drosophila* males that do not undergo reciprocal recombination (Morgan 1912), meiotic chromosomes associate with the nuclear periphery. Chromosome decondensation and a NE-associated sequence of events transforms premeiotic homolog pairing into high levels of meiotic pairing during an extended G_2 phase to MI (Vazquez et al. 2001, 2002). Thus, chromosome/NE contacts in male *Drosophila* meiosis may represent a structural analogy to recombinational meiosis where homolog pairing occurs also in the context of NE attachment, which is albeit reduced to telomeres.

In summary, it appears that in the vast majority of species meiotic telomeres are firmly attached to the inner nuclear membrane through telomere repeats and their associated proteins. The solid attachment is likely mediated by the numerous short telomere DNA loops at the AE end, which create an increased surface for telomere/NE contact and manifest as an attachment plate. This raises the question: Why is such an attachment necessary in meiosis? Telomere attachment is likely required to restrict the mobility of meiotic chromosome ends to two dimensions, which facilitates directed chromosome movements and reduces the complexity of homology search (Rhoades 1961; Moses 1968; Dorninger et al. 1995; Carlton et al. 2003) and provides a strong

Table 1. Mutants and species with aberrant or unusual conditions of meiotic telomere distribution

Species	Mutant/ condition	Telomeres at nucl. periphery	Telomere clustering	DSB repair[a]/ recombination	Synapsis	Function of resp. protein	Ref.
Saccharomyces cerevisiae	*ndj1/tam1Δ*	defective	–	reduced	reduced	meiotic telomere protein	Chua & Roeder 1997; Conrad et al. 1997; Trelles-Sticken et al. 2000
	sir3Δ	+	+	nd	+	silencing protein	Trelles-Sticken et al. 2003
	kar3Δ	+	+ +	reduced	aberrant	motor protein	Trelles-Sticken et al. 2003
	set1Δ	+	–	defective	defective	histone methylase	E. Trelles-Sticken et al. 2005b
	mlp1Δ, mlp2Δ	+	+	nd	+	NUP extension	E. Trelles-Sticken and H. Scherthan, (unpubl.)
	spo11Δ	+	+ +	–	–	transesterase	Trelles-Sticken et al. 1999
	rad50SΔ	+	+ +	defective	–	DNA repair	Trelles-Sticken et al. 1999
	rec8Δ	+	+ + +	defective	–	meiotic cohesin	Trelles-Sticken et al. 2005a
	haploid	+	+	aberrant	non-homologous	na[b]	Trelles-Sticken et al. 2003
Schizosaccharomyces pombe	*taz1⁻*	–	–	reduced	na	telomere repeat binding	Cooper et al. 1998; Nimmo et al. 1998
	rap1⁻	–	–	reduced	na	telomere protein interacting	Chikashige and Hiraoka 2001; Kanoh and Ishikawa 2001

(Continued)

Table 1. (*continued*)

Species	Mutant/ condition	Telomeres at nucl. periphery	Telomere clustering	DSB repair[a]/ recombination	Synapsis	Function of resp. protein	Ref.
	$kms1^-$	+	–	reduced	na	SPB component	Shimanuki et al. 1997; Niwa et al. 2000
	$rik1^-$	–	–	nd	na	heterochromatin formation	Tuzon et al. 2004
	$rad3^-\ tel1^-$	–	–	+	na	kinase	Naito et al. 1998
	mei4 (dot4)	compromised	compromised	nd	na	transcription factor	Jin et al. 2002
	mei4-636	compromised	compromised	nd	na	transcription factor	Jin et al. 2002
Mus musculus	$Atm^{-/-}$	+	+++	defective[c]	residual	serine-threonine kinase	Pandita et al. 1999; Scherthan et al. 2000a
	$H2ax^{-/-}$	+	++	defective	+	histone H2 variant	Fernandez-Capetillo et al. 2003
	$Sycp3^{-/-}$	+	++	defective	–	AE protein	Liebe et al. 2004
	$Spo11^{-/-}$	+	++	–	residual	transesterase	B. Liebe et al. (in prep.)
	$Smc1\beta^{-/-}$	reduced	(-)	defective	+	meiotic cohesin	Revenkova et al. 2004
	$mGcl1^{-/-}$	+	+	+	+	inner nuclear membrane protein	B. Liebe et al. (in prep.)
	$Lmna^{-/-}$ ♂	+	(-)	nd	+, defective for XY	A-type lamin	Alsheimer et al. 2004
	scid	+	+	nd	+	DNA-PKcs	B. Liebe et al. (in prep.)

Caenorhabditis elegans	wild type	+ (only one end involved)	+ (only one end involved)	+	+	na	Goldstein and Slaton 1982
	chk2	nd	chromosome polarization[d] −	defective	−	checkpoint protein kinase	MacQueen and Villeneuve 2001
	syp1	nd	chromosome polarization ++	defective	−	SC protein	MacQueen et al. 2002
Zea mays (maize)	*pam1*	+	−	altered	−	?	Golubovskaya et al. 2002
	dy	lost prematurely	+	altered	−		Bass et al. 2003
	dsy1	incomplete	±	altered	aberrant	?	Bass et al. 2003
Secale cereale (rye)	sy1	+	±	compromised	−	?	Mikhailova et al. 2001
Sordaria	*spo11*	+	+++	−	−	transesterase	Storlazzi et al. 2003
Bombyx mori	♀	+	+	−	+	na	Maeda 1939; Rasmussen 1977b
	triploid ♀	+	+	−	+	na	Rasmussen 1977a
Drosophila	♀	~at pachytene	likely absent	+	+	na	Carpenter 1975
	♂	?	−	−	−	na	Rasmussen 1973

Telomere clustering: +, wild type; ++, extended; +++, persistent; ±, relaxed in tightness; −, absent; (-), reduced. nd, not determined. SPB, spindle pole body; ?, not known; AE, axial element; SC, synaptonemal complex.

[a]Any type of recombinational repair defect.

[b]na, not applicable since *S. pombe* lacks SC.

[c]Fidelity of recombination in multicellular eukaryotes is usually assessed by alterations in repair protein localization in meiocytes.

[d]Only chromosome polarization tested.

point at the NE that supports chromosomes during a phase of vigorous movement.

TELOMERE CLUSTERING AND MOVEMENTS

Attached meiotic telomeres undergo movements along the inner nuclear membrane, which culminates in their clustering at a restricted sector of the nuclear periphery defined by the position of the cytoplasmic MTOC in animals and fungi (Fig. 1A,B). In plants, which lack a localized MTOC (Sheldon et al. 1988; Schmit et al. 1996), telomere clustering occurs in a microtubule-poor region opposing clustered nuclear pores (Cowan et al. 2002). In some species like *C. elegans* or coccids (intracellular parasites), telomere clustering involves only one end per chromosome (Reichenow 1927; Goldstein and Slaton 1982). The nuclear topology conferred by telomere clustering, known as chromosomal bouquet, is observed for a limited period of prophase I, where it is usually restricted to the leptotene/ zygotene transition, which is the case, e.g., in budding yeast, maize, and male mammals (Belar 1928; Scherthan et al. 1996; Trelles-Sticken et al. 1999; Bass et al. 2000; Pfeifer et al. 2001). The bouquet is known to persist from zygotene to pachytene in most mammalian oocytes (Fig. 1C) (Pfeifer et al. 2003; Roig et al. 2004; Tankimanova et al. 2004), flatworm (Gelei 1921), salamander (Fig. 1A) (Eisen 1900; Kezer and Macgregor 1971), flea beetle (Virkki 1974), and wheat meiosis (Martinez-Perez et al. 1999).

In the asynaptic meiosis (without SC) of *S. pombe* and the ciliate *T. thermophila*, telomere clustering is maintained to the end of prophase I when recombination has been completed (Chikashige et al. 1994; Ding et al. 2004; Loidl and Scherthan 2004). The meiotic nucleus of live *S. pombe* meiocytes undergoes sweeping movements during entire prophase I, which are led by the SPB and attached telomeres and confer a horse-tail-like appearance to the nucleus (Chikashige et al. 1994). This nuclear motility depends on the dynein heavy chain motor protein and cytoplasmic microtubules along which the nucleus is dragged forth and back (Svoboda et al. 1995; Ding et al. 1998; Yamamoto et al. 1999). Nuclear movements, but not telomere clustering, are absent in mutants that lack the dynein heavy chain motor protein Dhc1 and the dynein light chain family protein Dlc1 (Miki et al. 2002). Nuclear movements are also absent in meiotic cultures treated with MT drugs (Chikashige et al. 1994; Svoboda et al. 1995; Ding et al. 1998). A special type of prophase I "motility" occurs in the ciliate *Tetrahymena* where telomeres cluster at the head pole of the micronucleus (Fig. 4E) (Loidl and Scherthan 2004) that is dramatically elongated during prophase I by extension NE-associated MT bundles

(Wolfe et al. 1976). This tubelike nuclear transformation likely brings and holds homologous DNA strands into register until recombination is completed (Loidl and Scherthan 2004).

Chromosome and nuclear movements have also been observed in live leptotene and zygotene meiocytes of insects (Vazquez et al. 2002) and mammals (Parvinen and Soderstrom 1976) and are sensitive to colcemid, a colchicine derivative that dissociates MTs but also damages the meiotic NE (Salonen et al. 1982). Treatment of rye anther cultures with concentrations of colchicine that do not disrupt cytoplasmic MTs inhibits telomere clustering (Cowan and Cande 2002). The nature of the colchicine-sensitive target in the meiotic NE remains to be unveiled (Cowan and Cande 2002). In budding yeast, meiotic telomere clustering is only moderately reduced in the presence of MT inhibitors. Telomere clustering occurs at a higher frequency in the absence of the MT motor protein Kar3 (Trelles-Sticken et al. 2003) required for turnover of MTs (Saunders et al. 1997) and recombination (Bascom-Slack and Dawson 1997).

It may be speculated that the cytoplasmic microtubule cytoskeleton, which embraces the NE in higher eukaryotes (Franke 1971; Cherry and Hsu 1984; Cowan et al. 2002), plays a supportive, but not essential, role for nuclear and telomere movements, at least in some species. In agreement with this assumption, we recently observed in yeast meiosis that telomere clustering depends on actin polymerization—actin inhibitor treatment prevented telomere clustering or induced perinuclear dispersion of clustered Rap1-tagged telomere cluster in live *S. cerevisiae* bouquet meiocytes (Trelles-Sticken et al. 2005a). Whether actin plays a role for telomere clustering in other organisms remains to be explored.

MEIOTIC TELOMERES AND HOMOLOG PAIRING

Triploid meiosis of silkworm and rainbow trout displays persistent telomere clustering that involves the unpaired chromosomes (Rasmussen 1977a; Oliveira et al. 1995). The temporal correlation of meiotic telomere clustering and presence of unpaired chromosomes has led to the suggestion that bouquet formation may facilitate homology search and pairing (see Zickler and Kleckner 1998). A contribution of telomere clustering to homology search has been bolstered by significantly delayed homolog pairing in the telomere clustering-deficient *ndj1Δ* mutant of *S. cerevisiae* (Chua and Roeder 1997; Conrad et al. 1997; Trelles-Sticken et al. 2000), as well as by greatly reduced recombination rates and homolog interactions in *taz1⁻* and *rap1⁻ S. pombe* mutants that lack telomere attachment and clustering at the SPB (Cooper et al. 1998;

Nimmo et al. 1998; Chikashige and Hiraoka 2001; Kanoh and Ishikawa 2001). Furthermore, both telomere clustering and homolog pairing/synapsis are compromised in the *pam1* mutant of maize (Golubovskaya et al. 2002).

In fission yeast meiosis, telomere clustering and nuclear movements contribute differently to homolog pairing: Full levels of recombination and homologous associations along chromosome arms depend on telomere clustering, whereas nuclear movements mediate pairing of centromeric regions (Ding et al. 2004). Because *S. pombe* meiosis lacks a synaptonemal complex (Olson et al. 1978; Bahler et al. 1993) that fortifies homolog interactions, it seems to require both nuclear movements and telomere clustering to mediate and maintain close homolog association throughout prophase I (Ding et al. 2004). Rapid nuclear, bivalent, and telomere movements are observed during budding yeast meiosis at a time when telomere clustering has been resolved (White et al. 2004; C. Adelfalk and H. Scherthan, unpubl.). An increased telomere-independent motility late in prophase I may contribute to the late onset of homolog pairing in telomere-clustering-deficient *ndj1Δ* meiosis (Trelles-Sticken et al. 2000). Chromosome/nuclear motility in this mutant may increase the overall probability of encounters and may explain the increased recombinogenic interactions between homologous sequences at nonhomologous (ectopic) positions in the extended prophase I of *ndj1Δ* strains (Goldman and Lichten 2000). Similarly, disruption of the bouquet in *S. pombe* was shown to increase ectopic versus homologous recombination (Niwa et al. 2000), suggesting that telomere clustering facilitates homology search by generating a nuclear topology that reduces the complexity of chromosome interactions.

Chromosome arm painting and 3D microscopy in structurally preserved human spermatocytes have shown extensive intermingling of nonhomologous (and by inference homologous) DNA sequences between leptotene chromosomes (Scherthan et al. 1998), which is absent in somatic cells (for review, see Cremer and Cremer 2001). In grasshopper meiosis, the presence of homologous AEs of unequal length has been shown to prevent synapsis near attached ends (Moens et al. 1989; del Cerro and Santos 1997). Therefore, meiotic telomere clustering appears to confer a favorable topology at the nuclear periphery that may support homology search by bringing AEs and chromosome axes in close register and by pushing the DNA loops of leptotene chromosomes in a lampbrush-like fashion through each other. This will facilitate DNA loop/AE contacts that are considered requisite for synaptic homolog pairing, especially in recombinogenic meiosis (for an elaboration, see

Storlazzi et al. 2003). In some conditions the persistence or recurrence of telomere clustering may increase the likelihood of contacts for chromosomes that have failed to associate in a first round of homology search or when surplus chromosomes are present or homologs absent (Barlow et al. 2002; Trelles-Sticken et al. 2003; Roig et al. 2005). Telomere clustering may thereby facilitate hetero- or self-synapsis that is common under such conditions (see Loidl et al. 1991).

TELOMERE CLUSTERING IS INDEPENDENT OF RECOMBINATION

Cytological and genetic analyses have shown that telomere clustering occurs irrespective of the presence of homologs, synapsis, and/or recombination (Table 1). This has been observed, e.g., in haploid rye (Levan 1942; Santos et al. 1994) and wheat (Wang 1988), haploid yeast (Trelles-Sticken et al. 2003), the achiasmate (nonrecombinant) female meiosis of the dipteran *Bombyx mori* (Rasmussen 1977b), and recombination-deficient mutants of the ascomycete fungus *Sordaria*, mouse, and budding yeast (Trelles-Sticken et al. 1999; Storlazzi et al. 2003; B. Liebe et al., in prep.). In zygotic *S. pombe* meiosis, telomeres cluster at the SPB of haploid nuclei before these move toward each other, fuse, and form a zygote nucleus that undergoes meiosis (Chikashige et al. 1994). In *Arabidopsis*, meiotic telomere clustering establishes prior to entry in prophase I at the nucleolus and is relaxed during leptotene when recombination is initiated (Armstrong et al. 2001), whereas in polyploid plants homologous centromeres associate prior to telomere clustering that contributes to tight pairing and chromatin change during onset of prophase I (Martinez-Perez et al. 2000; Prieto et al. 2004). Collectively, these observations indicate that meiotic telomere movement and clustering are independent of the initiation of recombination.

MEIOTIC TELOMERE CLUSTER RESOLUTION IS COORDINATED WITH PROGRESS OF RECOMBINATION

In most species, telomere redistribution occurs during a time period of prophase I when meiotic chromosomes carry double-stranded DNA breaks. The latter initiate recombination (for review, see Roeder 1997; Keeney 2001) and contribute to homolog alignment and pairing in most species (for review, see Lichten 2001; Villeneuve and Hillers 2001). FISH analysis of mutants with defects in initiation or progression in recombinational repair has revealed that under such conditions telomere clustering is maintained for a longer period, as deduced from increased frequencies of bouquet cells in cell suspensions of sporulating cultures or reproductive organs (Table 1). This is the case in *spo11* mutants of budding yeast (Trelles-Sticken et al. 1999), *Sordaria* (Storlazzi et al. 2003), and mouse (B. Liebe et al., in prep.) that all fail to make meiotic double-strand breaks (DSBs) (Keeney et al. 1997; Baudat et al. 2000; Romanienko and Camerini-Otero 2000), suggesting that initiation and progress in recombination contribute to rapid exit from the bouquet stage. Prolonged chromosome polarization has also been noted in the recombination-defective *syp-1* mutant of *C. elegans* (MacQueen et al. 2002). In mammals, bouquet duration is extended in the sex with longer synaptonemal complex length (Pfeifer et al. 2003; Roig et al. 2004; Tankimanova et al. 2004). This raises the possibility that chromosome/axial compaction may contribute to telomere cluster resolution. This, at least in mammalian meiosis, seems unlikely as increased levels of telomere clustering have been observed in mice with normal ($H2ax^{-/-}$) and defective ($Sycp3^{-/-}$) bivalent condensation (Fernandez-Capetillo et al. 2003; Liebe et al. 2004; B. Liebe and H. Scherthan, unpubl.).

In some mutant conditions, persistence of the bouquet stage seems to be the consequence of a generally slowed transition through prophase I as cells take longer to pass a particular substage (Storlazzi et al. 2003). Differences in the duration of telomere clustering have been known from a variety of species (Belar 1928) and a sex-specific bouquet duration has been noted in cattle and human prophase I, with telomere clustering lasting longer in the female (Pfeifer et al. 2003; Roig et al. 2004; Tankimanova et al. 2004). In the human, where both genders display similar prophase I duration (Garcia et al. 1987), prolonged telomere clustering in oocytes coincides with delayed transition through intermediate steps of recombination and relative to progression of synapsis as assayed by colocalization of γ-H2AX and RPA recombination markers in pachytene oocytes (Roig et al. 2004). These observations and prolonged telomere clustering in mutants that form DSBs but fail in early steps of DSB processing like end resection or strand invasion (Table 1) (Trelles-Sticken et al. 1999; Storlazzi et al. 2003) suggest that the resolution of telomere clustering is influenced to some extent by progress of recombinational repair.

In agreement, aberrantly high frequencies of spermatocytes with clustered telomeres are detected in male mice lacking the ATM kinase (Pandita et al. 1999; Scherthan et al. 2000a), which is mutated in the chromosomal instability disorder ataxia telangiectasia, and its downstream target histone H2AX (Fernandez-Capetillo et al. 2003) that becomes phosphorylated in response to ectopic and meiotic DSBs (Rogakou et al. 1998; Mahadevaiah et al. 2001). The increased frequency of bouquet spermatocytes induced by

ATM disruption can be reverted to nearly wild-type values by Spo11 inactivation (B. Liebe et al., in prep.), which suggests that telomere movements and clustering are regulated by the machinery that senses and controls the repair of DSBs to which ATM is central (Abraham 2001; Pandita 2001). Interestingly, ATM has recently been found to physically interact with TRF2 telomere-binding protein (Karlseder et al. 2004), an interaction that provides a direct link between meiotic telomeres and ATM. Furthermore, it has been observed in *S. cerevisiae* meiosis that the dissolution of telomere clustering requires a functional cohesin complex (Trelles-Sticken et al. 2005). Since ATM phosphorylates the cohesin subunit Smc1 in response to DNA damage in vegetative cells (Kim et al. 2002; Yazdi et al. 2002), it will be interesting to see whether this DNA damage signaling branch is involved in regulating the exit from meiotic telomere clustering.

TELOMERE DISTRIBUTION IN POSTMEIOTIC CELLS

After meiocytes have passed both meiotic divisions, the resulting haploid cells undergo a complex differentiation that involves a dramatic condensation of the nucleus in spores or sperm. Telomere positioning has been studied in sperm, because the condensed nucleus of this cell type is still amenable to cytological analysis. Sperm chromatin condensation is associated with the replacement of most histones by transition proteins and finally by protamines (for review, see Zalenskaya and Zalensky 2002). Sperm telomeres remain associated with specialized histones (Churikov et al. 2004) and TRF proteins (H. Scherthan, unpubl.) and locate to the nuclear periphery where they form miniclusters or pairs (Fig. 4), which is the case in cattle, mouse, rat, and human sperm (Haaf and Ward 1995a; Zalensky et al. 1997; Meyer-Ficca et al. 1998). The peculiar architecture of the sperm nucleus seems to be the consequence of end-to-end associations of proximal and distal telomeres and may render a protective conformation to the male chromosomes, which facilitates their unpacking and decondensation in the zygote (Zalensky et al. 1995; Zalenskaya and Zalensky 2004). The maintenance of a special chromatin composition at sperm telomeres may protect the telomere complex and t-loops from disruption by protamine/histone replacement during nuclear condensation.

CONCLUSIONS

In summary, it appears that in the majority of sexually reproducing species, meiotic telomeres, unlike their somatic counterparts, become attached to the NE after cells have passed premeiotic S phase. This attachment seems to reduce chromosome entanglement and facilitates motility during homology search. Accumulation of all chromosome ends in a restricted nuclear subcompartment also provides a unique opportunity for setting a uniform telomere length among ends, e.g., by telomere rapid deletion or other recombination-mediated mechanisms (Joseph et al. 2005). Meiotic telomere clustering also takes up a role as matchmaker for meiotic chromosomes by bringing about physical proximity of chromosomes that failed to associate in a first round of homology search.

This predicts that meiotic telomere clustering should be dispensable in species where homolog pairing is present in premeiotic cells, does not rely on DSBs, and lacks the need for looped organization of DNA fibers and chromosome cores. Such a case may have developed in the meiosis of *Drosophila* males where homologs are juxtaposed prior to entry into prophase (Metz 1916; Hiraoka et al. 1993) and where recombination (Morgan 1912) and axial elements (Meyer 1960; Rasmussen 1973) are absent. Precise homologous pairing in male flies is associated with euchromatin decondensation at the NE (Vazquez et al. 2002) and special pairing sites (McKee and Karpen 1990; McKee et al. 1992; Merrill et al. 1992). The NE association of meiotic chromosomes may be required for stabilizing decondensed prophase I chromatin that seems to be instrumental for homology recognition. In the presence of DSBs and/or a large genome, the surface of the NE may be a limiting factor for chromosome stabilization. Extensive chromosome/NE interactions will impair chromosome movements and consequently homology search, especially in species with large genomes and/or high chromosome numbers. These restrictions do not apply when chromatin loops are attached to AEs and chromosome/NE attachment is limited to the ends of the cores, as is the case in the meiosis of most species. Interestingly, lateral elements and SC develop in the recombinogenic meiosis of *Drosophila* females (Carpenter 1975).

Overall, it seems plausible that meiotic chromosome cores and telomere/NE attachment have evolved to stabilize the relaxed chromatin of meiotic chromosomes when these undergo extensive movements and carry DSBs that are instrumental for homology search. Telomere clustering reduces the complexity of homology search, thereby improving the fidelity of meiosis. Finally, it is likely that meiotic telomere clustering has been evolutionarily conserved as a recombination-independent mechanism that is capable of mediating homolog association in karyotypes with rearranged chromosomes, because permutations in chromosome number and structure are the hallmarks of karyotypic evolution.

ACKNOWLEDGMENTS

I am grateful to C. Adelfalk, B. Liebe, and E. Trelles-Sticken for communication of results prior to their publication and to J. Loidl, University of Vienna, and C. Adelfalk for critical comments on the manuscript. I apologize to those whose work could be cited only through reviews. Support from the Deutsche Forschungsgemeinschaft and H.-H. Ropers, MPI for Molecular Genetics, Berlin, is acknowledged. The author is not aware of any conflicts of interest.

REFERENCES

Abraham R.T. 2001. Cell cycle checkpoint signaling through the ATM and ATR kinases. *Genes Dev.* **15:** 2177–2196.

Achi M.V., Ravindranath N., and Dym M. 2000. Telomere length in male germ cells is inversely correlated with telomerase activity. *Biol. Reprod.* **63:** 591–598.

Ahmed S. and Hodgkin J. 2000. MRT-2 checkpoint protein is required for germline immortality and telomere replication in *C. elegans. Nature* **403:** 159–164.

Alexander M.K. and Zakian V.A. 2003. Rap1p telomere association is not required for mitotic stability of a C(3)TA(2) telomere in yeast. *EMBO J.* **22:** 1688–1696.

Alsheimer M., von Glasenapp E., Hock R., and Benavente R. 1999. Architecture of the nuclear periphery of rat pachytene spermatocytes: Distribution of nuclear envelope proteins in relation to synaptonemal complex attachment sites. *Mol. Biol. Cell* **10:** 1235–1245.

Alsheimer M., Liebe B., Sewell L., Stewart C.L., Scherthan H., and Benavente R. 2004. Disruption of spermatogenesis in mice lacking A-type lamins. *J. Cell Sci.* **117:** 1173–1178.

Armstrong S.J., Franklin F.C., and Jones G.H. 2001. Nucleolus-associated telomere clustering and pairing precede meiotic chromosome synapsis in *Arabidopsis thaliana. J. Cell Sci.* **114:** 4207–4217.

Armstrong, S.J., Caryl A.P., Jones G.H., and Franklin F.C. 2002. Asy1, a protein required for meiotic chromosome synapsis, localizes to axis-associated chromatin in *Arabidopsis* and *Brassica. J. Cell Sci.* **115:** 3645–3655.

Ashley T., Walpita D., and de Rooij D.G. 2001. Localization of two mammalian cyclin dependent kinases during mammalian meiosis. *J. Cell Sci.* **114:** 685–693.

Bahler J., Wyler T., Loidl J., and Kohli J. 1993. Unusual nuclear structures in meiotic prophase of fission yeast: A cytological analysis. *J. Cell Biol.* **121:** 241–256.

Barlow A.L., Tease C., and Hulten M.A. 2002. Meiotic chromosome pairing in fetal oocytes of trisomy 21 human females. *Cytogenet. Genome Res.* **96:** 45–51.

Bascom-Slack C.A. and Dawson D.S. 1997. The yeast motor protein, Kar3p, is essential for meiosis I. *J. Cell Biol.* **139:** 459–467.

Bass H.W., Bordoli S.J., and Foss E.M. 2003. The desynaptic (dy) and desynaptic1 (dsy1) mutations in maize (*Zea mays* L) cause distinct telomere-misplacement phenotypes during meiotic prophase. *J. Exp. Bot.* **54:** 39–46.

Bass H.W., Riera-Lizarazu O., Ananiev E.V., Bordoli S.J., Rines H.W., Phillips R.L., Sedat J.W., Agard D.A., and Cande W.Z. 2000. Evidence for the coincident initiation of homolog pairing and synapsis during the telomere-clustering (bouquet) stage of meiotic prophase. *J. Cell Sci.* **113:** 1033–1042.

Baudat F., Manova K., Yuen J.P., Jasin M., and Keeney S. 2000. Chromosome synapsis defects and sexually dimorphic meiotic progression in mice lacking Spo11. *Mol. Cell* **6:** 989–998.

Belar K. 1928. Chromosomenreduktion. Die cytologischen Grundlagen der Vererbung. In *Handbuch der Vererbungswissenschaft* (ed. M. Hartmann), pp. 168–201. Geb. Borntraeger, Berlin.

Biessmann H., Mason J.M., Ferry K., d'Hulst M., Valgeirsdottir K., Traverse K.L., and Pardue M.L. 1990. Addition of telomere-associated HeT DNA sequences "heals" broken chromosome ends in *Drosophila*. *Cell* **61:** 663–673.

Blasco M.A., Lee H.W., Hande M.P., Samper E., Lansdorp P.M., DePinho R.A., and Greider C.W. 1997. Telomere shortening and tumor formation by mouse cells lacking telomerase RNA. *Cell* **91:** 25–34.

Boa S., Coert C., and Patterton H.G. 2003. *Saccharomyces cerevisiae* Set1p is a methyltransferase specific for lysine 4 of histone H3 and is required for efficient gene expression. *Yeast* **20:** 827–835.

Bojko M. 1983. Human meiosis VIII. Chromosome pairing and formation of the synaptonemal complex in oocytes. *Carlsberg Res. Commun.* **48:** 457–483.

———. 1990. Synaptic adjustment of inversion loops in *Neurospora crassa*. *Genetics* **124:** 593–598.

Boveri T. 1904. *Ergebnisse über die Konstitution der chromatischen Substanz des Zellkerns.* Gustav Fischer, Jena.

Briggs S.D., Bryk M., Strahl B.D., Cheung W.L., Davie J.K., Dent S.Y., Winston F., and Allis C.D. 2001. Histone H3 lysine 4 methylation is mediated by Set1 and required for cell growth and rDNA silencing in *Saccharomyces cerevisiae*. *Genes Dev.* **15:** 3286–3295.

Byers B. and Goetsch L. 1975. Electron microscopic observations on the meiotic karyotype of diploid and tetraploid *Saccharomyces cerevisiae*. *Proc. Natl. Acad. Sci.* **72:** 5056–5060.

Carlton P.M. and Cande W.Z. 2002. Telomeres act autonomously in maize to organize the meiotic bouquet from a semipolarized chromosome orientation. *J. Cell Biol.* **157:** 231–242.

Carlton P.M., Cowan C.R., and Cande W.Z. 2003. Directed motion of telomeres in the formation of the meiotic bouquet revealed by time course and simulation analysis. *Mol. Biol. Cell* **14:** 2832–2843.

Carpenter A.T. 1975. Electron microscopy of meiosis in *Drosophila melanogaster* females. I. Structure, arrangement, and temporal change of the synaptonemal complex in wild-type. *Chromosoma* **51:** 157–182.

Cherry L. and Hsu T. 1984. Antitubulin immunofluorescence studies of spermatogenesis in the mouse. *Chromosoma:* **90:** 265–274.

Chikashige Y. and Hiraoka Y. 2001. Telomere binding of the Rap1 protein is required for meiosis in fission yeast. *Curr. Biol.* **11:** 1618–1623.

Chikashige Y., Ding D.Q., Funabiki H., Haraguchi T., Mashiko S., Yanagida M., and Hiraoka Y. 1994. Telomere-led premeiotic chromosome movement in fission yeast. *Science* **264:** 270–273.

Chua P.R. and Roeder G.S. 1997. Tam1, a telomere-associated meiotic protein, functions in chromosome synapsis and crossover interference. *Genes Dev.* **11:** 1786–1800.

Churikov D., Zalenskaya I.A., and Zalensky A.O. 2004. Male germline-specific histones in mouse and man. *Cytogenet. Genome Res.* **105:** 203–214.

Conrad M.N., Dominguez A.M., and Dresser M.E. 1997. Ndj1p, a meiotic telomere protein required for normal chromosome synapsis and segregation in yeast. *Science* **276:** 1252–1255.

Cooper J.P., Watanabe Y., and Nurse P. 1998. Fission yeast Taz1 protein is required for meiotic telomere clustering and recombination. *Nature* **392:** 828–831.

Couteau F., Nabeshima K., Villeneuve A., and Zetka M. 2004. A component of *C. elegans* meiotic chromosome axes at the interface of homolog alignment, synapsis, nuclear reorganization, and recombination. *Curr. Biol.* **14:** 585–592.

Cowan C.R. and Cande W.Z. 2002. Meiotic telomere clustering is inhibited by colchicine but does not require cytoplasmic microtubules. *J. Cell Sci.* **115:** 3747–3756.

Cowan C.R., Carlton P.M., and Cande W.Z. 2002. Reorganization and polarization of the meiotic bouquet-stage cell can be uncoupled from telomere clustering. *J. Cell Sci.* **115:** 3757–3766.

Cremer T. and Cremer C. 2001. Chromosome territories, nuclear architecture and gene regulation in mammalian cells. *Nat. Rev. Genet.* **2:** 292–301.

de Lange T. 1992. Human telomeres are attached to the nuclear matrix. *EMBO J.* **11:** 717–724.

del Cerro L. and Santos J. 1997. Chiasma redistribution in the presence of different sized supernumerary segments in a grasshopper: Dependence on nonhomologous synapsis. *Genome* **40:** 682–688.

Ding D.Q., Chikashige Y., Haraguchi T., and Hiraoka Y. 1998. Oscillatory nuclear movement in fission yeast meiotic prophase is driven by astral microtubules, as revealed by continuous observation of chromosomes and microtubules in living cells. *J. Cell Sci.* **111:** 701–712.

Ding D.Q., Yamamoto A., Haraguchi T., and Hiraoka Y. 2004. Dynamics of homologous chromosome pairing during meiotic prophase in fission yeast. *Dev. Cell* **6:** 329–341.

Dobson M.J., Pearlman R.E., Karaiskakis A., Spyropoulos B., and Moens P.B. 1994. Synaptonemal complex proteins: Occurrence, epitope mapping and chromosome disjunction. *J. Cell Sci.* **107:** 2749–2760.

Dong F. and Jiang J. 1998. Non-Rabl patterns of centromere and telomere distribution in the interphase nuclei of plant cells. *Chromosome Res.* **6:** 551–558.

Dorninger D., Karigl G., and Loidl J. 1995. Simulation of chromosomal homology searching in meiotic pairing. *J. Theor. Biol.* **176:** 247–260.

Egel-Mitani M., Olson L.W., and Egel R. 1982. Meiosis in *Aspergillus nidulans*: Another example for lacking synaptonemal complexes in the absence of crossover interference. *Hereditas* **97:** 179–187.

Eijpe M., Offenberg H., Jessberger R., Revenkova E., and Heyting C. 2003. Meiotic cohesin REC8 marks the axial elements of rat synaptonemal complexes before cohesins SMC1beta and SMC3. *J. Cell Biol.* **160:** 657–670.

Eisen G. 1900. The spermatogenesis of *Batrachoseps*. *J. Morphol.* **17:** 1–117.

Esponda P. and Gimenez-Martin G. 1972. The attachment of the synaptonemal complex to the nuclear envelope. An ultrastructural and cytochemical analysis. *Chromosoma* **38:** 405–417.

Fawcett D.W. 1956. The fine structure of chromosomes in the meiotic prophase of vertebrate spermatocytes. *J. Biophys. Biochem. Cytol.* **2:** 403.

Fernandez-Capetillo O., Liebe B., Scherthan H., and Nussenzweig A. 2003. H2AX regulates meiotic telomere clustering. *J. Cell Biol.* **163:** 15–20.

Flemming W. 1895. Über die Wirkung von Chrom-Osmium-Essigsäure auf Zellkerne. *Arch. Mikrosk. Anat.* **45:** 162–166.

Flory M.R., Carson A.R., Muller E.G., and Aebersold R. 2004. An SMC-domain protein in fission yeast links telomeres to the meiotic centrosome. *Mol. Cell* **16:** 619–630.

Franco S., Alsheimer M., Herrera E., Benavente R., and Blasco M.A. 2002. Mammalian meiotic telomeres: Composition and ultrastructure in telomerase-deficient mice. *Eur. J. Cell Biol.* **81:** 335–340.

Franke W. 1971. Relationship of nuclear membranes with filaments and microtubules. *Protoplasma* **73:** 263–292.

Funabiki H., Hagan I., Uzawa S., and Yanagida M. 1993. Cell cycle-dependent specific positioning and clustering of centromeres and telomeres in fission yeast. *J. Cell Biol.* **121:** 961–976.

Fussell C.P. 1987. The Rabl orientation: A prelude to synapsis. In *Meiosis* (ed. P.B. Moens), pp. 275–299. Academic Press, Orlando, FL.

Galy V., Olivo-Marin J.C., Scherthan H., Doye V., Rascalou N., and Nehrbass U. 2000. Nuclear pore complexes in the organization of silent telomeric chromatin. *Nature* **403:** 108–112.

Garcia M., Dietrich A.J., Freixa L., Vink A.C., Ponsa M., and Egozcue J. 1987. Development of the first meiotic prophase stages in human fetal oocytes observed by light and electron microscopy. *Hum. Genet.* **77:** 223–232.

Gelei J. 1921. Weitere Studien über die Oogenese des Dendrocoelum lacteum. III. Die Konjugation der Chromosomen in der Literatur und meine Befunde. *Arch. Zellforsch.* **16:** 300–365.

Gillies C.B. 1975. Synaptonemal complex and chromosome structure. *Annu. Rev. Genet.* **9:** 91–109.

Goldman A.S. and Lichten M. 2000. Restriction of ectopic recombination by interhomolog interactions during *Saccharomyces cerevisiae* meiosis. *Proc. Natl. Acad. Sci.* **97:** 9537–9542.

Goldstein P. and Slaton D.E. 1982. The synaptonemal complexes of *Caenorhabditis elegans:* Comparison of wild-type and mutant strains and pachytene karyotype analysis of wild-type. *Chromosoma* **84:** 585–597.

Golubovskaya I.N., Harper L.C., Pawlowski W.P., Schichnes D., and Cande W.Z. 2002. The pam1 gene is required for meiotic bouquet formation and efficient homologous synapsis in maize (*Zea mays* L.). *Genetics* **162:** 1979–1993.

Haaf T. and Ward D.C. 1995a. Higher order nuclear structure in mammalian sperm revealed by in situ hybridization and extended chromatin fibers. *Exp. Cell Res.* **219:** 604–611.

———. 1995b. Rabl orientation of CENP-B box sequences in *Tupaia belangeri* fibroblasts. *Cytogenet. Cell Genet.* **70:** 258–262.

Hall I.M., Noma K., and Grewal S.I. 2003. RNA interference machinery regulates chromosome dynamics during mitosis and meiosis in fission yeast. *Proc. Natl. Acad. Sci.* **100:** 193–198.

Hausmann M., Liebe B., Perner B., Jerratsch M., Greulich K.O., and Scherthan H. 2003. Imaging of human meiotic chromosomes by scanning near-field optical microscopy (SNOM). *Micron* **34:** 441–447.

Hecht A., Laroche T., Strahl-Bolsinger S., Gasser S.M., and Grunstein M. 1995. Histone H3 and H4 N-termini interact with SIR3 and SIR4 proteins: A molecular model for the formation of heterochromatin in yeast. *Cell* **80:** 583–592.

Hemann M.T., Rudolph K.L., Strong M.A., DePinho R.A., Chin L., and Greider C.W. 2001. Telomere dysfunction triggers developmentally regulated germ cell apoptosis. *Mol. Biol. Cell* **12:** 2023–2030.

Heng H.H., Chamberlain J.W., Shi X.M., Spyropoulos B., Tsui L.C., and Moens P.B. 1996. Regulation of meiotic chromatin loop size by chromosomal position. *Proc. Natl. Acad. Sci.* **93:** 2795–2800.

Heyting C. 1996. Synaptonemal complexes: Structure and function. *Curr. Opin. Cell Biol.* **8:** 389–396.

Hiraoka T. 1952. Observational and experimental studies of meiosis with special reference to the bouquet stage. XIV. Some considerations on a probable mechanism of the bouquet formation. *Cytologia* **17:** 292–299.

Hiraoka Y., Dernburg A.F., Parmelee S.J., Rykowski M.C., Agard D.A., and Sedat J.W. 1993. The onset of homologous chromosome pairing during *Drosophila melanogaster* embryogenesis. *J. Cell Biol.* **120:** 591–600.

Hochstrasser M., Mathog D., Gruenbaum Y., Saumweber H., and Sedat J.W. 1986. Spatial organization of chromosomes in the salivary gland nuclei of *Drosophila melanogaster*. *J. Cell Biol.* **102:** 112–123.

Honigberg S.M. and Purnapatre K. 2003. Signal pathway integration in the switch from the mitotic cell cycle to meiosis in yeast. *J. Cell Sci.* **116:** 2137–2147.

Ishikawa F. and Naito T. 1999. Why do we have linear chromosomes? A matter of Adam and Eve. *Mutat. Res.* **434:** 99–107.

James R.D., Schmiesing J.A., Peters A.H., Yokomori K., and Disteche C.M. 2002. Differential association of SMC1α and SMC3 proteins with meiotic chromosomes in wild-type and SPO11-deficient male mice. *Chromosome Res.* **10:** 549–560.

Jessberger R. 2002. The many functions of SMC proteins in chromosome dynamics. *Nat. Rev. Mol. Cell Biol.* **3:** 767–778.

Jin Q.W., Fuchs J., and Loidl J. 2000. Centromere clustering is a major determinant of yeast interphase nuclear organization. *J. Cell Sci.* **113:** 1903–1912.

Jin Q., Trelles-Sticken E., Scherthan H., and Loidl J. 1998. Yeast nuclei display prominent centromere clustering that is reduced in nondividing cells and in meiotic prophase. *J. Cell Biol.* **141:** 21–29.

Jin Y., Uzawa S., and Cande W.Z. 2002. Fission yeast mutants affecting telomere clustering and meiosis-specific spindle pole body integrity. *Genetics* **160:** 861–876.

John B. 1990. *Meiosis.* Cambridge University Press, Cambridge, New York, Sidney.

Joseph I., Jia D., and Lustig A.J. 2005. Ndj1p-dependent epigenetic resetting of telomere size in yeast meiosis. *Curr. Biol.* **15:** 231–237.

Kanoh J. and Ishikawa F. 2001. spRap1 and spRif1, recruited to telomeres by Taz1, are essential for telomere function in fission yeast. *Curr. Biol.* **11:** 1624–1630.

Karlseder J., Hoke K., Mirzoeva O.K., Bakkenist C., Kastan M.B., Petrini J.H., and Lange Td. T. 2004. The telomeric protein TRF2 binds the ATM kinase and can inhibit the ATM-dependent DNA damage response. *PLoS Biol.* **2:** E240.

Keeney S. 2001. Mechanism and control of meiotic recombination initiation. *Curr. Top. Dev. Biol.* **52:** 1–53.

Keeney S., Giroux C.N., and Kleckner N. 1997. Meiosis-specific DNA double-strand breaks are catalyzed by Spo11, a member of a widely conserved protein family. *Cell* **88:** 375–384.

Kezer J. and Macgregor H.C. 1971. A fresh look at meiosis and centromeric heterochromatin in the red-backed salamander, *Plethodon cinereus cinereus* (Green). *Chromosoma* **33:** 146–166.

Kim S-T., Xu B., and Kastan M.B. 2002. Involvement of the cohesin protein, Smc1, in Atm-dependent and -independent responses to DNA damage. *Genes Dev.* **16:** 560–570.

Kimura T., Ito C., Watanabe S., Takahashi T., Ikawa M., Yomogida K., Fujita Y., Ikeuchi M., Asada N., Matsumiya K., Okuyama A., Okabe M., Toshimori K., and Nakano T. 2003. Mouse germ cell-less as an essential component for nuclear integrity. *Mol. Cell. Biol.* **23:** 1304–1315.

Klein F., Laroche T., Cardenas M.E., Hofmann J.F., Schweizer D., and Gasser S.M. 1992. Localization of RAP1 and topoisomerase II in nuclei and meiotic chromosomes of yeast. *J. Cell Biol.* **117:** 935–948.

Klein F., Mahr P., Galova M., Buonomo S.B., Michaelis C., Nairz K., and Nasmyth K. 1999. A central role for cohesins in sister chromatid cohesion, formation of axial elements, and recombination during yeast meiosis. *Cell* **98:** 91–103.

Lachner M. and Jenuwein T. 2002. The many faces of histone lysine methylation. *Curr. Opin. Cell Biol.* **14:** 286–298.

Lammers J.H., Offenberg H.H., van Aalderen M., Vink A.C., Dietrich A.J., and Heyting C. 1994. The gene encoding a major component of the lateral elements of synaptonemal complexes of the rat is related to X-linked lymphocyte-regulated genes. *Mol. Cell. Biol.* **14:** 1137–1146.

Lee H.W., Blasco M.A., Gottlieb G.J., Horner J.W., 2nd, Greider C.W., and DePinho R.A. 1998. Essential role of mouse telomerase in highly proliferative organs. *Nature* **392:** 569–574.

Levan A. 1942. Studies on the mechanism of haploid rye. *Hereditas* **28:** 177–211.

Lichten M. 2001. Meiotic recombination: Breaking the genome to save it. *Curr. Biol.* **11:** R253–256.

Liebe B., Alsheimer M., Hoog C., Benavente R., and Scherthan H. 2004. Telomere attachment, meiotic chromosome condensation, pairing, and bouquet stage duration are modified in spermatocytes lacking axial elements. *Mol. Biol. Cell* **15:** 827–837.

Liu L., Franco S., Spyropoulos B., Moens P.B., Blasco M.A., and Keefe D.L. 2004. Irregular telomeres impair meiotic synapsis and recombination in mice. *Proc. Natl. Acad. Sci.* **101:** 6496–6501.

Loidl J. and Scherthan H. 2004. Organization and pairing of meiotic chromosomes in the ciliate *Tetrahymena thermophila*. *J. Cell Sci.* **117:** 5791–5801.

Loidl J., Nairz K., and Klein F. 1991. Meiotic chromosome synapsis in a haploid yeast. *Chromosoma* **100:** 221–228.

Lombard D.B. and Guarente L. 2000. Nijmegen breakage syndrome disease protein and MRE11 at PML nuclear bodies and meiotic telomeres. *Cancer. Res.* **60:** 2331–2334.

Longtine M.S., Wilson N.M., Petracek M.E., and Berman J. 1989. A yeast telomere binding activity binds to two related telomere sequence motifs and is indistinguishable from RAP1. *Curr. Genet.* **16:** 225–239.

Luderus M.E., van Steensel B., Chong L., Sibon O.C., Cremers F.F., and de Lange T. 1996. Structure, subnuclear distribution, and nuclear matrix association of the mammalian telomeric complex. *J. Cell Biol.* **135:** 867–881.

Mackay S. 2000. Gonadal develoment in mammals at the cellular and molecular levels. *Int. Rev. Cytol.* **200:** 47–99.

MacQueen A.J. and Villeneuve A.M. 2001. Nuclear reorganization and homologous chromosome pairing during meiotic prophase require *C. elegans chk-2*. *Genes Dev.* **15:** 1674–1687.

MacQueen A.J., Colaiacovo M.P., McDonald K., and Villeneuve A.M. 2002. Synapsis-dependent and -independent mechanisms stabilize homolog pairing during meiotic prophase in *C. elegans*. *Genes Dev.* **16:** 2428–2442.

Maddar H., Ratzkovsky N., and Krauskopf A. 2001. Role for telomere cap structure in meiosis. *Mol. Biol. Cell* **12:** 3191–3203.

Maeda T. 1939. Chiasma studies in the silkworm *Bombyx mori* L. *Jap. J. Genet.* **15:** 118–127.

Mahadevaiah S.K., Turner J.M., Baudat F., Rogakou E.P., de Boer P., Blanco-Rodriguez J., Jasin M., Keeney S., Bonner W.M., and Burgoyne P.S. 2001. Recombinational DNA double-strand breaks in mice precede synapsis. *Nat. Genet.* **27:** 271–276.

Manheim E.A. and McKim K.S. 2003. The synaptonemal complex component C(2)M regulates meiotic crossing over in *Drosophila. Curr. Biol.* **13:** 276–285.

Marston A.L. and Amon A. 2004. Meiosis: Cell-cycle controls shuffle and deal. *Nat. Rev. Mol. Cell Biol.* **5:** 983–997.

Martinez-Perez E., Shaw P., Reader S., Aragon-Alcaide L., Miller T., and Moore G. 1999. Homologous chromosome pairing in wheat. *J. Cell Sci.* **112:** 1761–1769.

Martinez-Perez E., Shaw P.J., and Moore G. 2000. Polyploidy induces centromere association. *J. Cell Biol.* **148:** 233–238.

McKee B.D. 2004. Homologous pairing and chromosome dynamics in meiosis and mitosis. *Biochim. Biophys. Acta* **1677:** 165–180.

McKee B.D. and Karpen G.H. 1990. *Drosophila* ribosomal RNA genes function as an X-Y pairing site during male meiosis. *Cell* **61:** 61–72.

McKee B.D., Habera L., and Vrana J.A. 1992. Evidence that intergenic spacer repeats of *Drosophila melanogaster* rRNA genes function as X-Y pairing sites in male meiosis, and a general model for achiasmatic pairing. *Genetics* **132:** 529–544.

Merrill C.J., Chakravarti D., Habera L., Das S., Eisenhour L., and McKee. B.D. 1992. Promoter-containing ribosomal DNA fragments function as X-Y meiotic pairing sites in *D. melanogaster* males. *Dev. Genet.* **13:** 468–484.

Metz C.W. 1916. Chromosome studies on the Diptera: II. The paired association of chromosomes in the Diptera and its significance. *J. Exp. Zool.* **21:** 213–279.

Meuwissen R.L., Offenberg H.H., Dietrich A.J., Riesewijk A., van Iersel M., and Heyting C. 1992. A coiled-coil related protein specific for synapsed regions of meiotic prophase chromosomes. *EMBO J.* **11:** 5091–5100.

Meyer G.F. 1960. The fine structure of spermatocyte nuclei of *Drosophila melanogaster. Proc. Eur. Reg. Conf. Electron Micros.* **2:** 951–954.

Meyer-Ficca M., Muller-Navia J., and Scherthan H. 1998. Clustering of pericentromeres initiates in step 9 of spermiogenesis of the rat (*Rattus norvegicus*) and contributes to a well defined genome architecture in the sperm nucleus. *J. Cell Sci.* **111:** 1363–1370.

Mikhailova E.I., Sosnikhina S.P., Kirillova G.A., Tikholiz O.A., Smirnov V.G., Jones R.N., and Jenkins G. 2001. Nuclear dispositions of subtelomeric and pericentromeric chromosomal domains during meiosis in asynaptic mutants of rye (*Secale cereale* L.). *J. Cell Sci.* **114:** 1875–1882.

Miki F., Okazaki K., Shimanuki M., Yamamoto A., Hiraoka Y., and Niwa O. 2002. The 14-kDa dynein light chain-family protein Dlc1 is required for regular oscillatory nuclear movement and efficient recombination during meiotic prophase in fission yeast. *Mol. Biol. Cell.* **13:** 930–946.

Moens P.B. 1969a. The fine structure of meiotic chromosome pairing in the triploid, *Lilium tigrinum. J. Cell Biol.* **40:** 273–279.

———. 1969b. The fine structure of meiotic chromosome polarization and pairing in *Locusta migratoria* spermatocytes. *Chromosoma* **28:** 1–25.

Moens P.B. and Pearlman R.E. 1990. Telomere and centromere DNA are associated with the cores of meiotic prophase chromosomes. *Chromosoma* **100:** 8–14.

Moens P., Bernelot-Moens C., and Spyropulos B. 1989. Chromosome core attachment to the nuclear envelope regulates synapsis in *Choelatis* (Orthoptera). *Genome* **32:** 601–610.

Molnar M., Doll E., Yamamoto A., Hiraoka Y., and Kohli J. 2003. Linear element formation and their role in meiotic sister chromatid cohesion and chromosome pairing. *J. Cell Sci.* **116:** 1719–1731.

Morgan T.H. 1912. Complete linkage in the second chromosome of the male of *Drosophila*. *Science* **36:** 719–720.

Moses M.J. 1956. Chromosomal structures in crayfish spermatocytes. *J. Biophys. Biochem. Cytol.* **2:** 215–218.

———. 1968. Synaptonemal complex. *Annu. Rev. Genet.* **2:** 363–412.

Naito T., Matsuura A., and Ishikawa F. 1998. Circular chromosome formation in a fission yeast mutant defective in two ATM homologs. *Nat. Genet.* **20:** 203–206.

Nasmyth K. 2002. Segregating sister genomes: The molecular biology of chromosome separation. *Science* **297:** 559–565.

Nili E., Cojocaru G.S., Kalma Y., Ginsberg D., Copeland N.G., Gilbert D.J., Jenkins N.A., Berger R., Shaklai S., Amariglio N., Brok-Simoni F., Simon A.J., and Rechavi G. 2001. Nuclear membrane protein LAP2β mediates transcriptional repression alone and together with its binding partner GCL (germ-cell-less). *J. Cell Sci.* **114:** 3297–3307.

Nimmo E.R., Pidoux A.L., Perry P.E., and Allshire R.C. 1998. Defective meiosis in telomere-silencing mutants of *Schizosaccharomyces pombe*. *Nature* **392:** 825–828.

Niwa O., Shimanuki M., and Miki F. 2000. Telomere-led bouquet formation facilitates homologous chromosome pairing and restricts ectopic interaction in fission yeast meiosis. *EMBO J.* **19:** 3831–3840.

Offenberg H.H., Schalk J.A., Meuwissen R.L., van Aalderen M., Kester H.A., Dietrich A.J., and Heyting C. 1998. SCP2: A major protein component of the axial elements of synaptonemal complexes of the rat. *Nucleic Acids Res* **26:** 2572–2579.

Oliveira C., Foresti F., Rigolino M., and Tabata Y. 1995. Synaptonemal complex formation in spermatocytes of the autotriploid rainbow trout, *Oncorhynchus mykiss* (Pisces, Salmonidae). *Hereditas* **123:** 215–220.

Olson L., Eden U., Egel-Mitani M., and Egel R. 1978. Asynaptic meiosis in fission yeast? *Hereditas* **89:** 189–199.

Page S.L. and Hawley R.S. 2003. Chromosome choreography: The meiotic ballet. *Science* **301:** 785–789.

———. 2004. The genetics and molecular biology of the synaptonemal complex. *Annu. Rev. Cell Dev. Biol.* **20:** 525–558.

Pandita T.K. 2001. The role of ATM in telomere structure and function. *Radiat. Res.* **156:** 642–647.

Pandita T.K., Westphal C.H., Anger M., Sawant S.G., Geard C.R., Pandita R.K., and Scherthan H. 1999. Atm inactivation results in aberrant telomere clustering during meiotic prophase. *Mol. Cell. Biol.* **19:** 5096–5105.

Parvinen M. and Soderstrom K.O. 1976. Chromosome rotation and formation of synapsis. *Nature* **260:** 534–535.

Pasierbek P., Fodermayr M., Jantsch V., Jantsch M., Schweizer D., and Loidl J. 2003. The *Caenorhabditis elegans* SCC-3 homolog is required for meiotic synapsis and for proper chromosome disjunction in mitosis and meiosis. *Exp. Cell Res.* **289:** 245–255.

Pelttari J., Hoja M.R., Yuan L., Liu J.G., Brundell E., Moens P., Santucci-Darmanin S., Jessberger R., Barbero J.L., Heyting C., and Hoog C. 2001. A meiotic chromosomal core consisting of cohesin complex proteins recruits DNA recombination proteins and promotes synapsis in the absence of an axial element in mammalian meiotic cells. *Mol. Cell. Biol.* **21:** 5667–5677.

Peters A.H., O'Carroll D., Scherthan H., Mechtler K., Sauer S., Schofer C., Weipoltshammer K., Pagani M., Lachner M., Kohlmaier A., Opravil S., Doyle M., Sibilia M., and Jenuwein T. 2001. Loss of the Suv39h histone methyltransferases impairs mammalian heterochromatin and genome stability. *Cell* **107:** 323–337.

Petronczki M., Siomos M.F., and Nasmyth K. 2003. Un menage a quatre: The molecular biology of chromosome segregation in meiosis. *Cell* **112:** 423–440.

Pezzi N., Prieto I., Kremer L., Perez Jurado L.A., Valero C., Del Mazo J., Martinez A.C., and Barbero J.L. 2000. STAG3, a novel gene encoding a protein involved in meiotic chromosome pairing and location of STAG3-related genes flanking the Williams-Beuren syndrome deletion. *FASEB J.* **14:** 581–592.

Pfeifer C., Scherthan H., and Thomsen P.D. 2003. Sex-specific telomere redistribution and synapsis initiation in cattle oogenesis. *Dev. Biol.* **255:** 206–215.

Pfeifer C., Thomsen P.D., and Scherthan H. 2001. Centromere and telomere redistribution precedes homologue pairing and terminal synapsis initiation during prophase I of cattle spermatogenesis. *Cytogenet. Cell Genet.* **93:** 304–314.

Platner G. 1885. Ueber die Entstehung des Nebenkerns und seine Beziehung zur Kerntheilung. *Arch. Mikrosk. Anatomie* **26:** 343–369.

Podgornaya O.I., Bugaeva E.A., Voronin A.P., Gilson E., and Mitchell A.R. 2000. Nuclear envelope associated protein that binds telomeric DNAs. *Mol. Reprod. Dev.* **57:** 16–25.

Prieto I., Suja J.A., Pezzi N., Kremer L., Martinez A.C., Rufas J.S., and Barbero J.L. 2001. Mammalian STAG3 is a cohesin specific to sister chromatid arms in meiosis I. *Nat. Cell Biol.* **3:** 761–766.

Prieto I., Pezzi N., Buesa J.M., Kremer L., Barthelemy I., Carreiro C., Roncal F., Martinez A., Gomez L., Fernandez R., Martinez A.C., and Barbero J.L. 2002. STAG2 and Rad21 mammalian mitotic cohesins are implicated in meiosis. *EMBO Rep.* **3:** 543–550.

Prieto P., Shaw P., and Moore G. 2004. Homologue recognition during meiosis is associated with a change in chromatin conformation. *Nat. Cell Biol.* **6:** 906–908.

Rabl C. 1885. Über Zelltheilung. *Morphol. Jahrbuch* **10:** 214–330.

Rasmussen S.W. 1973. Ultrastructural studies of spermatogenesis in *Drosophila melanogaster* Meigen. *Z. Zellforsch. Mikrosk. Anat.* **140:** 125–144.

———. 1977a. Chromosome pairing in triploid females of *Bombyx mori* analyzed by three dimensional reconstructions of synaptonemal complexes. *Carlsberg Res. Commun.* **42:** 163–197.

———. 1977b. Meiosis in *Bombyx mori* females. *Philos. Trans. R. Soc. Lond. B Biol. Sci.* **277:** 343–350.

Ravindranath N., Dalal R., Solomon B., Djakiew D., and Dym M. 1997. Loss of telomerase activity during male germ cell differentiation. *Endocrinology* **138:** 4026–4029.

Reichenow E. 1927. Ergebnisse der Nuklaerfärbung bei Protozoen. Vorbemerkungen und 1. Teil: Die Entwicklungsgeschichte von Caryolysus. *Archiv. Protistenkd* **61:** 144–166.

Revenkova E., Eijpe M., Heyting C., Hodges C.A., Hunt P.A., Liebe B., Scherthan H., and Jessberger R. 2004. Cohesin SMC1β is required for meiotic chromosome dynamics, sister chromatid cohesion and DNA recombination. *Nat. Cell Biol.* **6:** 555–562.

Rhoades M. 1961. Meioses. In *The cell: Biochemistry, physiology, and morphology* (ed. A. Mirsky), pp. 1–75. Academic Press, New York.

Rockmill B. and Roeder G.S. 1998. Telomere-mediated chromosome pairing during meiosis in budding yeast. *Genes Dev.* **12:** 2574–2586.

Roeder G.S. 1997. Meiotic chromosomes: It takes two to tango. *Genes Dev.* **11:** 2600–2621.

Rogakou E.P., Pilch D.R., Orr A.H., Ivanova V.S., and Bonner W.M. 1998. DNA double-stranded breaks induce histone H2AX phosphorylation on serine 139. *J. Biol. Chem.* **273:** 5858–5868.

Roig I., Liebe B., Egozcue J., Cabero L., Garcia M., and Scherthan H. 2004. Female-specific features of recombinational double-stranded DNA repair in relation to synapsis and telomere dynamics in human oocytes. *Chromosoma* **113:** 22–33.

Roig I., Robles P., Garcia R., Martínez-Flores I., Cabero L., Egozcue J., Liebe B., Scherthan H., and Garcia M. 2005. Chromosome 18 pairing behavior in human trisomic oocytes. Presence of an extra chromosome extends bouquet stage. *Reproduction* **129:** 565–575.

Romanienko P.J. and Camerini-Otero R.D. 2000. The mouse Spo11 gene is required for meiotic chromosome synapsis. *Mol. Cell* **6:** 975–987.

Sadaie M., Naito T., and Ishikawa F. 2003. Stable inheritance of telomere chromatin structure and function in the absence of telomeric repeats. *Genes Dev.* **17:** 2271–2282.

Salonen K., Paranko J., and Parvinen M. 1982. A colcemid-sensitive mechanism involved in regulation of chromosome movements during meiotic pairing. *Chromosoma* **85:** 611–618.

Santos J., Jimenez M., and Diaz M. 1994. Meiosis in haploid rye: Extensive synapsis and low chiasma frequency. *Heredity* **73:** 580–588.

Saunders W., Hornack D., Lengyel V., and Deng C. 1997. The *Saccharomyces cerevisiae* kinesin-related motor Kar3p acts at preanaphase spindle poles to limit the number and length of cytoplasmic microtubules. *J. Cell Biol.* **137:** 417–431.

Scherthan H., Jerratsch M., Dhar S., Wang Y.A., Goff S.P., and Pandita T.K. 2000a. Meiotic telomere distribution and Sertoli cell nuclear architecture are altered in Atm- and Atm-p53-deficient mice. *Mol. Cell. Biol.* **20:** 7773–7783.

Scherthan H., Weich S., Schwegler H., Heyting C., Harle M., and Cremer T. 1996. Centromere and telomere movements during early meiotic prophase of mouse and man are associated with the onset of chromosome pairing. *J. Cell Biol.* **134:** 1109–1125.

Scherthan H., Eils R., Trelles-Sticken E., Dietzel S., Cremer T., Walt H., and Jauch A. 1998. Aspects of three-dimensional chromosome reorganization during the onset of human male meiotic prophase. *J. Cell Sci.* **111:** 2337–2351.

Scherthan H., Jerratsch M., Li B., Smith S., Hulten M., Lock T., and de Lange T. 2000b. Mammalian meiotic telomeres: Protein composition and redistribution in relation to nuclear pores. *Mol. Biol. Cell* **11:** 4189–4203.

Schreiner A. and Schreiner K.E. 1905. Über die Entwicklung der männlichen Geschlechtszellen von *Myxine glutinosa* (L). *Arch. Biol.* **21:** 183–314.

Schmit A.C., Endle M.C., and Lambert A.M. 1996. The perinuclear microtubule-organizing center and the synaptonemal complex of higher plants share a common antigen: Its putative transfer and role in meiotic chromosomal ordering. *Chromosoma* **104:** 405–413.

Sheldon J., Wilson C., and Dickenson H.G. 1988. Interaction between the nucleus and cytoskeleton during the pairing stages of male meiosis in flowering plants. In *Kew Chromosome Conference III* (ed. P.E. Brandham), pp. 27–35. Her Majesty's Stationery Office, London.

Shimanuki M., Miki F., Ding D.Q., Chikashige Y., Hiraoka Y., Horio T., and Niwa O. 1997. A novel fission yeast gene, kms1+, is required for the formation of meiotic prophase-specific nuclear architecture. *Mol. Gen. Genet.* **254:** 238–249.

Siomos M.F., Badrinath A., Pasierbek P., Livingstone D., White J., Glotzer M., and Nasmyth K. 2001. Separase is required for chromosome segregation during meiosis I in *Caenorhabditis elegans. Curr. Biol.* **11:** 1825–1835.

Smith A.V. and Roeder G.S. 1997. The yeast Red1 protein localizes to the cores of meiotic chromosomes. *J. Cell Biol.* **136:** 957–967.

Smith K.N. and Nicolas A. 1998. Recombination at work for meiosis. *Curr. Opin. Genet. Dev.* **8:** 200–211.

Sollier J., Lin W., Soustelle C., Suhre K., Nicolas A., Geli V., and De La Roche Saint-Andre C. 2004. Set1 is required for meiotic S-phase onset, double-strand break formation and middle gene expression. *EMBO J.* **23:** 1957–1967.

Sperling K. and Ludtke E.K. 1981. Arrangement of prematurely condensed chromosomes in cultured cells and lymphocytes of the Indian muntjac. *Chromosoma* **83:** 541–553.

Stack S. and Clark C. 1974. Chromosome polarization and nuclear rotation in *Allium cepa* roots. *Cytologia* **275:** 632–637.

Storlazzi A., Tesse S., Gargano S., James F., Kleckner N., and Zickler D. 2003. Meiotic double-strand breaks at the interface of chromosome movement, chromosome remodeling, and reductional division. *Genes Dev.* **17:** 2675–2687.

Strambio-de-Castillia C., Blobel G., and Rout M.P. 1999. Proteins connecting the nuclear pore complex with the nuclear interior. *J. Cell Biol.* **144:** 839–855.

Sturtevant A.H. 1915. No crossing over in the female of the silk worm moth. *Am. Nat.* **49:** 42–44.

Suphapeetiporn K., Greally J.M., Walpita D., Ashley T., and Bale A.E. 2002. MEN1 tumor-suppressor protein localizes to telomeres during meiosis. *Genes Chromosomes Cancer* **35:** 81–85.

Svoboda A., Bahler J., and Kohli J. 1995. Microtubule-driven nuclear movements and linear elements as meiosis-specific characteristics of the fission yeasts *Schizosaccharomyces versatilis* and *Schizosaccharomyces pombe. Chromosoma* **104:** 203–214.

Sym M., Engebrecht J.A., and Roeder G.S. 1993. ZIP1 is a synaptonemal complex protein required for meiotic chromosome synapsis. *Cell* **72:** 365–378.

Tankimanova M., Hulten M.A., and Tease C. 2004. The initiation of homologous chromosome synapsis in mouse fetal oocytes is not directly driven by centromere and telomere clustering in the bouquet. *Cytogenet. Genome Res.* **105:** 172–181.

Thomas J.B. and Kaltsikes P.J. 1976. A bouquet-like attachment plate for telomeres in leptotene of rye revealed by heterochromatin staining. *Heredity* **36:** 155–162.

Trelles-Sticken E., Dresser M.E., and Scherthan H. 2000. Meiotic telomere protein Ndj1p is required for meiosis-specific telomere distribution, bouquet formation and efficient homologue pairing. *J. Cell Biol.* **151:** 95–106.

Trelles-Sticken E., Loidl J., and Scherthan H. 1999. Bouquet formation in budding yeast: Initiation of recombination is not required for meiotic telomere clustering. *J. Cell Sci.* **112:** 651–658.

———. 2003. Increased ploidy and KAR3 and SIR3 disruption alter the dynamics of meiotic chromosomes and telomeres. *J. Cell Sci.* **116:** 2431–2442.

Trelles-Sticken E., Adelfalk C., Loidl J., and Scherthan H. 2005a. Meiotic telomere clustering requires actin for its formation and cohesin for its resolution. *J. Cell Biol.* **170:** 213–223.

Trelles-Sticken E., Bonfils S., Sollier J., Géli V., Scherthan H., and de La Roche Saint-André C. 2005b. Setl- and Clb5-deficiencies disclose the differential regulation of centromere and telomere dynamics in *Saccharomyces cerevisiae* meiosis. *J. Cell Sci.* (in press).

Tuzon C.T., Borgstrom B., Weilguny D., Egel R., Cooper J.P., and Nielsen O. 2004. The fission yeast heterochromatin protein Rik1 is required for telomere clustering during meiosis. *J. Cell Biol.* **165:** 759–765.

Uetz P., Giot L., Cagney G., Mansfield T.A., Judson R.S., Knight J.R., Lockshon D., Narayan V., Srinivasan M., Pochart P., Qureshi-Emili A., Li Y., Godwin B., Conover D., Kalbfleisch T., Vijayadamodar G., Yang M., Johnston M., Fields S., and Rothberg J.M. 2000. A comprehensive analysis of protein-protein interactions in *Saccharomyces cerevisiae*. *Nature* **403:** 623–627.

Vazquez J., Belmont A.S., and Sedat J.W. 2001. Multiple regimes of constrained chromosome motion are regulated in the interphase *Drosophila* nucleus. *Curr. Biol.* **11:** 1227–1239.

———. 2002. The dynamics of homologous chromosome pairing during male *Drosophila* meiosis. *Curr. Biol.* **12:** 1473–1483.

Villeneuve A.M. and Hillers K.J. 2001. Whence meiosis? *Cell* **106:** 647–650.

Virkki N. 1974. The bouquet. *J. Agric. Univ. P. R.* **58:** 479–485.

Voet T., Liebe B., Labaere C., Marynen P., and Scherthan H. 2003. Telomere-independent homologue pairing and checkpoint escape of accessory ring chromosomes in male mouse meiosis. *J. Cell Biol.* **162:** 795–807.

von Wettstein D. 1984. The synaptonemal complex and genetic segregation. *Symp. Soc. Exp. Biol.* **38:** 195–231.

Vourc'h C., Taruscio D., Boyle A.L., and Ward D.C. 1993. Cell cycle-dependent distribution of telomeres, centromeres, and chromosome-specific subsatellite domains in the interphase nucleus of mouse lymphocytes. *Exp. Cell Res.* **205:** 142–151.

Wandall A. and Svendsen A. 1985. Transition from somatic to meiotic pairing and progressional changes of the synaptonemal complex. *Chromosoma* **92:** 254–264.

Wang X. 1988. Chromosome pairing analysis in haploid wheat by spreading of meiotic nuclei. *Carlsberg Res. Commun.* **53:** 135–166.

Watanabe Y. and Nurse P. 1999. Cohesin Rec8 is required for reductional chromosome segregation at meiosis. *Nature* **400:** 461–464.

Weierich C., Brero A., Stein S., von Hase J., Cremer C., Cremer T., and Solovei I. 2003. Three-dimensional arrangements of centromeres and telomeres in nuclei of human and murine lymphocytes. *Chromosome Res* **11:** 485–502.

West J.A. and Daley G.Q. 2004. In vitro gametogenesis from embryonic stem cells. *Curr. Opin. Cell Biol.* **16:** 688–692.

White E.J., Cowan C., Cande W.Z., and Kaback D.B. 2004. In vivo analysis of synaptonemal complex formation during yeast meiosis. *Genetics* **167:** 51–63.

Wilson E.B. 1925. *The cell in heredity and development.* MacMillan, New York.

Wolfe J., Hunter B., and Adair W.S. 1976. A cytological study of micronuclear elongation during conjugation in *Tetrahymena*. *Chromosoma* **55:** 289–308.

Wolgemuth D.J. 2003. Insights into regulation of the mammalian cell cycle from studies on spermatogenesis using genetic approaches in animal models. *Cytogenet. Genome Res.* **103:** 256–266.

Woollam D.H., Ford E.H., and Millen J.W. 1966. The attachment of pachytene chromosomes to the nuclear membrane in mammalian spermatocytes. *Exp. Cell Res.* **42:** 657–661.

Yamamoto A., West R.R., McIntosh J.R., and Hiraoka Y. 1999. A cytoplasmic dynein heavy chain is required for oscillatory nuclear movement of meiotic prophase and efficient meiotic recombination in fission yeast. *J. Cell Biol.* **145:** 1233–1249.

Yazdi P.T., Wang Y., Zhao S., Patel N., Lee E.Y., and Qin J. 2002. SMC1 is a downstream effector in the ATM/NBS1 branch of the human S-phase checkpoint. *Genes Dev.* **16:** 571–582.

Yu H.-G. and Koshland D.E. 2003. Meiotic condensin is required for proper chromosome compaction, SC assembly, and resolution of recombination-dependent chromosome linkages. *J. Cell Biol.* **163:** 937–947.

Yuan L., Liu J.G., Zhao J., Brundell E., Daneholt B., and Hoog C. 2000. The murine SCP3 gene is required for synaptonemal complex assembly, chromosome synapsis, and male fertility. *Mol. Cell* **5:** 73–83.

Zalenskaya I.A. and Zalensky A.O. 2002. Telomeres in mammalian male germline cells. *Int. Rev. Cytol.* **218:** 37–67.

———. 2004. Non-random positioning of chromosomes in human sperm nuclei. *Chromosome Res.* **12:** 163–173.

Zalensky A.O., Tomilin N.V., Zalenskaya I.A., Teplitz R.L., and Bradbury E.M. 1997. Telomere-telomere interactions and candidate telomere binding protein(s) in mammalian sperm cells. *Exp. Cell Res.* **232:** 29–41.

Zalensky A.O., Allen M.J., Kobayashi A., Zalenskaya I.A., Balhorn R., and Bradbury E.M. 1995. Well-defined genome architecture in the human sperm nucleus. *Chromosoma* **103:** 577–590.

Zetka M.C., Kawasaki I., Strome S., and Muller F. 1999. Synapsis and chiasma formation in *Caenorhabditis elegans* require HIM-3, a meiotic chromosome core component that functions in chromosome segregation. *Genes Dev.* **13:** 2258–2270.

Zickler D. 1977. Development of the synaptonemal complex and the "recombination nodules" during meiotic prophase in the seven bivalents of the fungus *Sordaria macrospora* Auersw. *Chromosoma* **61:** 289–316.

Zickler D. and Kleckner N. 1998. The leptotene-zygotene transition of meiosis. *Annu. Rev. Genet.* **32:** 619–697.

———. 1999. Meiotic chromosomes: Integrating structure and function. *Annu. Rev. Genet.* **33:** 603–754.

Zickler D. and Olson L.W. 1975. The synaptonemal complex and the spindle plaque during meiosis in yeast. *Chromosoma* **50:** 1–23.

10

Telomere Position Effect: Silencing Near the End

Michelle A. Mondoux and Virginia A. Zakian
Department of Molecular Biology, Lewis Thomas Laboratories
Princeton University
Princeton, New Jersey 08544-1014

HETEROCHROMATIC SILENCING IS DISTINCT FROM TRANSCRIPTIONAL GENE REPRESSION

The organization of genomes into heterochromatic and euchromatic regions is a global method of gene regulation, in contrast to single-gene transcriptional control. Originally defined cytologically in higher organisms as dark-staining chromatin, heterochromatin is now characterized by a collection of molecular markers including repetitive DNA, regular nucleosome spacing, low gene density, late S phase replication, and histone tail modifications including hypoacetylation (for review, see Henikoff 2000; Richards and Elgin 2002). Heterochromatin is often found at centromeric and telomeric loci, and active genes that are translocated to heterochromatin are heritably silenced, often in a mosaic or variegated pattern from cell to cell. This variegation is thought to reflect a stochastic heterochromatin assembly at a formerly euchromatic locus. Termed "facultative heterochromatin," these genes are silent in only a subset of the cells, as opposed to "constitutive heterochromatin," where silencing is stable in all cells (for review, see Richards and Elgin 2002).

Heterochromatic transcriptional silencing is distinct from transcriptional repression seen at individual promoters that occurs at particular times in the cell cycle or during organismal development. Unlike the mechanisms of repression at specific promoters, heterochromatic silencing is generally promoter-nonspecific, such that almost all promoters are silenced by heterochromatin. Heterochromatin constitutes a repressive environment for gene expression over a large distance, in contrast to gene-specific transcriptional repression that occurs over the relatively short range of the individual gene. Certain histone modifications associated with reduced transcription are found in both heterochromatic silencing and gene-specific transcriptional repression (see, e.g., Snowden et al. 2002; Kouskouti et al. 2004; Peinado et al. 2004). Although these histone modifications are shared, heterochromatin silencing and gene-specific repression are distinct in terms of specificity, timing, and regulation.

Telomere position effect (TPE) is one heterochromatic silencing phenomenon. TPE refers to the silencing of genes near telomeres and is classically defined as a continuous spread of heterochromatin from the telomere inward, although discontinuous heterochromatin formation has also been observed at telomeres in different organisms. TPE has been observed in diverse organisms, including baker's yeast (*Saccharomyces cerevisiae*), fission yeast (*Schizosaccharomyces pombe*), *Drosophila melanogaster*, the sleeping sickness parasite *Trypanosoma brucei*, the malaria parasite *Plasmodium falciparum*, plants, and humans (Hazelrigg et al. 1984; Levis et al. 1985; Gottschling et al. 1990; Matzke et al. 1994; Nimmo et al. 1994; Horn and Cross 1995; Scherf et al. 1998; Baur et al. 2001). Although first discovered in *Drosophila*, TPE is best understood in the yeast *S. cerevisiae*. Therefore, we use budding yeast as a paradigm, but we also discuss mechanisms of silencing in other organisms. We review how common silencing mechanisms, including subtelomeric structure, histone modifications, and nuclear localization, contribute to TPE. Finally, we discuss the potential biological roles of TPE.

THE DISCOVERY OF TELOMERE POSITION EFFECT

TPE was first observed 20 years ago in *D. melanogaster* (Gehring et al. 1984; Hazelrigg et al. 1984; Levis et al. 1985). Position effects resulting in the inactivation of genes had already been observed in flies as a result of natural or X-ray-induced chromosomal rearrangements that brought genes close to centromeric heterochromatin. This positional silencing is termed position effect variegation (PEV). The advent of P-element transposon DNA transformation allowed genes to be introduced into many distinct sites. When a P element carrying the *white* gene is introduced into the genome, it usually produces a red (wild-type) eye color. However, some fly stocks are obtained where the eyes have a mutant, mosaic phenotype (Hazelrigg et al. 1984). In these flies, the *white* gene is inserted near centromeric heterochromatin or near an autosomal telomere (Fig. 1A). This mosaic phenotype is not due to mutations in the *white*

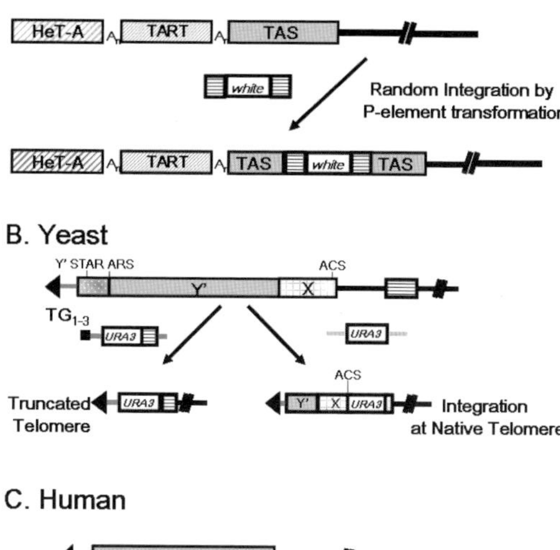

Figure 1. Telomere structure and TPE analysis in (*A*) *Drosophila*, (*B*) yeast, and (*C*) humans. (*A*) *Drosophila* telomeres consist of tandem arrays of the 6-kb *HeT-A* transposable element and the 12-kb *TART* transposable element. The number and arrangement of transposable elements vary from telomere to telomere. Proximal to the telomere is the subtelomeric telomere-associated sequence (TAS) array, tandem repeats that also vary in sequence and number between telomeres. TPE is assayed in *Drosophila* via P-element transformation of the *white* gene, which integrates randomly into the genome. P elements containing *white* that integrate within or adjacent to the TAS array are subject to TPE. (*B*) Natural yeast telomeres bear middle repetitive elements. The X element is heterogeneous, ranging from 0.3 to 3.75 kb in size. The "core X" element, containing an ARS consensus sequence (ACS), is 300–500 bp and found at most telomeres. The X element may also contain subtelomeric repeats (STRs) A–D in variable number and arrangement. The Y′ element is more conserved, with two classes of Y′ elements—long (6.7 kb) and short (5.2 kb). A given telomere may have 0–4 copies of Y′. TPE can be assayed at truncated or native yeast telomeres. Truncated telomeres are created via integration of a reporter gene and telomere seed at an upstream segment of unique DNA, eliminating the subtelomeric X and Y′ elements. TPE has also been assayed at native telomeres by integrating a reporter at the X-ACS site. (*C*) Human telomeres have highly variable subtelomeric repeats (Srpts). These repeats range from 1 kb to more than 200 kb in size. Some repeats are present at only one telomere and others are shared between several telomeres. TPE can be assayed in human cells via the random integration of a luciferase gene and telomere seed. When these integrations occur near a telomere, a new "healed" telomere is formed, analogous to the truncation events used to study yeast TPE.

gene, as mobilizing the P element produces flies with wild-type eye color (Levis et al. 1985).

Drosophila telomeres, which consist of transposable elements, are distinct from most other eukaryotic telomeres, which consist of highly repetitive telomerase-generated DNA (reviewed in Chapter 14). However, TPE is widespread among eukaryotes and is not unique to the unusual *Drosophila* telomeres. TPE was also discovered serendipitously in *S. cerevisiae* (Gottschling et al. 1990). In this case, a marker gene, *URA3*, was introduced next to a single telomere so that this unique sequence tag could be used to study the chromatin structure of an individual telomere (Fig. 1B). *URA3*, along with an adjacent seed of TG_{1-3} telomeric DNA, integrated at *ADH4*, the gene closest to the VII-L telomere. This integration event deletes the terminal 15 kb of the chromosome, including the subtelomeric middle repetitive X and Y′ elements; a new telomere is formed at the TG_{1-3} seed. As a result, the *URA3* transcription start site is positioned ~1.1 kb from the newly formed telomere. As expected, the cells bearing the telomeric *URA3* gene grow in the absence of uracil, indicating that *URA3* is expressed. Unexpectedly, however, 20–60% of the cells grew in the presence of 5-fluoro-orotic acid (5-FOA), a drug toxic to cells expressing *URA3* (Gottschling et al. 1990). As in flies, this effect is not due to mutation or inactivation of *URA3*. In fact, TPE at this truncated yeast telomere is reversible, as the FOA-grown colonies can be restreaked to plates lacking uracil, where *URA3* expression is required for growth. TPE was shown to function at multiple promoters, because expression of *ADE2* is also repressed when it is placed at the telomere. Red (*ade2⁻*) and white (*ade2⁺*) sectored colonies result, indicating that both the silent and the expressed states are stable through multiple cell cycles (Fig. 2). TPE is not specific to telomere VII-L: It can occur at other truncated telomeres as well as at some natural telomeres that bear the subtelomeric X and Y′ repeats (Gottschling et al. 1990; Renauld et al. 1993).

The *ADE2* reporter offers a visual demonstration of the variegated, stochastic nature of TPE at truncated telomeres. Because a strain with *ADE2* at a truncated telomere produces largely red (*ade2⁻*) or largely white (*ade2⁺*) colonies, both the silent and expressed states are stable through multiple cell divisions. However, red colonies contain white sectors and white colonies contain red sectors, indicating that cells retain the ability to switch expression states (Fig. 2). The regulation of expression state switching differs from telomere to telomere. This difference has been observed using a *URA3-GFP* fusion as a TPE reporter at both truncated and native telomeres (E. Louis, pers. comm.). At native telomeres, all cells in the culture seem to have a characteristic expression level for *GFP*,

A

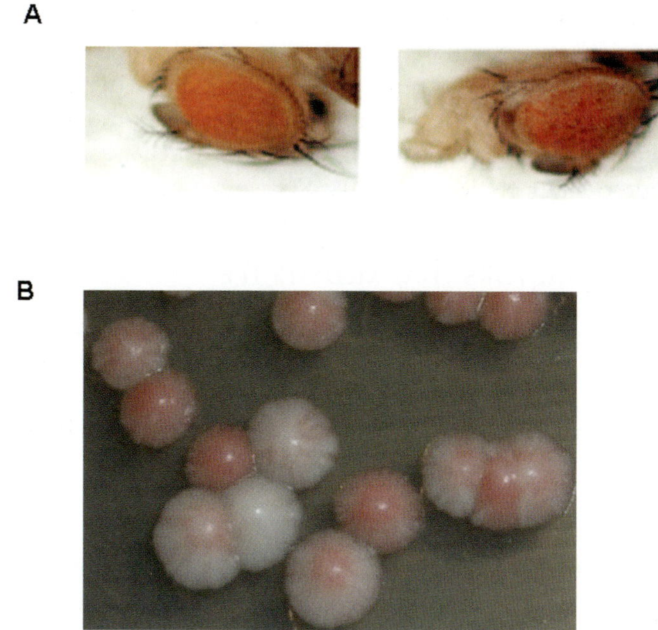

B

Figure 2. Variegation in (*A*) *Drosophila* and (*B*) yeast TPE. (*A*) TPE can be metastable at *Drosophila* telomeres. Complete expression of the *white* gene results in a red eye color; complete silencing results in a white eye color. TPE can be variegated, such that some cells maintain silencing and others lose silencing at the telomere, resulting in a mosaic phenotype, shown here at telomeres 3R (*left*) and the long arm of chromosome 4 (*right*). (*B*) Variegation at yeast telomeres. When *ADE2* is integrated at truncated telomere VII-L, cells appear red when the gene is silenced and white when the gene is expressed (Gottschling et al. 1990). The presence of sectored colonies indicates that although silencing is metastable, both silencing states are stable through multiple cell cycles. (*A*, Reprinted, with permission, from Shanower et al. 2005 [©GSA].)

which can be low or high depending on the telomere being assayed. At truncated telomeres, each cell in a culture can display a distinct expression level, from essentially no *GFP* expression to high *GFP* expression, even though each cell has the same reporter at the same truncated telomere.

Early attempts to detect TPE in human cells were unsuccessful, leading to early doubts as to whether this form of epigenetic silencing occurs in mammals (Sprung et al. 1996; Ofir et al. 1999). Ultimately, TPE was observed in human cells by constructing a linear plasmid containing the gene for puromycin resistance and a luciferase reporter gene adjacent to a 1.6-kb telomere repeat seed and transfecting the plasmid into telomerase-positive HeLa cells (Baur et al. 2001). When this construct integrates into the chromosome, a new telomere is formed at the telomere seed, analogous to the strains first used to observe TPE at truncated yeast telomeres (Fig. 1C). TPE in humans (hTPE) has many of the molecular hallmarks observed for yeast TPE, including variegation, length dependence, and a role for histone deacetylation, all of which are discussed later in this chapter (Baur et al. 2001, 2004; Koering et al. 2002).

TRANS-ACTING FACTORS THAT MEDIATE TPE

Yeast TPE was discovered by placing *URA3* next to the truncated left telomere of chromosome VII (Gottschling et al. 1990). Almost all of the early analyses to determine requirements and features of TPE were done with this chromosome. However, as are discussed later in this chapter, the TPE properties of different telomeres are not necessarily the same. In this section, requirements for silencing are based mainly on analysis of the truncated VII-L telomere.

Yeast TPE requires the double-stranded TG_{1-3} binding protein Rap1p (repressor activator protein 1), the end-binding heterodimeric Ku complex, and the multiprotein Sir2-4 (silent information regulator) complex, which does not bind DNA but is brought to telomeric chromatin via protein–protein interactions (for review, see Huang 2002; Tham and Zakian 2002; Perrod and Gasser 2003). Deleting any one of the Sir proteins or either Ku subunit essentially eliminates TPE. Likewise, deleting the carboxy-terminal domain of Rap1p eliminates TPE without impairing the essential function of Rap1p (Kyrion et al. 1993). All of these proteins likely act directly to affect TPE, as each is telomere bound in vivo (Conrad et al. 1990; Bourns et al. 1998; Gravel et al. 1998). Rap1p, Sir proteins, and Ku all have roles in processes other than TPE, including telomere length regulation, transcriptional repression at the silent mating type loci, rDNA silencing, and DNA repair (Ivy et al. 1986; Lustig et al. 1990; Mages et al. 1996; Porter et al. 1996; Smith and Boeke 1997). In addition to Rap1p, the Ku complex, and the Sir complex, there are more than 50 yeast genes that affect TPE. Many of these genes have relatively modest effects on the level of TPE, and probably many of them do not act directly. Table 1 presents a list of genes that affect TPE in *S. cerevisiae*. Not all of these genes are discussed in this chapter.

Based on the number of binding sites per telomere, Rap1p is thought to be present in ~10–20 copies at most yeast telomeres (Gilson et al. 1993). Sir4p and Sir3p both bind Rap1p, and mutations that eliminate this binding reduce TPE (Moretti et al. 1994; Moretti and Shore 2001). Sir4p also interacts with Ku (Tsukamoto et al. 1997). In the absence of Ku,

Table 1. Genes involved in Sir-mediated telomere position effect in *Saccharomyces cerevisiae*

Gene	Protein function	TPE phenotype in deletion mutant[a]
RAP1	binds to duplex $C_{1-3}A$ telomere repeats; binds Sir4p, Rif1p, and Rif2p; essential for role as transcription factor	carboxy-terminal deletion eliminates[b] TPE (Kyrion et al. 1993)
SIR2	histone deacetylase; binds Sir4p; required for *HM* and rDNA silencing	eliminated (Aparicio et al. 1991)
SIR3	spreading protein of silencing complex; binds Sir4p and histones H3 and H4; required for *HM* silencing	eliminated (Aparicio et al. 1991)
SIR4	binds Sir2p, Sir3p, Esc1p, and histones H3 and H4; required for *HM* silencing	eliminated (Aparicio et al. 1991)
YKU70 *YKU80*	as complex, promotes telomere lengthening via interaction with *TLC1*; required for peripheral localization of some telomeres; DNA repair	eliminated (Boulton and Jackson 1998)
RIF1 *RIF2*	compete with Sirs for Rap1p binding at telomere; antagonize telomere lengthening	increased (Kyrion et al. 1993)
SIR1	binds ORC; required for establishment of *HM* silencing	reduced at native telomere XI-L (Pryde and Louis 1999)
ORC2 *ORC5*	essential subunits of the origin recognition complex	partial loss-of-function alleles reduce TPE (Fox et al. 1997)
MCM5	essential helicase; component of MCM complex that primes origins of DNA replication	reduced at native telomeres[c] (Dziak et al. 2003)
ABF1	essential transcription factor; required for *HM* silencing; binds some ARSs	binding site mutation reduces TPE at native telomere XI-L (Pryde and Louis 1999)
ESC1	binds Sir4p; required for peripheral localization of some telomeres	reduced (Andrulis et al. 2002)
ESC8	binds Sir2p	reduced (Cuperus and Shore 2002)
HHT1 *HHT2*	histone H3; binds Sir3p and Sir4p	amino-terminal deletions eliminate TPE (Mann and Grunstein 1992; Thompson et al. 1994)
HHF1 *HHF2*	histone H4; binds Sir3p and Sir4p	amino-terminal deletions eliminate TPE (Aparicio et al. 1991; Mann and Grunstein 1992; Thompson et al. 1994)

(Continued)

Table 1. (*continued*)

Gene	Protein function	TPE phenotype in deletion mutant[a]
HTA1 *HTA2*	histone H2A	amino-terminal deletions reduce TPE (Wyatt et al. 2003)
HTZ1	histone-variant H2AZ; prevents spread of silent heterochromatin	increased at native telomeres[c] (Meneghini et al. 2003)
SAS2	histone H4K16 acetyltransferase; global histone acetylation	eliminated (Reifsnyder et al. 1996); increased at native telomere VI-R[c] (Kimura et al. 2002; Suka et al. 2002)
SAS4 *SAS5*	components of Sas2p complex	eliminated (Xu et al. 1999)
DOT1	histone H3K79 methyltransferase	reduced; overexpression also reduces TPE (Singer et al. 1998)
UBP10	ubiquitin-specific protease; may regulate silencing by acting on Sir4p	reduced; overexpression also reduces TPE (Singer et al. 1998)
DOT5	nuclear thiol peroxidase	overexpression reduces TPE (Singer et al. 1998)
CAC1 *CAC2* *CAC3*	subunits of CAF-I chromatin assembly complex; contribute to localization of Rap1p	reduced (Enomoto et al. 1997; Kaufman et al. 1997; Monson et al. 1997)
HIR1 *HIR2*	corepressors of histone gene transcription	increased; eliminated in concert with *cacΔ* (Kaufman et al. 1998)
NAT1 *ARD1*	subunit of amino-terminal acetyltransferase NatA	eliminated (Aparicio et al. 1991)
RAD6	E2 ubiquitin conjugating enzyme	eliminated (Huang et al. 1997)
BRE1	E3 ubiquitin ligase for Rad6p; required for the ubiquitination of histone H2B	eliminated (Wood et al. 2003)
HAT1 *HAT2* *HIF1*	subunits of the HAT-B histone acetyltransferase complex that acetylates free histone H4K12	reduced in concert with H3K14R mutation (Kelly al. 2000; Poveda et al. 2004)
ASF1	histone chaperone/chromatin assembly factor	overexpression reduces TPE (Le et al. 1997)
RPD3 *SAP30* *SIN3*	histone deacetylase complex	increased (Rundlett et al. 1996; Sun and Hampsey 1999)
SHG1 *SDC1* *SWD1*	components of the COMPASS complex required for H3K4 methylation	eliminated or reduced (Nislow et al. 1997; Krogan et al. 2003)

Table 1. (*continued*)

Gene	Protein function	TPE phenotype in deletion mutant[a]
SWD2		
SWD3		
SPP1		
BRE2		
SET1		
HST2	cytosolic member of Sir2 family of NAD$^+$-dependent deacetylases	overexpression eliminates TPE (Perrod et al. 2001)
HST3	members of Sir2 family of NAD$^+$-dependent deacetylases	double mutant eliminates TPE (Brachmann et al. 1995)
HST4		
NPT1	nicotinate phosphoribosyltransferase; NAD$^+$ biosynthesis	eliminated (Sandmeier et al. 2002)
PNC1	pyrazinamidase and nicotinamidase; NAD$^+$ biosynthesis	reduced (Sandmeier et al. 2002)
SUM1	suppressor of *sir* mutations	*SUM1-1* allele increases TPE (Chi and Shore 1996)
PKC1	MAP kinase cascade; serine/threonine MAP kinase Slt2p phosphorylates Sir3p	*slt2Δ* eliminates TPE (Ray et al. 2003)
BCK1		
MKK1		
SLT2		
STE7	MAP kinase cascade that results in hyperphosphorylation of Sir3p	activated *STE11-4* allele increases TPE (Stone and Pillus 1996)
STE11		
STE12		
FUS3		
YAF9	component of NuA4 histone H4 acetyltransferase complex	increased at some native telomeres[c] (Zhang et al. 2004)
EPL1	essential component of NuA4 histone H4 acetyltransferase complex; homologous to *Drosophila* Enhancer of Polycomb	partial loss-of-function alleles eliminate or reduce TPE (Boudreault et al. 2003)
POL1	catalytic subunit of the DNA polymerase α-primase complex; essential	partial loss-of-function allele eliminates TPE (Adams Martin et al. 2000)
POL2	DNA polymerase II; essential	partial loss-of-function allele reduces TPE (Iida and Araki 2004)
DPB3	subunits of DNA polymerase II complex	reduced (Iida and Araki 2004)
DPB4		
RTF1	subunit of the RNA polymerase II-associated Paf1 complex	eliminated (Krogan et al. 2003; Ng et al. 2003)

(Continued)

Table 1. (*continued*)

Gene	Protein function	TPE phenotype in deletion mutant[a]
PCNA	essential protein; functions as the sliding clamp for DNA polymerase delta	partial loss-of-function alleles reduce TPE (Zhang et al. 2000)
SPT4	mediates activation and inhibition of transcription elongation; pre-mRNA processing; kinetochore function	eliminated (Crotti and Basrai 2004)
ISW2 *DLS1*	subunits of ISW2/yCHRAC chromatin accessibility complex	increased (Iida and Araki 2004)
RRM3	DNA helicase; promotes replication fork progression at nonhistone protein-DNA complexes	reduced (Ivessa et al. 2002)
MEC1	essential ATR kinase; checkpoint control; viability restored in *mec1Δ sml1Δ*	reduced in *mec1Δ sml1Δ* (Craven and Petes 2000)
SCS2	suppressor of choline sensitivity; regulation of *INO1*	reduced (Craven and Petes 2001)
MEC3	DNA repair; checkpoint control	increased (Corda et al. 1999)
MRC1	S-phase checkpoint protein required for DNA replication	reduced (Hu et al. 2001)
ELG1	Required for S-phase progression and telomere length regulation; forms an alternative replication factor C	increased (Smolikov et al. 2004)
GAL11	component of the RNA polymerase II holoenzyme; acts as target of activators and repressors	reduced (Suzuki and Nishizaw 1994)
SNF2	component of Swi/Snf complex; transcriptional regulator	eliminated (Dror and Winston 2004)
WTM1 *WTM2* *WTM3*	WD-repeat-containing transcriptional regulators	increased (Pemberton and Blobel 1997)
BDF1	promotes transcription initiation at TATA-containing promoters	increased at several native telomeres[c] (Ladurner et al. 2003)
PBP2 *HEK1*	RNA-binding proteins with similarity to mammalian heterogeneous nuclear RNP K protein	increased (Denisenko and Bomsztyk 2002)
UPF2	component of the nonsense-mediated mRNA decay pathway	reduced (Lew et al. 1998)
RPT4 *RPT6*	essential components of the 26S proteasome	eliminated in temperature sensitive mutants (Ezhkova and Tansey 2004)

Table 1. (*continued*)

Gene	Protein function	TPE phenotype in deletion mutant[a]
ZDS1	interact with Sir proteins and Rap1p by	overexpression increases TPE
ZDS2	two-hybrid; involved in YAC stability	(Roy and Runge 1999)
IFH1	essential protein; potential Cdc28p substrate	overexpression reduces TPE (Singer et al. 1998)

[a]Refers to reporter gene at truncated telomere VII-L and/or truncated telomere V-R unless otherwise noted; point mutations in some of these genes may also affect TPE.

[b]Eliminated: less than 1% TPE; increased: twofold or greater increase relative to wild type; reduced: twofold or greater decrease relative to wild type, but greater than 1%.

[c]In genome-wide study.

Sir4p binding to the VI-R telomere is reduced to ~40% of wild-type levels, and Sir complex binding to subtelomeric chromatin is lost (Martin et al. 1999; Luo et al. 2002). These data suggest that the Sir complex can be brought to the telomere by its interaction with either Rap1p or Ku. Unlike Sir2 or Sir3p, Sir4p is still telomere bound at both the VII-L and VI-R telomeres in the absence of the other Sir proteins (Bourns et al. 1998; Luo et al. 2002). Thus, silencing is thought to initiate by Sir4p binding to the telomere followed by recruitment of Sir2p and Sir3p. The Sir complex spreads into adjacent subtelomeric chromatin by its ability to interact with histone tails (Hecht et al. 1995; Strahl-Bolsinger et al. 1997). This spreading extends ~3 kb from the telomere. The enzymatic activity of Sir2p, which is a histone deacetylase (HDAC), is required for this spreading (Hoppe et al. 2002; interactions between Sir proteins and histones and the activity of Sir2p are discussed in more detail in the section TPE and Histone Modifications).

Telomere-binding proteins that mediate TPE have been identified in only some of the organisms where TPE has been observed. In *Drosophila*, HP1 (heterochromatin protein 1), first discovered for its role in PEV in centromeric heterochromatin, is also needed for TPE at the largely heterochromatic chromosome 4 (for review, see Wallrath and Elgin 1995; Mason et al. 2000). However, HP1 is not required for TPE at the other *Drosophila* telomeres. HP1 is conserved in many eukaryotes but is not found in baker's yeast. The HP1 homolog in *S. pombe*, Swi6, is involved in TPE and is believed to be the structural protein that mediates the spread of silent chromatin in a manner analogous to Sir3p in *S. cerevisiae* (Allshire et al. 1995; for review, see Huang 2002). The conservation of HP1 may account for the lack of Sir3p and Sir4p homologs in other eukaryotes. Other conserved telomere-binding proteins do not mediate TPE in other organisms. The Ku complex, for example, is required for telomere maintenance in trypanosomes (Conway et al. 2002) and fission yeast (Manolis et al. 2001), and for telomere accessibility and length regulation in *Drosophila* (Melnikova et al. 2005), but there is no evidence that Ku has a role in TPE in these organisms. The yeast telomeric DNA-binding protein Rap1p, which recruits the Sir complex to telomeres, is conserved in fission yeast (spRap1) and humans (hRap1), although spRap1 and hRap1 do not bind directly to the telomeric DNA, but rather are recruited by other telomere-binding proteins (Li et al. 2000; Kanoh and Ishikawa 2001). In *S. pombe*, this mediator protein, Taz1p, is required for TPE, as is spRap1 (Nimmo et al. 1998; Park et al. 2002). hRap1's function in telomere length regulation is conserved from yeast to humans (Li et al. 2000). The roles of the hRap1 and its mediator protein, TRF2 (the duplex human $\underline{T}_2\underline{A}\underline{G}_3$ $\underline{r}$epeat binding $\underline{f}$actor) in human TPE are unknown. However, another human telomere-binding protein, TRF1, has been shown to negatively regulate TPE when overexpressed (Koering et al. 2002). It is unclear if TRF1 binding to the telomere inhibits TPE or if its overexpression titrates another factor away from the telomere that is itself necessary for human TPE.

THE ROLE OF TELOMERE LENGTH IN TPE

Telomere Length and TPE in Yeast: Length Is Only Part of the Story

Because Rap1p is a sequence-specific duplex DNA-binding protein that initiates TPE by recruiting the Sir complex, long telomeres should have higher TPE than short telomeres. Indeed, in otherwise wild-type yeast, a gene adjacent to a long telomere shows higher and more stable repression than a gene next to a short telomere (Kyrion et al. 1993). However, if telomeres are made longer by mutations that affect the establishment or maintenance of silencing, TPE is reduced. For example, *rap1^t* mutations eliminate a region in the carboxyl terminus of Rap1p that is necessary for both Sir complex and Rif1p/Rif2p binding (Rap1p interacting factor 1 and 2; Hardy et al. 1992; Moretti et al. 1994; Wotton and Shore 1997). Because Rif1p and Rif2p inhibit telomerase-mediated telomere lengthening (Teng et al. 2000), *rap1^t* cells have very long telomeres. However, the *rap1^t* mutation also prevents Sir binding, so, despite its long telomeres, this strain has no TPE (Kyrion et al. 1992, 1993). The Sir complex and the Rif proteins, which bind to the same portion of Rap1p, compete for Rap1p binding (Moretti et al. 1994; Wotton and Shore 1997). A *rif1Δ rif2Δ* strain, which has very long telomeres, also has elevated TPE (Kyrion et al. 1993). This higher level of silencing is due not only to more

Rap1p sites per telomere but also to reduced competition for Sir binding to telomeres.

Some of the mutations that eliminate TPE, such as *sir3Δ*, *sir4Δ*, and *yku70Δ* also cause telomere shortening. However, the *sir3Δ* and *sir4Δ* mutants have only modest effects on telomere length (Palladino et al. 1993), so it is unlikely that the complete loss of TPE in these strains is due to shortened telomeres. Certainly short telomeres are not sufficient to eliminate TPE: *tel1* cells have very short telomeres of <100 bp (Lustig and Petes 1986; Greenwell et al. 1995), yet TPE is normal in these cells (Gottschling et al. 1990; Runge and Zakian 1996).

Telomeres are also very short in *yku70Δ* cells. In this strain, Sir4p recruitment is impaired not only by the smaller number of Rap1p binding sites but also by the absence of the alternative Ku-mediated pathway for Sir4p recruitment. As predicted, Sir4p binding is reduced at telomeres and subtelomeric chromatin in *ykuΔ* strains (Martin et al. 1999; Luo et al. 2002). Ku does not appear to provide any direct activity that is essential for silencing, because in an otherwise wild-type *yku70* strain, a truncated VII-L telomere with a long telomeric repeat tract exhibits normal TPE (Mishra and Shore 1999). Moreover, relatively modest telomere lengthening (~100 bp) can restore telomere silencing to the truncated VII-L telomere in *yku70 rif1* strain, where the Sir complex does not have to compete with Rif1p for Rap1p binding. Thus, the loss of TPE in Ku-deficient cells is best explained by reduced Sir4p binding, caused partly by short telomeres and partly by loss of the Rap1-independent pathway for Sir4p recruitment. Ku has additional effects on telomeres and these might also influence TPE. For example, *ykuΔ* telomeres bear long single-strand G-tails throughout the cell cycle, not just in late S/G_2 phase, as seen in wild-type cells (Wellinger et al. 1993b; Gravel et al. 1998). Telomerase recruitment is also defective in the absence of Ku (Fisher et al. 2004), and some telomeres require Ku to target them to the nuclear periphery (Hediger et al. 2002b; discussed in detail in the section Nuclear Localization and TPE). It is not known if these changes contribute to the loss of silencing that is characteristic of Ku-deficient cells.

Telomere Length and TPE in Humans

Telomere length in human cells, as in yeast, follows the general trend of longer telomeres correlating with higher gene silencing. HeLa cell telomeres that are elongated from ~5 kb to ~14 kb by overexpression of the human telomerase reverse transcriptase (hTERT) display a twofold to tenfold increase in TPE at a luciferase reporter gene (Baur et al. 2001). As in yeast, however, there is not a simple, absolute correlation between telomere length and TPE levels. A study of telomere reporter gene expression in a different cell type found no correlation between resting telomere length and TPE (Koering et al. 2002). It is possible that relatively small differences in telomere length are not sufficient to produce a detectable change in TPE, whereas large changes, such as the 9-kb extension that accompanies hTERT overexpression, are required for measurable effects.

Telomere Length in *Drosophila* Contributes Negatively to TPE

Telomere length in *Drosophila* also contributes to TPE; however, the trend is opposite to that in yeast and humans—in *Drosophila*, longer telomeres display decreased TPE compared to shorter telomeres (Mason et al. 2003). This inverse correlation is probably a result of the unique telomere structure in flies. Unlike most eukaryotes, whose telomeres are composed of simple repetitive, noncoding DNA, *Drosophila* telomeres are made up of *HeT-A* and *TART* transposable elements. A new transposable element, *TAHRE*, has also recently been found at *Drosophila* telomeres (Abad et al. 2004). One, two, or all of these elements can be present at a given telomere, although *HeT-A* seems to be the most abundant (Levis et al. 1993; Pardue and DeBaryshe 1999). The negative effect of the transposable elements on TPE may be due to their transcription, as the 3′ end of *HeT-A* has promoter activity when tested with a *lacZ* reporter (Danilevskaya et al. 1997). Internal to the transposable elements, *Drosophila* telomeres bear telomere-associated sequence (TAS) repeats, which are similar to the subtelomeric repeats found on yeast and human chromosomes. When TPE is monitored at telomeres with the same TAS array but with different numbers of telomeric transposable elements *in cis*, the shortest telomere has high TPE while the reporter gene is not repressed as much in the long telomere strain (Mason et al. 2003).

In contrast to the telomeric transposons, the *Drosophila* subtelomeric TAS elements act positively to promote TPE. Indeed, TPE is detected only when transgenes are inserted within or adjacent to TAS arrays (Karpen and Spradling 1992; Levis et al. 1993; Cryderman et al. 1999). Further evidence for the importance of TAS elements is that their deletion at telomere 2L greatly reduces TPE of a reporter gene on the other copy of 2L (Golubovsky et al. 2001). Moreover, a deficiency screen for dominant suppressors of TPE at 2L found several positive loci that mapped back to the 2L TAS (Mason et al. 2004). The ability of TAS elements to mediate silencing is not limited to the telomere. When the TAS array from telomere 2L, along with a reporter gene, is moved into a nontelomeric site, the transgene is still silenced and its level of silencing is dependent on the length of the TAS array (Kurenova et al. 1998).

How the *Drosophila* TAS array mediates heterochromatin formation is unclear, as binding sites for *trans*-acting silencing proteins in the TAS elements have not been identified. Polycomb group proteins localize to telomeres, and there is a putative polycomb-response element (PRE) in the TAS on the left telomere of the X chromosome (Boivin et al. 2003). Direct binding of polycomb group proteins to the TAS has yet to be demonstrated. TAS elements may function by recruiting polycomb group or other yet to be identified silencing proteins. Alternatively, or in addition, silencing may be mediated by homolog pairing or nuclear localization. A role for higher-order structure in *Drosophila* TPE is suggested by the fact that deletions in the 2L TAS array have effects on TPE not only at telomere 2L but also at telomere 3R (Golubovsky et al. 2001).

TELOMERE IDENTITY AND EFFECTS ON TPE

Different Yeast Telomeres Have Different Subtelomeric Structures and Different TPE Phenotypes

Middle repetitive subtelomeric sequences are immediately proximal to the telomere repeats in many species, including yeast, *Drosophila*, and humans, and in several organisms have been shown to affect telomere length (Craven and Petes 1999; Figueiredo et al. 2002; Jacob et al. 2004). In yeast, there are two types of subtelomeric repeats, X and Y′ (for review, see Louis 1995). X elements are found at essentially all yeast telomeres and range in size from ~300 bp to 3 kb. Although X elements are quite heterogeneous, each contains a "core X" repeat, consisting of an ARS (autonomously replicating sequence) consensus sequence (ACS) and a binding site for the protein Abf1p, a transcription factor that, like Rap1p, also functions at the *HM* silencers (Diffley and Stillman 1989; Kurtz and Shore 1991; for review, see Haber 1998). Some X repeats also contain STR (<u>s</u>ub<u>t</u>elomeric <u>r</u>epeat) elements A–D. STR elements have recognition sites for Reb1p, which also functions in RNA polymerase I gene transcription and termination (Morrow et al. 1989; Wang et al. 1990), and Tbf1p (T_2AG_3 binding factor), an essential protein of unknown function (Brigati et al. 1993) that can act as a boundary element in subtelomeric DNA (Fourel et al. 1999, 2001). The combination, number, and arrangement of STR elements vary from telomere to telomere.

In contrast to the ubiquitous X element, the Y′ element is found at only one-half to two-thirds of yeast telomeres. When present, Y′ is distal to X and is found in up to four tandem copies (Chan and Tye 1983). Short tracts of TG_{1-3} telomeric sequence are found between some tandem Y' elements and at some X-Y' junctions (Walmsley et al. 1984). Because internal TG_{1-3} tracts can recruit Rap1p and other silencing proteins (Stavenhagen and Zakian 1994; Bourns et al. 1998), their presence may bolster TPE. Like X, Y' contains an ARS that can bind ORC and possibly Abf1p. Y' also contains sequences that counter the spread of TPE. The Y' STAR (subtelomeric antisilencing repeat) element has additional Tbf1p and Reb1p binding sites and functions as a boundary element, able to block the spread of heterochromatin into adjacent DNA (Fourel et al. 1999). Y' elements also contain two open reading frames (ORFs), one of which codes for a putative RNA helicase.

Although all yeast telomeres end in TG_{1-3} repeats and there are only two classes of subtelomeric repeats, the X and Y' elements are sufficiently diverse that each telomere is unique. Thus, the 32 yeast telomeres can be thought of as having a "barcode" that confers a distinct identity on each telomere. This barcode is not immutable, as the X and Y' content of a given chromosome end varies among different strains (Zakian and Blanton 1988).

What is the effect of the different subtelomeric structures on TPE? Early TPE studies in yeast were done using reporter genes integrated immediately adjacent to the telomeric TG_{1-3} tract, such that these truncated telomeres lack both the X and Y' elements (Gottschling et al. 1990). Although there are no reports of a truncated telomere that lacks TPE, the level of silencing at different truncated telomeres can vary widely. For example, in the same strain background, silencing at truncated telomere V-R is about tenfold lower ($\sim$4%) than at truncated VII-L ($\sim$33%; Gottschling et al. 1990). The source of TPE differences at truncated telomeres is not known, but they might reflect differences in the identity and transcription rate of internal sequences, nuclear localization, telomere length, or specific telomere–telomere associations.

TPE at "native" telomeres has been studied in two ways: (1) inserting *URA3* or another reporter gene into the X-ACS, keeping the Y' and X elements largely intact (Pryde and Louis 1999), and (2) observing expression levels of subtelomeric genes, either at individual unmodified telomeres (Vega-Palas et al. 1997, 2000) or on a genome-wide scale (Wyrick et al. 1999, 2001). Results from this second method are discussed below in the section Biological Functions of TPE.

Because individual telomeres have different subtelomeric structures that have binding sites for proteins that can promote or limit TPE, it is not surprising that the level of silencing at native telomeres varies considerably from telomere to telomere. In reporter gene assays, TPE is virtually nonexistent (<1%) at many native telomeres, occurring at only 6 of the 17 "native" telomeres that have been tested by insertion of reporter genes (Pryde and Louis 1999; M.A. Mondoux and V.A. Zakian, unpubl.; E. Louis, pers. comm.). The X-only telomeres III-R and IV-L, which do not support TPE, also do not bind Rap1p by a chromatin immunoprecipitation assay (Lieb et al. 2001). This result was unexpected, as Rap1p was assumed to bind to all yeast telomeres and to be the major determinant of telomeric chromatin. No or even very low Rap1p binding is sufficient to explain the lack of TPE at these telomeres. In contrast, some native X-only telomeres have very high TPE, comparable to TPE levels at truncated telomeres, with silencing seen in ~30–60% of cells (Pryde and Louis 1999). TPE is even higher, occurring in essentially all cells, at the core X-only telomere VI-R (M.A. Mondoux and V.A. Zakian, unpubl.). Given that the X and Y' content of a given telomere can vary from strain to strain, the TPE phenotype of a given chromosome end is probably also not fixed.

In addition to TPE levels, requirements for silencing proteins are different at different telomeres. Sir1p, which acts in the establishment of silencing at the yeast silent mating type loci (Pillus and Rine 1989), is not necessary for TPE at truncated telomeres (Aparicio et al. 1991). Nevertheless, if Sir1p is tethered to the truncated VII-L telomere, silencing increases (Chien et al. 1993). In contrast, *sir1Δ* reduces TPE at the native telomere XI-L by approximately twofold (Fourel et al. 1999; Pryde and Louis 1999). Although the ACS sites located in core X elements (X-ACS) do not seem to be active as origins of replication, the X-ACS does bind the origin recognition complex (ORC; Wyrick et al. 2001). Mutation of the X-ACS reduces TPE at telomere XI-L ~100-fold (Pryde and Louis 1999). Sir1p could play a role in native telomere silencing by binding to ORC at the X telomeres, just as it does at the silent mating type loci (Foss et al. 1993; Triolo and Sternglanz 1996). Indeed, the core X element can improve silencing at a weakened *HML* locus that lacks most of the *HML-E* silencer (Lebrun et al. 2001). The core X element also contains a binding site for Abf1p, another protein necessary for *HM* silencing. Mutation of the Abf1p-binding site at telomere XI-L reduces TPE at this telomere by approximately tenfold (Pryde and Louis 1999). Clearly, the subtelomeric DNA, which recruits different *trans*-acting factors to different telomeres, makes a strong contribution to the silencing profile of individual telomeres.

Unlike the X element, which can have positive effects on TPE, the Y' element has antisilencing properties. These effects are observed when Y's are located between the telomere and a reporter gene, or between an *HML* silencer and a reporter gene, consistent with their functioning as heterochromatin boundaries (Fourel et al. 1999, 2001). Although Reb1p is a transcriptional activator at other sites in the genome (Wang et al. 1990), in subtelomeric DNA, Reb1p acts as a boundary protein as does Tbf1p. These two Y′ binding proteins are sufficient for the Y′ boundary activity, as multiple binding sites for either protein recapitulate the Y′ STAR boundary activity at truncated telomere VII-L (Fourel et al. 1999).

At truncated telomere VII-L, TPE can be detected inward from the telomere, but TPE levels decrease exponentially with distance from the telomere (Gottschling et al. 1990). In contrast, at native telomeres, TPE does not decrease in a regular manner as a function of distance from the telomere. Rather, at native ends, there are zones of transcriptional repression that are punctuated by regions of normal gene expression (Fourel et al. 1999; Pryde and Louis 1999). The discontinuous nature of silencing at native ends is a consequence of the subtelomeric X and Y′ elements, which have binding sites for proteins that promote TPE and for boundary proteins that limit heterochromatin spread.

The unique identities of native telomeres in terms of sequence content and *trans*-acting binding proteins make it difficult to generalize telomere behavior from the study of one or even a few telomeres. This issue is especially troublesome when several different telomere phenotypes are being monitored, such as TPE, telomere length, or telomere position within the nucleus. Given the diversity from telomere to telomere, even between different truncated telomeres, it is critical that all behaviors be measured at the same telomere. These considerations are not unique to yeast, because, as described below, the subtelomeric regions of *Drosophila* and human chromosomes are also diverse.

Drosophila Subtelomeric Structure Also Leads to Different TPE Requirements and Phenotypes

Each of the eight *Drosophila* telomeres is distinguished by the length and sequence of its TAS array, with different telomeres bearing ~10 to ~20 kb of these repeats. The 2L and 3L TAS arrays have a highly similar sequence composition (Abad et al. 2004), but they are not similar to the XL, 2R, or 3R telomeres by the criterion of in situ hybridization (Mechler et al. 1985; Walter et al. 1995). The presence or absence of a TAS array on a particular chromosome end can vary by *Drosophila* stock. These TAS differences may explain differences in the requirements for TPE at different telomeres. For example, TPE on the long arm of the fourth chromosome, which is largely heterochromatic and has a much shorter distance from the centromere to the telomere, requires many of the classic suppressors of position effect variegation (*Su(var)*s), including HP1

(Wallrath and Elgin 1995). In contrast, neither telomere on the second or third chromosomes requires the *Su(var)*s or HP1 for TPE. Rather, TPE at these telomeres is sensitive to mutations in the Polycomb group genes, which do not affect TPE at chromosome 4.

Given the small size of the fourth chromosome, the TPE phenotype of chromosome 4 telomeres may be influenced by their proximity to pericentric heterochromatin. When the telomeric region of the *Drosophila* fourth chromosome is translocated to telomere 2L or 2R, there is a dramatic loss of TPE at the newly formed 2-4 hybrid telomere. This residual TPE is still regulated by the chromosome 4 TPE modifier HP1, not by the chromosome 2 modifier *Su(z)2* (Cryderman et al. 1999). This continued dependence on HP1 at this new location, far from a centromere, argues that genetic dependencies for TPE at chromosome 4 telomeres are inherent to them. When the 2R telomere is translocated to chromosome 4, there is no change in its TPE level, and the 2R transgene is still regulated by the chromosome 2 modifiers (Cryderman et al. 1999). Thus, the local TAS environment dictates the requirements for TPE, although chromosomal context also plays a role.

Even when *Drosophila* telomeres share requirements for TPE, their response to mutations in those genes can vary. For example, *grappa*, the H3K79 methyltransferase (*DOT1* in yeast), is an essential gene in flies that is required for TPE but not for pericentric PEV (Shanower et al. 2005). Three partial loss-of-function alleles of *grappa* have recently been identified that eliminate TPE at all *Drosophila* telomeres except 2L, which is only weakly affected by these alleles. Other alleles with differential effects on TPE were also identified, including gpp^{94A}, which reduces TPE at 3R but has weak or no effects at the other telomeres tested (Shanower et al. 2005). Grappa is the first example of a protein that modulates TPE on the long arm of the fourth chromosome that does not affect PEV. The lack of a role for *grappa* in PEV is further evidence that silencing at the chromosome 4 telomeres is a form of TPE, not an extension of PEV.

Human Subtelomeric Repeats and Links to Genetic Diseases

Human subtelomeric repeats are less well-studied than in yeast or *Drosophila* and represent one of the last frontiers in the human genome project, as their highly repetitive nature makes them difficult to sequence and assemble. Human subtelomeric repeats seem to have undergone evolutionarily recent duplications and deletions, as their number and chromosomal location can vary in different individuals (for review, see Mefford and Trask 2002). A recent sequencing effort (Riethman et al.

2004) found that human subtelomeric repeats range from 1 kb to more than 200 kb in size. Some of these subtelomeric repeats (Srpts) are present at only one telomere, whereas others are common to several telomeres (Riethman et al. 2004). Another component of human subtelomeric DNA, the X region, is a restriction-endonuclease-resistant 2–4-kb segment. The length of the X region has been calculated by comparing apparent telomere lengths on gels and quantitative fluorescent in situ hybridization (Hultdin et al. 1998; Steinert et al. 2004). The size of the X region varies, even at a specific telomere, as a direct function of telomere length (Steinert et al. 2004). The cause of the restriction-endonuclease refractory nature of the X region is unknown, but it does not seem to be due to DNA methylation (Steinert et al. 2004).

One rare example of a well-studied human subtelomeric repeat is the 3.3-kb D4Z4 repeat. This element is of particular interest because of its linkage to facioscapulohumeral muscular dystrophy (FSHD). In the human population, D4Z4 is present in varying numbers at telomeres 4q and 10q. FSHD patients have only 1–10 copies of the repeat at the subtelomere of one of the chromosome 4 homologs. In contrast, unaffected individuals have 11–150 copies of the repeat at both homologs (van Deutekom et al. 1993; Lemmers et al. 2001). Severity of the disease seems to correlate with decreased numbers of D4Z4 repeats. Portions of the D4Z4 repeat are highly homologous to known heterochromatic repeats on other chromosomes (1q, 21p, and 22p; Meneveri et al. 1993; Hewitt et al. 1994). An attractive model, then, is that as the D4Z4 repeat number decreases, there is a loss of TPE and increased expression at the nearby FSHD candidate genes that result in the disease state. In support of this model, three genes upstream of the D4Z4 repeats are inappropriately expressed in muscle samples from affected individuals, with the biggest effect at the gene closest to the telomere (Gabellini et al. 2002). Further analysis showed that there is a D4Z4 binding element (DBE) that binds a multiprotein repressor complex. The DBE is capable of acting as a silencer, with multiple copies of the DBE resulting in greater repression of the reporter gene (Gabellini et al. 2002). However, using a different quantitation technique, a second group did not observe increased expression in 4q subtelomeric genes in FSHD patients (Jiang et al. 2003). In addition, the second group argued that the 4q region was not heterochromatic based on histone H4 acetylation levels. This group proposes a "looping model," as opposed to a heterochromatin spreading model, as the mechanism for D4Z4's role in FSHD. These models are not necessarily mutually exclusive, as telomeres are known to loop in mammals, ciliates, trypanosomes, and yeast (Griffith et al. 1999; Murti and Prescott 1999; de Bruin et al. 2000; Muñoz-Jordan et al. 2001). This looping could bring silencing factors into closer proximity to the target genes. In this way, the subtelomeric repeats could play a role in silencing that is distinct from the classic definition of TPE that resembles the looping model for boundary activity (for review, see Schedl and Broach 2003).

TPE AND HISTONE MODIFICATIONS

Early Evidence That Chromatin Structure Determines TPE Status

The early cytological studies that defined heterochromatin identified the telomeres and centromeres as regions associated with condensed, darkly staining DNA. What "heterochromatin" meant in molecular terms, and how this influenced gene expression, was unknown. Although a correlation between increased histone acetylation and increased transcription was first observed 40 years ago (Allfrey et al. 1964), histone modifications and their effects on heterochromatin have only recently been explored.

Chromatin was first implicated in yeast gene silencing at the silent mating type loci, *HML* and *HMR*. Point mutations in the amino-terminal "tail" of histone H4 eliminate silencing at *HML* (Johnson et al. 1990; Megee et al. 1990; Park and Szostak 1990). Three of the six mutations that affect *HML* silencing, H4-K16Q, H4-K16G, and H4-R17G, also do not support TPE at truncated telomere VII-L (Aparicio et al. 1991). Alleles of *SIR3* were identified that partially suppress the *HM* silencing defect of histone H4 point mutants (Johnson et al. 1990). However, these *sir3* alleles do not restore TPE in cells with the same H4 mutations (Aparicio et al. 1991). This finding is one of several that indicates that the requirements for silencing at the *HM* loci and telomeres are similar but not identical. Further mutational analysis defined domains of both histones H3 and H4 that are completely contained within the amino-terminal tails of these histones that are essential for TPE, but not for cell viability (Mann and Grunstein 1992; Thompson et al. 1994). Amino-terminal deletions of H2B have no effect on either *HM* silencing or TPE (Kayne et al. 1988; Thompson et al. 1994). However, amino-terminal deletions of H2A have recently been shown to reduce TPE (Wyatt et al. 2003). Both the H3 and H4 tails contain several highly conserved lysine residues. Because lysines are subject to posttranslational acetylation, and acetylation had already been linked to transcriptional activity, the individual H3 lysines were mutated either to arginine to mimic the unacetylated state or to glycine or glutamine to mimic the acetylated state. Some of the lysine mutants that mimic acetylation, like K16Q, eliminate TPE (Aparicio et al. 1991; Thompson et al. 1994). Other lysine residues have no phenotype when mutated alone, but mutating several of these lysine residues affects TPE (Thompson et al. 1994).

In biochemical experiments, the H3 and H4 amino termini interact with Sir3p and Sir4p but not with Rap1p or Sir2p (Hecht et al. 1995). These interactions are lost when the tail domains of H3 and H4, known to be necessary in vivo for TPE, are deleted. Furthermore, specific point mutations known to eliminate TPE, like H4-K16Q, also disrupt the interactions between the Sir proteins and the histone tails (Hecht et al. 1995). Amino-terminal deletions of histones H3 or H4 or the H4-K16Q mutation also eliminate the foci of silencing proteins at the nuclear periphery (Hecht et al. 1995). Thus, the requirement of the H3 and H4 amino-terminal tails for silencing can be explained by their role in Sir protein recruitment.

Although Sir3p and Sir4p binding to histone tails is thought to play a structural role that makes subtelomeric chromatin less accessible to the transcription apparatus, Sir2p plays an enzymatic role in the formation of telomeric heterochromatin. Overexpression of *SIR2* reduces histone acetylation in vivo, leading to the suggestion that Sir2p could be an HDAC (Braunstein et al. 1993). Subsequently, several biochemical studies showed that Sir2p is an NAD^+-dependent lysine deacetylase (Tanny et al. 1999; Imai et al. 2000; Landry et al. 2000; Smith et al. 2000). The HDAC activity of Sir2p is required for TPE, *HM* silencing, and rDNA silencing (for review, see Gasser and Cockell 2001; Moazed 2001). After being recruited to the telomere by Sir4p, Sir2p can deacetylate adjacent histone lysine residues, creating binding sites for the structural heterochromatin component Sir3p. The repetition of this process is thought to propagate heterochromatin toward the centromere, with the spreading being limited by the concentration of Sir3p (Renauld et al. 1993).

In addition to the formation of heterochromatin, the balance of HDAC and histone acetyltransferase (HAT) activities is also important in setting up boundaries between heterochromatin and euchromatin. At telomeres, the HAT that counteracts Sir2p activity is Sas2p (something about silencing 2). Sas2p, a HAT required for the global acetylation of H4-K16, is also involved in boundary formation at *HMR* (Reifsnyder et al. 1996; Ehrenhofer-Murray et al. 1997; Kimura et al. 2002; Suka et al. 2002). In a wild-type cell, there is a gradient of acetylated H4-K16 from hyperacetylated at internal chromosomal regions to hypoacetylated at telomeric regions (Kimura et al. 2002). This gradient of acetylation corresponds to an inverse gradient of increasing Sir3p binding to chromatin (Kimura et al. 2002). In a *sas2Δ* strain, Sir3p spreads further from the telomere, increasing the span of Sir3-mediated heterochromatin from ~3 kb at telomere VI-R to ~15 kb, and concomitantly reducing the span of H4 lysine acetylation (Kimura et al. 2002; Suka et al. 2002). This change in chromatin state is correlated with a significant reduction in the transcription level of subtelomeric ORFs below that seen in wild-type cells (Kimura et al. 2002; Suka et al. 2002). This reduction is Sir3p dependent, as a *sas2Δ sir3Δ* strain has wild-type expression levels of the subtelomeric ORFs (Kimura et al. 2002; Suka et al. 2002).

Histone Modifications Other Than Acetylation That Affect Yeast TPE

In addition to histone acetylation, there are other, more recently identified, posttranslational histone modifications that affect TPE, including methylation, phosphorylation, ubiquitination, and ADP-ribosylation (for review, see Berger 2002; Richards and Elgin 2002). These individual histone tail modifications act in a combinatorial manner to mediate gene expression or silencing. For example, histone H3-S10 phosphorylation stimulates the acetylation of H3-K9, leading to gene activation (Lo et al. 2000). This connection is just one of many examples that led to the "histone code" hypothesis, which proposes that different patterns of histone tail modifications lead to distinct binding patterns for both activator and repressor proteins. These proteins, in turn, facilitate the formation of a euchromatic or heterochromatic DNA structure that is permissive or restrictive for gene expression (for review, see Strahl and Allis 2000; Rice and Allis 2001; Khan and Hampsey 2002; Richards and Elgin 2002).

The histone code model is useful for thinking about the role of Dot1p, a protein of previously unknown function that is necessary for TPE. *DOT1* was originally identified in a screen for high-copy disruptors of telomeric silencing (Singer et al. 1998). *DOT1* overexpression eliminates or greatly reduces TPE at truncated telomeres VII-L and V-R and also decreases *HM* and rDNA silencing (Singer et al. 1998). Intriguingly, *dot1Δ* also eliminates TPE and *HM* silencing but has no effect on rDNA silencing (Singer et al. 1998). *dot1Δ* mutants also fail to form wild-type foci of silencing proteins and demonstrate a decrease in Sir2p and Sir3p association with subtelomeric DNA (San-Segundo and Roeder 2000; Ng et al. 2002; van Leeuwen et al. 2002).

How could the same protein function to relieve silencing both when deleted and on overexpression? At the time of the initial screen, Dot1p showed no sequence similarity to any known proteins. Using a computational approach, the Dot1p secondary structure was shown to match the structure of known S-adenosyl-L-methionine methyltransferases (SAM-MTs; Dlakic 2001). This in silico experiment was validated by subsequent biochemical and genetic analyses that demonstrate that Dot1p is indeed a histone methyltransferase that methylates ~90% of the H3-K79 residues (Briggs et al. 2002; Lacoste et al. 2002; Ng et al. 2002; van Leeuwen et al. 2002). These data led to a model whereby methylation of H3-K79 prevents Sir complex binding. In this model, in wild-type cells, methylation is absent at the heterochromatic subtelomeric regions. In *dot1Δ* cells, TPE is disrupted because the resulting lack of H3-K79 methylation in nontelomeric regions allows the Sir complex to bind chromatin away from the telomere, resulting in decreased Sir binding at the telomeres. In a *DOT1* overexpression strain, TPE is disrupted because the H3-K79 methylation extends into the subtelomeric heterochromatin, preventing Sir complex binding to this region.

Unlike most of the other posttranslational histone modifications that have been identified, histone H3-K79 is located in the nucleosome core, not on an amino-terminal tail. The methylation of H3-K79 is dependent on the ubiquitination of histone H2B, suggesting that the histone code also extends beyond the tails (Briggs et al. 2002). Recently, a series of ten amino acids within the core region, including K79, that are necessary for TPE at truncated telomere VII-L and for *HM* silencing was identified (Thompson et al. 2003). Modification of the core domain is also important for heterochromatin formation in higher eukaryotes, as H3-K79 seems to be methylated by Dot1p homologs in flies and humans (Feng et al. 2002; Schubeler et al. 2004; Shanower et al. 2005). The *Drosophila DOT1* homolog, *grappa*, is also necessary for TPE (Shanower et al. 2005).

NUCLEAR LOCALIZATION AND TPE

Some or all telomeres are localized to the nuclear periphery in yeasts, flies, humans, and in the pathogenic protozoa *T. brucei* and *P. falciparum*. Localization of the protozoan telomeres is discussed later in this chapter in the section on Biological Functions of TPE.

Peripheral Localization of Yeast Telomeres and TPE

Localization within the nucleus has long been thought to contribute to heterochromatin formation and gene silencing. Specifically, the nuclear periphery seems to be a region of the nucleus that is conducive to silencing. For example, the human inactive X chromosome, which localizes to the nuclear periphery (Bourgeois et al. 1985), and the *Drosophila* chromocenter, a cluster of the pericentric heterochromatin from all three autosomes, is found at the nuclear periphery (Mathog et al. 1984;

Hochstrasser and Sedat 1987). Perhaps the most compelling experiment arguing that this peripheral localization has functional significance comes from yeast, where tethering a weakened *HMR* silencer lacking its ORC- and Rap1p-binding sites to the nuclear periphery increases the fraction of cells that exhibit silencing (Andrulis et al. 1998).

In baker's yeast, many of the proteins required for TPE localize in foci at the nuclear periphery. Rap1p, Sir2, Sir3, Sir4p, and the Ku complex colocalize in three to six foci at the nuclear periphery (Klein et al. 1992; Palladino et al. 1993; Gotta et al. 1996; Laroche et al. 1998). Deleting any one of the Sir or Ku proteins or the carboxy-terminal Sir-interaction domain of Rap1p eliminates TPE and disperses these foci (Aparicio et al. 1991; Hecht et al. 1995; Laroche et al. 1998). Therefore, all of these key silencing proteins are required not only for TPE but also for the integrity of these foci. Whether the foci themselves contribute to TPE is not clear.

Yeast chromosomes are quite small and therefore not easily visualized by fluorescent hybridization (FISH) in intact nuclei, unless probes for multicopy sequences are used. Using FISH, ~70% of Y$'$ sequences localize to the foci of silencing proteins (Gotta et al. 1996). Deletion of the Sirs or Rap1p-C has no effect on the Y$'$ localization (Gotta et al. 1996), whereas deleting Ku leads to more Y$'$ foci, some of which are no longer at the nuclear periphery (Laroche et al. 1998). It is difficult to generalize about telomere behavior from FISH, especially in terms of TPE, as the Y$'$ signal detects only the subset of telomeres that have this element, most of which do not exhibit TPE (Pryde and Louis 1999). However, the position of individual telomeres can be determined by inserting a *lac* operator array (LacO) near a telomere and using *lac* repressor–green fluorescent protein (LacI-GFP) fusions, which bind the array in vivo, to visualize the position of the telomere (Fig. 3) (Robinett et al. 1996; Michaelis et al. 1997). The nuclear envelope can be detected in fixed cells using an antibody to a nuclear pore protein or in vivo by expressing a nuclear pore–GFP fusion protein. With the LacO visualization system, localization and silencing can be monitored at the same telomere in a population of cells.

This type of analysis was first performed in fixed cells for truncated telomere VII-L (Tham et al. 2001). As anticipated from the Y$'$ analysis, the truncated telomere VII-L is often at the periphery. However, even under conditions where TPE is high (80% repression), the telomere is at the periphery in only a subset of cells (66%). The fraction of telomeres at the periphery varies throughout the cell cycle, with the highest level of association being in G_1 phase. Nonetheless, even in G_1 phase, the truncated VII-L telomere is away from the periphery in many cells. Late in the cell cycle, the association with the periphery is low. Localization of the

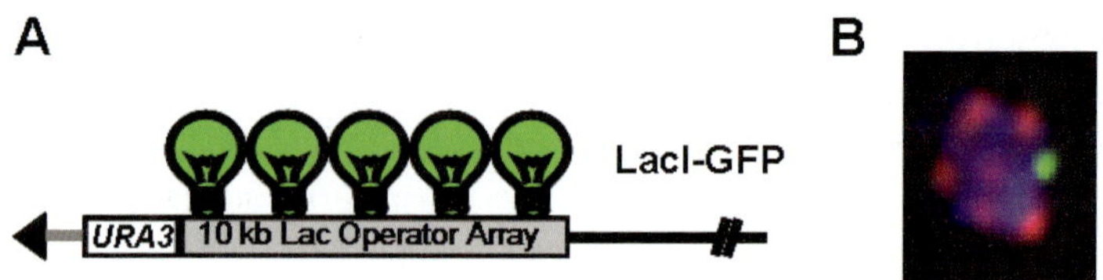

Figure 3. Visualization of single telomeres in the yeast nucleus. (*A*) Schematic of the system to measure TPE and nuclear localization of the truncated VII-L telomere. A *URA3* reporter gene is integrated adjacent to the telomere, along with a *lac* operator array, containing 256 copies of the LacO repeat (as in Tham et al. 2001). In the same strain, a fusion protein between the *lac* repressor and GFP is expressed. LacI-GFP binds to the LacO array, which can be visualized under the microscope as in *B*. (*B*) Visualization of the VII-L telomere and the Rap1p-silencing foci. In this haploid cell, the truncated VII-L telomere is visualized via the LacI-GFP protein. Antibody staining for Rap1p in the characteristic peripheral silencing foci (Gotta et al. 1996) is shown in *red*; DAPI is shown in *blue*. Projection of 3D deconvolution image taken under 100× power. (M.A. Mondoux, unpubl.)

truncated VII-L telomere is not reduced in *sir3Δ* or *yku70Δ* cells, even though TPE is eliminated at this telomere in these strains. Thus, neither Sir3p nor Ku is essential for localization of truncated telomere VII-L to the nuclear periphery.

The LacO assay has also been used to study the behavior of several "native" telomeres in living cells (Hediger et al. 2002b). As found for truncated VII-L, the X-only VI-R telomere, as well as the X-Y′ XIV-L and VIII-L telomeres, is at the periphery in some but not all cells (~50–60%). As with VII-L, this fraction varies with position in the cell cycle, but not in exactly the same way for each telomere. Unlike truncated VII-L, localization of the VI-R and VIII-L telomeres is Ku dependent. Native VI-R also requires Sir proteins to localize to the periphery. In contrast, truncated VI-R, like truncated VII-L, does not require Ku to bind the periphery. Ku independence is not limited to truncated telomeres as the association of the XIV-L telomere is Ku-independent in S phase and only partially Ku-dependent in G_1 phase. This telomere also requires Sir4p for association with the periphery in G_1.

It is difficult to put the data on telomere localization into a simple coherent picture. What is clear (at least for the subset of telomeres that have been examined) is that telomeres are at the periphery in many, but by no means all, cells, and this association decreases late in the cell cycle. Different telomeres show different dependencies on Ku and Sir proteins for localization to the periphery. With the limited data available, it is impossible to attribute a given pattern of genetic dependencies to the presence or absence of specific subtelomeric repeats. In this way, telomere behaviors in terms of TPE status and nuclear localization are similar. Different telomeres have different personalities.

Localization of a telomere to the periphery presumably requires at least two players, a telomere-bound protein and a protein associated with the nuclear periphery or nuclear envelope. From the data described above, the Sir complex, Ku complex, and perhaps other telomere-binding proteins can provide the telomere link. What provides the connection between telomeric chromatin and the nuclear periphery?

There are several candidates for the protein bridge between yeast telomeres and the nuclear envelope. Mlp1p (myosinlike protein 1) was identified in a screen to find nonnucleoporin proteins that associate with the nuclear envelope. Mlp1p and Mlp2p, a related protein, both localize to a filamentous structure that stretches from the inner nuclear envelope toward the nucleoplasm (Strambio-de-Castillia et al. 1999). Mlp2p also associates with the nucleoporin Nic96p (Kosova et al. 2000) and coimmunoprecipitates with yKu70p (Galy et al. 2000). These cell-biological approaches suggested a link between Mlps and telomere localization. However, these data are controversial. One group reported that the number of telomere foci detected by FISH increases, and some of the foci are less peripheral in *mlp1Δ mlp2Δ* cells (Galy et al. 2000), similar to the FISH results in *yku70Δ* cells (Laroche et al. 1998). However, the disruption in nuclear architecture that is seen in *mlp1Δ mlp2Δ* cells (Hediger et al. 2002a) makes it difficult to interpret these changes in Y′ foci. Moreover, using the LacO system, the localization of the X-only telomere VI-R is not altered in the absence of Mlp proteins (Hediger et al. 2002b).

The effects of Mlp proteins on TPE are also disputed. One group reported that *mlp1Δ mlp2Δ* cells show an ~1000-fold reduction in TPE at truncated telomere VII-L (Galy et al. 2000). This result is compromised by the fact that the wild-type strain in this study inexplicably showed poor growth on plates lacking uracil, a result not seen in earlier studies from multiple labs with the same truncated telomere. Two other groups found normal levels of TPE in *mlp1Δ mlp2Δ* cells, both at the truncated VII-L telomere and at the V-R telomere (Andrulis et al. 2002; Hediger et al. 2002a). This result was true in two strain backgrounds and using two reporter genes (Hediger et al. 2002a). Taken together, there is not compelling evidence that Mlp proteins function in TPE or telomere positioning. Mlp1p plays a role in export of unspliced mRNAs, so it is possible that under some conditions, TPE is affected indirectly in *mlp* mutants (Galy et al. 2004).

Are there other candidates for proteins that serve as the link between telomeres and the nuclear envelope? *ESC1* (establishes silent chromatin) was identified in a screen for proteins that could mediate silencing when tethered to the truncated VII-L telomere in a *rap1ΔC* background, in which the mutated Rap1p cannot recruit the Sir complex (Andrulis et al. 2002). Because Esc1p localizes to the inner nuclear membrane (but not to the nuclear pores or the foci of silencing proteins), it is in the right place to be part of the telomere tethering apparatus (Andrulis et al. 2002; Taddei and Gasser 2004). Moreover, by two-hybrid analysis, Esc1p interacts with a portion of Sir4p known as the partitioning and anchoring domain (PAD; Andrulis et al. 2002; Taddei and Gasser 2004), a region that is able to promote partitioning of unstable ARS plasmids in mitosis (Ansari and Gartenberg 1997). Both the carboxyl terminus of Esc1p and the Sir4p PAD are capable of repositioning an internal chromosomal locus to the nuclear periphery when tethered to the locus via LexA, as expected if the two proteins cooperate to bring sequences to the nuclear envelope (Taddei and Gasser 2004).

Taken together, these data are consistent with a model in which telomere-bound Sir4p interacts with Esc1p at the inner nuclear membrane to hold telomeres at the periphery. However, the peripheral localization of both truncated telomere VI-R and the X–Y′ telomere XIV-L is not perturbed in an *esc1Δ* strain (Taddei and Gasser 2004). These telomeres are also localized to the nuclear periphery in a *yku70Δ* strain (Hediger et al. 2002b). Both of these telomeres show a random distribution within the nucleus in an *esc1Δ yku70Δ* double mutant or a *sir4Δ yku70Δ* double mutant (Taddei and Gasser 2004). Thus, at least for these two telomeres, there are two redundant telomere localization pathways, one mediated by Esc1p/Sir4p and the other by the Ku complex in concert with an as-yet-unidentified nuclear-membrane-associated protein. The presence of redundant targeting pathways can also explain the behavior of truncated VII-L, which is localized to the nuclear periphery in both Sir- and Ku-deficient cells (Tham et al. 2001).

It is tempting to speculate that the association of telomeres with the nuclear periphery promotes TPE by bringing telomeres into close proximity to the foci of silencing proteins. In addition (or alternatively), placement near the nuclear periphery may promote TPE in other ways. For example, constraining telomere mobility by an association with the nuclear periphery may make it more difficult for RNA polymerase to transcribe through a telomere-linked gene. Telomere positioning at the periphery is clearly not sufficient for TPE. For example, several telomeres remain at the periphery when their silencing is eliminated, as does the VII-L telomere in *sir3* and *yku70* strains (Tham et al. 2001). Also, Y′ telomeres are often localized both to the nuclear periphery and to foci of silencing proteins (Palladino et al. 1993) despite their having very low or no TPE (Pryde and Louis 1999). If telomere positioning is important for TPE, then mechanisms that eliminate tethering should also eliminate TPE. Indeed, both Ku and Sir4p are essential for TPE. However, the loss of TPE seen in their absence is not due to loss of telomere positioning, as truncated telomeres are still at the nuclear periphery in Ku- and Sir-deficient cells (Tham et al. 2001; Hediger et al. 2002b). What about Esc1p? The foci of silencing proteins are not perturbed, and TPE is only modestly reduced at truncated telomeres VII-L and VI-R in *esc1Δ* cells (Andrulis et al. 2002; Taddei and Gasser 2004). However, these telomeres are still associated with the nuclear periphery in the absence of Esc1p, presumably via the redundant Ku-mediated pathway. A critical test of the importance of tethering for TPE could be the identification of a telomere whose association with the periphery occurs solely by the Esc1/Sir4 pathway, but none of the small subset of telomeres that have been tested is tethered in this way.

Another way of gauging if TPE and peripheral localization are causally linked is to ask if the fraction of telomeres associated with the periphery decreases when TPE is low and increases when TPE is high. The truncated VII-L telomere has the same high level of association with the nuclear periphery (Tham et al. 2001) and with silencing foci (M.A. Mondoux and V.A. Zakian, unpubl.) when wild-type cells are grown in medium that requires expression of the telomere-linked *URA3* gene. Thus, even in silencing competent cells, where foci of silencing proteins are not disturbed, telomeres can be actively transcribed yet still show high association with the periphery and with foci of silencing proteins.

Another experiment tested telomere localization in cells where TPE is increased by deleting the histone deacetylase Rpd3p, which increases silencing ~25-fold at truncated telomere V-R and confers TPE on the otherwise TPE-deficient native X-Y' V-R telomere (Rundlett et al. 1996). Because TPE is associated with hypoacetylated histones, the effects of Rpd3p on TPE are probably not due to its effects on subtelomeric chromatin. Rather, its HDAC activity may normally act to reduce expression of a gene that is needed for TPE. If the level of TPE at a given telomere correlates with telomere localization, then one expects that peripheral localization would increase in *rpd3Δ* cells. The results were mixed. The X-only VI-R telomere is more often at the nuclear periphery in mid-late S phase in *rpd3Δ* (more than 70%) compared to wild-type cells (~45%), but there is no change in the peripheral localization of telomere VI-R in G_1 or early S phase (Hediger et al. 2002b). Although the effects of Rpd3p on TPE at telomere VI-R were not determined in this study, in another study a reporter gene inserted at the native VI-R telomere was 100% silenced in wild-type cells (M.A. Mondoux and V.A. Zakian, unpubl.). Thus, it is difficult to correlate telomere position with silencing. In considering the current data, there is still no compelling evidence that TPE and peripheral positioning are causally linked. Nonetheless, although a variety of genes and conditions have been tested, other than late in the cell cycle (Tham et al. 2001), there is no case where telomeres are away from the periphery and silenced. Thus, localization to the nuclear periphery may be necessary for TPE.

Nuclear Localization and TPE in *Drosophila*

As in yeast, *Drosophila* telomeres are localized to the nuclear periphery; however, the exact pattern of localization is not the same for each telomere. The chromocenter, which is itself at the periphery, is formed by the clustering of the centromeric regions of all the chromosomes. The telomeres of the X, second, and third chromosomes are also at the nuclear periphery but are positioned on the opposite side of the nucleus from the chromocenter (the so-called Rabl orientation; Mathog et al. 1984; Hochstrasser et al. 1986). The chromosome 4 telomeres colocalize with the chromocenter. At the periphery, the same telomeres are not always next to each other, and the arrangements of different telomeres can change during development (Dernburg et al. 1996).

As described in the earlier section on *Drosophila* subtelomeric structure, requirements for TPE on the fourth chromosome are different from the requirements for TPE on the second and third chromosomes. When the telomere of the fourth chromosome is translocated to the end of the right arm of chromosome 2, silencing is reduced at this telomere, but the protein requirements for TPE at this new telomere mimic those for TPE at chromosome 4 (Cryderman et al. 1999). However, the translocated fourth telomere shows a localization pattern typical of the 2R telomere. That is, the fourth telomere fused to the right arm of chromosome 2 localizes to the nuclear periphery opposite the chromocenter (Cryderman et al. 1999). Compared to the chromocenter, the nuclear periphery has a lower concentration of HP1 (Kellum et al. 1995), which probably explains why the translocated fourth telomere, whose TPE is HP1 dependent, shows lowered silencing (Kellum et al. 1995; Cryderman et al. 1999). The reciprocal translocation of the 2R telomere to the fourth chromosome also results in its relocation. In this case, the telomere relocates to the HP1-rich chromocenter, but shows no change in TPE levels (Cryderman et al. 1999). It is possible that the proteins required for TPE at 2R are evenly distributed throughout the nucleus, or are as abundant in the chromocenter as they are in 2R's normal location near the periphery, and therefore TPE at this hybrid telomere is not reduced. Although TPE requirements are mediated by subtelomeric structure, localization seems to be mediated by chromosomal context as translocated telomeres retain their characteristic protein dependencies for TPE but lose their characteristic localization patterns. As in yeast, it is not clear if silencing requires localization to the nuclear periphery.

Peripheral Localization of a Particular Human Telomere May Play a Role in Silencing

Human telomeres are usually not localized to the nuclear periphery. Rather, telomeres are distributed throughout the nucleus in HeLa tissue culture cells, primary myoblasts, and fibroblasts (Luderus et al. 1996; Tam et al. 2004). Nonetheless, certain telomeres are specifically localized to heterochromatin at the nuclear and nucleolar peripheries. Although only 17% of telomeres localize to the nuclear periphery in primary fibroblasts, telomere 4q, which contains the D4Z4 repeats associated with FSHD, localizes with the nuclear periphery in primary fibroblasts in 65% of cells (Tam et al. 2004). This localization to heterochromatin compartments is even more pronounced (90%) in muscle precursor cells, the cell type where the disease is manifest (Tam et al. 2004). In addition, translocating the 4q tip onto the human active X chromosome significantly increases peripheral localization of the active X and decreases peripheral localization of the hybrid chromosome 4q (Tam et al. 2004).

Is there a link between D4Z4 subtelomeric structure, nuclear localization, and TPE? A plausible model is that as the D4Z4 subtelomeric repeats are lost, localization to the peripheral heterochromatin and thus TPE are reduced, resulting in expression of genes close to the telomere and the FSHD disease phenotype. To test this model, 4q localization was examined in heterozygous FSHD patient cell lines. Although there is a slight decrease in peripheral localization of the mutant allele compared to the normal allele, this difference is not statistically significant and the association of the mutant allele with the heterochromatin compartment remains high in all cell lines tested (Tam et al. 2004).

Mutant alleles all contain at least one copy of the D4Z4 repeat, which may be sufficient to direct localization of the telomere to the nuclear and nucleolar periphery. Telomere 10q also contains D4Z4 repeats, although the 10q repeats are not associated with FSHD. Telomere 10q also localizes to the nuclear periphery in ~65% of myoblasts (Tam et al. 2004). Thus, it is possible that the presence of the D4Z4 repeats mediates peripheral localization to heterochromatin, permitting and facilitating TPE at telomere 4q and, perhaps, at other telomeres. In this model, repeat loss leads to a loss of TPE through a mechanism other than loss of peripheral localization, for example, loss of subtelomeric binding proteins that promote heterochromatin formation. An alternate hypothesis, given the controversial subtelomeric gene expression data, is that D4Z4 acts as an insulator to block the spread of heterochromatin at telomere 4q. In this model, the peripheral nuclear localization is a consequence of insulator or boundary activity. In both models, subtelomeric elements and the nuclear periphery play a role in human gene regulation.

TPE AND THE CELL CYCLE

It is well established that heterochromatic regions, such as the pericentric heterochromatin of *Drosophila* and the mammalian inactive X chromosome, are late replicating (for review, see Gilbert 2002). Yeast telomeres also replicate late in S phase (McCarroll and Fangman 1988; Raghuraman et al. 2001). Late replication of heterochromatin could be explained by two general models. First, active origins may not be present in heterochromatin, either because origins are absent or because the compact chromatin structure of the region makes existing origins inaccessible to the replication machinery. Alternatively, heterochromatin may contain active origins, but these origins may not fire until late S phase. In baker's yeast, the X and Y′ subtelomeric repeats contain origins that are active on plasmids (Chan and Tye 1983) but that are rarely activated in their normal chromosomal context (Dubey et al. 1991; Vujcic et al. 1999). However, these origins are functional in the chromosomal context, because if replication fork progression is slowed by mutations in replication factors, these origins fire, although their efficiency of use is still relatively low (Ivessa et al. 2002, 2003). Other origins, such as ARS501, which is ~25 kb from the chromosome end and is located within unique DNA, are active but late firing (Ferguson et al. 1991).

Reduced or late origin activation is conferred by proximity to the telomere, not by the sequence of the origin. For example, when the late-firing origin ARS501 is placed on a circular plasmid, it activates early (Ferguson and Fangman 1992). This change in origin timing is not due to its placement on a small plasmid because the same origin on a linear plasmid activates in late S phase. Likewise, if the normally early-firing origin ARS1 is placed in the subtelomeric region of chromosome V-R or the early-firing 2-μm ARS is placed on a linear plasmid, both are still active as origins, but they now fire in late S phase (Ferguson and Fangman 1992; Wellinger et al. 1993a,b). Thus, yeast telomeres exert a position effect on origin activation, in some cases inhibiting and in other cases delaying origin activation.

Are the telomere effects on replication exerted through the same heterochromatic chromatin structure that limits transcription? The Ku complex, which is critical for TPE, is also necessary for late replication of telomeric regions (Cosgrove et al. 2002). In a *yku70Δ* strain, origins that normally fire in late S phase, such as ARS501, fire much earlier in S phase. This effect on replication appears to be limited to telomeres as the absence of Ku does not affect replication timing in other parts of the chromosome. However, Ku's effects on replication timing are not easily explained by its TPE function because ARS501 fires only slightly earlier in a *sir3Δ* strain compared to wild type (Stevenson and Gottschling 1999; Cosgrove et al. 2002). Likewise, the origin near the III-L telomere, which is inactive in wild-type cells, is also inactive in *sir2*, *sir3*, or *sir4* cells, although this origin is used when replication fork progression is slowed (Ivessa et al. 2003). Thus, Ku's function in setting replication timing is unlikely to be due to its role in TPE.

Human telomeres replicate throughout S phase. One model for studying whether heterochromatin and replication timing are linked at human telomeres is a naturally occurring truncation of chromosome 22q that results in the deletion of 130 kb. The region of the truncation break-point, which is now closer to the telomere, is late replicating. In contrast, the same piece of DNA on the intact allele replicates in mid S phase (Ofir et al. 1999). This change in replication timing may be due to the deletion of active origins in that region. Nevertheless, late replication does not seem to be accompanied by imposition of a heterochromatic-like chromatin structure: DNase I sensitivity, methylation, and expression of the closest gene (~54 kb from the breakpoint) are indistinguishable at the mutant and wild-type alleles (Ofir et al. 1999). Taken together, the data from yeast and humans suggest that the telomere's effects on tran-scription and replication timing are separable phenomena.

Cell Cycle Requirements for the Establishment of Silencing

S phase progression plays a role in the establishment of heterochromatin. Silencing can be established at the *HM* loci in cells that have passed from G_1 to M phase, but not in cells that are arrested in early S phase. In contrast, derepression can occur in both cell populations (Miller and Nasmyth 1984). These results were interpreted as indicating that S-phase progres-sion, and specifically DNA replication, are needed to establish the silent state. Several lines of evidence confirmed that S-phase progression is, in fact, crucial for establishment of silencing, but suggested that replication is not the critical event (Ehrenhofer-Murray et al. 1995; Fox et al. 1997). For example, silencing can be established on a nonreplicating, plasmid-borne *HMR* locus, as long as cells carrying the plasmid progress through S phase (Kirchmaier and Rine 2001; Li et al. 2001). Moreover, because Sir protein recruitment and spreading do not occur robustly until G_2/M phase, the complete repression of the *HM* loci occurs after S phase (Lau et al. 2002).

How do cell cycle progression and DNA replication affect TPE? When *URA3* is adjacent to the truncated VII-L telomere, its transactivator, Ppr1p, has different effects on *URA3* expression at different points in the cell cycle. Ppr1p is able to activate transcription in cells that have been arrested at G_2/M, but cannot activate transcription in cells arrested in stationary phase, G_1, or early S phase (Aparicio and Gottschling 1994). The time at which a telomeric *URA3* gene can be activated corresponds to the time in the cell cycle when the VII-L telomere moves away from the nuclear periphery (Tham et al. 2001). Moreover, if the truncated VII-L telomere is not associated with the nuclear periphery, TPE can be eliminated efficiently in G_1 phase cells (Tham et al. 2001). These data suggest that although S-phase progression is usually required for telomeres to switch from a silent to an active transcription state, the event that allows switching is probably not DNA replication. Rather, the key event may be movement of the telomere away from the periphery, which occurs normally after DNA replication (Tham et al. 2001). Cell cycle control of TPE is not limited to baker's yeast, as TPE at the *P. falciparum* subtelomeric *var* genes is also regulated by cell cycle progression and is established during S phase (Deitsch et al. 2001).

BIOLOGICAL FUNCTIONS OF TPE

Yeast TPE as a Mechanism for Metabolism, Stress Response, and Adaptation

TPE was discovered serendipitously in both *Drosophila* and yeast when genes that are normally far from telomeres were positioned adjacent to a chromosome end and found to be transcriptionally repressed (Levis et al. 1985; Gottschling et al. 1990). Although there are considerable data that bear on the mechanism by which telomeres affect transcription, the in vivo importance of TPE has been more difficult to assess.

The early experiments on TPE in yeast demonstrated that it was regulated by many of the key genes needed for silencing at the silent mating type loci (Aparicio et al. 1991). Thus, one way to determine if TPE is a bona fide mechanism of transcriptional regulation is to ask if genes that are naturally near telomeres are expressed at low levels in a Sir-dependent manner. As a further test, these telomere-linked genes can be moved to a nontelomeric site. If the gene's low-level expression at its telomeric site is due to TPE, then its transcription should increase at an internal site, even in a *SIR*-proficient strain.

Early attempts to address the question of the biological relevance of TPE examined the expression of individual, telomere-linked yeast genes by conventional methods. One study found that transcription of the Ty-5 transposon near the III-L telomere is low in wild-type cells, and this level increases in a *sir3* strain (Vega-Palas et al. 1997). Likewise, an ORF near the VI-R telomere is transcriptionally repressed and this repression is relieved in Sir-deficient strains (Vega-Palas et al. 2000). This same study found three other telomere-linked genes whose transcription is not increased when *SIR* genes are deleted.

The introduction of techniques for genome-wide transcription analysis made it possible to compare expression levels of all telomere-linked genes to that of the remainder of the genome. Consistent with TPE being a biologically relevant phenomenon, genes near telomeres are generally expressed at lower levels than nontelomeric genes (Wyrick et al. 1999). The 267 yeast genes that are located within 20 kb of a telomere are represented by an average of 0.5 RNA molecules per cell, a fivefold lower level than the average for nontelomeric genes. However, expression of only 20 of these genes is increased in the absence of Sir proteins. Similar effects on transcription are seen in *sir2*, *sir3*, and *sir4* strains. Most of the Sir-sensitive genes are within 8 kb of a telomere. Thus, by the classical definition, some yeast genes are regulated by TPE, but their number is relatively small, and they are very close to a chromosome end. Moreover, none of these genes has been moved to a nontelomeric site to determine if their Sir-mediated low-level transcription is telomere dependent. Without this test, it could be that genes that are expressed at low levels tend to accumulate near telomeres, but their low-level expression does not require telomere proximity.

Genome-wide transcription analysis is one of two ways used to assess TPE at "native" telomeres; the other method involves inserting the *URA3* reporter at the X-ACS in a manner that does not delete any of the telomeric repeats (see the section Telomere Identity and Effects on TPE and Fig. 1B for description). Both methods indicate that TPE functions at only a subset of "native" telomeres, but the two methods do not completely agree. Genome-wide analysis suggests that TPE regulates Sir-dependent gene expression at telomeres I-R, III-L, IV-R, V-L, VI-L, VI-R, VII-L, VIII-L, IX-R, X-R, XIII-R, XIV-L, XIV-R, XV-L, and XVI-R (Wyrick et al. 1999). Although not all of the telomeres have been tested via insertion of *URA3* at the X-ACS, of those that have, four telomeres containing genes up-regulated in the absence of *SIR3* do not support TPE when *URA3* is inserted at the X-ACS (VI-L, X-R, XIV-L, and XV-L; Wyrick et al. 1999; E. Louis, pers. comm.). In these cases, the ORFs naturally affected by TPE are very close to or within the subtelomeric repeats and are thus closer to the telomere than the reporter gene inserted at the X-ACS. It is intriguing that these ORFs within the subtelomeric repeats, some of which are within Y', may be naturally regulated by TPE, even though *URA3* reporter genes inserted within the Y' repeat are not subject to TPE (Pryde and Louis 1999). Two telomeres that support TPE by the *URA3* integration assay do not contain any natural ORFs that are up-regulated in the absence of *SIR3* (II-R and XI-L; Wyrick et al. 1999; E. Louis, pers. comm.). In fact, one of the ORFs close to telomere II-R is actually down-regulated in the absence of *SIR3* (Wyrick et al. 1999). These results emphasize that the specific subtelomeric context has a strong impact on TPE. This DNA sequence and chromatin conformation context may not be identical at the same telomere in different strains.

The effects of other mutations suggest that there are also Sir-independent transcriptional repression mechanisms that act preferentially on telomere proximal regions. Depletion of histone H4 increases gene expression at ∼15% of yeast genes, most of which are located near a telomere (Wyrick et al. 1999). Under these conditions, ∼50% of the 267 genes that are within 20 kb of the telomeres are derepressed, many more than the number affected by Sir depletion. Likewise, genome-wide HDAC activity maps reveal that Hda1p, which specifically deacetylates lysines on H3 and H2B (Wu et al. 2001), plays a role in deacetylation of histones in repressed domains that are ∼10–25 kb away from the telomeres (Robyr et al. 2002). These regions, which are termed HAST (Hda1-affected subtelomeric) domains are distinct from the immediately sub-telomeric zones that are the targets of Sir-protein-mediated TPE. These experiments suggest that we should expand our definition of TPE to include genes whose transcription is reduced by proximity to the telo-mere in a Sir-independent, Hda1p-dependent fashion. Again, none of the HAST domain genes has been tested to see if they are expressed at higher levels at a nontelomeric site.

What can we learn about functions of Sir-dependent and Sir-independent TPE from the identities of the low-expression telomere-linked genes? Rapamycin treatment, pheromone treatment, nutrient starvation, and heat shock all lead to hyperphosphorylation of Sir3p, which is correlated with decreased TPE at truncated telomere VII-L (Stone and Pillus 1996; Ai et al. 2002). In concert with the loss of TPE, upon Sir3p hyperphosphorylation, there is an up-regulation of at least some of the subtelomeric *PAU* genes, which encode cell wall proteins (Ai et al. 2002). Overexpression of the *PAU* genes confers partial rapamycin resistance (Ai et al. 2002). In addition to the *PAU* genes, there are several other subtelomeric genes, such as the *MAL* and *SUC* families, that regulate growth and stress response. Thus, in yeast, Sir-dependent TPE may be a mechanism for responding to nutrient deprivation and environmental stress.

HAST domains contain genes necessary for gluconeogenesis, growth in nonglucose carbon sources, and stress response, again genes that are not expressed under normal growth conditions (Robyr et al. 2002). For example, four of the five members of the *FLO* gene family are subtelomeric, located from 15 to 35 kb from chromosome ends. The four subtelomeric *FLO* genes are transcriptionally silent, whereas *FLO11*, which is ~50 kb from a telomere, is the only *FLO* gene expressed in the Σ1278b strain (Guo et al. 2000). When expressed, the subtelomeric *FLO* genes enhance cell-to-cell adherence, promote adhesion to surfaces, and promote pseudohyphal growth. Although expression of the subtelomeric *FLO* genes is not Sir2p-dependent, they are regulated by the Sir2p homologs Hst1p and Hst2p (homolog of Sir two). *FLO11*, which is further from a telomere than most HAST domain genes, is nonetheless regulated by Hda1p (Robyr et al. 2002; Halme et al. 2004).

In *S. pombe*, deletion of the Hda1p ortholog Clr3 results in the up-regulation of a large class of subtelomeric genes up to 50 kb from the telomere (Hansen et al. 2005). Many of these genes are involved in the response to nitrogen starvation. Cell wall proteins are affected by TPE in the yeast *Candida glabrata*, an opportunistic human pathogen. In *Candida*, the *EPA* genes, which encode cell wall proteins that regulate cell-to-cell adherence and surface adhesion, are located in two gene clusters adjacent to the telomeres. Although *EPA1* is expressed, the *EPA2-5* genes are transcriptionally silent in a Sir- and Rap-dependent manner (De Las Peñas et al. 2003). These data suggest that the organization of genes for stress and alternative growth strategies into subtelomeric domains is evolutionarily conserved.

Telomere Localization and Gene Expression in Eukaryotic Parasites

T. brucei is a single-celled protozoan that causes sleeping sickness. TPE exists in *Trypanosoma* because an active, nontelomeric promoter becomes transcriptionally inert when moved next to a telomere (Horn and Cross 1995). One mechanism this organism has evolved to evade the host immune system is a regular changing of its surface coat proteins, termed antigenic variation (for review, see Borst and Ulbert 2001). In *T. brucei*, there are approximately 1000 VSG (variant surface glycoprotein) genes that encode the coat proteins. However, there are only approximately 20 sites from which these genes can be expressed, all of which are subtelomeric. In *T. brucei*, only 1 of the 20 subtelomeric VSG genes is active at a time and the rest are silent, prompting an attractive model in which TPE prevents VSG expression at all but one subtelomeric expression site. Thus, one very interesting question is to determine how the active subtelomeric expression site escapes TPE and how it differs from the inactive expression sites in the same cell (Chaves et al. 1999). By FISH analysis, the 20 subtelomeric expression sites are randomly distributed in the trypanosome nucleus, without any evident clustering or organizational pattern (Chaves et al. 1998; Navarro and Gull 2001). However, these experiments could not distinguish the position of the single active subtelomeric expression site from the inactive sites. Unlike most protein-coding genes, the VSGs are transcribed by RNA polymerase I (Shea et al. 1987; Brown et al. 1992). In *T. brucei* nuclei, RNA polymerase I is found in the nucleolus, as expected for its role in rDNA transcription. There is also a transcriptionally active, extranucleolar structure that contains RNA polymerase I that is not associated with rDNA transcription (Navarro and Gull 2001). Using the *lac* operator system to visualize different expression sites, the active expression site is observed to associate with this RNA polymerase I body, whereas a silent expression site does not (Navarro and Gull 2001). This nuclear structure was thus renamed the expression site body (ESB). A new model for antigenic variegation at subtelomeric genes, then, is that the active expression site associates with the ESB, and switching occurs not as a loss or gain of conventional TPE, but as a relocalization to or away from the ESB. It is also possible that the localization of the active ES with the ESB is a secondary consequence of its association with RNA polymerase I.

Antigenic variation also occurs in the malaria parasite *P. falciparum*, modulating the transcription of the subtelomeric *var* genes, whose products are expressed on the surface of the infected erythrocyte (for review, see Wahlgren et al. 1999). As in *Trypanosoma*, only one of the *var* genes is expressed at a time, suggesting that their expression might also be regulated by TPE. The 28 *Plasmodium* telomeres cluster into four to seven foci at the nuclear periphery in a manner that is affected by their subtelomeric DNA content (Freitas-Junior et al. 2000; Figueiredo et al. 2002). Telomeres lacking the subtelomeric TAS array fail to localize to telomere clusters, though they still localize to the nuclear periphery (Figueiredo et al. 2002). Telomere clustering may play a role in the frequent ectopic recombination events between the *var* genes (Freitas-Junior et al. 2000; Figueiredo et al. 2002).

TPE and Aging in Yeast and Metazoan Cells

Because human telomeres shorten with age, it is possible that TPE is developmentally regulated in a manner that impacts the aging process. For example, as human telomeres shorten over time, the expression of TPE-regulated subtelomeric genes might change, leading to the expression of genes that either contribute to or combat the aging process. Alternatively, as telomeres shorten, key silencing proteins that are normally sequestered at the telomere might be freed, enabling them to silence nontelomeric genes that inhibit aging. Although attractive, there is currently no direct evidence to support a role for TPE in human aging. For example, no human genes that contribute to aging as we understand it are known to be located at telomeres and to increase expression over time.

There are, however, connections between the conserved Sir2 HDAC and aging in several organisms. Aging in baker's yeast is defined by how many times mother cells can bud to give rise to daughter cells. Using this metric, *sir2Δ* yeast have decreased life spans (~50%) and cells overexpressing Sir2p have increased life spans (30–40%; Kaeberlein et al. 1999; Roy and Runge 2000). Aging in *C. elegans* is defined by the decay of the nondividing cells of the soma. Despite this being a very different metric for aging, as in yeast, increased dosage of the *C. elegans sir-2.1* gene extends life span (Tissenbaum and Guarente 2001). However, there is no evidence in either organism that the effects of Sir2p on life span are connected to its role in TPE. Yeast Sir2p has multiple functions: It inhibits transcription not only at telomeres but also at the silent mating type loci and in the rDNA. Yeast Sir2p also inhibits activation of origins of DNA replication, both in the rDNA and elsewhere (Pasero et al. 2002; Pappas et al. 2004). The replication function of Sir2p has been implicated in aging through its role in preventing the accumulation of extrachromosomal rDNA circles, although the association of rDNA circles with aging has been challenged (Sinclair et al. 1997; Kaeberlein et al. 1999; Roy and Runge 2000; Falcon and Aris 2003). Sir2p's function as an NAD$^+$ deacetylase may link its role in aging to metabolism. Caloric restriction increases life span in yeast, worms, and humans (for review, see Tissenbaum and Guarente 2002). In yeast and worms, mutating *SIR2* is argued to eliminate the increased life span associated with caloric restriction (Lin et al. 2000, 2002). However, more recent evidence suggests that although caloric restriction and Sir2p both function in the regulation of yeast life span, they act in independent pathways (Jiang et al. 2002; Kaeberlein et al. 2004).

Sir2p is the only one of the yeast Sir proteins that is conserved among diverse organisms (Brachmann et al. 1995). Moreover, there are multiple members of the Sir2 family in most organisms, including baker's yeast, which has four HST genes. Compounds that activate Sir2 homologs promote longevity in yeast, human cells, *Drosophila*, and *C. elegans* (Howitz et al. 2003; Wood et al. 2004). In mammalian cells, the effects of caloric restriction on life span are argued to be mediated by the closest of the Sir2p homologs, SIRT1 (Cohen et al. 2004). In addition, the human SIRT3 gene, which also encodes a NAD-dependent deacetylase, is located close to the 11p telomere. This region of the chromosome contains five genes that have been linked to aging (De Luca et al. 2001; Bonafe et al. 2002; Tan et al. 2002).

Sir2p and its homologs are not the only HDACs with correlations to aging. In yeast, the HDAC Rpd3p contributes negatively to longevity (Jiang et al. 2002). A partial reduction in Rpd3 levels also results in increased life span in flies (*Drosophila* Rpd3 is essential). Because *Drosophila* Sir2 expression is increased twofold in the *rpd3* mutants, the effects of Rpd3 on life span may be indirect (Rogina et al. 2002). In yeast, deleting the HDAC *HDA1*, required to inhibit transcription of HAST domain genes, has no effect by itself on longevity but its deletion acts synergistically with caloric restriction to increase life span (Jiang et al. 2002). Is the role of Sir2p and other histone deacetylases in aging related to their roles in TPE and Sir-independent subtelomeric gene regulation? The most likely link seems to be metabolism and stress response, as gene families involved in these processes are located in subtelomeric regions and are subject to TPE or Sir-independent regulation by acetylation. However, at this point, there is at best a tenuous connection between telomeric gene expression and aging.

CONCLUDING REMARKS

In diverse organisms, telomeres exert position effects on the expression of subtelomeric genes. TPE is classically defined in *S. cerevisiae* as the Sir-protein-mediated spread of heterochromatin from the telomere inward. However, there is recent evidence in yeast for Sir-independent transcriptional inhibition of genes that are near telomeres but outside of the Sir-associated domain. These larger subtelomeric domains are still regulated by histone modifications, but they are silenced via the Hda1p or Hst1/2p, not the Sir2p, HDAC. One model for TPE at human telomere 4q involves the looping of the telomere to bring silencing factors in proximity to target genes, as opposed to a continuous spread of heterochromatin. We should therefore expand our definition of TPE to include all gene silencing that is mediated by the telomeric and subtelomeric repeats, not just that which is Sir-mediated or continuous.

Different species have evolved similar mechanisms to regulate TPE, including conserved or analogous telomere-binding proteins, telomere length regulation, nuclear localization, and cell cycle regulation. Although different species may use similar mechanisms to silence subtelomeric genes, different telomeres within a species can have differing abilities to support TPE, levels of TPE, and even requirements for TPE. Some, but not all, of these differences can be explained by the varied subtelomeric repeat elements found at different telomeres.

TPE is a biologically relevant phenomenon, regulating rarely used genes for growth and stress response in *S. cerevisiae* and *S. pombe*, cell adhesion in the opportunistic pathogen *C. glabrata*, immune system evasion in the parasites *T. brucei* and *P. falciparum*, and potentially a form of muscular dystrophy in humans. The study of telomere position effect thus not only serves as a model for epigenetic silencing and regulation, but also provides the potential for understanding wide-ranging biological problems, including genome organization, survival mechanisms, pathogenesis, and disease.

ACKNOWLEDGMENTS

We thank our colleagues who sent manuscripts that helped us write this paper. We also thank Ed Louis, Art Lustig, Jim Mason, Lorraine Pillus, Kurt Runge, and Woody Wright for their comments and the NIH for support of research in our laboratory.

REFERENCES

Abad J.P., De Pablos B., Osoegawa K., De Jong P.J., Martin-Gallardo A., and Villasante A. 2004. *TAHRE*, a novel telomeric retrotransposon from *Drosophila melanogaster*, reveals the origin of *Drosophila* telomeres. *Mol. Biol. Evol.* **21:** 1620–1624.

Adams Martin A., Dionne I., Wellinger R.J., and Holm C. 2000. The function of DNA polymerase α at telomeric G tails is important for telomere homeostasis. *Mol. Cell. Biol.* **20:** 786–796.

Ai W., Bertram P.G., Tsang C.K., Chan T.F., and Zheng X.F. 2002. Regulation of subtelomeric silencing during stress response. *Mol. Cell* **10:** 1295–1305.

Allfrey V.G., Faulkner R., and Mirsky A.E. 1964. Acetylation and methylation of histones and their possible role in the regulation of RNA synthesis. *Proc. Natl. Acad. Sci.* **51:** 786–794.

Allshire R.C., Nimmo E.R., Ekwall K., Javerzat J.P., and Cranston G. 1995. Mutations derepressing silent centromeric domains in fission yeast disrupt chromosome segregation. *Genes Dev.* **9:** 218–233.

Andrulis E.D., Neiman A.M., Zappulla D.C., and Sternglanz R. 1998. Perinuclear localization of chromatin facilitates transcriptional silencing. *Nature* **394:** 592–595.

Andrulis E.D., Zappulla D.C., Ansari A., Perrod S., Laiosa C.V., Gartenberg M.R., and Sternglanz R. 2002. Esc1, a nuclear periphery protein required for Sir4-based plasmid anchoring and partitioning. *Mol. Cell. Biol.* **22:** 8292–8301.

Ansari A. and Gartenberg M.R. 1997. The yeast silent information regulator Sir4p anchors and partitions plasmids. *Mol. Cell. Biol.* **17:** 7061–7068.

Aparicio O.M. and Gottschling D.E. 1994. Overcoming telomeric silencing: A *trans*-activator competes to establish gene expression in a cell cycle-dependent way. *Genes Dev.* **8:** 1133–1146.

Aparicio O.M., Billington B.L., and Gottschling D.E. 1991. Modifiers of position effect are shared between telomeric and silent mating-type loci in *S. cerevisiae*. *Cell* **66:** 1279–1287.

Baur J.A., Shay J.W., and Wright W.E. 2004. Spontaneous reactivation of a silent telomeric transgene in a human cell line. *Chromosoma* **112:** 240–246.

Baur J.A., Zou Y., Shay J.W., and Wright W.E. 2001. Telomere position effect in human cells. *Science* **292:** 2075–2077.

Berger S.L. 2002. Histone modifications in transcriptional regulation. *Curr. Opin. Genet. Dev.* **12:** 142–148.

Boivin A., Gally C., Netter S., Anxolabehere D., and Ronsseray S. 2003. Telomeric associated sequences of *Drosophila* recruit polycomb-group proteins *in vivo* and can induce pairing-sensitive repression. *Genetics* **164:** 195–208.

Bonafe M., Barbi C., Olivieri F., Yashin A., Andreev K.F., Vaupel J.W., De Benedictis G., Rose G., Carrieri G., Jazwinski S.M., and Franceschi C. 2002. An allele of HRAS1 3′ variable number of tandem repeats is a frailty allele: Implication for an evolutionarily-conserved pathway involved in longevity. *Gene* **286:** 121–126.

Borst P. and Ulbert S. 2001. Control of VSG gene expression sites. *Mol. Biochem. Parasitol.* **114:** 17–27.

Boudreault A.A., Cronier D., Selleck W., Lacoste N., Utley R.T., Allard S., Savard J., Lane W.S., Tan S., and Cote J. 2003. Yeast enhancer of polycomb defines global Esa1-dependent acetylation of chromatin. *Genes Dev.* **17:** 1415–1428.

Boulton S.J. and Jackson S.P. 1998. Components of the Ku-dependent non-homologous end-joining pathway are involved in telomeric length maintenance and telomeric silencing. *EMBO J.* **17:** 1819–1828.

Bourgeois C.A., Laquerriere F., Hemon D., Hubert J., and Bouteille M. 1985. New data on the in-situ position of the inactive X chromosome in the interphase nucleus of human fibroblasts. *Hum. Genet.* **69:** 122–129.

Bourns B.D., Alexander M.K., Smith A.M., and Zakian V.A. 1998. Sir proteins, Rif proteins and Cdc13p bind *Saccharomyces* telomeres *in vivo*. *Mol. Cell. Biol.* **18:** 5600–5608.

Brachmann C.B., Sherman J.M., Devine S.E., Cameron E.E., Pillus L., and Boeke J.D. 1995. The SIR2 gene family, conserved from bacteria to humans, functions in silencing, cell cycle progression, and chromosome stability. *Genes Dev.* **9:** 2888–2902.

Braunstein M., Rose A.B., Holmes S.G., Allis C.D., and Broach J.R. 1993. Transcriptional silencing in yeast is associated with reduced nucleosome acetylation. *Genes Dev.* **7:** 592–604.

Brigati C., Kurtz S., Balderes D., Vidali G., and Shore D. 1993. An essential yeast gene encoding a TTAGGG repeat-binding protein. *Mol. Cell. Biol.* **13:** 1306–1314.

Briggs S.D., Xiao T., Sun Z.W., Caldwell J.A., Shabanowitz J., Hunt D.F., Allis C.D., and Strahl B.D. 2002. Gene silencing: *Trans*-histone regulatory pathway in chromatin. *Nature* **418:** 498.

Brown S.D., Huang J., and Van der Ploeg L.H. 1992. The promoter for the procyclic acidic repetitive protein (PARP) genes of *Trypanosoma brucei* shares features with RNA polymerase I promoters. *Mol. Cell. Biol.* **12:** 2644–2652.

Chan C.S.M. and Tye B.-K. 1983. Organization of DNA sequences and replication origins at yeast telomeres. *Cell* **33:** 563–573.

Chaves I., Rudenko G., Dirks-Mulder A., Cross M., and Borst P. 1999. Control of variant surface glycoprotein gene-expression sites in *Trypanosoma brucei*. *EMBO J.* **18:** 4846–4855.

Chaves I., Zomerdijk J., Dirks-Mulder A., Dirks R.W., Raap A.K., and Borst P. 1998. Subnuclear localization of the active variant surface glycoprotein gene expression site in *Trypanosoma brucei*. *Proc. Natl. Acad. Sci.* **95:** 12328–12333.

Chi M.H. and Shore D. 1996. SUM1-1, a dominant suppressor of SIR mutations in *Saccharomyces cerevisiae*, increases transcriptional silencing at telomeres and HM mating-type loci and decreases chromosome stability. *Mol. Cell. Biol.* **16:** 4281–4294.

Chien C.T., Buck S., Sternglanz R., and Shore D. 1993. Targeting of SIR1 protein establishes transcriptional silencing at *HM* loci and telomeres in yeast. *Cell* **75:** 531–541.

Cohen H.Y., Miller C., Bitterman K.J., Wall N.R., Hekking B., Kessler B., Howitz K.T., Gorospe M., de Cabo R., and Sinclair D.A. 2004. Calorie restriction promotes mammalian cell survival by inducing the SIRT1 deacetylase. *Science* **305:** 390–392.

Conrad M.N., Wright J.H., Wolf A.J., and Zakian V.A. 1990. RAP1 protein interacts with yeast telomeres *in vivo*: Overproduction alters telomere structure and decreases chromosome stability. *Cell* **63:** 739–750.

Conway C., McCulloch R., Ginger M.L., Robinson N.P., Browitt A., and Barry J.D. 2002. Ku is important for telomere maintenance, but not for differential expression of telomeric VSG genes, in African trypanosomes. *J. Biol. Chem.* **277:** 21269–21277.

Corda Y., Schramke V., Longhese M.P., Smokvina T., Paciotti V., Brevet V., Gilson E., and Geli V. 1999. Interaction between Set1p and checkpoint protein Mec3p in DNA repair and telomere functions. *Nat. Genet.* **21:** 204–208.

Cosgrove A.J., Nieduszynski C.A., and Donaldson A.D. 2002. Ku complex controls the replication time of DNA in telomere regions. *Genes Dev.* **16:** 2485–2490.

Craven R.J. and Petes T.D. 1999. Dependence of the regulation of telomere length on the type of subtelomeric repeat in the yeast *Saccharomyces cerevisiae*. *Genetics* **152:** 1531–1541.

———. 2000. Involvement of the checkpoint protein Mec1p in silencing of gene expression at telomeres in *Saccharomyces cerevisiae*. *Mol. Cell. Biol.* **20:** 2378–2384.

———. 2001. The *Saccharomyces cerevisiae* suppressor of choline sensitivity (SCS2) gene is a multicopy suppressor of mec1 telomeric silencing defects. *Genetics* **158:** 145–154.

Crotti L.B. and Basrai M.A. 2004. Functional roles for evolutionarily conserved Spt4p at centromeres and heterochromatin in *Saccharomyces cerevisiae*. *EMBO J.* **23:** 1804–1814.

Cryderman D.E., Morris E.J., Biessmann H., Elgin S.C., and Wallrath L.L. 1999. Silencing at *Drosophila* telomeres: Nuclear organization and chromatin structure play critical roles. *EMBO J.* **18:** 3724–3735.

Cuperus G. and Shore D. 2002. Restoration of silencing in *Saccharomyces cerevisiae* by tethering of a novel Sir2-interacting protein, Esc8. *Genetics* **162:** 633–645.

Danilevskaya O.N., Arkhipova I.R., Traverse K.L., and Pardue M.L. 1997. Promoting in tandem: The promoter for telomere transposon HeT-A and implications for the evolution of retroviral LTRs. *Cell* **88:** 647–655.

de Bruin D., Kantrow S.M., Liberatore R.A., and Zakian V.A. 2000. Telomere folding is required for the stable maintenance of telomere position effects in yeast. *Mol. Cell. Biol.* **20:** 7991–8000.

Deitsch K.W., Calderwood M.S., and Wellems T.E. 2001. Cooperative silencing elements in *var* genes. *Nature* **412:** 875–876.

De Las Peñas A., Pan S.J., Castano I., Alder J., Cregg R., and Cormack B.P. 2003. Virulence-related surface glycoproteins in the yeast pathogen *Candida glabrata* are encoded in subtelomeric clusters and subject to *RAP1*- and *SIR*-dependent transcriptional silencing. *Genes Dev.* **17:** 2245–2258.

De Luca M., Rose G., Bonafe M., Garasto S., Greco V., Weir B.S., Franceschi C., and De Benedictis G. 2001. Sex-specific longevity associations defined by Tyrosine Hydroxylase-Insulin-Insulin Growth Factor 2 haplotypes on the 11p15.5 chromosomal region. *Exp. Gerontol.* **36:** 1663–1671.

Denisenko O. and Bomsztyk K. 2002. Yeast hnRNP K-like genes are involved in regulation of the telomeric position effect and telomere length. *Mol. Cell. Biol.* **22:** 286–297.

Dernburg A.F., Broman K.W., Fung J.C., Marshall W.F., Philips J., Agard D.A., and Sedat J.W. 1996. Perturbation of nuclear architecture by long-distance chromosome interactions. *Cell* **85:** 745–759.

Diffley J.F. and Stillman B. 1989. Similarity between the transcriptional silencer binding proteins ABF1 and RAP1. *Science* **246:** 1034–1038.

Dlakic M. 2001. Chromatin silencing protein and pachytene checkpoint regulator Dot1p has a methyltransferase fold. *Trends Biochem. Sci.* **26:** 405–407.

Dror V. and Winston F. 2004. The Swi/Snf chromatin remodeling complex is required for ribosomal DNA and telomeric silencing in *Saccharomyces cerevisiae*. *Mol. Cell. Biol.* **24:** 8227–8235.

Dubey D.D., Davis L.R., Greenfeder S.A., Ong L.Y., Zhu J.G., Broach J.R., Newlon C.S., and Huberman J.A. 1991. Evidence suggesting that the *ARS* elements associated with silencers of the yeast mating-type locus HML do not function as chromosomal DNA replication origins. *Mol. Cell. Biol.* **11:** 5346–5355.

Dziak R., Leishman D., Radovic M., Tye B.K., and Yankulov K. 2003. Evidence for a role of MCM (mini-chromosome maintenance)5 in transcriptional repression of sub-telomeric and *Ty*-proximal genes in *Saccharomyces cerevisiae*. *J. Biol. Chem.* **278:** 27372–27381.

Ehrenhofer-Murray A.E., Rivier D.H., and Rine J. 1997. The role of Sas2, an acetyltransferase homologue of *Saccharomyces cerevisiae*, in silencing and ORC function. *Genetics* **145:** 923–934.

Ehrenhofer-Murray A.E., Gossen M., Pak D.T., Botchan M.R., and Rine J. 1995. Separation of origin recognition complex functions by cross-species complementation. *Science* **270:** 1671–1674.

Enomoto S., McCune-Zierath P.D., Gerami-Nejad M., Sanders M.A., and Berman J. 1997. *RLF2*, a subunit of yeast chromatin assembly-I, is required for telomeric chromatin function in vivo. *Genes Dev.* **11:** 358–370.

Ezhkova E. and Tansey W.P. 2004. Proteasomal ATPases link ubiquitylation of histone H2B to methylation of histone H3. *Mol. Cell* **13:** 435–442.

Falcon A.A. and Aris J.P. 2003. Plasmid accumulation reduces life span in *Saccharomyces cerevisiae*. *J. Biol. Chem.* **278:** 41607–41617.

Feng Q., Wang H., Ng H.H., Erdjument-Bromage H., Tempst P., Struhl K., and Zhang Y. 2002. Methylation of H3-lysine 79 is mediated by a new family of HMTases without a SET domain. *Curr. Biol.* **12:** 1052–1058.

Ferguson B.M. and Fangman W.L. 1992. A position effect on the time of replication origin activation in yeast. *Cell* **68:** 333–339.

Ferguson B.M., Brewer B.J., Reynolds A.E., and Fangman W.L. 1991. A yeast origin of replication is activated late in S phase. *Cell* **65:** 507–515.

Figueiredo L.M., Freitas-Junior L.H., Bottius E., Olivo-Marin J.C., and Scherf A. 2002. A central role for *Plasmodium falciparum* subtelomeric regions in spatial positioning and telomere length regulation. *EMBO J.* **21:** 815–824.

Fisher T.S., Taggart A.K.P., and Zakian V.A. 2004. Cell cycle-dependent regulation of yeast telomerase by Ku. *Nat. Struct. Mol. Biol.* **11:** 1198–1205.

Foss M., McNally F.J., Laurenson P., and Rine J. 1993. Origin recognition complex (ORC) in transcriptional silencing and DNA replication in *S. cerevisiae*. *Science* **262:** 1838–1844.

Fourel G., Revardel E., Koering C.E., and Gilson E. 1999. Cohabitation of insulators and silencing elements in yeast subtelomeric regions. *EMBO J.* **18:** 2522–2537.

Fourel G., Boscheron C., Revardel E., Lebrun E., Hu Y.F., Simmen K.C., Muller K., Li R., Mermod N., and Gilson E. 2001. An activation-independent role of transcription factors in insulator function. *EMBO Rep.* **2:** 124–132.

Fox C.A., Ehrenhofer-Murray A.E., Loo S., and Rine J. 1997. The origin recognition complex, *SIR1*, and the S phase requirement for silencing. *Science* **276:** 1547–1551.

Freitas-Junior L.H., Bottius E., Pirrit L.A., Deitsch K.W., Scheidig C., Guinet F., Nehrbass U., Wellems T.E., and Scherf A. 2000. Frequent ectopic recombination of virulence factor genes in telomeric chromosome clusters of *P. falciparum*. *Nature* **407:** 1018–1022.

Gabellini D., Green M.R., and Tupler R. 2002. Inappropriate gene activation in FSHD: A repressor complex binds a chromosomal repeat deleted in dystrophic muscle. *Cell* **110:** 339–348.

Galy V., Gadal O., Fromont-Racine M., Romano A., Jacquier A., and Nehrbass U. 2004. Nuclear retention of unspliced mRNAs in yeast is mediated by perinuclear Mlp1. *Cell* **116:** 63–73.

Galy V., Olivo-Marin J.C., Scherthan H., Doye V., Rascalou N., and Nehrbass U. 2000. Nuclear pore complexes in the organization of silent telomeric chromatin. *Nature* **403:** 108–112.

Gasser S.M. and Cockell M.M. 2001. The molecular biology of the SIR proteins. *Gene* **279:** 1–16.

Gehring W.J., Klemenz R., Weber U., and Kloter U. 1984. Functional analysis of the white+ gene of *Drosophila* by P-factor-mediated transformation. *EMBO J.* **3:** 2077–2085.

Gilbert D.M. 2002. Replication timing and transcriptional control: Beyond cause and effect. *Curr. Opin. Cell Biol.* **14:** 377–383.

Gilson E., Roberge M., Giraldo R., Rhodes D., and Gasser S.M. 1993. Distortion of the DNA double helix by RAP1 at silencers and multiple telomeric binding sites. *J. Mol. Biol.* **231:** 293–310.

Golubovsky M.D., Konev A.Y., Walter M.F., Biessmann H., and Mason J.M. 2001. Terminal retrotransposons activate a subtelomeric white transgene at the 2L telomere in *Drosophila*. *Genetics* **158:** 1111–1123.

Gotta M., Laroche T., Formenton A., Maillet L., Scherthan H., and Gasser S.M. 1996. The clustering of telomeres and colocalization with Rap1, Sir3, and Sir4 proteins in wild-type *Saccharomyces cerevisiae. J. Cell Biol.* **134:** 1349–1363.

Gottschling D.E., Aparicio O.M., Billington B.L., and Zakian V.A. 1990. Position effect at *S. cerevisiae* telomeres: Reversible repression of Pol II transcription. *Cell* **63:** 751–762.

Gravel S., Larrivee M., Labrecque P., and Wellinger R.J. 1998. Yeast Ku as a regulator of chromosomal DNA end structure. *Science* **280:** 741–744.

Greenwell P.W., Kronmal S.L., Porter S.E., Gassenhuber J., Obermaier B., and Petes T.D. 1995. *TEL1*, a gene involved in controlling telomere length in *S. cerevisiae*, is homologous to the human ataxia telangiectasia gene. *Cell* **82:** 823–829.

Griffith J.D., Comeau L., Rosenfield S., Stansel R.M., Bianchi A., Moss H., and de Lange T. 1999. Mammalian telomeres end in a large duplex loop. *Cell* **97:** 503–514.

Guo B., Styles C.A., Feng Q., and Fink G.R. 2000. A *Saccharomyces* gene family involved in invasive growth, cell-cell adhesion, and mating. *Proc. Natl. Acad. Sci.* **97:** 12158–12163.

Haber J.E. 1998. Mating-type gene switching in *Saccharomyces cerevisiae. Annu. Rev. Genet.* **32:** 561–599.

Halme A., Bumgarner S., Styles C., and Fink G.R. 2004. Genetic and epigenetic regulation of the FLO gene family generates cell-surface variation in yeast. *Cell* **116:** 405–415.

Hansen K.R., Burns G., Mata J., Volpe T.A., Martienssen R.A., Bahler J., and Thon G. 2005. Global effects on gene expression in fission yeast by silencing and RNA interference machineries. *Mol. Cell. Biol.* **25:** 590–601.

Hardy C.F., Sussel L., and Shore D. 1992. A RAP1-interacting protein involved in transcriptional silencing and telomere length regulation. *Genes Dev.* **6:** 801–814.

Hazelrigg T., Levis R., and Rubin G.M. 1984. Transformation of *white* locus DNA in *Drosophila*: Dosage compensation, *zeste* interaction, and position effects. *Cell* **36:** 469–481.

Hecht A., Laroche T., Strahl-Bolsinger S., Gasser S.M., and Grunstein M. 1995. Histone H3 and H4 N-termini interact with SIR3 and SIR4 proteins: A molecular model for the formation of heterochromatin in yeast. *Cell* **80:** 583–592.

Hediger F., Dubrana K., and Gasser S.M. 2002a. Myosin-like proteins 1 and 2 are not required for silencing or telomere anchoring, but act in the Tel1 pathway of telomere length control. *J. Struct. Biol.* **140:** 79–91.

Hediger F., Neumann F.R., Van Houwe G., Dubrana K., and Gasser S.M. 2002b. Live imaging of telomeres: yKu and Sir proteins define redundant telomere-anchoring pathways in yeast. *Curr. Biol.* **12:** 2076–2089.

Henikoff S. 2000. Heterochromatin function in complex genomes. *Biochim. Biophys. Acta.* **1470:** 1–8.

Hewitt J.E., Lyle R., Clark L.N., Valleley E.M., Wright T.J., Wijmenga C., van Deutekom J.C., Francis F., Sharpe P.T., Hofker M., et al. 1994. Analysis of the tandem repeat locus D4Z4 associated with facioscapulohumeral muscular dystrophy. *Hum. Mol. Genet.* **3:** 1287–1295.

Hochstrasser M. and Sedat J.W. 1987. Three-dimensional organization of *Drosophila melanogaster* interphase nuclei. I. Tissue-specific aspects of polytene nuclear architecture. *J. Cell Biol.* **104:** 1455–1470.

Hochstrasser M., Mathog D., Gruenbaum Y., Saumweber H., and Sedat J.W. 1986. Spatial organization of chromosomes in the salivary gland nuclei of *Drosophila melanogaster. J. Cell Biol.* **102:** 112–123.

Hoppe G.J., Tanny J.C., Rudner A.D., Gerber S.A., Danaie S., Gygi S.P., and Moazed D. 2002. Steps in assembly of silent chromatin in yeast: Sir3-independent binding of a

Sir2/Sir4 complex to silencers and role for Sir2-dependent deacetylation. *Mol. Cell. Biol.* **22:** 4167–4180.

Horn D. and Cross G.A. 1995. A developmentally regulated position effect at a telomeric locus in *Trypanosoma brucei. Cell* **83:** 555–561.

Howitz K.T., Bitterman K.J., Cohen H.Y., Lamming D.W., Lavu S., Wood J.G., Zipkin R.E., Chung P., Kisielewski A., Zhang L.L., Scherer B., and Sinclair D.A. 2003. Small molecule activators of sirtuins extend *Saccharomyces cerevisiae* lifespan. *Nature* **425:** 191–196.

Hu F., Alcasabas A.A., and Elledge S.J. 2001. Asf1 links Rad53 to control of chromatin assembly. *Genes Dev.* **15:** 1061–1066.

Huang H., Kahana A., Gottschling D.E., Prakash L., and Liebman S.W. 1997. The ubiquitin-conjugating enzyme Rad6 (Ubc2) is required for silencing in *Saccharomyces cerevisiae. Mol. Cell. Biol.* **17:** 6693–6699.

Huang Y. 2002. Transcriptional silencing in *Saccharomyces cerevisiae* and *Schizosaccharomyces pombe. Nucleic Acids Res.* **30:** 1465–1482.

Hultdin M., Gronlund E., Norrback K., Eriksson-Lindstrom E., Just T., and Roos G. 1998. Telomere analysis by fluorescence *in situ* hybridization and flow cytometry. *Nucleic Acids Res.* **26:** 3651–3656.

Iida T. and Araki H. 2004. Noncompetitive counteractions of DNA polymerase ε and ISW2/yCHRAC for epigenetic inheritance of telomere position effect in *Saccharomyces cerevisiae. Mol. Cell. Biol.* **24:** 217–227.

Imai S., Armstrong C.M., Kaeberlein M., and Guarente L. 2000. Transcriptional silencing and longevity protein Sir2 is an NAD-dependent histone deacetylase. *Nature* **403:** 795–800.

Ivessa A.S., Lenzmeier B.A., Bessler J.B., Goudsouzian L.K., Schnakenberg S.L., and Zakian V.A. 2003. The *Saccharomyces cerevisiae* helicase Rrm3p facilitates replication past nonhistone protein-DNA complexes. *Mol. Cell* **12:** 1525–1536.

Ivessa A.S., Zhou J.-Q., Schulz V.P., Monson E.M., and Zakian V.A. 2002. *Saccharomyces* Rrm3p, a 5′ to 3′ DNA helicase that promotes replication fork progression through telomeric and sub-telomeric DNA. *Genes Dev.* **16:** 1383–1396.

Ivy J.M., Klar A.J.S., and Hicks J.B. 1986. Cloning and characterization of four *SIR* genes of *Saccharomyces cerevisiae. Mol. Cell. Biol.* **6:** 688–702.

Jacob N.K., Stout A.R., and Price C.M. 2004. Modulation of telomere length dynamics by the subtelomeric region of tetrahymena telomeres. *Mol. Biol. Cell.* **15:** 3719–3728.

Jiang G., Yang F., van Overveld P.G., Vedanarayanan V., van der Maarel S., and Ehrlich M. 2003. Testing the position-effect variegation hypothesis for facioscapulohumeral muscular dystrophy by analysis of histone modification and gene expression in subtelomeric 4q. *Hum. Mol. Genet.* **12:** 2909–2921.

Jiang J.C., Wawryn J., Shantha Kumara H.M., and Jazwinski S.M. 2002. Distinct roles of processes modulated by histone deacetylases Rpd3p, Hda1p, and Sir2p in life extension by caloric restriction in yeast. *Exp. Gerontol.* **37:** 1023–1030.

Johnson L.M., Kayne P.S., Kahn E.S., and Grunstein M. 1990. Genetic evidence for an interaction between SIR3 and histone H4 in the repression of the silent mating loci in *Saccharomyces cerevisiae. Proc. Natl. Acad. Sci.* **87:** 6286–6290.

Kaeberlein M., McVey M., and Guarente L. 1999. The SIR2/3/4 complex and SIR2 alone promote longevity in *Saccharomyces cerevisiae* by two different mechanisms. *Genes Dev.* **13:** 2570–2580.

Kaeberlein M., Kirkland K.T., Fields S., and Kennedy B.K. 2004. Sir2-independent life span extension by calorie restriction in yeast. *PLoS Biol.* **2:** E296.

Kanoh J. and Ishikawa F. 2001. spRap1 and spRif1, recruited to telomeres by Taz1, are essential for telomere function in fission yeast. *Curr. Biol.* **11:** 1624–1630.

Karpen G.H. and Spradling A.C. 1992. Analysis of subtelomeric heterochromatin in the *Drosophila* minichromosome *Dp1187* by single P element insertional mutagenesis. *Genetics* **132:** 737–753.

Kaufman P.D., Cohen J.L., and Osley M.A. 1998. Hir proteins are required for position-dependent gene silencing in *Saccharomyces cerevisiae* in the absence of chromatin assembly factor I. *Mol. Cell. Biol.* **18:** 4793–4806.

Kaufman P.D., Kobayashi R., and Stillman B. 1997. Ultraviolet radiation sensitivity and reduction of telomeric silencing in *Saccharomyces cerevisiae* cells lacking chromatin assembly factor-I. *Genes Dev.* **11:** 345–357.

Kayne P.S., Kim U.-J., Han M., Mullen J., Yoshizaki F., and Grunstein M. 1988. Extremely conserved histone H4 N-terminus is dispensable for growth but essential for repressing the silent mating loci in yeast. *Cell* **55:** 27–39.

Kellum R., Raff J.W., and Alberts B.M. 1995. Heterochromatin protein 1 distribution during development and during the cell cycle in *Drosophila* embryos. *J. Cell Sci.* **108:** 1407–1418.

Kelly T.J., Qin S., Gottschling D.E., and Parthun M.R. 2000. Type B histone acetyltransferase Hat1p participates in telomeric silencing. *Mol. Cell. Biol.* **20:** 7051–7058.

Khan A.U. and Hampsey M. 2002. Connecting the DOTs: Covalent histone modifications and the formation of silent chromatin. *Trends Genet.* **18:** 387–389.

Kimura A., Umehara T., and Horikoshi M. 2002. Chromosomal gradient of histone acetylation established by Sas2p and Sir2p functions as a shield against gene silencing. *Nat. Genet.* **32:** 370–377.

Kirchmaier A.L. and Rine J. 2001. DNA replication-independent silencing in *S. cerevisiae*. *Science* **291:** 646–650.

Klein F., Laroche T., Cardenas M.E., Hofmann J.F., Schweizer D., and Gasser S.M. 1992. Localization of RAP1 and topoisomerase II in nucleic and meiotic chromosomes of yeast. *J. Cell. Biol.* **117:** 935–948.

Koering C.E., Pollice A., Zibella M.P., Bauwens S., Puisieux A., Brunori M., Brun C., Martins L., Sabatier L., Pulitzer J.F., and Gilson E. 2002. Human telomeric position effect is determined by chromosomal context and telomeric chromatin integrity. *EMBO Rep.* **3:** 1055–1061.

Kosova B., Pante N., Rollenhagen C., Podtelejnikov A., Mann M., Aebi U., and Hurt E. 2000. Mlp2p, a component of nuclear pore attached intranuclear filaments, associates with nic96p. *J. Biol. Chem.* **275:** 343–350.

Kouskouti A., Scheer E., Staub A., Tora L., and Talianidis I. 2004. Gene-specific modulation of TAF10 function by SET9-mediated methylation. *Mol. Cell* **14:** 175–182.

Krogan N.J., Dover J., Wood A., Schneider J., Heidt J., Boateng M.A., Dean K., Ryan O.W., Golshani A., Johnston M., Greenblatt J.F., and Shilatifard A. 2003. The Paf1 complex is required for histone H3 methylation by COMPASS and Dot1p: Linking transcriptional elongation to histone methylation. *Mol. Cell* **11:** 721–729.

Kurenova E., Champion L., Biessmann H., and Mason J.M. 1998. Directional gene silencing induced by a complex subtelomeric satellite from *Drosophila*. *Chromosoma* **107:** 311–320.

Kurtz S. and Shore D. 1991. RAP1 protein activates and silences transcription of mating-type genes in yeast. *Genes Dev.* **5:** 616–628.

Kyrion G., Boakye K.A., and Lustig A.J. 1992. C-terminal truncation of RAP1 results in the deregulation of telomere size, stability, and function in *Saccharomyces cerevisiae*. *Mol. Cell. Biol.* **12:** 5159–5173.

Kyrion G., Liu K., Liu C., and Lustig A.J. 1993. RAP1 and telomere structure regulate telomere position effects in *Saccharomyces cerevisiae*. *Genes Dev.* **7:** 1146–1159.

Lacoste N., Utley R.T., Hunter J.M., Poirier G.G., and Cote J. 2002. Disruptor of telomeric silencing-1 is a chromatin-specific histone H3 methyltransferase. *J. Biol. Chem.* **277:** 30421–30424.

Ladurner A.G., Inouye C., Jain R., and Tjian R. 2003. Bromodomains mediate an acetyl-histone encoded antisilencing function at heterochromatin boundaries. *Mol. Cell* **11:** 365–376.

Landry J., Sutton A., Tafrov S.T., Heller R.C., Stebbins J., Pillus L., and Sternglanz R. 2000. The silencing protein SIR2 and its homologs are NAD-dependent protein deacetylases. *Proc. Natl. Acad. Sci.* **97:** 5807–5811.

Laroche T., Martin S.G., Gotta M., Gorham H.C., Pryde F.E., Louis E.J., and Gasser S.M. 1998. Mutation of yeast Ku genes disrupts the subnuclear organization of telomeres. *Curr. Biol.* **8:** 653–656.

Lau A., Blitzblau H., and Bell S.P. 2002. Cell-cycle control of the establishment of mating-type silencing in *S. cerevisiae*. *Genes Dev.* **16:** 2935–2945.

Le S., Davis C., Konopka J.B., and Sternglanz R. 1997. Two new S-phase-specific genes from *Saccharomyces cerevisiae*. *Yeast* **13:** 1029–1042.

Lebrun E., Revardel E., Boscheron C., Li R., Gilson E., and Fourel G. 2001. Protosilencers in *Saccharomyces cerevisiae* subtelomeric regions. *Genetics* **158:** 167–176.

Lemmers R.J.L., de Kievit P., van Geel M., van der Wielen M.J., Bakker E., Padberg G.W., Frants R.R., and van der Maarel S.M. 2001. Complete allele information in the diagnosis of facioscapulohumeral muscular dystrophy by triple DNA analysis. *Ann. Neurol.* **50:** 816–819.

Levis R., Hazelrigg T., and Rubin G.M. 1985. Effects of genomic position on the expression of transduced copies of the *white* gene of *Drosophila*. *Science* **229:** 558–561.

Levis R.W., Ganesan R., Houtchens K., Tolar L.A., and Sheen F.-M. 1993. Transposons in place of telomeric repeats at a *Drosophila* telomere. *Cell* **75:** 1083–1093.

Lew J.E., Enomoto S., and Berman J. 1998. Telomere length regulation and telomeric chromatin require the nonsense-mediated mRNA decay pathway. *Mol. Cell. Biol.* **18:** 6121–6130.

Li B., Oestreich S., and de Lange T. 2000. Identification of human Rap1: Implications for telomere evolution. *Cell* **101:** 471–483.

Li Y.C., Cheng T.H., and Gartenberg M.R. 2001. Establishment of transcriptional silencing in the absence of DNA replication. *Science* **291:** 650–653.

Lieb J.D., Liu X., Botstein D., and Brown P.O. 2001. Promoter-specific binding of Rap1 revealed by genome-wide maps of protein-DNA association. *Nat. Genet.* **28:** 327–334.

Lin S.J., Defossez P.A., and Guarente L. 2000. Requirement of NAD and SIR2 for life-span extension by calorie restriction in *Saccharomyces cerevisiae*. *Science* **289:** 2126–2128.

Lin S.J., Kaeberlein M., Andalis A.A., Sturtz L.A., Defossez P.A., Culotta V.C., Fink G.R., and Guarente L. 2002. Calorie restriction extends *Saccharomyces cerevisiae* lifespan by increasing respiration. *Nature* **418:** 344–348.

Lo W.S., Trievel R.C., Rojas J.R., Duggan L., Hsu J.Y., Allis C.D., Marmorstein R., and Berger S.L. 2000. Phosphorylation of serine 10 in histone H3 is functionally linked in vitro and in vivo to Gcn5-mediated acetylation at lysine 14. *Mol. Cell* **5:** 917–926.

Louis E.J. 1995. The chromosome ends of *Saccharomyces cerevisiae*. *Yeast* **11:** 1553–1574.

Luderus M.E., van Steensel B., Chong L., Sibon O.C., Cremers F.F., and de Lange T. 1996. Structure, subnuclear distribution, and nuclear matrix association of the mammalian telomeric complex. *J. Cell Biol.* **135:** 867–881.

Luo K., Vega-Palas M.A., and Grunstein M. 2002. Rap1-Sir4 binding independent of other Sir, yKu, or histone interactions initiates the assembly of telomeric heterochromatin in yeast. *Genes Dev.* **16:** 1528–1539.

Lustig A.J. and Petes T.D. 1986. Identification of yeast mutants with altered telomere structure. *Proc. Natl. Acad. Sci.* **83:** 1398–1402.

Lustig A.J., Kurtz S., and Shore D. 1990. Involvement of the silencer and UAS binding protein RAP1 in regulation of telomere length. *Science* **250:** 549–553.

Mages G.J., Feldmann H.M., and Winnacker E.L. 1996. Involvement of the *Saccharomyces cerevisiae* HDF1 gene in DNA double-strand break repair and recombination. *J. Biol. Chem.* **271:** 7910–7915.

Mann R.K. and Grunstein M. 1992. Histone H3 N-terminal mutations allow hyperactivation of the yeast GAL1 gene *in vivo*. *EMBO J.* **11:** 3297–3306.

Manolis K.G., Nimmo E.R., Hartsuiker E., Carr A.M., Jeggo P.A., and Allshire R.C. 2001. Novel functional requirements for non-homologous DNA end joining in *Schizosaccharomyces pombe*. *EMBO J.* **20:** 210–221.

Martin S.G., Laroche T., Suka N., Grunstein M., and Gasser S.M. 1999. Relocalization of telomeric Ku and SIR proteins in response to DNA strand breaks in yeast. *Cell* **97:** 621–633.

Mason J., Ransom J., and Konev A. 2004. A deficiency screen for dominant suppressors of telomeric silencing in *Drosophila*. *Genetics* **168:** 1353–1370.

Mason J.M., Konev A.Y., Golubovsky M.D., and Biessmann H. 2003. *Cis-* and *trans*-acting influences on telomeric position effect in *Drosophila melanogaster* detected with a subterminal transgene. *Genetics* **163:** 917–930.

Mason J.M., Haoudi A., Konev A.Y., Kurenova E., Walter M.F., and Biessmann H. 2000. Control of telomere elongation and telomeric silencing in *Drosophila melanogaster*. *Genetica* **109:** 61–70.

Mathog D., Hochstrasser M., Gruenbaum Y., Saumweber H., and Sedat J. 1984. Characteristic folding pattern of polytene chromosomes in *Drosophila* salivary gland nuclei. *Nature* **308:** 414–421.

Matzke M.A., Moscone E.A., Park Y.D., Papp I., Oberkofler H., Neuhuber F., and Matzke A.J. 1994. Inheritance and expression of a transgene insert in an aneuploid tobacco line. *Mol. Gen. Genet.* **245:** 471–485.

McCarroll R.M. and Fangman W.L. 1988. Time of replication of yeast centromeres and telomeres. *Cell* **54:** 505–513.

Mechler B.M., McGinnis W., and Gehring W.J. 1985. Molecular cloning of lethal(2)giant larvae, a recessive oncogene of *Drosophila melanogaster*. *EMBO J.* **4:** 1551–1557.

Mefford H.C. and Trask B.J. 2002. The complex structure and dynamic evolution of human subtelomeres. *Nat. Rev. Genet.* **3:** 91–102.

Megee P.C., Morgan B.A., Mittman B.A., and Smith M.M. 1990. Genetic analysis of histone H4: Essential role of lysines subject to reversible acetylation. *Science* **247:** 841–845.

Melnikova L., Biessmann H., and Georgiev P. 2005. The Ku protein complex is involved in length regulation of *Drosophila* telomeres. *Genetics* **170:** 221–235.

Meneghini M.D., Wu M., and Madhani H.D. 2003. Conserved histone variant H2A.Z protects euchromatin from the ectopic spread of silent heterochromatin. *Cell* **112:** 725–736.

Meneveri R., Agresti A., Marozzi A., Saccone S., Rocchi M., Archidiacono N., Corneo G., Della Valle G., and Ginelli E. 1993. Molecular organization and chromosomal location of human GC-rich heterochromatic blocks. *Gene* **123:** 227–234.

Michaelis C., Ciosk R., and Nasmyth K. 1997. Cohesins: Chromosomal proteins that prevent premature separation of sister chromatids. *Cell* **91:** 35–45.

Miller A.M. and Nasmyth K.A. 1984. Role of DNA replication in the repression of silent mating type loci in yeast. *Nature* **312:** 247–251.

Mishra K. and Shore D. 1999. Yeast Ku protein plays a direct role in telomeric silencing and counteracts inhibition by rif proteins. *Curr. Biol.* **9:** 1123–1126.

Moazed D. 2001. Enzymatic activities of Sir2 and chromatin silencing. *Curr. Opin. Cell Biol.* **13:** 232–238.

Monson E.K., de Bruin D., and Zakian V.A. 1997. The yeast Cacl protein is required for the stable inheritance of transcriptionally repressed chromatin at telomeres. *Proc. Natl. Acad. Sci.* **94:** 13081–13086.

Moretti P. and Shore D. 2001. Multiple interactions in Sir protein recruitment by Rap1p at silencers and telomeres in yeast. *Mol. Cell. Biol.* **21:** 8082–8094.

Moretti P., Freeman K., Coodly L., and Shore D. 1994. Evidence that a complex of SIR proteins interacts with the silencer and telomere-binding protein RAP1. *Genes Dev.* **8:** 2257–2269.

Morrow B.E., Johnson S.P., and Warner J.R. 1989. Proteins that bind to the yeast rDNA enhancer. *J. Biol. Chem.* **264:** 9061–9068.

Muñoz-Jordan J.L., Cross G.A., de Lange T., and Griffith J.D. 2001. t-loops at trypanosome telomeres. *EMBO J.* **20:** 579–588.

Murti K.G. and Prescott D.M. 1999. Telomeres of polytene chromosomes in a ciliated protozoan terminate in duplex DNA loops. *Proc. Natl. Acad. Sci.* **96:** 14436–14439.

Navarro M. and Gull K. 2001. A pol I transcriptional body associated with VSG mono-allelic expression in *Trypanosoma brucei*. *Nature* **414:** 759–763.

Ng H.H., Dole S., and Struhl K. 2003. The Rtf1 component of the Paf1 transcriptional elongation complex is required for ubiquitination of histone H2B. *J. Biol. Chem.* **278:** 33625–33628.

Ng H.H., Feng Q., Wang H., Erdjument-Bromage H., Tempst P., Zhang Y., and Struhl K. 2002. Lysine methylation within the globular domain of histone H3 by Dot1 is important for telomeric silencing and Sir protein association. *Genes Dev.* **16:** 1518–1527.

Nimmo E.R., Cranston G., and Allshire R.C. 1994. Telomere-associated chromosome breakage in fission yeast results in variegated expression of adjacent genes. *EMBO J.* **13:** 3801–3811.

Nimmo E.R., Pidoux A.L., Perry P.E., and Allshire R.C. 1998. Defective meiosis in telomere-silencing mutants of *Schizosaccharomyces pombe*. *Nature* **392:** 825–828.

Nislow C., Ray E., and Pillus L. 1997. *SET1*, a yeast member of the *trithorax* family, functions in transcriptional silencing and diverse cellular processes. *Mol. Biol. Cell* **8:** 2421–2436.

Ofir R., Wong A.C., McDermid H.E., Skorecki K.L., and Selig S. 1999. Position effect of human telomeric repeats on replication timing. *Proc. Natl. Acad. Sci.* **96:** 11434–11439.

Palladino F., Laroche T., Gilson E., Axelrod A., Pillus L., and Gasser S.M. 1993. SIR3 and SIR4 proteins are required for the positioning and integrity of yeast telomeres. *Cell* **75:** 543–555.

Pappas D.L., Jr., Frisch R., and Weinreich M. 2004. The NAD^+-dependent Sir2p histone deacetylase is a negative regulator of chromosomal DNA replication. *Genes Dev.* **18:** 769–781.

Pardue M.L. and DeBaryshe P.G. 1999. *Drosophila* telomeres: Two transposable elements with important roles in chromosomes. *Genetica* **107:** 189–196.

Park E.-C. and Szostak J.W. 1990. Point mutations in the yeast histone H4 gene prevent silencing of the silent mating type locus *HML. Mol. Cell. Biol.* **10:** 4932–4934.

Park M.J., Jang Y.K., Choi E.S., Kim H.S., and Park S.D. 2002. Fission yeast Rap1 homolog is a telomere-specific silencing factor and interacts with Taz1p. *Mol. Cell* **13:** 327–333.

Pasero P., Bensimon A., and Schwob E. 2002. Single-molecule analysis reveals clustering and epigenetic regulation of replication origins at the yeast rDNA locus. *Genes Dev.* **16:** 2479–2484.

Peinado H., Ballestar E., Esteller M., and Cano A. 2004. Snail mediates E-cadherin repression by the recruitment of the Sin3A/histone deacetylase 1 (HDAC1)/HDAC2 complex. *Mol. Cell. Biol.* **24:** 306–319.

Pemberton L.F. and Blobel G. 1997. Characterization of the Wtm proteins, a novel family of *Saccharomyces cerevisiae* transcriptional modulators with roles in meiotic regulation and silencing. *Mol. Cell. Biol.* **17:** 4830–4841.

Perrod S. and Gasser S.M. 2003. Long-range silencing and position effects at telomeres and centromeres: Parallels and differences. *Cell. Mol. Life Sci.* **60:** 2303–2318.

Perrod S., Cockell M.M., Laroche T., Renauld H., Ducrest A.L., Bonnard C., and Gasser S.M. 2001. A cytosolic NAD-dependent deacetylase, Hst2p, can modulate nucleolar and telomeric silencing in yeast. *EMBO J.* **20:** 197–209.

Pillus L. and Rine J. 1989. Epigenetic inheritance of transcriptional states in *S. cerevisiae. Cell* **59:** 637–647.

Porter S.E., Greenwell P.W., Ritchie K.B., and Petes T.D. 1996. The DNA-binding protein Hdf1p (a putative Ku homologue) is required for maintaining normal telomere length in *Saccharomyces cerevisiae. Nuc. Acids Res.* **24:** 582–585.

Poveda A., Pamblanco M., Tafrov S., Tordera V., Sternglanz R., and Sendra R. 2004. Hif1 is a component of yeast histone acetyltransferase B, a complex mainly localized in the nucleus. *J. Biol. Chem.* **279:** 16033–16043.

Pryde F.E. and Louis E.J. 1999. Limitations of silencing at native yeast telomeres. *EMBO J.* **18:** 2538–2550.

Raghuraman M.K., Winzeler E.A., Collingwood D., Hunt S., Wodicka L., Conway A., Lockhart D.J., Davis R.W., Brewer B.J., and Fangman W.L. 2001. Replication dynamics of the yeast genome. *Science* **294:** 115–121.

Ray A., Hector R.E., Roy N., Song J.H., Berkner K.L., and Runge K.W. 2003. Sir3p phosphorylation by the Slt2p pathway effects redistribution of silencing function and shortened lifespan. *Nat. Genet.* **33:** 522–526.

Reifsnyder C., Lowell J., Clarke A., and Pillus L. 1996. Yeast *SAS* silencing genes and human genes associated with AML and HIV-1 Tat interactions are homologous with acetyltransferases. *Nat. Genet.* **14:** 42–49.

Renauld H., Aparicio O.M., Zierath P.D., Billington B.L., Chhablani S.K., and Gottschling D.E. 1993. Silent domains are assembled continuously from the telomere and are defined by promoter distance and strength, and by *SIR3* dosage. *Genes Dev.* **7:** 1133–1145.

Rice J.C. and Allis C.D. 2001. Histone methylation versus histone acetylation: New insights into epigenetic regulation. *Curr. Opin. Cell Biol.* **13:** 263–273.

Richards E.J. and Elgin S.C. 2002. Epigenetic codes for heterochromatin formation and silencing: Rounding up the usual suspects. *Cell* **108:** 489–500.

Riethman H., Ambrosini A., Castaneda C., Finklestein J., Hu X.L., Mudunuri U., Paul S., and Wei J. 2004. Mapping and initial analysis of human subtelomeric sequence assemblies. *Genome Res.* **14:** 18–28.

Robinett C.C., Straight A., Li G., Willhelm C., Sudlow G., Murray A., and Belmont A.S. 1996. In vivo localization of DNA sequences and visualization of large-scale chromatin organization using lac operator/repressor recognition. *J. Cell Biol.* **135:** 1685–1700.

Robyr D., Suka Y., Xenarios I., Kurdistani S.K., Wang A., Suka N., and Grunstein M. 2002. Microarray deacetylation maps determine genome-wide functions for yeast histone deacetylases. *Cell* **109:** 437–446.

Rogina B., Helfand S.L., and Frankel S. 2002. Longevity regulation by *Drosophila* Rpd3 deacetylase and caloric restriction. *Science* **298:** 1745.

Roy N. and Runge K.W. 1999. The ZDS1 and ZDS2 proteins require the Sir3p component of yeast silent chromatin to enhance the stability of short linear centromeric plasmids. *Chromosoma* **108:** 146–161.

———. 2000. Two paralogs involved in transcriptional silencing that antagonistically control yeast life span. *Curr. Biol.* **10:** 111–114.

Rundlett S.E., Carmen A.A., Kobayashi R., Bavykin S., Turner B.M., and Grunstein M. 1996. HDA1 and RPD3 are members of distinct yeast histone deacetylase complexes that regulate silencing and transcription. *Proc. Natl. Acad. Sci.* **93:** 14503–14508.

Runge K.W. and Zakian V.A. 1996. *TEL2,* an essential gene required for telomere length regulation and telomere position effect in *Saccharomyces cerevisiae. Mol. Cell. Biol.* **16:** 3094–3105.

Sandmeier J.J., Celic I., Boeke J.D., and Smith J.S. 2002. Telomeric and rDNA silencing in *Saccharomyces cerevisiae* are dependent on a nuclear NAD$^+$ salvage pathway. *Genetics* **160:** 877–889.

San-Segundo P.A. and Roeder G.S. 2000. Role for the silencing protein Dot1 in meiotic checkpoint control. *Mol. Biol. Cell* **11:** 3601–3615.

Schedl P. and Broach J.R. 2003. Making good neighbors: The right fence for the right job. *Nat. Struct. Biol.* **10:** 241–243.

Scherf A., Hernandez-Rivas R., Buffet P., Bottius E., Benatar C., Pouvelle B., Gysin J., and Lanzer M. 1998. Antigenic variation in malaria: *In situ* switching, relaxed and mutually exclusive transcription of *var* genes during intra-erythrocytic development in *Plasmodium falciparum. EMBO J.* **17:** 5418–5426.

Schubeler D., MacAlpine D.M., Scalzo D., Wirbelauer C., Kooperberg C., van Leeuwen F., Gottschling D.E., O'Neill L.P., Turner B.M., Delrow J., Bell S.P., and Groudine M. 2004. The histone modification pattern of active genes revealed through genome-wide chromatin analysis of a higher eukaryote. *Genes Dev.* **18:** 1263–1271.

Shanower G.A., Muller M., Blanton J.L., Honti V., Gyurkovics H., and Schedl P. 2005. Characterization of the *grappa* gene, the *Drosophila* histone H3 lysine 79 methyltransferase. *Genetics* **169:** 173–184.

Shea C., Lee M.G., and Van der Ploeg L.H. 1987. VSG gene 118 is transcribed from a cotransposed pol I-like promoter. *Cell* **50:** 603–612.

Sinclair D.A., Mills K., and Guarente L. 1997. Accelerated aging and nucleolar fragmentation in yeast sgs1 mutants. *Science* **277:** 1313–1316.

Singer M.S., Kahana A., Wolf A.J., Meisinger L.L., Peterson S.E., Goggin C., Mahowald M., and Gottschling D.E. 1998. Identification of high-copy disruptors of telomeric silencing in *Saccharomyces cerevisiae. Genetics* **150:** 613–632.

Smith J.S. and Boeke J.D. 1997. An unusual form of transcriptional silencing in yeast ribosomal DNA. *Genes Dev.* **11:** 241–254.

Smith J.S., Brachmann C.B., Celic I., Kenna M.A., Muhammad S., Starai V.J., Avalos J.L., Escalante-Semerena J.C., Grubmeyer C., Wolberger C., and Boeke J.D. 2000. A phylogenetically conserved NAD$^+$-dependent protein deacetylase activity in the Sir2 protein family. *Proc. Natl. Acad. Sci.* **97:** 6658–6663.

Smolikov S., Mazor Y., and Krauskopf A. 2004. ELG1, a regulator of genome stability, has a role in telomere length regulation and in silencing. *Proc. Natl. Acad. Sci.* **101:** 1656–1661.

Snowden A.W., Gregory P.D., Case C.C., and Pabo C.O. 2002. Gene-specific targeting of H3K9 methylation is sufficient for initiating repression in vivo. *Curr. Biol.* **12:** 2159–2166.

Sprung C.N., Sabatier L., and Murnane J.P. 1996. Effect of telomere length on telomeric gene expression. *Nucleic Acids Res.* **24:** 4336–4340.

Stavenhagen J.B. and Zakian V.A. 1994. Internal tracts of telomeric DNA act as silencers in *Saccharomyces cerevisiae*. *Genes Dev.* **8:** 1411–1422.

Steinert S., Shay J.W., and Wright W.E. 2004. Modification of subtelomeric DNA. *Mol. Cell. Biol.* **24:** 4571–4580.

Stevenson J. and Gottschling D. 1999. Telomeric chromatin modulates replication timing near chromosome ends. *Genes Dev.* **15:** 146–151.

Stone E.M. and Pillus L. 1996. Activation of an MAP kinase cascade leads to Sir3p hyperphosphorylation and strengthens transcriptional silencing. *J. Cell Biol.* **135:** 571–583.

Strahl B.D. and Allis C.D. 2000. The language of covalent histone modifications. *Nature* **403:** 41–45.

Strahl-Bolsinger S., Hecht A., Luo K., and Grunstein M. 1997. SIR2 and SIR4 interactions differ in core and extended telomeric heterochromatin in yeast. *Genes Dev.* **11:** 83–93.

Strambio-de-Castillia C., Blobel G., and Rout M.P. 1999. Proteins connecting the nuclear pore complex with the nuclear interior. *J. Cell Biol.* **144:** 839–855.

Suka N., Luo K., and Grunstein M. 2002. Sir2p and Sas2p opposingly regulate acetylation of yeast histone H4 lysine16 and spreading of heterochromatin. *Nat. Genet.* **32:** 378–383.

Sun Z.W. and Hampsey M. 1999. A general requirement for the Sin3-Rpd3 histone deacetylase complex in regulating silencing in *Saccharomyces cerevisiae*. *Genetics* **152:** 921–932.

Suzuki Y. and Nishizaw M. 1994. The yeast GAL11 protein is involved in regulation of the structure and the position effect of telomeres. *Mol. Cell. Biol.* **14:** 3791–3799.

Taddei A. and Gasser S.M. 2004. Multiple pathways for telomere tethering: Functional implications of subnuclear position for heterochromatin formation. *Biochim. Biophys. Acta* **1677:** 120–128.

Tam R., Smith K.P., and Lawrence J.B. 2004. The 4q subtelomere harboring the FSHD locus is specifically anchored with peripheral heterochromatin unlike most human telomeres. *J. Cell. Biol.* **167:** 269–279.

Tan Q., Bellizzi D., Rose G., Garasto S., Franceschi C., Kruse T., Vaupel J.W., De Benedictis G., and Yashin A.I. 2002. The influences on human longevity by HUMTHO1.STR polymorphism (Tyrosine Hydroxylase gene). A relative risk approach. *Mech. Ageing Dev.* **123:** 1403–1410.

Tanny J.C., Dowd G.J., Huang J., Hilz H., and Moazed D. 1999. An enzymatic activity in the yeast Sir2 protein that is essential for gene silencing. *Cell* **99:** 735–745.

Teng S.-C., Chang J., McCowan B., and Zakian V.A. 2000. Telomerase-independent lengthening of yeast telomeres occurs by an abrupt Rad50p-dependent, Rif-inhibited recombinational process. *Mol. Cell.* **6:** 947–952.

Tham W.H. and Zakian V.A. 2002. Transcriptional silencing at *Saccharomyces* telomeres: Implications for other organisms. *Oncogene* **21:** 512–521.

Tham W.H., Wyithe J.S., Ferrigno P.K., Silver P.A., and Zakian V.A. 2001. Localization of yeast telomeres to the nuclear periphery is separable from transcriptional repression and telomere stability functions. *Mol. Cell* **8:** 189–199.

Thompson J.S., Ling X., and Grunstein M. 1994. Histone H3 amino terminus is required for telomeric and silent mating locus repression in yeast. *Nature* **369:** 245–247.

Thompson J.S., Snow M.L., Giles S., McPherson L.E., and Grunstein M. 2003. Identification of a functional domain within the essential core of histone H3 that is required for telomeric and HM silencing in *Saccharomyces cerevisiae*. *Genetics* **163:** 447–452.

Tissenbaum H.A. and Guarente L. 2001. Increased dosage of a *sir-2* gene extends lifespan in *Caenorhabditis elegans*. *Nature* **410:** 227–230.

———. 2002. Model organisms as a guide to mammalian aging. *Dev. Cell* **2:** 9–19.

Triolo T. and Sternglanz R. 1996. Role of interactions between the origin recognition complex and SIR1 in transcriptional silencing. *Nature* **381:** 251–253.

Tsukamoto Y., Kato J., and Ikeda H. 1997. Silencing factors participate in DNA repair and recombination in *Saccharomyces cerevisiae*. *Nature* **388:** 900–903.

van Deutekom J.C., Wijmenga C., van Tienhoven E.A., Gruter A.M., Hewitt J.E., Padberg G.W., van Ommen G.J., Hofker M.H., and Frants R.R. 1993. FSHD associated DNA rearrangements are due to deletions of integral copies of a 3.2 kb tandemly repeated unit. *Hum. Mol. Genet.* **2:** 2037–2042.

van Leeuwen F., Gafken P.R., and Gottschling D.E. 2002. Dot1p modulates silencing in yeast by methylation of the nucleosome core. *Cell* **109:** 745–756.

Vega-Palas M.A., Martin-Figueroa E., and Florencio F.J. 2000. Telomeric silencing of a natural subtelomeric gene. *Mol. Gen. Genet.* **263:** 287–291.

Vega-Palas M.A., Venditti S., and Di Mauro E. 1997. Telomeric transcriptional silencing in a natural context. *Nat. Genet.* **15:** 232–233.

Vujcic M., Miller C.A., and Kowalski D. 1999. Activation of silent replication origins at autonomously replicating sequence elements near the HML locus in budding yeast. *Mol. Cell. Biol.* **19:** 6098–6109.

Wahlgren M., Fernandez V., Chen Q., Svard S., and Hagblom P. 1999. Waves of malarial *var*-iations. *Cell* **96:** 603–606.

Wallrath L.L. and Elgin S.C. 1995. Position effect variegation in *Drosophila* is associated with an altered chromatin structure. *Genes Dev.* **9:** 1263–1277.

Walmsley R.M., Chan C.S.M., Tye B.-K., and Petes T.D. 1984. Unusual DNA sequences associated with the ends of yeast chromosomes. *Nature* **310:** 157–160.

Walter M.F., Jang C., Kasravi B., Donath J., Mechler B.M., Mason J.M., and Biessmann H. 1995. DNA organization and polymorphism of a wild-type *Drosophila* telomere region. *Chromosoma* **104:** 229–241.

Wang H., Nicholson P.R., and Stillman D.J. 1990. Identification of a *Saccharomyces cerevisiae* DNA-binding protein involved in transcriptional regulation. *Mol. Cell. Biol.* **10:** 1743–1753.

Wellinger R.J., Wolf A.J., and Zakian V.A. 1993a. Origin activation and formation of single-strand TG_{1-3} tails occur sequentially in late S phase on a yeast linear plasmid. *Mol. Cell. Biol.* **13:** 4057–4065.

————. 1993b. *Saccharomyces* telomeres acquire single-strand TG_{1-3} tails late in S phase. *Cell* **72:** 51–60.

Wood A., Krogan N.J., Dover J., Schneider J., Heidt J., Boateng M.A., Dean K., Golshani A., Zhang Y., Greenblatt J.F., Johnston M., and Shilatifard A. 2003. Bre1, an E3 ubiquitin ligase required for recruitment and substrate selection of Rad6 at a promoter. *Mol. Cell* **11:** 267–274.

Wood J.G., Rogina B., Lavu S., Howitz K., Helfand S.L., Tatar M., and Sinclair D. 2004. Sirtuin activators mimic caloric restriction and delay ageing in metazoans. *Nature* **430:** 686–689.

Wotton D. and Shore D. 1997. A novel Rap1p-interacting factor, Rif2p, cooperates with Rif1p to regulate telomere length in *Saccharomyces cerevisiae. Genes Dev.* **11:** 748–760.

Wu J., Suka N., Carlson M., and Grunstein M. 2001. TUP1 utilizes histone H3/H2B-specific HDA1 deacetylase to repress gene activity in yeast. *Mol. Cell* **7:** 117–126.

Wyatt H.R., Liaw H., Green G.R., and Lustig A.J. 2003. Multiple roles for *Saccharomyces cerevisiae* histone H2A in telomere position effect, Spt phenotypes and double-strand-break repair. *Genetics* **164:** 47–64.

Wyrick J.J., Aparicio J.G., Chen T., Barnett J.D., Jennings E.G., Young R.A., Bell S.P., and Aparicio O.M. 2001. Genome-wide distribution of ORC and MCM proteins in *S. cerevisiae:* High-resolution mapping of replication origins. *Science* **294:** 2357–2360.

Wyrick J.J., Holstege F.C., Jennings E.G., Causton H.C., Shore D., Grunstein M., Lander E.S., and Young R.A. 1999. Chromosomal landscape of nucleosome-dependent gene expression and silencing in yeast. *Nature* **402:** 418–421.

Xu E.Y., Kim S., and Rivier D.H. 1999. SAS4 and SAS5 are locus-specific regulators of silencing in *Saccharomyces cerevisiae. Genetics* **153:** 25–33.

Zakian V.A. and Blanton H.M. 1988. Distribution of telomere-associated sequences on natural chromosomes of *Saccharomyces cerevisiae. Mol. Cell. Biol.* **8:** 2257–2260.

Zhang H., Richardson D.O., Roberts D.N., Utley R., Erdjument-Bromage H., Tempst P., Cote J., and Cairns B.R. 2004. The Yaf9 component of the SWR1 and NuA4 complexes is required for proper gene expression, histone H4 acetylation, and Htz1 replacement near telomeres. *Mol. Cell. Biol.* **24:** 9424–9436.

Zhang Z., Shibahara K., and Stillman B. 2000. PCNA connects DNA replication to epigenetic inheritance in yeast. *Nature* **408:** 221–225.

11

The Structural Biology of Telomeres

Daniela Rhodes

MRC Laboratory of Molecular Biology
Cambridge CB2 2QH, United Kingdom

ALTHOUGH THE LAST 10 YEARS HAVE WITNESSED an explosion in telomere biology with the identification of a large number of protein components of yeast and human telomeres, structural information on telomeres and their subcomplexes is in its infancy. Despite this, the three-dimensional structural information that has been obtained provides important and detailed insights into the function and evolutionary relationship of telomeric proteins. The first three-dimensional structure of a telomeric protein was determined in 1996—the DNA-binding domain of *Saccharomyces cerevisiae* Rap1p in complex with its double-stranded telomeric DNA-binding site (Konig et al. 1996). This was followed in 1998 by the structure of the *Oxytricha nova* telomere end-binding protein (TEBPα-β) heterodimer in complex with the single-stranded G-overhang (Horvath et al. 1998). Since then, a number of other structures of telomeric proteins and their complexes have been determined. This chapter summarizes the three-dimensional structural information on telomeric DNA and telomeric protein/DNA complexes obtained from direct structural methods such as X-ray crystallography and nuclear magnetic resonance (NMR). Features of the overall architecture of telomeres that have emerged from biological and biochemical studies in combination with electron microscopy are also highlighted.

DNA G-QUADRUPLEX STRUCTURES

The sequence conservation of telomeric DNA repeats, organized in a region of double-stranded DNA and a single-stranded G-overhang, not only imparts on the DNA the ability to recruit telomere-binding proteins, but also gives telomeric DNA unusual properties. The composition of telomeric DNA repeats containing runs of three or four guanines imparts on the single-stranded G-overhang the inherent ability to form G-quadruplex structures at physiological salt conditions. Such structures could in themselves provide a telomeric cap and are attractive because they involve self-recognition. The ability of G-rich sequences to fold into G-quadruplexes and the resulting structures has been extensively documented by structural studies during the last 15 years and will not be reviewed here (for reviews, see Henderson 1995; Rhodes and Giraldo 1995). Over the years the running question has been: Do such G-quadruplex structures form in vivo and do they have a biological function? Recent observations provide the first direct evidence for their existence at telomeres in vivo.

The building blocks of G-quadruplex structures are G-quartets, which arise from the association of four guanines into a cyclic hydrogen-bonding arrangement. G-quartets stack on top of each other, giving rise to four-stranded helical structures that are stabilized by K^+ and Na^+ ions that bind in the core of the structure. Such structures can be formed from the intramolecular folding of one strand, as well as the intermolecular assembly of two or four strands. G-quadruplex structures are very polymorphic and can be classified into two main classes, parallel or antiparallel, depending on the orientation of the strands (for reviews, see Henderson 1995; Rhodes and Giraldo 1995). The most relevant question for biology is the kinetics of their formation, which is strongly dependent on the number of strands involved. For long G-overhangs such as those present at human telomeres (50–200 nucleotides) (McElligott and Wellinger 1997; Wright et al. 1997), the intramolecular folding of the single strand will occur without any kinetic barriers (since it is strand-concentration-independent). In fact, one of the roles of the proteins that bind to the telomeric G-overhang (see next section) might be to inhibit the folding of the G-overhang into these compact and stable structures. On the other hand, the formation of G-quadruplex structures from two or four G-overhangs might require the aid of protein chaperones to accelerate their formation (Fang and Cech 1993a; Giraldo et al. 1994).

Recently, the crystal structure of the intramolecularly folded K^+ form of a 22-nucleotide fragment, d[AGGG(TTAGGG)$_3$], of the human G-overhang was determined at 2.1 Å resolution (Parkinson et al. 2002). It reveals a striking propeller-like G-quadruplex structure in which the strands are in a parallel orientation. In this novel structure, the four G-tracts form a stack of three G-quartets with the K^+ ions sandwiched between them. The TTA loops protrude out like the blades in a propeller

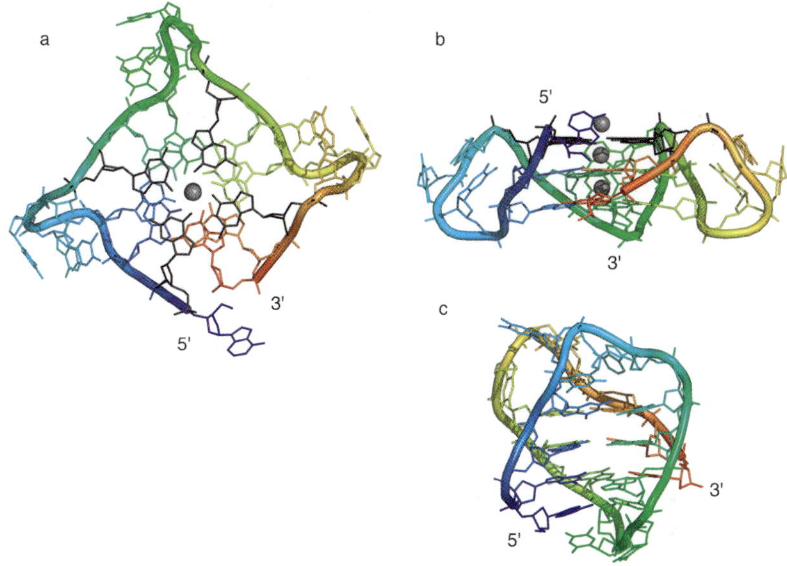

Figure 1. Human telomeric G-quadruplex structure. (*a,b*) End-on and side view of the crystal structure of the human telomeric sequence d[AGGG(TTAGGG)$_3$] folded in K$^+$ ions. The *gray spheres* represent the K$^+$ ions. (*c*) Side view of NMR structure of the same sequence folded in Na$^+$ ions. The DNA strand is color-coded *blue* to *yellow* going from the 5′ to the 3′ end.

(Fig. 1a). The structure formed is a flat disk (Fig. 1b), and consecutive folded 22-mers can stack on top of each other forming a continuous structure. The topology of this structure is such that it facilitates folding and unfolding (Parkinson et al. 2002). However, the NMR structure of the same sequence folded in Na$^+$ ions is topologically different, with the four strands in an antiparallel orientation (Fig. 1c) (Wang and Patel 1993). These two structures illustrate the polymorphic nature of G-quadruplex formation, but the K$^+$-coordinated G-quadruplex is the more likely structure to form in vivo since the intracellular K$^+$ concentration exceeds that of Na$^+$.

Since the finding that G-quadruplex structures inhibit telomerase-dependent telomere elongation in vitro (Zahler et al. 1991), by presumably sequestering the G-strand substrate and making it inaccessible to the telomerase, there have been considerable efforts to design small organic ligands that stabilize G-quadruplexes (prevent their unfolding). The premise for such ligands was that they may act as antitumor agents by impairing telomerase function. It has been shown that the use of highly G-quadruplex-specific ligands in telomerase-positive human cell lines results in senescence and telomere shortening, providing indirect evidence for the formation of a G-quadruplex structure (Riou et al. 2002; Kim et al. 2003). Very recently, G-quadruplexes have also been implicated in explaining the role of the *Rtel* gene, which is required for telomere elongation in murine cells (Ding et al. 2004). The *Rtel* gene encodes a helicase-like protein that has sequence homology with the *dog-1* helicase gene from *Caenorhabditis elegans*, which appears to be involved in resolving the secondary structure of G-tracts in vivo (Cheung et al. 2002). On the basis of this similarity, it was proposed that RTEL is involved in resolving G-quadruplex structures that could arise at telomeres during lagging-strand DNA replication (Ding et al. 2004). Evidence for G-quadruplex function also comes from the RNA world. It has been found that the mammalian Fragile-X syndrome protein binds to RNA quadruplexes present in mRNA (Darnell et al. 2001).

The most direct evidence for the presence of G-quadruplexes at telomeres in vivo has come from studies using antibodies (Schaffitzel et al. 2001). Using a library of in-vitro-generated antibodies, which are highly selective and can discriminate between parallel and antiparallel G-quadruplex structures, one was found that reacted specifically and with high affinity with the macronucleus, but not the micronucleous, of the ciliate *Stylonychia lemnae* (Schaffitzel et al. 2001). Furthermore, it was demonstrated that the structure recognized by the antibody is an antiparallel G-quadruplex. This observation is consistent with an old observation of the unusual behavior of *S. lemnae* telomeres (Lipps et al. 1982). During ciliate macronuclear differentiation and the accompanying genome fragmentation, the resulting gene-sized fragments are capped by short telomeres with a d[(TTTTGGGG)$_2$] G-overhang. When macronuclei were gently lysed, the chromosomes were found in long fibers, suggesting that telomeric G-overhangs are involved in end-to-end joining. It was suggested that a way to achieve telomeric end-to-end joining could be via the formation of an antiparallel G-quadruplex structure (Lipps et al. 1982), which is indeed the only type of G-quadruplex that can arise from the dimerization of two short G-overhangs containing two G-tracts each. Very recent experiments (Paesche et al. 2005), using RNA interference (RNAi) technology to silence gene expression of the two G-overhang-binding proteins TEBPα and TEBPβ, provide direct evidence that these telomeric proteins are involved in controlling the formation of G-quadruplexes in vivo. This observation is consistent with early experiments that showed that TEBPβ from the related ciliate *O. nova* accelerates G-quadruplex formation in vitro (Fang and Cech 1993a,b). Thus, the expectations based on the in vitro structural evidence and physical chemistry of G-quadruplex structure formation has been fulfilled by the demonstration of their presence in vivo. However, the precise role of G-quadruplexes at telomeres remains to be defined.

RECOGNITION OF THE SINGLE-STRANDED G-OVERHANG

The first protein to be identified that specifically recognizes and caps the single-stranded G-overhang was the ciliate *O. nova* TEBP (Gottschling and Zakian 1986; Price and Cech 1987). It is composed of two subunits, α and β. These two proteins can form two alternative complexes, an α-β heterodimer and an α-α homodimer, that bind specifically, but differently, to the telomeric overhang of the macronuclear chromosomes. A single copy of the α-β heterodimer is sufficient for binding the 16-nucleotide *O. nova* G-overhang in vitro (Fang and Cech 1993c). From these observations, it has been proposed that the different complexes might be involved in the assembly and disassembly of higher-order telomeric complexes (Peersen et al. 2002). Furthermore, the α-β heterodimer inhibits the action of telomerase, whereas the α-homodimer does so to a lesser extent. Hence, the two different complexes appear to have different functions at telomeres (Froelich-Ammon et al. 1998). In *Saccharomyces cerevisiae*, the single-stranded G-overhang is bound specifically by Cdc13p (Lin and Zakian 1996; Nugent et al. 1996). Cdc13p has two separate functions: It is involved in both chromosome end protection and the recruitment of the telomerase and hence in telomere replication (Nugent et al. 1996; Pennock et al. 2001). More recently, POT1, a widespread protein found in fission yeast, humans, and other species, was identified through a weak sequence similarity to the amino-terminal region of the *O. nova* TEBPα (Baumann and Cech 2001). Deletion of the *POT1* gene in fission yeast leads to rapid loss of telomeric DNA and chromosome circularization, providing evidence that POT1 has a crucial role in telomere capping (Baumann and Cech 2001). In human cells, POT1 appears to be involved in telomere length regulation through its interaction with both the G-overhang and the human TRF1 (hTRF1) complex (Colgin et al. 2003; Loayza and de Lange 2003).

The information emerging from the determination of the three-dimensional structures of the TEBP, Cdc13p, and POT1 end-binding proteins is that they bind the single-stranded G-overhang using a conserved DNA-binding motif, the OB fold. The OB fold is of bacterial origin and was originally identified as an oligonucleotide or oligosaccharide-binding motif. It is a structural domain of 70–180 amino acids in length with diverse functions (Murzin 1993). OB folds are difficult to recognize from amino acid sequence comparisons alone because of very low sequence identity, and consequently structural information is often required. The role of the OB fold in the recognition of the single-stranded telomeric G-overhang was first revealed by the crystal structure of the TEBP in complex with single-stranded DNA (see below) (Horvath et al. 1998). More recently, the NMR structure of the well-characterized DNA-binding domain of *S. cerevisiae* Cdc13p (Hughes et al. 2000) revealed that it consists of an OB fold (Mitton-Fry et al. 2002, 2004). As expected from sequence comparisons, the crystal structure of the amino-terminal region of *Schizosaccharomyces pombe* POT1 confirmed the presence of an OB fold (Lei et al. 2003), whereas the crystal structure of the DNA-binding domain of human POT1 contains two OB folds (Lei et al. 2004).

The superposition of the OB folds of *O. nova* TEBPα, *S. cerevisiae* Cdc13p, and *S. pombe* POT1 shows that the three OB-fold domains share a very similar central core consisting of a curved five-stranded antiparallel β-barrel and a helical extension at their carboxyl termini (Fig. 2a). However, there is considerable structural variability in the outer regions of the domain. The single-stranded DNA primarily binds in a groove formed by one face of the β-barrel and two flanking loops: On one side is the loop linking β-strands 1 and 2, and on the other side is the loop linking β-strands 4 and 5 (Fig. 2). The common feature is that these two loops provide interacting amino acid side chains that clamp the DNA strand in place. In all structures, the DNA strands bind with the same polarity and take up a more or less extended and irregular conformation, with the DNA ribose-phosphate backbones solvent-exposed and the bases partially or completely buried in the protein surface. Predominantly aromatic and basic amino acid side chains interact with the DNA bases and backbone. A recurring theme in the OB-fold domain–single-stranded DNA recognition is the stacking of aromatic and hydrophobic side chains onto bases, and bases with each other (Figs. 3b,d and 4b) (Horvath et al. 1998; Lei et al. 2003, 2004; Mitton-Fry et al. 2004). In detail, however, for each OB-fold domain, the mode of single-stranded DNA recognition is both versatile and different. The most evident difference is the length of G-overhang bound to each OB fold, which varies from 6 nucleotides by *S. pombe* POT1 to 11 nucleotides by *S. cerevisiae* Cdc13p. For TEBP, the binding of DNA to the first OB fold (Fig. 2b) is discontinuous since the 12-mer DNA is bound by three different OB folds in the context of a much larger heterodimeric complex, discussed below (Fig. 4a,b).

The NMR structure of the *S. cerevisiae* Cdc13p DNA-binding domain in complex with single-stranded DNA was determined with the minimal

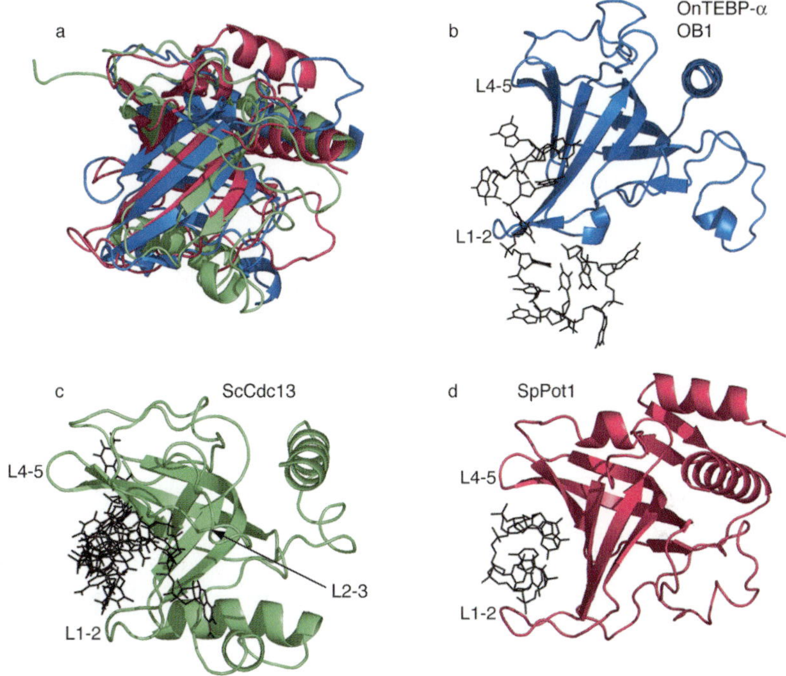

Figure 2. OB folds structure and single-stranded DNA recognition. (*a*) Superposition of the structures of the OB folds from *O. nova* TEBP, *S. cerevisiae* Cdc13p, and *S. pombe* POT1; (*b*) structure of TEBPα OB1 in complex with d(GGGGTTTTGGGG); (*c*) structure of Cdc13p OB fold in complex with d(GTGTGGGTGTG); (*d*) structure of POT1 OB fold in complex with d(GGTTAC).

11-mer binding site d(GTGTGGGTGTG) (Mitton-Fry et al. 2004). Recognition of this long binding site by a single OB-fold domain is achieved by an unusually long loop between β-strands 2 and 3 that packs on the protein surface, extending the DNA interaction surface of this OB fold (Fig. 2c).

The crystal structure (1.9 Å resolution) of the OB fold identified at the amino terminus of *S. pombe* POT1 was determined in complex with the 6-mer sequence d(GGTTAC) (Fig. 3a,b) (Lei et al. 2003). The DNA strand folds on the surface of the domain, and in addition to making a number of stacking and hydrogen bond interactions with amino acid side chains, it makes two intrastrand hydrogen bonds, between bases (G1 with T3 and G2 with T4) (Fig. 3b). This mode of DNA self-recognition adds specificity (Lei et al. 2003), but it is not a mechanism observed in other OB fold–single-stranded DNA complexes (Bochkarev et al. 1997;

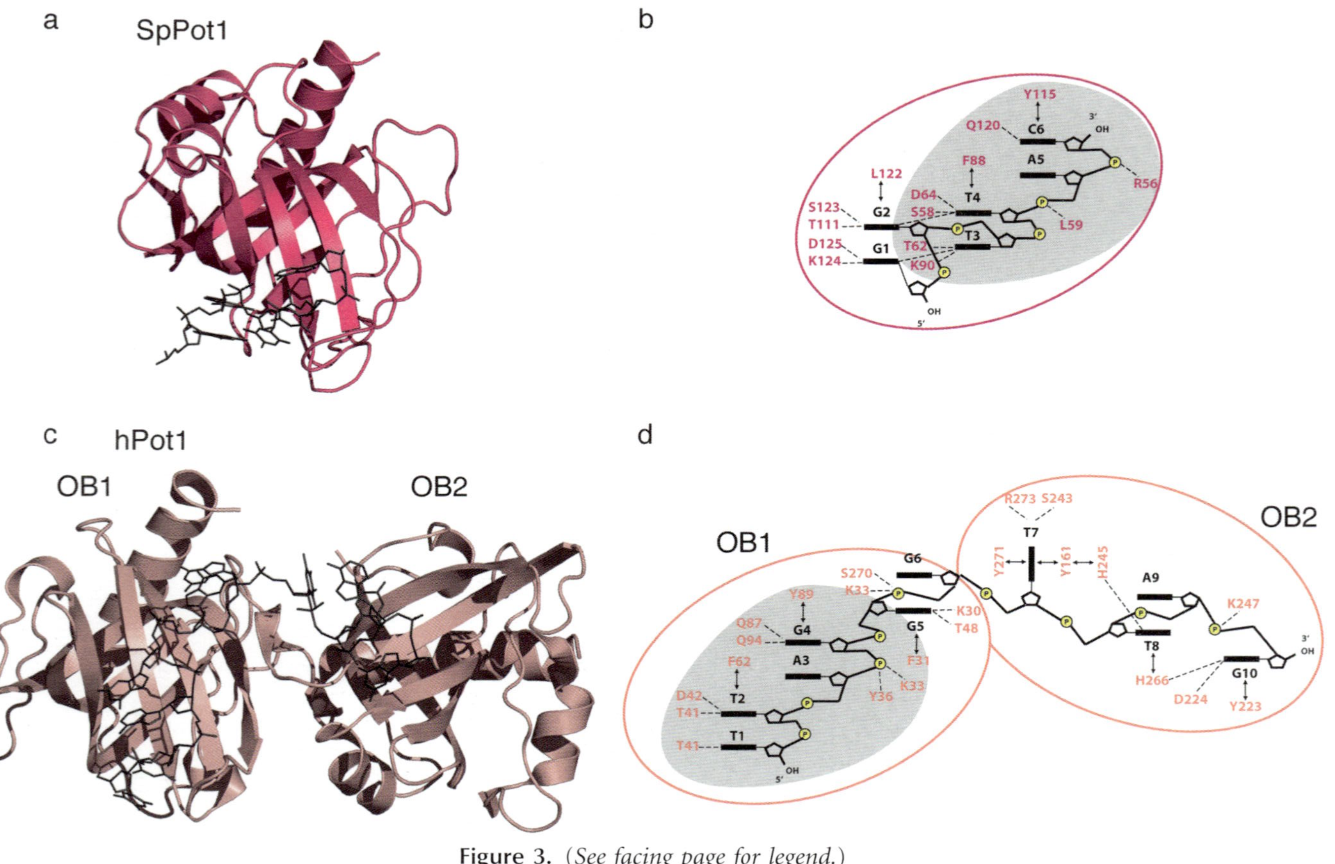

Figure 3. (*See facing page for legend.*)

Horvath et al. 1998; Mitton-Fry et al. 2004). Because the DNA-binding site was defined using the single OB fold (Lei et al. 2002), it remains to be demonstrated whether it constitutes the complete DNA-binding domain of the *S. pombe* POT1. Significantly, examination of the DNA-binding requirements of full-length human POT1 defined a minimal binding site that is significantly longer: the 9-mer sequence d(TAGGGT-TAG) (Loayza et al. 2004). The crystal structure (1.73 Å resolution) of the DNA-binding domain of human POT1 revealed two tandem OB folds bound to the 10-mer sequence d(TTAGGGTTAG) (Fig. 3c,d) (Lei et al. 2004). The two OB folds of human POT1 pack together so that the second OB fold (OB2) is tilted backward by about 90° and rotated relative to the first (OB1), resulting in the 10-mer single-stranded DNA binding in a more or less continuous path (Fig. 3c,d). The arrangement of the two OB folds relative to each other is different from both that in TEBPα (see Fig. 4) (Horvath et al. 1998; Peersen et al. 2002) and that in the replication protein A (RPA), which also uses two tandem OB folds for single-stranded DNA recognition (Bochkarev et al. 1997). For RPA, there is good evidence that the arrangement of the two OB folds is induced by DNA binding (Bochkareva et al. 2001). Of the 10-mer d(TTAGGGTTAG), T1 to G6 interact with the OB1 and G7 to G10 interact with OB2 (Fig. 3d). OB1 is more important in single-stranded DNA binding since it makes many more contacts with the G-overhang DNA than does OB2. Comparison between the *S. pombe* and human POT1 OB1 domains shows an overlap in DNA recognition. The binding of T1 to G4 by the human OB1 mimics that of T3 to C6 by the *S. pombe* OB fold: The stacking arrangement of the bases, with T stacking on the following T and A stacking on the following base, and the pattern of side chain interactions are the same and use the same protein interaction surface (Fig. 3b,d). However, the human OB1 recognizes two additional nucleotides (G5 and G6) as the single strand crosses to OB2. The two bases at the 5′ end of the *S. pombe* site, G1 and G2, are not part of the minimal human POT1-binding site (Fig. 3b,d). Importantly and contrary to the observation

Figure 3. Structures of the OB folds from *S. pombe* and human POT1. (*a*) *S. pombe* OB fold domain in complex with d(GGTTAC). (*b*) Schematic representation of amino acid side chain–single-stranded DNA interactions. (*Solid arrows*) Stacking interactions; (*dashed lines*) hydrogen bonds. (*c*) Human POT1 DNA-binding domain containing two tandem OB folds in complex with (TTAGGGTTAG). (*d*) Schematic representation of amino acid side chain–single-stranded DNA interactions. (*Solid arrows*) Stacking interactions; (*dashed lines*) hydrogen bonds. The *gray patch* identifies the conservation in recognition between the two POT1s.

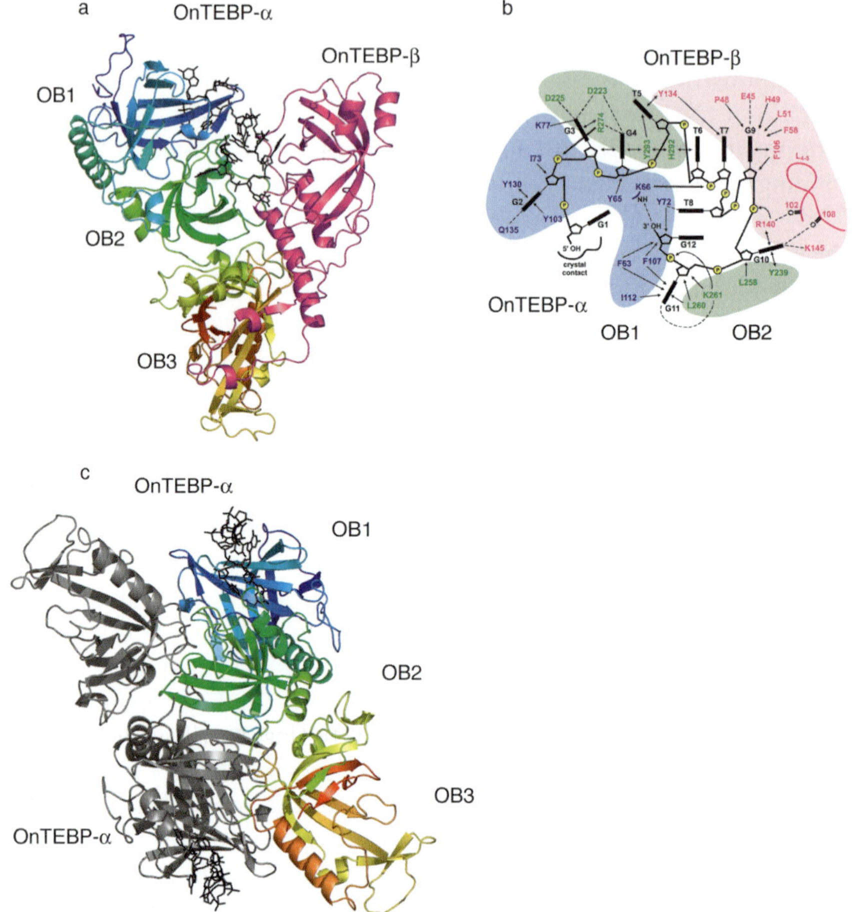

Figure 4. Multiple OB folds are involved in TEBP binding of the *O. nova* G-overhang. (*a*) Structure of TEBPα-TEBPβ heterodimer in complex with d(GGGGTTTTGGGG). The α subunit is color-coded from *blue* to *red*, going from the amino to the carboxyl terminus. (*Magenta*) β subunit. (*b*) Schematic representation of the protein–single-stranded DNA interaction in the TEBPα-TEBPβ–d(GGGGTTTTGGGG) complex. (*Solid arrows*) Stacking interactions; (*dashed lines*) hydrogen bonds. The patches indicate the approximate region of sequence recognized by each OB fold in the complex: *Blue* represents OB fold 1 and *green* represents OB fold 2 from the α-subunit fold; *pink* represents the OB fold from the β subunit. (*c*) Structure of TEBPα homodimer in complex with d(TTTTGGGG).

from the *S. pombe* POT1 OB fold–single-stranded DNA complex where the 3′ end of the DNA is solvent-accessible, the 3′ guanine (G10) in the human POT1 DNA-binding domain complex is buried deeply in a binding pocket, consistent with the observation that human POT1 acts as a negative regulator of telomerase in vitro (Kelleher et al. 2005).

Two crystal structures, those of the *O. nova* TEBPα-β heterodimer and the TEBPα homodimer in complex with the *O. nova* telomeric G-overhang, preceded the discovery that both Cdc13p and POT1 use OB folds for telomeric G-overhang recognition. The crystal structure (2.8 Å and 1.8 Å resolution) of the heterodimeric TEBPα-TEBPβ–d(GGGGTTTTGGGG) 1:1:1 complex (Horvath et al. 1998); (Horvath and Schultz 2001) showed that TEBPα consists of three tandem OB folds and TEBPβ contains one OB fold. The four OB folds have different functions: Three are used in single-stranded DNA binding and the fourth for protein–protein interactions (Fig. 4a). The three OB folds in the α subunit are arranged in two structural domains: an amino-terminal domain consisting of two tandem OB folds (OB1 and OB2) that bind DNA and a carboxy-terminal domain containing a third OB fold (OB3) that interacts with the β subunit. The single OB fold present in the TEBPβ also interacts with DNA. Together, the three DNA-binding OB folds form a deep DNA-binding cleft, which is likely to be formed by cofolding of protein and DNA (Fig. 4a) (Horvath et al. 1998). OB1 and OB2 of TEBPα pack tightly together, and the loops of the two domains cooperate to interact with G1 to T5, T8 and G10 to G12, whereas the OB fold of the β subunit interacts with T5 to G10 (Fig. 4b). The DNA strand adopts an irregular, folded back structure, so that both the 5′ and 3′ ends of the 12-mer G-overhang oligonucletide interact with OB1 of TEBPα. The ribose-phosphate backbone is solvent exposed and the bases are completely buried in the protein cleft. Because of this intimate mode of binding—more intimate than is seen for either Cdc13p or POT1 proteins—a wealth of different interactions are made, including the stacking of most of the bases with aromatic amino acid side chains or with another base of the single-stranded DNA (Fig. 4b). All of the guanines in the 12-mer single-stranded DNA sequence are involved in hydrogen-bonding interactions. Significantly, G12, the 3′-terminal guanine, is buried deep within the complex, making it inaccessible to telomerase, consistent with the finding that the TEBPα-TEBPβ complex blocks telomerase activity (Froelich-Ammon et al. 1998).

The crystal structure (2.7 Å resolution) of the TEBPα–d (TTTTGGGG) 2:2 complex (Fig. 4c) revealed a homodimeric head-to-tail arrangement, in which each α subunit binds one telomeric single-stranded DNA overhang on opposite sides of the dimer, suggesting a protein-mediated telomere–telomere association (Peersen et al. 2002). Comparison between the structures of the homodimeric and heterodimeric complexes (Fig. 4a,c) shows a reorientation of domains. In the heterodimeric complex, OB3 of the α subunit is tilted by about 45° and rotated clockwise by about 90° relative to its position in the homodimer. This is achieved by distorting the linker that connects OB2 with OB3 in the α subunit. This conformational rearrangement has consequences for nucleic acid recognition, because in the α homodimer, OB2—which in the heterodimeric complex is involved in DNA binding—is here primarily involved in protein–protein interactions. Despite the switch in function, the spatial arrangement of OB1 and OB2 in TEBPα is the same in the two complexes. Consequently, homodimerization of the α-subunit and DNA binding as seen in the α-β complex are mutually exclusive. Significantly and contrary to the situation in the heterodimeric complex, in the TEBPα homodimer, the 3′-hydroxyl group of the terminal guanine is solvent exposed and accessible to the telomerase, consistent with functional differences between the homodimeric and heterodimeric TEBP complexes (Froelich-Ammon et al. 1998).

The structural information on a number of TEBPs has revealed not only that recognition of the single-stranded G-overhang takes place via a conserved motif, the OB-fold domain, but also that the number of OB folds involved in single-stranded DNA binding varies from one to three, resulting in the recognition of G-overhangs from 5 to 12 nucleotides in length. Is this a true variability or is it possible that some of the structures are incomplete and contain only parts of the DNA-binding domains of these TEBPs? Recent sequence alignments using sequence profile methods have identified three tandem OB folds in POT1, suggesting a much closer structural homology with the ciliate TEBPα than had been thought (Simonsson 2003; Theobald and Wuttke 2004; Wei and Price 2004). The use of multiple OB folds for single-stranded DNA recognition is also a feature shared by the human heterotrimeric replication complex RPA (Bochkarev and Bochkareva 2004). The relationship between TEBPα and POT1 proteins is strengthened by the recent crystal structure showing that the DNA-binding domain of human POT1 DNA consists of two OB folds, OB1 and OB2 (Lei et al. 2004). This observation raises the question of whether *S. pombe* POT1 may recognize single-stranded DNA in the same way. In fact, although it has been reported that the amino-terminal OB fold of *S. pombe* POT1 binds to telomeric DNA with higher affinity than the full-length protein, suggesting it represents the entire DNA-binding domain of the protein (Lei et al. 2002, 2003), no experimental evidence supporting this assumption is available. If the structural homology between the ciliate, the TEBPα, and the metazoan POT1 proteins holds true, this would suggest a possible functional homology with the well-characterized homodimeric and heterodimeric TEBP complexes (Fig. 4) (Horvath et al. 1998; Peersen et al. 2002), where multiple OB folds are used for the recognition of long regions of single-stranded DNA, as well as in the binding of protein partners. The budding yeast Cdc13p contains two OB folds (Theobald and Wuttke 2004), but these are arranged differently from those in POT1 and TEBPα and are located in the carboxy-terminal half of the protein. Since the DNA-binding domain of Cdc13p is well characterized and consists of the first OB fold (Hughes et al. 2000), it would seem that Cdc13p belongs to a family of G-overhang binding proteins different from those of POT1.

RECOGNITION OF DOUBLE-STRANDED TELOMERIC DNA

S. cerevisiae Rap1p was the first protein to be discovered that binds sequence-specifically to double-stranded telomeric DNA repeats. Rap1p was originally identified as a regulator of transcription and only later was it discovered that it binds the irregular telomeric sequence $d(GTG_{1-3})$ of *S. cerevisiae* (Longtine et al. 1989; for review, see Shore 1994). Rap1p is a negative regulator of telomere length (Conrad et al. 1990; Lustig et al. 1990), and the number of bound Rap1p molecules at a telomere seems to constitute a telomere-length-measuring mechanism (Krauskopf and Blackburn 1996; Marcand et al. 1997). The double-stranded d(TTAGGG) repeats of mammals are bound sequence-specifically by two related proteins, TRF1 (Chong et al. 1995) and TRF2 (Bilaud et al. 1997; Broccoli et al. 1997). The two TRFs have somewhat different functions at telomeres. TRF1, like Rap1p, is a negative regulator of telomere length (van Steensel and de Lange 1997). Although hTRF2 is also involved in telomere length regulation (Smogorzewska et al. 2000), its primary role appears to be in capping and protecting chromosome ends (Ancelin et al. 1998). It has also been implicated in the formation of a telomeric higher-order structure, the t-loop (for review, see de Lange 2002; Smogorzewska and de Lange 2004). Similarly, fission yeast telomeres are protected from end-to-end fusions by the TRF-related protein, TAZ1 (Cooper et al. 1997).

Both the TRFs and *S. cerevisiae* Rap1p have multidomain structures, identified through amino acid sequence conservation and functional studies. Such domains have specific and apparently separable functions and are used in telomeric DNA recognition and in protein–protein interactions with other proteins in the telomeric complex (for review, see Shore 1994; de Lange 2002; Smogorzewska and de Lange 2004). Whereas *S. cerevisiae* Rap1p binds to DNA as a monomer (Gilson et al. 1993), TRF1 and TRF2 bind to telomeric repeats as preformed homodimers (Bianchi et al. 1997; Broccoli et al. 1997). Dimerization of the TRFs involves the TRF-homology (TRFH) domain, a region of about 200 amino acids in the amino-terminal half of the two proteins (Bianchi et al. 1997; Li et al. 2000). Located at the carboxyl termini of both TRF1 and TRF2 are the Myb/homeodomain-like DNA-binding domains, which are linked to the dimerization domains via linkers with no apparent inherent structure.

S. pombe TAZ1 is likely to share the architecture of the TRFs (Li et al. 2000; Fairall et al. 2001). Proteolysis studies to identify structural domains within TRF1, together with the observation that TRF1 not only binds to adjacent telomeric repeats, but also binds to binding sites spaced far apart (Bianchi et al. 1999) and even to two different DNA molecules (Griffith et al. 1998), provide evidence of a highly flexible structure, a feature that is likely to be relevant for the dynamic structure of telomeres. Unfortunately, flexible protein architectures preclude the determination of the three-dimensional structure of the full-length proteins, so that the structural information available is that of the various folded subdomains determined separately.

The crystal structures of the TRFH dimerization domain from both human TRF1 and TRF2 (2.9 Å and 2.1 Å resolution) revealed that they have almost identical and entirely α-helical structures (Fairall et al. 2001). Each dimer is formed by two monomers interacting in an antiparallel manner forming a symmetrical dimer (Fig. 5). The overall structure resembles a twisted horseshoe, providing a large protein interaction surface (see Fig. 8). Despite the structural conservation, the sequence divergence between TRF1 and TRF2 TRFH domains (Bianchi et al. 1997; Li et al. 2000) results in the presentation of different interaction surfaces, imparting on them the ability to interact with different protein partners. Each monomer consists of a bundle of nine α helices with a novel topology (Fairall et al. 2001). The dimer interface is formed by six α helices, helices 1, 2, and 9, from each of the two monomers (Fig. 5). At the interface,

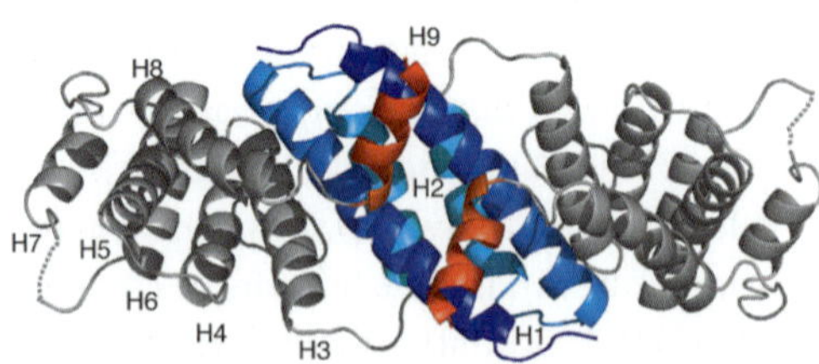

Figure 5. Structure of the TRFH dimerization domain of hTRF1. The view shows the dimerization interface consisting of helices 1, 2, and 9 from each monomer. The helices are numbered. Another view representing a 90° rotation can be seen in Fig. 8.

helices 1 and 2 from one monomer pack against helices 1 and 2 from the other monomer, forming a symmetrical antiparallel four-helix bundle. The four-helix bundle is stabilized by the two helix 9s packing against each other and perpendicularly to the two helix 1s, forming a cross-brace at the top and the bottom of the dimer interface (Fig. 5). The mode of dimerization by TRF1 and TRF2 is used by many other proteins and is most similar to the plasmid replication protein Rop (Banner et al. 1987). The large and hydrophobic nature of the dimer interface explains why TRF1 and TRF2 only exist as dimeric proteins and not as monomeric proteins. The biological importance of a dimeric structure for the TRFs was highlighted by a mutational analysis which showed that point mutations at the helix 1 protein–protein interface prevent telomere localization in vivo (Fairall et al. 2001).

The three-dimensional information from the crystal structures of the *S. cerevisiae* Rap1p DNA-binding domain (2.2 Å resolution) (Konig et al. 1996; Taylor et al. 2000) and the hTRF1 and hTRF2 DNA-binding domains (2.0 Å and 1.8 Å resolution) (Court et al. 2005), in complex with their respective telomeric DNA-binding sites, as well as the NMR structure of the TRF1 DNA-binding domain in complex with a telomeric DNA site (Nishikawa et al. 1998, 2001), shows that recognition of double-stranded telomeric DNA repeats is via homeodomains (Kissinger et al. 1990; Konig et al. 1998). The DNA-binding motif present in TRF1 (Chong et al. 1995) and TRF2 was originally identified (Bilaud et al. 1996, 1997; Broccoli et al. 1997; Konig and Rhodes 1997) through its high sequence similarity to the Myb DNA-binding motif (Kanei-Ishii et al. 1990) and was so named. Structurally, the Myb and homeodomain DNA-binding motifs are closely related, and both belong to the helix-turn-helix (H-T-H) family of DNA-recognition motifs found in many transcription regulators, including bacterial DNA-binding proteins (Pabo and Sauer 1992). However, there is one important difference in the way Myb and homeodomains bind DNA. The superposition of the homeodomains from TRF1 and TRF2, and the core structure of the homeodomains of Rap1p in complex with DNA (Fig. 6a), shows that the DNA-binding domains of the three proteins have the same overall architecture consisting of a bundle of three α helices, in which the second and third helices form the H-T-H motif. The third α helix is the DNA-recognition helix, which makes base-specific contacts in the major groove of DNA. In addition, all three motifs have an amino-terminal arm that makes base-specific contacts in the minor groove of DNA (Fig. 6a). The amino-terminal arm, which is a feature of homeodomains and not Myb domains, extends the sequence-specific recognition from 2 bp for a single Myb domain (Ogata et al. 1994)

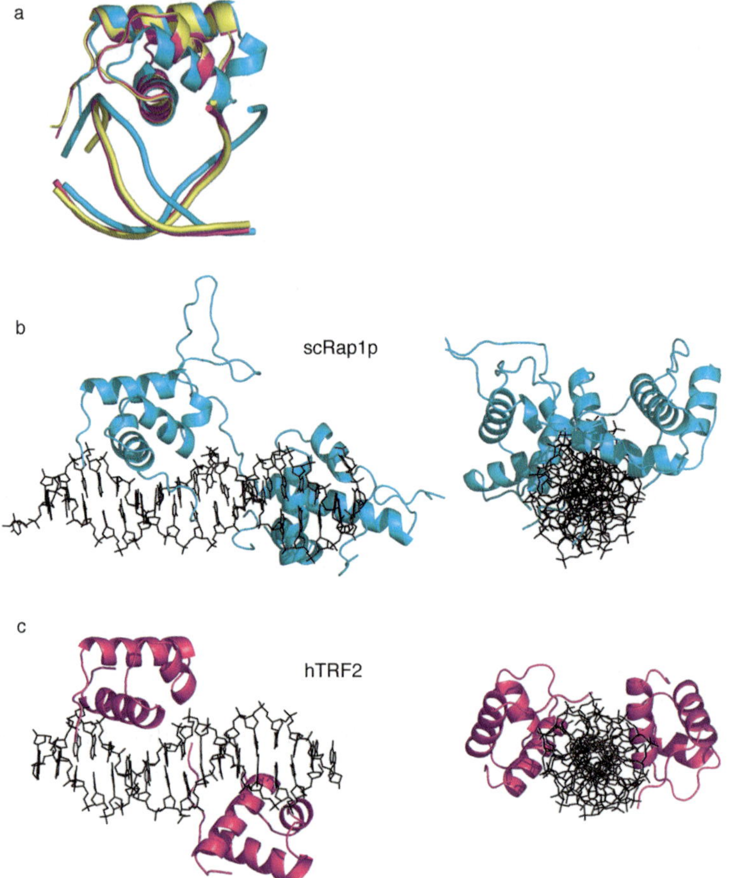

Figure 6. Homeodomain structure and double-strand recognition. (*a*) Superposition of the homeodomain structures from Rap1 domain 1 (*yellow*), TRF1 (*blue*), and TRF (*magenta*). (*b*) Two views of the DNA-binding domain of Rap1p in complex with the double-stranded d(TGGTGTGTGGGTGTG). (*c*) Two views of the structures of two TRF2 homeodomains in complex with the double-stranded d(TTAGGGTTAGGGTTA).

to 5 bp, and hence increases both stability and specificity. Consequently, the mode of DNA-binding classifies both Rap1p and TRF DNA-binding domains as homeodomains.

The crystal structure of the complex of *S. cerevisiae* Rap1p DNA-binding domain with the high-affinity binding site d(TGGTGT-GTGGGTGTG) revealed that it had a bipartite structure consisting of two homeodomains (Fig. 6b) (Konig et al. 1996; Taylor et al. 2000). Both Rap1p homeodomains are augmented by additional structural elements, but these are not involved in DNA binding. The two domains bind DNA in a tandem orientation, so that each is aligned and making very similar interactions with sites centered on the two G-tracts present in the binding site (Fig. 6b). Each domain makes contact with an 8-bp site: Domain 1 binds to GTGGGTGT and domain 2 binds to CTGGTGTG. Because the two domains do not form any protein–protein interactions, their relative positioning on the DNA must be determined upon DNA binding. The domains bind with an 8-bp spacing, but wrap around the DNA in a left-handed path (opposite to the DNA double helix) with a rotation of about 90°. This arrangement of the two domains permits adjacent Rap1p molecules to bind with a spacing smaller than the size of the binding site (Konig et al. 1996), which is consistent with footprinting data (Gilson et al. 1993), and would result in the DNA being engulfed by the protein (Fig. 6b).

The crystal structure of the hTRF1 and hTRF2 DNA-binding domains in complex with human telomeric repeats was determined with two homeodomains bound in each complex (Fig. 6c) (Court et al. 2005). The binding site present in the crystals was defined from footprinting and SELEX data on full-length dimeric hTRF1 and contained 2.5 TTAGGG repeats (Konig et al. 1998; Bianchi et al. 1999). Each DNA-binding domain contacts a 7-bp site: TAGGGTA. In each complex, two DNA-binding domains are bound to two adjacent binding sites, presumably mimicking the geometry of binding of the DNA-binding domains in the full-length dimeric protein. Overall, the structures of the TRF1 and TRF2 complexes are very similar and differ only in minor details (Court et al. 2005). As illustrated by the hTRF2DBD-DNA complex (Fig. 6c), the two DNA-binding domains are bound on the DNA with a 6-bp spacing, or 205° rotation, and hence are bound on almost opposite faces of the DNA double helix, similar to Rap1p.

Not only is the type of DNA-binding domain conserved, but the homeodomains of Rap1p, TRF1, and TRF2 align on the telomeric DNA repeats so that the pattern of protein-DNA contacts is largely conserved (Fig. 7). Significantly, the guanine clusters that characterize telomeric DNA repeats have a central role in the recognition. The patterns of sequence-specific contacts made by the TRFs and Rap1p are almost identical: arginine and lysine side chains from similar positions on the DNA-recognition helix make hydrogen bond contacts in the major groove to guanines in the G-strand and also to a cytosine in the C-strand (Fig. 7). The amino-terminal arm in all four domains crosses the ribose-phosphate backbone into the minor groove, and an arginine or lysine side chain contacts primarily adenines in the C-strand. The binding is further stabilized by multiple interactions to the ribose-phosphate backbone, and also an unusually large number of water molecules at the protein–DNA

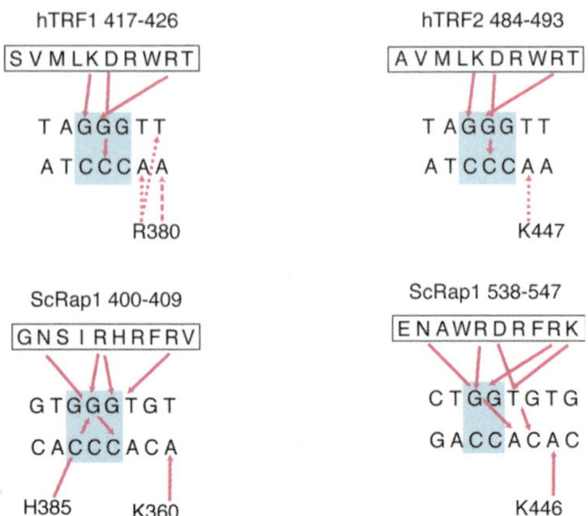

Figure 7. Comparison of the contacts made by the two homeodomains in Rap1p and the TRF1 and TRF2 homeodomains with double-stranded DNA. Only the contacts made by the DNA-recognition helix are shown. (*Pink arrows*) Hydrogen bonds; (*dashed arrows*) alternative contacts.

interface mediate protein–DNA interactions in the TRF complexes (Court et al. 2005). Figure 8 shows the possible architecture of the TRF proteins bound to double-stranded DNA. The structure is a composite built from the crystal structure of the TRF2 dimerization domain (Fairall

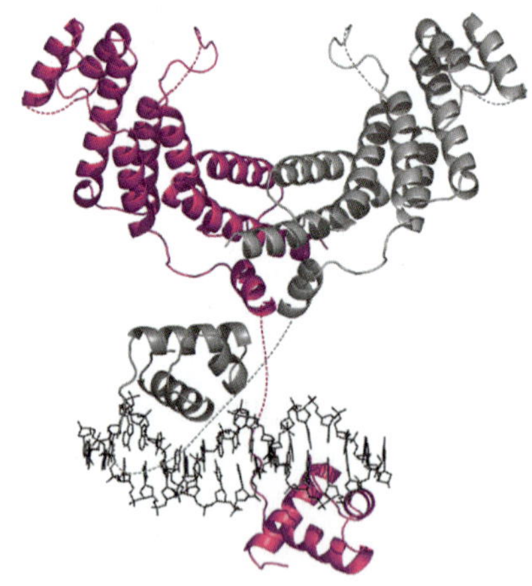

Figure 8. Structure of TRF2 bound to double-stranded telomeric DNA. The structure is a composite using the crystal structure of the TRFH dimerization domain (Fig. 5) and the homeodomain-DNA complex (Fig. 6c). One monomer is shown in *magenta* and the other in *gray*. (*Dashed line*) The linkers linking the dimerization domain with the DNA-binding domains and for which there is no structural information are shown as *dotted lines*.

et al. 2001) and the structure of the two TRF2 homeodomains in complex with double-stranded telomeric DNA repeats (Court et al. 2004).

Besides the conservation of double-stranded telomeric DNA recognition by Myb/homeodomains, which belong to the widespread family of H-T-H DNA-binding motifs found in both prokaryotic and eukaryotic DNA-binding proteins (Pabo and Sauer 1992), the recognition of double-stranded telomeric DNA appears to require the juxtaposition of two DNA-binding domains. In the mammalian TRFs and very likely in the fission yeast TAZ1 (Cooper et al. 1997), this requirement is fulfilled by homodimerization of monomers each carrying a single DNA-binding domain (Bianchi et al. 1997; Broccoli et al. 1997). In contrast to the TRFs, the budding yeast Rap1p binds telomeric DNA as a monomer (Gilson et al. 1993), and here this requirement is fulfilled by the tandem arrangement of two DNA-binding domains within the monomeric protein (Konig et al. 1996). Comparison of the telomeric components of budding and fission yeast with mammals suggests that during the course of evolution, budding yeast might have lost the TRF-like proteins, resulting in the acquisition of a direct DNA-binding activity by Rap1p (Kanoh and Ishikawa 2001). This was probably through gene duplication of the Myb motif present in a single copy in mammalian *RAP1* and partly duplicated in *S. pombe RAP1p*, neither of which binds sequence-specifically to DNA (Li 2000; Kanoh and Ishikawa 2001). Geometrically, the use of two tandemly arranged DNA-binding domains to recognize tandem DNA repeats is a more efficient solution than homodimerization and is also likely to facilitate recognition of the more irregular sequence composition of budding yeast telomeric DNA.

TELOMERE HIGHER-ORDER STRUCTURE

Evidence from a large number of studies suggests that both yeast and mammalian telomeres exist in higher-order structures. A protein-free, lasso-like structure called the t-loop was isolated from human telomeres from DNA cross-linking carried out in vivo and subsequently visualized by electron microscopy (Griffith et al. 1999). These t-loops contain thousands of base pairs of TTAGGG repeats. The t-loop is proposed to form by the insertion of the G-overhang into the double-stranded region of telomeric DNA (Griffith et al. 1999). Such a structure can be recapitulated by incubating pure hTRF2 with double-stranded telomeric DNA ending in a G-overhang (Stansel et al. 2001). Subsequently, t-loop structures have been found at the end of telomeres of other organisms, and hence appear to be an evolutionarily conserved feature of telomere structure (for re-

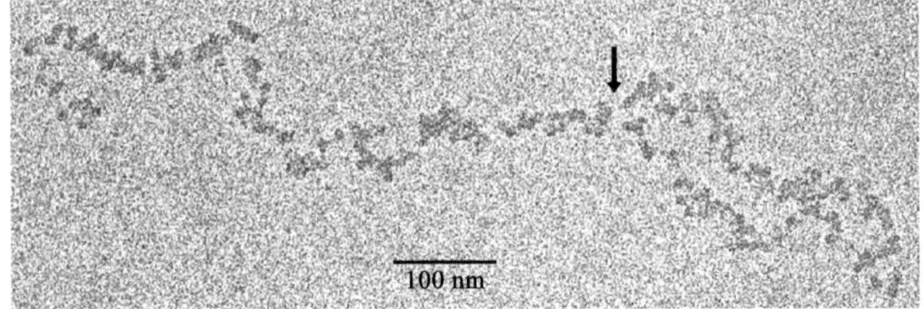

Figure 9. t-loops contain nucleosomes. Positively stained telomeric chromatin shows an open beads-on-a-string conformation typical of nucleosome arrays. The arrow indicates the t-loop junction. The illustration was kindly provided by Chris Woodcock (University of Massachusetts, Amherst).

view, see de Lange 2004). However, the abundance of these t-loop structures remains unknown. More recently, both chicken and mouse telomeres have been isolated in a native state and electron microscopy analysis shows the presence of a closed chromatin loop conformation (Fig. 9) (Nikitina and Woodcock 2004). Significantly, although the characteristic beads-on-a-string appearance of nucleosome arrays is observed in low salt, when the salt was raised, the nucleosome array compacted into a 30-nm chromatin fiber, and hence telomeric chromatin behaves and folds like bulk chromatin (Nikitina and Woodcock 2004). The observation of nucleosomes covering the telomeric DNA repeats is consistent with microccocal nuclease digestion studies of rat liver telomeric chromatin, which revealed regularly spaced nucleosomes in vertebrates and invertebrates. However, telomeric chromatin may have special features because it is more nuclease-sensitive than bulk chromatin, and the nucleosome repeat length is significantly shorter than that of the corresponding bulk chromatin (Makarov et al. 1993; Tommerup et al. 1994; Lejnine et al. 1995). On the other hand, functional studies on hTRF2 suggest that as many as thousands of TRF2 molecules could be present at each mammalian chromosome end, a sufficiently large number to coat the thousands of base pairs of double-stranded telomeric DNA (Karlseder et al. 2004). These two observations raise the question of interplay or competition between the specific proteins that bind telomeric DNA and nucleosomes.

In terms of nucleosome dynamics, in vitro analysis of nucleosome formation on telomeric DNA repeats shows that the free energies involved in the association of telomeric DNA with histones are the highest among a variety of biological DNA sequences. By comparison to mixed sequence DNA, the energy required to form nucleosomes on telomeric

DNA is 10–15 times higher, resulting in less stable and more mobile nucleosomes on telomeres than bulk chromatin (Rossetti et al. 1998). The source of this instability is the G-rich nature of telomeric DNA and a sequence periodicity of 6–8 bp, which is out of phase with the helical repeat of DNA, disfavoring the wrapping of DNA around the histone octamer (Rossetti et al. 1998; Widom 2001). This observation provides an explanation for the irregular micrococcal nuclease digestion pattern obtained from probing telomeres in vivo (Tommerup et al. 1994). Since telomeric double-stranded binding proteins bind with very high affinities (Buchman et al. 1988; Konig et al. 1998), it is likely they would compete with nucleosome formation in a concentration-dependent manner (Rossetti et al. 2001).

The first evidence for a physical interaction between telomeric components and chromatin came from studies in budding yeast (for review, see Grunstein 1997). These studies led to the notion of a telomere fold-back structure to explain silencing of a gene placed close to telomeres. In this model, for which there is experimental evidence, Rap1 molecules bound to the terminal telomeric DNA repeats recruit the Sir2/3/4p complex (Moretti et al. 1994). This complex then folds back, allowing Sir3 and Sir4 to interact with the H3 and H4 histone tails (Hecht et al. 1995) of internal nucleosomes, forming a protective heterochromatic complex. Significantly, Sir2 is a histone deacetylase (Imai et al. 2000), providing the first indication that histone modifications are involved in the modulation of chromatin structure at telomeres. Recent observations strengthen this view with the report that histone methylation is also involved in the regulation of telomere length in mammalian cells (Garcia-Cao et al. 2004).

SUMMARY

The information on the three-dimensional structures of telomeric proteins and their complexes has provided important insights into the architecture of these proteins, the conservation in their structures, and the precise function in single-stranded and double-stranded telomeric DNA recognition. The elusive role of G-quadruplex structures at telomeres has become a reality by the demonstration of the presence of such structures in vivo. Future structural information on various telomeric multiprotein complexes will be required to understand fully the function of various telomeric proteins in the assembly and regulation of telomere architecture. Besides structural information on specific complexes, overall structural information on t-loops and telomeric chromatin needs to be investigated in order to understand the likely interplay between specific telomeric proteins and chromatin.

ACKNOWLEDGMENTS

I am indebted to Louise Fairall for preparation of Figures 1–8. I thank Chris Woodcock for providing Figure 9 and Lynda Chapman for critical reading of the manuscript. The British Medical Research Council and a Human Frontiers Science Program grant supported the research on telomeres in my laboratory.

REFERENCES

Ancelin K., Brun C., and Gilson E. 1998. Role of the telomeric DNA-binding protein TRF2 in the stability of human chromosome ends. *Bioessays* **20:** 879–883.

Banner D.W., Kokkinidis M., and Tsernoglu D. 1987. Structure of ColE1 Rop protein at 1.7 Å resolution. *J. Mol. Biol.* **196:** 657–675.

Baumann P. and Cech T.R. 2001. Pot1, the putative telomere end-binding protein in fission yeast and humans. *Science* **292:** 1171–1175.

Bianchi A., Smith S., Chong L., Elias P., and de Lange T. 1997. TRF1 is a dimer and bends telomeric DNA. *EMBO J.* **16:** 1785–1794.

Bianchi A., Stansel R.M., Fairall L., Griffith J.D., Rhodes D., and de Lange T. 1999. TRF1 binds a bipartite telomeric site with extreme spatial flexibility. *EMBO J.* **18:** 5735–5744.

Bilaud T., Brun C., Ancelin K., Koering C.E., Laroche T., and Gilson E. 1997. Telomeric localization of TRF2, a novel human telobox protein. *Nat. Genet.* **17:** 236–239.

Bilaud T., Koering C.E., Binet-Brasselet E., Ancelin K., Pollice A., Gasser S.M., and Gilson E. 1996. The telobox, a Myb-related telomeric DNA binding motif found in proteins from yeast, plants and human. *Nucleic Acids Res.* **24:** 1294–1303.

Bochkarev A. and Bochkareva E. 2004. From RPA to BRCA2: Lessons from single-stranded DNA binding by the OB-fold. *Curr. Opin. Struct. Biol.* **14:** 36–42.

Bochkarev A., Pfuetzner R.A, Edwards A.M., and Frappier L. 1997. Structure of the single-stranded-DNA-binding domain of replication protein A bound to DNA. *Nature* **385:** 176–181.

Bochkareva E., Belegu V., Korolev S., and Bochkarev A. 2001. Structure of the major single-stranded DNA-binding domain of replication protein A suggests a dynamic mechanism for DNA binding. *EMBO J.* **20:** 612–618.

Broccoli D., Smogorzewska A., Chong L., and de Lange T. 1997. Human telomeres contain two distinct Myb-related proteins, TRF1 and TRF2. *Nat. Genet.* **17:** 231–235.

Buchman A.R., Kimmerly W.J., Rine J., and Kornberg R.D. 1988. Two DNA-binding factors recognize specific sequences at silencers, upstream activating sequences, autonomously replicating sequences, and telomeres in *Saccharomyces cerevisiae*. *Mol. Cell. Biol.* **8:** 210–225.

Cheung I., Schertzer M., Rose A., and Lansdorp P.M. 2002. Disruption of dog-1 in *Caenorhabditis elegans* triggers deletions upstream of guanine-rich DNA. *Nat. Genet.* **31:** 405–409.

Chong L., van Steensel B., Broccoli D., Erdjument-Bromage H., Hanish J., Tempst P., and de Lange T. 1995. A human telomeric protein. *Science* **270:** 1663–1667.

Colgin L.M., Baran K., Baumann P., Cech T.R., and Reddel R.R. 2003. Human POT1 facilitates telomere elongation by telomerase. *Curr. Biol.* **13:** 942–946.

Conrad M.N., Wright J.H., Wolf A.J., and Zakian V.A. 1990. RAP1 protein interacts with yeast telomeres in vivo: Overproduction alters telomere structure and decreases chromosome stability. *Cell* **63:** 739–750.

Cooper J.P., Nimmo E.R., Allshire R.C., and Cech T.R. 1997. Regulation of telomere length and function by a Myb-domain protein in fission yeast. *Nature* **385:** 744–747.

Court R., Chapman L., Fairall L., and Rhodes D. 2005. How the human telomeric proteins TRF1 and TRF2 recognise telomeric DNA: A view from high-resolution crystal structures. *EMBO Rep.* **6:** 191.

Darnell J.C., Jensen K.B., Jin P., Brown V., Warren S.T., and Darnell R.B. 2001. Fragile X mental retardation protein targets G quartet mRNAs important for neuronal function. *Cell* **107:** 489–499.

de Lange T. 2002. Protection of mammalian telomeres. *Oncogene* **21:** 532–540.

———. 2004. T-loops and the origin of telomeres. *Nat. Rev. Mol. Cell. Biol.* **5:** 323–329.

Ding H., Schertzer M., Wu X., Gertsenstein M., Selig S., Kammori M., Pourvali R., Poon S., Vulto I., Chavez E., Tam P.P., Nagy A., and Lansdorp P.M. 2004. Regulation of murine telomere length by Rtel: An essential gene encoding a helicase-like protein. *Cell* **117:** 873–886.

Fairall L., Chapman L., Moss H., de Lange T., and Rhodes D. 2001. Structure of the TRFH dimerization domain of the human telomeric proteins TRF1 and TRF2. *Mol. Cell* **8:** 351–361.

Fang G. and Cech T.R. 1993a. The beta subunit of *Oxytricha* telomere-binding protein promotes G-quartet formation by telomeric DNA. *Cell* **74:** 875–885.

———. 1993b. Characterization of a G-quartet formation reaction promoted by the beta-subunit of the *Oxytricha* telomere-binding protein. *Biochemistry* **32:** 11646–11657.

———. 1993c. *Oxytricha* telomere-binding protein: DNA-dependent dimerization of the alpha and beta subunits. *Proc. Natl. Acad. Sci.* **90:** 6056–6060.

Froelich-Ammon S.J., Dickinson B.A., Bevilacqua J.M., Schultz S.C., and Cech T.R. 1998. Modulation of telomerase activity by telomere DNA-binding proteins in *Oxytricha*. *Genes Dev.* **12:** 1504–1514.

Garcia-Cao M., O'Sullivan R., Peters A.H., Jenuwein T., and Blasco M.A. 2004. Epigenetic regulation of telomere length in mammalian cells by the Suv39h1 and Suv39h2 histone methyltransferases. *Nat. Genet.* **36:** 94–99.

Gilson E., Roberge M., Giraldo R., Rhodes D., and Gasser S.M. 1993. Distortion of the DNA double helix by RAP1 at silencers and multiple telomeric binding sites. *J. Mol. Biol.* **231:** 293–310.

Giraldo R., Suzuki M., Chapman L., and Rhodes D. 1994. Promotion of parallel DNA quadruplexes by a yeast telomere binding protein: A circular dichroism study. *Proc. Natl. Acad. Sci.* **91:** 7658–7662.

Gottschling D.E. and Zakian V.A. 1986. Telomere proteins: Specific recognition and protection of the natural termini of *Oxytricha* macronuclear DNA. *Cell* **47:** 195–205.

Griffith J., Bianchi A., and de Lange T. 1998. TRF1 promotes parallel pairing of telomeric tracts in vitro. *J. Mol. Biol.* **278:** 79–88.

Griffith J.D., Comeau L., Rosenfield S., Stansel R.M., Bianchi A., Moss H., and de Lange T. 1999. Mammalian telomeres end in a large duplex loop [see comments]. *Cell* **97:** 503–514.

Grunstein M. 1997. Molecular model for telomeric heterochromatin in yeast. *Curr. Opin. Cell Biol.* **9:** 383–387.

Hecht A., Laroche T., Strahl-Bolsinger S., Gasser S.M., and Grunstein M. 1995. Histone H3 and H4 N-termini interact with SIR3 and SIR4 proteins: A molecular model for the formation of heterochromatin in yeast. *Cell* **80:** 583–592.

Henderson E. 1995. Telomere DNA structure. In *Telomeres* (ed. E.H. Blackburn and C.W. Greider), pp. 11–34. Cold Spring Harbor Laboratory Press, Cold Spring Harbor, New York.

Horvath M.P. and Schultz S.C. 2001. DNA G-quartets in a 1.86 Å resolution structure of an *Oxytricha nova* telomeric protein-DNA complex. *J. Mol. Biol.* **310:** 367–377.

Horvath M.P., Schweiker V.L., Bevilacqua J.M., Ruggles J.A., and Schultz S.C. 1998. Crystal structure of the *Oxytricha nova* telomere end binding protein complexed with single strand DNA. *Cell* **95:** 963–974.

Hughes T.R., Weilbaecher R.G., Walterscheid M., and Lundblad V. 2000. Identification of the single-strand telomeric DNA binding domain of the *Saccharomyces cerevisiae* Cdc13 protein. *Proc. Natl. Acad. Sci.* **97:** 6457–6462.

Imai S., Johnson F.B., Marciniak R.A., McVey M., Park P.U., and Guarente L. 2000. Sir2: An NAD-dependent histone deacetylase that connects chromatin silencing, metabolism, and aging. *Cold Spring Harbor Symp. Quant. Biol.* **65:** 297–302.

Kanei-Ishii C., Sarai A., Sawazaki T., Nakagoshi H., He D.N., Ogata K., Nishimura Y., and Ishii S. 1990. The tryptophan cluster: A hypothetical structure of the DNA-binding domain of the *myb* protooncogene product. *J. Biol. Chem.* **265:** 19990–19995.

Kanoh J. and Ishikawa F. 2001. spRap1 and spRif1, recruited to telomeres by Taz1, are essential for telomere function in fission yeast. *Curr. Biol.* **11:** 1624–1630.

Karlseder J., Hoke K., Mirzoeva O.K., Bakkenist C., Kastan M.B., Petrini J.H., and de Lange T. 2004. The telomeric protein TRF2 binds the ATM kinase and can inhibit the ATM-dependent DNA damage response. *PLoS Biol.* **2:** E240.

Kelleher C., Kurth I., and Lingner J. 2005. Human protection of telomeres 1 (POT1) is a negative regulator of telomerase activity in vitro. *Mol. Cell. Biol.* **25:** 808–818.

Kim M.Y., Gleason-Guzman M., Izbicka E., Nishioka D., and Hurley L.H. 2003. The different biological effects of telomestatin and TMPyP4 can be attributed to their selectivity for interaction with intramolecular or intermolecular G-quadruplex structures. *Cancer Res.* **63:** 3247–3256.

Kissinger C.R., Liu B.S., Martin-Blanco E., Kornberg T.B., and Pabo C.O. 1990. Crystal structure of an engrailed homeodomain-DNA complex at 2.8 Å resolution: A framework for understanding homeodomain-DNA interactions. *Cell* **63:** 579–590.

Konig P. and Rhodes D. 1997. Recognition of telomeric DNA. *Trends Biochem. Sci.* **22:** 43–47.

Konig P., Fairall L., and Rhodes D. 1998. Sequence-specific DNA recognition by the myb-like domain of the human telomere binding protein TRF1: A model for the protein-DNA complex. *Nucleic Acids Res.* **26:** 1731–1740.

Konig P., Giraldo R., Chapman L., and Rhodes D. 1996. The crystal structure of the DNA-binding domain of yeast RAP1 in complex with telomeric DNA. *Cell* **85:** 125–136.

Krauskopf A. and Blackburn E.H. 1996. Control of telomere growth by interactions of RAP1 with the most distal telomeric repeats. *Nature* **383:** 354–357.

Lei M., Baumann P., and Cech T.R. 2002. Cooperative binding of single-stranded telomeric DNA by the Pot1 protein of *Schizosaccharomyces pombe*. *Biochemistry* **41:** 14560–14568.

Lei M., Podell E.R., and Cech T.R. 2004. Structure of human POT1 bound to telomeric single-stranded DNA provides a model for chromosome end-protection. *Nat. Struct. Mol. Biol.* **11:** 1223–1229.

Lei M., Podell E.R., Baumann P., and Cech T.R. 2003. DNA self-recognition in the structure of Pot1 bound to telomeric single-stranded DNA. *Nature* **426:** 198–203.

Lejnine S., Makarov V.L., and Langmore J.P. 1995. Conserved nucleoprotein structure at the ends of vertebrate and invertebrate chromosomes. *Proc. Natl. Acad. Sci.* **92:** 2393–2397.

Li B., Oestreich S., and de Lange T. 2000. Identification of human Rap1: Implications for telomere evolution. *Cell* **101:** 471–483.

Lin J.J. and Zakian V.A. 1996. The *Saccharomyces CDC13* protein is a single-strand TG1–3 telomeric DNA-binding protein *in vitro* that affects telomere behavior *in vivo*. *Proc. Natl. Acad. Sci.* **93:** 13760–13765.

Lipps H.J., Gruissem W., and Prescott D.M. 1982. Higher order DNA structure in macronuclear chromatin of the hypotrichous ciliate *Oxytricha nova*. *Proc. Natl. Acad. Sci.* **79:** 2495–2499.

Loayza D. and de Lange T. 2003. POT1 as a terminal transducer of TRF1 telomere length control. *Nature* **424:** 1013–1018.

Loayza D., Parsons H., Donigian J., Hoke K., and de Lange T. 2004. DNA binding features of human POT1: A nonamer 5′-TAGGGTTAG-3′ minimal binding site, sequence specificity, and internal binding to multimeric sites. *J. Biol. Chem.* **279:** 13241–13248.

Longtine M.S., Wilson N.M., Petracek M.E., and Berman J. 1989. A yeast telomere binding activity binds to two related telomere sequence motifs and is indistinguishable from RAP1. *Curr. Genet.* **16:** 225–239.

Lustig A.J., Kurtz S., and Shore D. 1990. Involvement of the silencer and UAS binding protein RAP1 in regulation of telomere length. *Science* **250:** 549–553.

Makarov V.L., Lejnine S., Bedoyan J., and Langmore J.P. 1993. Nucleosomal organization of telomere-specific chromatin in rat. *Cell* **73:** 775–787.

Marcand S., Wotton D., Gilson E., and Shore D. 1997. Rap1p and telomere length regulation in yeast. *Ciba Found. Symp.* **211:** 76–93.

McElligott R. and Wellinger R.J. 1997. The terminal DNA structure of mammalian chromosomes. *EMBO J.* **16:** 3705–3714.

Mitton-Fry R.M., Anderson E.M., Hughes T.R., Lundblad V., and Wuttke D.S. 2002. Conserved structure for single-stranded telomeric DNA recognition. *Science* **296:** 145–147.

Mitton-Fry R.M., Anderson E.M., Theobald D.L., Glustrom L.W., and Wuttke D.S. 2004. Structural basis for telomeric single-stranded DNA recognition by yeast Cdc13. *J. Mol. Biol.* **338:** 241–255.

Moretti P., Freeman K., Coodly L., and Shore D. 1994. Evidence that a complex of SIR proteins interacts with the silencer and telomere-binding protein RAP1. *Genes Dev.* **8:** 2257–2269.

Murzin A.G. 1993. OB(oligonucleotide/oligosaccharide binding)-fold: Common structural and functional solution for non-homologous sequences. *EMBO J.* **12:** 861–867.

Nikitina T. and Woodcock C.L. 2004. Closed chromatin loops at the ends of chromosomes. *J. Cell Biol.* **166:** 161–165.

Nishikawa T., Nagadoi A., Yoshimura S., Aimoto S., and Nishimura Y. 1998. Solution structure of the DNA-binding domain of human telomeric protein, hTRF1. *Structure* **6:** 1057–1065.

Nishikawa T., Okamura H., Nagadoi A., Koig P., Rhodes D., and Nishimura Y. 2001. Structure of the DNA-binding domain of human telomeric protein, TRF1 and its interaction with telomeric DNA. *Nucleic Acids Res. Suppl.* **2001:** 273–274.

Nugent C.I., Hughes T.R., Lue N.F., and Lundblad V. 1996. Cdc13p: A single-strand telomeric DNA-binding protein with a dual role in yeast telomere maintenance. *Science* **274:** 249–252.

Ogata K., Morikawa S., Nakamura H., Sekikawa A., Inoue T., Kanai H., Sarai A., Ishii S., and Nishimura Y. 1994. Solution structure of a specific DNA complex of the Myb DNA-binding domain with cooperative recognition helices. *Cell* **79:** 639–648.

Pabo C.O. and Sauer R.T. 1992. Transcription factors: Structural families and principles of DNA recognition. *Annu. Rev. Biochem.* **61:** 1053–1095.

Paeschke K., Simonsson T., Postberg J., Rhodes D., and Lipps H.-J. 2005. Telomere end-binding proteins control the formation of G-quadruplex DNA structure in vivo. *Nat. Struct. Mol. Biol.* (in press).

Parkinson G.N., Lee M.P., and Neidle S. 2002. Crystal structure of parallel quadruplexes from human telomeric DNA. *Nature* **417:** 876–808.

Peersen O.B., Ruggles J.A., and Schultz S.C. 2002. Dimeric structure of the *Oxytricha nova* telomere end-binding protein α-subunit bound to ssDNA. *Nat. Struct. Biol.* **9:** 182–187.

Pennock E., Buckley K., and Lundblad V. 2001. Cdc13 delivers separate complexes to the telomere for end protection and replication. *Cell* **104:** 387–396.

Price C.M. and Cech T.R. 1987. Telomeric DNA-protein interactions of *Oxytricha* macronuclear DNA. *Genes Dev.* **1:** 783–793.

Rhodes D. and Giraldo R. 1995. Telomere structure and function. *Curr. Opin. Struct. Biol.* **5:** 311–322.

Riou J.F., Guittat L., Mailliet P., Laoui A., Renou E., Petitgenet O., Megnin-Chanet F., Helene C., and Mergny J.L. 2002. Cell senescence and telomere shortening induced by a new series of specific G-quadruplex DNA ligands. *Proc. Natl. Acad. Sci.* **99:** 2672–2677.

Rossetti L., Cacchione S., Fua M., and Savino M. 1998. Nucleosome assembly on telomeric sequences. *Biochemistry* **37:** 6727–6737.

Rossetti L., Cacchione S., De Menna A., Chapman L., Rhodes D., and Savino M. 2001. Specific interactions of the telomeric protein Rap1p with nucleosomal binding sites. *J. Mol. Biol.* **306:** 903–913.

Schaffitzel C., Berger I., Postberg J., Hanes J., Lipps H.J., and Plückthun A. 2001. In vitro generated antibodies specific for telomeric guanine-quadruplex DNA react with *Stylonychia lemnae* macronuclei. *Proc. Natl. Acad. Sci.* **98:** 8572–8577.

Shore D. 1994. RAP1: A protean regulator in yeast. *Trends Genet.* **10:** 408–412.

Simonsson T. 2003. A substrate for telomerase. *Trends Biochem. Sci.* **28:** 632–638.

Smogorzewska A. and de Lange T. 2004. Regulation of telomerase by telomeric proteins. *Annu. Rev. Biochem.* **73:** 177–208.

Smogorzewska A., van Steensel B., Bianchi A., Oelmann S., Schaefer M.R., Schnapp G., and de Lange T. 2000. Control of human telomere length by TRF1 and TRF2. *Mol. Cell. Biol.* **20:** 1659–1668.

Stansel R.M., de Lange T., and Griffith J.D. 2001. T-loop assembly in vitro involves binding of TRF2 near the 3′ telomeric overhang. *EMBO J.* **20:** 5532–5540.

Taylor H.O., O'Reilly M., Leslie A.G., and Rhodes D. 2000. How the multifunctional yeast Rap1p discriminates between DNA target sites: A crystallographic analysis. *J. Mol. Biol.* **303:** 693–707.

Theobald D.L. and Wuttke D.S. 2004. Prediction of multiple tandem OB-fold domains in telomere end-binding proteins Pot1 and Cdc13. *Structure* **12:** 1877–1879.

Tommerup H., Dousmanis A., and de Lange T. 1994. Unusual chromatin in human telomeres. *Mol. Cell. Biol.* **14:** 5777–5785.

van Steensel B. and de Lange T. 1997. Control of telomere length by the human telomeric protein TRF1. *Nature* **385:** 740–743.

Wang Y. and Patel D.J. 1993. Solution structure of the human telomeric repeat d[AG3(T2AG3)3] G-tetraplex. *Structure* **1:** 263–282.

Wei C. and Price C.M. 2004. Cell cycle localization, dimerization, and binding domain architecture of the telomere protein cPot1. *Mol. Cell. Biol.* **24:** 2091–2102.

Widom J. 2001. Role of DNA sequence in nucleosome stability and dynamics. *Q. Rev. Biophys.* **34:** 269–324.

Wright W.E., Tesmer V.M., Huffman K.E., Levene S.D., and Shay J.W. 1997. Normal human chromosomes have long G-rich telomeric overhangs at one end. *Genes Dev.* **11:** 2801–2809.

Zahler A.M., Williamson J.R., Cech T.R., and Prescott D.M. 1991. Inhibition of telomerase by G-quartet DNA structures. *Nature* **350:** 718–720.

12

Budding Yeast Telomeres

Vicki Lundblad

The Molecular and Cell Biology Laboratory
The Salk Institute
La Jolla, California 92037

YEAST TELOMERE BIOLOGY EMERGED IN THE EARLY 1980s with a seminal study by Szostak and Blackburn (1982). Combining their respective interests in yeast artificial chromosomes and ciliate telomeres, they constructed a trans-kingdom chromosome: a linear yeast episome with ciliate telomeric DNA on its ends. When introduced into budding yeast, these evolutionarily distant ciliate termini were elongated by the direct addition of yeast telomeric DNA. This lent impetus to the hypothesis that cells contained a telomere-specific enzymatic activity, analogous to terminal transferase, that was capable of extending the ends of chromosomes (see the Appendix). Of more immediate importance to the nascent yeast telomere biology field, however, was the molecular cloning of yeast telomeres, which provided a crucial molecular reagent that could be used to monitor the behavior of yeast chromosome termini (Shampay et al. 1984; Walmsley et al. 1984). This prompted a flurry of studies that probed the dynamic nature of chromosome ends in yeast (Dunn et al. 1984; Horowitz et al. 1984; Pluta et al. 1984) and also initiated the process of identifying the genes required for telomere length regulation (Carson and Hartwell 1985; Lustig and Petes 1986). Over the past 20 years, studies from numerous laboratories have greatly expanded our knowledge of the details of yeast telomere structure, as well as the gene products that maintain yeast chromosome ends. This chapter discusses some of the main insights that have resulted from these studies and highlights several major unanswered questions.

YEAST TELOMERE STRUCTURE

Telomeric DNA: An Irregular Repeat Sequence with a Conserved Core

The irregular telomeric repeat array that characterizes *Saccharomyces cerevisiae* telomeres is often denoted as $G_{1-3}T$, although a more accurate consensus is $G_{2-3}(TG)_{1-6}$ (Shampay et al. 1984; Wang and Zakian 1990). This divergence from the more common theme of homogeneous telomeric repeat sequences, found at ciliate or metazoan chromosome ends, is due to degenerate copying of the template region of the yeast telomerase RNA (called TLC1; Singer and Gottschling 1994). The sequence pattern observed at budding yeast telomeres appears to be due to alignment of telomeres in multiple registers along the 17-nucleotide-long TLC1 template (3'-ACACACACCCACACCAC-5'), possibly combined with abortive synthesis at several points along the template (Prescott and Blackburn 1997a,b; Forstemann et al. 2000; Forstemann and Lingner 2001, 2005). Other processes, such as nucleolytic degradation and/or incomplete semi-conservative DNA replication of the duplex tract of the telomere, may also contribute to the degree of sequence diversity.

This sequence degeneracy is not a quirk that is unique to *S. cerevisiae.* Analysis of telomeres cloned from multiple yeast species from within the *Saccharomyces* genus, as well as from more distantly related yeasts such as *Kluyveromyces lactis* and *Candida albicans*, has shown that telomeric repeat variation is a characteristic shared by many fungal species (McEachern and Hicks 1993; McEachern and Blackburn 1994; Cohn et al. 1998), including even the evolutionarily distant fission yeast (Sugawara 1988). Telomeres can also be composed of a much longer repeat sequence in a subset of budding yeasts, with repeat lengths up to 26 bp in length (Cohn et al. 1998). Examination of the template region of *TLC1* genes from a dozen yeast species reveals the source of both the variation in the repeat length and the sequence diversity (Fig. 1). However, even these degenerate telomere sequences retain a reiterated conserved core sequence of ~6 nucleotides, which can also be detected in the alignment of TLC1 templates shown in Figure 1. Blackburn and colleagues have argued that this core preserves a binding site for the essential duplex telomere-binding protein Rap1 (Cohn et al. 1998), and indeed the Rap1 consensus corresponds closely to the conserved telomeric repeat core. There may also be a selection for the binding site of other telomere-binding proteins: Localization of the Cdc13 protein to yeast telomeres, for example, also relies on sequence-specific recognition of single-stranded telomeric DNA (Mitton-Fry et al. 2004).

S. cerevisiae	3′-cauuguaa**ACACACACCCACACCAC**uaccaucc-5′
S. mikitae	uauuauaguu**ACACACCCACACCA**uuaccaucu
S. bayanus	aauuacaag**CACACACCCACACCA**uuaccaucu
S. castellii	uuuauaug**ACACAGACCCACAC**uaguccauuuc
S. kluyveri	auaugguu**CCCACCUGUACGCAUGACACUCCAGACCCACCUGUA**uaugguuu
K. lactis	ccaaaacu**AAACUAAUCCAUACACCACAUGCCUAAACU**caugguggu
K. nonfermentans	uuccuaac**UAACUAAUCCAUACACCACAUGCCUAAACU**auguggac
K. dobzhanskii	uuucgcca**AAACUAAUCCAUACACCACAUGCCUAAACU**cacgguac
K. marxianus	cuucgcca**AAACUAAUCAAUACACCACAUGCCUAAACU**uucaggug
K. wickerhamii	uuccgcca**AAACUAAUCCAUACACCACAUGCCUAAAUA**ucaugguu
K. aestuarii	uuccuaac**UAACUAAUCCAUACACCACAUGCCUAAACU**auguggac

Figure 1. Alignment of the template from various *Saccharomcyes* and *Kluyveromyces* telomerase RNAs; template sequences are indicated in uppercase, bracketed by non-templating sequences in lowercase. Despite the high degree of sequence divergence, a conserved core (highlighted in *gray*) is present in all 11 template sequences; this partially overlaps with the Rap1 consensus binding sequence (Konig and Rhodes 1997).

The degenerate nature of *S. cerevisiae* telomeric repeats has turned out to be a feature that can be exploited experimentally, which has revealed that the extreme termini of yeast chromosomes are highly dynamic (Wang and Zakian 1990; Forstemann et al. 2000). Detailed sequence analysis of clonally derived telomeres has shown that telomeres have a centromere-proximal region that displays little sequence variation, whereas the more terminal 40–100 nucleotides exhibit much more divergence. The variation in the most distal telomeric repeats, which is abolished in a telomerase-defective background, suggests a highly dynamic turnover at the extreme terminus that is a consequence of a balance between telomere-shortening processes and telomerase-mediated extension. As pointed out by Lingner and colleagues, the length of the distal variable domain is inconsistent, however, with an exact balance between these two activities, whereby telomerase acts on every telomere in every cell cycle (Forstemann et al. 2000). Instead, telomerase acts preferentially only on a subset of telomeres in a given cell cycle (Teixeira et al. 2004), a point that is discussed more extensively later in this chapter.

This turnover at the distal region of chromosomes reflects a second feature of yeast telomeres, which is that the $G_{1-3}T$ repeat tract is heterogeneous in length among a population of cells, even for a single chromosomal telomere (Horowitz et al. 1984; Walmsley et al. 1984; Lustig and Petes 1986; Shampay and Blackburn 1988). Most wild-type *S. cerevisiae* strains possess ~350–500 bp of $G_{1-3}T$ repeats at their termini, although there can be substantial strain-to-strain variation (a reflection, at least in part, of the consequence that there is no single "wild type" parental *S. cerevisiae* strain). This telomere-length heterogeneity may not be dictated by a single genome-wide mechanism that coordinately regulates the shortening and lengthening activities at every terminus: Subtle differences in telomere-length regulation can be observed among different classes of telomeres, at least in response to perturbations of telomere gene function (Craven and Petes 1999; A. Salinger and V. Lundblad, unpubl.). Clonal analysis of the length of multiple individual telomeres also argues for an imperfectly regulated balance between shortening and lengthening activities that varies at individual termini (Shampay and Blackburn 1988).

The Subtelomeric Repeat Region

The ~25-kb subtelomeric region of yeast chromosomes, located immediately centromere-proximal to the G-rich tract, is composed of a mosaic of middle repetitive repeat sequences. These repeated elements can exhibit substantial variation in distribution among different chromosome ends within a single strain, as well as differences among the different parental yeast strains used by different laboratories. Historically, these repeat elements were originally grouped into two discrete categories, dubbed X and Y elements (Chan and Tye 1983a,b). However, subsequent analysis has revealed far more complexity, particularly among the X element category. All chromosome ends have a so-called "core" X element, which may be accompanied by a heterogeneous array of smaller subtelomeric repeat (STR) A, B, C, and D elements (Louis 1995). Y′ subtelomeric elements, in contrast, are typically located at ~1/2 to 2/3 of chromosome ends, embedded within the $G_{1-3}T$ telomeric repeat tract. They are often present in tandem arrays of up to four copies that are separated by short stretches of $G_{1-3}T$ telomeric repeats. Although X and Y′ elements have received more attention (because of historical precedence), recent analysis has uncovered additional repeat sequence families present in the subtelomeric regions of yeast chromosomes (Louis and Vershinin 2005).

The contribution that subtelomeric repeats make to telomere function in wild-type yeast strains has not yet been fully defined, which perhaps is a reflection of the polymorphic nature of these repeat elements. The 473-bp core X element possesses binding sites for the yeast origin recognition complex (ORC) as well as the Abf1 transcriptional factor, suggesting a functional role for this element in chromosome maintenance. However, artificial chromosomes that completely lack X or Y′ elements nevertheless are transmitted normally during both mitosis and meiosis (Murray and Szostak 1983), and both X and Y′ elements are so poorly conserved that they are not detected in the genomes of even closely related yeasts. The selection pressure for such elements at yeast chromosome termini, therefore, may be the consequence of a function(s) that is not specific to telomere biology per se; for example, these regions have been proposed to protect the genome from deleterious ectopic recombination or to provide a region that is permissive for gene amplification as a rapid response to environmental perturbations (Pryde and Louis 1999).

There are two well-characterized situations in which STRs can have a pronounced biological effect, however. First, the presence and specific arrangement of STR elements can modulate the expression of telomere-localized genes. Genes adjacent to telomeres are subject to telomere position effect (TPE), whereby gene expression becomes metastable (see Chapter 10). The degree to which native telomeres exhibit TPE can be highly variable, presumably a reflection of the wide variation in subtelomeric structure (Pryde and Louis 1997).

Global subtelomeric rearrangements can also contribute to cell viability by alleviating the lethal effects of a telomerase deficiency. Movement of Y′ elements among different termini normally occurs at extremely low frequencies in wild-type yeast strains (Horowitz and Haber 1985). However, both the copy number and distribution of Y′ elements can change dramatically in response to a telomere replication defect (Lundblad and Blackburn 1993). As described in Chapter 8, recombination can replace telomerase as an alternative means of replenishing the G-rich repeats found at chromosome termini. One version of this recombination-dependent telomere maintenance pathway is characterized by extensive amplification and expansion of tandem arrays of Y′ elements, with an accompanying expansion of the short $G_{1-3}T$ tracts interspersed between these elements. Restoration of telomere function could be the consequence of the reappearance of G-rich telomeric repeat DNA, rather than the amplification of Y′ elements, at these under-replicated termini. In addition, the now tandemly repeated Y′ sequences could form heterochromatin that stabilizes the telomeres, as proposed for *Drosophila* telomeres (Chapter 14).

The Single-Strand G-rich 3′ Extension

The presence of a single-strand extension of the G-strand of the telomere—often called a G-tail—appears to be a ubiquitous feature of chromosome ends in all species that rely on telomerase-maintained termini. Disruption of this structure, due to either loss of the G-tail itself or exposure of the C-strand to nucleolytic attack, is lethal. Thus, careful maintenance of this structure is a crucial task for cells. In budding yeast, this overhang is bound and protected by the essential single-strand DNA-binding protein Cdc13, but numerous other activities also contribute to the preservation of this terminal structure.

The extent of the G-rich overhang is highly regulated through the budding yeast cell cycle, in sharp contrast to the situation in human cells. In late S phase, after semi-conservative replication is complete, yeast chromosome termini exhibit G-tails that are estimated to be more than 30 nucleotides (Wellinger et al. 1993), whereas the G-rich overhang is only 12–14 nucleotides throughout the rest of the cell cycle (Larrivee et al. 2004). This regulation of overhang length implies that there is a resection activity(s) that generates the S-phase-specific structure, as well as a pathway to convert the more extended overhang present at the end of S phase to the 12–14-nucleotide version that prevails through the rest of the cell cycle. The mechanism by which the overhang is generated is still unclear, but protein complexes that may contribute to this process are discussed in several places in this chapter.

In humans, this G-rich overhang is crucial to the formation of t-loops, which are thought to be generated as the result of TRF2-mediated invasion of the 3′ single-stranded overhang into the duplex telomeric tract (Griffith et al. 1999). The presence of t-loops at chromosome ends provides an elegant architectural solution for telomere capping, as sequestration of the 3′ OH terminus in a t-loop can mask the G-tail from a variety of detrimental activities (see Chapter 13). These t-loop structures have been observed at telomeres in a wide range of species. Whether t-loops similarly play a functional role at budding yeast telomeres has been a matter of some debate, however, in part because the sequence heterogeneity of *S. cerevisiae* telomeres suggests insufficient base-pairing between an invading single-stranded terminus and the duplex tract. This, combined with the absence of an obvious TRF2 ortholog at *S. cerevisiae* telomeres, has been used as an argument against a role for t-loops in budding yeast. However, even though fission yeast telomeres also exhibit some degree of sequence heterogeneity, the fission yeast ortholog of TRF2, Taz1, is able to promote the formation of t-loop-like structures in vitro with model telomere substrates, suggestive of a role for t-loops in this species (Tomaska et al. 2004). Thus, the reiterated short core consensus sequence motifs present at both fission and budding yeast telomeres may be sufficient (potentially with a stabilizing factor) for t-loop formation.

IDENTIFICATION OF GENES THAT MEDIATE YEAST TELOMERE FUNCTION: HOW CLOSE ARE WE TO ACQUIRING THE COMPLETE GENE SET?

A comprehensive picture of telomere function in any organism will depend on both the identification of all factors that contribute to telomere function, and a unified picture of how these factors perform their respective biochemical activities. Studies in *S. cerevisiae* have made many contributions toward this goal over the past two decades, beginning with pioneering studies in the mid-1980s that identified the first genes that promote telomere function (Carson and Hartwell 1985; Lustig and Petes 1986; Shore and Nasmyth 1987). Since these initial studies, diverse approaches from numerous laboratories have uncovered dozens of genes that contribute to various aspects of telomere biology in yeast. These gene identification efforts have been followed by more careful mechanistic analysis of specific activities such as the telomerase RNP, the protein complex that binds to duplex telomeric DNA, the single-strand telomere-binding protein Cdc13, the Ku heterodimer, the Tel1 and Mec1 kinases, as well as other factors too numerous to list here. Later sections in this chapter will highlight major conclusions resulting from detailed investigations of a subset of these protein complexes.

These efforts, which have driven numerous models about how various aspects of telomere function are performed, may have only scratched the surface, however. In a major advance, McEachern and colleagues have examined the 4800 viable yeast deletion strains for alterations in telomere length (Askree et al. 2004). Their analysis uncovered more than 150 nonessential genes that had not been previously appreciated for their effects on telomere length, indicating that as much as 3% of the yeast genome contributes, either directly or indirectly, to telomere length regulation. Not surprisingly, many of these new genes affect cell processes, such as DNA replication, chromatin remodeling, and nucleotide metabolism, that would be expected to influence telomere length homeostasis. Less obvious is the recovery of a large number of genes that are required for certain other activities, such as vacuolar sorting, mitochondrial function, or ribosome structure. However, these additional classes of genes may point the way to novel new regulatory mechanisms. For example, as Askree et al. (2004) point out in their discussion, the large number of genes in their collection that are required for yeast vacuole (the analog of the mammalian lysosome) function suggests that one or more telomere proteins may be regulated by vacuole-dependent degradation.

The above genome-wide study, of necessity, only examined the viable yeast deletion set. There are, however, several notable examples of essential genes that contribute to telomere length regulation (Carson and Hartwell 1985; Lustig et al. 1990; Nugent et al. 1996; Runge and Zakian 1996; Grandin et al. 1997, 2001; Grossi et al. 2004). Thus, there are undoubtedly further genes, among the ~1200 essential yeast genes, that are yet to be discovered for their effects on telomere length regulation. Tools recently developed to monitor alterations in essential gene activity, using titratable promoter alleles, may facilitate the identification of these hypothesized additional genes (Mnaimneh et al. 2004). The Askree et al. (2004) study also only monitored disruptions in telomere length homeo-stasis. Similar genome-wide strategies that probe alterations in the single-strand overhang, or other telomere-specific properties, will undoubtedly uncover additional new factors that make pivotal contributions to telo-mere biology.

This wealth of data provides a challenge for the field: How can this new information be rapidly incorporated into the ongoing analyses of the approximately two dozen genes that encode subunits of well-studied complexes? Certainly one answer will come from integration of the data from many other genome-wide studies, such as the analysis shown in Figure 2, which has the potential to uncover unexpected physical or genetic interactions. In parallel, many laboratories are using SGA technology to identify those viable deletion strains that display synthetic interactions with a particular mutation of interest. One limitation of these high-throughput genetic analyses, however, is the pleiotrophic nature of strains bearing deletions of those telomere-associated proteins that possess mul-tiple, discrete biochemical activities. As illustrated several times in this

Figure 2. Interactions among the 172 genes identified by Askree et al. (2004) which, when deleted, confer a short or long telomere length phenotype. Strains with long telomeres are indicated by orange gene names, and strains with short telomeres by white gene names. Genes are represented by circles and are colored according to their Gene Ontology classification (*upper left*). The types of interactions among the genes are represented by colored lines, according to the appropriate experimental system, as indicated in the legend (*lower left*). (Modifed, with permission, from Edmonds et al. 2004 [©National Academy of Sciences U.S.A.]; updated information from Breitkreutz et al., unpubl.)

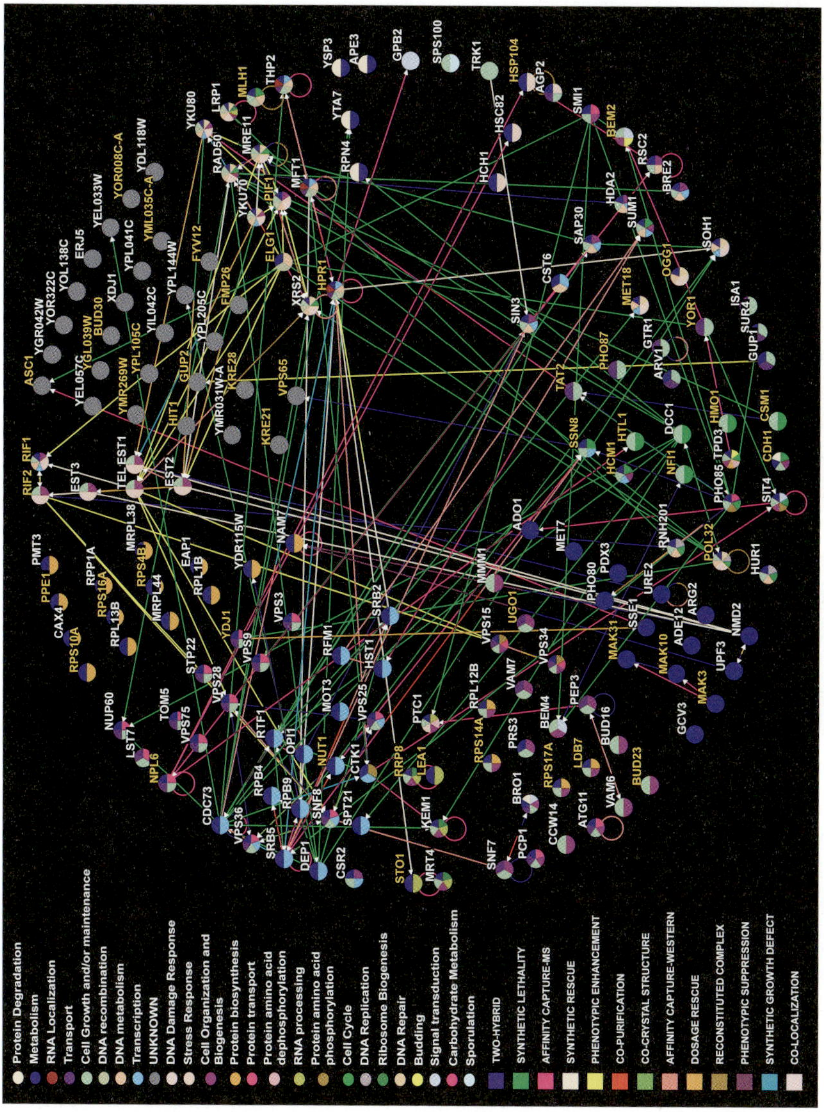

Figure 2. (*See facing page for legend.*)

chapter, often a missense mutation that surgically eliminates a single activity has been key in elucidating a particular pathway.

PROTEIN COMPLEXES THAT INTERACT WITH THE SINGLE-STRANDED G-RICH OVERHANG

As described in the opening section of this chapter, yeast chromosomes terminate with a single-strand 3′ extension of the G-rich strand. The length of the overhang is well under 100 nucleotides, and yet this extremely small region of the genome is the target of an impressive number of complexes. This abundance of attending factors is perhaps not so surprising, given the consequences for the cell if this unique region of the chromosome becomes unprotected. Loss of any of a number of factors that contribute to telomere capping can expose these termini to unregulated resection or end-to-end fusions, which are genotoxic for the cell (Garvik et al. 1995; Booth et al. 2001; Hackett et al. 2001; Craven et al. 2002; Chan and Blackburn 2003; Hackett and Greider 2003; Mieczkowski et al. 2003). The 3′ terminus of the G-rich strand is also the substrate for telomerase; therefore, both telomerase and at least a subset of the activities that regulate telomerase interact with the single-stranded overhangs. This section highlights several key protein complexes that interact with this single-stranded overhang and contribute to the processes of telomere replication and capping.

The Cdc13 Protein

First discovered in Hartwell's legendary screen for *cdc* mutants, the Cdc13 protein plays a central role in coordinating multiple events at yeast chromosome termini. Cdc13 associates with telomeres through its ability to recognize, with high affinity and high sequence specificity, the single-stranded G-strand extension present at chromosome termini. While resident at telomeres, Cdc13 contributes to both telomere capping and telomere length regulation. Although the full spectrum of Cdc13 activities is not fully understood, numerous studies have provided several insights into this important yeast telomere protein.

CDC13 performs its essential function at telomeres by protecting chromosome ends from degradation. Loss of *CDC13* function exposes these termini to immediate extensive resection of the C-strand of telomeres, with the resulting exposed single-stranded region leading to a *RAD9*-mediated arrest (Weinert and Hartwell 1993; Garvik et al. 1995; Lydall and Weinert 1995; Booth et al. 2001). Despite the essential aspect of this role, the molecular mechanism by which Cdc13 protects termini from resection is poorly understood, as is the identity of the resecting nuclease(s). Two essential Cdc13-associated proteins, called Stn1 and Ten1, also contribute to this capping activity, based on the increased resection that also occurs in strains impaired for either *STN1* or *TEN1* function (Grandin et al. 1997, 2001). In fact, the lethality of a *cdc13-Δ* strain can be rescued if the Stn1 protein is ectopically delivered to telomeres, through a fusion of the Stn1 protein to the Cdc13 DNA-binding domain (DBD); neither the DBD alone, nor the DBD fused to other telomere-specific proteins, is sufficient to rescue a *cdc13-Δ* strain, indicating that Cdc13's primary role in end protection might be to deliver Stn1 to chromosome termini (Pennock et al. 2001).

Cdc13 also plays two functionally discrete roles as a regulator of telomere length (Evans and Lundblad 2000). First, Cdc13 acts as a positive regulator of telomere replication. The elucidation of this process has relied heavily on the characterization of a particular separation-of-function allele of *CDC13*, called *cdc13-2*, which confers a severe telomere replication defect without affecting the end protection activity of *CDC13* (Lendvay et al. 1996; Nugent et al. 1996). This analysis has shown that Cdc13 promotes telomere replication through a recruitment activity that resides in a 15-kD domain of the amino terminus of the protein (Evans and Lundblad 1999; Pennock et al. 2001; Bianchi et al. 2004). As described in more detail in the next subsection, Cdc13 recruits telomerase to its site of action through an interaction with the Est1 subunit of telomerase. This requirement for a recruitment mechanism argues that telomerase does not passively find its substrate; instead, telomerase access to the telomere is actively regulated.

In addition to positively regulating telomerase action at telomeres, Cdc13 has a separate, more poorly characterized role as a negative regulator of telomeric DNA elongation, which it performs in conjunction with Stn1 and Ten1. Strains bearing any of a number of different temperature-sensitive alleles of *CDC13*, *STN1*, or *TEN1* exhibit elongated telomeres at the semi-permissive temperature (Grandin et al. 1997, 2001; A. Chandra and V. Lundblad, unpubl.), which initially suggested that the essential function and the negative length regulation activity resulted from the same biochemical property. However, an allele of *CDC13*, called *cdc13-5*, separates these two activities: Telomere length is increased by more than 1000 bp in the *cdc13-5* strain as a result of unregulated elongation by telomerase, but the essential end protection function of *CDC13* is unimpaired (Chandra et al. 2001). The *cdc13-5* defect is the consequence of a carboxy-terminal deletion that truncates the Cdc13 protein by

230 amino acids, pointing to a possible negative regulatory domain of Cdc13.

The *cdc13-2* and *cdc13-5* mutations thus appear to define two discrete steps for Cdc13 in telomere replication: Telomerase is recruited to chromosome ends, followed by a second Cdc13-dependent mechanism that regulates the extent of elongation by telomerase. The molecular basis for this second step remains elusive, although a number of observations point to an interaction between Cdc13 and DNA polymerase α that mediates coordination of the synthesis of the two strands of the telomere. This model is discussed more extensively later in this chapter.

The 23.5-kD DBD of Cdc13 utilizes a single OB fold, with an extended interaction surface provided by an unusually large loop, for DNA recognition (Mitton-Fry et al. 2002, 2004). This DBD employs the same protein fold as that of the multimeric *O. nova* telomere end-binding protein (*On*TEBP) complex (Mitton-Fry et al. 2002) and also exhibits weak sequence similarity to the amino-terminal OB-fold found in the Pot1 protein (Theobald et al. 2003). In vitro studies with the full-length Cdc13 protein have established that binding does not depend on an exposed 3′ terminus (Nugent et al. 1996), which correlates with the fact that telomeric single-stranded DNA wraps around the surface of the Cdc13 DBD OB-fold (Mitton-Fry et al. 2004). This is in apparent contrast to the structure of *On*TEPB, in which the 3′ end of the single-stranded DNA is buried deep within the complex (Horvath et al. 1998). Whether this difference reflects a true biological difference or not is an interesting and unresolved question. The ternary TEBP complex (composed of the full-length 56-kD α subunit, an amino-terminal 28-kD region of the β subunit, and a 12-nucleotide single-stranded DNA) possesses three OB-folds that form a deep cleft that binds single-stranded DNA, providing an elegant mechanism for telomere capping (see Chapter 11). Whether Cdc13, possibly along with additional proteins (such as Stn1 and Ten1), will form a similar structure that employs multiple OB-folds to sequester chromosome termini awaits further structural analysis. Notably, an additional potential OB-fold in the carboxy-terminal domain of Cdc13 has been predicted on the basis of a sensitive sequence-based profile-profile analysis (Theobald and Wuttke 2004), which corresponds to the same domain that is deleted in the *cdc13-5* mutation described above.

The lack of a requirement for an exposed 3′ terminus also influences models about the specific in vivo location of Cdc13 at telomeres. Cdc13 association with yeast telomeres, as assessed by chromatin immunoprecipitation (ChIP), is enhanced during late S phase (Taggart et al. 2002;

Schramke et al. 2004), which coincides with the period of the cell cycle when Cdc13 would be expected to mediate both chromosome end protection and telomerase recruitment. This peak of increased Cdc13 association has been widely interpreted to be due to binding of Cdc13 to the extended single-strand G-tails that appear at chromosome termini during late S phase. However, this period of the cell cycle also coincides with replication forks passing through the terminal ~0.5 kb of telomeric duplex DNA; thus, the increased detection of Cdc13 at chromosome termini may instead be the consequence of the single-stranded telomeric DNA that becomes exposed during conventional replication of the terminal 0.5 kb of the chromosome. In fact, the peak of Cdc13 association shifts when the timing of semi-conservative DNA replication at the ends of chromosomes is altered by the absence of Ku function (Fisher et al. 2004). Because of these caveats and the difficulty of using ChIP to accurately assess the number of proteins bound to a specific locus, how much Cdc13 is bound during the late S phase, and where the protein is localized, remains an open question.

As mentioned above, the Cdc13 DBD shows both structural parallels and weak sequence similarity to the DNA-binding domains utilized by telomere-binding proteins from other species. Furthermore, loss of the fission yeast Pot1 telomere end-binding protein results in near-lethality and extensive loss of terminal sequences (Baumann and Cech 2001), which is somewhat similar to the defects observed in budding yeast when Cdc13 function is impaired. Since conditional alleles of the fission yeast *pot1* gene have not yet been isolated, it has not been possible to determine whether the extensive C-strand resection observed in the *S. cerevisiae cdc13-1* strain also occurs immediately following loss of *pot1* function in *Schizosaccharomyces pombe*. However, the roughly similar consequences following loss of either Cdc13 or Pot1 function suggest that these two telomere-binding proteins may perform analogous functions, at least with regard to chromosome end protection, despite their somewhat disparate domain structure and lack of notable sequence relationship.

The Yeast Telomerase RNP

Identification of components of the ciliate telomerase complex has relied almost exclusively on biochemical approaches (Greider and Blackburn 1989; Lingner and Cech 1996; Aigner et al. 2000; Witkin and Collins 2004). In contrast, genetic strategies have been the key to uncovering components of the yeast telomerase complex. Protein subunits of the yeast telomerase RNP were identified from mutant hunts based on the predicted behavior—called the Est phenotype—of a telomerase-defective strain (Lundblad and Szostak 1989; Lendvay et al. 1996). The Est phenotype (for *ever shorter telomeres*) is characterized by two properties: continuous telomere shortening and an accompanying decline in cell viability, also referred to as senescence. In contrast, the gene encoding the RNA subunit, called TLC1, was recovered using an approach that screened for genes which, when expressed in high copy, disrupted telomeric silencing (Singer and Gottschling 1994; Singer et al. 1998).

Despite the fact that the protein components and the RNA were recovered from two unrelated strategies, strains deleted for *EST1*, *EST2*, *EST3*, or the *TLC1* RNA exhibit indistinguishable senescence and telomere-shortening phenotypes (Lundblad and Szostak 1989; Singer and Gottschling 1994; Lendvay et al. 1996). All four deletion strains also appear to be equally defective in an in vivo telomere addition assay that measures the extent of telomerase-mediated elongation at a de novo telomere (Diede and Gottschling 1999). Consistent with these phenotypic properties, the three Est proteins are tightly associated subunits of the telomerase RNP (Hughes et al. 2000; Friedman et al. 2003). These observations indicate that these four gene products function in vivo solely as subunits of the telomerase complex, with no roles for individual Est proteins outside the complex. However, careful measurements of the shortening rate of a single marked telomere have revealed a slightly higher rate of sequence loss in an *est1*-Δ strain, when compared to a *tlc1*-Δ strain (Marcand et al. 1999). This suggests that Est1 may be able to protect telomeres from degradation through a pathway that is independent of its role as a component of the telomerase complex. Differences in the rates of telomere shortening have also been observed in telomerase-defective derivatives of the opportunistic pathogen *C. albicans* (Singh et al. 2002). Confusingly, loss of the *CaEST1* gene has a more modest effect on telomere-shortening rates, relative to loss of the catalytic *CaTERT* subunit, in apparent contradiction to the observations by Marcand et al. (1999); this inconsistency, however, might be a consequence of the complexities of constructing isogenic homozygous null mutant strains in *C. albicans*, which is naturally diploid.

The catalytic core of the budding yeast enzyme comprises the Est2 protein (the TERT subunit) and the TLC1 telomerase RNA. Analyses of these two *S. cerevisiae* subunits have exploited the advantages of studying telomerase in a genetically tractable system, which has facilitated the rapid dissection of functional domains of both components. Such studies have shown, for example, that TLC1 serves as a scaffold upon which the RNP assembles, with discrete stem-loop structures that bind individual telomerase-interacting proteins. Similarly, analysis of specific missense mutations in *EST2*-encoded reverse transcriptase have yielded insights into the polymerase mechanism. The reader is referred to Chapters 2 and 3 for in-depth discussions of Est2 and TLC1.

These efforts have been somewhat hampered, however, by the fact that telomerase activity recovered from *S. cerevisiae* extracts is very inefficient at translocating the RNP, relative to the DNA primer: The enzyme remains tightly bound to its substrate after the addition of only a single telomeric repeat (Prescott and Blackburn 1997a,b). The functional significance of this in vitro behavior remains unclear, since single termini can be elongated in vivo by more than 100 nucleotides during a single S phase (Teixeira et al. 2004); thus, the enzyme is either much more processive in vivo, or there is a mechanism that promotes rapid turnover of the enzyme. An additional limitation of *S. cerevisiae* telomerase studies has been the fact that in vitro reconstitution of yeast telomerase activity has not yet been established, potentially because of the difficulties of correctly folding the large 1.3-kb TLC1 RNA. Finally, low levels of each protein subunit have dictated the use of epitope tagging, where alleles of Est1, Est2, or Est3 bear long tandem arrays of myc or HA epitopes. Although these tagging strategies have greatly facilitated analysis, they are not completely benign, as strains bearing tagged alleles of telomerase subunits have at least a modest alteration in telomere length (see, e.g., Hughes et al. 2000; Taggart et al. 2002).

Est1 and Est3, In Vivo Regulatory Subunits of the Telomerase RNP

In contrast to Est2 and TLC1, Est1 and Est3 are telomerase-associated proteins that are not essential for catalysis in either *S. cerevisiae* (Cohn and Blackburn 1995; Lingner et al. 1997) or *C. albicans* (Singh et al. 2002). The experiments in *S. cerevisiae* relied on assessments of partially fractionated extracts prepared from *est1-Δ* and *est3-Δ* mutant strains, using assay conditions where primer was in excess. Thus, the possibility that either subunit contributed to some aspect of the polymerization activity of telomerase—such as the processivity of the nucleotide addition reaction, or the translocation step on the template—could not be excluded. In fact, as described in more detail below, the *C. albicans* Est1 telomerase subunit appears to help promote elongation of certain primers across the telomerase RNA template. Whether the Est3 protein similarly contributes to some aspect of telomerase activity also bears careful reexamination.

The highly basic 82-kD Est1 protein, which has been the subject of investigation by numerous laboratories, possesses three functionally distinct biochemical activities. Est1 associates with the telomerase RNP, presumably through a direct interaction with the TLC1 RNA. In addition to its RNA-binding activity, Est1 binds to the telomere-binding protein Cdc13 and also interacts with single-stranded telomeric DNA. Each of these three biochemical properties of Est1 is discussed in more detail in the following paragraphs.

Est1 interacts with telomerase by binding to a bulged stem loop in the TLC1 RNA (Seto et al. 2002). Despite several efforts, the domain of Est1 that is responsible for RNA binding has not yet been unambiguously defined. Recombinant Est1 protein that has been partially purified from *Escherichia coli* exhibits an in vitro RNA-binding activity, which maps to a carboxy-terminal 130-amino acid domain (Virta-Pearlman et al. 1996). A short sequence that exhibits weak sequence similarity to an RNA recognition motif (RRM) has also been reported in this carboxy-terminal region (Zhou et al. 2000). However, a mutant Est1 protein that is deleted for this RRM-like sequence still retains roughly wild-type levels of association with a catalytically active telomere RNP (Evans and Lundblad 2002), arguing against the conclusion that Est1 employs a canonical RNA recognition sequence to bind RNA. Recently, alignment of Est1 homologs from multiple species has revealed a conserved amino-terminal domain with the features of a tetratricopeptide repeat (TPR) domain (Clissold and Ponting 2000; Beernink et al. 2003; Reichenbach et al. 2003; Snow et al. 2003), a structural motif that has often been implicated in protein–protein interactions (for review, see D'Andrea and Regan 2003). Notably, several alleles of *EST1* that map to this TPR region are defective in their ability to associate with the telomerase RNP, as assessed by immunoprecipitation experiments (Evans and Lundblad 2002; K. Buckley and V. Lundblad, unpubl.).

The primary regulatory role for the Est1 protein is to recruit telomerase to its site of action, through an electrostatic interaction with the single-strand telomere DNA-binding protein Cdc13. Elucidation of this recruitment activity has stemmed from a detailed characterization of a particular class of mutations in *CDC13* and *EST1*. Strains bearing either *cdc13-2* or *est1-60* alleles display the Est phenotype described above, but without a detectable effect on telomerase activity (Nugent et al. 1996: Lingner et al. 1997; Pennock et al. 2001). Strikingly, telomere length maintenance can be fully restored in a *cdc13-2 est1-60* double mutant strain (Pennock et al. 2001). This reciprocal cosuppression argues for a direct interaction between Est1 and Cdc13, and demonstrates that telomere replication fails if these two proteins cannot interact. Furthermore, the *est1-60* strain is completely inactive for telomere addition, as assessed by an assay that monitors telomerase elongation at roughly single-nucleotide resolution (Teixeira et al. 2004), which is inconsistent with the idea that Est1 is acting solely as a processivity or "activation" factor, as had been proposed by the Zakian group (Taggart et al. 2002). As a rigorous test of the recruitment model, Bianchi et al. (2004) examined the consequence of placing Cdc13 or Est1 at non-telomeric sites. Relocalization of the catalytic Est2 subunit to ectopic sites such as a double-strand break depended on the Cdc13-Est1 interaction, even in the absence of telomere DNA synthesis. The Est2 catalytic subunit could even follow Est1 to an internal chromosome location, far from chromosome termini (Bianchi et al. 2004). This experimental approach has established that the function of Cdc13 and Est1, as defined by the *cdc13-2* and *est1-60* mutations, is to recruit telomerase.

In addition, Est1 may perform a second step in telomere replication that occurs subsequent to recruitment. The Est1 protein exhibits a weak, but sequence-specific, affinity for single-stranded telomeric DNA, which maps to a 130-amino acid carboxy-terminal region of the protein (Virta-Pearlman et al. 1996). Notably, binding requires a free 3' terminus. This suggests that Est1 mediates recognition of the end of the chromosome by telomerase, although an assessment of the contribution of this activity to telomere replication awaits identification of missense mutations that specifically eliminate DNA binding. Work in *C. albicans*, however, also supports the postulated contribution of the Est1 subunit in substrate recognition, as telomerase activity from *CaEST1*-deleted *C. albicans* strains exhibits an altered primer specificity (Singh and Lue 2003). The sequence of the 3' termini of CaEst1-responsive primers corresponds to two points in the ultra-long template of the *C. albicans* telomerase RNA (see Fig. 1), suggesting that the CaEst1 protein may help promote synthesis across barriers present in this long template.

The analysis of the Cdc13-dependent recruitment and DNA-binding properties suggests a model in which the Est1 telomerase subunit performs two discrete, successive steps in telomere replication (Fig. 3). In the first step, telomerase is recruited to its site of action, as the result of the protein–protein interaction between Est1 and Cdc13. The proposed second step occurs when Est1 binds to telomeric DNA, presumably to aid in the polymerization reaction. Est1 could promote telomere synthesis through its ability to recognize the 3' terminus of the telomere, thereby facilitating delivery of the telomere DNA substrate to the enzyme active site. Alternatively, as described for the *C. albicans* enzyme, Est1 may relieve an allosteric barrier to synthesis across the template. The role for Est1 in telomerase recruitment has been well established, but evidence for a DNA-binding-dependent step that occurs subsequent to telomerase

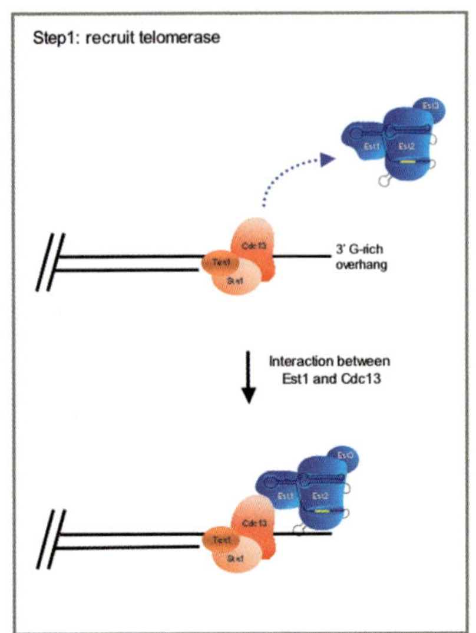

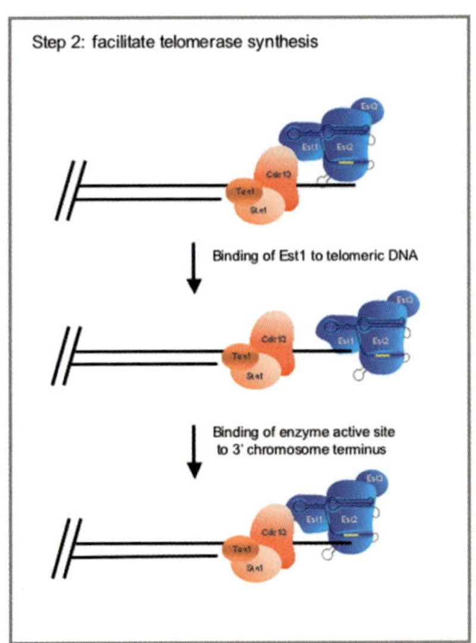

Figure 3. Proposed two successive steps for the Est1 telomerase subunit in telomere replication. In step 1, telomerase is recruited to the chromosome terminus, as the result of the protein–protein interaction between Est1 and Cdc13. Step 2, which is proposed to occur subsequent to telomerase recruitment, occurs when Est1 binds to telomeric DNA, which promotes the polymerization reaction.

recruitment is more circumstantial. However, a class of mutant *est1* alleles with properties that distinguish them from recruitment-defective *est1* mutations supports a second activity for Est1 in telomere replication (Evans and Lundblad 2002).

The discussion of the budding yeast telomerase proteins concludes with a brief discussion of the 19-kD Est3 protein, which is the smallest subunit of the telomerase complex (Morris and Lundblad 1997; Hughes et al. 2000). No biochemical activity(s) has yet been described for this telomerase protein, which has not been subjected to the same level of scrutiny as Est1. The Est3 protein sequence itself does not have any motifs that have revealed a possible function, although phylogenetic comparison of Est3 proteins from a dozen species has uncovered a conserved 70-amino acid domain (D. Morris and V. Lundblad, unpubl.). Association of Est3 with the telomerase RNP requires an intact catalytic core (Hughes et al. 2000), presumably due to a physical association between the Est2 and Est3 proteins (Friedman et al. 2003). In *S. cerevisiae*, synthesis of the full-length Est3 protein is dependent on a programmed +1 ribosome frameshift (Morris and Lundblad 1997). This, along with the relationship of the Est2 protein to reverse transcriptases (Eickbush 1997), furthered the proposed evolutionary link between telomerase and retrotransposition. However, the requirement for +1 frameshifting appears to be restricted to *S. cerevisiae* and closely related species, as synthesis of the full-length Est3 protein from other species does not depend on translation frameshifting (Singh et al. 2002; A. Chappell and V. Lundblad, unpubl.).

Are there additional components of the yeast telomerase complex yet to be uncovered? Neither of the genetic approaches that yielded *TLC1* or the *EST* genes was exhaustive, which left open the possibility of more components yet to be discovered. Notably, the genome-wide screen of viable yeast deletion strains, described above, did not uncover additional genes that, when deleted, exhibited the Est phenotype characteristic of a telomerase deficiency (Askree et al. 2004). However, such genome-wide screens are subject to the caveat that the collection of viable yeast deletion strains may be incomplete, due to losses during propagation of the strain set. Furthermore, the possibility that the yeast telomerase complex could contain one or more protein subunits encoded by essential genes cannot be excluded. Notably absent has been a biochemical approach to this problem, through purification of the yeast telomerase holoenzyme, comparable to the successful studies performed with the *Euplotes* and *Tetrahymena* enzymes (Lingner and Cech 1996; Witkin and Collins 2004).

The Ku Heterodimer

The Ku heterodimer, composed of 70-kD and 80-kD subunits, is an abundant heterodimeric complex that binds with high affinity to double-stranded DNA ends in a sequence-independent fashion (Downs and Jackson 2004). In both yeast and higher eukaryotes, Ku is required for the accurate and efficient re-ligation of double-strand breaks, through the so-called nonhomologous end-joining pathway. Paradoxically, despite the fact that telomeres are normally refractory to such end-to-end fusions, the Ku heterodimer nevertheless is also present at telomeres, and its absence affects a surprisingly diverse array of telomere-specific activities. Ku mediates telomere length by modulating telomerase function (Boulton and Jackson 1996; Porter et al. 1996; Peterson et al. 2001; Stellwagen et al. 2003) and contributes to the regulation of the process(es) that generates the single-strand G-rich overhang (Gravel et al. 1998; Polotnianka et al. 1998); both of these activities place Ku in proximity of the single-stranded terminus. As an inferred component of duplex telomeric chromatin, Ku influences the expression of telomere-localized genes (Boulton and Jackson 1998) and contributes to the clustering of telomeres to the nuclear periphery (Laroche et al. 1998; Hediger et al. 2002).

This surfeit of roles argues that the Ku heterodimer makes several biochemically distinct contributions at yeast telomeres, which could occur at physically discrete locations. The Ku heterodimer forms a quasi-symmetrical ring-like structure through which DNA can be threaded, with the DNA end minimally exposed (Walker et al. 2001). This ability of the Ku heterodimer to bind to DNA ends is certainly consistent with the premise that Ku localizes to the extreme termini of yeast chromosomes, although biochemical studies would suggest that Ku most likely resides at the transition from single-strand DNA to duplex DNA. Chromatin immunoprecipitation (ChIP) has also demonstrated that Ku is a constitutive component of telomeres (Gravel et al. 1998; Fisher et al. 2004; Schramke et al. 2004). However, ChIP does not provide enough resolution to distinguish between localization at the single-stranded terminus versus association with duplex telomeric chromatin. With this caveat in mind, this subsection considers the contribution of two properties of the Ku heterodimer that are presumed to require interaction of Ku with the extreme terminus, whereas Ku's presumed separate role as a component of duplex chromatin is considered in a later section of this chapter.

The multiple activities of the Ku heterodimer at telomeres have begun to be teased apart, through the isolation of separation-of-function alleles that specifically eliminate a single function. Two such alleles, which map to the *YKU80* subunit and the *TLC1* gene (*yku80-135i* and *tlc1-Δ48*, respectively), have been instrumental in characterizing an interaction between the Ku heterodimer and a 48-nucleotide stem loop of TLC1 (Peterson et al. 2001; Stellwagen et al. 2003). In the *yku80-135i* strain, the resulting defect in the interaction with TLC1 results in telomere shortening, although the other functions performed by the Ku heterodimer, at both telomeres and double-strand breaks, remain intact. However, although telomere length regulation is compromised in the *yku80-135i* and *tlc1-Δ48* strains, telomere addition can still occur, albeit at reduced kinetics, as assessed by an in vivo assay that measures de novo telomere addition (Stellwagen et al. 2003). Consistent with this, telomere length in the *yku80-135i* and *tlc1-Δ48* strains is stably short, in contrast to the more severe telomere replication defect displayed by telomerase null strains.

Stellwagen et al. (2003) also analyzed the effect of the *yku80-135i* defect on another class of DNA ends, those that arise as a result of chromosome breakage. As expected, such ends are poor substrates for telomerase, although at extremely low frequencies, telomerase can even elongate these breaks. These telomerase-mediated additions usually depend on the presence of a short "seed" sequence that resembles telomeric G-rich DNA (Kramer and Haber 1993; Kolodner et al. 2002), although termini that completely lack these telomere-like sequences can also be extended (Chen and Kolodner 1999; Myung et al. 2001). In the *yku80-135i* strain, telomere addition is specifically abolished at the particular class of double-strand breaks that lack short telomeric seed sequences. Thus, the Ku–TLC1 interaction appears to promote a sequence-independent pathway for telomerase access at double-strand breaks, whereas this interaction is dispensable at breaks that retain a more telomere-like character. Curiously, however, this interaction is clearly not expendable at natural telomeres, which possess a full complement of telomeric DNA. These slightly contradictory observations lead to two possible models to explain the role of the Ku–TLC1 interaction. The first possibility is that Ku, like Cdc13, contributes to telomerase recruitment. Consistent with this, when Ku70 is tethered to an internal chromosomal site, Est2 is weakly recruited to this site, through a process that is dependent on the 48-nucleotide stem loop of TLC1 that binds Ku (Bianchi et al. 2004); this could reflect Ku's ability to localize telomerase to telomeric chromatin during the G_1 phase of the cell cycle (Fisher et al. 2004). Alternatively, Ku may associate with telomerase to promote an activity that occurs subsequent to enzyme recruitment, such as stabilizing base-pairing between the TLC1 template and a DNA end in order to promote the elongation reaction; this latter model might help explain the sequence dependency of the *yku80-135i* strain when healing de novo breaks. These two possible models may be distinguished by analysis of the *yku80-135i* and *tlc1-Δ48* mutations with assays that can differentiate between a defect in recruitment and a defect in elongation (such as that described by Teixeira et al. 2004).

In parallel with its role in telomere length regulation, the Ku heterodimer participates in another terminus-specific process. In the absence of Ku function, the regulation of the structure of the G-rich overhang is altered, suggesting that Ku contributes, either directly or indirectly, to the process(es) that generates single-strand termini. In wild-type cells, an extended G-rich overhang is only observed in S phase (Wellinger et al. 1993), whereas an exaggerated G-tail is present throughout the cell cycle in *yku70-Δ* or *yku80-Δ* strains (Gravel et al. 1998; Polotnianka et al. 1998). This altered terminal structure is at least in part due to increased exposure of the C-strand to resection by nucleolytic activities, as the single-strandedness of these abnormal termini can be at least partially rescued by a mutation in the *EXO1* gene (Bertuch and Lundblad 2004). *EXO1*-dependent resection extends even into the subtelomeric regions in Ku-defective strains that are grown under more stringent conditions (Maringele and Lydall 2002). These observations certainly support the conclusion that Ku plays a role in blocking aberrant access of certain nucleases, such as Exo1, to telomeres. However, since an *exo1-Δ* strain does not have any detectable telomere phenotypes (Moreau et al. 2001; Bertuch and Lundblad 2003), it is hard to argue that Exo1 plays a role in normal telomere processing. Whether Ku directly regulates the activity of a nuclease that is in fact required for telomere function in wild-type cells is an open question.

As with the Ku–TLC1 interaction, insight into this particular function of Ku may be revealed by analysis of a second set of *YKU80* separation-of-function alleles, called the *yku80^{tel}* missense mutations (Bertuch and Lundblad 2003). The *yku80^{tel}* strains exhibit altered chromosome end structure, although repair of double-strand breaks is unimpaired in these mutant strains. A subset of the *yku80^{tel}* missense mutations target residues that are conserved between the yeast and human Ku heterodimer, arguing strongly that insights to come from these yeast mutants will instruct studies with the human Ku heterodimer.

An additional property of *yku70-Δ* and *yku80-Δ* strains may be very relevant to the above discussion, which is that the timing of semi-conservative DNA replication at the ends of chromosomes is altered by the absence of Ku function (Cosgrove et al. 2002). Replication origins located adjacent to the ends of chromosomes normally initiate late, but these origins are activated much earlier in mutants lacking Ku function, with a consequent effect on S-phase progression. Notably, the processing step that generates the single-stranded G-strand extensions present at chromosome termini in late S phase also depends on passage of a replication fork through telomeres (Dionne and Wellinger 1998). This suggests that disruption of the mechanistic link between semi-conservative DNA replication and telomere processing could alter chromosome end structure, as an indirect consequence of the altered replication timing of telomeres in *yku70-Δ* and *yku80-Δ* strains. This model, therefore, proposes that Ku contributes to chromosome end structure indirectly, by coupling replication fork passage and telomere processing events, rather than through direct regulation of a telomere-specific nuclease. This also implies that Ku might modulate end structure as a component of telomeric chromatin, as opposed to a factor that is resident at the extreme terminus. If so, this could explain why the *yku80^{tel}* strains, which are severely compromised for chromosome end structure, also exhibit properties indicative of a defect in telomere heterochromatin (Bertuch and Lundblad 2003).

The MRX Complex

Like the Ku heterodimer, the evolutionarily conserved MRX complex (composed of three subunits, encoded by the *MRE11*, *RAD50*, and *XRS2* genes) makes a multifaceted set of contributions both to repair of double-strand breaks and to telomere function. This complex promotes the repair of double-strand breaks, through either homologous recombination or nonhomologous end-joining. Numerous enzymatic activities are also associated with the MRX complex: single-stranded endonuclease, single-strand and double-strand exonuclease, and DNA-unwinding activities all have been reported from in vitro studies. A further paradox arises from an assessment of the in vivo consequences for an MRX defect, which results in impaired 5′ to 3′ resection of double-strand breaks, even though the Mre11-associated exonuclease activity has the opposite polarity in vitro. Mre11 also functions as a DNA damage sensor, and finally, the MRX complex promotes recombination between sister chromatids. (For a review and primary literature references, see Stracker et al. 2004 and Lichten 2005.)

This abundance of data, from multiple experimental approaches, has led to the comment that "as many questions as answers regarding the in vivo roles of the complex, and the mechanisms by which they are mediated" have arisen (Stracker et al. 2004). Indeed, the situation at telomeres is not any clearer. This complex plays an important, but mechanistically undefined, role in the telomerase pathway for telomere length maintenance (Kironmai and Muniyappa 1997; Nugent et al. 1998; Ritchie and Petes 2000; Diede and Gottschling 2001; Tsukamoto et al. 2001; Lewis et al. 2002). It is also a key regulator in a telomerase-independent pathway for telomere length control called telomere rapid deletion (Li and Lustig 1996; Bucholc et al. 2001). It also contributes to, but is not essential for, the maintenance of the single-stranded G-rich overhang at telomeres (Larrivee et al. 2004; Takata et al. 2005). These seemingly unrelated properties may nevertheless stem from one proposed function of the MRX complex, which is to load other proteins onto the telomere. In particular, in *mre11-Δ* strains, the localization of Cdc3 at both de novo telomeres (Diede and Gottschling 2001) and native chromosome ends (Takata et al. 2005) is diminished; reduced Cdc13 protein levels at telomeres may therefore explain the pleiotrophic telomere-related phenotypes of MRX-defective strains.

PROTEIN COMPLEXES THAT INTERACT WITH THE DUPLEX TELOMERIC REPEAT TRACT

A number of proteins contribute to the heterochromatic structure that forms at yeast telomeres, with consequent effects on the expression of telomere-adjacent genes (TPE) as well as effects on telomere length control. For a detailed discussion of the role of these proteins in TPE, see Chapter 10.

The Rap1 Protein and Associated Factors

In budding yeast, duplex telomeric DNA is bound by the Rap1 protein through two tandem myb-like DNA-binding domains (Konig et al. 1996; Konig and Rhodes 1997; Chapter 11). While bound to telomeres, Rap1 contributes to both telomeric silencing and telomere length control, through a single domain at its carboxyl terminus and the use of separate sets of Rap1-interacting proteins. *RAP1* is also an essential gene, but this appears to be a consequence of its role in transcriptional regulation of gene expression, rather than a telomere-specific function that is required for viability.

Several early studies, which analyzed the effects of either overexpression of the carboxyl terminus or the properties of an allele of *RAP1* that was truncated at its carboxyl terminus (*rap1^t*), implicated this region of Rap1 in negative length regulation (Conrad et al. 1990; Kyrion et al. 1992). An elegant study by Shore and colleagues uncovered the basis for this carboxy-terminal-mediated length regulation: They demonstrated that tethering this carboxyl terminus immediately adjacent to the telomeric repeat tract caused an effect on telomere length, in *cis*, that was proportional to the number of tethered molecules (Marcand et al. 1997). This observation can be explained by the so-called "counting model," which proposes that the length of a given telomere is dictated by the number of Rap1 molecules bound to that telomere, thereby serving as a readout of telomere length. Those telomeres with fewer Rap1 molecules bound would be more likely to be elongated by telomerase, whereas longer telomeres, with more bound Rap1 protein, would become refractory to telomerase action. The immediate effectors of this length regulation are Rif1 and Rif2, which both interact with the carboxyl terminus of Rap1: Tethering these two Rap1-interacting proteins to telomeres also shortened telomere length in *cis*, in a manner that is also roughly proportional to the number of tethered molecules (Levy and Blackburn 2004). Further discussion of this negative feedback mechanism, and how Rap1 might transmit information about telomere length to telomerase, appears in the next section.

The Ku Heterodimer as a Component of Duplex Heterochromatin

As discussed earlier, the Ku heterodimer may localize to telomeres at the transition from single-strand DNA to duplex DNA, at the extreme chromosome terminus. However, Ku, like Rap1, is required for TPE (Boulton and Jackson 1998; Laroche et al. 1998; Nugent et al. 1998), suggesting that

Ku also resides on the duplex portion of the telomere, as a component of telomeric heterochromatin. Ku influences telomeric silencing through its interaction with the Sir4 protein (Tsukamoto et al. 1997; Mishra and Shore 1999; Roy et al. 2004), whose recruitment to telomeres is essential for both the initiation and spreading of telomeric heterochromatin into subtelomeric regions. This interaction with Sir4 dictates the outcome of a competition between Sir and Rif proteins for access to the Rap1 protein, thereby contributing, at least in part, to the balance between telomeric silencing and telomere length control (Mishra and Shore 1999). Ku may also mediate telomere length control by another mechanism, by sequestering the telomerase catalytic core during the G_1 phase of the cell cycle (Fisher et al. 2004).

REGULATION OF TELOMERE LENGTH

Telomere length homeostasis is the convergence of numerous regulatory mechanisms that confer stable telomere length in telomerase-proficient cells. Stable telomere length ultimately is the result of a balance between activities, such as telomerase, that elongate chromosome ends, and activities, such as nucleases or incomplete semi-conservative DNA replication, that result in telomere shortening. Much of the recent focus in yeast telomere length regulation has been directed at certain mechanisms that are responsible for regulation of telomerase. Other obvious points of regulation, such as telomerase assembly, have received far less attention. Very little is known about the counterbalancing shortening activities that must act on telomeres, which is addressed in the following section.

Telomeres Switch between Nonextendible and Extendible States

As discussed in the previous section, one level of telomere length regulation is achieved via a negative feedback loop, in which the number of Rap1 molecules bound to the duplex region of the telomere dictates whether further elongation by telomerase will occur, in *cis*. However, left unanswered was whether this regulation was achieved by modulating the frequency with which the telomerase accessed telomeres, or the extent of telomerase-mediated synthesis. This issue was elegantly addressed through the development of a powerful assay that provides nucleotide-level resolution of the frequency and extent of telomerase-mediated elongation events within a single cell cycle (Teixeira et al. 2004). This assay has revealed that telomerase is regulated at the initial initiation event: In wild-type cells, only ~10% of telomeres are elongated by telomerase during a single cell cycle, and this low level of telomerase action is preferentially targeted to short telomeres. In contrast, the extent of elongation is less regulated: The number of nucleotides added to a telomere in a single cell cycle varies from a few to ≥ 100 nucleotides, and this variation is independent of telomere length. When this assay was applied to strains that lacked the Rap1-interacting factors, Rif1 and Rlf2, the frequency of elongation events increased, so that a larger proportion of telomeres within a given cell cycle was elongated, when this negative feedback loop was relieved.

This led Lingner and colleagues to propose that long telomeres are in a "telomerase-nonextendible state," whereas short telomeres are "telomerase-extendible" (Teixeira et al. 2004). The implication of this model is that there must be some mechanism which marks the shortest telomeres as preferred substrates for telomerase. The basis for this preference has not been elucidated, but two obvious possibilities come to mind. The first is that telomerase is preferentially recruited to short telomeres. The alternative is that the single-stranded G-rich substrate itself must be converted from a nonextendible form (a t-loop, perhaps?) to a form that can be elongated by telomerase.

Telomerase Elongation Is Tightly Coupled to Lagging-Strand DNA Synthesis

Elongation of the G-strand by telomerase is also tightly coordinated at another level. Using a sensitive assay that monitors de novo telomere synthesis, Diede and Gottschling (1999) have shown that telomere addition by telomerase can be completely abolished if the lagging-strand polymerase (DNA pol α or pol δ) is inactive. This indicates that elongation of the G-strand by yeast telomerase is tightly coupled to synthesis of the corresponding C-strand by the conventional DNA replication machinery (Fig. 4). Comparable observations in the ciliate *Euplotes* also support the premise that synthesis of the two strands is coordinately regulated: Partial inhibition of DNA polymerases α and δ during de novo telomere synthesis in *Euplotes* not only alters the length of the telomeric C-strand, but also causes an increase in the length and heterogeneity of the G-strand (Fan and Price 1997; for more discussion, see Chapter 15).

The specific mechanism by which synthesis of these two strands is coupled is unknown, but several lines of evidence point to Cdc13 and its associated proteins Stn1 and Ten1. Strains bearing defects in subunits of the lagging-strand machinery have long been known to show both longer, and more heterogeneous, telomere lengths when these strains

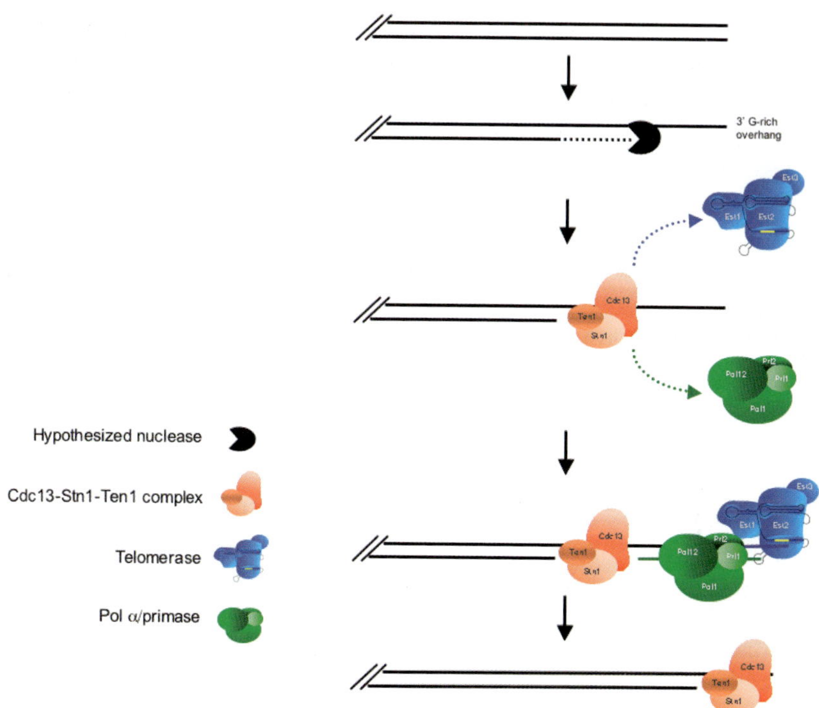

Figure 4. Proposed series of steps that occur on the presumed blunt terminus that is created as a result of leading-strand synthesis, following completion of semi-conservative DNA replication in late S phase. Resection of the C-strand by a hypothesized nuclease creates a G-rich single-stranded overhang, which will be bound by the Cdc13 telomere-binding protein, potentially in a complex with Stn1 and Ten1. Once bound, the Cdc13–Stn1–Ten1 complex is responsible for recruiting telomerase and may also mediate access of the DNA pol α/primase complex. Elongation of the G-strand by telomerase is coordinated with fill-in synthesis of the C-strand by DNA pol α/primase, to generate the shorter 12–14-nucleotide G-rich overhang that persists throughout the cell cycle.

were propagated at a semi-permissive temperature (Carson and Hartwell 1985: Adams and Holm 1996; Adams Martin et al. 2000). Similarly, many mutations in *CDC13*, *STN1*, or *TEN1* result in the same unregulated telomere elongation (Grandin et al. 1997, 2001; A. Chandra and V. Lundblad, unpubl.). Notably, both Cdc13 and its interacting protein Stn1 appear to directly contact subunits of the DNA polymerase α/primase complex, suggesting a direct mechanistic link between these two complexes (Qi and Zakian 2000; Grossi et al. 2004). Such a link is supported by in vivo interactions: Increased expression of the Stn1 protein suppresses the telomere elongation phenotype of DNA polymerase α mutant strains (Chandra et al. 2001), whereas loss of the ability of Stn1 to interact with the DNA polymerase α/primase complex confers telomere elongation (Grossi et al. 2004). Since Cdc13 is also responsible for recruiting telomerase, a simple model proposes that the Cdc13–Stn1–Ten1 complex is responsible for recruiting both telomerase and the DNA polymerase α/primase complex and for coordinating synthesis of the G- and C-strands, by coupling the action of these two sets of polymerases (Fig. 4).

A suggestive implication of the coupling model is that telomerase must be in a complex with the lagging-strand machinery in order to be active. Thus, as pointed out by Diede and Gottschling (1999), recruitment of DNA pol α and pol δ to telomeres may be as crucial to telomere length maintenance as recruitment of telomerase. This premise expands the possible list of gene products that, when mutant, would give rise to the Est phenotype: Mutations that disrupt the coupling between telomerase and the lagging-strand machinery, or that inhibit recruitment of DNA pol α and pol δ, would be expected to confer ever shorter telomeres and senescence. This model prompts a possible reinterpretation of the proposed role of replication protein A (RPA) in telomere replication (Schramke et al. 2004). An *rfa2-Δ40* strain, bearing an amino-terminal deletion of the middle subunit of RPA, exhibits a severe telomere replication defect. Strikingly, the Est1 protein also exhibits decreased association with telomeres in the *rfa2-Δ40* strain, as assessed by ChIP. This led Geli and colleagues to conclude that RPA is responsible for loading Est1 onto telomeres, based on the assumption that the change in occupancy of Est1 that they observed was due to a decrease in the "on rate" for Est1. However, ChIP cannot distinguish whether a change in the steady-state occupancy of a particular protein at telomeres is due to a change in the "on rate" or the "off rate." Thus, an equally plausible, and highly intriguing, alternative interpretation is that RPA fulfills its role in telomere replication by stabilizing the interaction between telomerase and the lagging-strand machinery, in a manner such that when this interaction is destabilized, Est1 cannot be retained at telomeres.

Telomeres Are Elongated in Late S Phase

Telomerase only acts on ~10% of telomeres in every cell cycle (Teixeira et al. 2004), indicating that telomere elongation is not an obligatory step in cell cycle progression. Consistent with this, loss of telomerase does not confer an inability to progress through the cell cycle, at least before telomeres begin to resemble double-strand breaks. Nevertheless, elongation of telomeres by telomerase is restricted to late S phase (Marcand et al. 2000). This restriction is not at the level of telomerase assembly, as enzyme activity can be detected in extracts prepared during other phases of the cell cycle (Diede and Gottschling 1999). The mechanism that restricts telomerase to this period of the cell cycle has not yet been determined, in part because several valid possibilities exist. One possibility is simple substrate accessibility: The long G-strand overhangs present in late S phase might be expected to be preferred substrates for telomerase. However, in the de novo telomere addition assay, in which telomerase is quite efficient, there is no evidence for a long single-stranded structure (Diede and Gottschling 2001). An alternative possibility stems from the fact that telomere addition is tightly coupled to the activity of DNA polymerases α and δ, as discussed above. Finally, the telomere localization of both Cdc13 and Est1, which are responsible for recruiting the catalytic core to chromosome ends, is restricted to late S phase.

Regulated Association of Telomerase Components with Telomeric Chromatin

Association of the Est1 and Est2 telomerase subunits with telomeric chromatin, using chromatin immunoprecipitation (ChIP), has been assessed in a number of different laboratories (Taggart et al. 2002; Smith et al. 2003; Bianchi et al. 2004; Schramke et al. 2004; Takata et al. 2005). These studies, some of which use a subtelomeric probe to detect sequences located $\sim$350–600 bp from the end of the chromosome, have uncovered several surprising differences in the pattern of association of the Est1 and Est2 proteins across the cell cycle.

The Est2 catalytic subunit localizes to telomeric chromatin throughout the cell cycle, until it is displaced during mitosis (Taggart et al. 2002; Smith et al. 2003). This pattern of association was unexpected, since telomere elongation by telomerase is restricted to S phase. Examination of Est2 association in different mutant strains has revealed that there are two distinct pathways which mediate Est2 localization to telomeres, during two discrete periods of the cell cycle. During S phase, Est2 telomere association is dramatically reduced in the presence of the recruitment-defective *cdc13-2* mutation (Taggart et al. 2002); Est2 association is predicted to be also reduced during S phase in an *est1-60* strain (as well as in an *est1-Δ* strain), although this has not been tested. The Cdc13-dependent localization of Est2 during S phase, along with other observations discussed elsewhere in this chapter, supports the idea that recruitment of telomerase to telomeres in late S phase is mediated by an interaction between Cdc13 and the Est1 telomerase subunit, in order to bring the catalytic core to its site of action at the chromosome terminus.

Association of the Est2 subunit to telomeric chromatin during the G_1 phase period of the cell cycle is not dependent on Cdc13, however, as G_1-specific levels of Est2 are unaffected by the *cdc13-2* mutation (Taggart et al. 2002). Instead, the Ku heterodimer appears to mediate G_1-dependent localization of Est2, and this localization is dependent on the Ku–TLC1 interaction described earlier in this chapter (Fisher et al. 2004). This suggests that the catalytic subunit of telomerase is sequestered in a subtelomeric locale, until it is recruited to the 3′ terminus of the G-strand. Assessing the functional significance of this proposed sequestration, however, will be difficult, as loss of the Ku–TLC1 interaction also has a twofold effect on Est2 telomere association during S phase (Fisher et al. 2004; Schramke et al. 2004), which may account for the modest telomere-shortening defects displayed by the *yku80-135i* and *tlc1-Δ48* strains. Also, it is worth noting that Geli and colleagues did not observe a noticeable decrease in Est2 association during the G_1 phase in the Ku-defective strain (Schramke et al. 2004).

In contrast to Est2 protein levels, which do not change appreciably through the cell cycle, the steady-state levels of the Est1 protein are low in the G_1 phase of the cell cycle, and they increase by ∼2- to 3-fold by S phase (Taggart et al. 2002). This suggests that Est1 protein synthesis may be regulated; alternatively, free Est1 protein that is unbound from its Cdc13 partner during G_1 may be subject to enhanced degradation. Not surprisingly, Est1 protein levels correlate with the pattern of association of Est1 with telomeres, which is also low in the G_1 phase of the cell cycle but enhanced in late S phase (Taggart et al. 2002; Schramke et al. 2004). In initial experiments, Est1 exhibited 2.5- to 5-fold increased association with telomeres in S phase, relative to G_1; surprisingly, this S-phase-dependent association was not eliminated by the *cdc13-2* mutation, which was a strong argument against the recruitment model (Taggart et al. 2002). More recent experiments, which have refined the sensitivity of detection of Est1 at telomeres, have reported a more robust 15- to 30-fold enhancement of Est1 localization to telomeres during S phase (Fisher et al. 2004; Schramke et al. 2004). The Cdc13 dependency of this Est1–telomere association, using this more robust detection method, has not yet been reported.

One discrepancy in the observations about Est2 localization with telomeres through the cell cycle has recently been resolved. Both Taggart et al. (2002) and Schramke et al. (2004) reported two peaks of Est2 association, in G_1 and late S phase, with a reproducible ∼2-fold reduction in Est2 association in early S phase. In contrast, Blackburn and colleagues observed continuous association of Est2 through the cell cycle, until mitosis (Smith et al. 2003). The association of Est2 interpreted by Taggart et al. and Schramke et al. as an early-S-phase reduction now appears to have been due to the fact that the nontelomeric sequence used to normalize the telomeric signal replicates earlier in S phase than does the telomeric sequence, thereby briefly skewing the normalization calculations (Fisher et al. 2004).

Telomeric Recombination in Telomere Length Control

As described above, telomerase acts preferentially on short telomeres, in order to restore these shortened termini to a wild-type length. Telomeric rapid deletion (TRD), discovered in the Lustig laboratory, is a recombination-dependent process by which overelongated telomeres are restored to wild-type lengths (Li and Lustig 1996; Bucholc et al. 2001; see Chapter 8). Dissection of this pathway has relied on the analysis of experimentally elongated telomeres that are generated in an otherwise wild-type yeast strain. These elongated telomeres can return to wild-type telomere length as the result of gradual loss of telomeric repeat sequences occurring during many population doublings, presumably reflective of shortening activities (either nucleases or incomplete semi-conservative replication) that are not counterbalanced by telomerase-mediated synthesis. However, the length of these elongated telomeres can also be reset by TRD through a single-step intra-chromatid deletion event. This therefore defines a pathway for telomere length homeostasis that occurs in wild-type cells, via a mechanism that is distinct from the telomerase-mediated pathway.

TRD is proposed to occur via invasion of the single-stranded G-rich overhang into the duplex telomeric repeat tract to create a recombination intermediate that leads to excision of the portion that is looped out (thereby forming a structure that bears a striking resemblance to a t-loop). This process is notable because of its precision: Elongated telomeres that undergo TRD are returned to a length that is dictated by the size of the majority of the telomeres in the cell. How the precision of this *trans*-acting process is regulated is an intriguing question, although it may rely on sub-nuclear localization of telomeres, such as to the nuclear periphery. Efficient TRD requires two subunits of the MRX complex, Mre11 and Rad50 (Bucholc et al. 2001). The role of Mre11 is likely to be quite complex, as different missense mutations of MRE11 implicate it in both positive and negative regulation of TRD (Williams et al. 2005). Notably, the nuclease activity of Mre11 is dispensable for this process (Williams et al. 2005).

Tel1 Kinase in Telomere Length Control

A discussion of telomere length regulation in yeast would not be complete without a consideration of the role of the Tel1 kinase. *TEL1* was one of the first genes identified for its role in telomere length maintenance (Lusting and Petes 1986) and the first PIKK kinase recognized to participate in telomere length maintenance. However, the target of Tel1 action has still not been identified. Tel1, often referred to as the yeast equivalent of the human ATM protein, and Mec1 (the ATR homolog), are two phosphoinositide (PI)-3-kinase-related kinases that mediate the cascade of events that occurs in response to DNA damage. Current models suggest that Tel1 (along with the NHEJ proteins) is among the first proteins to localize to a double-strand break in order to initiate DNA repair, but is subsequently replaced by Mec1 (and the HR proteins), in a transition that is dependent on resection of one strand of the double-strand break (for review, see Garber et al. 2005).

The Tel1 protein also makes a pivotal contribution to telomere length control: *tel1* strains have extremely short telomeres (roughly comparable to that of a telomerase null strain), although telomeres are not so short that a *tel1* strain exhibits the senescence growth characteristics of telomerase-defective strains (Lustig and Petes 1986). The Mec1 protein also makes a very slight contribution to telomere length maintenance, and not surprisingly, combining a *tel1* and *mec1* mutation in the same strain results in an additive effect on telomere length, resulting in further telomere shortening, as well as a growth defect that is roughly analogous to the senescence phenotype of telomerase-defective strains (Ritchie et al. 1999). Tel1 presumably modulates telomere length by phosphorylating a target protein: This target protein is also an intriguing candidate for the factor that marks short telomeres for elongation by telomerase.

Matsuura and colleagues have pursued the role of Tel1 and Mec1 in telomere function through a careful examination of the association of these two proteins with telomeres through the cell cycle (Takata et al. 2004). Both proteins exhibit cell-cycle-specific localization, in a mutually exclusive manner. Mec1 associates with telomeres in late S phase, and intriguingly, association is preferentially with short telomeres; Takata et al. propose, therefore, that Mec1 acts as a sensor for structural abnormalities at chromosome ends. Tel1 exhibits a reciprocal association: It is present at telomeres in G_1 and early S phase, but curiously, Tel1 is absent from telomeres in late S phase. This latter observation is not easily reconciled with Tel1's role in telomere length regulation, unless telomeres are "marked" for telomere elongation by Tel1 earlier in the cell cycle. The pattern of association of Mec1 and Tel1 was also examined in various mutant backgrounds, which showed that even in genetic perturbations that alter the association of one of these two PIKK proteins, compensation for the altered association occurs, such that association of the other PIKK protein changes in a reciprocal pattern. This latter set of observations shows that either Mec1 or Tel1 will always be present at the telomere, providing an elegant molecular view of how partially redundant proteins perform their overlapping functions.

REGULATION OF THE TELOMERIC END STRUCTURE

A critical feature of chromosome ends is the single-stranded G-rich overhangs, which presumably are present at every chromosome terminus, in order to promote chromosome end protection. In yeast, the length of this structure is highly regulated: The overhang is considerably longer in late S phase, following completion of semi-conservative replication, than the 12–14-nucleotide extensions that are present during the rest of the cell cycle (Wellinger et al. 1993; Larrivee et al. 2004). Estimating the exact length of the overhang is a substantial technical challenge; the overhang during late S phase is at least 30 nucleotides long, but it could be even longer (Wellinger et al. 1993). Although it is obviously difficult to assess whether all native chromosome ends have such structures, examination of both termini of a linear plasmid reveals the presence of G-tails (Wellinger et al. 1996), consistent with the idea that this feature is essential for end protection.

Information about the factors required for the formation of these overhangs is still preliminary. Although one obvious mechanism invokes synthesis by telomerase, this has been ruled out by the fact that overhangs are still present in telomerase-defective cells (Wellinger et al. 1996). In hindsight, this is perhaps not such an unexpected result, given that telomerase only elongates a subset of telomeres in every cell cycle (Teixeira et al. 2004). Furthermore, immediate loss of telomerase is not lethal (Lundblad and Szostak 1989; Singer and Gottschling 1994), as would be predicted if telomerase were essential for maintaining the cap structure at the ends of chromosomes. A separate set of studies has revealed that the MRX complex contributes to this process: The extent of the G-tail overhang is reduced, but not abolished, in MRX mutants (Larrivee et al. 2004; Takata et al. 2005). Other studies have examined the extent of resection, and the genetic dependency of the observed resection, in strains in which chromosome end protection is impaired (Lydall and Weinert 1995; Booth et al. 2001; Maringele and Lydall 2002; Bertuch and Lundblad 2004; Jia et al. 2004; Zubko et al. 2004). However, as discussed earlier in this chapter, it is possible that such studies are primarily monitoring the aberrant access of certain nucleases, which are not normally required for telomere end processing.

Collectively, these studies suggest that the resection activity that processes chromosome termini immediately following the completion of conventional DNA replication has yet to be identified. The premise for such an activity stems, at least in part, from a consideration of the structure of chromosome termini produced by semi-conservative DNA replication: The immediate product of leading-strand replication is a blunt terminus that lacks a G-strand overhang (for a more detailed discussion, see Chapter 1). These blunt-ended termini, if left unprocessed, would presumably be repaired as double-strand breaks, with lethal consequences. Thus, many models postulate an activity that acts following DNA replication in order to expose the G-strand (Fig. 4), presumably by resection of the C-strand (although note that other activities, such as helicases, cannot be ruled out). The resulting exposure of the G-strand would create a substrate for telomere-specific end-binding factors, such as the Cdc13 DNA-binding protein. Once bound, Cdc13 and associated factors could promote telomere-specific activities, such as telomerase-mediated elongation, as well as shielding this terminus from recombination and repair activities. Efforts to identify this hypothesized resection activity are ongoing in multiple laboratories.

SUMMARY AND PERSPECTIVE

The last 20 years of yeast telomere analysis have yielded a number of crucial insights about how yeast telomeres are maintained: Subunits of the telomerase complex have been discovered, a number of pathways that regulate telomere length have been worked out at a mechanistic level, and a wealth of structural, biochemical, and genetic information has been acquired about proteins that bind the duplex and single-strand regions of the chromosome. Future investigations should provide a working model for how more than 175 genes collaborate to regulate telomere length, address the complexities of multifunctional proteins such as the Ku and MRX complexes, and identify missing activities such as the hypothesized resecting nuclease described above.

ACKNOWLEDGMENTS

I gratefully acknowledge the many conversations with numerous colleagues, which have influenced and shaped the ideas presented here, as well as funding from the National Institutes of Health and Ellison Medical Foundation for research in my laboratory. My thanks are also extended to the folks at Cold Spring Harbor Laboratory Press, who patiently waited for a ligament in my spine to cooperate enough so that I could complete this chapter.

REFERENCES

Adams A.K. and Holm C. 1996. Specific DNA replication mutations affect telomere length in *Saccharomyces cerevisiae. Mol. Cell. Biol.* **16:** 4614–4620.

Adams Martin A., Dionne I., Wellinger R.J., and Holm C. 2000. The function of DNA polymerase α at telomeric G tails is important for telomere homeostasis. *Mol. Cell. Biol.* **20:** 786–796.

Aigner S., Lingner J., Goodrich K.J., Grosshans C.A., Shevchenko A., Mann M., and Cech T.R. 2000. *Euplotes* telomerase contains an La motif protein produced by apparent translational frameshifting. *EMBO J.* **19:** 6230–6239.

Askree S.H., Yehuda T., Smolikov S., Gurevich R., Hawk J., Coker C., Krauskopf A., Kupiec M., and McEachern M.J. 2004. A genome-wide screen for *Saccharomyces cerevisiae* deletion mutants that affect telomere length. *Proc. Natl. Acad. Sci.* **101:** 8658–8663.

Baumann P. and Cech T.R. 2001. Pot1, the putative telomere end-binding protein in fission yeast and humans. *Science* **292:** 1171–1175.

Beernink H.T., Miller K., Deshpande A., Bucher P., and Cooper J.P. 2003. Telomere maintenance in fission yeast requires an Est1 ortholog. *Curr. Biol.* **13:** 575–580.

Bertuch A.A. and Lundblad V. 2003. The Ku heterodimer performs separable activities at double strand breaks and chromosome termini. *Mol. Cell. Biol.* **23:** 8202–8215.

———. 2004. *EXO1* contributes to telomere maintenance in both telomerase-proficient and telomerase-deficient *Saccharomyces cerevisiae. Genetics* **166:** 1651–1659.

Bianchi A., Negrini S., and Shore D. 2004. Delivery of yeast telomerase to a DNA break depends on the recruitment functions of Cdc13 and Est1. *Mol. Cell* **16:** 139–146.

Booth C., Griffith E., Brady G., and Lydall D. 2001. Quantitative amplification of single-stranded DNA (QAOS) demonstrates that *cdc13-1* mutants generate ssDNA in a telomere to centromere direction. *Nucleic Acids Res.* **20:** 4414–4422.

Boulton S.J. and Jackson S.P. 1996. Identification of a *Saccharomyces cerevisiae* Ku80 homologue: Roles in DNA double strand break rejoining and in telomeric maintenance. *Nucleic Acids Res.* **24:** 4639–4648.

———. 1998. Components of the Ku-dependent non-homologous end-joining pathway are involved in telomeric length maintenance and telomeric silencing. *EMBO J.* **17:** 1819–1828.

Bucholc M., Park Y., and Lustig A.J. 2001. Intrachromatid excision of telomeric DNA as a mechanism for telomere size control in *Saccharomyces cerevisiae. Mol. Cell. Biol.* **21:** 6559–6573.

Carson M. and Hartwell L. 1985. CDC17: An essential gene that prevents telomere elongation in yeast. *Cell* **42:** 249–257.

Chan C.S. and Tye B.K. 1983a. A family of *Saccharomyces cerevisiae* repetitive autonomously replicating sequences that have very similar genomic environments. *J. Mol. Biol.* **168:** 505–523.

———. 1983b. Organization of DNA sequences and replication origins at yeast telomeres. *Cell* **33:** 563–573.

Chan S.W. and Blackburn E.H. 2003. Telomerase and ATM/Tel1p protect telomeres from nonhomologous end joining. *Mol. Cell* **11:** 1379–1387.

Chandra A., Hughes T.R., Nugent C.I., and Lundblad V. 2001. Cdc13 both positively and negatively regulates telomere replication. *Genes Dev.* **15:** 404–414.

Chen C. and Kolodner R.D. 1999. Gross chromosomal rearrangements in *Saccharomyces cerevisiae* replication and recombination defective mutants. *Nat. Genet.* **23:** 81–85.

Clissold P.M. and Ponting C.P. 2000. PIN domains in nonsense-mediated mRNA decay and RNAi. *Curr. Biol.* **10:** R888–R890.

Cohn M. and Blackburn E.H. 1995. Telomerase in yeast. *Science* **269:** 396–400.

Cohn M., McEachern M.J., and Blackburn E.H. 1998. Telomeric sequence diversity within the genus *Saccharomyces*. *Curr. Genet.* **33:** 83–91.

Conrad M.N., Wright J.H., Wolf A.J., and Zakian V.A. 1990. RAP1 protein interacts with yeast telomeres in vivo: Overproduction alters telomere structure and decreases chromosome stability. *Cell* **63:** 739–750.

Cosgrove A.J., Nieduszynski C.A., and Donaldson A.D. 2002. Ku complex controls the replication time of DNA in telomere regions. *Genes Dev.* **16:** 2485–2490.

Craven R.J. and Petes T.D. 1999. Dependence of the regulation of telomere length on the type of subtelomeric repeat in the yeast *Saccharomyces cerevisiae*. *Genetics* **152:** 1531–1541.

Craven R.J., Greenwell P.W., Dominska M., and Petes T.D. 2002. Regulation of genome stability by *TEL1* and *MEC1*, yeast homologs of the mammalian ATM and ATR genes. *Genetics* **161:** 493–507.

D′Andrea L.D. and Regan L. 2003. TPR proteins: The versatile helix. *Trends Biochem. Sci.* **26:** 655–662.

Diede S.J. and Gottschling D.E. 1999. Telomerase-mediated telomere addition in vivo requires DNA primase and DNA polymerases alpha and delta. *Cell* **99:** 723–733.

———. 2001. Exonuclease activity is required for sequence addition and Cdc13p loading at a de novo telomere. *Curr. Biol.* **11:** 1336–1340.

Dionne I. and Wellinger R.J. 1998. Processing of telomeric DNA ends requires the passage of a replication fork. *Nucleic Acids Res.* **26:** 5365–5371.

Downs J.A. and Jackson S.P. 2004. A means to a DNA end: The many roles of Ku. *Nat. Rev. Mol. Cell Biol.* **5:** 367–378.

Dunn B., Szauter P., Pardue M.L., and Szostak J.W. 1984. Transfer of yeast telomeres to linear plasmids by recombination. *Cell* **39:** 191–201.

Edmonds D., Breitkreutz B.J., and Harrington L. 2004. A genome-wide telomere screen in yeast: The long and short of it all. *Proc. Natl. Acad. Sci.* **101:** 9515–9516.

Eickbush T.H. 1997. Telomerase and retrotransposons: which came first? *Science* **277:** 911–912.

Evans S.K. and Lundblad V. 1999. Est1 and Cdc13 as comediators of telomerase access. *Science* **286:** 117–120.

———. 2000. Positive and negative regulation of telomerase access to the telomere. *J. Cell Sci.* **113:** 3357–3364.

———. 2002. The Est1 subunit of *Saccharomyces cerevisiae* telomerase makes multiple contributions to telomere length maintenance. *Genetics* **162:** 1101–1115.

Fan X. and Price C.M. 1997. Coordinate regulation of G- and C-strand length during new telomere synthesis. *Mol. Biol. Cell* **8:** 2145–2155.

Fisher T.S., Taggart A.K., and Zakian V.A. 2004. Cell cycle-dependent regulation of yeast telomerase by Ku. *Nat. Struct. Mol. Biol.* **11:** 1198–1205.

Forstemann K. and Lingner J. 2001. Molecular basis for telomere repeat divergence in budding yeast. *Mol. Cell. Biol.* **21:** 7277–7286.

———. 2005. Telomerase limits the extent of base pairing between template RNA and telomeric DNA. *EMBO Rep.* **6:** 361–366.

Forstemann K., Hoss M., and Lingner J. 2000. Telomerase-dependent repeat divergence at the 3′ ends of yeast telomeres. *Nucleic Acids Res.* **28:** 2690–2694.

Friedman K.L., Heit J.J., Long D.M., and Cech T.R. 2003. N-terminal domain of yeast telomerase reverse transcriptase: Recruitment of Est3p to the telomerase complex. *Mol. Biol. Cell.* **14:** 1–13.

Garber P.M., Vidanes G.M., and Toczyski D.P. 2005. Damage in transition. *Trends Biochem. Sci.* **30:** 63–66.

Garvik B., Carson M., and Hartwell L. 1995. Single-stranded DNA arising at telomeres in *cdc13* mutants may constitute a specific signal for the *RAD9* checkpoint. *Mol. Cell. Biol.* **15:** 6128–6138.

Grandin N., Damon C., and Charbonneau M. 2001. Ten1 functions in telomere end protection and length regulation in association with Stn1 and Cdc13. *EMBO J.* **20:** 1173–1183.

Grandin N., Reed S.I., and Charbonneau M. 1997. Stn1, a new *Saccharomyces cerevisiae* protein, is implicated in telomere size regulation in association with Cdc13. *Genes. Dev.* **11:** 512–527.

Gravel S., Larrivee M., Labrecque P., and Wellinger R.J. 1998. Yeast Ku as a regulator of chromosomal DNA end structure. *Science* **280:** 741–745.

Greider C.W. and Blackburn E.H. 1989. A telomeric sequence in the RNA of *Tetrahymena* telomerase required for telomere repeat synthesis. *Nature* **337:** 331–337.

Griffith J.D., Comeau L., Rosenfield S., Stansel R.M., Bianchi A., Moss H., and de Lange T. 1999. Mammalian telomeres end in a large duplex loop. *Cell* **97:** 503–514.

Grossi S., Puglisi A., Dmitriev P.V., Lopes M., and Shore D. 2004. Pol12, the B subunit of DNA polymerase α, functions in both telomere capping and length regulation. *Genes Dev.* **18:** 992–1006.

Hackett J.A. and Greider C.W. 2003. End resection initiates genomic instability in the absence of telomerase. *Mol. Cell. Biol.* **23:** 8450–8461.

Hackett J.A., Feldser D.M., and Greider C.W. 2001. Telomere dysfunction increases mutation rate and genomic instability. *Cell* **106:** 275–286.

Hediger F., Neumann F.R., Van Houwe G., Dubrana K., and Gasser S.M. 2002. Live imaging of telomeres: yKu and Sir proteins define redundant telomere-anchoring pathways in yeast. *Curr. Biol.* **12:** 2076–2089.

Horowitz H. and Haber J.E. 1985. Identification of autonomously replicating circular subtelomeric Y′ elements in *Saccharomyces cerevisiae*. *Mol. Cell. Biol.* **5:** 2369–2380.

Horowitz H., Thorburn P., and Haber J.E. 1984. Rearrangements of highly polymorphic regions near telomeres of *Saccharomyces cerevisiae*. *Mol. Cell. Biol.* **4:** 2509–2517.

Horvath M.P., Schweiker V.L., Bevilacqua J.M., Ruggles J.A., and Schultz S.C. 1998. Crystal structure of the *Oxytricha nova* telomere end binding protein complexed with single strand DNA. *Cell* **95:** 963–974.

Hughes T.R., Evans S.K., Weilbaecher R.G., and Lundblad V. 2000. The Est3 protein is a subunit of yeast telomerase. *Curr. Biol.* **10:** 809–812.

Jia X., Weinert T., and Lydall D. 2004. Mec1 and Rad53 inhibit formation of single-stranded DNA at telomeres of *Saccharomyces cerevisiae cdc13-1* mutants. *Genetics* **166:** 753–764.

Kironmai K.M. and Muniyappa K. 1997. Alteration of telomeric sequences and senescence caused by mutations in *RAD50* of *Saccharomyces cerevisiae*. *Genes Cells* **2:** 443–455.

Kolodner R.D., Putnam C.D., and Myung K. 2002. Maintenance of genome stability in *Saccharomyces cerevisiae*. *Science* **297:** 552–557.

Konig P. and Rhodes D. 1997. Recognition of telomeric DNA. *Trends Biochem. Sci.* **22:** 43–47.

Konig P., Giraldo R., Chapman L., and Rhodes D. 1996. The crystal structure of the DNA-binding domain of yeast RAP1 in complex with telomeric DNA. *Cell* **85:** 125–136.

Kramer K.M. and Haber J. E. 1993. New telomeres in yeast are initiated with a highly selected subset of TG1-3 repeats. *Genes Dev.* **7:** 2345–2356.

Kyrion G., Boakye K.A., and Lustig A.J. 1992. C-terminal truncation of RAP1 results in the deregulation of telomere size, stability, and function in *Saccharomyces cerevisiae*. *Mol. Cell. Biol.* **12:** 5159–5173.

Laroche T., Martin S.G., Gotta M., Gorham H.C., Pryde F.E., Louis E.J., and Gasser S.M. 1998. Mutation of yeast Ku genes disrupts the subnuclear organization of telomeres. *Curr. Biol.* **8:** 653–656.

Larrivee M., LeBel C., and Wellinger R.J. 2004. The generation of proper constitutive G-tails on yeast telomeres is dependent on the MRX complex. *Genes Dev.* **18:** 1391–1396.

Lendvay T.S., Morris D.K., Sah J., Balasubramanian B., and Lundblad V. 1996. Senescence mutants of *Saccharomyces cerevisiae* with a defect in telomere replication identify three additional *EST* genes. *Genetics* **144:** 1399–1412.

Levy D.L. and Blackburn E.H. 2004. Counting of Rif1p and Rif2p on *Saccharomyces cerevisiae* telomeres regulates telomere length. *Mol. Cell. Biol.* **24:** 10857–10867.

Lewis L.K., Karthikeyan G., Westmoreland J.W., and Resnick M.A. 2002. Differential suppression of DNA repair deficiencies of yeast *rad50*, *mre11* and *xrs2* mutants by *EXO1* and *TLC1* (the RNA component of telomerase). *Genetics* **160:** 49–62.

Li B. and Lustig A.J. 1996. A novel mechanism for telomere size control in *Saccharomyces cerevisiae*. *Genes Dev.* **10:** 1310–1326.

Lichten M. 2005. Rad50 connects by hook or by crook. *Nat. Struct. Mol. Biol.* **12:** 392–393.

Lingner J. and Cech T.R. 1996. Purification of telomerase from *Euplotes aediculatus*: Requirement of a primer 3′ overhang. *Proc. Natl. Acad. Sci.* **93:** 10712–10717.

Lingner J., Cech T.R., Hughes T.R., and Lundblad V. 1997. Three Ever Shorter Telomere (*EST*) genes are dispensable for in vitro yeast telomerase activity. *Proc. Natl. Acad. Sci.* **94:** 11190–11195.

Louis E.J. 1995. The chromosome ends of *Saccharomyces cerevisiae*. *Yeast* **11:** 1553–1573.

Louis E.J. and Vershinin A.V. 2005. Chromosome ends: Different sequences may provide conserved functions. *Bioessays* **27:** 685–697.

Lundblad V. and Blackburn E.H. 1993. An alternative pathway for yeast telomere maintenance rescues *est1⁻* senescence. *Cell* **73:** 347–360.

Lundblad V. and Szostak J.W. 1989. A mutant with a defect in telomere elongation leads to senescence in yeast. *Cell* **57:** 633–643.

Lustig A. and Petes T.D. 1986. Identification of yeast mutants with altered telomere structure. *Proc. Natl. Acad. Sci.* **83:** 1398–1402.

Lustig A.J., Kurtz S., and Shore D. 1990. Involvement of the silencer and UAS binding protein RAP1 in regulation of telomere length. *Science* **250:** 549–553.

Lydall D. and Weinert T. 1995. Yeast checkpoint genes in DNA damage processing: Implications for repair and arrest. *Science* **270:** 1488–1491.

Marcand S., Brevet V., and Gilson E. 1999. Progressive *cis*-inhibition of telomerase upon telomere elongation. *EMBO J.* **18:** 3509–3519.

Marcand S., Gilson E., and Shore D. 1997. A protein-counting mechanism for telomere length regulation in yeast. *Science* **275:** 986–990.

Marcand S., Brevet V., Mann C., and Gilson E. 2000. Cell cycle restriction of telomere elongation. *Curr. Biol.* **10:** 487–490.

Maringele L. and Lydall D. 2002. Exol-dependent single-stranded DNA at telomeres activates subsets of DNA damage and spindle checkpoint pathways in budding yeast *yku70Δ* mutants. *Genes Dev.* **16:** 1919–1933.

McEachern M.J. and Blackburn E.H. 1994. A conserved sequence motif within the exceptionally diverse telomeric sequences of budding yeasts. *Proc. Natl. Acad. Sci.* **91:** 3453–3457.

McEachern M.J. and Hicks J.B. 1993. Unusually large telomeric repeats in the yeast *Candida albicans. Mol. Cell. Biol.* **13:** 551–560.

Mieczkowski P.A., Mieczkowska J.O., Dominska M., and Petes T.D. 2003. Genetic regulation of telomere-telomere fusions in the yeast *Saccharomyces cerevisiae. Proc. Natl. Acad. Sci.* **100:** 10854–10859.

Mishra K. and Shore D. 1999. Yeast Ku protein plays a direct role in telomeric silencing and counteracts inhibition by rif proteins. *Curr. Biol.* **9:** 1123–1126.

Mitton-Fry R.M., Anderson E.M., Hughes T.R., Lundblad V., and Wuttke D.S. 2002. Conserved structure for single-stranded telomeric DNA recognition. *Science* **296:** 145–147.

Mitton-Fry R.M., Anderson E.M., Theobald D.L., Glustrom L.W., and Wuttke D.S. 2004. Structural basis for telomeric single-stranded DNA recognition by yeast Cdc13. *J. Mol. Biol.* **338:** 241–255.

Mnaimneh S., Davierwala A.P., Haynes J., Moffat J., Peng W.T., Zhang W., Yang X., Pootoolal J., Chua G., Lopez A., Trochesset M., Morse D., Krogan N.J., Hiley S.L., Li Z., Morris Q., Grigull J., Mitsakakis N., Roberts C.J., Greenblatt J.F., Boone C., Kaiser C.A., Andrews B.J., and Hughes T.R. 2004. Exploration of essential gene functions via titratable promoter alleles. *Cell* **118:** 31–44.

Moreau S., Morgan E.A., and Symington L.S. 2001. Overlapping functions of the *Saccharomyces cerevisiae* Mre11, Exo1 and Rad27 nucleases in DNA metabolism. *Genetics* **159:** 1423–1433.

Morris D.K. and Lundblad V. 1997. Programmed translational frameshifting in a gene required for yeast telomere replication. *Curr. Biol.* **7:** 969–976.

Murray A.W. and Szostak J.W. 1983. Construction of artificial chromosomes in yeast. *Nature* **305:** 189–193.

Myung K., Datta A., and Kolodner R.D. 2001. Suppression of spontaneous chromosomal rearrangements by S phase checkpoint functions in *Saccharomyces cerevisiae. Cell* **104:** 397–408.

Nugent C.I., Hughes T.R., Lue N.F., and Lundblad V. 1996. Cdc13p: A single-strand telomeric DNA-binding protein with a dual role in yeast telomere maintenance. *Science* **274:** 249–252.

Nugent C.I., Bosco G., Ross L.O., Evans S.K., Salinger A.P., Moore J.K., Haber J.E., and Lundblad V. 1998. Telomere maintenance is dependent on activities required for end repair of double-strand breaks. *Curr. Biol.* **8:** 657–660.

Pennock E., Buckley K., and Lundblad V. 2001. Cdc13 delivers separate complexes to the telomere for end protection and replication. *Cell* **104:** 387–396.

Peterson S.E., Stellwagen A.E., Diede S.J., Singer M.S., Haimberger Z.W., Johnson C.O., Tzoneva M., and Gottschling D.E. 2001. The function of a stem-loop in telomerase RNA is linked to the DNA repair protein Ku. *Nat. Genet.* **27:** 64–67.

Pluta A.F., Dani G.M., Spear B.B., and Zakian V.A. 1984. Elaboration of telomeres in yeast: Recognition and modification of termini from *Oxytricha* macronuclear DNA. *Proc. Natl. Acad. Sci.* **81:** 1475–1479.

Polotnianka R.M., Li J., and Lustig A.J. 1998. The yeast Ku heterodimer is essential for protection of the telomere against nucleolytic and recombinational activities. *Curr. Biol.* **8:** 831–834.

Porter S.E., Greenwell P.W., Ritchie K.B., and Petes T.D. 1996. The DNA-binding protein Hdf1p (a putative Ku homologue) is required for maintaining normal telomere length in *Saccharomyces cerevisiae*. *Nucleic Acids Res.* **24:** 582–585.

Prescott J. and Blackburn E.H. 1997a. Functionally interacting telomerase RNAs in the yeast telomerase complex. *Genes Dev.* **11:** 2790–2800.

———. 1997b. Telomerase RNA mutations in *Saccharomyces cerevisiae* alter telomerase action and reveal nonprocessivity in vivo and in vitro. *Genes Dev.* **11:** 528–540.

Pryde F.E. and Louis E.J. 1997. *Saccharomyces cerevisiae* telomeres. A review. *Biochemistry* **62:** 1232–1241.

———. 1999. Limitations of silencing at native yeast telomeres. *EMBO J.* **18:** 2538–2550.

Qi H. and Zakian V.A. 2000. The *Saccharomyces* telomere-binding protein Cdc13p interacts with both the catalytic subunit of DNA polymerase α and the telomerase-associated Est1 protein. *Genes Dev.* **14:** 1777–1788.

Reichenbach P., Hoss M., Azzalin C.M., Nabholz M., Bucher P., and Lingner J. 2003. A human homolog of yeast Est1 associates with telomerase and uncaps chromosome ends when overexpressed. *Curr. Biol.* **13:** 568–574.

Ritchie K.B. and Petes T.D. 2000. The Mre11p/Rad50p/Xrs2p complex and the Tel1p function in a single pathway for telomere maintenance in yeast. *Genetics* **155:** 475–479.

Ritchie K.B., Mallory J.C., and Petes T.D. 1999. Interactions of *TLC1* (which encodes the RNA subunit of telomerase), *TEL1*, and *MEC1* in regulating telomere length in the yeast *Saccharomyces cerevisiae*. *Mol. Cell. Biol.* **19:** 6065–6075.

Roy R., Meier B., McAinsh A.D., Feldmann H.M., and Jackson S.P. 2004. Separation-of-function mutants of yeast Ku80 reveal a Yku80p-Sir4p interaction involved in telomeric silencing. *J. Biol. Chem.* **279:** 86–94.

Runge K.W. and Zakian V.A. 1996. *TEL2*, an essential gene required for telomere length regulation and telomere position effect in *Saccharomyces cerevisiae*. *Mol. Cell. Biol.* **16:** 3094–3105.

Schramke V., Luciano P., Brevet V., Guillot S., Corda Y., Longhese M.P., Gilson E., and Geli V. 2004. RPA regulates telomerase action by providing Est1p access to chromosome ends. *Nat. Genet.* **36:** 46–54.

Seto A.G., Livengood A.J., Tzfati Y., Blackburn E.H., and Cech T.R. 2002. A bulged stem tethers Est1p to telomerase RNA in budding yeast. *Genes Dev.* **16:** 2800–2812.

Shampay J. and Blackburn E.H. 1988. Generation of telomere-length heterogeneity in *Saccharomyces cerevisiae*. *Proc. Natl. Acad. Sci.* **85:** 534–538.

Shampay J., Szostak J.W., and Blackburn E.H. 1984. DNA sequences of telomeres maintained in yeast. *Nature* **310:** 154–157.

Shore D. and Nasmyth K. 1987. Purification and cloning of a DNA binding protein from yeast that binds to both silencer and activator elements. *Cell.* **51:** 721–732.

Singer M.S. and Gottschling D.E. 1994. TLC1: Template RNA component of *Saccharomyces cerevisiae* telomerase. *Science* **266:** 404–409.

Singer M.S., Kahana A., Wolf A.J., Meisinger L.L., Peterson S.E., Goggin C., Mahowald M., and Gottschling D.E. 1998. Identification of high-copy disruptors of telomeric silencing in *Saccharomyces cerevisiae*. *Genetics* **150**: 613–632.

Singh S.M. and Lue N.F. 2003. Ever shorter telomere 1 (*EST1*)-dependent reverse transcription by *Candida* telomerase in vitro: Evidence in support of an activating function. *Proc. Natl. Acad. Sci.* **100**: 5718–5723.

Singh S.M., Steinberg-Neifach O., Mian I.S., and Lue N.F. 2002. Analysis of telomerase in *Candida albicans*: Potential role in telomere end protection. *Eukaryot. Cell.* **1**: 967–977.

Smith C.D., Smith D.L., DeRisi J.L., and Blackburn E.H. 2003. Telomeric protein distributions and remodeling through the cell cycle in *Saccharomyces cerevisiae*. *Mol. Biol. Cell.* **14**: 556–570.

Snow B.E., Erdmann N., Cruickshank J., Goldman H., Gill R.M., Robinson M.O., and Harrington L. 2003. Functional conservation of the telomerase protein Est1p in humans. *Curr. Biol.* **13**: 698–704.

Stellwagen A.E., Haimberger Z.W., Veatch J.R., and Gottschling D.E. 2003. Ku interacts with telomerase RNA to promote telomere addition at native and broken chromosome ends. *Genes Dev.* **17**: 2384–2395.

Stracker T.H., Theunissen J.W., Morales M., and Petrini J.H. 2004. The Mre11 complex and the metabolism of chromosome breaks: The importance of communicating and holding things together. *DNA Repair* **3**: 845–854.

Sugawara N. 1998. "DNA sequences at the telomeres of the fission yeast *S. pombe*." Ph.D. thesis, Harvard University, Cambridge, Massachusetts.

Szostak J.W. and Blackburn E.H. 1982. Cloning yeast telomeres on linear plasmid vectors. *Cell* **29**: 245–255.

Taggart A.K., Teng S.C., and Zakian V.A. 2002. Est1p as a cell cycle-regulated activator of telomere-bound telomerase. *Science* **297**: 1023–1026.

Takata H., Tanaka Y., and Matsuura A. 2005. Late S phase-specific recruitment of Mre11 complex triggers hierarchical assembly of telomere replication proteins in *Saccharomyces cerevisiae*. *Mol. Cell* **17**: 573–583.

Takata H., Kanoh Y., Gunge N., Shirahige K., and Matsuura A. 2004. Reciprocal association of the budding yeast ATM-related proteins Tel1 and Mec1 with telomeres in vivo. *Mol. Cell* **14**: 515–522.

Teixeira M.T., Arneric M., Sperisen P., and Lingner J. 2004. Telomere length homeostasis is achieved via a switch between telomerase-extendible and -nonextendible states. *Cell* **117**: 323–335.

Theobald D.L. and Wuttke D.S. 2004. Prediction of multiple tandem OB-fold domains in telomere end-binding proteins Pot1 and Cdc13. *Structure* **12**: 1877–1879.

Theobald D.L., Cervantes R.B., Lundblad V., and Wuttke D.S. 2003. Homology among telomeric end-protection proteins. *Structure* **11**: 1049–1050.

Tomaska L., Willcox S., Slezakova J., Nosek J., and Griffith J.D. 2004. Taz1 binding to a fission yeast model telomere: Formation of telomeric loops and higher order structures. *J. Biol. Chem.* **279**: 50764–50772.

Tsukamoto Y., Kato J., and Ikeda H. 1997. Silencing factors participate in DNA repair and recombination in *Saccharomyces cerevisiae*. *Nature* **388**: 900–903.

Tsukamoto Y., Taggart A.K., and Zakian V.A. 2001. The role of the Mre11-Rad50-Xrs2 complex in telomerase-mediated lengthening of *Saccharomyces cerevisiae* telomeres. *Curr. Biol.* **11**: 1328–1335.

Virta-Pearlman V., Morris D.K., and Lundblad V. 1996. Est1 has the properties of a single-stranded telomere end-binding protein. *Genes Dev.* **10**: 3094–3104.

Walker J.R., Corpina R.A., and Goldberg J. 2001. Structure of the Ku heterodimer bound to DNA and its implications for double-strand break repair. *Nature* **412:** 607–614.

Walmsley R.W., Chan C.S., Tye B.K., and Petes T.D. 1984. Unusual DNA sequences associated with the ends of yeast chromosomes. *Nature* **310:** 157–160.

Wang S.S. and Zakian V.A. 1990. Sequencing of *Saccharomyces* telomeres cloned using T4 DNA polymerase reveals two domains. *Mol. Cell. Biol.* **10:** 4415–4419.

Weinert T.A. and Hartwell L.H. 1993. Cell cycle arrest of *cdc* mutants and specificity of the *RAD9* checkpoint. *Genetics* **134:** 63–80.

Wellinger R.J., Wolf A.J., and Zakian V.A. 1993. Saccharomyces telomeres acquire single-strand TG1-3 tails late in S phase. *Cell* **72:** 51–60.

Wellinger R.J., Ethier K., Labrecque P., and Zakian V.A. 1996. Evidence for a new step in telomere maintenance. *Cell* **85:** 423–433.

Williams B., Bhattacharyya M.K., and Lustig A.J. 2005. Mre11p nuclease activity is dispensable for telomeric rapid deletion. *DNA Repair* **15:** 994–1005.

Witkin K.L. and Collins K. 2004. Holoenzyme proteins required for the physiological assembly and activity of telomerase. *Genes Dev.* **18:** 1107–1118.

Zhou J., Hidaka K., and Futcher B. 2000. The Est1 subunit of yeast telomerase binds the Tlc1 telomerase RNA. *Mol. Cell. Biol.* **20:** 1947–1955.

Zubko M.K., Guillard S., and Lydall D. 2004. Exo1 and Rad24 differentially regulate generation of ssDNA at telomeres of *Saccharomyces cerevisiae cdc13-1* mutants. *Genetics* **168:** 103–115.

13

Mammalian Telomeres

Titia de Lange
Laboratory of Cell Biology and Genetics
The Rockefeller University
New York, New York 10021-6399

As IN MOST OTHER EUKARYOTES, THE ENDS of mammalian chromosomes are protected by the combined action of telomeric DNA, telomere-associated proteins, and telomerase. In the decade since the last Cold Spring Harbor Laboratory Telomere Monograph was published, substantial progress has been made on each aspect of mammalian telomere biology. Important facets of the DNA component, including the t-loop configuration and the structure of the telomere terminus, have been illuminated; a telomere-specific protein complex, now referred to as shelterin, has been identified; and several DNA-damage response and repair factors have been implicated in telomere function. Studies of telomere pathology, resulting from shelterin inhibition or other insults, have revealed the fate of dysfunctional telomeres and their impact on chromosomes and cells. Furthermore, the principles of telomere length homeostasis and the role of shelterin in controlling telomere elongation by telomerase have emerged. This chapter focuses on the DNA and protein components of mammalian telomeres and the mechanisms of telomere function. The details of mammalian telomerases are discussed in Chapters 2 and 3 and aspects of telomere function that relate to cancer and aging are covered in Chapters 4–6.

TELOMERIC DNA

The Telomeric TTAGGG Repeat Array

Mammals and all other vertebrates have telomeres made up of tandem TTAGGG repeats (Fig. 1) (Moyzis et al. 1988; Meyne et al. 1989). This

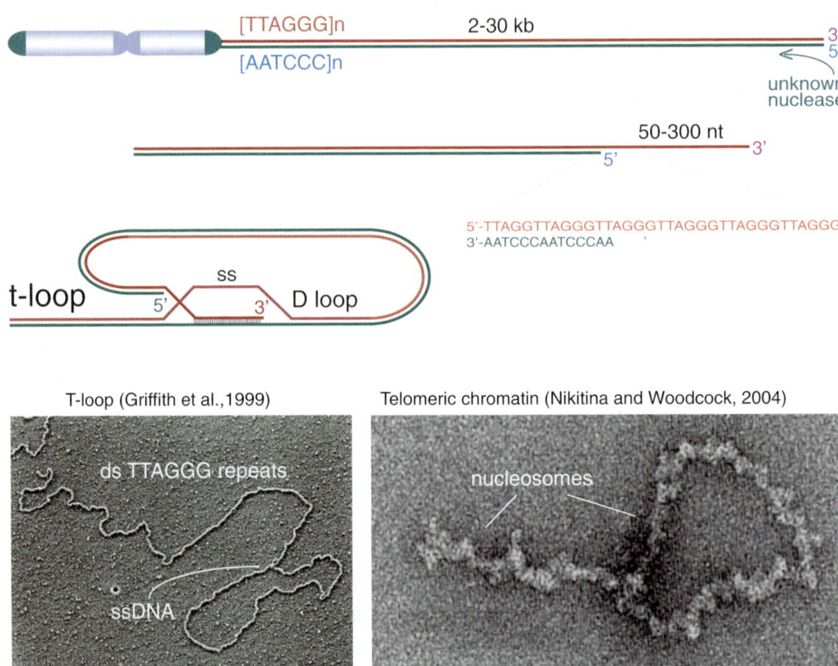

Figure 1. Telomere structure. (*Top*) Schematic of human telomeres in the "open" and t-loop states. (*Bottom left*) Electron microscopic micrographs of mouse telomeric DNA in the t-loop configuration in naked isolated DNA. (*Bottom right*) Electron microscopic micrograph of telomeric chromatin from chicken erythrocytes. (*Bottom left*, Reprinted, with permission, from Griffith et al. 1999 [©Elseveier]; *bottom right*, modified, with permission, from Nikitina and Woodcock 2004 [©The Rockefeller University Press].)

sequence is dictated by telomerase and may be the oldest telomeric repeat sequence because it is also found in several fungi, protozoa, and plants. The length of the TTAGGG repeat tract is an inherited trait with length settings that vary from species to species (Kipling and Cooke 1990; Starling et al. 1990; Zhu et al. 1998; Hathcock et al. 2002). The longest telomeres are found in rats and some strains of *Mus musculus*, which have TTAGGG repeat arrays of up to 150 kb (Kipling and Cooke 1990; Starling et al. 1990; Makarov et al. 1993). Human telomeres are typically about 10 kb at birth and gradually shorten with cell divisions (Cooke and Smith 1986; Allshire et al. 1988, 1989; de Lange et al. 1990; Harley et al. 1990). The minimum tract length required for mammalian telomere function is not known. New telomeres can be formed with telomere seeds that have as little as 400 bp of telomeric DNA (Farr et al. 1991; Barnett et al. 1993; Hanish et al. 1994), but such newly formed telomeres always have telomeric DNA added to them.

Telomerase-positive human cell lines have widely varying telomere length settings, ranging from 2 to 30 kb. As discussed later in this chapter, telomere length is governed by a homeostasis mechanism that is established by telomere-associated proteins. Although telomere length homeostasis keeps telomeres within a set range, each telomere is heterogeneous in size. The mechanism by which this heterogeneity arises is poorly understood, but overexpression of certain *Rap1* alleles can either sharpen or broaden the size distribution, indicating some level of regulation of this process (Li and de Lange 2003).

A region made up of ΨTTAGGG repeats, including TTGGGG, TGAGGG, and TCAGGG units, is found proximal to the TTAGGG repeat array of human telomeres (Baird et al. 1995; Varley et al. 2002). Further inward, the subtelomeric regions of human chromosomes are highly repetitive (Riethman et al. 2004) and undergo frequent deletions and translocations (Brown et al. 1990; Wilkie et al. 1991).

TTAGGG-like sequences are not limited to chromosome ends. They occur at pericentric sites in certain rodents, fruit bats, lemurs, chickens, tree frogs, and other vertebrates and can represent up to 5% of the genome (Southern 1970; Meyne et al. 1990). Even the streamlined genome of the Japanese pufferfish (*Fugu rubripes*) contains TTAGGG repeat sequences among its minisatellites (Edwards et al. 1998). Fortunately, the molecular analysis of human and mouse telomeres is simplified by the absence of extensive interstitial telomeric DNA in these species.

The Telomere Terminus

Like many other eukaryotes, mammals have telomeres that end in a 3′ overhang (Fig. 1) (Makarov et al. 1997). The telomeric overhang is composed of TTAGGG repeats and can be several hundred nucleotides in length (Wright et al. 1997; Huffman et al. 2000). It is anticipated that the overhang is crucial for the function of telomeres, because it is involved in the formation of t-loops and because it binds the single-stranded DNA-binding protein POT1. Since telomerase is not required for overhang formation (Hemann and Greider 1999; Huffman et al. 2000), some other pathway must be responsible. A 3′ overhang could be generated in the course of lagging-strand DNA synthesis, but this process cannot account for the presence of the 3′ overhang at the telomere end created by leading-strand DNA synthesis. Most likely, a 5′ exonuclease or a combination of an endonuclease and a helicase are needed to process telomere termini (Makarov et al. 1997).

The trimming of telomere termini by a nuclease could explain the high rate of shortening of human telomeres. The rate of shortening predicted from the "end-replication" problem is approximately 3 bp/end/cell division (Chapter 2), which is indeed observed in budding yeast lacking telomerase. However, human cells lacking telomerase lose 50–300 bp/end/population doubling (PD) (Huffman et al. 2000). Perhaps this rapid attrition is due to the nuclease that creates the 3' overhang. Given that telomere shortening is a proposed tumor suppressor pathway (see Chapters 4 and 5), it will be important to understand this process in greater detail.

The 3' nucleotide of the telomere can represent each position within the TTAGGG repeat (Sfeir et al. 2005). When telomerase is present, a modest prevalence of TAG-3' ends is observed, consistent with the in vitro products of telomerase (see Chapters 2 and 3). In contrast to the variability of the 3' end, the 5' end of the telomere is nearly always ATC-5' (Sfeir et al. 2005). Human cells with diminished POT1 levels do not show this precision at the 5' ends of their chromosomes, hinting at a processing reaction that is controlled by shelterin (Hockemeyer et al. 2005).

The t-loop

Electron microscopy of purified human and mouse telomeres revealed a lariat configuration, referred to as the telomeric loop or t-loop (Fig. 1) (Griffith et al. 1999). t-loops are large duplex loops formed through the strand invasion of the 3' overhang into the duplex part of the telomere. The size distribution of the circle part of the t-loops is correlated with the length of the telomeric repeat array and compatible with more or less random strand invasion (Griffith et al. 1999). The crucial aspect of t-loop formation is assumed to be the sequestration of the chromosome end.

In vivo, telomeric proteins presumably maintain the t-loop configuration, but once protein-free DNA is isolated, the invaded overhang can be displaced by the D-loop, resolving the lariat. Visualization of t-loops in naked DNA by electron microscopy therefore requires interstrand cross-linking with psoralen/UV. Using this method, up to 20–40% of the telomeric DNA can be recovered in the t-loop state, but the actual frequency of t-loops may be higher as t-loops could unfold because of insufficient cross-linking or might break during their isolation. t-loop-like structures were also observed in whole telomeric chromatin isolated from chicken erythrocytes and mouse splenocytes (Nikitina and Woodcock 2004).

THE PROTEIN COMPONENTS OF MAMMALIAN TELOMERES

Three types of factors associate with telomeric DNA: nucleosomes (Fig. 1), the six proteins that make up shelterin (see Fig. 2), and a number of other chromosome transaction factors that also function elsewhere (see Table 1). The current knowledge about each of these components of the telomeric chromatin is summarized below.

Telomeric Nucleosomes

Studies on the long telomeres of rat cells showed that most of the TTAGGG repeat tract is packaged in nucleosomal chromatin (Makarov et al. 1993). The telomeric nucleosomes are closely spaced and the chromatin contains H1 (Makarov et al. 1993; Nikitina and Woodcock 2004). Electron microscopic analysis of t-loop chromatin shows nucleosomes throughout the tail and the circle, and the chromatin resembles canonical nucleosomal arrays (Nikitina and Woodcock 2004). One unusual aspect of telomeric nucleosomes is that the core particle is hypersensitive to MNase I (Tommerup et al. 1994). This may be due to the fact that TTAGGG repeats are a suboptimal sequence for the trajectory around the nucleosome core. Telomeric chromatin contains histone H3 that is di- and trimethylated on lysine 9, and this modification is required for the association with HP1 isoforms and also to maintain normal telomere length (Garcia-Cao et al. 2004).

Telomeric Silencing, but No Clustering, Peripheral Position, or Late Replication

In many eukaryotes, telomeres have the ability to silence adjacent genes (see Chapter 10). Similarly in human cells, genes placed next to a telomere are suppressed, and the suppression increases with telomere length (Baur et al. 2001, 2004; Koering et al. 2002). Inhibition of histone deacetylases can alleviate the silencing, implicating chromatin modification in this process. In contrast to telomeric silencing, several other well-known features of telomeric chromatin in unicellular organisms do not occur in mammalian cells. Budding yeast telomeres form aggregates of approximately eight chromosome ends that are positioned close to the nuclear envelope (see Chapter 12), whereas mammalian telomeres are not obviously clustered or close to the nuclear envelope (Vourc'h et al. 1993; Luderus et al. 1996). Furthermore, whereas yeast telomeres replicate late, replication of human telomeric

DNA can be detected throughout S phase (Ten Hagen et al. 1990; Wright et al. 1999).

Shelterin

During the past 10 years, it has become clear that human telomeres contain a complex of at least six telomere-specific proteins (Fig. 2). Two of these proteins, TRF1 and TRF2, bind to double-stranded telomeric DNA and a third, POT1, binds to TTAGGG repeats in single-stranded form. These three DNA-binding proteins, held together by protein interactions, give the telomeric complex exquisite specificity for the sequence and structure of telomeric DNA. This complex is referred to as shelterin because it protects chromosome ends. Shelterin is implicated in the formation of t-loops, affects the structure of the telomere terminus, and controls the synthesis of telomeric DNA by telomerase. These effects of shelterin on the structure of the telomeric DNA are proposed to help protect chromosome ends from DNA damage surveillance and repair pathways.

Whereas the six components of shelterin can be found together in a single complex in fractionated nuclear extracts (Liu et al. 2004a; Ye et al. 2004b), subcomplexes lacking one or two shelterin subunits have also been observed (Fig. 2) (Liu et al. 2004a; Mattern et al. 2004; Ye et al. 2004b). These variants of shelterin are likely to be relevant to telomere function, but the exact stoichiometry of the shelterin components and subcomplexes on the telomere is not known and the distribution of shelterin on the telomeric repeat array needs to be determined. It is unlikely that any shelterin component has a singular position along telomeres because their abundance on chromosome ends increases as telomeres lengthen (Smogorzewska et al. 2000; Loayza and de Lange 2003).

Shelterin is ubiquitously expressed and abundant. Estimates suggest that hundreds of copies of shelterin are bound along the duplex telomeric repeat array of each chromosome end. Shelterin is at telomeres throughout the cell cycle, it does not accumulate elsewhere in the nucleus, and its function is limited to telomeres. On the basis of these criteria, shelterin proteins are distinguished from several DNA-damage-processing proteins and various other factors found at telomeres. These additional telomere-associated proteins can have crucial roles at telomeres, but they also have nontelomeric functions. They often accumulate elsewhere in the cell, they are less abundant at telomeres than shelterin, and several are at telomeres transiently. The contribution of these additional factors to telomere function is discussed following the sections on the shelterin subunits.

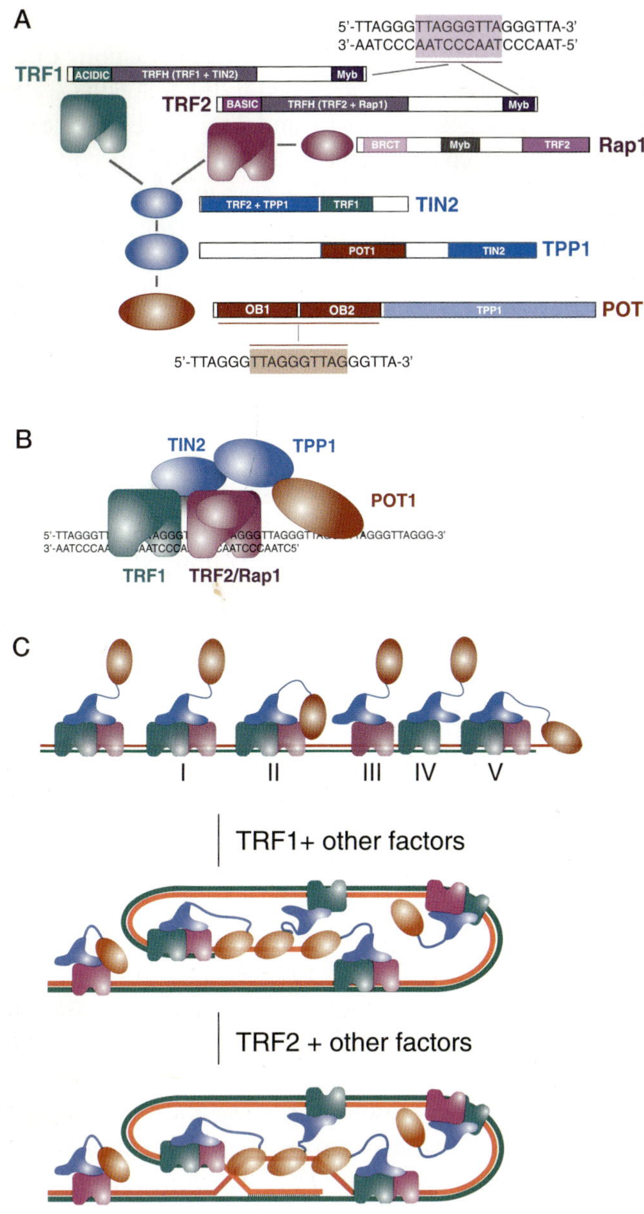

Figure 2. Shelterin. (*A*) Schematic of the subunits of shelterin and their interactions (not shown is the interaction between TRF2 and POT1 [Yang et al. 2005]). (*B*) Schematic showing the overall composition of shelterin. (*C*) Alternative shelterin complexes and telomere configurations. (*I–V*) Five alternative shelterin complexes for which in vitro and/or in vivo data are available. Telomeres are drawn in three distinct configurations: linear, folded, and t-loop. TRF1 and TRF2 have in vitro activities that could mediate telomere folding and formation of t-loops.

The Subunits of Shelterin: TRF1, TRF2, TIN2, Rap1, TPP1, POT1

The identification, characteristics, and protein interactions of each of the six known shelterin subunits are described in the following sections. Initial functional studies revealed that TRF1 and TIN2 regulate telomere length, whereas TRF2 protects chromosome ends from fusions and prevents activation of a DNA-damage response. However, emerging evidence indicates that the shelterin subunits act together to control telomere length and protect chromosome ends. Therefore, these functional aspects are not discussed for each subunit. Instead, the function of shelterin is discussed in separate sections on telomere protection and telomere length homeostasis.

TRF1

The first component of shelterin to be identified, TRF1, was detected in HeLa cell nuclear extracts as a DNA-binding activity that associates specifically with duplex TTAGGG repeats (Zhong et al. 1992) and hence its name, TTAGGG-repeat-binding factor 1. TRF1 is a 439-amino-acid protein with three domains: an amino-terminal acidic region, a dimerization domain, and a carboxy-terminal DNA-binding domain (Fig. 2A) (Chong et al. 1995; Bianchi et al. 1997).

The TRF1 DNA-binding domain is an approximately 50-amino-acid region of the SANT/Myb-type that is also structurally related to the homeodomain (see Chapter 11). TRF1 is a dimer in solution and binds DNA predominantly in this form (Bianchi et al. 1997; Griffith et al. 1998). Dimerization of TRF1 is mediated by the TRF homology (TRFH) domain, the hallmark of the TRF family, which includes TRF2, *Schizosaccharomyces pombe* Taz1, and the TRF from trypanosomes (Bianchi et al. 1997; Broccoli et al. 1997a,b; Li et al. 2000, 2005; Fairall et al. 2001). The linker region between the Myb domain and the TRFH domain appears to be flexible and diverges rapidly in mammalian evolution (Broccoli et al. 1997b). As a result of alternative splicing, human TRF1 is expressed as two closely migrating proteins that differ by 20 amino acids in the linker region (Shen et al. 1997; van Steensel and de Lange 1997). These two forms of human TRF1 are functionally identical, and only one form of TRF1 is expressed in mouse cells (Broccoli et al. 1997b).

SELEX (systematic evolution of ligands by exponential enrichment) revealed that the TRF1 consensus sequence is a half-site of 5'-YTAGGGTTR-3' and that single-base changes in the AGGGTT core are not tolerated (Fig. 2A) (Bianchi et al. 1999). TRF1 will bind these YTAGGGTTR sites independently of the surrounding sequences or the proximity of a DNA end. In vivo, TRF1 also binds TTAGGG repeats independently of their location within the chromosome (Smogorzewska et al. 2000; Krutilina et al. 2001). As expected from its dimeric nature, the TRF1-binding substrates that emerge from SELEX usually contain two YTAGGGTTR half-sites (Bianchi et al. 1999). The simultaneous binding of two Myb domains endows TRF1 dimers with an approximately tenfold greater affinity for DNA. The half-sites can be in an overlapping arrangement (5′-YTAGGGTTAGGGTTR-3′), but TRF1 can also bind two half-sites at a distance and in opposite orientation. From these studies, the view emerged that TRF1 has extreme spatial flexibility, presumably because of an unstructured linker region in between the TRFH and Myb domain.

Immunofluorescence and chromatin immunoprecipitation (ChIP) data show that TRF1 is specifically localized at human and mouse telomeres at all stages of the mitotic and meiotic cell cycle (Chong et al. 1995; Broccoli et al. 1997b; van Steensel and de Lange 1997; Scherthan et al. 2000; Crabbe et al. 2004).

Two-hybrid screens identified the shelterin subunit TIN2 as well as the tankyrases as TRF1-interacting proteins (Smith et al. 1998; Kim et al. 1999; Kaminker et al. 2001). These proteins are discussed in detail below. Other telomere-associated proteins that bind to TRF1 directly are the Ku70/80 heterodimer and the BLM (Bloom's Syndrome) helicase (Hsu et al. 2000; Lillard-Wetherell et al. 2004; Opresko et al. 2004). TRF1 also binds to the carboxyl terminus of the ATM (ataxia telangiectasia–mutated) kinase and can be phosphorylated by ATM on Ser-219 (Kishi et al. 2001). Finally, the nucleotide diphosphate kinase nm23-H2 (Nosaka et al. 1998), the rRNA maturation factor PINX1 (Zhou and Lu 2001), components of the mitotic spindle (Nakamura et al. 2001), and the transcriptional repressor SALL1 (Netzer et al. 2001) have been reported to interact with TRF1, but the significance of these interactions has not yet been established.

In telomerase-positive human cells, TRF1 acts as a negative regulator of telomere length. Diminished TRF1 loading on telomeres leads to telomere elongation, but no telomere deprotection phenotype is observed. However, TRF1 is essential in the mouse, suggesting that TRF1 contributes to the protection of telomeres (Karlseder et al. 2003; Iwano et al. 2004). $Terf1^{-/-}$ embryos die early in development at E5–E6 and the inner cell mass of the blastocysts has a severe growth defect accompanied by apoptosis. The early embryonic phenotype is partially suppressed by deletion of p53, consistent with its being due to telomere deprotection. Conditional deletion of TRF1 will be required to explore this possibility further.

TRF2

TRF2 was identified on the basis of sequence similarity to the Myb domain of TRF1 (Bilaud et al. 1997; Broccoli et al. 1997a). Cloning of the full-length cDNA revealed a 500-amino-acid protein with a TRFH domain, and a short basic amino-terminal domain that is rich in arginine residues (Broccoli et al. 1997a). The structures of TRFH domains of TRF1 and TRF2 are very similar, but a steric clash prevents heterodimerization and distinct proteins bind to the TRFH domains of TRF1 and TRF2 (Fairall et al. 2001; see Chapter 11).

Although the DNA-binding features of TRF2 have not been fully worked out, there are strong parallels with TRF1 (Broccoli et al. 1997a; A. Bianchi et al., unpubl.). TRF1 and TRF2 have the same sequence specificity and spatial flexibility, but TRF2 has a greater tendency to form higher-order oligomers and does not have telomeric DNA-pairing activity (Stansel et al. 2001).

Two shelterin subunits, Rap1 and TIN2, interact with TRF2 (Fig. 2A) (Li et al. 2000; Houghtaling et al. 2004; Liu et al. 2004a; Ye et al. 2004b). TRF2 has also been reported to interact with POT1 (Yang et al. 2005), the WRN RecQ helicases (Opresko et al. 2002), Ku70 (Song et al. 2000), and the ATM kinase (Karlseder et al. 2004).

Like TRF1, TRF2 is essential, and $TRF2^{-/-}$ mice die early in embryogenesis (Celli and de Lange 2005). A floxed TRF2 allele ($TRF2^{F}$) was created, allowing conditional deletion of TRF2 in mouse embryo fibroblasts (MEFs). The deletion of $TRF2$ results in p53-dependent senescence, genome-wide chromosome end fusions, and a DNA-damage response (Celli and de Lange 2005). These phenotypes explain the early embryonic lethality of $TRF2$ null mice and account for the death of embryos lacking TRF1 or TIN2, as these proteins stabilize TRF2 on telomeres (see below).

TIN2

Although this component of shelterin was identified in a two-hybrid screen as the second TRF1-interacting nuclear protein, TIN2 (Kim et al. 1999), it also binds to TRF2 and to TPP1 (Houghtaling et al. 2004; Kim et al. 2004; Liu et al. 2004a; Ye et al. 2004a,b). TIN2 is a small (354 amino acid/40 kD) factor that binds the TRFH domain of TRF1 using its central region (amino acids 195–284). Its amino terminal half (amino acids 1–195) binds TRF2 and TPP1. TIN2 can be recovered in association with TRF1, TRF2, and TPP1 in vivo and accumulates at telomeres as determined by immunofluorescence and ChIP analyses (Kim et al. 1999; Loayza and de Lange 2003).

TIN2 is the central component of shelterin. Because of its protein interactions, TIN2 connects the three telomeric DNA-binding proteins, TRF1, TRF2, and (through TPP1) POT1 (Fig. 2). TIN2 can bind to TRF1 and TRF2 simultaneously, and in vivo studies indicate that this stabilizes TRF2 on telomeres (Kim et al. 2004; Ye et al. 2004b). In general, interactions between two DNA-binding proteins will increase their affinity for a DNA molecule when both binding sites are accessible. This effect may be particularly important with regard to TRF1 and TRF2, which have a high off-rate in vitro (T. de Lange, unpubl.). As noted above, the ability of TRF1 and TIN2 to stabilize TRF2 on telomeres can explain the lethal phenotype of TRF1 and TIN2 deficiency in the mouse (Karlseder et al. 2003; Chiang et al. 2004; Iwano et al. 2004). In agreement, mutant versions of TIN2 that disrupt the TRF2–TIN2–TRF1 connection show a telomere deprotection phenotype (Kim et al. 2004). TIN2 also acts to stabilize shelterin by protecting TRF1 from tankyrase 1 (Ye and de Lange 2004) (see below).

Rap1

The mammalian ortholog of yeast Rap1 emerged unexpectedly from searches for TRF2-interacting proteins (Li et al. 2000; Zhu et al. 2000). Staining of purified endogenous TRF2/Rap1 complexes suggests that these two proteins are present in a 1:1 stoichiometry. Human Rap1 is a 399-amino-acid (~49 kD) protein with three domains: a BRCT domain, a single Myb domain, a coiled region, and the Rap1 carboxy-terminal (RCT) domain (Fig. 2A). The RCT region mediates the interaction with TRF2 as well as homotypic interactions. The Rap1 carboxyl terminus also contains a putative nuclear localization signal (NLS).

Mammalian Rap1 is highly diverged from budding yeast Rap1, which binds to telomeric DNA directly by virtue of having two Myb domains and does not require a protein interaction to localize to telomeres. Mammalian Rap1 has only one Myb domain that does not bind to DNA (Li et al. 2000). Its Myb domain does not have the overall positive charge necessary for interaction with DNA and is more likely to mediate protein–protein interactions with partners that are still at large (Hanaoka et al. 2001). Similarly, no interaction partner is known for the BRCT domain that is expected to recruit a factor necessary for the maintenance of telomere length heterogeneity (Li and de Lange 2003). When TRF2 is absent, mammalian Rap1 does not bind to telomeres, and the Rap1 protein level declines (Celli and de Lange 2005). In this regard, mammalian Rap1 is more similar to its ortholog in *S. pombe*, which also localizes to telomeres through a TRF2-like protein, Taz1p (Chikashige and Hiraoka 2001; Kanoh and Ishikawa 2001; Chapter 16). Mass spectrometry has identified Ku70/80, PARP1, Mre11, and Rad50 in the Rap1 complex (O'Connor et al. 2004; Ye et al. 2004b). It is not clear whether any of these proteins bind Rap1 directly.

Rap1 has a role in telomere length regulation and affects telomere length heterogeneity (Li et al. 2000; Li and de Lange 2003). Mice lacking Rap1 are not viable (M. van Overbeek and T. de Lange, unpubl.), suggesting that Rap1 may also be important for the protective activity of shelterin.

TPP1

The most recently identified component of shelterin, TPP1, emerged from mass spectrometry of TRF1/TIN2-associated proteins and a two-hybrid screen with TIN2. TPP1 was reported as PTOP, PIP1, and TINT1 (Houghtaling et al. 2004; Liu et al. 2004b; Ye et al. 2004a) before a single name was agreed on. The carboxy-terminal 60 amino acids of TPP1 binds to the amino-terminal half of TIN2, and a central 100-amino-acid region in TPP1 binds to the carboxy-terminal half of POT1 (Fig. 2A). Other than a serine-rich region separating these two interaction domains, the TPP1 sequence does not reveal notable features.

TPP1 is important for the recruitment of POT1 to telomeres, and the phenotype of TPP1 inhibition is consistent with this recruitment. Specifically, short hairpin RNA (shRNA) knockdown of TPP1 or expression of defective mutants results in inappropriate telomere elongation, a phenotype associated with diminished POT1 loading (Liu et al. 2004b; Ye et al. 2004a).

A splice defect in the TPP1 gene was recently reported to result in a severe developmental phenotype in the ACD (adrenocortical dysplasia) mouse (Keegan et al. 2005). The ACD mutation arose spontaneously in the Jackson Laboratory. The severity of the developmental abnormalities, which include a defect in the organs derived from the urogenital ridge, depends on the genetic background. In some settings, acd/acd mice are born and live to adulthood. This is surprising because all other components of shelterin are required for early embryonic development. It has not been ruled out, however, that the ACD mutation represents a hypomorphic allele of TPP1.

POT1

Initial biochemical searches for a single-stranded telomeric DNA-binding protein in mammals or other vertebrates failed to deliver the desired factor (McKay and Cooke 1992; Cardenas et al. 1993; Ishikawa et al. 1993). Eventually, this protein, POT1 (protection of telomeres 1) was identified in the database through its homology with the DNA-binding domain of TEBPα, a ciliate telomere terminus factor (Baumann and Cech 2001; see Chapters 11 and 15). Like TEBPα, POT1 binds single-stranded DNA with two oligonucleotide/oligosaccharide-binding (OB) folds (Lei et al. 2004). Human POT1 is highly specific for the sequence 5′-(T)TAGGGTTAG-3′, which it can bind at a 3′ end or within a longer single-stranded region (Lei et al. 2004; Loayza et al. 2004). The crystal structure of a POT1-DNA complex suggests that POT1 could protect the 3′ telomere terminus if it ends on TAG-3′ (Lei et al. 2004). Such TAG-3′ ends occur in vivo, but other 3′ ends are not uncommon (Sfeir et al. 2005). According to the biochemical and structural data, POT1 can also bind to any site along a single-stranded TTAGGG repeat array, regardless of the terminal sequence. In addition, recent biochemical data suggest that POT1 binds to its site when it is positioned close to a double/single-stranded transition, especially in the natural configuration at the telomere in which a POT1 site is 2 nucleotides from the ATC-5′ end (Fig. 1) (F. Ishikawa, pers. comm.; D. Hockemeyer and T. de Lange, unpubl.). This preferred binding may explain why the specificity of the 5′ end nucleotide is lost upon POT1 inhibition (Hockemeyer et al. 2005).

A region in the carboxy-terminal half of POT1 binds to TPP1 (Fig. 2A). This interaction is important for the localization of POT1 to telomeres, whereas the DNA-binding domain of POT1 is not (Loayza and de Lange 2003; Liu et al. 2004b). ChIP data indicate that longer telomeres contain more POT1, arguing that most POT1 is bound to the other shelterin subunits on the double-stranded TTAGGG repeat array. Given the looped structure of telomeres, POT1 bound along the duplex part of the telomere might still interact with the single-stranded telomeric DNA, including the D-loop (see Fig. 2C).

Human cells express two forms of POT1 from alternatively spliced mRNAs: one full-length form and a shorter form (POT1-55) that lacks the first of the two OB folds (OB1) required for single-stranded DNA binding (Loayza et al. 2004; Hockemeyer et al. 2005). POT1-55 retains the TPP1 interaction domain that is responsible for the recruitment of POT1 to telomeres. The abundance of POT1-55 is tenfold lower than full-length POT1 and its function is not known. The mouse genome contains two separate genes encoding full-length POT1 proteins (POT1a and POT1b) (D. Hockemeyer and T. de Lange, unpubl). Although POT1a and POT1b are nearly identical and associate with telomeres, mice lacking POT1a die early in embryogenesis, indicating that these two versions of POT1 are not functionally redundant (D. Hockemeyer and T. de Lange, unpubl.). RNA interference (RNAi)-mediated inhibition of POT1 also leads to a growth defect in primary human cells (Veldman et al. 2004; Hockemeyer et al. 2005).

POT1 has a crucial function in telomere length homeostasis, acting as the terminal transducer of telomere length control (Loayza and de Lange 2003; Liu et al. 2004b; Ye et al. 2004a) (see below). It also contributes to the protection of chromosome ends, since partial knockdown of POT1 with RNAi results in a DNA-damage response at telomeres, reduction in the single-stranded telomeric DNA, changes in the 5′ end of the chromosome, and a mild telomere fusion phenotype (Veldman et al. 2004; Hockemeyer et al. 2005; Yang et al. 2005).

In Vitro DNA Remodeling Activities of Shelterin Subunits

The emerging view of shelterin is that it changes the structure of the telomeric DNA in order to protect chromosome ends. This view is based on the in vitro DNA remodeling activities of TRF1 and TRF2. TRF2 has the ability to form t-loop-like structures in vitro (Griffith et al. 1999; Stansel et al. 2001). When purified recombinant TRF2 is incubated with a duplex TTAGGG repeat array ending in a 3′ overhang, 10–15% of the resulting complexes have a DNA loop of the size of the TTAGGG repeat array. The loops can be preserved after removal of TRF2 if the DNA is cross-linked with psoralen/UV, consistent with a strand-invasion event. Optimal substrates have TTAGGG sequences at the double/single-stranded junction and at least one TTAGGG repeat in the 3′ tail. The sequence of the 3′-terminal nucleotide is not important. It remains to be determined how TRF2 assembles t-loops. TRF2 does not have a helicase domain, and the reaction does not require an NTP cofactor. An intriguing feature of TRF2 in this regard is its propensity to form a higher-order complex at the end of the DNA and at the strand-invasion point, suggesting a preference for double/single-stranded junctions that might stabilize the t-loop.

The architectural activities of TRF1 include DNA bending, looping, and pairing. TRF1 induces a shallow bend upon binding to sites with 3–12 tandem TTAGGG repeats (Bianchi et al. 1997). When provided with two half-sites spaced 200 bp apart, TRF1 can form a DNA loop (Bianchi et al. 1999). The most dramatic effect of TRF1 is the pairing of telomeric DNA (Griffith et al. 1998). Electron microscopic studies showed that TRF1 can bind at high density all along a telomeric tract, forming a tightly packed protein filament. These filaments are often paired and occasionally clustered. The synaptic complexes are not stabilized by DNA cross-linking, indicating that they are held together by TRF1. Interactions between TRF1 dimers could facilitate the association of telomeric repeat arrays, or alternatively, the two Myb domains of a TRF1 dimer might engage two different DNA molecules. The bending, looping, and pairing activity of TRF1 could facilitate the folding of telomeres (Fig. 2C), whereas TRF2 might be instrumental in facilitating the strand invasion that forms the t-loop.

Now that most of the other components of shelterin have been identified, it will be important to establish how this complex remodels telomeric DNA in vitro and in vivo. The only TRF1- or TRF2-interacting factors studied in this regard are Rap1 and TIN2. Rap1 can bind to TRF2 when it is associated with DNA but has no obvious effect on the complex (Li et al. 2000). TIN2 has multiple effects on the binding of TRF1 to telomeric DNA in vitro (Kim et al. 1999, 2003). Addition of the carboxy-terminal half of TIN2 (which includes the TRF1-binding domain) to TRF1 gel-shift reactions results in a large complex that contains both proteins bound to DNA. Full-length TIN2 can stimulate the ability of TRF1 to aggregate telomeric DNA in a "probe-clustering" assay, but in other settings, it inhibits the DNA-binding activity of TRF1 (Kim et al. 1999, 2003).

Telomere-associated Proteins That Are Not Part of Shelterin

PARPs at Telomeres

The sequence RGCADG in the amino terminus of TRF1 interacts with tankyrase 1 and 2, two closely related proteins that were discovered in two-hybrid screens with TRF1 (Smith et al. 1998; Kaminker et al. 2001; Sbodio and Chi 2002). Tankyrase 1 can be detected at telomeres by immunofluorescence and ChIP analyses, but its abundance at telomeres is much less than that of the shelterin subunits (Smith et al. 1998; Loayza and de Lange 2003). The tankyrases were named as *TRF1-interacting ankyrin*-related (ADP-ribose) polymer*ases* (Smith et al. 1998), reflecting the presence of an ankyrin domain and the catalytic domain of the poly(ADP-ribose) polymerases (PARPs). The ankyrin domain contains 24 ankyrin repeats of which a subset mediates TRF1 binding (Seimiya and Smith 2002; Seimiya et al. 2004). The PARPs use βNAD^+ as a precursor to add long branched chains of poly(ADP-ribose) (PAR) to the carboxyl group of a glutamic acid residue in a protein acceptor. In vitro, the tankyrases have the ability to PARsylate themselves as well as TRF1 (Smith et al. 1998; Cook et al. 2002). Tankyrase 1 and 2 also have a SAM

Table 1. Examples of proteins at human telomeres that are not part of shelterin

Protein complex	Nontelomeric function	Effects at telomeres	Telomere recruitment
Mre11/Rad50/ Nbs1	recombinational repair, DNA damage sensor	t-loop formation/ resolution? required for t-loop HR	associated with shelterin
ERCC1/XPF	NER, cross-link repair, 3′ flap endonuclease	deficiency leads to formation of TDMs; implicated in overhang processing after TRF2 loss	associated with shelterin
WRN helicase	branch migration G4 DNA resolution	deficiency results in loss of lagging-strand telomeres	binds to TRF2
BLM helicases	branch migration	not known	binds to TRF2
DNA-PK	NHEJ	deficiency leads to mild fusion phenotype	associated with shelterin
PARP-2	BER	not known	binds to TRF2
Tankyrases	role in mitosis, Golgi	positive regulator of telomere length through inhibition of TRF1	binds to TRF1
Rad51D	unknown (HR?)	deficiency leads to mild fusion phenotype	unknown

(sterile α module) motif, which has been shown to mediate oligomerization in chicken tankyrase 1 (De Rycker et al. 2003). The two tankyrases also can interact with each other (Sbodio et al. 2002).

PARsylation of TRF1 inhibits its ability to bind telomeric DNA in vitro, and enforced expression of tankyrases removes TRF1 from telomeres (Smith et al. 1998; Smith and de Lange 2000). The amount of TRF1 protein is also reduced in this setting, and in vitro data suggest that the dislodged TRF1 is degraded through a ubiquitin/proteasome pathway (Chang et al. 2003). However, when TIN2 is associated with TRF1 in vitro or in vivo, the modification of TRF1 by tankyrase 1 is strongly reduced, and the stability of TRF1 on telomeres is improved (Ye and de Lange 2004).

The TRF1-tankyrase 1-TIN2 interplay is involved in telomere length regulation. Overexpression of tankyrase 1 in the nucleus leads to excessive telomere elongation as if telomerase has liberal access to telomeres (Smith and de Lange 2000). Conversely, inhibition of tankyrase 1 with shRNA or a presumed dominant-negative allele leads to telomere shortening

(Seimiya et al. 2005; J. Donigian and T. de Lange, unpubl.). Furthermore, partial knockdown of TIN2 elongates telomeres, consistent with increased modification of TRF1 by tankryase 1 (Ye and de Lange 2004). The working model is that tankyrase 1 removes some of the shelterin from telomeres to allow telomerase access to the chromosome end. Because tankyrases are likely to have nontelomeric functions (Smith and de Lange 1999; Chi and Lodish 2000; Sbodio and Chi 2002; Sbodio et al. 2002; Dynek and Smith 2004), it will be a challenge to dissect their role at telomeres. An additional confounding aspect is that mouse TRF1 lacks the tankyrase-binding site and is resistant to enforced tankyrase overexpression (Sbodio and Chi 2002; J. Donigian and T. de Lange, unpubl.).

In addition to the tankyrases, PARP-1 and PARP-2 have been suggested to have a role at telomeres. PARP-2 binds to the Myb domain of TRF2, PARsylates the TRFH domain, and inhibits the binding of TRF2 to telomeric DNA in vitro (Dantzer et al. 2004). Interestingly, the TRF2 Myb also binds to PAR itself and this might block its ability to bind to TTAGGG repeats. Perhaps the binding of TRF2 to PAR could explain the curious observation that TRF2 localizes to regions with massive DNA damage induced by laser light/Hoechst treatment (Bradshaw et al. 2005). Although PARP-2 has been observed at the specialized telomeric clusters in ALT cells, PARP-2-deficient mice have no overt telomere defect (Dantzer et al. 2004). Therefore, the role of PARP-2 at telomeres remains to be established. Similarly, the phenotype of PARP-1 deficiency has not provided unequivocal evidence for a role at telomeres (d'Adda di Fagagna et al. 1999; Samper et al. 2001), although PARP-1 is recovered in association with TRF2/Rap1 (O'Connor et al. 2004).

The WRN and BLM RecQ Helicases

Werner's syndrome (WS) patients suffer from several premature aging symptoms caused by a mutation in the WRN RecQ helicase. The biological function of WRN is not fully understood, but it is likely that this helicase is important for the resolution of aberrant DNA structures formed during recombination and replication (for review, see Ozgenc and Loeb 2005). Modeling in the mouse suggests that telomere dysfunction can contribute to the phenotypes of WRN deficiency (see Chapter 5), and several lines of evidence implicate WRN in telomere maintenance in human cells.

Although WS cells lose telomeric DNA at a normal rate, primary WS fibroblasts senesce sooner than normal human fibroblasts, and senescent WS cells have longer telomeres than senescent control cells

(Schulz et al. 1996; Baird et al. 2004). As expression of human telomerase reverse transcriptase (hTERT) can immortalize primary WS cells, aberrantly short telomeres are the likely cause for their premature growth arrest (Johnson et al. 2001). Using immunofluorescence, the WRN protein is only detectable at the large telomeric foci in ALT-associated PML (promyelocytic leukemia) bodies (Johnson et al. 2001; see Chapter 7). However, ChIP analysis on primary and telomerase-positive human cells shows WRN on "normal" telomeres in S phase (Crabbe et al. 2004). Consistent with its presence at telomeres, WRN binds to TRF2 in vitro (Opresko et al. 2002, 2004; Machwe et al. 2004).

Expression of a helicase-deficient form of WRN induces stochastic telomere losses in human cells (Bai and Murnane 2003b). Some of the events affect one of the two sister telomeres, indicating that the deletions take place during or after DNA replication (Bai and Murnane 2003b; Crabbe et al. 2004). Furthermore, CO-FISH (chromosome orientation–fluorescence in situ hybridization; see Chapter 7) revealed that the deletions preferentially affect the sister telomere generated by lagging-strand DNA synthesis (Crabbe et al. 2004). Since the RecQ helicases have the ability to resolve G-quadruplex DNA, WRN may be required to remove these structures from the TTAGGG repeat template of lagging-strand DNA synthesis (Sun et al. 1998; Huber et al. 2002). Persistence of the G quadruplexes in WRN-deficient cells could lead to fork stalling. How such stalled forks would give rise to deletions of only the lagging-strand sister telomere remains to be resolved.

A second RecQ helicase, BLM, mutated in Bloom's syndrome patients, may also have a role at mammalian telomeres. BLM interacts with TRF1 and can be detected at the ALT-associated PML bodies (APBs) of ALT cells (Yankiwski et al. 2000; Stavropoulos et al. 2002; Lillard-Wetherell et al. 2004). Whether BLM has a role at "normal" telomeres is not yet clear.

RAD51D

Like the other five mammalian RAD51 paralogs, RAD51D (also called RAD51L3) is thought to have a role in homology-directed repair of double-strand breaks (DSBs). Recently, a small amount of RAD51D, but not the other RAD51 paralogs, has been detected at mammalian telomeres (Tarsounas et al. 2004). Telomere fusions are observed at low frequency in mouse cells lacking both RAD51D and p53, suggesting a modest contribution of RAD51D to telomere protection. How RAD51D is recruited to chromosome ends is not known.

NHEJ Factors

The chromosome end fusions that are pathognomonic of telomere dysfunction are generated by the NHEJ (non homologous end-joining) pathway (see Fig. 4) (Smogorzewska et al. 2002; Celli and de Lange 2005). Paradoxically, components of the NHEJ machinery are associated with telomeres and have a role in the protection of chromosome ends. Several observations indicate that the Ku heterodimer is associated with shelterin and mammalian telomeres: the Ku70/80 heterodimer interacts with TRF1 and TRF2 in vitro (Hsu et al. 2000; Song et al. 2000); Ku is recovered in the TRF2/Rap1 complex (O'Connor et al. 2004); and Ku as well as the DNA–protein kinase catalytic subunit (DNA-PKcs) can be detected at telomeres by ChIP (Hsu et al. 1999; d'Adda di Fagagna et al. 2001).

Ku appears to be essential in human cells, and its function at telomeres is therefore difficult to establish (Li et al. 2002). However, even partial reduction of human Ku80 results in a telomere deprotection phenotype that includes loss of telomeric DNA and occasional chromosome end fusions (Jaco et al. 2004; Myung et al. 2004). In mouse cells, lack of Ku or DNA-PKcs function increases the frequency of telomere fusions slightly without obvious loss of telomeric DNA (Bailey et al. 1999; Hsu et al. 2000; Samper et al. 2000; d'Adda di Fagagna et al. 2001; Gilley et al. 2001; Goytisolo et al. 2001). Furthermore, embryonic stem (ES) cells lacking Artemis, the NHEJ-associated nuclease involved in V(D)J recombination, show a mild telomere fusion phenotype (Rooney et al. 2003). Although significant, the telomeric phenotypes of Ku, DNA-PKcs, or Artemis deficiency are minor compared to shelterin loss, and whereas shelterin is essential in the mouse, DNA-PK null mice are viable. This suggests that the protective role of NHEJ factors at telomeres is largely redundant with other pathways. For instance, protection by NHEJ factors may only be necessary when an unusual replication intermediate is formed at telomeres. Another possibility is that the telomere fusions in NHEJ-deficient cells are not covalent and can be resolved in mitosis, thereby avoiding chromosome breakage or nondisjunction.

CO-FISH analyses to distinguish the telomeres generated by leading- and lagging-strand DNA synthesis show frequent telomere sister chromatid exchanges (T-SCEs) when TRF2 is deleted from Ku70$^{-/-}$ MEFs (G. Celli et al., unpubl.). T-SCEs are not significantly increased in Ku70$^{-/-}$ cells or in cells that lack TRF2 and DNA ligase IV, indicating a redundant role for Ku70 and TRF2 in repressing this type of homologous recombination at telomeres. The role of the Ku heterodimer in repression of homologous recombination at telomeres is discussed further in the sections on telomere protection.

The Mre11 Complex

The Mre11 complex, composed of Mre11, Rad50, and Nbs1, is recruited to telomeres through TRF2 (Zhu et al. 2000). The details of the interactions remain unclear, but indirect evidence suggests that Rap1 has a role in bringing Rad50 to telomeres (O'Connor et al. 2004). Although only a small fraction of the Mre11 complex is located at chromosome ends, its components can be detected at telomeres by immunofluorescence and ChIP. Mre11 and Rad50 are present at telomeres throughout the cell cycle, but Nbs1 is only detectable at telomeres in S phase (Zhu et al. 2000; Loayza and de Lange 2003).

The Mre11 complex accumulates at sites of DNA damage, acts both downstream from and upstream of the ATM kinase, and has multiple functions in the maintenance of genome integrity (for review, see Petrini and Stracker 2003). When telomeres are deprotected, the amount of associated Mre11 complex increases sharply and this binding is independent of ATM signaling or the action of other PI-3-kinase-related kinase (PIKKs) (Takai et al. 2003). Because the Mre11 complex is essential in mammals, it has been difficult to establish its function at telomeres; the available hypomorphic mutations are not associated with obvious telomere phenotype. In support of a telomeric function of the Mre11 complex, a mutant allele of Nbs1 induces telomere loss in some settings (Bai and Murnane 2003a). In addition, Nbs1 contributes to the telomerase-mediated telomere elongation pathway (Ranganathan et al. 2001), and the Mre11 complex is important for the ALT pathway (Chapter 7).

ERCC1/XPF

Mass spectrometric identification of proteins associated with TRF2 revealed the presence of the ERCC1/XPF heterodimer (Zhu et al. 2003). ERCC1/XPF is a 3′ flap endonuclease that functions in nucleotide excision repair (NER), cross-link repair, and some aspects of homologous recombination. A small fraction of the cellular ERCC1/XPF is found at telomeres throughout the cell cycle. Observations on ERCC1-deficient mouse cells have hinted at a function for ERCC1/XPF at telomeres. These cells generate double-minute chromosomes composed of telomeric and nontelomeric DNA and are referred to as telomeric DNA containing double minutes (TDMs). It was proposed that the TDMs arise through illegitimate recombination between telomeres and interstitial telomeric DNA (see Fig. 6). The implication is that shelterin-associated ERCC1/XPF is important to prevent this type of recombination. This model is discussed further below in the section on the protective function of shelterin.

As is the case for several of the shelterin-associated DNA repair factors, ERCC1/XPF can also have a detrimental effect on telomeres as it is implicated in the removal of the 3′ overhang (Zhu et al. 2003). When TRF2 is inhibited with a dominant-negative allele, the amount of single-stranded DNA at chromosome ends is gradually reduced to approximately 50% of the original level. However, in human XPF cells that lack normal ERCC1/XPF function, this overhang loss does not occur. Indirect evidence suggests that the NHEJ complex is involved in the recruitment of ERCC1/XPF to deprotected telomeres or that the NHEJ machinery activates this nuclease. This view arose from conditional deletion of TRF2 from MEFs (Celli and de Lange 2005). When TRF2 is deleted from cells that also lack DNA ligase IV, the 3′ overhang remains intact, implying that ERCC1/XPF is incapable of removing the single-stranded protrusion in the absence of a functional NHEJ pathway. Recruitment by the NHEJ machinery could explain how ERCC1/XPF can associate with telomeres when TRF2 is not there.

Est1, Rif1, Tel2, and Other Orthologs of Yeast Telomeric Proteins

The orthologs of several budding yeast telomeric proteins recently surfaced as a consequence of the human genome sequencing effort. There are three human genes with similarity to the Est1 telomerase recruitment factor of budding yeast (see Chapter 12). Est1A and B both associate with telomerase (Reichenbach et al. 2003; Snow et al. 2003; Chapter 2). The interaction of Est1C with telomerase has not been studied. These proteins all share a 14-3-3 phosphopeptide-binding fold and have been implicated in nonsense-mediated RNA decay (Conti and Izaurralde 2005; Fukuhara et al. 2005). Overexpression of a truncated form of Est1A has a telomere deprotection phenotype that includes formation of fused chromosome ends and cell cycle arrest (Reichenbach et al. 2003). This might indicate that Est1A has a protective function at telomeres or that its overexpression removes one of the protecting proteins. Human Rif1 is not associated with telomeres, and mammalian cells with diminished or absent Rif1 have no obvious defect in telomere function (Silverman et al. 2004; Xu and Blackburn 2004; S. Buonomo and T. de Lange, unpubl.). Human Tel2 has been reported to have a slight effect on telomere length in human cells (Jiang et al. 2003), but Tel2$^{-/-}$ mouse cells do not show an obvious telomeric phenotype (H. Takai and T. de Lange, unpubl.). A human ortholog of the Cdc13-interacting factor Stn1 was recently identified

(F. Ishikawa, pers. comm.), but orthologs of other budding yeast telomeric proteins, such as Rif2 and Ten1 are still at large.

PREVENTING THE DNA-DAMAGE RESPONSE AT NATURAL CHROMOSOME ENDS

Eukaryotic cells have the ability to distinguish natural chromosome ends from DSBs and other forms of DNA damage. When shelterin function is compromised, chromosome ends lose this privileged status, and the deprotected telomeres activate the canonical DNA-damage response. Experiments in which the function of TRF2, TIN2, or POT1 was curtailed have shown that dysfunctional telomeres can activate a DNA-damage signaling pathway, become associated with DNA-damage response factors, and inhibit cell cycle progression (Fig. 3). Similarly, telomeres that have become dysfunctional due to attrition of the TTAGGG repeats induce a DNA-damage response. The following sections discuss the telomere-damage response to shelterin inhibition, the possible mechanisms by which shelterin acts to repress this response, and the relevance of these findings to telomere attrition.

The Telomere-Damage Response

The protective function of shelterin can be wracked by conditional deletion of TRF2 from mouse cells or through overexpression of a dominant-negative allele of TRF2, TRF2$^{\Delta B \Delta M}$ (van Steensel et al. 1998). This TRF2 allele lacks the amino-terminal basic domain and the carboxy-terminal DNA-binding domain but retains the TRFH domain, the NLS, and most protein–protein interaction interfaces (see Fig. 2). TRF2$^{\Delta B \Delta M}$ forms an inactive heterodimer with the endogenous TRF2, effectively stripping TRF2, Rap1, and a portion of POT1 from telomeres (van Steensel et al. 1998; Loayza and de Lange 2003). TRF1 and TIN2 are largely unaffected. Similarly, conditional deletion of TRF2 results in complete loss of Rap1 from telomeres, whereas (some of) TRF1 remains (Celli and de Lange 2005).

Upon TRF2 inhibition, telomeres become associated with DNA-damage response factors such as 53BP1, Nbs1, phosphorylated ATM, and Rad17, Rif1, and MDC1 (Fig. 3) (d'Adda di Fagagna et al. 2003; Takai et al. 2003; Silverman et al. 2004; N. Dimitrova and TdL, unpubl.). These so-called telomere dysfunction-induced foci (TIFs) extend well beyond the telomeres, in some cases forming elongated structures suggestive of a DNA-damage domain that stretches far into the chromosome (Takai et al. 2003).

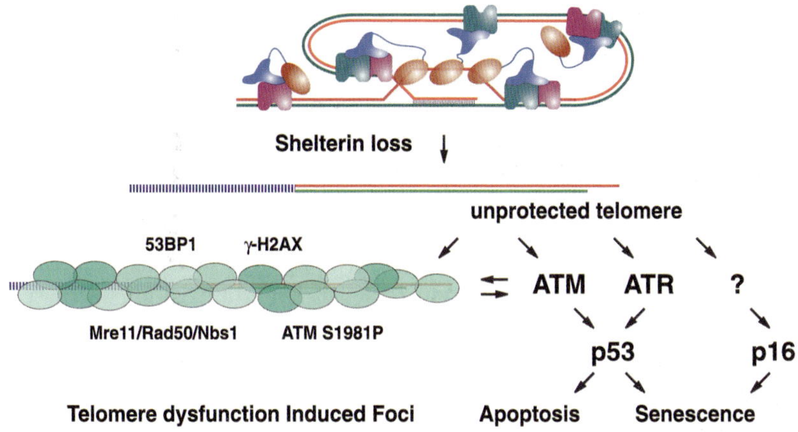

Figure 3. The telomere damage signaling pathway. Upon loss of shelterin, telomeres are detected by the canonical DNA-damage response and numerous DNA-damage response factors associate with the chromosome end, generating so-called telomere dysfunction-induced foci (TIFs). Activation of the ATM kinase pathway leads to up-regulation of p53 and induction of senescence. In the absence of ATM, ATR can become activated. The p16 pathway can also be activated and induce cell cycle arrest, but the details of this process are not known. Shelterin loss can lead to a reduction in the single-stranded DNA at telomeres (not shown), but this overhang degradation is not required for the DNA-damage response. When telomeres shorten, a similar set of events is observed.

Likewise, a "ChIP on chip" assay on senescent cells detected γ-H2AX hundreds of kilobases from the chromosome ends (d'Adda di Fagagna et al. 2003). When TRF2 is deleted, the majority of the telomeres accumulate TIFs, indicating that most (if not all) telomeres resemble sites of DNA damage when their protection is compromised (Celli and de Lange 2005).

When ERCC1/XPF is present and when the NHEJ machinery is intact, TRF2 inhibition leads to loss of approximately 50% of the single-stranded TTAGGG repeats (van Steensel et al. 1998). However, this degradation of the overhang is not required for the formation of TIFs or other aspects of the telomere-damage response (Celli and de Lange 2005). The response is not diminished in NHEJ-deficient cells even though the overhang persists apparently intact.

The telomere-damage signal is transduced by caffeine-sensitive PIKKs (Takai et al. 2003). The ATM kinase is a major transducer of the signal as evident from the diminished TIF formation in A-T cells (Takai et al. 2003) and from the fact that the ATM-responsive protein Rif1 is present in TIFs (Silverman et al. 2004). In absence of ATM, ATR (ataxia telangiectasia/Rad3-related) can respond to dysfunctional telomeres, and ATR interacting protein (ATRIP) is observed at dysfunctional telomeres in A-T cells (Herbig et al. 2004). Inhibition of both kinases with caffeine results in the rapid disappearance of TIFs, indicating that PIKK signaling is required for their persistence (Takai et al. 2003). Among the DNA-damage factors at deprotected telomeres, the Mre11 complex behaves differently, being unaffected by caffeine or ATM status (Takai et al. 2003). Similarly, the Mre11 complex does not require ATM signaling to accumulate at random sites of DNA damage. These findings are consistent with the idea that the Mre11 complex contributes to an upstream sensing step in the DNA-damage response (Petrini and Stracker 2003).

Activation of the ATM kinase by dysfunctional telomeres leads to the expected downstream events, including phosphorylation of ATM substrates (e.g., Nbs1 and Chk2), delays in cell cycle progression, and apoptosis or senescence (van Steensel et al. 1998; Karlseder et al. 1999; d'Adda di Fagagna et al. 2003; Takai et al. 2003). The cellular outcome is in part dependent on the cell type. Apoptosis is prominent in some epithelial cells and in lymphocytes, whereas fibroblasts respond with senescence to telomere dysfunction (Karlseder et al. 1999). This response pattern is also seen after ionizing radiation of these cell types.

The cell cycle arrest upon telomere damage is primarily mediated by p53 (Chin et al. 1999; Karlseder et al. 1999; Smogorzewska and de Lange 2002). Stabilization and activation of p53 result in increased levels of p21, an inhibitor of Cdk2, and an arrest before entry into S phase. Other mechanisms of Cdk2 inhibition (e.g., degradation of Cdc25A) have not been investigated. In addition to p21, telomere damage induces p16, an inhibitor of Cdk4/6 (Smogorzewska and de Lange 2002; Jacobs and de Lange 2004). By displacing p21 from Cdk4/6, p16 can increase the pool of p21 available for Cdk2 inhibition. Although p53 is crucial for the cell cycle arrest in response to dysfunctional telomeres, when p53 is impaired, up-regulation of p16 can act as a second mechanism to block S-phase entry upon telomere damage (Jacobs and de Lange 2004). The telomere-damage signaling pathway is different in mouse cells where p16, although induced by telomere dysfunction, does not induce a G_1/S arrest (Smogorzewska and de Lange 2002).

As expected from the important role of TIN2 in holding shelterin together, the telomere-damage response is activated by inhibition of this connecting factor. Specifically, overexpression of TIN2 mutants that lack the ability to bind to both TRF1 and TRF2 result in loss of TRF2 from telomeres and a telomere-damage response marked by TIFs (Kim et al. 2004).

The formation of TIFs is also a prominent phenotype of POT1 inhibition (Hockemeyer et al. 2005). In human cells, knockdown of POT1 with shRNA results in the presence of TIFs at most telomeres. In contrast to TIFs generated by TRF2 inhibition, the response to knockdown of POT1 is largely confined to G_1, and the TIFs disappear by the time cells enter S phase. Presumably, the telomere damage inflicted by POT1 loss is repaired during G_1, explaining why the cells do not arrest in the cell cycle. As TIFs resulting from TRF2 inhibition are seen throughout interphase, they are probably caused by a different type of telomere damage.

Repression of the Telomere-Damage Response by Shelterin

The challenge is to understand why telomeres containing functional shelterin are ignored by the ATM kinase pathway. As shelterin is implicated in the formation of t-loops, shelterin's repression of the telomere-damage response could be due to this structure. Why would the t-loop structure make the chromosome end undetectable? One possibility is that the t-loop simply hides the chromosome end. This strategy would foil a DNA-damage sensor that depends on a free DNA end to detect DSBs. A second explanation is that the t-loop creates a nucleosomal organization similar to that of intact chromosome internal DNA. If the DNA damage surveillance senses an altered chromatin structure that is specific to the free DNA end, preventing the formation of this structure by securing the terminus in the t-loops could be an effective way to mask the chromosome end. Indirect evidence has suggested that changes in nucleosomal chromatin indeed contribute to the mechanism by which the ATM kinase is activated (Bakkenist and Kastan 2003; Huyen et al. 2004).

A second mechanism by which shelterin might repress the DNA-damage response was recently suggested by the finding that TRF2 can inhibit the ATM kinase in certain settings (Karlseder et al. 2004). TRF2 can bind directly to the ATM kinase in vitro and in vivo, but it is not clear whether this interaction occurs at telomeres. When TRF2 is highly overexpressed so that it is present throughout the nucleoplasm, the activation of the ATM kinase by DSBs is blunted. The autophosphorylation of ATM on S1981 is diminished by TRF2 overexpression, and ATM-dependent cell cycle checkpoints are largely abrogated. Thus, in this setting, TRF2 appears to act as an inhibitor of the ATM signaling pathway. If telomere-bound TRF2 similarly can repress ATM activation, it might prevent or dampen the DNA-damage response at telomeres. The inhibition of ATM would be limited to chromosome ends where TRF2 is abundant.

The Response to Telomere Attrition

The programmed shortening of human telomeres limits the proliferation of primary human cells (see Chapter 4). The resulting cell cycle arrest has the hallmarks of a DNA-damage response: DNA damage foci are detectable in (newly) senescent cells, the ATM kinase pathway is activated, and p53 enforces a G_1/S arrest (d'Adda di Fagagna et al. 2003; Bakkenist et al. 2004; Herbig et al. 2004). Similarly, p53 is the main effector of the response to shortened telomeres in the telomerase knockout mouse (Chapter 5).

It has been difficult to determine the details of the response to telomere attrition, because the dysfunction of a single (or a few) telomere is sufficient to induce cell cycle arrest (Hemann et al. 2001; Zou et al. 2004). Therefore, only a fraction of the telomeres of senescent cells can be expected to be in a deprotected state. A second confounding aspect is that inappropriate growth conditions can induce p16-dependent senescence in human and mouse cells (Sherr and DePinho 2000; Ramirez et al. 2001). Despite these hurdles, the available data indicate that the shortest telomeres induce a DNA-damage response similar to (or the same as) the response to telomeres lacking protection by shelterin. The major question that remains to be answered is why critically short telomeres lose protection. Perhaps very short telomeres simply contain too little shelterin, but other possibilities have not been excluded.

NHEJ AND HOMOLOGOUS RECOMBINATION AT DYSFUNCTIONAL TELOMERES

The fate of dysfunctional mammalian telomeres, primarily surmised from shelterin-inhibition experiments, is very similar to the fate of DSBs. Two main pathways are involved in processing the exposed DNA ends, NHEJ (Fig. 4) and homologous recombination (Figs. 5 and 6). NHEJ is responsible for telomere fusions, formation of dicentric chromosomes, and the associated genome instability. Homologous recombination can delete large segments of telomeric DNA and mediate exchange of sequences between sister telomeres. Repression of NHEJ by shelterin is proposed to involve the formation of t-loops. How shelterin controls homologous recombination is unclear.

NHEJ

Chromosome end fusions are a prominent phenotype of telomere dysfunction. They can occur when telomeres are shortened, when shelterin components are inhibited, and upon loss of certain other telomere-associated proteins. Chromosome end fusions are distinguished from end associations on the basis of the DNA-staining pattern in metaphase spreads. If the staining is not contiguous, the event is referred to as an association. However, the absence of a staining gap is not proof of a co-valent ligation. It is not known whether two telomeres held together by protein interactions, a strand invasion event, or G-G base-pairing between their 3′ overhangs will appear different from telomere–telomere ligations in metaphase spreads.

Upon expression of TRF2$^{\Delta B \Delta M}$ or conditional deletion of TRF2 from mouse cells, telomere fusions are rapidly induced (Fig. 4) (van Steensel et al. 1998; Celli and de Lange 2005). These fusions are so frequent that they are detectable as a new class of TTAGGG-repeat-containing restriction frag-

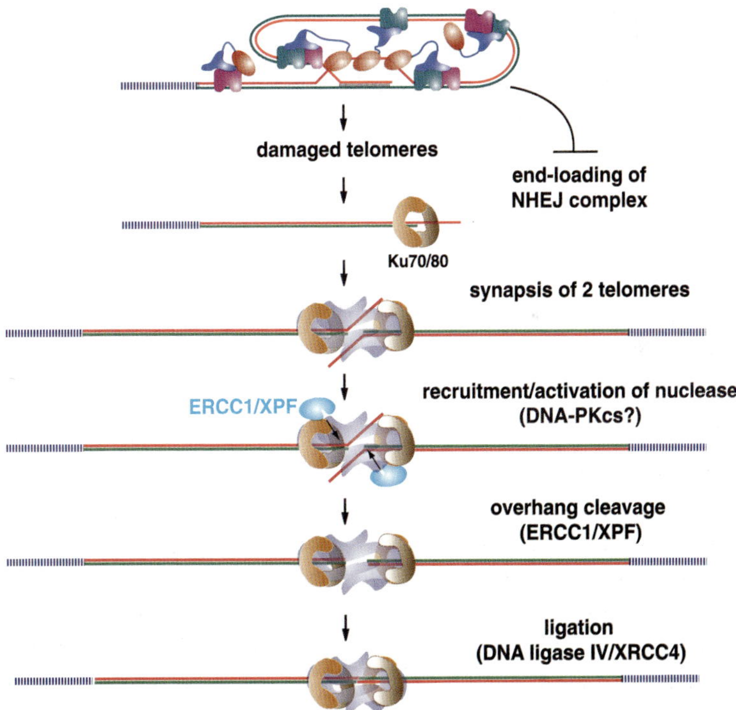

Figure 4. NHEJ of damaged telomeres. When shelterin is compromised, NHEJ is responsible for the formation of telomere fusions. The ERCC1/XPF NER endonuclease is implicated in the removal of the 3′ overhang. In absence of the NHEJ machinery, the overhang is not removed, suggesting that ERCC1/XPF is recruited or activated by this complex. The t-loop is proposed to block the NHEJ pathway at functional telomeres.

ments with a median size approximately twice that of the original telomeres. The fusion products are also detectable in denatured DNA, consistent with ligation of the G-rich strand of one telomere to the C-rich strand of another (Smogorzewska et al. 2002). In cells lacking DNA ligase IV or Ku70, the fusions are largely abrogated, indicating that they are mediated by NHEJ (Smogorzewska et al. 2002; Celli and de Lange 2005; G. Celli et al., unpubl.). NHEJ of deprotected telomeres can take place before DNA replication, giving rise to a chromosome-type fusion in metaphase spreads. Fusions can also occur after DNA replication, yielding chromatid-type unions. Post-replicative fusions more often involve the sister telomere replicated by leading-strand DNA synthesis (Bailey et al. 2001). The reason for this is not known.

The processing of deprotected telomeres by NHEJ involves the removal of the 3′ overhang. The amount of single-stranded TTAGGG repeats decreases after TRF2 inhibition, and the ERCC1/XPF nuclease appears to be required for this process (van Steensel et al. 1998; Zhu et al. 2003). As discussed above, NHEJ and 3′ overhang processing are linked. ERCC1$^{-/-}$ MEFs have a diminished ability to join deprotected telomeres after introduction of TRF2$^{\Delta B \Delta M}$ (Zhu et al. 2003). Conversely, DNA ligase-IV-deficient mouse cells show persistence of the overhang when TRF2 is deleted (Celli and de Lange 2005).

The t-loop is proposed to be a simple architectural method to prevent telomere fusions (Fig. 4). The NHEJ machinery is not expected to gain access to the telomere end when the 3′ overhang is paired with the duplex part of the telomere. The presumed ability of shelterin to remodel telomeres into t-loops could therefore protect the chromosome end from this type of inappropriate repair. Given that NHEJ and overhang cleavage are linked, the ability of shelterin to generate t-loops would also serve to protect the overhang from nucleolytic attack.

Homologous Recombination between Telomeres: Telomere Sister-chromatid Exchanges

CO-FISH analysis on ALT cells has recently revealed that their unusual telomeres frequently exchange sequences after DNA replication (see Fig. 5 in Chapter 7) (Bailey et al. 2004; Bechter et al. 2004; Londono-Vallejo et al. 2004). As sister-chromatid exchanges are thought to be mediated by homologous recombination (Sonoda et al. 1999), the frequent T-SCEs may be due to a general derepression of homologous recombination at ALT cell telomeres. How are T-SCEs repressed in non-ALT cells? So far, two factors have been implicated: TRF2 and Ku70. Loss of TRF2 alone (in a NHEJ-deficient background) or absence of the Ku heterodimer has no effect, but T-SCEs are very frequent in TRF2$^{-/-}$Ku70$^{-/-}$ MEFs (G. Celli et al., unpubl.). One possibility is that T-SCEs are normally not formed because any free end generated in the TTAGGG repeat array is processed to form a t-loop. However, this would require that the strand invasion only takes place in *cis*. In this regard, it is interesting that when TRF2 generates t-loop like structures in vitro, the assocation of the end with the duplex part of the repeat array always occurs in *cis* (Stansel et al. 2001). Through what mechanism the Ku heterodimer represses homologous recombination at telomeres is also not known, but it is well established that Ku (and not NHEJ per se) inhibits homologous recombination at other DNA ends (Adachi et al. 2001; Clikeman et al. 2001; Fukushima et al. 2001; Pierce et al. 2001; Kooistra et al. 2004).

Homologous-Recombination-mediated t-Loop Deletion

The t-loop resembles a structure formed during homologous recombination. When a DSB is processed by homologous recombination, a 3′ overhang is generated that executes the search for a homologous donor (e.g., the sister chromatid). The next step is the formation of a Holliday junction (HJ) through branch migration. HJs can be resolved by HJ resolvases, and when a double HJ is present, this resolution can lead to a crossover or a noncrossover product. If a single or double HJ is formed at the base of the t-loop, HJ resolution could lead to deletion of the telomeric DNA distal to the strand invasion point (Fig. 5).

Work with a mutant form of TRF2 lacking the amino-terminal basic domain, TRF2$^{\Delta B}$, showed that this so-called t-loop homologous recombination reaction can take place (Wang et al. 2004). TRF2$^{\Delta B}$ lacks the arginine-rich amino terminus, but it has the ability to bind to telomeres and interacts with Rap1 and TIN2. When this form of TRF2 displaces the endogenous TRF2 from telomeres, NHEJ continues to be repressed and the 3′ overhang is not degraded. However, large segments of telomeric DNA are rapidly deleted. These deletions can be detected as a reduction in the telomeric repeat signal in genomic blots, and they are visible in metaphase spreads. Often, the telomere loss takes place on one of the two sister telomeres, indicating that the deletion occurs after DNA replication. Several observations indicate that the TRF2$^{\Delta B}$-induced deletions are due to homologous recombination. The telomere-loss phenotype is abrogated in cells that lack XRCC3, a RAD51 paralog that is associated with RAD51C and HJ resolvase activity. Human cells that lack a functional Mre11 complex due to a mutation in Nbs1 also fail to show

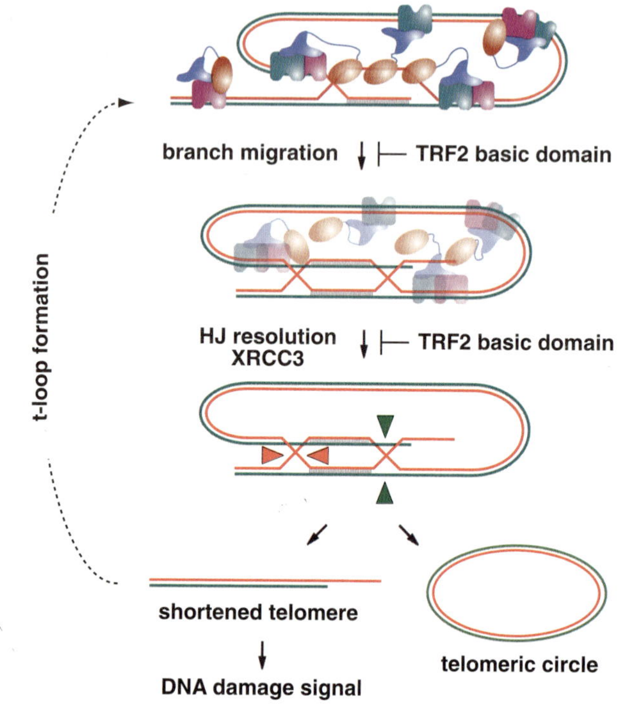

Figure 5. T-loop homologous recombination. Cells overexpressing a form of TRF2 that lacks the basic domain display t-loop homologous recombination. Presumably, the t-loop structure is altered to form a single or double HJ that is resolved by HJ resolvase. The products are a telomeric circle and shortened telomeres that could induce a DNA-damage signal or reform a t-loop. The basic domain of TRF2 might repress either the branch migration step that forms the HJ or the resolution of the HJ. XRCC3, a component of a complex with HJ resolvase activity, and the Mre11 complex have been shown to be required for t-loop homologous recombination.

loss of telomeric DNA after TRF2$^{\Delta B}$ expression. Furthermore, two-dimensional gels show that TRF2$^{\Delta B}$ induces the formation of circular telomeric DNAs with a size distribution similar to that of the circle part of the t-loops (Fig. 5).

The current model for t-loop homologous recombination (Fig. 5) is similar to that proposed for TRDs in yeast (Chapter 8). In this model, branch migration occurs at the base of the t-loop leading to formation of two HJs as proposed by Haber (2004). The second step would be the resolution of the double HJ, and in one orientation, this would release the telomeric loop as a circle and shorten the telomere. The basic domain of TRF2 is implicated in repression of either the branch migration step or resolution of the HJ with crossover. There are no known interacting factors of this part of TRF2, and how it functions is not yet known. As discussed in Chapter 7, telomeric circles are frequent in ALT cells, suggesting that this type of homologous recombination is derepressed at ALT cell telomeres.

Recombination of Telomeres with Interstitial Telomeric DNA

In human cells, telomeric sequences are largely confined to the chromosome end, but in many other vertebrates, the genome is littered with these sequences, especially at pericentric sites (see above). Recombination of telomeres with interstitial TTAGGG repeats can lead to major chromosome aberrations, including inversions, deletions, and translocations. It is assumed that this type of recombination is repressed by the same processes that control other forms of homologous recombination at telomeres. As discussed above, ERCC1$^{-/-}$ MEFs show signs of aberrant recombination between telomeres and interstitial telomeric DNA, implicating the ERCC1/XPF nuclease in repressing these events (Zhu et al. 2003). The TDM chromosomes observed in ERCC1$^{-/-}$ cells are proposed to be the product of a recombination between a telomere and a stretch of telomeric DNA proximal on the same chromosome (Fig. 6). In some cases, the TDMs contain centromeric heterochromatin, but most are

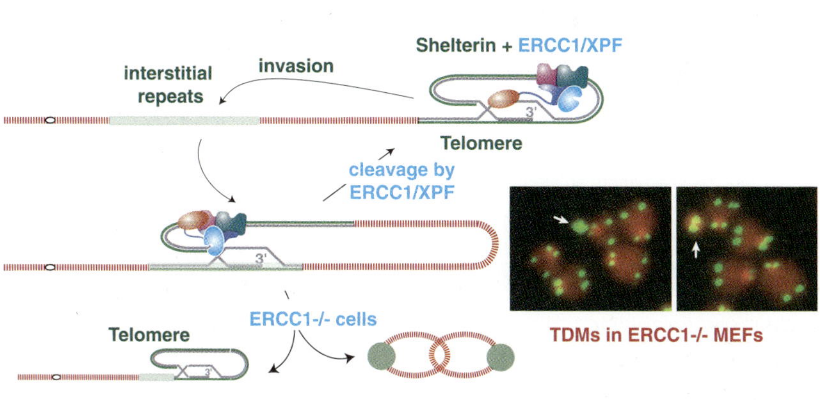

Figure 6. Model for the formation of TDMs through recombination of a telomere with interstitial telomeric DNA. (*Left*) Proposed series of events that lead to the formation of TDMs and the role of ERCC1/XPF in repressing this type of recombination by cleaving a recombination intermediate. (*Right*) TDMs in ERCC1$^{-/-}$ MEFs. (*Red*) DNA (DAPI); (*green*) telomeric DNA (FISH). The arrows point to TDMs. (Adapted, with permission, from Zhu et al. 2003 [©Elsevier].)

acentric and presumably unstable. The model in Figure 6 predicts that ERCC1/XPF cleaves an intermediate in the recombination releasing the strand-invasion event. If ERCC1/XPF is absent, recombination could generate a TDM or an inversion, depending on the orientation of the interstitial TTAGGG repeats.

REGULATION OF TELOMERE LENGTH

As in other eukaryotes, the length of mammalian telomeres can be controlled by a homeostasis pathway. Presumably, this system functions to maintain telomere length in the germ line, but telomere length control is also responsible for the stable telomere length settings of human tumor cell lines. Telomere length homeostasis is achieved through a negative feedback loop established by a negative regulator that binds to the product of telomerase. The negative regulation is provided by shelterin which senses telomere length through its ability to accumulate along the duplex telomeric repeat array. As a telomere becomes elongated, it associates with more shelterin, which decreases the chance of further elongation of that telomere by telomerase (Fig. 7). The resulting steady-state length of all telomeres in a cell clone depends on the activity of telomerase, the rate of telomere shortening, and the level of shelterin.

The mechanism of human telomere length maintenance has been studied in cultured cells, primarily in the HTC75 subclone of the human fibrosarcoma tumor cell line HT1080. Initial experiments established that inhibition of TRF1 resulted in elongation of telomeres, whereas increased loading of TRF1 on telomeres shortened them (van Steensel and de Lange

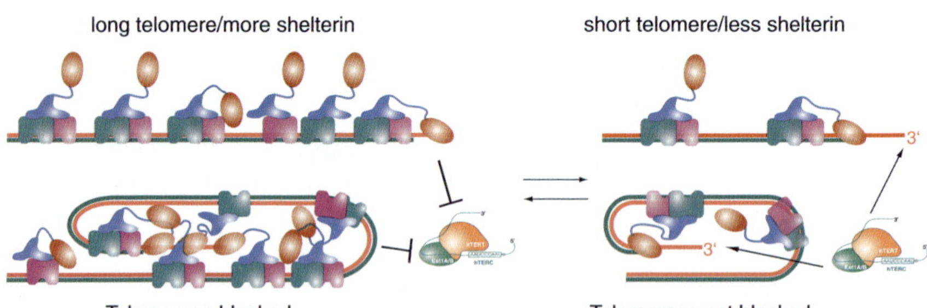

Figure 7. Model for telomere length homeostasis. Long telomeres contain more shelterin. The increased amount of shelterin increases the chance that POT1 will block telomerase at the 3′ end of the chromosome. Short telomeres or inhibition of shelterin will lead to the situation (*right*) in which telomerase will have a greater chance of gaining access to the 3′ end.

1997). These length changes occurred without effects on telomerase activity or the shortening rate of the telomeres, indicating that TRF1 affects the ability of telomerase to elongate telomeres. The amount of TRF1 at telomeres increases with the length of the TTAGGG repeat array (Smogorzewska et al. 2000; Loayza and de Lange 2003), consistent with a "protein counting" model for telomere length control (Marcand et al. 1997; van Steensel and de Lange 1997). This concept was confirmed through tethering experiments that demonstrated the *cis*-acting effect of TRF1 (Ancelin et al. 2002). Other components of shelterin were found to behave similarly, including TIN2, TPP1, and POT1 (Kim et al. 1999; Loayza and de Lange 2003; Houghtaling et al. 2004; Liu et al. 2004b; Ye et al. 2004a). TRF2 also increases at telomeres as they become longer, and overexpression of TRF2 leads to telomere shortening, suggesting that TRF2 can affect telomere length control as well (Smogorzewska et al. 2000; Loayza and de Lange 2003). Finally, indirect evidence suggests that Rap1 levels at the telomere modulate telomere elongation (Li et al. 2000; Li and de Lange 2003; O'Connor et al. 2004). Thus, telomere length homeostasis appears to be affected by all components of shelterin.

Telomerase acts on the telomere terminus and hence shelterin must exert its negative control at this site. How does shelterin, bound along the duplex repeat array, affect telomerase at the telomere terminus? The crucial player is POT1, the only component of shelterin that can bind the single-stranded DNA. A mutant of POT1 that lacks OB1 induces rapid and extensive telomere elongation without affecting the other components of shelterin (Loayza and de Lange 2003). The inappropriately elongated telomeres in cells with altered POT1 function still accumulate large amounts of TRF1, TIN2, Rap1, and TRF2, yet telomerase is no longer inhibited (Loayza and de Lange 2003). This indicates that POT1 is required for the effect of shelterin on telomerase, acting downstream from TRF1, TIN2, and TPP1.

The current model is that POT1 acts as a terminal transducer that converts information on telomere length into a regulatory signal at the telomere terminus (Fig. 7). As telomeres get longer, more shelterin is loaded along the telomeric repeat array. As a result, telomeres will contain more POT1, increasing the chance that POT1 will associate with the single-stranded telomeric DNA. The binding of POT1 to the single-stranded telomeric DNA is proposed to block telomerase. A simple mechanism would be that POT1 loading occludes the 3′ terminus, thus blocking telomerase's access to its primer. In vitro, POT1 loading on the 3′ end of a telomerase substrate diminishes the ability of the enzyme to add TTAGGG repeats (Kelleher et al. 2005; Lei et al. 2005). The competition for the 3′ end would occur at some point during S phase when telomerase is thought to act and when DNA replication is expected to lead to t-loop resolution. This model does not preclude a positive role for POT1 in the telomerase pathway, in keeping with a similarly dual role for Cdc13, the POT1-like factor of budding yeast (Chapter 12). In vitro, when POT1 is positioned near, but not at, the 3′ end of a telomerase primer, a positive effect is noted that could reflect a recruitment function (Lei et al. 2005).

As noted above, tankyrase affects modulated shelterin by modifying TRF1 and causing its release from telomeric DNA. As expected, tankyrase 1 overexpression also lowers the amount of POT1 on telomeres (Loayza and de Lange 2003), and this can explain the telomere elongation induced by excess tankyrase 1 activity in the nucleus. TIN2 has a complex role in this interplay since it functions both to tether TPP1/POT1 to telomeres and to inhibit tankyrase 1 (Ye and de Lange 2004). Exactly what determines whether tankyrase 1 will have the ability to remove shelterin and allow telomere elongation is not clear.

Several nonshelterin factors affect telomere length. In human cells, Nbs1 and Est1A mediate telomere elongation (Ranganathan et al. 2001; Snow et al. 2003). Rtel is an essential helicase and is one of several factors implicated in the species-specific telomere length setting in mice (Zhu et al. 1998; Manning et al. 2002; Ding et al. 2004). Mouse cells lacking all three retinoblastoma (Rb) family proteins have unusually long telomeres (Garcia-Cao et al. 2002), and altered telomere length settings have been reported in mice with deficiencies for DNA–PK (Hande et al. 1999; d'Adda di Fagagna et al. 2001; Espejel et al. 2002), PARP1 (d'Adda di Fagagna et al. 1999), Rad54 (Jaco et al. 2003), Rad51D (Tarsounas et al. 2004), Suv39h1/2 (Garcia-Cao et al. 2002), and heterogeneous nuclear RNP (hnRNP) A1/UP1 (LaBranche et al. 1998).

SUMMARY

Mammalian telomeres have emerged as a perplexing puzzle full of paradoxes and unexpected twists. Although lacking any evidence for intelligent design, telomeres have evolved effective mechanisms to fend off the pathways that threaten the integrity of linear chromosomes. They block NHEJ, restrain several forms of homologous recombination, prevent activation of a DNA damage signal, and control telomerase. Some hints as to how telomeres execute these tasks have come from the finding of t-loops, the subunits of shelterin, and the additional players on shelterin's "tool belt." The future challenge is to understand the mechanism by which these components collaborate to execute the functions of telomeres. It is becoming clear that telomeres use multiple tactics, each tailored to a different threat. A deeper understanding of how telomeres work will illuminate a crucial part of mammalian chromosomes and will have ramifications for several human diseases. In addition, insights into the mechanisms by which telomeres control DNA repair and DNA damage surveillance are expected to provide important clues about the function of the DNA-damage response.

ACKNOWLEDGMENTS

Although this chapter aims to give a complete overview of the function of mammalian telomeres, it is not a comprehensive review of the literature, and I apologize to colleagues whose work I was unable to cite. David Sealey and Lea Harrington are thanked for comments on this manuscript. I am very grateful for the continued support for our research from the National Institutes of Health (G.M., A.G.) and the National Cancer Institute (C.A.). Members of my laboratory are thanked for fact checking. Mike and Harold provided valuable advice on mice.

REFERENCES

Adachi N., Ishino T., Ishii Y., Takeda S., and Koyama H. 2001. DNA ligase IV-deficient cells are more resistant to ionizing radiation in the absence of Ku70: Implications for DNA double-strand break repair. *Proc. Natl. Acad. Sci.* **98:** 12109–12113.

Allshire R.C., Dempster M., and Hastie N.D. 1989. Human telomeres contain at least three types of G-rich repeat distributed non-randomly. *Nucleic Acids Res.* **17:** 4611–4627.

Allshire R.C., Gosden J.R., Cross S.H., Cranston G., Rout D., Sugawara N., Szostak J.W., Fantes P.A., and Hastie N.D. 1988. Telomeric repeat from *T. thermophila* cross hybridizes with human telomeres. *Nature* **332:** 656–659.

Ancelin K., Brunori M., Bauwens S., Koering C.E., Brun C., Ricoul M., Pommier J.P., Sabatier L., and Gilson E. 2002. Targeting assay to study the *cis* functions of human telomeric proteins: Evidence for inhibition of telomerase by TRF1 and for activation of telomere degradation by TRF2. *Mol. Cell. Biol.* **22:** 3474–3487.

Bai Y. and Murnane J.P. 2003a. Telomere instability in a human tumor cell line expressing NBS1 with mutations at sites phosphorylated by ATM. *Mol. Cancer Res.* **1:** 1058–1069.

———. 2003b. Telomere instability in a human tumor cell line expressing a dominant-negative WRN protein. *Hum. Genet.* **113:** 337–347.

Bailey S.M., Brenneman M.A., and Goodwin E.H. 2004. Frequent recombination in telomeric DNA may extend the proliferative life of telomerase-negative cells. *Nucleic Acids Res.* **32:** 3743–3751.

Bailey S.M., Cornforth M.N., Kurimasa A., Chen D.J., and Goodwin E.H. 2001. Strand-specific postreplicative processing of mammalian telomeres. *Science* **293:** 2462–2465.

Bailey S.M., Meyne J., Chen D.J., Kurimasa A., Li G.C., Lehnert B.E., and Goodwin E.H.

1999. DNA double-strand break repair proteins are required to cap the ends of mammalian chromosomes. *Proc. Natl. Acad. Sci.* **96:** 14899–14904.

Baird D.M., Jeffreys A.J., and Royle N.J. 1995. Mechanisms underlying telomere repeat turnover, revealed by hypervariable variant repeat distribution patterns in the human Xp/Yp telomere. *EMBO J.* **14:** 5433–5443.

Baird D.M., Davis T., Rowson J., Jones C.J., and Kipling D. 2004. Normal telomere erosion rates at the single cell level in Werner syndrome fibroblast cells. *Hum. Mol. Genet.* **13:** 1515–1524.

Bakkenist C.J. and Kastan M.B. 2003. DNA damage activates ATM through intermolecular autophosphorylation and dimer dissociation. *Nature* **421:** 499–506.

Bakkenist C.J., Drissi R., Wu J., Kastan M.B., and Dome J.S. 2004. Disappearance of the telomere dysfunction-induced stress response in fully senescent cells. *Cancer Res.* **64:** 3748–3752.

Barnett M.A., Buckle V.J., Evans E.P., Porter A.C., Rout D., Smith A.G., and Brown W.R. 1993. Telomere directed fragmentation of mammalian chromosomes. *Nucleic Acids Res.* **21:** 27–36.

Baumann P. and Cech T.R. 2001. Pot1, the putative telomere end-binding protein in fission yeast and humans. *Science* **292:** 1171–1175.

Baur J.A., Shay J.W., and Wright W.E. 2004. Spontaneous reactivation of a silent telomeric transgene in a human cell line. *Chromosoma* **112:** 240–246.

Baur J.A., Zou Y., Shay J.W., and Wright W.E. 2001. Telomere position effect in human cells. *Science* **292:** 2075–2077.

Bechter O.E., Zou Y., Walker W., Wright W.E., and Shay J.W. 2004. Telomeric recombination in mismatch repair deficient human colon cancer cells after telomerase inhibition. *Cancer Res.* **64:** 3444–3451.

Bianchi A., Smith S., Chong L., Elias P., and de Lange T. 1997. TRF1 is a dimer and bends telomeric DNA. *EMBO J.* **16:** 1785–1794.

Bianchi A., Stansel R.M., Fairall L., Griffith J.D., Rhodes D., and de Lange T. 1999. TRF1 binds a bipartite telomeric site with extreme spatial flexibility. *EMBO J.* **18:** 5735–5744.

Bilaud T., Brun C., Ancelin K., Koering C.E., Laroche T., and Gilson E. 1997. Telomeric localization of TRF2, a novel human telobox protein. *Nat. Genet.* **17:** 236–239.

Bradshaw P.S., Stavropoulos D.J., and Meyn M.S. 2005. Human telomeric protein TRF2 associates with genomic double-strand breaks as an early response to DNA damage. *Nat. Genet.* **37:** 193–197.

Broccoli D., Smogorzewska A., Chong L., and de Lange T. 1997a. Human telomeres contain two distinct Myb-related proteins, TRF1 and TRF2. *Nat. Genet.* **17:** 231–235.

Broccoli D., Chong L., Oelmann S., Fernald A.A., Marziliano N., van Steensel B., Kipling D., Le Beau M.M., and de Lange T. 1997b. Comparison of the human and mouse genes encoding the telomeric protein, TRF1: Chromosomal localization, expression and conserved protein domains. *Hum. Mol. Genet.* **6:** 69–76.

Brown W.R., MacKinnon P.J., Villasante A., Spurr N., Buckle V.J., and Dobson M.J. 1990. Structure and polymorphism of human telomere-associated DNA. *Cell* **63:** 119–132.

Cardenas M.E., Bianchi A., and de Lange T. 1993. A *Xenopus* egg factor with DNA-binding properties characteristic of terminus-specific telomeric proteins. *Genes Dev.* **7:** 883–894.

Celli G. and de Lange T. 2005. DNA processing not required for ATM-mediated telomere damage response after TRF2 deletion. *Nat. Cell Biol.* **7:** 712–718.

Chang W., Dynek J.N., and Smith S. 2003. TRF1 is degraded by ubiquitin-mediated proteolysis after release from telomeres. *Genes Dev.* **17:** 1328–1333.

Chi N.W. and Lodish H.F. 2000. Tankyrase is a golgi-associated mitogen-activated protein kinase substrate that interacts with IRAP in GLUT4 vesicles. *J. Biol. Chem.* **275:** 38437–38444.

Chiang Y.J., Kim S.H., Tessarollo L., Campisi J., and Hodes R.J. 2004. Telomere-associated protein TIN2 is essential for early embryonic development through a telomerase-independent pathway. *Mol. Cell. Biol.* **24:** 6631–6634.

Chikashige Y. and Hiraoka Y. 2001. Telomere binding of the Rap1 protein is required for meiosis in fission yeast. *Curr. Biol.* **11:** 1618–1623.

Chin L., Artandi S.E., Shen Q., Tam A., Lee S.L., Gottlieb G.J., Greider C.W., and DePinho R.A. 1999. p53 deficiency rescues the adverse effects of telomere loss and cooperates with telomere dysfunction to accelerate carcinogenesis. *Cell* **97:** 527–538.

Chong L., van Steensel B., Broccoli D., Erdjument-Bromage H., Hanish J., Tempst P., and de Lange T. 1995. A human telomeric protein. *Science* **270:** 1663–1667.

Clikeman J.A., Khalsa G.J., Barton S.L., and Nickoloff J.A. 2001. Homologous recombinational repair of double-strand breaks in yeast is enhanced by MAT heterozygosity through yKU-dependent and -independent mechanisms. *Genetics* **157:** 579–589.

Conti E. and Izaurralde E. 2005. Nonsense-mediated mRNA decay: Molecular insights and mechanistic variations across species. *Curr. Opin. Cell Biol.* **17:** 316–325.

Cook B.D., Dynek J.N., Chang W., Shostak G., and Smith S. 2002. Role for the related poly(ADP-Ribose) polymerases tankyrase 1 and 2 at human telomeres. *Mol. Cell. Biol.* **22:** 332–342.

Cooke H.J. and Smith B.A. 1986. Variability at the telomeres of the human X/Y pseudoautosomal region. *Cold Spring Harbor Symp. Quant. Biol.* **51:** 213–219.

Crabbe L., Verdun R.E., Haggblom C.I., and Karlseder J. 2004. Defective telomere lagging strand synthesis in cells lacking WRN helicase activity. *Science* **306:** 1951–1953.

d'Adda di Fagagna F., Hande M.P., Tong W.M., Lansdorp P.M., Wang Z.Q., and Jackson S.P. 1999. Functions of poly(ADP-ribose) polymerase in controlling telomere length and chromosomal stability. *Nat. Genet.* **23:** 76–80.

d'Adda di Fagagna F., Hande M.P., Tong W.M., Roth D., Lansdorp P.M., Wang Z.Q., and Jackson S.P. 2001. Effects of DNA nonhomologous end-joining factors on telomere length and chromosomal stability in mammalian cells. *Curr. Biol.* **11:** 1192–1196.

d'Adda di Fagagna F., Reaper P.M., Clay-Farrace L., Fiegler H., Carr P., Von Zglinicki T., Saretzki G., Carter N.P., and Jackson S.P. 2003. A DNA damage checkpoint response in telomere-initiated senescence. *Nature* **426:** 194–198.

Dantzer F., Giraud-Panis M.J., Jaco I., Ame J.C., Schultz I., Blasco M., Koering C.E., Gilson E., Menissier-de Murcia J., de Murcia G., and Schreiber V. 2004. Functional interaction between poly(ADP-Ribose) polymerase 2 (PARP-2) and TRF2: PARP activity negatively regulates TRF2. *Mol. Cell. Biol.* **24:** 1595–1607.

de Lange T., Shiue L., Myers R.M., Cox D.R., Naylor S.L., Killery A.M., and Varmus H.E. 1990. Structure and variability of human chromosome ends. *Mol. Cell. Biol.* **10:** 518–527.

De Rycker M., Venkatesan R.N., Wei C., and Price C.M. 2003. Vertebrate tankyrase domain structure and sterile α motif (SAM)-mediated multimerization. *Biochem. J.* **372:** 87–96.

Ding H., Schertzer M., Wu X., Gertsenstein M., Selig S., Kammori M., Pourvali R., Poon S., Vulto I., Chavez E., Tam P.P., Nagy A., and Lansdorp P.M. 2004. Regulation of murine telomere length by *Rtel*: An essential gene encoding a helicase-like protein. *Cell* **117:** 873–886.

Dynek J.N. and Smith S. 2004. Resolution of sister telomere association is required for progression through mitosis. *Science* **304:** 97–100.

Edwards Y.J., Elgar G., Clark M.S., and Bishop M.J. 1998. The identification and characterization of microsatellites in the compact genome of the Japanese pufferfish, *Fugu rubripes*: Perspectives in functional and comparative genomic analyses. *J. Mol. Biol.* **278:** 843–854.

Espejel S., Franco S., Sgura A., Gae D., Bailey S.M., Taccioli G.E., and Blasco M.A. 2002. Functional interaction between DNA-PKcs and telomerase in telomere length maintenance. *EMBO J.* **21:** 6275–6287.

Fairall L., Chapman L., Moss H., de Lange T., and Rhodes D. 2001. Structure of the TRFH dimerization domain of the human telomeric proteins TRF1 and TRF2. *Mol. Cell* **8:** 351–361.

Farr C., Fantes J., Goodfellow P., and Cooke H. 1991. Functional reintroduction of human telomeres into mammalian cells. *Proc. Natl. Acad. Sci.* **88:** 7006–7010.

Fukuhara N., Ebert J., Unterholzner L., Lindner D., Izaurralde E., and Conti E. 2005. SMG7 is a 14-3-3-like adaptor in the nonsense-mediated mRNA decay pathway. *Mol. Cell* **17:** 537–547.

Fukushima T., Takata M., Morrison C., Araki R., Fujimori A., Abe M., Tatsumi K., Jasin M., Dhar P.K., Sonoda E., Chiba T., and Takeda S. 2001. Genetic analysis of the DNA-dependent protein kinase reveals an inhibitory role of Ku in late S-G$_2$ phase DNA double-strand break repair. *J. Biol. Chem.* **276:** 44413–44418.

Garcia-Cao M., Gonzalo S., Dean D., and Blasco M.A. 2002. A role for the Rb family of proteins in controlling telomere length. *Nat. Genet.* **32:** 415–419.

Garcia-Cao M., O'Sullivan R., Peters A.H., Jenuwein T., and Blasco M.A. 2004. Epigenetic regulation of telomere length in mammalian cells by the Suv39h1 and Suv39h2 histone methyltransferases. *Nat. Genet.* **36:** 94–99.

Gilley D., Tanaka H., Hande M.P., Kurimasa A., Li G.C., Oshimura M., and Chen D.J. 2001. DNA-PKcs is critical for telomere capping. *Proc. Natl. Acad. Sci.* **98:** 15084–15088.

Goytisolo F.A., Samper E., Edmonson S., Taccioli G.E., and Blasco M.A. 2001. The absence of the DNA-dependent protein kinase catalytic subunit in mice results in anaphase bridges and in increased telomeric fusions with normal telomere length and G-strand overhang. *Mol. Cell. Biol.* **21:** 3642–3651.

Griffith J., Bianchi A., and de Lange T. 1998. TRF1 promotes parallel pairing of telomeric tracts in vitro. *J. Mol. Biol.* **278:** 79–88.

Griffith J.D., Comeau L., Rosenfield S., Stansel R.M., Bianchi A., Moss H., and de Lange T. 1999. Mammalian telomeres end in a large duplex loop. *Cell* **97:** 503–514.

Haber J.E. 2004. Telomeres thrown for a loop. *Mol. Cell* **16:** 502–503.

Hanaoka S., Nagadoi A., Yoshimura S., Aimoto S., Li B., de Lange T., and Nishimura Y. 2001. NMR structure of the hRap1 Myb motif reveals a canonical three-helix bundle lacking the positive surface charge typical of Myb DNA-binding domains. *J. Mol. Biol.* **312:** 167–175.

Hande P., Slijepcevic P., Silver A., Bouffler S., van Buul P., Bryant P., and Lansdorp P. 1999. Elongated telomeres in *scid* mice. *Genomics* **56:** 221–223.

Hanish J.P., Yanowitz J.L., and de Lange T. 1994. Stringent sequence requirements for the formation of human telomeres. *Proc. Natl. Acad. Sci.* **91:** 8861–8865.

Harley C.B., Futcher A.B., and Greider C.W. 1990. Telomeres shorten during ageing of human fibroblasts. *Nature* **345:** 458–460.

Hathcock K.S., Hemann M.T., Opperman K.K., Strong M.A., Greider C.W., and Hodes R.J. 2002. Haploinsufficiency of mTR results in defects in telomere elongation. *Proc. Natl. Acad. Sci.* **99:** 3591–3596.

Hemann M.T. and Greider C.W. 1999. G-strand overhangs on telomeres in telomerase-deficient mouse cells. *Nucleic Acids Res.* **27:** 3964–3969.

Hemann M.T., Strong M.A., Hao L.Y., and Greider C.W. 2001. The shortest telomere, not average telomere length, is critical for cell viability and chromosome stability. *Cell* **107:** 67–77.

Herbig U., Jobling W.A., Chen B.P., Chen D.J., and Sedivy J.M. 2004. Telomere shortening triggers senescence of human cells through a pathway involving ATM, p53, and p21[(CIP1)], but not p16[(INK4a)]. *Mol. Cell* **14:** 501–513.

Hockemeyer D., Sfeir A.J., Shay J.W., Wright W.E., and de Lange T. 2005. POT1 protects telomeres from a transient DNA damage response and determines how human chromosomes end. *EMBO J.* **24:** 2667–2678.

Houghtaling B.R., Cuttonaro L., Chang W., and Smith S. 2004. A dynamic molecular link between the telomere length regulator TRF1 and the chromosome end protector TRF2. *Curr. Biol.* **14:** 1621–1631.

Hsu H.L., Gilley D., Blackburn E.H., and Chen D.J. 1999. Ku is associated with the telomere in mammals. *Proc. Natl. Acad. Sci.* **96:** 12454–12458.

Hsu H.L., Gilley D., Galande S.A., Hande M.P., Allen B., Kim S.H., Li G.C., Campisi J., Kohwi-Shigematsu T., and Chen D.J. 2000. Ku acts in a unique way at the mammalian telomere to prevent end joining. *Genes Dev.* **14:** 2807–2812.

Huber M.D., Lee D.C., and Maizels N.P. 2002. G4 DNA unwinding by BLM and Sgs1p: Substrate specificity and substrate-specific inhibition. *Nucleic Acids Res.* **30:** 3954–3961.

Huffman K.E., Levene S.D., Tesmer V.M., Shay J.W., and Wright W.E. 2000. Telomere shortening is proportional to the size of the G-rich telomeric 3′-overhang. *J. Biol. Chem.* **275:** 19719–19722.

Huyen Y., Zgheib O., Ditullio R.A., Jr., Gorgoulis V.G., Zacharatos P., Petty T.J., Sheston E.A., Mellert H.S., Stavridi E.S., and Halazonetis T.D. 2004. Methylated lysine 79 of histone H3 targets 53BP1 to DNA double-strand breaks. *Nature* **432:** 406–411.

Ishikawa F., Matunis M.J., Dreyfuss G., and Cech T.R. 1993. Nuclear proteins that bind the pre-mRNA 3′ splice site sequence r(UUAG/G) and the human telomeric DNA sequence d(TTAGGG)$_n$. *Mol. Cell. Biol.* **13:** 4301–4310.

Iwano T., Tachibana M., Reth M., and Shinkai Y. 2004. Importance of TRF1 for functional telomere structure. *J. Biol. Chem.* **279:** 1442–1448.

Jaco I., Munoz P., and Blasco M.A. 2004. Role of human Ku86 in telomere length maintenance and telomere capping. *Cancer Res.* **64:** 7271–7278.

Jaco I., Munoz P., Goytisolo F., Wesoly J., Bailey S., Taccioli G., and Blasco M.A.P. 2003. Role of mammalian Rad54 in telomere length maintenance. *Mol. Cell. Biol.* **23:** 5572–5580.

Jacobs J.J. and de Lange T. 2004. Significant role for p16[(INK4a)] in p53-independent telomere-directed senescence. *Curr. Biol.* **14:** 2302–2308.

Jiang N., Benard C.Y., Kebir H., Shoubridge E.A., and Hekimi S. 2003. Human CLK2 links cell cycle progression, apoptosis, and telomere length regulation. *J. Biol. Chem.* **278:** 21678–21684.

Johnson F.B., Marciniak R.A., McVey M., Stewart S.A., Hahn W.C., and Guarente L. 2001. The *Saccharomyces cerevisiae* WRN homolog Sgs1p participates in telomere maintenance in cells lacking telomerase. *EMBO J.* **20:** 905–913.

Kaminker P.G., Kim S.H., Taylor R.D., Zebarjadian Y., Funk W.D., Morin G.B., Yaswen P., and Campisi J. 2001. TANK2, a new TRF1-associated PARP, causes rapid induction of cell death upon overexpression. *J. Biol. Chem.* **276:** 35891–35899.

426 de Lange

Kanoh J. and Ishikawa F. 2001. spRap1 and spRif1, recruited to telomeres by Taz1, are essential for telomere function in fission yeast. *Curr. Biol.* **11:** 1624–1630.

Karlseder J., Broccoli D., Dai Y., Hardy S., and de Lange T. 1999. p53- and ATM-dependent apoptosis induced by telomeres lacking TRF2. *Science* **283:** 1321–1325.

Karlseder J., Hoke K., Mirzoeva O.K., Bakkenist C., Kastan M.B., Petrini J.H., and de Lange T. 2004. The telomeric protein TRF2 binds the ATM kinase and can inhibit the ATM-dependent DNA damage response. *PLoS Biol.* **2:** E240.

Karlseder J., Kachatrian L., Takai H., Mercer K., Hingorani S., Jacks T., and de Lange T. 2003. Targeted deletion reveals an essential function for the telomere length regulator Trf1. *Mol. Cell. Biol.* **23:** 6533–6541.

Keegan C.E., Hutz J.E., Else T., Adamska M., Shah S.P., Kent A.E., Howes J.M., Beamer W.G., and Hammer G.D. 2005. Urogenital and caudal dysgenesis in adrenocortical dysplasia (*acd*) mice is caused by a splicing mutation in a novel telomeric regulator. *Hum. Mol. Genet.* **14:** 113–123.

Kelleher C., Kurth I., and Lingner J. 2005. Human protection of telomeres 1 (POT1) is a negative regulator of telomerase activity in vitro. *Mol. Cell. Biol.* **25:** 808–818.

Kim S.H., Kaminker P., and Campisi J. 1999. TIN2, a new regulator of telomere length in human cells. *Nat. Genet.* **23:** 405–412.

Kim S.H., Han S., You Y.H., Chen D.J., and Campisi J. 2003. The human telomere-associated protein TIN2 stimulates interactions between telomeric DNA tracts in vitro. *EMBO Rep.* **4:** 685–691.

Kim S.H., Beausejour C., Davalos A.R., Kaminker P., Heo S.J., and Campisi J. 2004. TIN2 mediates functions of TRF2 at human telomeres. *J. Biol. Chem.* **279:** 43799–43804.

Kipling D. and Cooke H.J. 1990. Hypervariable ultra-long telomeres in mice. *Nature* **347:** 400–402.

Kishi S., Zhou X.Z., Ziv Y., Khoo C., Hill D.E., Shiloh Y., and Lu K.P. 2001. Telomeric protein Pin2/TRF1 as an important ATM target in response to double strand DNA breaks. *J. Biol. Chem.* **276:** 29282–29291.

Koering C.E., Pollice A., Zibella M.P., Bauwens S., Puisieux A., Brunori M., Brun C., Martins L., Sabatier L., Pulitzer J.F., and Gilson E. 2002. Human telomeric position effect is determined by chromosomal context and telomeric chromatin integrity. *EMBO Rep.* **3:** 1055–1061.

Kooistra R., Hooykaas P.J., and Steensma H.Y. 2004. Efficient gene targeting in *Kluyveromyces lactis. Yeast* **21:** 781–792.

Krutilina R.I., Oei S., Buchlow G., Yau P.M., Zalensky A.O., Zalenskaya I.A., Bradbury E.M., and Tomilin N.V. 2001. A negative regulator of telomere-length protein TRF1 is associated with interstitial (TTAGGG)n blocks in immortal Chinese hamster ovary cells. *Biochem. Biophys. Res. Commun.* **280:** 471–475.

LaBranche H., Dupuis S., Ben-David Y., Bani M.R., Wellinger R.J., and Chabot B. 1998. Telomere elongation by hnRNP A1 and a derivative that interacts with telomeric repeats and telomerase. *Nat. Genet.* **19:** 199–202.

Lei M., Podell E.R., and Cech T.R. 2004. Structure of human POT1 bound to telomeric single-stranded DNA provides a model for chromosome end-protection. *Nat. Struct. Mol. Biol.* **11:** 1223–1229.

Lei M., Zaug A.J., Podell E.R., and Cech T.R. 2005. Switching human telomerase on and off with hPOT1 protein in vitro. *J. Biol. Chem.* **280:** 20449–20456.

Li B. and de Lange T. 2003. Rap1 affects the length and heterogeneity of human telomeres. *Mol. Biol. Cell* **14:** 5060–5068.

Li B., Espinal A., and Cross G.A. 2005. Trypanosome telomeres are protected by a homologue of mammalian TRF2. *Mol. Cell. Biol.* **25:** 5011–5021.

Li B., Oestreich S., and de Lange T. 2000. Identification of human Rap1: Implications for telomere evolution. *Cell* **101:** 471–483.

Li G., Nelsen C., and Hendrickson E.A. 2002. Ku86 is essential in human somatic cells. *Proc. Natl. Acad. Sci.* **99:** 832–837.

Lillard-Wetherell K., Machwe A., Langland G.T., Combs K.A., Behbehani G.K., Schonberg S.A., German J., Turchi J.J., Orren D.K., and Groden J. 2004. Association and regulation of the BLM helicase by the telomere proteins TRF1 and TRF2. *Hum. Mol. Genet.* **13:** 1919–1932.

Liu D., O'Connor M.S., Qin J., and Songyang Z. 2004a. Telosome, a mammalian telomere-associated complex formed by multiple telomeric proteins. *J. Biol. Chem.* **279:** 51338–51342.

Liu D., Safari A., O'Connor M.S., Chan D.W., Laegeler A., Qin J., and Songyang Z. 2004b. PTOP interacts with POT1 and regulates its localization to telomeres. *Nat. Cell Biol.* **6:** 673–680.

Loayza D. and de Lange T. 2003. POT1 as a terminal transducer of TRF1 telomere length control. *Nature* **424:** 1013–1018.

Loayza D., Parsons H., Donigian J., Hoke K., and de Lange T. 2004. DNA binding features of human POT1: A nonamer 5'-TAGGGTTAG-3' minimal binding site, sequence specificity, and internal binding to multimeric sites. *J. Biol. Chem.* **279:** 13241–13248.

Londono-Vallejo J.A., Der-Sarkissian H., Cazes L., Bacchetti S., and Reddel R.R. 2004. Alternative lengthening of telomeres is characterized by high rates of telomeric exchange. *Cancer Res.* **64:** 2324–2327.

Luderus M.E., van Steensel B., Chong L., Sibon O.C., Cremers F.F., and de Lange T. 1996. Structure, subnuclear distribution, and nuclear matrix association of the mammalian telomeric complex. *J. Cell Biol.* **135:** 867–881.

Machwe A., Xiao L., and Orren D.K. 2004. TRF2 recruits the Werner syndrome (WRN) exonuclease for processing of telomeric DNA. *Oncogene* **23:** 149–156.

Makarov V.L., Hirose Y., and Langmore J.P. 1997. Long G tails at both ends of human chromosomes suggest a C strand degradation mechanism for telomere shortening. *Cell* **88:** 657–666.

Makarov V.L., Lejnine S., Bedoyan J., and Langmore J.P. 1993. Nucleosomal organization of telomere-specific chromatin in rat. *Cell* **73:** 775–787.

Manning E.L., Crossland J., Dewey M.J., and Van Zant G. 2002. Influences of inbreeding and genetics on telomere length in mice. *Mamm. Genome* **13:** 234–238.

Marcand S., Gilson E., and Shore D. 1997. A protein-counting mechanism for telomere length regulation in yeast. *Science* **275:** 986–990.

Mattern K.A., Swiggers S.J., Nigg A.L., Lowenberg B., Houtsmuller A.B., and Zijlmans J.M. 2004. Dynamics of protein binding to telomeres in living cells: Implications for telomere structure and function. *Mol. Cell. Biol.* **24:** 5587–5594.

McKay S.J. and Cooke H. 1992. A protein which specifically binds to single stranded TTAGGGn repeats. *Nucleic Acids Res.* **20:** 1387–1391.

Meyne J., Ratliff R.L., and Moyzis R.K. 1989. Conservation of the human telomere sequence (TTAGGG)n among vertebrates. *Proc. Natl. Acad. Sci.* **86:** 7049–7053.

Meyne J., Baker R.J., Hobart H.H., Hsu T.C., Ryder O.A., Ward O.G., Wiley J.E., Wurster-Hill D.H., Yates T.L., and Moyzis R.K. 1990. Distribution of non-telomeric sites of the (TTAGGG)n telomeric sequence in vertebrate chromosomes. *Chromosoma* **99:** 3–10.

428   de Lange

Moyzis R.K., Buckingham J.M., Cram L.S., Dani M., Deaven L.L., Jones M.D., Meyne J., Ratliff R.L., and Wu J.R. 1988. A highly conserved repetitive DNA sequence, (TTAGGG)n, present at the telomeres of human chromosomes. *Proc. Natl. Acad. Sci.* **85:** 6622–6626.

Myung K., Ghosh G., Fattah F.J., Li G., Kim H., Dutia A., Pak E., Smith S., and Hendrickson E.A. 2004. Regulation of telomere length and suppression of genomic instability in human somatic cells by Ku86. *Mol. Cell. Biol.* **24:** 5050–5059.

Nakamura M., Zhou X.Z., Kishi S., Kosugi I., Tsutsui Y., and Lu. K.P. 2001. A specific interaction between the telomeric protein Pin2/TRF1 and the mitotic spindle. *Curr. Biol.* **11:** 1512–1516.

Netzer C., Rieger L., Brero A., Zhang C.D., Hinzke M., Kohlhase J., and Bohlander S.K. 2001. *SALL1*, the gene mutated in Townes-Brocks syndrome, encodes a transcriptional repressor which interacts with TRF1/PIN2 and localizes to pericentromeric heterochromatin. *Hum. Mol. Genet.* **10:** 3017–3024.

Nikitina T. and Woodcock C.L. 2004. Closed chromatin loops at the ends of chromosomes. *J. Cell Biol.* **166:** 161–165.

Nosaka K., Kawahara M., Masuda M., Satomi Y., and Nishino H. 1998. Association of nucleoside diphosphate kinase nm23-H2 with human telomeres. *Biochem. Biophys. Res. Commun.* **243:** 342–348.

O'Connor M.S., Safari A., Liu D., Qin J., and Songyang Z. 2004. The human Rap1 protein complex and modulation of telomere length. *J. Biol. Chem.* **279:** 28585–28591.

Opresko P.L., Von Kobbe C., Laine J.P., Harrigan J., Hickson I.D., and Bohr V.A. 2002. Telomere binding protein TRF2 binds to and stimulates the Werner and Bloom syndrome helicases. *J. Biol. Chem.* **277:** 41110–41119.

Opresko P.L., Otterlei M., Graakjaer J., Bruheim P., Dawut L., Kolvraa S., May A., Seidman M.M., and Bohr V.A. 2004. The Werner syndrome helicase and exonuclease cooperate to resolve telomeric D loops in a manner regulated by TRF1 and TRF2. *Mol. Cell* **14:** 763–774.

Ozgenc A. and Loeb L.A. 2005. Current advances in unraveling the function of the Werner syndrome protein. *Mutat. Res.* **577:** 237–251.

Petrini J. and Stracker T.H. 2003. The cellular response to DNA double strand breaks: Defining the sensors and mediators. *Trends Cell Biol.* **13:** 458–462.

Pierce A.J., Hu P., Han M., Ellis N., and Jasin M. 2001. Ku DNA end-binding protein modulates homologous repair of double-strand breaks in mammalian cells. *Genes Dev.* **15:** 3237–3242.

Ramirez R.D., Morales C.P., Herbert B.S., Rohde J.M., Passons C., Shay J.W., and Wright W.E. 2001. Putative telomere-independent mechanisms of replicative aging reflect inadequate growth conditions. *Genes Dev.* **15:** 398–403.

Ranganathan V., Heine W.F., Ciccone D.N., Rudolph K.L., Wu X., Chang S., Hai H., Ahearn I.M., Livingston D.M., Resnick I., Rosen F., Seemanova E., Jarolim P., DePinho R.A., and Weaver D.T. 2001. Rescue of a telomere length defect of Nijmegen breakage syndrome cells requires NBS and telomerase catalytic subunit. *Curr. Biol.* **11:** 962–966.

Reichenbach P., Hoss M., Azzalin C.M., Nabholz M., Bucher P., and Lingner J. 2003. A human homolog of yeast Est1 associates with telomerase and uncaps chromosome ends when overexpressed. *Curr. Biol.* **13:** 568–574.

Riethman H., Ambrosini A., Castaneda C., Finklestein J., Hu X.L., Mudunuri U., Paul S., and Wei J. 2004. Mapping and initial analysis of human subtelomeric sequence assemblies. *Genome Res.* **14:** 18–28.

Rooney S., Alt F.W., Lombard D., Whitlow S., Eckersdorff M., Fleming J., Fugmann S., Ferguson D.O., Schatz D.G., and Sekiguchi J. 2003. Defective DNA repair and increased genomic instability in Artemis-deficient murine cells. *J. Exp. Med.* **197:** 553–565.

Samper E., Goytisolo F.A., Slijepcevic P., van Buul P.P., and Blasco M.A. 2000. Mammalian Ku86 protein prevents telomeric fusions independently of the length of TTAGGG repeats and the G-strand overhang. *EMBO Rep.* **1:** 244–252.

Samper E., Goytisolo F.A., Menissier-de Murcia J., Gonzalez-Suarez E., Cigudosa J.C., de Murcia G., and Blasco M.A. 2001. Normal telomere length and chromosomal end capping in poly(ADP-ribose) polymerase-deficient mice and primary cells despite increased chromosomal instability. *J. Cell Biol.* **154:** 49–60.

Sbodio J.I. and Chi N.W. 2002. Identification of a tankyrase-binding motif shared by IRAP, TAB182, and human TRF1 but not mouse TRF1. NuMA contains this RXXPDG motif and is a novel tankyrase partner. *J. Biol. Chem.* **277:** 31887–31892.

Sbodio J.I., Lodish H.F., and Chi N.W. 2002. Tankyrase-2 oligomerizes with tankyrase-1 and binds to both TRF1 (telomere-repeat-binding factor 1) and IRAP (insulin-responsive aminopeptidase). *Biochem. J.* **361:** 451–459.

Scherthan H., Jerratsch M., Li B., Smith S., Hulten M., Lock T., and de Lange T. 2000. Mammalian meiotic telomeres: Protein composition and redistribution in relation to nuclear pores. *Mol. Biol. Cell* **11:** 4189–4203.

Schulz V.P., Zakian V.A., Ogburn C.E., McKay J., Jarzebowicz A.A., Edland S.D., and Martin G.M. 1996. Accelerated loss of telomeric repeats may not explain accelerated replicative decline of Werner syndrome cells. *Hum. Genet.* **97:** 750–754.

Seimiya H. and Smith S. 2002. The telomeric poly(ADP-ribose) polymerase, tankyrase 1, contains multiple binding sites for telomeric repeat binding factor 1 (TRF1) and a novel acceptor, 182-kDa tankyrase-binding protein (TAB182). *J. Biol. Chem.* **277:** 14116–14126.

Seimiya H., Muramatsu Y., Ohishi T., and Tsuruo T. 2005. Tankyrase 1 as a target for telomere-directed molecular cancer therapeutics. *Cancer Cell* **7:** 25–37.

Seimiya H., Muramatsu Y., Smith S., and Tsuruo T. 2004. Functional subdomain in the ankyrin domain of tankyrase 1 required for poly(ADP-ribosyl)ation of TRF1 and telomere elongation. *Mol. Cell. Biol.* **24:** 1944–1955.

Sfeir A.J., Chai W., Shay J.W., and Wright W.E. 2005. Telomere-end processing the terminal nucleotides of human chromosomes. *Mol. Cell* **18:** 131–138.

Shen M., Haggblom C., Vogt M., Hunter T., and Lu K.P. 1997. Characterization and cell cycle regulation of the related human telomeric proteins Pin2 and TRF1 suggest a role in mitosis. *Proc. Natl. Acad. Sci.* **94:** 13618–13623.

Sherr C.J. and DePinho R.A. 2000. Cellular senescence: Mitotic clock or culture shock? *Cell* **102:** 407–410.

Silverman J., Takai H., Buonomo S.B., Eisenhaber F., and de Lange T. 2004. Human Rif1, ortholog of a yeast telomeric protein, is regulated by ATM and 53BP1 and functions in the S-phase checkpoint. *Genes Dev.* **18:** 2108–2119.

Smith S. and de Lange T. 1999. Cell cycle dependent localization of the telomeric PARP, tankyrase, to nuclear pore complexes and centrosomes. *J. Cell Sci.* **112:** 3649–3656.

———. 2000. Tankyrase promotes telomere elongation in human cells. *Curr. Biol.* **10:** 1299–1302.

Smith S., Giriat I., Schmitt A., and de Lange T. 1998. Tankyrase, a poly(ADP-ribose) polymerase at human telomeres. *Science* **282:** 1484–1487.

Smogorzewska A. and de Lange T. 2002. Different telomere damage signaling pathways in human and mouse cells. *EMBO J.* **21:** 4338–4348.

Smogorzewska A., Karlseder J., Holtgreve-Grez H., Jauch A., and de Lange T. 2002. DNA ligase IV-dependent NHEJ of deprotected mammalian telomeres in G1 and G2. *Curr. Biol.* **12:** 1635–1644.

Smogorzewska A., van Steensel B., Bianchi A., Oelmann S., Schaefer M.R., Schnapp G., and de Lange T. 2000. Control of human telomere length by TRF1 and TRF2. *Mol. Cell. Biol.* **20:** 1659–1668.

Snow B.E., Erdmann N., Cruickshank J., Goldman H., Gill R.M., Robinson M.O., and Harrington L. 2003. Functional conservation of the telomerase protein Est1p in humans. *Curr. Biol.* **13:** 698–704.

Song K., Jung D., Jung Y., Lee S.G., and Lee I. 2000. Interaction of human Ku70 with TRF2. *FEBS Lett.* **481:** 81–85.

Sonoda E., Sasaki M.S., Morrison C., Yamaguchi-Iwai Y., Takata M., and Takeda S. 1999. Sister chromatid exchanges are mediated by homologous recombination in vertebrate cells. *Mol. Cell. Biol.* **19:** 5166–5169.

Southern E.M. 1970. Base sequence and evolution of guinea-pig alpha-satellite DNA. *Nature* **227:** 794–798.

Stansel R.M., de Lange T., and Griffith J.D. 2001. T-loop assembly in vitro involves binding of TRF2 near the 3′ telomeric overhang. *EMBO J.* **20:** 5532–5540.

Starling J.A., Maule J., Hastie N.D., and Allshire R.C. 1990. Extensive telomere repeat arrays in mouse are hypervariable. *Nucleic Acids Res.* **18:** 6881–6888.

Stavropoulos D.J., Bradshaw P.S., Li X., Pasic I., Truong K., Ikura M., Ungrin M., and Meyn M.S. 2002. The Bloom syndrome helicase BLM interacts with TRF2 in ALT cells and promotes telomeric DNA synthesis. *Hum. Mol. Genet.* **11:** 3135–3144.

Sun H., Karow J.K., Hickson I.D., and Maizels N. 1998. The Bloom's syndrome helicase unwinds G4 DNA. *J. Biol. Chem.* **273:** 27587–27592.

Takai H., Smogorzewska A., and de Lange T. 2003. DNA damage foci at dysfunctional telomeres. *Curr. Biol.* **13:** 1549–1556.

Tarsounas M., Muñoz P., Claas A., Smiraldo P.G., Pittman D.L., Blasco M.A., and West S.C. 2004. Telomere maintenance requires the RAD51D recombination/repair protein. *Cell* **117:** 337–347.

Ten Hagen K.G., Gilbert D.M., Willard H.F., and Cohen S.N. 1990. Replication timing of DNA sequences associated with human centromeres and telomeres. *Mol. Cell. Biol.* **10:** 6348–6355.

Tommerup H., Dousmanis A., and de Lange T. 1994. Unusual chromatin in human telomeres. *Mol. Cell. Biol.* **14:** 5777–5785.

van Steensel B. and de Lange T. 1997. Control of telomere length by the human telomeric protein TRF1. *Nature* **385:** 740–743.

van Steensel B., Smogorzewska A., and de Lange T. 1998. TRF2 protects human telomeres from end-to-end fusions. *Cell* **92:** 401–413.

Varley H., Pickett H.A., Foxon J.L., Reddel R.R., and Royle N.J. 2002. Molecular characterization of inter-telomere and intra-telomere mutations in human ALT cells. *Nat. Genet.* **30:** 301–305.

Veldman T., Etheridge K.T., and Counter C.M. 2004. Loss of hPot1 function leads to telomere instability and a *cut*-like phenotype. *Curr. Biol.* **14:** 2264–2270.

Vourc'h C., Taruscio D., Boyle A.L., and Ward D.C. 1993. Cell cycle-dependent distribution of telomeres, centromeres, and chromosome-specific subsatellite domains in the interphase nucleus of mouse lymphocytes. *Exp. Cell Res.* **205:** 142–151.

Wang R.C., Smogorzewska A., and de Lange T. 2004. Homologous recombination generates T-loop-sized deletions at human telomeres. *Cell* **119:** 355–368.

Wilkie A.O., Higgs D.R., Rack K.A., Buckle V.J., Spurr N.K., Fischel-Ghodsian N., Ceccherini I., Brown W.R., and Harris P.C. 1991. Stable length polymorphism of up to 260 kb at the tip of the short arm of human chromosome 16. *Cell* **64:** 595–606.

Wright W.E., Tesmer V.M., Liao M.L., and Shay J.W. 1999. Normal human telomeres are not late replicating. *Exp. Cell Res.* **251:** 492–499.

Wright W.E., Tesmer V.M., Huffman K.E., Levene S.D., and Shay J.W. 1997. Normal human chromosomes have long G-rich telomeric overhangs at one end. *Genes Dev.* **11:** 2801–2809.

Xu L. and Blackburn E.H. 2004. Human Rif1 protein binds aberrant telomeres and aligns along anaphase midzone microtubules. *J. Cell Biol.* **167:** 819–830.

Yang Q., Zheng Y.L., and Harris C.C. 2005. POT1 and TRF2 cooperate to maintain telomeric integrity. *Mol. Cell. Biol.* **25:** 1070–1080.

Yankiwski V., Marciniak R.A., Guarente L., and Neff N.F. 2000. Nuclear structure in normal and Bloom syndrome cells. *Proc. Natl. Acad. Sci.* **97:** 5214–5219.

Ye J.Z. and de Lange T. 2004. TIN2 is a tankyrase 1 PARP modulator in the TRF1 telomere length control complex. *Nat. Genet.* **36:** 618–623.

Ye J.Z., Hockemeyer D., Krutchinsky A.N., Loayza D., Hooper S.M., Chait B.T., and de Lange T. 2004a. POT1-interacting protein PIP1: A telomere length regulator that recruits POT1 to the TIN2/TRF1 complex. *Genes Dev.* **18:** 1649–1654.

Ye J.Z., Donigian J.R., Van Overbeek M., Loayza D., Luo Y., Krutchinsky A.N., Chait B.T., and de Lange T. 2004b. TIN2 binds TRF1 and TRF2 simultaneously and stabilizes the TRF2 complex on telomeres. *J. Biol. Chem.* **279:** 47264–47271.

Zhong Z., Shiue L., Kaplan S., and de Lange T. 1992. A mammalian factor that binds telomeric TTAGGG repeats in vitro. *Mol. Cell. Biol.* **12:** 4834–4843.

Zhou X.Z. and Lu K.P. 2001. The Pin2/TRF1-interacting protein PinX1 is a potent telomerase inhibitor. *Cell* **107:** 347–359.

Zhu L., Hathcock K.S., Hande P., Lansdorp P.M., Seldin M.F., and Hodes R.J. 1998. Telomere length regulation in mice is linked to a novel chromosome locus. *Proc. Natl. Acad. Sci.* **95:** 8648–8653.

Zhu X.D., Kuster B., Mann M., Petrini J.H., and de Lange T. 2000. Cell-cycle-regulated association of RAD50/MRE11/NBS1 with TRF2 and human telomeres. *Nat. Genet.* **25:** 347–352.

Zhu X.D., Niedernhofer L., Kuster B., Mann M., Hoeijmakers J.H., and de Lange T. 2003. ERCC1/XPF removes the 3′ overhang from uncapped telomeres and represses formation of telomeric DNA-containing double minute chromosomes. *Mol. Cell* **12:** 1489–1498.

Zou Y., Sfeir A., Gryaznov S.M., Shay J.W., and Wright W.E. 2004. Does a sentinel or a subset of short telomeres determine replicative senescence? *Mol. Biol. Cell* **15:** 3709–3718.

14

Drosophila Telomeres

Sergio Pimpinelli

Istituto Pasteur
Fondazione Cenci Bolognetti and
Dipartimento di Genetica e Biologia molecolare
Università "La Sapienza"
00185 Roma, Italy

THE DNA ENDS IN *DROSOPHILA*

Despite the fact that they were first defined in *Drosophila*, fruit fly telomeres are dramatically different from those of other organisms, at least regarding replication functions. This species lacks telomerase; instead, peculiar transposable elements seem to act as buffers against telomere loss during DNA replication. The *Drosophila* telomeres contain arrays of two non-LTR (long terminal repeat) retrotransposon elements called *HeT-A* and *TART* (Fig. 1) (Rubin 1978; Young et al. 1983; Levis et al. 1993; for reviews, see Mason and Biessmann 1995; Pardue 1995; Pardue and DeBaryshe 2003). Like many other non-LTR retroelements, *TART* encodes the Gag and Pol proteins, whereas *HeT-A* encodes only the Gag protein; both have been placed in the *jockey* clade (Eickbush 1994; Malik et al. 1999) (for a description of their structure see Fig. 2). The elements in these arrays are in a head-to-tail arrangement with their 5' ends always oriented toward the terminus. Many of the elements in these telomeric arrays are truncated to varying degrees at their 5' ends. The probable mechanism of telomere elongation is the addition of *HeT-A* and *TART* copies generated from RNA templates by the activity of a reverse transcriptase (Fig. 3) (Mason and Biessmann 1995; Pardue et al. 1996). Their targeting seems to be independent of specific terminal DNA sequences (Traverse and Pardue 1988; Biessmann et al. 1990, 1992; Sheen and Levis 1994). *HeT-A* is transcribed in the normal 5' to 3' direction from a promoter located at the 3' UTR (untranslated region) of another

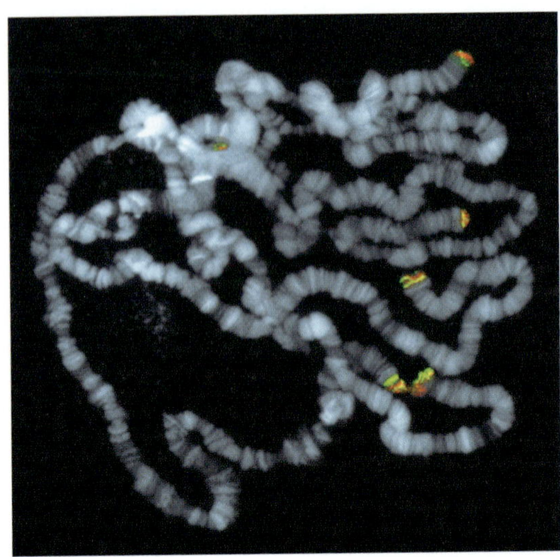

Figure 1. Localization of *HeT-A* (*red signals*) and *TART* (*green signals*) retrotransposons on polytene chromosomes of *D. melanogaster*. Both elements appear to be located at the ends of all the chromosomes with partially overlapped patterns.

upstream element (Danilevskaya et al. 1997), whereas *TART* is transcribed in both sense and antisense directions from a still unknown promoter (Danilevskaya et al. 1999). The bidirectional transcription of *TART* allows the formation of a double-stranded RNA that could be involved in regulating *HeT-A* expression via RNA interference (RNAi). Alternatively, *TART* could be autoregulated by RNAi for producing a limited amount

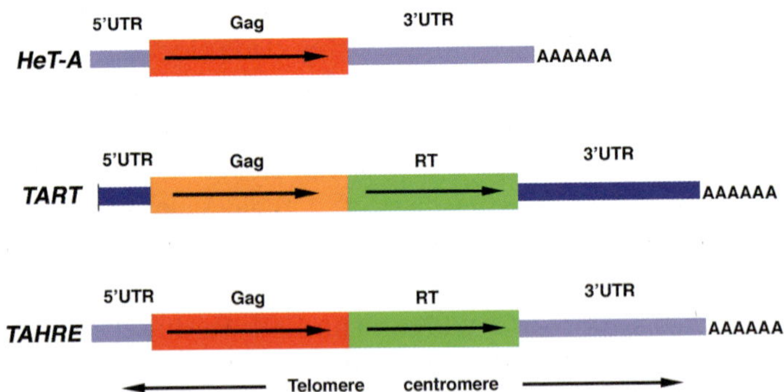

Figure 2. Diagram of the molecular structure of the RNA transposition intermediate of *Drosophila HeT-A*, *TART*, and *TAHRE* telomeric retrotransposons. The ORFs and the 3'- and 5'-noncoding sequences are indicated. As indicated by the colored boxes, the comparative sequence analysis has shown that the 5' UTR, ORF1, and 3' UTR of *TAHRE* are very similar to the corresponding sequences of *HeT-A*, whereas ORF2 is more similar to that of *TART* (see Pardue et al. 1996; Abad et al. 2004).

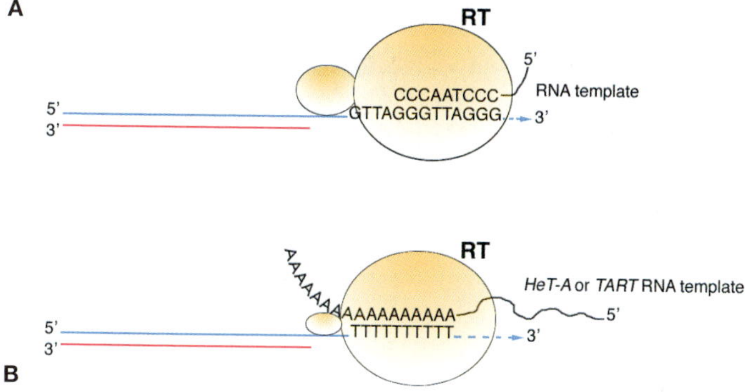

Figure 3. Mechanisms of telomere elongation by telomerase and the proposed mechanism of telomere elongation of *Drosophila* telomeres by the retransposons *HeT-A* and *TART*. (*A*) The alignment of the telomerase RNA component with the 3′ end permits copying several times a short central sequence to be attached to the chromosome. (*B*) In *Drosophila*, the 3′ end of the retrotransposons associates with the DNA ends, and by a reverse transcriptase activity, all the transposon is copied as DNA producing a telomere elongation (see Pardue et al. 1996).

of its reverse transcriptase that in turn could regulate the rate of telomeric elongation of both retrotransposons. However, if so, the control would be posttranscriptional and on the basal transcription because both elements are transcriptionally repressed by the capping protein HP1 (heterochromatin protein 1) (Perrini et al. 2004) and by a still unknown protein encoded by the *Tel* gene (Siriaco et al. 2002). Several studies have suggested a model for telomere elongation (see Fig. 4) (Rashkova et al. 2002a,b). After leaving the nucleus, *HeT-A* and *TART* transcripts would be used as mRNA for translating the two Gag polypeptides and the TART reverse transcriptase. The Gag protein encoded by *HeT-A* would mediate the reentry into the nucleus of both types of RNAs (suggested by the observation that this protein contains three nuclear localization signals). It is not known if *HeT-A* and *TART* RNAs recognize the telomeres for reverse transcription by an interaction of Gag proteins with some component of the telomere-capping complex or by another mechanism.

Analyses of their expression patterns during development have shown that *HeT-A* and *TART* are transcribed in dividing diploid cells (George and Pardue 2003; Walter and Biessmann 2004). The correlation of *HeT-A* and *TART* activity with cell proliferation parallels the correlation of telomerase activity in human proliferating cells during early development (for review, see Collins and Mitchell 2002) and would seem to suggest that proliferating tissues in *Drosophila* also require telomere elongation during development. However, several lines of evidence confute

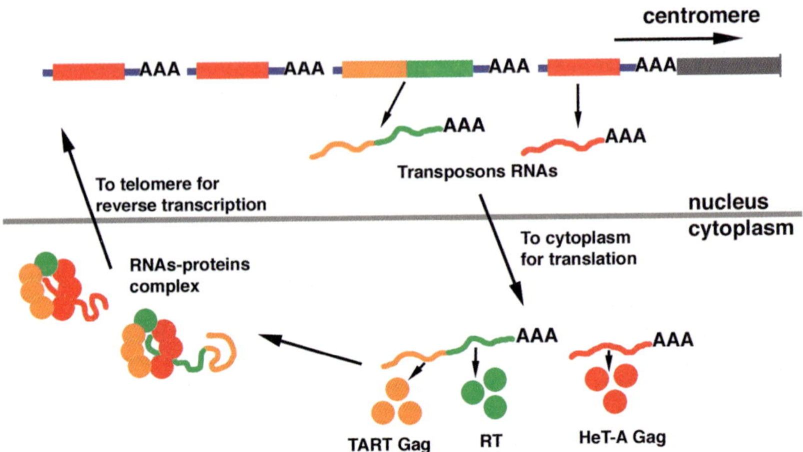

Figure 4. Diagrammatic representation of the model proposed by Rashkova et al. (2002a) for telomere elongation by *HeT-A* and *TART* transposition in *Drosophila*. Above, *HeT-A* (*red*) and *TART* (*orange/green*) arrays at a telomere. One of the members of each retrotransposon is transcribed and transferred to the cytoplasm for translation. The GAG (*red*) encoded by *HeT-A* would then make a complex with the transposition template and both GAG (*orange*) and RT (*green*) encoded by *TART* to target the transposition template to the chromosome ends (see Rashkova et al. 2002b).

the likelihood of this requirement. *Drosophila* can live for generations with a terminal deletion lacking all the telomeric transposons (Mason et al. 1984; Levis 1989). Because a normal telomere contains more than 20 kb of dispensable transposon sequences, and the predicted rate of terminal loss is about 2.5 kb per year, the incomplete replication of chromosomal ends should not present a problem for many years. Moreover, in addition to *HeT-A* and *TART* transposition, recombination-based mechanisms, such as gene conversion, seem to be important in regulating telomere length in *Drosophila* (Mikhailovsky et al. 1999; Kahn et al. 2000). At least one gene, called *E(tc)* (*Enhancer of terminal gene conversion*), specifically controls this mechanism (Melnikova and Georgiev 2002). Thus, the combination of the low-frequency *HeT-A* and *TART* telomeric transposition and the slow rate of transposon attrition is sufficient to resolve the end-replication problem. Previous studies have shown that *HeT-A*- and *TART*-like sequences are also located at nontelomeric sites in the autosomal and Y-chromosome mitotic heterochromatin (Traverse and Pardue 1989; Danilevskaya et al. 1993; Losada et al. 1997, 1999; Agudo et al. 1999), suggesting some structural or evolutionary link among telomeres, heterochromatin, and centromeres (Agudo et al. 1999). Several types of tandem repeats containing *TART*- and *HeT-A*-related sequences have been characterized in clones from these nontelomeric sites (Danilevskaya et al. 1993; Losada et al. 1997, 1999;

Agudo et al. 1999). The *TART-* and *HeT-A*-like sequences in these nontelomeric tandem repeats differ from the telomeric elements in at least two respects: They are interspersed with other repetitive sequences and they are not always oriented in the same direction.

TART and *HeT-A* elements have been found in other *Drosophila* species, suggesting that these species share a common mechanism of telomeric elongation (for review, see Pardue and DeBaryshe 2003). Consistent with this, probes containing *TART* and *HeT-A* sequences have been shown to hybridize to the telomeres of the polytene chromosomes of several *Drosophila* species (Young et al. 1983; Danilevskaya et al. 1998; Casacuberta and Pardue 2002, 2003a). Recent in situ hybridization experiments, using both *TART* and *HeT-A* probes, on mitotic and polytene chromosomes of *Drosophila simulans*, *Drosophila sechellia*, *Drosophila mauritiana*, *Drosophila yakuba*, and *Drosophila teissieri* have shown that in all of these species, cross-hybridizing sequences are located not only at the telomeres, but, although with different patterns, also at nontelomeric sites on heterochromatic Y chromosomes (Berloco et al. 2005).

The evolutionary origins of both *HeT-A* and *TART* remain controversial. It has been suggested that the close relationship of their Gag proteins derives from convergent evolution, rather than originating from a common ancestor (Danilevskaya et al. 1999; Pardue and DeBaryshe 2002; Casacuberta and Pardue 2003b). This conceptual frame implies that *HeT-A* was derived from telomerase and not from a preexisting retroelement that lost the *pol* gene (Pardue et al. 1996; Pardue and DeBaryshe 1999, 2002, 2003). Recent data have described a novel telomere-specific retroelement *TAHRE* (*telomere-associated and HeT-A-related element*) (Fig. 2). The analysis of their phylogenetic relationships has shown that *TAHRE* and *TART* probably arose from a common ancestor and that *HeT-A* was derived from *TAHRE* by a loss of its reverse transcriptase (Abad et al. 2004).

Telomeres in Other Insects

The information on telomere organization in other insects is recent and still incomplete. However, as shown in Figure 5, the emerging general picture is that at least three telomere types are present in this large class. The most widespread telomere motif is a pentanucleotide $(TTAGG)_n$ repeat, identified for the first time in the silkworm *Bombyx mori* (Okazaki et al. 1993; Meyne et al. 1995; Sahara et al. 1999). In several species that are phylogenetically distantly related, these sequences are associated with the presence of telomerase activity; however, in *Bombyx* itself, such activity could not be detected (Sasaki and Fujiwara 2000). In some orders, there are species that lack the pentanucleotide repeats, and these repeats seem to be completely

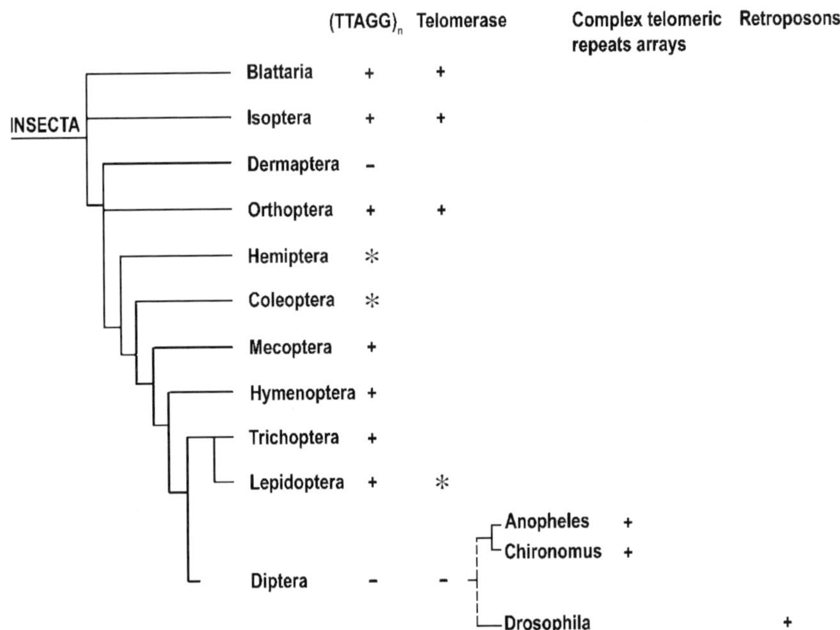

Figure 5. Patterns of distribution of different types of telomeric DNA sequences and telomerase among insects. The (TTAGG)$_n$ repeats were assayed by cross-hybridization and FISH analyses. Plus indicates presence and minus indicates absence in all tested species of each order. Asterisk indicates orders that have some species with and others without the telomeric (TTAGG)$_n$ repeats or telomerase (see Sasaki and Fujiwara 2000).

absent in the order Diptera. Several studies conducted in different species of this order have shown that beside the presence of specific telomeric retroposons in the genus *Drosophila*, long arrays of complex telomeric repeats have been found in the genus *Chironomus* (for review, see Kamnert et al. 1997) and in the mosquito *Anopheles gambiae* (Biessmann et al. 1998). Although in *Drosophila* telomere elongation is mainly based on transposon activity, in *Chironomus* and *Anopheles* it seems to depend on homologous recombination or gene conversion between the telomeric complex repeats (Cohn and Edstrom 1992; Lopez et al. 1996; Roth et al. 1997). The wide distribution of the pentanucleotide motif among insects suggests that it has been generally conserved, but sporadically lost, in some species or orders, such as Diptera, during insect evolution.

Proteins Involved in Telomeric Capping in *Drosophila*

Although in organisms with telomerase activity the DNA telomeric sequences function as "caps," in *Drosophila* the *HeT-A* and *TART* elements do not seem to be required for telomere stability. In contrast to the situation in yeast (Sandell and Zakian 1993), chromosomes carrying terminal deletions in *Drosophila* have been recovered (Mason et al. 1984; Levis 1989). Their molecular analysis has shown that their ends in many cases completely lack all of the normal telomere elements and that these chromosomes continue to lose terminal DNA sequences. Nevertheless, these broken chromosomes are stably transmitted through many generations as normal capped chromosomes (Levis 1989; Biessmann et al. 1990). Only occasionally do the *HeT-A* and *TART* elements transpose to the receding ends of the broken chromosomes (Biessmann et al. 1992, 1994; Sheen and Levis 1994). Apparently, these elements in *Drosophila* are essential for telomere elongation but dispensable for chromosome stability. Thus, unlike other organisms with telomerase-dependent telomeres, it appears that the replication functions of the telomere in *Drosophila* depend on specific DNA sequences, whereas the capping function is attributable to protein complexes.

The recovery of viable terminal deletions seems to be in discrepancy with the failure of Muller (1938, 1940) to recover terminally deleted chromosomes after X-irradiation in *Drosophila*. He observed that the recovered broken chromosomes were always capped by other chromosome fragments, suggesting that the chromosome ends were specialized structures that prevent chromosome damage. In accord with Muller's conclusions, there is a mechanism in *Drosophila* that can detect and react to a single broken chromosome (Ahmad and Golic 1999). A freshly generated nontelomeric end, caused by breakage of a dicentric Y chromosome during somatic anaphase, induces mitotic arrest and apoptosis along with developmental defects. Because the Y chromosome is not required for any somatic function, the data strongly suggest that it is the induction of a double-strand break, and not aneuploidy, that activates a cell cycle checkpoint in mitotic cells. However, as mentioned above, viable terminal deletions have been obtained in *Drosophila*; they have been recovered among progeny of X-ray-irradiated females carrying a mutation at the *mu-2* (*mutator-2*) locus (Mason et al. 1984, 1997; Biessmann et al. 1990) or by the destabilization of a P-element transposon inserted in a subtelomeric region (Levis 1989). The apparent discrepancy between these data and the Muller data seems to depend on the different experimental approaches in inducing chromosome breaks. Several observations have suggested that *mu-2* mutations affect the processing of double-strand breaks in oocytes and that the induced breaks remain unrepaired for a long time (Mason et al. 1997). It has been suggested that cell cycle checkpoint control is inactive in mature oocytes (Kasravi et al. 1999). Because there is evidence that a cell cycle checkpoint control is not present in early embryos until nuclear migration (Sullivan et al. 1990, 1993; Fogarty et al. 1997), the double-strand breaks induced in postrecombination oocytes would not cause arrest before syncytial blastoderm. Therefore, the broken chromosomes should be "healed" after fertilization and during preblastoderm stage by capping factors already present in the cytoplasm (Kasravi et al. 1999). If so, the terminal deletions induced by the destabilization of the P transposon should also be recovered, because subterminal breakages do not trigger a cell cycle arrest due to inactivity of the surveillance mechanism in both mature oocytes and early embryos (Levis 1989). Importantly, these explanations again implicate the requirement of proteins for telomere capping rather than DNA sequences. As illustrated below, proteins involved in telomere capping have in fact been found.

Heterochromatin Protein 1

Heterochromatin Protein 1 (HP1) is a conserved chromosomal protein first discovered in *Drosophila melanogaster* and now known to be associated with the heterochromatin of many organisms (James and Elgin 1986; James et al. 1989; Singh et al. 1991; Kellum et al. 1995; Fanti et al. 1998). In *Drosophila*, HP1 is a 206-amino-acid protein encoded by a modifier of position-effect variegation, the *Su(var)205* locus (Fig. 6) (Eissenberg et al. 1990). It has two conserved sequence motifs important for chromatin binding: the chromodomain, which is also present in the negative developmental regulator POLYCOMB protein (Paro and Hogness 1991), and the chromoshadow domain (Aasland and Stewart 1995) (Fig. 6). Both domains are likely involved in protein–protein interactions (Paro and Hogness 1991; Aasland and Stewart 1995), and there is also evidence for direct binding of the chromodomain to RNA (Akhtar et al. 2000;

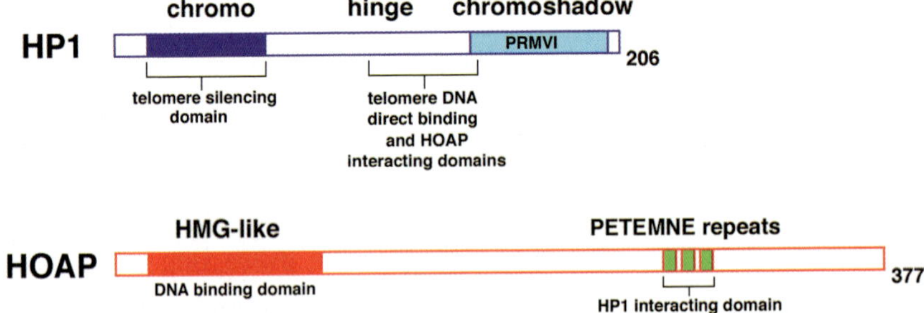

Figure 6. Diagrammatic representation of the structure of *Drosophila* HP1 and HOAP proteins. Only the regions involved in telomere metabolism and in HP1/HOAP interaction are shown.

Piacentini et al. 2003). The nuclear targeting activity of the protein depends on a portion of the carboxyl terminus, whereas the amino- and carboxy-terminal halves have independent capacities to bind heterochromatin (Powers and Eissenberg 1993; Platero et al. 1995).

A detailed cytogenetic analysis of HP1 distribution along the chromosomes has shown that this protein, besides its heterochromatic location, is also a stable component of all the telomeres in *Drosophila*, including the ends of stable terminal deletions lacking the telomeric transposons (for an example, see Fig. 7) (Fanti et al. 1998). Mutations in HP1 cause multiple telomere–telomere fusions in mutant cells, resulting in a striking spectrum of abnormal chromosome configurations (Fanti et al. 1998) (for examples, see Figs. 8 and 9). Analyses of metaphase and anaphase configurations have shown that the telomeric fusions result in the formation

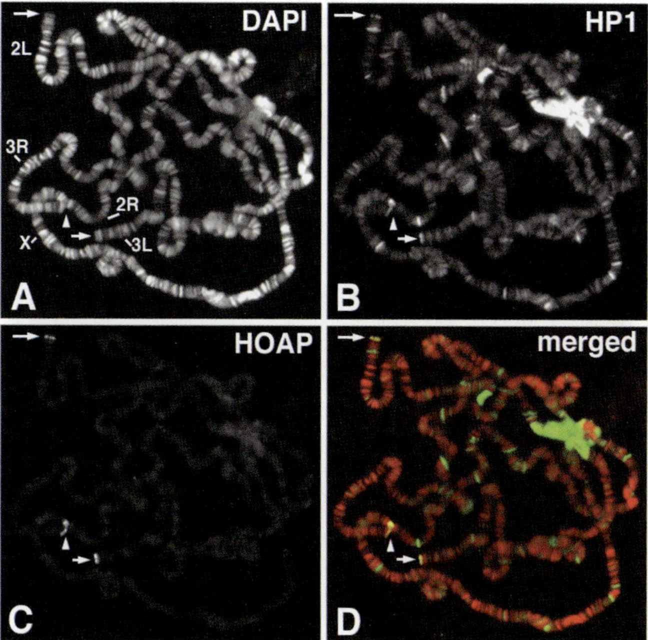

Figure 7. Immunolocalization of HP1 and HOAP on polytene chromosomes of *Drosophila* larval salivary glands. (*A*) DAPI staining pattern. The numbers indicate the different chromosomes and the letters the left (L) and the right (R) arms. The arrows indicate the telomeres and the arrowhead the three associated telomeres. (*B*) Immunostaining by a specific HP1 antibody reveals that the protein is located on the chromocenter, many euchromatic sites, and all the telomeres. (*C*) The immunopattern of HOAPs shows that such protein is mainly located on all the telomeres. (*D*) The merged figure clearly shows that the two proteins overlap at the telomeres.

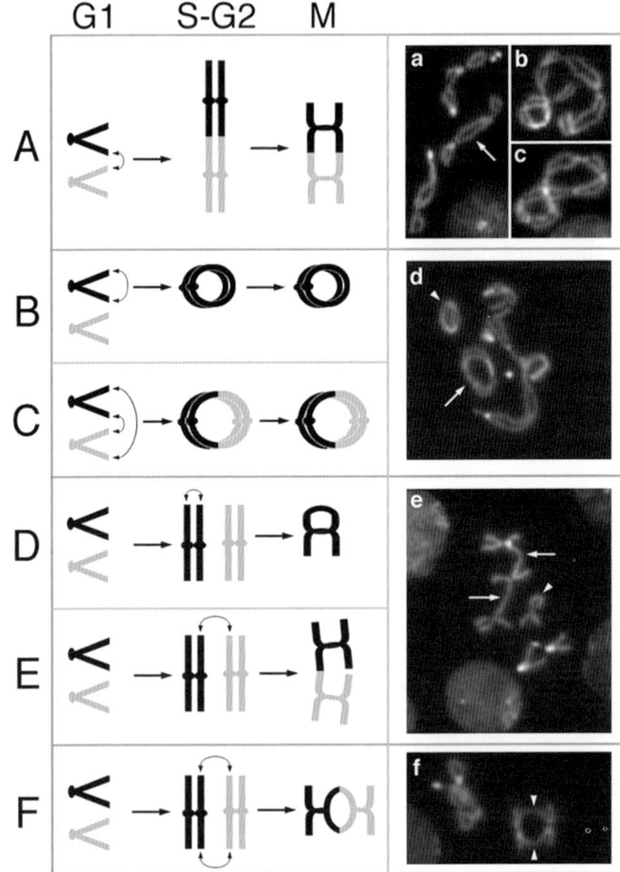

Figure 8. Diagrammatic representation of different types of telomere fusions, before (*A–C*) and after (*D–E*) the replication phase, when the telomeres are "uncapped," and the consequent chromosome configurations as illustrated by the DAPI-stained metaphases. (*A*) A fusion of two telomeres of different chromosomes before the replication phase produces a double-telomere attachment (DTA) after DNA replication, resulting in a dicentric chromosome in metaphase as shown by an arrow in *a*. Multiple fusions produce long chromosomal chains that can be linear (*b*) or circular (*c*). (*B*) The fusion of the two telomeres of the same chromosome before the replication phase produces a single chromosome ring in metaphase as indicated by the arrowhead in *d*. (*C*) The fusion of all the telomeres of two chromosomes also produces a ring as shown by the arrow in *d*. (*D*) After the replication phase, the sister telomeres can fuse as shown by the arrowhead in *e*. (*E*) Single or multiple fusion of chromatid telomeres produces single-telomere attachments (STAs) that could involve two or more chromosomes as shown by the arrows in *e*. (*F*) The fusion of all the telomeres of two nonsister chromatids produces the chromosome configuration indicated by arrowheads in *f*.

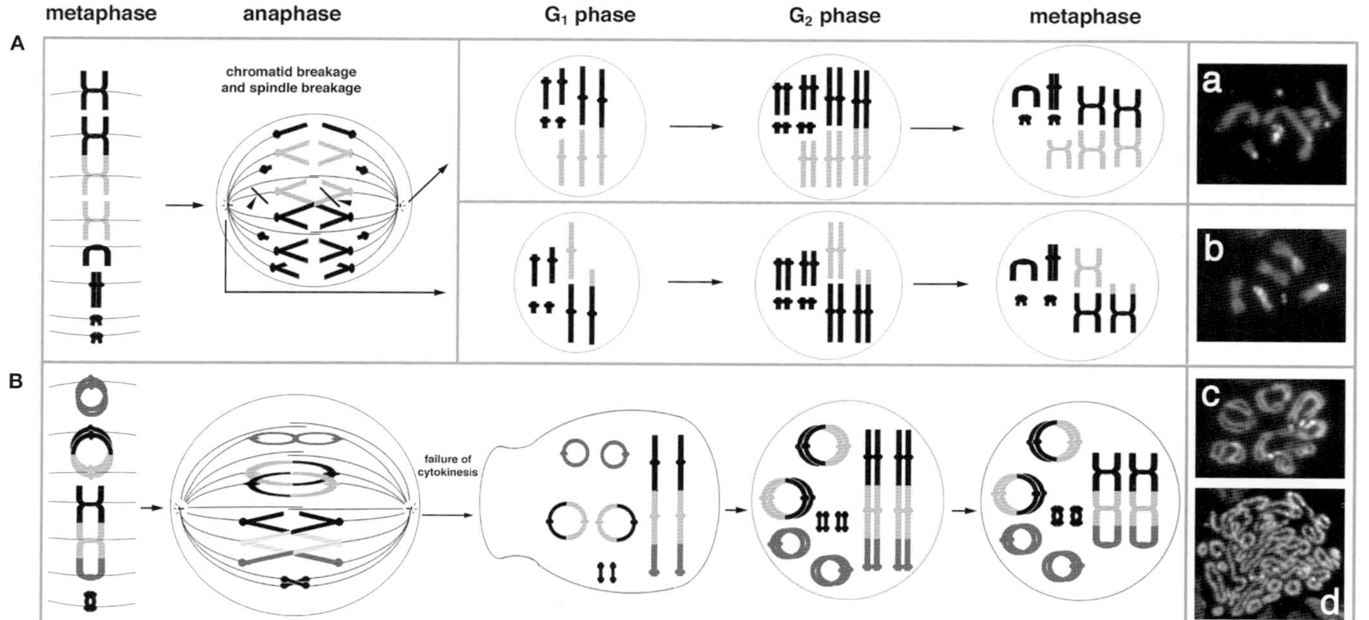

Figure 9. Aneuploid and polyploid nuclei are frequently present in HP1 mutant cells. (*A*) Diagrammatic representation of the generation of a nucleus with an additional centric fragment (*a*) and another lacking one chromosome (*b*) due to a chromatid and spindle breakage (*arrowheads*) during anaphase of a cell with a telomeric fusion. (*B*) Diagrammatic representation of multiple bridges during the anaphase of a cell with multiple telomeric fusions causing a failure of cytokinesis. The result is the generation of a G_1 nucleus with duplicated chromosomes that after replication will produce a polyploid cell (*c*). This type of cell after other rounds of replication will generate highly polyploid nuclei (*d*).

of chromosome bridges during anaphase, causing extensive chromosome breakage. This chromosome breakage strongly suggests that the telomeric fusions involve DNA–end fusions rather than proteinaceous bridges and that such chromosome breakages initiate a breakage-fusion-bridge cycle (BFB cycle) (Fanti et al. 1998). Analysis of the different types of telomere fusions produced by the lack of HP1 has shown that the double-telomere attachments (DTAs) are much more frequent than single-telomere attachments (STAs). DTAs are also more frequently involved in producing intra- and interchromosomal rings and long chains of multiple fused chomosomes. Because DTAs arose from telomere fusions before DNA replication and STAs arose after DNA replication, this suggests that all of the telomeres are closer to each other in G_1 phase than in other phases of the cell cycle and supports the model of their Rabl orientation (Rabl 1885) in this phase (Cenci et al. 1997).

Taken together, these observations implicate HP1 as a telomeric "cap" protein, essential for telomere stability independent of the sequences at the chromosome termini (Fanti et al. 1998). Interestingly, lesions in the chromodomain do not affect HP1 telomere localization and do not induce telomere fusions, suggesting that the chromodomain is dispensable for HP1 telomeric localization and its function in maintaining telomere stability (Fanti et al. 1998). According to the most general model in different species, the formation of the heterochromatin depends on histone methyltransferase enzymes (HMTases) that methylate histone H3 at lysine 9 (H3-MeK9), creating selective binding sites for themselves and for the chromodomain of HP1 (for review, see Jenuwein 2001). However, although it has been shown that histone H3 trimethylated at lysine 9 (H3-Me3K9) is present at the *Drosophila* telomeres (Cowell et al. 2002), the lack of requirement for the HP1 chromodomain makes it unlikely that the telomeric localization of HP1 is solely mediated by such modified histone.

Using a cross-linking assay with *cis*-diaminedichloroplatinum (*cis*-DDP) in intact nuclei purified from *Drosophila* larvae and an immunoprecipitation assay (X-ChIP assay), Perrini et al. (2004) have shown that the HP1 telomere-capping function depends on its direct binding to telomeric DNA, independent of specific sequences. Gel-shift experiments, using different HP1 fragments, have also shown that the hinge region of this protein is capable of binding double-stranded and, with greater affinity, single-stranded telomeric DNA (Perrini et al. 2004).

Most recently, it has been shown that HP1 is involved not only in the capping, but also in the replication of chromosome ends. In heterozygous HP1 mutant stocks, over time, the telomeres become elongated and the transcription of both TART and HeT-A significantly increases (Savitsky et al. 2002; Perrini et al. 2004). The transcription of the telomeric transposons is much more abundant in mutant cells completely lacking HP1 (Perrini et al. 2004). It is also relevant that a mutation disrupting the HP1 chromodomain significantly increases the quantity of telomeric transposon transcripts. Thus, a functional HP1 chromodomain, although dispensable for HP1 telomeric localization and telomere stability, is necessary for the silencing of telomere transposons and probably also for telomere elongation. Intriguingly, it has been found that a functional HP1 chromodomain is also required for the methylation of H3-K9 at the telomeres, thus suggesting that the HP1 chromodomain and H3-Me3K9, although not required for telomere capping, are, however, necessary for the control of telomeric transposon transcription and telomere elongation (Perrini et al. 2004).

In conclusion, it appears that the telomere-capping and the replication functions of HP1 depend on two different types of binding of this protein to telomeres. Telomere capping depends on its direct binding to telomeric sequences, whereas the transcriptional control of such sequences depends on the interaction of the chromodomain with H3-Me3K9. The observation that H3-K9 methylation depends on the presence of HP1 at the telomeres also suggests a previous interaction of HP1 with a specific HMTase. However, known HMTases such as E(Z) and SU(VAR)3-9 have been tested for their possible involvement in H3-K9 methylation at the telomeres with negative results. Finally, it has also been shown that the RNAi mechanism is probably not involved in recruiting HP1 and a putative HMTase to the telomeres (Perrini et al. 2004).

Interestingly, because there is evidence for a telomeric localization of the different HP1 isoforms on mammalian metaphase chromosomes (Aagaard et al. 2000; Koering et al. 2002: Garcia-Cao et al. 2004), it is possible that the role of HP1 at telomeres is conserved between *Drosophila* and humans.

HP1-ORC-associated Protein

HP1-ORC-associated protein (HOAP) is a DNA-binding protein identified as a component of a multiprotein complex containing HP1 and ORC (origin-recognition complex) subunits that is present in the cytoplasm of *Drosophila* early embryos (Shareef et al. 2001). Sequence analysis has shown that the amino terminus of HOAP shares similarity with the HMG box of sequence-specific HMG proteins and that the carboxyl terminus contains three copies of a novel PETEMNE repeated sequence (Fig. 6)

(Shareef et al. 2001). This protein is able to bind in vitro specific satellite sequences and a telomere-associated sequence; it is weakly localized to the heterochromatin and is able to suppress heterochromatin-induced gene silencing (Shareef et al. 2001; Badugu et al. 2003). Importantly, immunostaining with a specific antibody against HOAP has shown that this protein is mainly located at the telomeres of both mitotic and polytene chromosomes. A recent study has shown that HOAP is encoded by the *caravaggio* (*cav*) gene and that a mutation in this gene causes extensive telomeric fusions in larval neuroblasts with a pattern similar to that produced by HP1 mutations (Cenci et al. 2003). Although HOAP can be found in a complex with ORC (Pak et al. 1997), ORC is not bound to telomeres (Pak et al. 1997), does not protect against telomere fusions (Cenci et al. 2003), and does not affect the telomere localization of HP1 (Huang et al. 1998) or HOAP (Cenci et al. 2003). Therefore, the HOAP-ORC association is unlikely to be important for the telomeric function of HOAP. To test if HOAP contributes to telomere chromatin repression, we looked for differences in telomeric transposon transcription in HOAP mutant cells, but we found little effect (L. Piacentini et al., unpubl.). We also observed that the localization of both HP1 and H3-Me3K9 at the telomeres of polytene chromosomes does not seem to be completely affected by the lack of HOAP (L. Fanti et al., unpubl.).

These data have opened the possibility that HP1 and HOAP form a complex that is required for telomere capping and have stimulated studies on the nature of HP1–HOAP interaction.

The Relationship between HP1 and HOAP in Telomere Capping

Badugu et al. (2003) have shown that HP1 and HOAP interact directly and that this interaction depends on HP1 dimerization. The domains responsible for the interaction have been mapped on both the proteins. The carboxyl terminus of HOAP can independently interact with both the hinge and the chromoshadow domains of HP1. The interaction with the hinge region depends on a repeated PETEMNE motif found in the HOAP carboxyl terminus, and a specific sequence of the HP1 hinge domain. It has also been shown that HOAP binding to the HP1 chromoshadow domain depends on a PRMVI peptide motif that is present in this HP1 domain but absent in HOAP. Because the PRMVI motif is present in several HP1-interacting proteins and is involved in HP1 homodimerization, the absence of this motif in HOAP strongly suggests that its requirement for HOAP–HP1 interaction reflects the requirement for HP1 dimerization.

Although not proven, the above-described HP1–HOAP interaction could be relevant for its capping function at the telomeres. However, it is important to note that the absence of HP1 in *Su(var)2-5* mutant mitotic cells does not prevent HOAP localization at the telomeres. In metaphases of wild-type cells, 98% of telomeres show immunosignals after immunostaining with a HOAP antibody, whereas in metaphases of HP1 mutant cells, 70% of telomeres not involved in telomere fusions have a clear HOAP signal, and only 15% of telomeres involved in telomere fusions have a HOAP signal (Cenci et al. 2003). Thus, HP1 does not have a major role in HOAP localization to the telomeres of mitotic chromosomes. Unfortunately, the diffuse HP1 immunopattern along mitotic chromosomes makes it much more difficult to determine with sufficient confidence whether or not the lack of HOAP influences HP1 telomeric localization.

Lacking this information, we must consider at least two possible models for HP1–HOAP interactions in telomeric capping. If HOAP mutations have no effects on HP1 telomere localization, this would suggest that the proteins are probably independently recruited to the telomeres and then interact to make a stable capping complex. Alternatively, an alteration of HP1 telomere localization by HOAP mutations would suggest a role of HOAP in HP1 telomeric targeting. In the latter case, it would be reasonable to suppose that HP1 is the cap protein and that the telomere fusions observable in HOAP mutant cells are due to the absence of HP1. Few observations seem to support the first model. As mentioned above, the lack of HOAP does not seem to strongly affect the transcription of telomere transposons and does not completely prevent the telomere localization of both HP1 and H3-Me3K9 in polytene chromosomes, suggesting that HP1 could be present at the mitotic telomeres in HOAP mutant cells. It is interesting that in HP1 mutant mitotic cells, HOAP is present at the unfused telomeres while prevalently absent at fused telomeres (Cenci et al. 2003). If this should turn out to be true also for the localization of HP1 to the telomere in HOAP mutant mitotic cells, it would suggest that both proteins can individually cap the telomeres, although with a low efficiency. This strongly reduced efficiency would suggest that the stability of HP1 and HOAP telomere binding is reciprocally reinforced by their interaction in making a capping complex.

The DNA Repair Components MRE11, RAD50, and ATM

Recent studies from several labs have shown that components of the DNA-repair MRN complex, which function in maintaining telomeres in yeast and mammals (for review, see Ferreira et al. 2004), have similar roles in *Drosophila*. Mutations in genes such as *Rad50*, *Mre11*, and the ataxia-telangiectasia-mutated (*ATM*) checkpoint kinase, encoded by the *telomere fusion* (*tefu*) gene (Queiroz-Machado et al. 2001), cause multiple telomeric fusions, anaphase bridges, extensive chromosome breakage, and chromosome aneuplody and polyploidy in larval proliferative tissues (Queiroz-Machado et al. 2001; Bi et al. 2004; Ciapponi et al. 2004; Oikemus et al. 2004; Silva et al. 2004; Song et al. 2004). Analyses of the effects of double-mutant combinations have shown that all of these genes act in the same pathway in controlling the stability of *Drosophila* telomeres (Bi et al. 2004; Ciapponi et al. 2004). In addition, telomere fusions induced by *ATM* mutations are partially dependent on DNA ligase IV, thus suggesting an involvement of a nonhomologous end-joining (NHEJ) mechanism (Bi et al. 2004).

Three independent studies have reported the effects of mutations in the *ATM*, *Rad50*, and *Mre11* genes on the telomere localization of HP1 and HOAP in both mitotic and polytene chromosomes. *ATM* mutations do not seem to strongly affect the localization of HOAP to the telomeres of mitotic chromosomes (Bi et al. 2004), but they do reduce the level of the protein at the telomeres of polytene chromosomes (Oikemus et al. 2004). However, the level of HOAP at the telomeres of both mitotic and polytene chromosomes is reduced by *Mre11* (Bi et al. 2004; Ciapponi et al. 2004) and *Rad50* (Ciapponi et al. 2004) mutations. In addition, mutations in *ATM*, *Mre11*, and *Rad50* reduce the level of HP1 at the telomeres of polytene chromosomes (Ciapponi et al. 2004; Oikemus et al. 2004). These results suggest that the MRN complex and ATM kinase are necessary for the targeting of HP1 and HOAP to the telomeres.

Although mutations in any of the components of the MRN complex produce uncapped chromosomes because of the absence of HP1 and HOAP localization, this defective repair system may not be completely efficient in producing telomeric fusions. Ciapponi et al. (2004) observed that the frequency of telomeric fusions in cells with *Mre11* and *Rad50* null mutations is much lower than the frequency caused by mutations in the HP1 (Fanti et al. 1998) or HOAP (Cenci et al. 2003) encoding genes. However, Bi et al. (2004) reported a higher frequency of telomeric fusions in cells with another *Mre11* null mutation. The lower frequency of telomeric fusions induced by *Mre11* and, above all, *Rad50* mutations could be explained by a residual activity of both HP1 and HOAP. However, it is also possible that the damage to the repair system caused by these mutations prevents the production of telomeric fusions that would normally be expected in uncapped chromosomes. The frequency of telomere fusions in double-mutant cells for either HP1 or HOAP and any of the MRN components should permit us to discriminate between the two possibilities. If the frequency observed in mutant cells for each of the MRN components does not significantly change in double-mutant cells, we can reasonably conclude that the lack of fusions is due to a defective repair system. Interestingly, it has been observed that the frequency of telomere fusions in ATM mutant cells is close to the frequency observed in HP1 and HOAP mutant cells (Bi et al. 2004; Oikemus et al. 2004). However, although ATM does not seem to strongly affect the telomeric localization of HOAP on mitotic chromosomes (Bi et al. 2004), it does clearly affect the telomeric localization of both HP1 and HOAP on polytene chromosomes (Oikemus et al. 2004). In addition, the effect of ATM on the telomeric localization of HP1 on mitotic chromosomes still needs to be elucidated. It is possible that ATM is involved in telomere localization of capping proteins, probably by its interaction with the MRN complex, and perhaps in their activation.

The role of the MRN complex in telomere metabolism in *Drosophila* resembles the role that it has in yeast, where it is also required for maintaining telomere length and for preventing telomere fusions (Nugent et al. 1998; Mieczkowski et al. 2003). In mammals, although the MRN complex is associated with both TRF1 and TRF2 (Wu et al. 2000; Zhu et al. 2000), it is not yet clear whether mutations in this complex induce telomere fusions. Telomeric fusions have not been observed in cells from ATDL patients (Hernandez et al. 1993; Klein et al. 1996). In mice, null mutations in *Mre11* and *Rad50* loci induce early embryonic lethality, preventing chromosome analysis (Xiao and Weaver 1997; Luo et al. 1999). However, telomere fusions have not been found in mice homozygous for a hypomorphic *Mre11* mutation (Theunissen et al. 2003). Interestingly, it has been proposed that the MRN complex might be the nuclease that generates 3′ overhangs which provide the substrates for telomerase and also allow telomeric loop (t-loop) formation (for detailed discussion, see de Lange and Petrini 2000; DuBois et al. 2000).

How does the MRN complex play this paradoxical part in telomere capping in *Drosophila*? In the simplest explanation, the MRN complex recognizes the telomeric ends as double-strand breaks and targets HP1 and HOAP to these ends. Perhaps some part of the MRN complex directly interacts with HP1 and HOAP. Alternatively, after the MRN complex binds to the telomeric ends, its nuclease activity might produce an extensive overhang that would be specifically recognized by HP1 and perhaps HOAP. The latter possibility seems to be supported by the evidence of a preferential affinity of HP1 for single-stranded telomeric

DNA, suggesting that HP1 could be recruited to telomeric DNA by specific recognition of the protruding telomeric ends. It is not unreasonable to think that this could also be true for HOAP recruitment.

The question of how MRN/ATM-HP1/HOAP could distinguish a telomere from a random break remains open. It could be based on a specific structure of telomeric ends, such as t-loops (Griffith et al. 1999), and/or specific histone modifications that would act as a stable telomeric marker, favoring the recruitment of capping proteins rather than repair enzymes. This appears to be reasonable considering the diverse fate of X-ray-induced double-strand breaks in *Drosophila* oocytes in Muller's experiments compared to the more recent experiments showing the recovery of viable terminal deletions. Muller recovered only broken chromosomes that were repaired by their fusion with another chromosome fragment containing a telomere. In contrast, terminal deletions induced in the oocytes of *mu-2* mutant females are "capped" and the capping is stably maintained through cell proliferation. This means that an induced subterminal break can be processed to assume a stable and heritable telomeric configuration independently on terminal DNA sequences.

UbcD1

Another interesting gene causing telomeric fusions has been described. Mutations in *UbcD1*, which encodes a class I ubiquitin-conjugating (E2) enzyme, cause frequent telomere–telomere associations during both mitosis and male meiosis in *Drosophila* (Cenci et al. 1997). However, the telomeric associations present in *UbcD1* mutants are resolved during mitotic anaphase and do not cause chromosome breakage. Therefore, such associations could be seen as a sort of telomere stickiness, due to proteinaceous bridges, rather than true telomeric associations involving DNA fusions as in HP1 or HOAP mutant cells. The most plausible explanation of these results is that during interphase, probably during the G_1 phase, the telomeres establish contact through *UbcD1* target proteins. In *UbcD1* mutants, the failure to degrade these proteins maintains the telomeric associations after interphase (Cenci et al. 1997). Intriguingly, also in this case, DTAs are more frequent than STAs, suggesting that the chromosomes maintain a Rabl orientation in G_1 phase (Cenci et al. 1997).

The lack of telomeric DNA fusions suggests that in *UbcD1* mutant cells, the telomeres are not exposed as uncapped ends and do not trigger the DNA repair mechanisms. Thus, there is a substantial functional difference between UbcD1 and the HP1 and HOAP capping proteins.

A possible explanation for the involvement of ubiquitin protein modification at the telomeres is suggested by the evidence that HP1 may also mediate the association of both the heterochromatin and telomeres with the inner nuclear membrane. It has recently been shown that the lamin B receptor (LBR), an integral protein of the inner nuclear membrane, interacts with *Drosophila* HP1 in a yeast two-hybrid assay (Shaffer et al. cited in Elgin 1996). The same LBR protein also interacts with a human chromodomain protein homologous to *Drosophila* HP1 (Ye and Worman 1996). There are also data suggesting a possible interaction of HP1 with Ku70 in human cells (Song et al. 2001), a protein that has been found to be associated with the telomeres of yeast (Gravel et al. 1998) and mammalian cells (Bianchi and de Lange 1999; Hsu et al. 1999) and with the nuclear matrix of human cells (Yu et al. 1998). It is not unreasonable to suppose that *UbcD1* is involved in the degradation of HP1 interacting proteins, like the LBR or Ku70, that could mediate the ordered interaction among telomeres and/or the interaction of telomeres with other structures such as the inner nuclear membrane or nuclear matrix.

A PRELIMINARY MODEL FOR TELOMERE CAPPING AND TELOMERE ELONGATION IN *DROSOPHILA*

Taken together, the existing data allow the formulation of a preliminary model for telomere capping and elongation in *Drosophila* (Fig. 10). According to the model, the MRN complex, with the probable involvement of ATM kinase, would recognize the natural ends as double-strand breaks, and by nuclease activity, it would make a 3′ overhanging strand that is directly bound by both HP1 and HOAP. After their direct binding, the two proteins would interact to form a stable capping complex. It is possible that ATM could be involved in such interaction. In addition, HP1 would also recruit HMTase that would methylate H3-K9, creating an additional binding site for HP1. The spreading of HP1, HMTase, and the H3-Me3K9 complex would form the telomeric repressive chromatin independently of HOAP. Clearly, this model only considers the known genes and proteins that contribute to telomere function and does not incorporate information on the structure of *Drosophila* telomeres. The open question is, in fact, if the *Drosophila* telomeres are also protected by the formation of t-loops, discovered by electron microscope analysis of telomeric DNA first in humans and then in other organisms (for review, see de Lange 2004). We have preliminary results that suggest the existence of these structures also in *Drosophila* (see Fig. 11) (B. Perrini et al., unpubl.).

Telomeric Position Effect

Several kilobases of complex satellite sequences, collectively called TASs (telomere-associated sequences), are located just proximal to the telomeric retrotransposon arrays in all the chromosomes. Although the TAS arrays share similarities, they have different lengths and are organized differently among the telomeres (Karpen and Spradling 1992; Levis et al. 1993; Walter et al. 1995). When a reporter gene, such as the mini-*white*[+] transgene, is inserted at the telomeres, it shows a variegating phenotype, suggesting that the chromosome ends have a sort of heterochromatic structure (for review, see Mason et al. 2003). This suggestion is reinforced

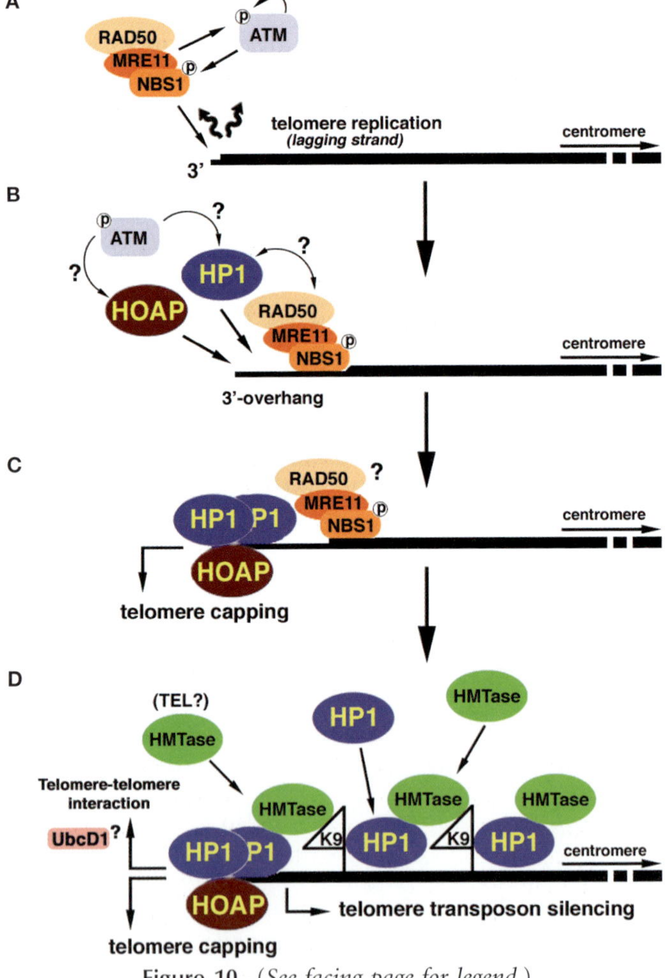

Figure 10. (*See facing page for legend.*)

by the presence of the HP1-H3-MeK9 complex that is typically involved in the formation of pericentromeric heterochromatin (for review, see Jenuwein 2001). This variegating phenotype resembles a well-known silencing caused by pericentromeric heterochromatin called position-effect variegation (PEV), first discovered in *Drosophila* by Muller (1930). When a normal euchromatic gene is relocated by a chromosome rearrangement in, or very close to, the heterochromatin, it undergoes a *cis*-heterochromatin inactivation in a portion of the cells during development, giving a mosaic phenotype (for review, see Weiler and Wakimoto 1995). Many genetic, cytogenetic, and molecular data have provided a general picture of PEV including the probable mechanism of gene inactivation by the heterochromatin. In *Drosophila melanogaster*, the euchromatic regions containing the variegating gene lose their normal morphology in polytene chromosomes and appear to be "heterochromatinized" (Grigliatti 1991); this is accompanied by a decrease in transcription (Rushlow et al. 1984). Moreover, mutations in certain genetic loci can cause an enhancement or suppression of PEV (for review, see Reuter and Spierer 1992).

By analogy with PEV, the induced telomeric variegating silencing has been termed "telomere position effect" (TPE) (Hazelrigg et al. 1984; Levis et al. 1985; for review, see Sandell and Zakian 1992; Shore 1996; see also Chapter 10 for further discussion on TPE). However, several data seem to suggest that TPE could be a phenomenon different from

Figure 10. Diagrammatic representation of a preliminary model for telomere capping and telomere silencing in *Drosophila*. Recent data suggest that, as in other organisms, the telomere ends are recognized by the MRN DNA-repair complex and the ATM kinase (wavy arrow) (*A*). The MRN complex produces a 3′ overhang recognized by the HP1 and HOAP capping proteins (*B*). It is not yet known if the recognition involves ATM activity and/or a direct interaction with some of the MRN components. However, it seems that these proteins independently bind the chromosome ends and then interact to make a stable telomere capping complex (*C*). It is not yet known if this interaction depends on some sort of activation involving ATM, nor is it yet known if the MRN complex is a stable part of the telomere or is released. In the following step, HP1 recruits a putative histone methyltranferase that trimethylates the lysine 9 of H3 histone, thus creating a binding site for the binding of the HP1 chromodomain (*D*). The reiteration of the second step creates repressive telomeric chromatin. In a variation of this model, HP1 binds the telomeric DNA as pre-established dimer with HMTase. Because the *Tel* gene also is involved in telomere elongation, it is not unreasonable to assume, as a working hypothesis, that this gene may encode a methyltransferase. UbcD1 seems to suggest that ubiquitination could be involved in the interactions among the telomeres and in the interactions of the telomeres with the inner nuclear membrane and/or nuclear matrix.

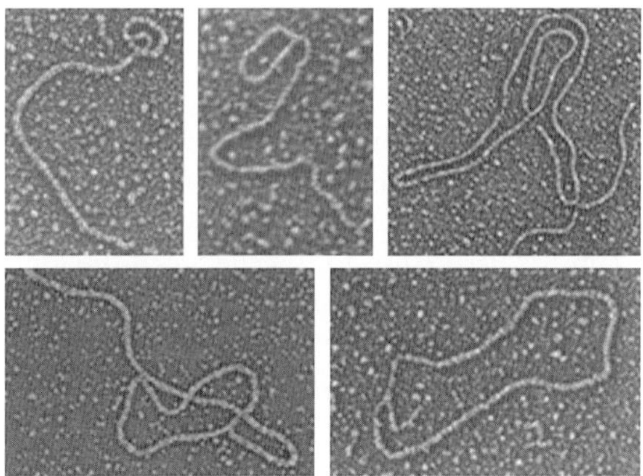

Figure 11. Electron micrographs of loop configurations of *HeT-A* DNA extracted from *Drosophila* salivary glands.

PEV. One of the stronger pieces of evidence is that the majority of the suppressors or enhancers of PEV, including the HP1-encoding gene, and mutation in several genes belonging to the Polycomb group (Pc-G) that have been tested, have no effect on TPE of *white*[+] transgenes except for the *Su(z)2* and *Psc* genes (Talbert et al. 1994; Roseman et al. 1995; Wallrath and Elgin 1995; Cryderman et al. 1999). Interestingly, all variegating *white*[+] transgenes so far studied were found to be inserted into, or very close to, the TAS sequences, suggesting that such sequences are responsible for the variegating silencing of the transgenes (Karpen and Spradling 1992; Levis et al. 1993; Wallrath and Elgin 1995; Cryderman et al. 1999; Golubovsky et al. 2001). This suggestion was confirmed by experiments with a transgene carrying TAS repeats located between the eye-specific enhancer and the mini-*white*[+] reporter gene, which found that this transgene is expressed in a variegating fashion independent of its chromosomal location (Kurenova et al. 1998). In addition, the strength of variegation is dependent on the length and the orientation of the TAS sequences and the activity of the *Su(z)2* gene. Studies on the variation in telomeric variegation strength over time were conducted on a *P(w*[var]*)* transgene carrying a complete *white*[+] genomic region and inserted between the TAS and the terminal retrotransposon array of the left arm of the second chromosome (Gehring et al. 1984; Golubovsky et al. 2001). The results have shown that the variation in variegation intensity was related to changes in the length of telomeric retrotransposon array: Loss or elongation of the array corresponded, respectively, to a decrease or increase of transgene expression (Golubovsky et al.

2001; Mason et al. 2003), the converse of the dependence on TAS array lengths.

An interesting feature of TPE is that the expression of subtelomeric transgenes is affected by the interaction among telomeres. The analysis of the eye color phenotype produced by *white*[+] reporter transgenes located in different telomeres has shown that a single dose produces almost colorless eyes and a double dose produces red eyes. The same phenotypes were observed in flies homozygous for either transgene or carrying the double heterozygote of the transgenes (Wallrath and Elgin 1995). However, the intensity of the phenotype cannot be wholly explained by dosage effect. Importantly, it has been observed that a single dose of the transgene can produce red eyes when the homologous telomere carries a terminal deletion including TAS. The expression of such transgenes may be due not only to a dosage effect, but also to the interaction among telomeres.

All of these results have suggested a model for TPE (Mason et al. 2000, 2003; Golubovsky et al. 2001). According to the model, the expression of a subtelomeric transgene would depend on the balance between the repressive effect of TAS and the inductive effect of telomeric retrotransposons. This expression would also be influenced by telomere interactions, which also implies that the telomeres would have contacts in interphasic nuclei, favoring the Rabl configuration model. However, it is important to consider that the variegating silencing of the transgenes inserted in TAS probably depends on chromatin conformation rather than DNA sequence organization. This is suggested by the observation that the product of *Su(z)2* gene is strongly accumulated at TAS (L. Fanti et al., unpubl.). In addition, the correlation of a decrease in transgene expression with the shortening of the retrotransposon array further supports the view that the chromosome ends have a particular structure with repressive properties, possibly because of the presence of the HP1-H3-MeK9 complex, although, as mentioned above, this complex does not seem to be directly involved in TPE.

CONCLUDING REMARKS

Although the telomeres of *Drosophila* seem to be different from those of many other organisms, they share several features. It is clear now that many proteins, such as the DNA repair complex, involved in telomere metabolism in other systems, are also present in *Drosophila*. On the other hand, there is growing evidence that some telomere components discovered in *Drosophila*, such as HP1, are also involved in telomere metabolism in other organisms including mammals. This suggests that the analysis of

Drosophila telomeres will give important information on telomere organization also in other organisms. Thus, the availability of sophisticated genetics and the ability to cytologically analyze chromosomal phenotypes in both mitotic and polytene cells again make *Drosophila* a powerful experimental system for studies on telomere biology.

ACKNOWLEDGMENTS

I thank all the members of my lab for helpful discussions. I particularly thank Maria Berloco, Laura Fanti, Barbara Perrini, and Lucia Piacentini for generously sharing unpublished data.

REFERENCES

Aagaard L., Schmid M., Warburton P., and Jenuwein T. 2000. Mitotic phosphorylation of SUV39H1, a novel component of active centromeres, coincides with transient accumulation at mammalian centromeres. *J. Cell Sci.* **113:** 817–829.

Aasland R. and Stewart A.F. 1995. The chromo shadow domain, a second chromo domain in heterochromatin-binding protein 1, HP1. *Nucleic Acids Res.* **23:** 3168–3173.

Abad J.P., de Pablos B., Osoegawa K., de Jong P.J., Martín-Gallardo A., and Villasante A. 2004. TAHRE, a novel telomeric retrotransposon from *Drosophila melanogaster*, reveals the origin of *Drosophila* telomeres. *Mol. Biol. Evol.* **21:** 1620–1624.

Agudo M., Losada A., Abad J.P., Pimpinelli S., Ripoll P., and Villasante A. 1999. Centromeres from telomeres? The centromeric region of the Y chromosome of *Drosophila melanogaster* contains a tandem array of the telomeric Het-A and TART-related sequences. *Nucleic Acids Res.* **27:** 3318–3324.

Ahmad K. and Golic K.G. 1999. Telomere loss in somatic cells of *Drosophila* causes cell cycle arrest and apoptosis. *Genetics* **151:** 1041–1051.

Akhtar A., Zink D., and Becker P.B. 2000. Chromodomains are protein-RNA interaction modules. *Nature* **407:** 405–409.

Badugu R., Shareef M.M., and Kellum R. 2003. Novel *Drosophila* heterochromatin protein 1 (HP1)/origin recognition complex-associated protein (HOAP) repeat motif in HP1/HOAP interactions and chromocenter associations. *J. Biol. Chem.* **5:** 34491–34498.

Berloco M., Fanti L., Sheen F., Levis R.W., and Pimpinelli S. 2005. Heterochromatic distribution of *HeT-A-* and *TART*-like sequences in several *Drosophila* species. *Cytogenet. Genome Res.* **110:** 124–133.

Bi X., Wei S., and Rong Y. 2004. Telomere protection without a telomerase: The role of *ATM* and *Mre11* in *Drosophila* telomere maintenance. *Curr. Biol.* **14:** 1348–1353.

Bianchi A. and de Lange T. 1999. Ku binds telomeric DNA *in vitro*. *J. Biol. Chem.* **274:** 21223–21227.

Biessmann H., Carter S.B., and Mason J.M. 1990. Chromosome ends in *Drosophila* without telomeric DNA sequences. *Proc. Natl. Acad. Sci.* **87:** 1758–1761.

Biessmann H., Kobeski F., Walter M.F., Kasravi A., and Roth C.W. 1998. DNA organization and length polymorphism at the 2L telomeric region of *Anopheles gambiae*. *Insect Mol. Biol.* **7:** 83–93.

Biessmann H., Champion L.E., O'Hair M., Ikenaga K., Kasravi B., and Mason J.M. 1992. Frequent transposition of *Drosophila melanogaster* HeT-A transposable elements to receding chromosome ends. *EMBO J.* **11:** 4459–4469.

Biessmann H., Mason J.M., Ferry C., d'Hulst M., Valgeirsdottir K., Traverse K.L., and Pardue M.L. 1994. Addition of telomere-associated HeT DNA sequences "heals" broken chromosome ends in *Drosophila. Cell* **61:** 663–673.

Casacuberta E. and Pardue M.L. 2002. Coevolution of the telomeric retrotransposons across *Drosophila* species. *Genetics* **161:** 1113–1124.

———. 2003a. Transposon telomeres are widely distributed in the *Drosophila* genus: TART elements in the virilis group. *Proc. Natl. Acad. Sci.* **100:** 3363–3368.

———. 2003b. HeT-A elements in *Drosophila virilis:* Retrotransposon telomeres are conserved across the *Drosophila* genus. *Proc. Natl. Acad. Sci.* **100:** 14091–14096.

Cenci G., Siriaco G., Raffa G.D., Kellum R., and Gatti M. 2003. The *Drosophila* HOAP protein is required for telomere capping. *Nat. Cell Biol.* **5:** 82–84.

Cenci G., Rawson R.B., Belloni G., Castrillon D.H., Tudor M., Petrucci R., Goldberg M.L., Wasserman S.A., and Gatti M. 1997. UbcD1, a *Drosophila* ubiquitin-conjugating enzyme required for proper telomere behavior. *Genes Dev.* **11:** 863–875.

Ciapponi L., Cenci G., Ducau J., Flores C., Johnson-Schlitz D., Gorski M.M., Engels W.R., and Gatti M. 2004. The *Drosophila* Mre11/Rad50 complex is required to prevent both telomeric fusion and chromosome breakage. *Curr. Biol.* **14:** 1360–1366.

Cohn M. and Edstrom J.E. 1992. Telomere-associated repeats in *Chironomus* form discrete subfamilies generated by gene conversion. *J. Mol. Evol.* **35:** 114–122.

Collins K. and Mitchell J.R. 2002. Telomerase in the human organism. *Oncogene* **21:** 564–579.

Cowell I.G., Aucott R., Mahadevaiah S.K., Burgonye P.S., Huskisson N., Bongiorni S., Prantera G., Fanti L., Pimpinelli S., Wu R., et al. 2002. Heterochromatin, HP1 and methylation at lysine 9 of histone H3 in animals. *Chromosoma* **111:** 22–36.

Cryderman D.E., Tang H., Bell C., Gilmour D.S., and Wallrath L.L. 1999. Heterochromatic silencing of *Drosophila* heat shock genes acts at the level of promoter potentiation. *Nucleic Acids Res.* **27:** 3364–3370.

Danilevskaya O.N., Arkhipova I.R., Traverse K.L., and Pardue M.L. 1997. Promoting in tandem: The promoter for telomere transposon HeT-A and implications for the evolution of retroviral LTRs. *Cell* **88:** 647–655.

Danilevskaya O., Lofsky A., Kurenova E.V., and Pardue M.L. 1993. The Y chromosome of *Drosophila melanogaster* contains a distinctive subclass of Het-A-related repeats. *Genetics* **134:** 531–543.

Danilevskaya O.N., Chen T., Wong J., Alibhai J., and Pardue M.L. 1998. Unusual features of the *Drosophila melanogaster* telomere transposable element HeT-A are conserved in *Drosophila yakuba* telomere elements. *Proc. Natl. Acad. Sci.* **95:** 3770–3775.

Danilevskaya O.N., Traverse K.L., Hogan N.C., DeBaryshe P.G., and Pardue M.L. 1999. The two *Drosophila* telomeric transposable elements have very different patterns of transcription. *Mol. Cell. Biol.* **19:** 873–881.

de Lange T. 2004. T-loops and the origin of telomeres. *Nat. Rev. Mol. Cell Biol.* **5:** 323–329.

de Lange T. and Petrini J.H. 2000. A new connection at human telomeres: Association of the Mre11 complex with TRF2. *Cold Spring Harbor Symp. Quant. Biol.* **65:** 265–273.

DuBois M.L., Diede S.J., Stellwagen A.E., and Gottschling D.E. 2000. All things must end: Telomere dynamics in yeast. *Cold Spring Harbor Symp. Quant. Biol.* **65:** 281–296.

Elgin S.C.R. 1996. Heterochromatin and gene regulation in *Drosophila. Curr. Opin. Genet. Dev.* **6:** 193–202.

Eickbush T.H. 1994. Origin and evolutionary relationships of retroelements. In *The Evolutionary biology of viruses* (ed. S.S. Morse), pp. 121–157. Raven Press, New York.

Eissenberg J.C., James T.C., Foster-Hartnett D.M., Hartnett T., Ngan V., and Elgin S.C.R. 1990. Mutation in a heterochromatin-specific chromosomal protein is associated with suppression of position effect variegation in *Drosophila melanogaster. Proc. Natl. Acad. Sci.* **87:** 9923–9927.

Fanti L., Giovinazzo M., Berloco G., and Pimpinelli S. 1998. The heterochromatin protein 1 prevents telomere fusions in *Drosophila. Mol. Cell* **2:** 1–20.

Ferreira M.G., Miller K.M., and Cooper J.P. 2004. Indecent exposure—When telomeres become uncapped. *Mol. Cell* **13:** 7–18.

Fogarty P., Campbell S.D., Abu-Shumays R., Desaint Phalle B., Yu K.R., Uy G.L., Goldberg M.L., and Sullivan W. 1997. The *Drosophila grapes* gene is related to checkpoint gene *chk1/rad27* and is required for late syncytial division fidelity. *Curr. Biol.* **7:** 418–426.

Garcia-Cao M., O'Sullivan R., Peters A.H., Jenuwein T., and Blasco M.A. 2004. Epigenetic regulation of telomere length in mammalian cells by the Suv39h1 and Suv39h2 histone methyltransferases. *Nat. Genet.* **36:** 94–99.

Gehring W.J., Klemenz R., Weber U., and Kloter U. 1984. Functional analysis of the *white*[+] gene of *Drosophila* by P-factor-mediated transformation. *EMBO J.* **3:** 2077–2085.

George J.A. and Pardue M.L. 2003. The promoter of the heterochromatic *Drosophila* telomeric retrotransposon, HeT-A, is active when moved into euchromatic locations. *Genetics* **163:** 625–635.

Golubovsky M.D., Konev A.Y., Walter M.F., Biessmann H., and Mason J.M. 2001. Terminal retrotransposons activate a subtelomeric *white* transgene at the 2L telomere in *Drosophila. Genetics* **158:** 1111–1123.

Gravel S., Larrivée M., Labrecque P., and Wellinger R. J. 1998. Yeast Ku as a regulator of chromosomal DNA end structure. *Science* **280:** 741–744.

Griffith J.D., Comeau L., Rosenfield S., Stansel R.M., Bianchi A., Moss H., and de Lange T. 1999. Mammalian telomeres end in a large duplex loop. *Cell* **97:** 503–514.

Grigliatti T. 1991. Position-effect variegation—An assay for nonhistone chromosomal proteins and chromatin assembly and modifying factors. *Methods Cell Biol.* **35:** 587–627.

Hazelrigg T., Levis R., and Rubin G.M. 1984. Transformation of *white* locus DNA in *Drosophila*: Dosage compensation, *zeste* interaction, and position effects. *Cell* **36:** 469–481.

Hernandez D., McConville C.M., Stacey M., Woods C.G., Brown M.M., Shutt P., Rysiecki G., and Taylor A.M. 1993. A family showing no evidence of linkage between the ataxia telangiectasia gene and chromosome 11q22–23. *J. Med. Genet.* **30:** 135–140.

Hsu H.-L., David G., Blackburn E.H., and Chen D.J. 1999. Ku is associated with the telomere in mammals. *Proc. Natl. Acad. Sci.* **96:** 12454–12458.

Huang D.W., Fanti L., Pak D.T.S., Botchan M.R., Pimpinelli S., and Kellum R. 1998. Distinct cytoplasmic and nuclear fractions of *Drosophila* heterochromatin protein 1: Their phosphorylation levels and associations with origin recognition complex proteins. *J. Cell Biol.* **142:** 307–318.

James T.C. and Elgin S.C. 1986. Identification of nonhistone chromosomal protein associated with heterochromatin in *Drosophila* and its gene. *Mol. Cell. Biol.* **6:** 3862–3872.

James T.C., Eissenberg J.C., Craig C., Dietrich V., Hobson A., and Elgin S.C.R. 1989. Distribution patterns of HP1, a heterochromatin-associated nonhistone chromosomal protein of *Drosophila*. *Eur. J. Cell Biol.* **50:** 170–180.

Jenuwein T. 2001. Re-SET-ting heterochromatin by histone methyltransferases. *Trends Cell Biol.* **11:** 266–273.

Kahn T., Savitsky M., and Georgiev P. 2000. Attachment of HeT-A sequences to chromosomal termini in *Drosophila melanogaster* may occur by different mechanisms. *Mol. Cell. Biol.* **20:** 7634–7642.

Kamnert I., Lopez C.C., Rosen M., and Edstrom J.E. 1997. Telomeres terminating with long complex tandem repeats. *Hereditas* **127:** 175–180.

Karpen G.H. and Spradling A.C. 1992. Analysis of subtelomeric heterochromatin in the *Drosophila* minichromosome *Dp 1187* by single-P element insertional mutagenesis. *Genetics* **132:** 737–753.

Kasravi A., Walter M.F., Brand S., Mason J.M., and Biessmann H. 1999. Molecular cloning and tissue-specific expression of the mutator2 gene (*mu2*) in *Drosophila melanogaster*. *Genetics* **152:** 1025–1035.

Kellum R., Raff J.W., and Alberts B. 1995. Heterochromatin protein 1 distribution during development and during cell cycle in *Drosophila* embryos. *J. Cell Sci.* **108:** 1407–1418.

Klein S., Zenvirth D., Dror V., Barton A.B., Kaback D.B., and Simchen G. 1996. Patterns of meiotic double-strand breakage on native and artificial yeast chromosomes. *Chromosoma* **105:** 276–284.

Koering C.E., Pollice A., Zibella M.P., Bauwens S., Puisieux A., Brunori M., Brun C., Martins L., Sabatier L., Pulitzer J.F., and Gilson E. 2002. Human telomeric position effect is determined by chromosomal context and telomeric chromatin integrity. *EMBO Rep.* **11:** 1055–1061.

Kurenova E., Champion L., Biessmann H., and Mason J.M. 1998. Directional gene silencing induced by a complex subtelomeric satellite from *Drosophila*. *Chromosoma* **107:** 311–320.

Levis R. 1989. Viable deletions of a telomere from a *Drosophila* chromosome. *Cell* **58:** 791–801.

Levis R., Hazelrigg T., and Rubin G.M. 1985. Effects of genomic position on the expression of transduced copies of the *white* gene of *Drosophila*. *Science* **229:** 558–561.

Levis R.W., Ganesan R., Houtchens K., Tolar L.A., and Sheen F.M. 1993. Transposons in place of telomeric repeats at a *Drosophila* telomere. *Cell* **75:** 1083–1093.

Lopez C.C., Nielsen L., and Edström J.E. 1996. Terminal long tandem repeats in chromosomes from *Chironomus pallidivittatus*. *Mol. Cell. Biol.* **16:** 3285–3290.

Losada A., Abad J.P., and Villasante A. 1997. Organization of DNA sequences near the centromere of the *Drosophila melanogaster* Y chromosome. *Chromosoma* **106:** 503–512.

Losada A., Agudo M., Abad J.P., and Villasante A. 1999. HeT-A telomere-specific retrotransposons in the centric heterochromatin of *Drosophila melanogaster* chromosome 3. *Mol. Gen. Genet.* **262:** 618–622.

Luo G., Yao M.S., Bender C.F., Mills M., Bladl A.R., Bradley A., and Petrini J.H. 1999. Disruption of mRad50 causes embryonic stem cell lethality, abnormal embryonic development, and sensitivity to ionizing radiation. *Proc. Natl. Acad. Sci.* **96:** 7376–7381.

Malik H.S., Burke W.D., and Eickbush T.H. 1999. The age and evolution of non-LTR retrotransposable elements. *Mol. Biol. Evol.* **16:** 793–805.

Mason J.M. and Biessmann H. 1995. The unusual telomeres of *Drosophila*. *Trends Genet.* **11:** 58–62.

Mason J.M., Champion L.E., and Hook G. 1997. Germ-line effects of a mutator, *mu-2*, in *Drosophila melanogaster*. *Genetics* **146**: 1381–1397.

Mason J.M., Konev A.Y., and Biessmann H. 2003. Telomeric position effect in *Drosophila melanogaster* reflects a telomere length control mechanism. *Genetica* **117**: 319–325.

Mason J.M., Strobel E., and Green M.M. 1984. *mu-2*: Mutator gene in *Drosophila* that potentiates the induction of terminal deficiencies. *Proc. Natl. Acad. Sci.* **81**: 6090–6094.

Mason J.M., Haoudi A., Konev A.Y., Kurenova E., Walter M.F., and Biessmann H. 2000. Control of telomere elongation and telomeric silencing in *Drosophila melanogaster*. *Genetica* **109**: 61–70.

Melnikova L. and Georgiev P. 2002. Enhancer of terminal gene conversion, a new mutation in *Drosophila melanogaster* that induces telomere elongation by gene conversion. *Genetics* **162**: 1301–1312.

Meyne J., Hirai H., and Imai H.T. 1995. FISH analysis of the telomere sequences of bulldog ants (*Myrmecia: Formicidae*). *Chromosoma* **104**: 14–18.

Mieczkowski P.A., Mieczkowska J.O., Dominska M., and Petes T.D. 2003. Genetic regulation of telomere-telomere fusions in the yeast *Saccharomyces cerevisiae*. *Proc. Natl. Acad. Sci.* **100**: 10854–10859.

Mikhailovsky S., Belenkaya T., and Georgiev P. 1999. Broken chromosomal ends can be elongated by conversion in *Drosophila melanogaster*. *Chromosoma* **108**: 114–120.

Muller H.J. 1930. Types of visible variations induced by X-rays in *Drosophila*. *J. Genet.* **22**: 299–334.

———. 1938. The remaking of chromosomes. *Collect. Net* **8**: 182–195.

———. 1940. An analysis of the process of structural change in chromosomes of *Drosophila*. *J. Genet.* **40**: 1–66.

Nugent C.I., Bosco G., Ross L.O., Evans S.K., Salinger A.P., Moore J.K., Haber J.E., and Lundblad V. 1998. Telomere maintenance is dependent on activities required for end repair of double-strand breaks. *Curr. Biol.* **8**: 657–660.

Oikemus S.R., McGinnis N., Queiroz-Machado J., Tukachinsky H., Takada S., Sunkel C.E., and Brodsky M.H. 2004. *Drosophila* atm/telomere fusion is required for telomeric localization of HP1 and telomere position effect. *Genes Dev.* **18**: 1850–1861.

Okazaki S., Tsuchida K., Maekawa H., Ishikawa H., and Fujiwara H. 1993. Identification of a pentanucleotide telomeric sequence, (TTAGG)n, in the silkworm *Bombyx mori* and in other insects. *Mol. Cell. Biol.* **13**: 1424–1432.

Pak D.T.S., Pflumm M., Chesnokov I., Huang D.W., Kellum R., Marr J., Romanowski P., and Botchan M.R. 1997. Association of the origin recognition complex with heterochromatin and HP1 in higher eukaryotes. *Cell* **91**: 311–323.

Pardue M.L. 1995. *Drosophila* telomeres: Another way to end it all. In *Telomeres* (ed. E.H. Blackburn and C.W. Greider), pp. 339–370. Cold Spring Harbor Laboratory Press, Cold Spring Harbor, New York.

Pardue M.L. and DeBaryshe P.G. 1999. Telomeres and telomerase: More than the end of the line. *Chromosoma* **108**: 73–82.

———. 2002. Telomeres and transposable elements. In *Mobile DNA II* (ed. N.L. Craig), pp. 870–887. ASM Press, Washington D.C.

———. 2003. Retrotransposons provide an evolutionarily robust non-telomerase mechanism to maintain telomeres. *Annu. Rev. Genet.* **37**: 485–511.

Pardue M.L., Danilevskaya O.N., Lowenhaupt K., Slot F., and Traverse K.L. 1996. *Drosophila* telomeres: New views on chromosome evolution. *Trends Genet.* **12**: 48–52.

Paro R. and Hogness D.S. 1991. The polycomb protein shares a homologous domain with a heterochromatin-associated protein of *Drosophila*. *Proc. Natl. Acad. Sci.* **88:** 263–267.

Perrini B., Piacentini L., Fanti L., Altieri F., Chichiarelli S., Berloco M., Turano C., Ferraro A., and Pimpinelli S. 2004. HP1 controls telomere capping, telomere elongation, and telomere silencing by two different mechanisms in *Drosophila*. *Mol. Cell* **15:** 467–476.

Piacentini L., Fanti L., Berloco M., Perrini B., and Pimpinelli S. 2003. Heterochromatin protein 1 (HP1) is associated with induced gene expression in *Drosophila* euchromatin. *J. Cell Biol.* **161:** 707–714.

Platero J.S., Hartnett T., and Eissenberg J.C. 1995. Functional analysis of the chromo domain of HP1. *EMBO J.* **14:** 3977–3986.

Powers J.A. and Eissenberg J.C. 1993. Overlapping domains of the heterochromatin-associated protein HP1 mediate nuclear localization and heterochromatin binding. *J. Cell Biol.* **120:** 291–299.

Queiroz-Machado J., Perdigao J., Simoes-Carvalho P., Herrmann S., and Sunkel C.E. 2001. *tef:* A mutation that causes telomere fusion and severe genome rearrangements in *Drosophila melanogaster*. *Chromosoma* **110:** 10–23.

Rabl C. 1885. Über Zelltheilung. *Morpholog. Jahrbuch* **10:** 214–330.

Rashkova S., Karam S.E., and Pardue M.L. 2002a. Element-specific localization of *Drosophila* retrotransposon Gag proteins occurs in both nucleus and cytoplasm. *Proc. Natl. Acad. Sci.* **99:** 3621–3626.

Rashkova S., Karam S.E., Kellum R., and Pardue M.L. 2002b. Gag proteins of the two *Drosophila* telomeric retrotransposons are targeted to chromosome ends. *J. Cell Biol.* **159:** 397–402.

Reuter G. and Spierer P. 1992. Position effect variegation and chromatin proteins. *BioEssays* **14:** 605–612.

Roseman R.R., Johnson E.A., Rodesch C.K., Bjerke M., Nagoshi R.N., and Geyer P.K. 1995. A P element containing suppressor of hairy-wing binding regions has novel properties for mutagenesis in *Drosophila melanogaster*. *Genetics* **141:** 1061–1074.

Roth C.W., Kobeski F., Walter M.F., and Biessmann H. 1997. Chromosome end elongation by recombination in the mosquito *Anopheles gambiae*. *Mol. Cell. Biol.* **17:** 5176–5183.

Rubin G.M. 1978. Isolation of a telomeric DNA sequence from *Drosophila melanogaster*. *Cold Spring Harbor Symp. Quant. Biol.* **42:** 1041–1046.

Rushlow C.A., Bender W., and Chovnick A. 1984. Studies on the mechanism of heterochromatic position effect at the *rosy* locus of *Drosophila melanogaster*. *Genetics* **108:** 603–615.

Sahara K., Marec F., and Traut W. 1999. TTAGG telomeric repeats in chromosomes of some insects and other arthropods. *Chromosome Res.* **7:** 449–460.

Sandell L.L. and Zakian V.A. 1992. Telomeric position effect in yeast. *Trends Cell Biol.* **2:** 10–14.

———. 1993. Loss of a yeast telomere: Arrest, recovery and chromosome loss. *Cell* **75:** 729–739.

Sasaki T. and Fujiwara H. 2000. Detection and distribution patterns of telomerase activity in insects. *Eur. J. Biochem.* **267:** 3025–3031.

Savitsky M., Kravchuk O., Melnikova L., and Georgiev P. 2002. Heterochromatin protein 1 is involved in control of telomere elongation in *Drosophila melanogaster*. *Mol. Cell. Biol.* **22:** 3204–3218.

Shareef M.M., King C., Damaj M., Badugu R.K., Huang D.W., and Kellum R. 2001. *Drosophila* heterochromatin protein 1 (HP1)/origin recognition complex (ORC) protein is associated with HP1 and ORC and functions in heterochromatin-induced silencing. *Mol. Biol. Cell* **12:** 1671–1685.

Sheen F.M. and Levis R.W. 1994. Transposition of the LINE-like retrotransposon TART to *Drosophila* chromosome termini. *Proc. Natl. Acad. Sci.* **91:** 12510–12514.

Shore D. 1996. The means to bind the ends. *Nat. Struct. Biol.* **3:** 491–493.

Silva E., Tiong S., Pedersen M., Homola E., Royou A., Fasulo B., Siriaco G., and Campbell S.D. 2004. ATM is required for telomere maintenance and chromosome stability during *Drosophila* development. *Curr. Biol.* **14:** 1341–1347.

Singh P.B., Miller J.R., Pearce J., Kothary R., Burton R.D., Paro R., James T.C., and Gaunt S.J. 1991. A sequence motif found in a *Drosophila* heterochromatin protein is conserved in animal and plants. *Nucleic Acids Res.* **19:** 789–794.

Siriaco G.M., Cenci G., Haoudi A., Champion L.E., Zhou C., Gatti M., and Mason J.M. 2002. Telomere elongation (Tel), a new mutation in *Drosophila melanogaster* that produces long telomeres. *Genetics* **160:** 235–245.

Song K., Jung Y., Jung D., and Lee I. 2001. Human Ku70 interacts with heterochromatin protein 1alpha. *J. Biol. Chem.* **276:** 8321–8327.

Song Y.-H., Mirey G., Betson M., Haber D.A., and Settleman J. 2004. The *Drosophila* ATM ortholog, dATM, mediates the response to ionizing radiation and to spontaneous DNA damage during development. *Curr. Biol.* **14:** 1354–1359.

Sullivan W., Minden J.S., and Alberts B.M. 1990. *daughterless-abo-like*, a *Drosophila* maternal-effect mutation that exhibits abnormal centrosome separation during the late blastoderm divisions. *Development* **110:** 311–323.

Sullivan W., Daily D.R., Fogarty P., Yook K.J., and Pimpinelli S. 1993. Delays in anaphase initiation occur in individual nuclei of the syncytial *Drosophila* embryo. *Mol. Biol. Cell* **4:** 885–896.

Talbert P.B., LeCiel C.D., and Henikoff S. 1994. Modification of the *Drosophila* heterochromatic mutation *brown Dominant* by linkage alterations. *Genetics* **136:** 559–571.

Theunissen J.W., Kaplan M.I., Hunt P.A., Williams B.R., Ferguson D.O., Alt F.W., and Petrini J.H. 2003. Checkpoint failure and chromosomal instability without lymphomagenesis in Mre11(ATLD1/ATLD1) mice. *Mol. Cell* **12:** 1511–1523.

Traverse K.L. and Pardue M.L. 1988. A spontaneously opened ring chromosome of *Drosophila melanogaster* has acquired He-T DNA sequences at both new telomeres. *Proc. Natl. Acad. Sci.* **85:** 8116–8120.

———. 1989. Studies of He-T DNA sequences in the pericentric regions of *Drosophila* chromosomes. *Chromosoma* **97:** 261–271.

Wallrath L.L. and Elgin S.C.R. 1995. Position effect variegation in *Drosophila* is associated with an altered chromatin structure. *Genes Dev.* **9:** 1263–1277.

Walter M.F. and Biessmann H. 2004. Expression of the telomeric retrotransposon HeT-A in *Drosophila melanogaster* is correlated with cell proliferation. *Dev. Genes Evol.* **214:** 211–219.

Walter M.F., Jang C., Kasravi B., Donath J., Mechler B.M., Mason J.M., and Biessmann H. 1995. DNA organization and polymorphism of wild-type *Drosophila* telomere region. *Chromosoma* **104:** 229–241.

Weiler K.S. and Wakimoto B.T. 1995. Heterochromatin and gene expression in *Drosophila*. *Annu. Rev. Genet.* **29:** 577–605.

Wu G., Lee W.H., and Chen P.L. 2000. NBS1 and TRF1 colocalize at promyelocytic leukemia bodies during late S/G2 phases in immortalized telomerase-negative cells. Implication of NBS1 in alternative lengthening of telomeres. *J. Biol. Chem.* **275:** 30618–30622.

Ye Q. and Worman H.J. 1996. Interaction between an integral protein of the nuclear envelope inner membrane and human chromo domain protein homologous to *Drosophila* HP1. *J. Biol. Chem.* **271:** 14653–14656.

Young B.S., Pession A., Traverse K.L., French C., and Pardue M.L. 1983. Telomere regions in *Drosophila* share complex DNA sequences with pericentric heterochromatin. *Cell* **34:** 85–94.

Yu E., Song K., Moon H., Maul G.G., and Lee I. 1998. Characteristic immunolocalization of Ku protein as nuclear matrix. *Hybridoma* **17:** 413–419.

Xiao Y. and Weaver D.T. 1997. Conditional gene targeted deletion by Cre recombinase demonstrates the requirement for the double-strand break repair Mre11 protein in murine embryonic stem cells. *Nucleic Acids Res.* **25:** 2985–2991.

Zhu X.D., Kuster B., Mann M., Petrini J.H., and de Lange T. 2000. Cell-cycle-regulated association of RAD50/MRE11/NBS1 with TRF2 and human telomeres. *Nat. Genet.* **25:** 347–352.

15

Ciliate Telomeres

Carolyn Price

Department of Molecular Genetics, Biochemistry, and Microbiology
University of Cincinnati
Cincinnati, Ohio 45267

CILIATED PROTOZOA HOLD A UNIQUE PLACE IN TELOMERE biology because so many key discoveries have been made with these organisms. The first telomeres to be sequenced were from *Tetrahymena* (Blackburn and Gall 1978) and the first telomere-binding protein was isolated from *Oxytricha* (*Sterkiella*[1]) (Gottschling and Zakian 1986; Price and Cech 1987). Telomerase was discovered in *Tetrahymena* (Greider and Blackburn 1985) and, after a decade-long search, the catalytic subunit was finally isolated from *Euplotes* and shown to be a reverse transcriptase (Lingner and Cech 1996; Lingner et al. 1997). In recent years, *Tetrahymena* and *Euplotes* have continued to provide important models for understanding telomerase biochemistry, telomere replication, and G-overhang processing (Collins 1999; Ray et al. 2002; Jacob et al. 2003), whereas the crystal structure of the *Oxytricha* telomere-binding protein has provided insight into G-overhang protection and telomere capping (Larson et al. 1987; Horvath et al. 1998; Lei et al. 2003). The reason ciliates have played such a central role in telomere biology lies in their unusual genomic organization, because it results in each cell containing literally thousands or millions of telomeres.

Ciliates have two functionally distinct nuclei: the germ-line micronucleus, which is transcriptionally silent and serves as a genetic repository during sexual development, and the somatic macronucleus, which is transcriptionally active and is used to support vegetative growth

---

[1] The class and species names of several spirotrichs have recently been changed. The old names are used here because they are better known, but the new names are presented in parentheses when the species or class is first mentioned.

(Prescott 1994; Jahn and Klobutcher 2002). The micronucleus is diploid and it divides by mitosis. It contains relatively few large chromosomes: 5 in *Tetrahymena* and ~120 in *Oxytricha*. In contrast, the macronucleus is highly polyploidy and undergoes amitotic division. It is generated from a copy of the micronucleus during a developmental process that takes place after cells mate. Differentiation of the macronucleus involves a series of genomic rearrangements that result in the large micronuclear chromosomes being broken into smaller units. Telomeres are added to the newly generated ends and the resulting macronuclear chromosomes are subject to multiple rounds of DNA replication. The process of macronuclear development generates ~40,000 telomeres in a single *Tetrahymena* macronucleus and ~48,000,000 telomeres in an *Oxytricha* or *Euplotes* macronucleus. This abundance of telomeres has greatly facilitated studies of telomere structure, telomere proteins, and telomerase.

The ciliates used most frequently to study telomere biology belong to two classes within the phylum Ciliophora: the class hypotrichea and the class oligohymenophorea. *Tetrahymena* and *Paramecium* species are oligohymenophoreans, whereas *Oxytricha*, *Euplotes*, and *Stylonychia* species are hypotrichs. Many aspects of telomere biology are conserved between the two classes, but they differ in certain features such as telomere length and the extent of chromosome fragmentation and telomere addition during macronuclear development. The macronuclear genomes of *Paramecium tetraurelia* and *Tetrahymena thermophila* have been completely sequenced and both micronuclear and macronuclear gene disruptions can be made in *T. thermophila*.

TELOMERE STRUCTURE

Telomeric DNA

The telomeres from micronuclear and macronuclear chromosomes are similar in sequence, but macronuclear telomeres are usually less than 1 kb in length whereas micronuclear telomeres are much longer. For example, *Tetrahymena* macronuclear telomeres are 250–350 bp in length and the micronuclear telomeres are 2.0–3.4 kb (Kirk and Blackburn 1995). *Paramecium* macronuclear telomeres are 100–500 bp (Gilley and Blackburn 1994). In *T. thermophila*, the micronuclear and macronuclear telomere are both composed of $G_4T_2 \cdot C_4A_2$ repeats, but the most internal 0.6–1.0 kb of the micronuclear telomeric tract contains homogeneous stretches of a variant $G_4T_3 \cdot C_4A_3$ repeat. Like mammalian and yeast chromosomes, the micronuclear chromosomes share regions of sequence homology in their telomere-associated regions; however, no such conservation has been observed in the subtelomeric sequences of macronuclear chromosomes. Macronuclear telomeres from spirotrichous ciliates such as *Oxytricha*, *Euplotes*, and *Stylonychia* are considerably shorter than those of *Tetrahymena* or *Paramecium* as they consist of only 28 or 20 bp of $G_4T_4 \cdot C_4A_4$ sequence (Klobutcher et al. 1981). The micronuclear telomeres are again longer, with those from *Oxytricha* estimated to be in the 3–20-kb range (Prescott 1994). The telomeres of most ciliates are composed of a homogeneous repeat sequence ($G_4T_2 \cdot C_4A_2$ or $G_4T_4 \cdot C_4A_4$); however, the macronuclear telomeres from *Paramecium* are unusual because they contain a mixture of $G_4T_2 \cdot C_4A_2$ and $G_3T_3 \cdot C_3A_3$ repeats (Baroin et al. 1987). As described below, both types of repeats seem to be synthesized by a single telomerase ribonucleoprotein particle (RNP) (McCormick-Graham and Romero 1996).

One interesting difference in the macronuclear telomeres from the spirotrichous and the oligohymenophorean ciliates lies in the precision of telomere length regulation. *Tetrahymena* and *Paramecium* telomeres are quite heterogeneous in length (Larson et al. 1987; Gilley and Blackburn 1994) such that in *Tetrahymena*, the telomeres from one particular macronuclear chromosome (e.g., the rDNA chromosome) can vary in length by as much as 100 bp (Jacob et al. 2004). Moreover, the telomeres from different chromosomes each exhibit a characteristic mean length that ranges from 250 to 450 bp. In contrast, the majority of telomeres from *Oxytricha*, *Euplotes*, and *Stylonychia* have exactly 20 bp (*Oxytricha* and *Stylonychia*) or 28 bp (*Euplotes*) of double-stranded telomeric sequence (Fig. 1) (Klobutcher et al. 1981). This almost complete lack of heterogeneity suggests spirotrichs have evolved a very precise mechanism

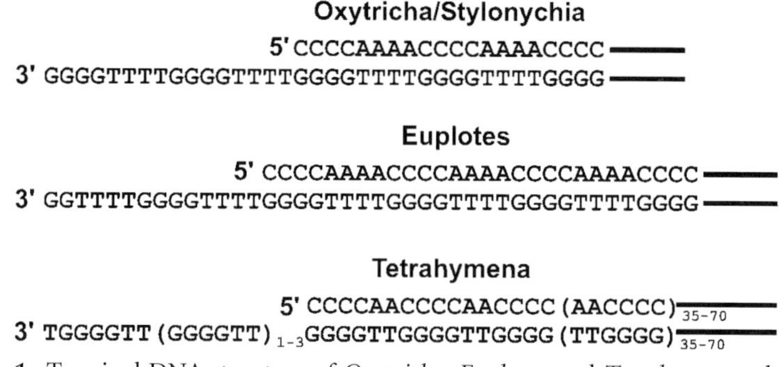

Figure 1. Terminal DNA structure of *Oxytricha*, *Euplotes*, and *Tetrahymena* telomeres. The lines represent nontelomeric DNA.

for telomere length regulation to ensure against catastrophic loss of an already minimal telomeric tract.

Telomeres of most, if not all, organisms terminate in a 3′ overhang on the G-rich strand (Wei and Price 2003). These overhangs are the substrate for specialized G-strand-binding proteins that serve to protect the end of the chromosome. In ciliates, G-strand overhangs are present on both ends of the macronuclear chromosomes, and they have a very precise structure (Fig. 1). Those from *Oxytricha* and *Stylonychia* are 16 nt in length and they terminate with the sequence $5′-T_4G_4$, whereas *Euplotes* overhangs are 14 nt long and terminate with $5′-T_4G_2$ (Klobutcher et al. 1981). The overhangs from *Tetrahymena* telomeres vary in length by multiples of 6 nt but most are 14–15- or 20–21-nt long and the majority end with the sequence $5′-G_4T$ (Henderson and Blackburn 1989; Jacob et al. 2001). The precise structure of ciliate G-strand overhangs and their presence on both telomeres provided some of the first evidence that G-overhangs are generated by DNA processing reactions (see below).

Telomeric Chromatin and Telomere Proteins

Ciliate macronuclear telomeres are packaged into protective DNA-protein complexes that encompass most of the telomeric tract and contain one or more specialized telomere-binding proteins. These telomeric complexes were first identified when *Tetrahymena, Oxytricha,* and *Euplotes* macronuclei were treated with micrococcal nuclease and the telomeric DNA was found to be released as part of a unique nonnucleosomal complex (Fig. 2A) (Blackburn and Chiou 1981; Gottschling and Cech 1984; Price 1990). In *Tetrahymena* the telomeric complex usually covers the full

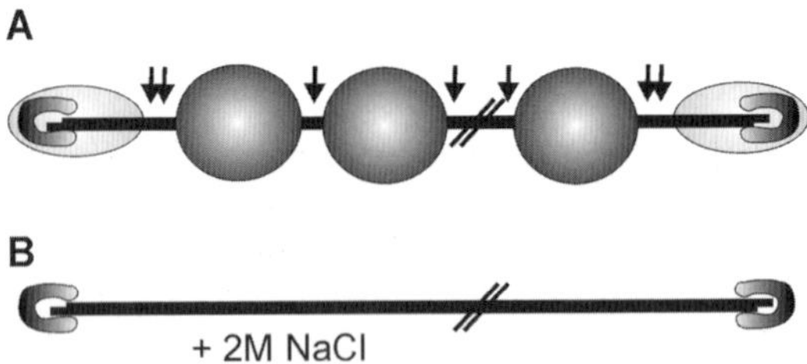

Figure 2. Telomeres exist as DNA-protein complexes. (*A*) MNase digestion of macronuclei releases nonnucleosomal telomeric complexes (*ovals*) from adjacent nucleosomes (*spheres*). *Arrows* indicate sites of digestion. (*B*) Purification of the onTEBP by treating macronuclei with 2 M NaCl. The histones are removed but the onTEBP remains bound to the G-strand overhang. The TEBP-DNA complexes can be purified by gel filtration or density gradient centrifugation.

250–350 bp of double-stranded telomeric DNA, but a few (10%) are protected by a smaller complex and an adjacent nucleosome (Budarf and Blackburn 1986; Cohen and Blackburn 1998). Even the largest complexes protect only a small amount (<50 bp) of subtelomeric DNA, and their size appears to be determined by the length of the telomeric tract as they become proportionately larger during telomere growth (Cohen and Blackburn 1998; Jacob et al. 2004). In *Oxytricha* and *Euplotes* the complex is smaller and contains a higher proportion of subtelomeric DNA as it encompasses the 20- or 28-bp telomeric tract and protects 50–100 bp of adjacent sequence (Price 1990). The chromatin structure of micronuclear telomeres has not been characterized in detail, but electron microscopy of DNA isolated from *Oxytricha* micronuclear chromosomes revealed the presence of large terminal loops (Murti and Prescott 1999) that are similar to the t-loops found at mammalian and plant telomeres (Griffith et al. 1999; Cesare et al. 2003). t-loops were not observed at *Oxytricha* macronuclear telomeres, which is not surprising given the short length of the telomeric tract. Whether they exist at the longer *Tetrahymena* macronuclear telomeres remains to be determined.

The complexes present at *Oxytricha* and *Euplotes* macronuclear telomeres seem to be composed of a single telomere end-binding protein (TEBP) that binds specifically to the G-strand overhang (Horvath et al. 1998; Wei and Price 2003). The *Oxytricha* TEBP is a heterodimer composed of a 56-kD α subunit and a 41-kD β subunit. The protein was first identified because it both was abundant and remained bound to telomeric DNA in 2 M NaCl, allowing a single-step purification (Fig. 2B) (Gottschling and Zakian 1986). It binds specifically to the $5'-G_4T_4G_4$ sequence that is present at the $3'$ end of the G-strand overhang, and the crystal structure of the $\alpha/\beta/$DNA ternary complex indicates that the 12 nt of $G_4T_4G_4$ sequence lie in a deep cleft between the α and β subunits (Horvath et al. 1998). The *Tetrahymena* telomeric complex almost certainly contains one or more homologs of the *Oxytricha* TEBP as the *Tetrahymena* genome encodes TEBP homologs (tPOT1a and tPOT1b), and disruption of one of these genes causes an increase in telomere length (C. Price and N. K. Jacob, unpubl.). Given the larger size of the telomeric tract, it is likely that the *Tetrahymena* complex also contains double-strand telomeric DNA-binding protein(s) that is analogous to *Saccharomyces cerevisiae* Rap1 or mammalian TRF1 or TRF2, but this remains to be identified.

Although *Oxytricha* macronuclear telomeres appear to be protected by the $\alpha\beta$ heterodimer throughout much of the cell cycle, the α subunit can bind DNA in the absence of the β subunit and an $\alpha_2/$DNA complex is proposed to be an intermediate in formation of the α/β/DNA ternary complex (Classen et al. 2001; Peersen et al. 2002). The DNA-binding surface on the *Oxyricha* α subunit is composed of two oligonucleotide/oligosaccharide-binding (OB) folds (Horvath et al. 1998). This OB-fold motif seems to be a conserved structural feature of all telomeric G-strand-binding proteins including those that share little sequence identity with the TEBP family (e.g., Cdc13 from *S. cerevisiae*) (Mitton-Fry et al. 2002). Although the sequences of G-overhang binding proteins are very divergent, the region corresponding to the first OB fold in the *Oxytricha* α subunit is conserved in members of the POT1 family, and this short region of conservation led to identification of the *Schizosaccharomyces pombe* and mammalian proteins (Baumann and Cech 2001; Wei and Price 2003). The β subunit differs from the α subunit in that it is not widespread in nature and neither the *Euplotes* TEBP nor members of the POT1 family have a β homolog. However, *Euplotes* has a second α subunit homolog (rTP or the replication telomere protein) that is expressed only during DNA replication and hence may substitute for the TEBP at this time (Skopp et al. 1996).

It has not been possible to explore the function of the *Oxytricha* or *Euplotes* TEBP in vivo because the tools necessary for genetic studies have not yet been developed for these ciliates. However, recombinant *Oxytricha* αβ heterodimer inhibits the in vitro extension of telomeric DNA by telomerase (Froelich-Ammon et al. 1998). This result, and the telomere elongation phenotype observed in the *Tetrahymena* POT1 knockout, indicates that ciliate TEBP/POT1 proteins regulate telomere length by restricting telomerase access to the telomere. In spirotrichs, binding of TEBP and formation of the telomeric complex may be responsible for all telomere length regulation. However, in *Tetrahymena* the situation is more complex as telomere length and telomere growth also are altered by changes in the organization of the subtelomeric chromatin (Jacob et al. 2004). Thus, the subtelomeric chromatin appears to cooperate with the telomeric complex to form a higher-order chromatin structure that controls telomere length. As the subtelomeric chromatin structure is different for each macronuclear chromosome, this explains why each telomere has a characteristic length.

TELOMERE MAINTENANCE

Macronuclear and micronuclear telomeres are both maintained by telomerase (Kirk et al. 1997). In ciliates, the telomerase holoenzyme contains the catalytic subunit telomerase reverse transcriptase (TERT), the telomerase RNA, plus one or more additional subunits that are important for ribonucleoprotein particle (RNP) biogenesis, stability, and processivity (see Chapters 2 and 3). *Tetrahymena* appears to have a single telomerase that is used to maintain both macronuclear and micronuclear telomeres (Miller and Collins 2000; Witkin and Collins 2004). However, *Euplotes* has multiple TERT genes, which suggests that different telomerase RNPs may be used for macronuclear and micronuclear telomere maintenance (Collins 1999; Karamysheva et al. 2003). The mixed repeat sequence found in the macronuclear telomeres of one species of *Paramecium*, *P. primaurelia*, initially suggested the existence of several telomerase RNAs that differed in their template sequence (Baroin et al. 1987; McCormick-Graham and Romero 1996). However, only one telomerase RNA gene has been found in *Paramecium*, *Euplotes*, and *Tetrahymena*. Thus, the mixed $G_4T_2{\cdot}C_4A_2$ and $G_3T_3{\cdot}C_3A_3$ repeats found at *P. primaurelia* telomeres appear to result from a single telomerase misincorporating deoxyadenosine triphosphate (dATP) opposite a specific C-residue in the RNA template (McCormick-Graham and Romero 1996; McCormick-Graham et al. 1997).

Although macronuclear telomeres are maintained by telomerase, the extreme end of the chromosome does not have the sequence that would be expected if it were generated by telomerase extending the 3′ terminus of the telomeric DNA to the end of the RNA template. In *Euplotes* and *Oxytricha*, extension to the end of the template would generate the sequence $5'–G_4T_4G$; however, the telomeres terminate with the sequence $5'–G_4T_4G_2$ or $5'–G_4T_4G_4$ (Klobutcher et al. 1981; Shippen-Lentz and Blackburn 1990; Melek et al. 1994). Likewise, the majority of *Tetrahymena* telomeres terminate with the sequence $5'–T_4G_4T$, whereas the telomerase RNA templates the sequence $5'–G_4T_4G$ (Greider and Blackburn 1989; Jacob et al. 2001). *Tetrahymena* telomerase is processive and generally copies all the way to the end of the RNA template (Collins 1999). Thus, this difference between the end of the chromosome and the sequence generated by telomerase suggested that *Tetrahymena* telomeres, and probably also *Euplotes* and *Oxytricha* telomeres, are subjected to some form of DNA processing. Subsequent studies with *Tetrahymena* showed that this is indeed the case, and the terminal DNA structure is created by a defined processing pathway that involves at least two different cleavage steps (Jacob et al. 2003). One step resects the C-strand to generate overhangs of defined length on both telomeres, whereas the second step trims the G-strand to generate the correct 3′ terminal sequence (Fig. 3). The first cleavage step is needed to make overhangs at the telomeres replicated by leading strand synthesis because this process is predicted to generate a blunt end. However, processing of the lagging strand telomere may also be required to give the correct length of overhang. The nucleases responsible

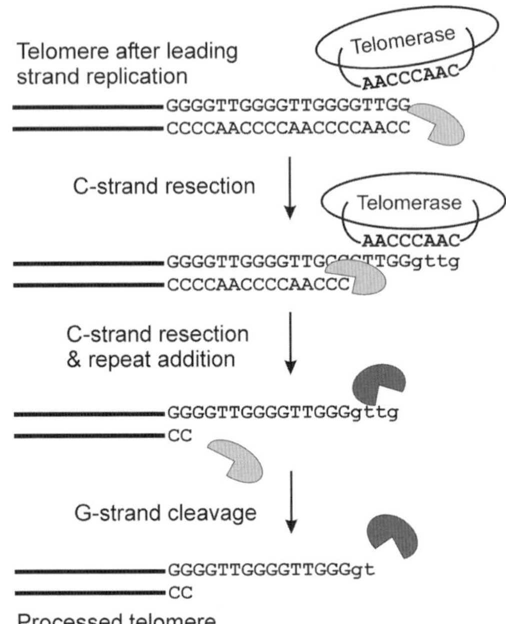

Figure 3. DNA processing steps required to generate the terminal DNA structure at *Tetrahymena* telomeres. The *filled structures* represent nucleases.

for the two processing reactions have not yet been identified, but various DNA repair proteins (Mre11, Exo1, ERCC1) are possible candidates.

TELOMERE BIOLOGY OF THE SPIROTRICHOUS CILIATES: *EUPLOTES, OXYTRICHA,* AND *STYLONYCHIA*

Genomic Organization and Macronuclear Development

In spirotrichs, the approximately 48 million macronuclear telomeres are generated by a unique but highly reproducible developmental program that involves extensive reorganization of the micronuclear genome. During this process of macronuclear development >95% of the micronuclear genome is eliminated and the remainder undergoes chromosome fragmentation to generate linear gene-sized molecules, de novo telomere synthesis, and endoreplication (Prescott 1994; Jahn and Klobutcher 2002). The various steps in macronuclear development have been studied in most detail for the marine ciliate *Euplotes crassus* (*MonoEuplotes crassus*) because this organism can be induced to mate synchronously after mixing of opposite mating types (Roth et al. 1985). However, the same events occur in other spirotrichs. The process of

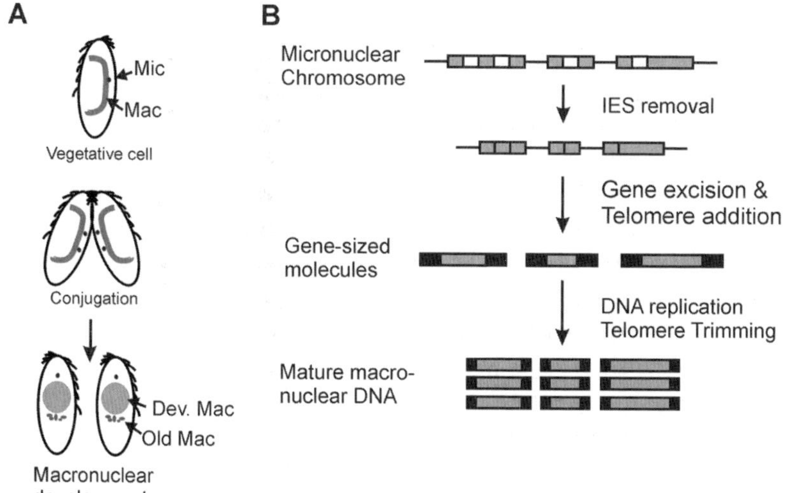

Figure 4. Macronuclear development. (*A*) General scheme showing nuclear duality and changes in nuclear content during and after conjugation. (Mic) Micronucleus; (Mac) macronucleus; (Dev Mac) developing macronucleus; (Old Mac) degenerating old macronucleus. (*B*) Stages in genomic reorganization through which micronuclear chromosomes are converted to mature macronuclear DNA molecules.

conjugation and macronuclear development usually starts when starvation induces cells of opposite mating types to form pairs and initiate meiosis (Fig. 4A). Only the micronucleus undergoes meiosis. The meiotic divisions produce haploid pronuclei that are exchanged between members of a mating pair; the exchanged nucleus then fuses with a resident nucleus to produce the new diploid zygotic nucleus. The zygotic nucleus divides mitotically, and one daughter is retained as a micronucleus while the other develops into a new macronucleus. As the new macronucleus develops, the old macronucleus ceases transcription and is eventually destroyed along with the unused meiotic products.

Conversion of micronuclear DNA into the macronuclear genome takes 3–4 days and occurs in several stages (Fig. 4B) (Prescott 1994; Jahn and Klobutcher 2002). The first stage involves removal of large amounts of interstitial sequence by DNA breakage and rejoining. The eliminated DNA ranges from short segments of <100 bp that interrupt the coding sequence (internal eliminated sequences [IESs]) to large transposon-like elements that mostly lie in intergenic regions. Because the IESs interrupt coding sequence, their excision must be very precise. This is achieved by marking the ends of the IESs with short direct repeats; these repeats are then used to define the sites of DNA cleavage (Prescott and DuBois 1996;

Doak et al. 1997; Prescott et al. 1998). In *E. crassus*, staggered cuts are made at each end of the IES, the IES is released, and the cut sites are religated to generate uninterrupted coding sequence that contains one copy of the repeat element (Klobutcher 1999). A similar mechanism is used to remove the transposon-like elements, TEC (transposon *Euplotes crassus*) elements (Jaraczewski and Jahn 1993). Although the direct repeats are required for IES and TEC removal, specialized chromatin structures are also thought to help define the cleavage site (Jahn 1999).

The second stage of genome rearrangement involves site-specific cleavage of the DNA to generate approximately 24,000 small linear DNA molecules (Fig. 4B) (Prescott 1994; Jahn and Klobutcher 2002). The majority of these macronuclear-destined sequences encode a single gene and they have an average size of ~2.0 kb. As described below, the position of cleavage may be defined by a conserved DNA sequence element (a chromosome breakage sequence [CBS]) or by the presence of an unusual DNA or chromatin structure adjacent to the breakage site. Cleavage of the DNA seems to be tightly coupled to telomere addition because macronuclear-destined molecules lacking telomeres cannot be detected by sensitive polymerase chain reaction (PCR)-based approaches (Klobutcher 1999). Telomere addition has been analyzed most extensively in *E. crassus* and has been found to involve both DNA synthesis and processing steps (Roth and Prescott 1985; Vermeesch et al. 1993; Vermeesch and Price 1994). The telomeres that are synthesized initially are heterogeneous in length and are on average 50 bp longer than mature macronuclear telomeres. Most of the G-strands have a 3′ overhang, but these differ from the overhangs at mature telomeres in that their length and terminal sequence are quite variable. Once telomere addition is complete, the macronuclear DNA is subject to several rounds of endoreplication. This results in most molecules being amplified to a copy number of ~1000. The long, heterogeneous telomeres persist throughout the rounds of replication and are replaced only by the mature telomere structure at the very end of macronuclear development. The short mature telomeres are made even if the final rounds of replication are inhibited, which indicates that they are generated by an active processing event rather than by gradual erosion of the telomeric tract during replication (Vermeesch et al. 1993).

Chromosome Fragmentation

In *Euplotes*, chromosome fragmentation is a very precise event that results in addition of telomeric sequence to the same sites each time a new macronucleus is generated. The chromosome breakage sites are marked by a 10-bp sequence (the E-CBS) that has a highly conserved 5-bp core element (5′ TTGAA 3′) (Klobutcher et al. 1998). The CBS is found either in the macronuclear-destined sequence where the 5′ end of the core element lies 17 bp from the site of telomere addition or in the eliminated sequence where it lies in an inverse orientation. The CBS seems to serve as a recognition and positioning sequence for the cleavage machinery, which then generates a staggered cut 17 bp from the 5′ end of the core element (Klobutcher 1999). This cut has a six-base 3′ overhang that serves as the substrate for telomerase. Although the non-macronuclear-destined sequences are eventually degraded, telomeres are added to both sides of the cleavage junction (Mollenbeck and Klobutcher 2002). Thus, DNA elimination is not triggered by the absence of a telomere. Analysis of the cleavage junctions indicated that the eliminated DNA may instead be marked by an alternative chromatin structure that makes it more susceptible to degradation. Although telomeric repeats are added directly to the cleavage site on macronuclear-destined sequences, the exact site of telomere addition is somewhat variable for the DNA that will be eliminated. This suggests that the eliminated DNA exists in a nuclease-sensitive chromatin structure that allows some degradation prior to telomere addition. In contrast, the macronuclear-destined sequences appear to be packaged into a protective structure that prevents DNA removal and/or promotes more rapid telomere addition. The different chromatin structures are likely to form prior to IES elimination when several development-specific histone variants are expressed (Jahn et al. 1997; Ray et al. 1999).

Although defined CBSs have been found in a variety of *Euplotes* species, an equivalent, well-defined sequence is not present near the chromosome breakage sites of other spirotrichs (Prescott 1994). Moreover, *Euplotes* is the only ciliate that lacks heterogeneity in the site of telomere addition. In *Oxytricha*, the site at which telomeres are added to a specific macronuclear DNA can vary by tens of base pairs, and in some cases a choice of fragmentation sites gives rise to molecules that vary in size by multiple kilobases (Williams et al. 2002). Interestingly, the macronuclear DNA molecules from several *Oxytricha* species display an unusual strand asymmetry in the purine to pyrimidine ratio within the subtelomeric region (Prescott and Dizick 2000), whereas in *Stylonychia* the subtelomeric regions of certain macronuclear-destined sequences are required for chromosome breakage (Jonsson et al. 2001). These findings suggest that chromosome fragmentation may be directed either by DNA or chromatin structure rather than by a conserved DNA element like the *Euplotes* CBS.

The Enzymology of New Telomere Synthesis

The DNA ends generated by chromosome fragmentation lack preexisting telomeric sequences in all ciliates examined to date. However, studies of telomere addition in *T. thermophila* have shown that the new telomeric DNA is synthesized by telomerase, and the available evidence indicates that this is also the case in other ciliates (Yu and Blackburn 1991; also see below). Because telomerase isolated from unmated, vegetative *Euplotes* and *Oxytricha* cells is unable to use nontelomeric DNA as a substrate (Zahler and Prescott 1988; Bednenko et al. 1997), the capacity to add telomeric repeats at the sites of chromosome breakage must result from changes associated with macronuclear development. Comparison of telomerase complexes isolated from mated and unmated *E. crassus* has revealed that these changes take the form of a dramatic modification in the telomerase holoenzyme such that the RNP architecture, holoenzyme composition, and substrate specificity are all altered (Greene and Shippen 1998).

When telomerase from unmated cells is fractionated by gel filtration, it separates as a complex of 280–400 kD (Greene and Shippen 1998). In contrast, telomerase from mated cells undergoing new telomere synthesis exists as a series of complexes of 550 kD, 1600 kD, and 5 mD. The telomerase in the 1600-kD and 5-mD complexes is more processive than the activity from the smaller complexes, and it is able to add repeats to primers that lack telomeric sequence at the 3′ end. This capacity to use nontelomeric primers is lost when the larger complexes are purified on glycerol gradients, but it can be restored by addition of gradient fractions that lack telomerase activity (Bednenko et al. 1997). These, and other findings discussed below, indicate that *Euplotes* telomerase undergoes a developmentally programmed switch from being an enzyme that can simply maintain telomeres during vegetative growth to one that can perform extensive de novo telomere addition in the absence of a telomeric DNA substrate.

Generation of a new telomere requires not only addition of the telomeric G-strand by telomerase, but also synthesis of the complementary C-strand. In *Euplotes*, new telomere synthesis also involves some form of length regulation, because, despite the heterogeneity in G-strand length, most of the C-strands are exactly 84 nt long (Vermeesch and Price 1994). Studies of newly synthesized telomeres have shown that G- and C-strand synthesis is coordinated, and this coordination is important for length regulation (Fan and Price 1997). If G- and C-strand synthesis is uncoupled by using aphidicolin to partially inhibit C-strand synthesis, the G-strands become longer and the C-strands become more heterogeneous. This finding suggests that initiation of C-strand synthesis by polα-primase limits further extension of the telomeric G-strand by telomerase.

Coordination of telomeric G- and C-strand synthesis has since been demonstrated in yeast (Chandra et al. 2001), and it is probably needed in all organisms where telomerase adds long stretches of G-strand in a single processive reaction. In *Euplotes*, coupling of G- and C-strand synthesis is achieved by incorporating DNA polα-primase into the developmentally regulated telomerase holoenzyme complexes (Ray et al. 2002). Analysis of the complexes has revealed that DNA primase is present in the 550-kD, 1600-kD, and 5-mD telomerase complexes, but not the 280-kD complex from vegetative cells. Presumably, this juxtaposition of the machinery needed to generate each strand of the telomere allows the very efficient telomere synthesis that is needed to create thousands of new telomeres within a short period of time (Fig. 5).

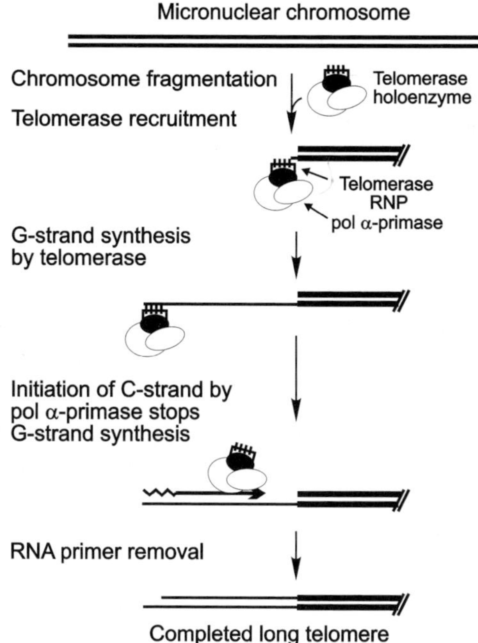

Figure 5. Coordination of G- and C-strand synthesis by the telomerase/polα-primase telomere synthesis complex.

Although all three developmentally regulated telomerase complexes contain DNA primase, they differ in their other components (Ray et al. 2002). Unlike the 5-mD complex, the 550-kD complex lacks proliferating cell nuclear antigen (PCNA), the processivity factor that associates with DNA polymerase δ and ε. The 550-kD complex also seems to lack factor(s) that promote telomerase processivity as it synthesizes shorter arrays of telomeric repeats than the two larger complexes. These observations suggest that the 550-kD and 1600-kD complexes may be assembly intermediates for a more complete 5-mD telomere synthesis complex that contains all the components needed to make a telomere. The large size of the 5-mD complex and the tight coupling between chromosome breakage and telomere addition (Klobutcher 1999) suggest that it may also contain the nucleases responsible for site-specific cleavage of the DNA.

Whereas the presence of replication factors such as polα-primase is one key difference between the vegetative and developmental forms of telomerase, a second fundamental difference lies in the components of the core telomerase RNP. Unlike higher eukaryotes, the *Euplotes* genome contains three different TERT genes that encode catalytic subunits that are only 70–80% identical (Karamysheva et al. 2003). Two of the genes are expressed during vegetative growth and hence must be responsible for macro- and micronuclear telomere maintenance. However, expression of these genes drops at the time of new telomere synthesis, and the third development-specific TERT is expressed instead. Thus, assembly of telomerase/polα-primase telomere synthesis complexes may well be triggered by the appearance of this developmental TERT. Intriguingly, the transcriptionally active form of the developmental TERT gene exists only for a short period of time following IES removal as the entire gene is later eliminated from the developing macronucleus. Presumably this elimination provides a fail-safe way to prevent unwanted de novo telomere addition at double-strand breaks during vegetative growth. Although developmental regulation of telomerase via the expression of different TERT genes is a unique mechanism that has thus far only been found in *Euplotes*, one wonders whether alternative splicing to generate different forms of TERT provides a similar regulatory mechanism during development in higher eukaryotes (Kilian et al. 1997; Ulaner et al. 1998).

Telomerase Biochemistry

Endogenous telomerase RNP (holoenzyme) was first purified from *E. aediculatus* by affinity chromatography using an antisense oligonucleotide complementary to the template region of the telomerase RNA

(Lingner and Cech 1996). The purified enzyme contained two proteins of 123 and 43 kD. Cloning and sequencing of the p123 gene revealed the presence of reverse transcriptase motifs (Lingner et al. 1997). Further analysis revealed that p123 shared sequence identity with Est2 (ever shorter telomeres), a protein known to be important for telomere maintenance in *S. cerevisiae*. Mutagenesis of conserved aspartic acid residues in the reverse transcriptase motifs of EST2 resulted in telomere shortening and senescence, thus providing the first conclusive evidence that telomerase uses a reverse transcriptase subunit in addition to the RNA template to catalyze synthesis of telomeric DNA.

The p43 subunit of *Euplotes* telomerase is a paralog of the La autoantigen, a nuclear protein that promotes biogenesis of pol III transcripts in a variety of eukaryotes (Aigner et al. 2000). However, p43 differs from La in that it is associated only with telomerase and not with primary transcripts of U6 or other small nuclear or nucleolar RNPs (Aigner et al. 2003). Moreover, p43 recognizes specific aspects of the telomerase RNA structure rather than the 3′ poly U tract that is characteristic of pol III transcripts and which is normally recognized by La. Recombinant p43 binds *Euplotes* and *Tetrahymena* telomerase RNA with almost the same K_d, and when it is included with the *Tetrahymena* RNA and TERT in a telomerase reconstitution reaction, it causes a slight increase in enzymatic activity and processivity (Aigner and Cech 2004). Thus, p43 appears to be a dedicated telomerase protein.

Unfortunately, studies of *Euplotes* telomerase have been hampered because in vitro reconstitution of catalytically active enzyme has not been possible and molecular genetic methods to introduce and express mutant genes in vivo are also unavailable. Nonetheless, analysis of the enzyme isolated from wild-type cells has made significant contributions to our understanding of telomerase biochemistry (see Chapters 2 and 3 for a detailed review). In particular, studies of telomerase from *E. aediculatus* have provided evidence that p123 (TERT) contacts the DNA substrate 20–22 nt upstream of the 3′ terminus (Hammond et al. 1997). These studies have also shown that the telomerase maintains a fairly constant number of base pairs between the RNA template and the primer during primer elongation and have revealed that deoxyguanine triphosphate (dGTP) promotes translocation and repositioning of the primer on the RNA template (Hammond and Cech 1997, 1998). Likewise, examination of telomerase from *E. crassus* has yielded novel information about the telomerase-associated nuclease, an activity that copurifies with the telomerase holoenzyme from organisms ranging from humans to yeast to ciliates (Melek et al. 1996; Greene et al. 1998). The activity from *Euplotes* is an endonuclease that can cleave at least 13 nt of nontelomeric sequence from the 3′ end of an otherwise telomeric primer. The primer is then extended by addition of canonical T_4G_4 repeats. The nuclease can also remove mismatches from telomeric DNA that is hybridized to the 5′ region of the template RNA. Thus, it may function as a proofreading activity that enhances telomerase fidelity.

TELOMERE BIOLOGY OF THE OLIGOHYMENORPHORANS: *TETRAHYMENA* AND *PARAMECIUM*

Genomic Organization and Macronuclear Development

Although *Tetrahymena* and *Paramecium* have fewer macronuclear telomeres than *Oxytricha* or *Euplotes*, these ciliates use a similar developmental program to form a new macronucleus after mating (Karrer 2000; Yao et al. 2002). When cells conjugate, the micronuclei undergo meiosis, pronuclear exchange, pronuclear fusion, and mitotic division. The macronucleus then develops from one or more of the zygotic nuclei, and this process again involves a series of DNA rearrangements including IES removal and chromosome fragmentation. However, the amount of DNA elimination, chromosome breakage, and telomere addition is less than in the spirotrichs. In *Tetrahymena*, only 10–20% of the genome is eliminated as a result of IES removal and chromosome fragmentation whereas >95% is eliminated in the spirotrichs.

About 6000 IESs and transposon-like elements are removed from the *Tetrahymena* genome by DNA breakage and rejoining, whereas about 50,000 are removed from the *Paramecium* genome (Yao et al. 2002). *Tetrahymena* IESs are 0.5 to >20 kb in length and they are flanked by short 1–8-bp direct repeats. IES recognition and removal occur by an RNAi-based mechanism that is guided by the old macronucleus (Mochizuki et al. 2002; Yao et al. 2003; Mochizuki and Gorovsky 2004). During this process the micronuclear genome is transcribed bidirectionally and the resulting double-stranded RNA is processed to small RNAs by a Dicer-like RNase. These small RNAs then go to the old macronucleus where RNAs that are homologous to the macronuclear DNA are degraded. The remaining RNAs then seem to be transferred to the developing macronucleus where they promote elimination of homologous DNA sequences.

Chromosome fragmentation and new telomere synthesis occur once IES removal is complete. In *Tetrahymena*, the five micronuclear chromosomes are fragmented into approximately 200 different macronuclear chromosomes of 100–1500 kbp, and telomeres are added to the newly generated ends (Yao et al. 2002). It is not known whether the newly synthesized telomeres resemble the preexisting micronuclear telomeres or the shorter macronuclear telomeres in terms of length or terminal DNA structure. The macronuclear chromosomes are then subject to several rounds of replication to produce the final copy number of >45. In *T. thermophila*, roughly half of the macronuclear telomeres are formed as the result of an unusual DNA rearrangement event that excises the single-copy ribosomal DNA gene and turns it into 21-kb minichromosome (Butler et al. 1995). During this process, the excised rDNA gene is duplicated into a palindrome to create a linear DNA molecule, containing two copies of the rRNA gene in a head-to-head orientation. Telomeres are added to each end of the rDNA molecule, and it is amplified to a copy number of ~9000. The abundance of the rDNA minichromosome and the ease with which it can be manipulated by molecular genetic techniques has made this molecule a prime tool for studying many aspects of *Tetrahymena* telomere biology.

Tetrahymena Molecular Genetics

Analysis of *Tetrahymena* telomeres and telomerase has been greatly facilitated by the existence of a robust transformation system (Turkewitz et al. 2002). Efficient transformation together with naturally high levels of homologous recombination allows disruption or replacement of both macronuclear and micronuclear genes, regulated protein expression, and generation of altered rDNA chromosomes. Gene disruptions and replacements have been made for TERT, TER, POT1, Ku70, and various exonucleases, whereas rDNA expression vectors have been used to generate telomerase RNAs that have mutations in template and nontemplate regions.

Disruption or replacement of macronuclear genes is particularly straightforward because gene-targeting constructs integrate into the homologous gene locus with high efficiency (Bruns and Cassidy-Hanley 2000; Gaertig and Kapler 2000). Moreover, *Tetrahymena* introns are quite small so it is easy design constructs that will delete or replace an entire gene locus. When a gene targeting construct is introduced into the macronucleus it will recombine with only a few of the macronuclear chromosomes carrying the targeted gene (Fig. 6). Thus, a construct carrying a selectable marker will initially confer a low level of drug resistance. Continuous culture with increasing drug concentration will then force phenotypic assortment so more and more chromosomes carry the gene disruption/replacement. If the targeted gene is nonessential, all the copies of the

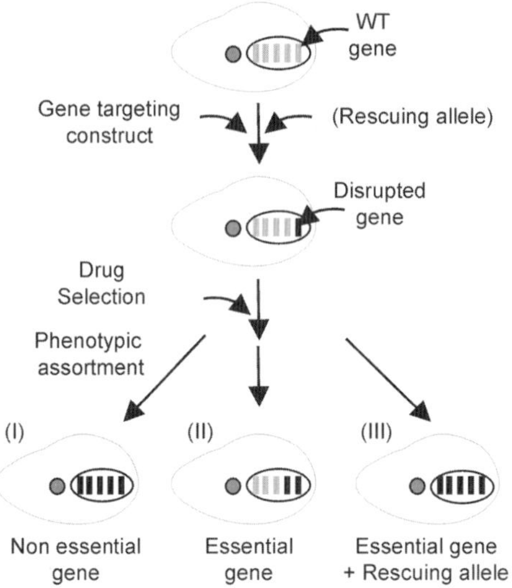

Figure 6. A full knockout can be obtained after drug selection if the gene is non-essential (*I*). If the gene is essential, a knockdown is obtained (*II*) unless a rescuing allele is introduced (*III*).

wild-type gene are eventually replaced. If the gene is essential, passaging in the highest tolerated drug concentration will result in reduction of the wild-type chromosomes to the minimum required for viability. In this situation, a complete macronuclear knockout can be generated by introducing a rescuing copy of the wild-type gene under control of a regulated promoter prior to performing the gene targeting.

Germ-line transformation to generate a micronuclear gene knockout is useful when the gene of interest is essential because expression of the wild-type macronuclear gene allows the cell to survive (Hai et al. 2000). The null phenotype can then be examined when the cells are mated and transformation with variant forms of the gene during mating allows mutant analysis (Butler et al. 1995).

Chromosome Fragmentation and New Telomere Synthesis

Fragmentation of *Tetrahymena* chromosomes is directed by a 15-bp sequence (the chromosome breakage sequence [CBS]) that is conserved in most species of *Tetrahymena* (Yao et al. 2002). The CBS is both necessary and sufficient for chromosome breakage, and mutations at most positions prevent breakage. Telomeres are added 5–25 bp away from each end of the CBS, which results in the loss of 25–65 bp from the breakage site (Fan and Yao 1996). The pattern of telomere addition suggests that the CBS is recognized by an endonuclease, or a protein that recruits an endonuclease, and this endonuclease then cuts the chromosome on either side of the CBS. Telomere addition is very efficient when chromosome breakage is directed by the CBS, but it does not occur at DNA ends generated by restriction digestion. Thus, it appears that chromosome breakage and telomere addition are tightly coupled.

Examination of the micronuclear DNA sequence adjacent to several sites of fragmentation revealed that telomeric repeats are absent from the breakage site and hence must be added de novo (Yao et al. 1987). Direct evidence that the new telomeric DNA is synthesized by telomerase was obtained by mating cells expressing both wild-type telomerase and a telomerase template mutant that generated an altered repeat sequence (Yu and Blackburn 1991). When the newly synthesized telomeres were cloned and sequenced, many were found to have mutant repeats directly adjacent to the unique sequence at the end of the chromosome (Fig. 7A). Thus, the new repeats must have been added directly to the broken end by telomerase from the parental cell. Surprisingly, the newly synthesized telomeres did not consist of homogeneous runs of either wild-type or mutant repeats but instead had the wild-type and mutant repeats interspersed. Thus, although *Tetrahymena* telomerase is highly processive in vitro, the enzyme appears to be much less processive in vivo.

Although chromosome fragmentation and telomere addition are tightly linked, *Tetrahymena* cells do not appear to form telomere synthesis complexes analogous to those found in *Euplotes*. Telomerase from mated and vegetative cells fractionates as a single complex, and the size of the complex does not increase during macronuclear development (Wang and Blackburn 1997). Although the lack of higher-order complexes could reflect problems associated with telomerase extraction, these results suggest that telomerase does not form an association with the chromosome breakage machinery. *Tetrahymena* telomerase also differs from the *Euplotes* enzyme in that telomerase from both vegetative and mated cells can add telomeric repeats to a DNA primer that lacks any preexisting telomeric sequence (Wang and Blackburn 1997; Wang et al. 1998). Thus, *Tetrahymena* may not need to form a specialized telomere synthesis complex during macronuclear development because the enzyme already has the ability to add telomeric sequence to a nontelomeric substrate. Also, the total number of telomeres that must be synthesized is much smaller than in *Euplotes*, so the process may not need to be as efficient.

A   rDNA telomere sequences

MIC.

.....GAAATTATTAAA<u>TTGGGGT</u>TTTAACTTATTTTAAAAA.....

MAC.

.....GAAATTATTAAA<u>TTGGGGTT</u>(**G**₅T₂)₆(G₄T₂)₄**G**₅T₂G₄T₂(**G**₅T₂)₄(G₄T₂)₂......

.....GAAATTATTAAA<u>TTGGGGTT</u>G₄T₂(**G**₅T₂)₂G₄T₂(**G**₅T₂)₂(G₄T₂)₄**G**₅T₂G₄T₂(**G**₅T₂)₃(G₄T₂)₃......

.....GAAATTATTAAA<u>TTGGGGTT</u>G₄T₂(**G**₅T₂)₂G₄T₂(**G**₅T₂)₂(G₄T₂)₄**G**₅T₂G₄T₂(**G**₅T₂)₃......

.....GAAATTATTAAA<u>TTGGGGTT</u>(**G**₅T₂)₃G₄T₂(**G**₅T₂)₄(G₄T₂)₆(**G**₅T₂)₃(G₄T₂)₄......

.....GAAATTATTAAA<u>TTGGGGTT</u>(**G**₅T₂)₆(G₄T₂)₅......

.....GAAATTATTAAA<u>TTGGGGTT</u>**G**₅T₂G₄T₂(**G**₅T₂)₃(G₄T₂)₈......

.....GAAATTATTAAA<u>TTGGGGTT</u>G₄TC**G**₄T₂G₄T₂G₄T₂......

B

	RNA template	Telomere repeat	Phenotype (growth)	Phenotype (telomere)
WT	3' 51—43 5' aaaAACCCCAACuu	5' GGGTTG	Wild type	Wild type
45+C	3' 51—43 5' aaaAACCCC**C**AAC uu	5' GGGG**G**TT	Monster cells & senescence	Lengthen
48T	3' 51—43 5' aaaAAC**T**CCA ACuu	5' G**T**GGTT	Monster cells & senescence	Shorten
44G	3' 51—43 5' aaaAACCCCA**G**Cuu	5' GGGGT**C**	Monster cells & senescence	Lengthen
43AA	3' 51—43 5' aaaAACCCCAAA**A**uu	5' GGGGTT**TT** 5' GGGGTT**T**	Monster cells & senescence, anaphase bridges	Lengthen

Figure 7. Phenotypes of telomerase template mutants. (*A*) Sequence of new telomeres synthesized in the presence of wild-type telomerase and the 45 + C template mutant. (Mic) Sequence in the telomere addition region of the micronuclear rDNA chromosome; (Mac) sequence of individual macronuclear telomeres. (*B*) Cell growth and telomere phenotypes of different template mutants.

Telomerase Biochemistry

Tetrahymena telomerase was discovered through biochemical studies designed to look for novel polymerase activities that could synthesize new telomeric repeats and hence circumvent the telomere shortening that arises due to incomplete replication by DNA polymerase (Greider and Blackburn 1985). *Tetrahymena* was chosen for this work because the large number of telomeres in each macronucleus made it likely that such an activity would be quite abundant. The *Tetrahymena* extracts were found to contain an unusual terminal transferase activity that could add T_2G_4

repeats to a telomeric DNA primer. The activity appeared to contain both RNA and protein as it was sensitive to proteinase K and RNase or micrococcal nuclease (Greider and Blackburn 1987). Fractionation of the activity revealed a copurifying RNA molecule that contained a short 9-nt region that was complementary to the G-strand of the telomeric DNA (Greider and Blackburn 1989). This region of the telomerase RNA was proposed to provide the template for repeat synthesis. Subsequent in vivo studies demonstrated that this was indeed the case as expression of mutant telomerase RNA (TER) with base changes in the template region resulted in incorporation of mutant repeats into the telomeric DNA (Fig. 7B) (Yu et al. 1990; Kirk et al. 1997). These experiments also demonstrated the essential role of telomeres because the mutant repeats led to defects in nuclear and cell division and senescence.

Both the TERT and TER subunits of *Tetrahymena* telomerase have been cloned, and active enzyme can be assembled in vitro from recombinant protein and RNA (Bryan et al. 1998; Collins and Gandhi 1998). Mutant or tagged telomerase holoenzyme can be generated in vivo by expressing versions of either subunit in *Tetrahymena* cells (Witkin and Collins 2004). Consequently, it has been possible to use a combination of in vivo and in vitro approaches to analyze the role of different regions in the TERT and TER molecules in repeat synthesis and hence to dissect the telomerase catalytic cycle. Phylogenetic comparison of TER from various species of *Tetrahymena* and other ciliates has led to a detailed model of the RNA secondary structure that has provided a critical tool for understanding TER function (Romero and Blackburn 1991; Lingner et al. 1994; McCormick-Graham and Romero 1995). Analysis of TER mutants has revealed that the RNA does much more than simply provide a template for DNA synthesis; it is instead an integral component of the holoenzyme that is involved in many aspects of the catalytic cycle (Gilley and Blackburn 1999; Licht and Collins 1999; Sperger and Cech 2001; Lai et al. 2002, 2003; Miller and Collins 2002; Mason et al. 2003; Cunningham and Collins 2005). Sequences directly 5′ and 3′ of the 9-nt template provide boundaries that ensure that only this region of the RNA is copied by TERT. TER is also important for enzyme processivity as segments of the RNA promote copying of the entire template and addition of successive telomeric repeats. Both high- and low-affinity TERT-binding sites have been mapped on the RNA. Likewise, analysis of TERT mutants has defined the domains required for RNA binding and catalytic activity (Bryan et al. 2000a,b; Miller et al. 2000; Lai et al. 2001, 2002). A detailed discussion of these and other aspects of *Tetrahymena* telomerase biochemistry can be found in Chapter 3.

Defining the precise components of the telomerase holoenzymes from different organisms has proved a difficult task because the RNP is labile and easily disrupted by conventional chromatographic techniques. However, addition of a TAP tag to the TERT carboxyl terminus has allowed the *Tetrahymena* holoenzyme to be isolated by a rapid and extremely gentle affinity purification protocol (Witkin and Collins 2004). The purified RNP was found to contain TERT, TER, and four novel proteins of 20, 45, 65, and 75 kD. It did not contain the 80- and 95-kD proteins that were originally found to cofractionate with telomerase activity using conventional chromatography. Subsequent studies have shown that p80 and p95 play a role in telomere length regulation, but they are unlikely to be components of the holoenzyme (Miller and Collins 2000; Mason et al. 2001). Interestingly, both the telomerase holoenzyme isolated using the TAP tag and the recombinant TERT + TER complex assembled in vitro fractionate as monomers of each subunit rather than dimers (Bryan et al. 2003; Witkin and Collins 2004). Both yeast and human telomerase can exist as dimers or higher-order complexes, and it has been suggested that dimerization may be essential for telomerase function (Kelleher et al. 2002). However, as the *Tetrahymena* enzyme shows full catalytic activity as a monomer of TERT + TER, dimerization is unlikely to be fundamental to the catalytic cycle.

The function of the p20 and p75 components of the telomerase holoenzyme components have not yet been investigated; however, both the p45 and p65 subunits are essential and their depletion causes telomere shortening (Witkin and Collins 2004). p65 resembles the p43 subunit from *Euplotes* telomerase in that it also contains a La motif and associates specifically with telomerase RNA rather than other PolIII transcripts. Depletion of p65 causes a decrease in the level of telomerase RNA, suggesting that it is required for RNP biogenesis and/or stability. It is striking that the telomerase holoenzymes from different species all seem to contain proteins that are important for RNP maturation (see Chapter 2), but in vertebrates and yeast these proteins (dyskerin or Sm) also play a role in the maturation of other small RNPs, whereas ciliates appear to have evolved telomerase-specific proteins.

CONCLUDING REMARKS

Over the years, ciliates have provided an excellent system for studying a wide range of topics in telomere biology. The role of any model organism is to provide insight into problems that are intractable in higher organisms. Ciliates with their thousands of telomeres and unusual biology have served this function admirably. As is often observed for the biology of single-cell eukaryotes, ciliate telomeres and telomerase seem to represent smaller simpler versions of their mammalian counterparts. This, together with the abundance of telomere and telomerase components, has made it possible to analyze telomere structure, composition, and telomerase biochemistry using techniques that were inadequate to analyze the larger, less abundant telomeres or telomerase from multicellular organisms. The discoveries made with ciliates have then provided guidelines for developing the more sophisticated techniques needed to perform similar analyses with mammalian telomeres. An additional benefit of working with ciliates is that they have revealed unexpected aspects of telomere biology and telomerase biochemistry such as chromosome healing by de novo telomere synthesis and developmental regulation of telomerase holoenzyme composition. The recent completion of the *Tetrahymena* and *Paramecium* genomes, coupled with excellent classic and molecular genetics, should ensure that studies with ciliates continue to fuel important advances and paradigm shifts within the telomere field.

ACKNOWLEDGMENTS

I thank Kathleen Collins and Dorothy Shippen for comments on the manuscript. The research conducted in the author's laboratory was supported by the National Institutes of Health (GM41803–14). The author declares that she has no competing financial interests.

REFERENCES

Aigner S. and Cech T.R. 2004. The *Euplotes* telomerase subunit p43 stimulates enzymatic activity and processivity in vitro. *RNA* **10:** 1108–1118.

Aigner S., Postberg J., Lipps H.J., and Cech T.R. 2003. The *Euplotes* La motif protein p43 has properties of a telomerase-specific subunit. *Biochemistry* **42:** 5736–5747.

Aigner S., Lingner J., Goodrich K.J., Grosshans C.A., Shevchenko A., Mann M., and Cech T.R. 2000. *Euplotes* telomerase contains an La motif protein produced by apparent translational frameshifting. *EMBO J.* **19:** 6230–6239.

Baroin A., Prat A., and Caron F. 1987. Telomeric site position heterogeneity in macronuclear DNA of *Paramecium primaurelia*. *Nucleic Acids Res.* **15:** 1717–1728.

Baumann P. and Cech T.R. 2001. Pot1, the putative telomere end-binding protein in fission yeast and humans. *Science* **292:** 1171–1175.

Bednenko J., Melek M., Greene E.C., and Shippen D.E. 1997. Developmentally regulated initiation of DNA synthesis by telomerase: Evidence for factor-assisted *de novo* telomere formation. *EMBO J.* **16:** 2507–2518.

Blackburn E.H. and Chiou S.S. 1981. Non-nucleosomal packaging of a tandemly repeated DNA sequence at termini of extrachromosomal DNA coding for rRNA in *Tetrahymena*. *Proc. Natl. Acad. Sci.* **78:** 2263–2267.

Blackburn E.H. and Gall J.G. 1978. A tandemly repeated sequence at the termini of the extrachromosomal ribosomal RNA genes in *Tetrahymena. J. Mol. Biol.* **120:** 33–53.

Bruns P.J. and Cassidy-Hanley D. 2000. Biolistic transformation of macro- and micronuclei. *Methods Cell Biol.* **62:** 501–512.

Bryan T.M., Goodrich K.J., and Cech T.R. 2000a. A mutant of *Tetrahymena* telomerase reverse transcriptase with increased processivity. *J. Biol. Chem.* **275:** 24199–24207.

———. 2000b. Telomerase RNA bound by protein motifs specific to telomerase reverse transcriptase. *Mol. Cell* **6:** 493–499.

———. 2003. *Tetrahymena* telomerase is active as a monomer. *Mol. Biol. Cell* **14:** 4794–4804.

Bryan T.M., Sperger J.M., Chapman K.B., and Cech T.R. 1998. Telomerase reverse transcriptase genes identified in *Tetrahymena thermophila* and *Oxytricha trifallax. Proc. Natl. Acad. Sci.* **95:** 8479–8484.

Budarf M.L. and Blackburn E.H. 1986. Chromatin structure of the telomeric region and 3'-nontranscribed spacer of *Tetrahymena* ribosomal RNA genes. *J. Biol. Chem.* **261:** 363–369.

Butler D.K., Yasuda L.E., and Yao M.C. 1995. An intramolecular recombination mechanism for the formation of the rRNA gene palindrome of *Tetrahymena thermophila. Mol. Cell. Biol.* **15:** 7117–7126.

Cesare A.J., Quinney N., Willcox S., Subramanian D., and Griffith J.D. 2003. Telomere looping in *P. sativum* (common garden pea). *Plant J.* **36:** 271–279.

Chandra A., Hughes T.R., Nugent C.I., and Lundblad V. 2001. Cdc13 both positively and negatively regulates telomere replication. *Genes Dev.* **15:** 404–414.

Classen S., Ruggles J.A., and Schultz S.C. 2001. Crystal structure of the N-terminal domain of *Oxytricha nova* telomere end-binding protein alpha subunit both uncomplexed and complexed with telomeric ssDNA. *J. Mol. Biol.* **314:** 1113–1125.

Cohen P. and Blackburn E.H. 1998. Two types of telomeric chromatin in *Tetrahymena thermophila. J. Mol. Biol.* **280:** 327–344.

Collins K. 1999. Ciliate telomerase biochemistry. *Annu. Rev. Biochem.* **68:** 187–218.

Collins K. and Gandhi L. 1998. The reverse transcriptase component of the *Tetrahymena* telomerase ribonucleoprotein complex. *Proc. Natl. Acad. Sci.* **95:** 8485–8490.

Cunningham D. and Collins K. 2005. Biological and biochemical functions of RNA in the *Tetrahymena* telomerase holoenzyme. *Mol. Cell. Biol.* **25:** 4442–4454.

Doak T.G., Witherspoon D.J., Doerder F.P., Williams K., and Herrick G. 1997. Conserved features of TBE1 transposons in ciliated protozoa. *Genetica* **101:** 75–86.

Fan Q. and Yao M. 1996. New telomere formation coupled with site-specific chromosome breakage in *Tetrahymena thermophila. Mol. Cell. Biol.* **16:** 1267–1274.

Fan X. and Price C.M. 1997. Coordinate regulation of G- and C-strand length during new telomere synthesis. *Mol. Biol. Cell* **8:** 2145–2155.

Froelich-Ammon S.J., Dickinson B.A., Bevilacqua J.M., Schultz S.C., and Cech T.R. 1998. Modulation of telomerase activity by telomere DNA-binding proteins in *Oxytricha. Genes Dev.* **12:** 1504–1514.

Gaertig J. and Kapler G. 2000. Transient and stable DNA transformation of *Tetrahymena thermophila* by electroporation. *Methods Cell Biol.* **62:** 485–500.

Gilley D. and Blackburn E.H. 1994. Lack of telomere shortening during senescence in *Paramecium. Proc. Natl. Acad. Sci.* **91:** 1955–1958.

———. 1999. The telomerase RNA pseudoknot is critical for the stable assembly of a catalytically active ribonucleoprotein. *Proc. Natl. Acad. Sci.* **96:** 6621–6625.

Gottschling D.E. and Cech T.R. 1984. Chromatin structure of the molecular ends of *Oxytricha* macronuclear DNA: Phased nucleosomes and a telomeric complex. *Cell* **38:** 501–510.

Gottschling D.E. and Zakian V.A. 1986. Telomere proteins: Specific recognition and protection of the natural termini of *Oxytricha* macronuclear DNA. *Cell* **47:** 195–205.

Greene E.C. and Shippen D.E. 1998. Developmentally programmed assembly of higher order telomerase complexes with distinct biochemical and structural properties. *Genes Dev.* **12:** 2921–2931.

Greene E.C., Bednenko J., and Shippen D.E. 1998. Flexible positioning of the telomerase-associated nuclease leads to preferential elimination of nontelomeric DNA. *Mol. Cell. Biol.* **18:** 1544–1552.

Greider C.W. and Blackburn E.H. 1985. Identification of a specific telomere terminal transferase activity in *Tetrahymena* extracts. *Cell* **43:** 405–413.

———. 1987. The telomere terminal transferase of *Tetrahymena* is a ribonucleoprotein enzyme with two kinds of primer specificity. *Cell* **51:** 887–898.

———. 1989. A telomeric sequence in the RNA of *Tetrahymena* telomerase required for telomere repeat synthesis. *Nature* **337:** 331–337.

Griffith J.D., Comeau L., Rosenfield S., Stansel R.M., Bianchi A., Moss H., and de Lange T. 1999. Mammalian telomeres end in a large duplex loop. *Cell* **97:** 503–514.

Hai B., Gaertig J., and Gorovsky M.A. 2000. Knockout heterokaryons enable facile mutagenic analysis of essential genes in *Tetrahymena*. *Methods Cell Biol.* **62:** 513–531.

Hammond P.W. and Cech T.R. 1997. dGTP-dependent processivity and possible template switching of *Euplotes* telomerase. *Nucleic Acids Res.* **25:** 3698–3704.

———. 1998. *Euplotes* telomerase: Evidence for limited base-pairing during primer elongation and dGTP as an effector of translocation. *Biochemistry* **37:** 5162–5172.

Hammond P.W., Lively T.N., and Cech T.R. 1997. The anchor site of telomerase from *Euplotes aediculatus* revealed by photo-cross-linking to single- and double-stranded DNA primers. *Mol. Cell. Biol.* **17:** 296–308.

Henderson E.R. and Blackburn E.H. 1989. An overhanging 3′ terminus is a conserved feature of telomeres. *Mol. Cell. Biol.* **9:** 345–348.

Horvath M.P., Schweiker V.L., Bevilacqua J.M., Ruggles J.A., and Schultz S.C. 1998. Crystal structure of the *Oxytricha nova* telomere end binding protein complexed with single strand DNA. *Cell* **95:** 963–974.

Jacob N.K., Kirk K.E., and Price C.M. 2003. Generation of telomeric G strand overhangs involves both G and C strand cleavage. *Mol. Cell* **11:** 1021–1032.

Jacob N.K., Skopp R., and Price C.M. 2001. G-overhang dynamics at *Tetrahymena* telomeres. *EMBO J.* **20:** 4299–4308.

Jacob N.K., Stout A.R., and Price C.M. 2004. Modulation of telomere length dynamics by the subtelomeric region of *Tetrahymena* telomeres. *Mol. Biol. Cell* **15:** 3719–3728.

Jahn C.L. 1999. Differentiation of chromatin during DNA elimination in *Euplotes crassus*. *Mol. Biol. Cell* **10:** 4217–4230.

Jahn C.L. and Klobutcher L.A. 2002. Genome remodeling in ciliated protozoa. *Annu. Rev. Microbiol.* **56:** 489–520.

Jahn C.L., Ling Z., Tebeau C.M., and Klobutcher L.A. 1997. An unusual histone H3 specific for early macronuclear development in *Euplotes crassus*. *Proc. Natl. Acad. Sci.* **94:** 1332–1337.

Jaraczewski J.W. and Jahn C.L. 1993. Elimination of Tec elements involves a novel excision process. *Genes Dev.* **7:** 95–105.

Jonsson F., Steinbruck G., and Lipps H.J. 2001. Both subtelomeric regions are required and sufficient for specific DNA fragmentation during macronuclear development in *Stylonychia lemnae*. *Genome Biol.* **2:** RESEARCH0005.

Karamysheva Z., Wang L., Shrode T., Bednenko J., Hurley L.A., and Shippen D.E. 2003. Developmentally programmed gene elimination in *Euplotes crassus* facilitates a switch in the telomerase catalytic subunit. *Cell* **113:** 565–576.

Karrer K.M. 2000. *Tetrahymena* genetics: Two nuclei are better than one. *Methods Cell Biol.* **62:** 127–186.

Kelleher C., Teixeira M.T., Forstemann K., and Lingner J. 2002. Telomerase: Biochemical considerations for enzyme and substrate. *Trends Biochem. Sci.* **27:** 572–579.

Kilian A., Bowtell D.L., Abud H.E., Hime G.R., Venter D.J., Keese P.K., Duncan E.L., Reddel R.R., and Jefferson R.A. 1997. Isolation of a candidate human telomerase catalytic subunit gene, which reveals complex splicing patterns in different cell types. *Hum. Mol. Genet.* **6:** 2011–2019.

Kirk K.E. and Blackburn E.H. 1995. An unusual sequence arrangement in the telomeres of the germ-line micronucleus in *Tetrahymena thermophila*. *Genes Dev.* **9:** 59–71.

Kirk K.E., Harmon B.P., Reichardt I.K., Sedat J.W., and Blackburn E.H. 1997. Block in anaphase chromosome separation caused by a telomerase template mutation. *Science* **275:** 1478–1481.

Klobutcher L.A. 1999. Characterization of in vivo developmental chromosome fragmentation intermediates in *E. crassus*. *Mol. Cell* **4:** 695–704.

Klobutcher L.A., Swanton M.T., Donini P., and Prescott D.M. 1981. All gene-sized DNA molecules in four species of hypotrichs have the same terminal sequence and an unusual 3′ terminus. *Proc. Natl. Acad. Sci.* **78:** 3015–3019.

Klobutcher L.A., Gygax S.E., Podoloff J.D., Vermeesch J.R., Price C.M., Tebeau C.M., and Jahn C.L. 1998. Conserved DNA sequences adjacent to chromosome fragmentation and telomere addition sites in *Euplotes crassus*. *Nucleic Acids Res.* **26:** 4230–4240.

Lai C.K., Miller M.C., and Collins K. 2002. Template boundary definition in *Tetrahymena* telomerase. *Genes Dev.* **16:** 415–420.

———. 2003. Roles for RNA in telomerase nucleotide and repeat addition processivity. *Mol. Cell* **11:** 1673–1683.

Lai C.K., Mitchell J.R., and Collins K. 2001. RNA binding domain of telomerase reverse transcriptase. *Mol. Cell. Biol.* **21:** 990–1000.

Larson D.D., Spangler E.A., and Blackburn E.H. 1987. Dynamics of telomere length variation in *Tetrahymena thermophila*. *Cell* **50:** 477–483.

Lei M., Podell E.R., Baumann P., and Cech T.R. 2003. DNA self-recognition in the structure of Pot1 bound to telomeric single-stranded DNA. *Nature* **426:** 198–203.

Licht J.D. and Collins K. 1999. Telomerase RNA function in recombinant *Tetrahymena* telomerase. *Genes Dev.* **13:** 1116–1125.

Lingner J. and Cech T.R. 1996. Purification of telomerase from *Euplotes aediculatus*: Requirement of a primer 3′ overhang. *Proc. Natl. Acad. Sci.* **93:** 10712–10717.

Lingner J., Hendrick L.L., and Cech T.R. 1994. Telomerase RNAs of different ciliates have a common secondary structure and a permuted template. *Genes Dev.* **8:** 1984–1998.

Lingner J., Hughes T.R., Shevchenko A., Mann M., Lundblad V., and Cech T.R. 1997. Reverse transcriptase motifs in the catalytic subunit of telomerase. *Science* **276:** 561–567.

Mason D.X., Autexier C., and Greider C.W. 2001. *Tetrahymena* proteins p80 and p95 are not core telomerase components. *Proc. Natl. Acad. Sci.* **98:** 12368–12373.

Mason D.X., Goneska E., and Greider C.W. 2003. Stem-loop IV of *Tetrahymena* telomerase RNA stimulates processivity in trans. *Mol. Cell. Biol.* **23:** 5606–5613.

McCormick-Graham M. and Romero D.P. 1995. Ciliate telomerase RNA structural features. *Nucleic Acids Res.* **23:** 1091–1097.

———. 1996. A single telomerase RNA is sufficient for the synthesis of variable telomeric DNA repeats in ciliates of the genus *Paramecium. Mol. Cell. Biol.* **16:** 1871–1879.

McCormick-Graham M., Haynes W.J., and Romero D.P. 1997. Variable telomeric repeat synthesis in *Paramecium tetraurelia* is consistent with misincorporation by telomerase. *EMBO J.* **16:** 3233–3242.

Melek M., Davis B.T., and Shippen D.E. 1994. Oligonucleotides complementary to the *Oxytricha nova* telomerase RNA delineate the template domain and uncover a novel mode of primer utilization. *Mol. Cell. Biol.* **14:** 7827–7838.

Melek M., Greene E.C., and Shippen D.E. 1996. Processing of nontelomeric 3′ ends by telomerase: Default template alignment and endonucleolytic cleavage. *Mol. Cell. Biol.* **16:** 3437–3445.

Miller M.C. and Collins K. 2000. The *Tetrahymena* p80/p95 complex is required for proper telomere length maintenance and micronuclear genome stability. *Mol. Cell* **6:** 827–837.

———. 2002. Telomerase recognizes its template by using an adjacent RNA motif. *Proc. Natl. Acad. Sci.* **99:** 6585–6590.

Miller M.C., Liu J.K., and Collins K. 2000. Template definition by *Tetrahymena* telomerase reverse transcriptase. *EMBO J.* **19:** 4412–4422.

Mitton-Fry R.M., Anderson E.M., Hughes T.R., Lundblad V., and Wuttke D.S. 2002. Conserved structure for single-stranded telomeric DNA recognition. *Science* **296:** 145–147.

Mochizuki K. and Gorovsky M.A. 2004. Conjugation-specific small RNAs in *Tetrahymena* have predicted properties of scan (scn) RNAs involved in genome rearrangement. *Genes Dev.* **18:** 2068–2073.

Mochizuki K., Fine N.A., Fujisawa T., and Gorovsky M.A. 2002. Analysis of a piwi-related gene implicates small RNAs in genome rearrangement in *Tetrahymena. Cell* **110:** 689–699.

Mollenbeck M. and Klobutcher L.A. 2002. De novo telomere addition to spacer sequences prior to their developmental degradation in *Euplotes crassus. Nucleic Acids Res.* **30:** 523–531.

Murti K.G. and Prescott D.M. 1999. Telomeres of polytene chromosomes in a ciliated protozoan terminate in duplex DNA loops. *Proc. Natl. Acad. Sci.* **96:** 14436–14439.

Peersen O.B., Ruggles J.A., and Schultz S.C. 2002. Dimeric structure of the *Oxytricha nova* telomere end-binding protein alpha-subunit bound to ssDNA. *Nat. Struct. Biol.* **9:** 182–187.

Prescott D.M. 1994. The DNA of ciliated protozoa. *Microbiol. Rev.* **58:** 233–267.

Prescott D.M. and Dizick S.J. 2000. A unique pattern of intrastrand anomalies in base composition of the DNA in hypotrichs. *Nucleic Acids Res.* **28:** 4679–4688.

Prescott D.M. and DuBois M.L. 1996. Internal eliminated segments (IESs) of *Oxytrichidae. J. Eukaryot. Microbiol.* **43:** 432–441.

Prescott J.D., DuBois M.L., and Prescott D.M. 1998. Evolution of the scrambled germline gene encoding alpha-telomere binding protein in three hypotrichous ciliates. *Chromosoma* **107:** 293–303.

Price C.M. 1990. Telomere structure in *Euplotes crassus*: Characterization of DNA-protein interactions and isolation of a telomere-binding protein. *Mol. Cell. Biol.* **10:** 3421–3431.

Price C.M. and Cech T.R. 1987. Telomeric DNA-protein interactions of *Oxytricha* macronuclear DNA. *Genes Dev.* **1:** 783–793.

Ray S., Jahn C., Tebeau C.M., Larson M.N., and Price C.M. 1999. Differential expression of linker histone variants in *Euplotes crassus*. *Gene* **231:** 15–20.

Ray S., Karamysheva Z., Wang L., Shippen D.E., and Price C.M. 2002. Interactions between telomerase and primase physically link the telomere and chromosome replication machinery. *Mol. Cell. Biol.* **22:** 5859–5868.

Romero D.P. and Blackburn E.H. 1991. A conserved secondary structure for telomerase RNA. *Cell* **67:** 343–353.

Roth M. and Prescott D.M. 1985. DNA intermediates and telomere addition during genome reorganization in *Euplotes crassus*. *Cell* **41:** 411–417.

Roth M., Lin M., and Prescott D.M. 1985. Large scale synchronous mating and the study of macronuclear development in *Euplotes crassus*. *J. Cell Biol.* **101:** 79–84.

Shippen-Lentz D. and Blackburn E.H. 1990. Functional evidence for an RNA template in telomerase. *Science* **247:** 546–552.

Skopp R., Wang W., and Price C. 1996. rTP: A candidate telomere protein that is associated with DNA replication. *Chromosoma* **105:** 82–91.

Sperger J.M. and Cech T.R. 2001. A stem-loop of *Tetrahymena* telomerase RNA distant from the template potentiates RNA folding and telomerase activity. *Biochemistry* **40:** 7005–7016.

Turkewitz A.P., Orias E., and Kapler G. 2002. Functional genomics: The coming of age for *Tetrahymena thermophila*. *Trends Genet.* **18:** 35–40.

Ulaner G.A., Hu J.F., Vu T.H., Giudice L.C., and Hoffman A.R. 1998. Telomerase activity in human development is regulated by human telomerase reverse transcriptase (hTERT) transcription and by alternate splicing of hTERT transcripts. *Cancer Res.* **58:** 4168–4172.

Vermeesch J.R. and Price C.M. 1994. Telomeric DNA sequence and structure following *de novo* telomere synthesis in *Euplotes crassus*. *Mol. Cell. Biol.* **14:** 554–566.

Vermeesch J.R., Williams D., and Price C.M. 1993. Telomere processing in *Euplotes*. *Nucleic Acids Res.* **21:** 5366–5371.

Wang H. and Blackburn E.H. 1997. De novo telomere addition by *Tetrahymena* telomerase in vitro. *EMBO J.* **16:** 866–879.

Wang H., Gilley D., and Blackburn E.H. 1998. A novel specificity for the primer-template pairing requirement in *Tetrahymena* telomerase. *EMBO J.* **17:** 1152–1160.

Wei C. and Price C.M. 2003. Protecting the terminus: T-loops and telomere end-binding proteins. *Cell. Mol. Life Sci.* **60:** 2283–2294.

Williams K.R., Doak T.G., and Herrick G. 2002. Telomere formation on macronuclear chromosomes of *Oxytricha trifallax* and *O. fallax*: Alternatively processed regions have multiple telomere addition sites. *BMC Genet.* **3:** 16.

Witkin K.L. and Collins K. 2004. Holoenzyme proteins required for the physiological assembly and activity of telomerase. *Genes Dev.* **18:** 1107–1118.

Yao M., Duharcourt S., and Chalker D.L. 2002. Genome-wide rearrangements of DNA in ciliates. In *Mobile DNA II* (ed. N.L. Craig et. al.), pp. 730–758. ASM Press, Washington, D.C.

Yao M.C., Fuller P., and Xi X. 2003. Programmed DNA deletion as an RNA-guided system of genome defense. *Science* **300:** 1581–1584.

Yao M.C., Zheng K., and Yao C.H. 1987. A conserved nucleotide sequence at the sites of developmentally regulated chromosomal breakage in *Tetrahymena*. *Cell* **48:** 779–788.

Yu G.L. and Blackburn E.H. 1991. Developmentally programmed healing of chromosomes by telomerase in *Tetrahymena. Cell* **67:** 823–832.

Yu G.L., Bradley J.D., Attardi L.D., and Blackburn E.H. 1990. *In vivo* alteration of telomere sequences and senescence caused by mutated *Tetrahymena* telomerase RNAs. *Nature* **344:** 126–132.

Zahler A.M. and Prescott D.M. 1988. Telomere terminal transferase activity in the hypotrichous ciliate *Oxytricha nova* and a model for replication of the ends of linear DNA molecules. *Nucleic Acids Res.* **16:** 6953–6972.

16

Fission Yeast Telomeres

Julia Promisel Cooper

Telomere Biology Laboratory
Cancer Research United Kingdom
London WC2A 3PX, United Kingdom

Yasushi Hiraoka

Kansai Advanced Research Center
National Institute of Information and Communications Technology
Kobe 651-2492, Japan

THE FISSION YEAST *SCHIZOSACCHAROMYCES POMBE* has a number of special characteristics that make it a convenient and illuminating model for studying telomere biology. First isolated from the cultures used to make beer in East Africa (*pombe* is a Swahili word for beer), *S. pombe* is a unicellular fungus that divides by medial fission. *S. pombe* is nearly as distant evolutionarily from budding yeast as it is from humans, having diverged from the budding yeast lineage more than 300 million years ago. The approximately 13.8-Mbp *S. pombe* genome contains about 4900 genes spread over only three chromosomes (Wood et al. 2002), housed in a nucleus that measures about 2 μm in diameter, centered in a 12-μm-long cell (Fig. 1). Facile transformation and integrative recombination provide an ease of genetic manipulation reminiscent of budding yeast. At the same time, fission yeast chromosome organization and dynamics are remarkably similar to those of humans. For example, fission yeast centromeres are large, complex, and heterochromatic, and fission yeast have a RNA interference (RNAi) pathway; both of these features, absent from *Saccharomyces cerevisiae*, are present in humans. Telomere protein organization is also highly reminiscent of that found in humans. This property, coupled with the small chromosome number that makes cytological studies and analysis of telomere fusions remarkably straightforward, has distinguished fission yeast as a paradigm for telomere research.

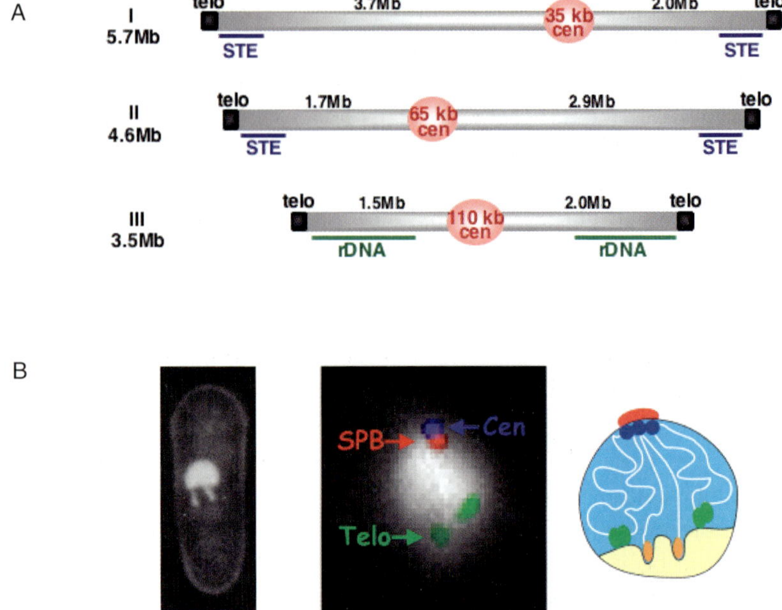

Figure 1. (*A*) Chromosome arrangement in *S. pombe.* (telo) The ~300 bp of telomere repeats (of the sequence shown in Fig. 3) found at each chromosome end; (STE) subtelomeric repeat elements (see text). (*B*) (*Left*) A fission yeast cell stained with DAPI. The nucleus is shaped like a plug, with the telomere-adjacent rDNA tracts at either end of chromosome III forming the "prongs" and extending into the nucleolus. During mitotic interphase, the centromeres cluster near the spindle pole body (SPB), and the telomeres form two to four clusters at the nuclear periphery (*middle*). (*Right*) An interpretive diagram of the fluorescence micrograph is shown.

TELOMERE COMPONENTS AND ARCHITECTURE

Telomeric and Subtelomeric Sequences

Each telomere contains approximately 300 bp of a degenerate repeat sequence in which the most frequently occurring motif is TTACAGG, with a consensus sequence of $TTAC(A)(C)G_{2-8}$. Inspection of several cloned telomeres (Fig. 2)(Sugawara 1988) shows that stretches of G·C longer than 2 bp are common, and indeed the only telomere-specific double-stranded DNA-binding protein characterized to date, Taz1, shows a preference for repeats containing three or more G·C bp (Cooper et al. 1997). As has been observed in budding yeast, an appreciable G-strand overhang signal is detectable by native gel analysis only during S phase in wild-type cells (Tomita et al. 2003). This analysis requires a chromosomal overhang of ≥30 nucleotides for detection. More sensitive techniques used to detect the shorter overhangs that persist throughout the cell cycle in budding yeast (Larrivee et al. 2004) have yet to be applied to fission yeast.

The subtelomeric regions at either end of chromosomes I and II contain at least 19 kbp of loosely conserved approximately 86 bp repeats, with modulation of the repeat sequence occurring inward toward the centromere (Sugawara 1988). This subtelomeric repeat region contains sequences homologous to the centromere repeats. Chromosome III harbors all of the rDNA repeats, which are located in two large blocks just proximal to either telomere of the chromosome, constituting about 1 Mbp of its 3.5 Mbp (Fig. 1A). Subtelomeric repeats may also be present in the region between the telomere and the rDNA at one end of chromosome III (Sugawara 1988). During interphase, all *S. pombe* telomeres localize to one to four clusters (usually one or two) at the nuclear periphery (Funabiki et al. 1993), whereas the centromeres cluster at the spindle pole body (SPB) (Fig. 1B). The telomeres of chromosome III reside at the border between the nucleolus and the nucleus, visible as "prongs" in the plug-shaped DAPI (4′,6-diamidino-2-phenylindole)-staining nucleus that penetrates the DAPI-excluding crescent containing the nucleolus (Fig. 1B).

Protein Components of the Telomere

Like chromosome ends in all eukaryotes, fission yeast telomeric DNA is bound by complexes of sequence-specific DNA-binding proteins and their interacting partners, as well as proteins that interact with DNA "ends" nonspecifically (Fig. 3). A current challenge in telomere biology is to understand how the activities of this latter group of proteins, which includes factors involved in general DNA breakage surveillance and repair, are both exploited and tightly controlled by the telomere-specific complex to promote telomere maintenance and protection, as discussed below.

The double-strand telomeric repeats bind Taz1, the only known ortholog of the mammalian telomere proteins TRF1 and TRF2 (Cooper et al. 1997; Li et al. 2000; Fairall et al. 2001). Like its higher eukaryotic counterparts, the 663-amino-acid Taz1 protein contains a carboxy-terminal helix-loop-helix motif that has homology with the Myb DNA-binding domain. These regions are commonly referred to as "Myb" domains, but they are even more homologous to homeodomain DNA-binding motifs, which contain not only recognition helices that contact the major groove of the telomere repeat, but also amino-terminal arms that penetrate the

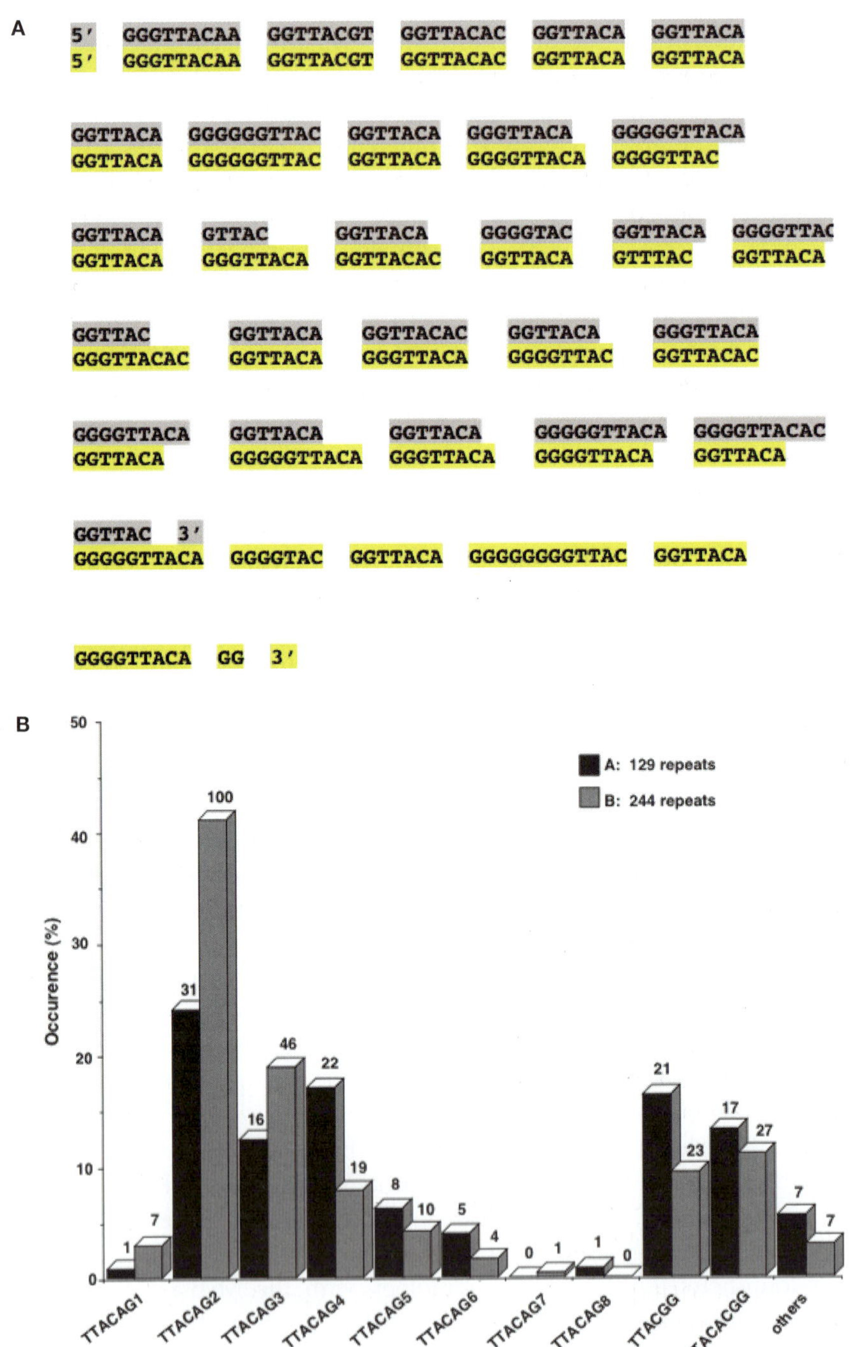

Figure 2. (*See facing page for legend.*)

minor groove of the T·A base pairs (Konig et al. 1998; Court et al. 2005). These homeodomains confer higher affinity than Myb domains because they make specific contacts with a longer sequence. Taz1 also shares a 227-amino-acid region of homology with the so-called telomere repeat factor-homology (TRFH) domain found in human TRF1 (hTRF1) and hTRF2 (Li et al. 2000); this region contains the dimerization domain of each protein, as well as regions that contact other proteins (Li et al. 2000). Taz1 is more homologous to TRF1 within both the 63-amino-acid Myb domain and the TRFH domain, but more homologous to TRF2 in the 207-amino-acid region between these domains (Fairall et al. 2001). Taz1 not only binds telomere repeats in vitro and in a one-hybrid assay (Cooper et al. 1997; Vassetzky et al. 1999; Spink et al. 2000), but also localizes to them in vivo (Chikashige and Hiraoka 2001; Kanoh and Ishikawa 2001; Nakamura et al. 2002; Hall et al. 2003; Tuzon et al. 2004) and controls a wide range of telomere functions (see below).

Two other open reading frames (ORFs) containing Myb-like domains have been identified in the *S. pombe* genome. One, termed SpX or Teb1, contains 2 Myb domains and displays specific binding in vitro to the human, but not the fission yeast, telomere repeat (Vassetzky et al. 1999). The other contains a single carboxy-terminal Myb domain and 28% homology (44% similarity) to budding yeast Tbf1 (Vassetzky et al. 1999; T. Simonson and D. Rhodes, in prep.); the functions of this protein are currently being analyzed.

Fission yeast Rap1 is recruited to telomeres by binding to Taz1. Like budding yeast and human Rap1, SpRap1 contains a BRCT (BRCA-1 carboxy-terminal) domain, but it lacks the so-called RCT (Rap1

Figure 2. Cloned telomere sequences from *S. pombe*. (*A*) The sequences of two cloned telomeres, one in *gray* and one in *yellow*. Note that the proximal 66 bp are conserved among different clones, whereas the distal telomere repeats vary in sequence. (*B*) Graph representing the occurrence (given as a number atop each bar) of each variant telomere repeat at cloned telomeres. *Black bars* labeled "A" are derived from the 129 repeats from five independent telomere clones (pSNU64, pSNU65, pSNU68, pSNU70, and pSNU71) (Sugawara 1988). *Gray bars* labeled "B" are derived from 244 repeats from two independent clones of *taz1*Δ telomeres of 754 bp and 1268 bp (Y. Chikashige and Y. Hiraoka, published in the public database DNA Data Bank of Japan [DDBJ]: accession numbers AB079786 and AB079787, respectively). These clones were generated from PCR products using genomic DNA from *taz1*Δ cells as template. The most frequent sequence is TTACAGG, with some variations in the number of Gs following TTACA. Note that the sequence TTAC or TTACAC always accompanies two G residues. (*A*, Reprinted, with permission, from Sugawara 1988.)

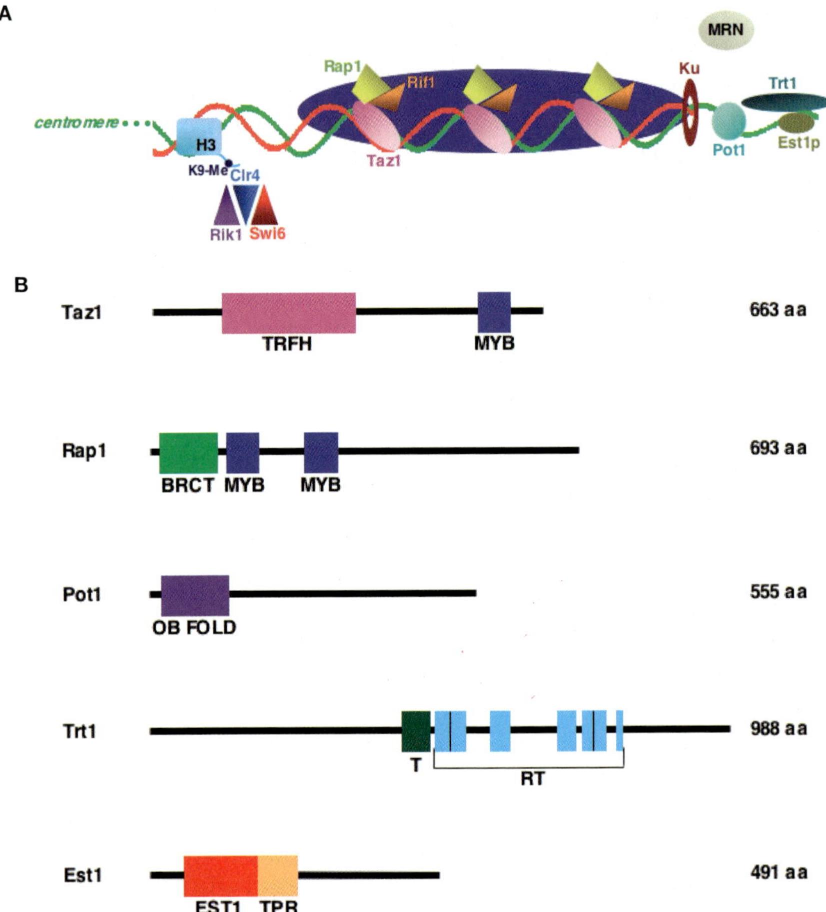

Figure 3. Components of fission yeast telomeres. (*A*) Diagram of fission yeast telomeres. Ku refers to the complex of SpKu70 and SpKu80. MRN refers to the complex of Rad32 (the fission yeast Mre11 homolog), Rad50, and Nbs1. The component of telomeres bound by MRN remains unknown. The heterochromatin proteins Swi6, Clr4, and Rik1 associate with subtelomeric repeats and may also associate with telomeres. (*B*) Domain organization of fission yeast telomere proteins. The locations of known domains within each protein are shown. (TRFH) TRF-homology domain; (Myb) a region homologous to the Myb-like DNA-binding domains, although those found in Rap1 do not appear to confer DNA-binding activity (see text for details); (BRCT) the Brca1 C-terminal motif; (RT) reverse trancriptase domains; (TPR) a tetratricopeptide repeat element.

carboxy-terminal) domain (Chikashige and Hiraoka 2001; Kanoh and Ishikawa 2001). BRCT motifs are found in numerous proteins involved in DNA repair and appear to serve as adaptors for protein–protein interactions. Unlike budding yeast Rap1, the two Myb-like motifs present in SpRap1 lack the ability to bind DNA. Instead, the recruitment of Rap1 to telomeres is largely dependent on Taz1, analogous to the situation in humans, in which Rap1 binds telomeres through protein contacts with TRF2 (Li et al. 2000; Li and de Lange 2003).

Like Rap1, fission yeast Rif1 was identified via homology with its budding yeast counterpart, but its behavior has diverged considerably from that of ScRif1 (Kanoh and Ishikawa 2001). The interaction of SpRif1 with telomeres is independent of Rap1, despite its acronym (*Rap1 interacting factor*). Rif1 can be detected at telomeres using chromatin immunoprecipitation assays, and this telomeric interaction depends on Taz1. When viewed by immunofluorescence, Rif1 appears diffusely localized through the nucleus in wild-type cells but is intensely localized to telomeres in cells lacking Rap1. This suggests that Rif1 and Rap1 may compete for binding to Taz1.

To date, the only known single-strand telomere-specific binding protein is Pot1, which contains an oligonucleotide/oligosaccharide binding (OB)-fold motif that has homology with those found in the *Oxytricha nova* telomere-binding protein and in budding yeast Cdc13 (Baumann and Cech 2001; Mitton-Fry et al. 2004). Biochemical and structural studies of the 185-amino-acid amino-terminal domain of Pot1 (Pot1^{DBD}) have shown that it binds single-stranded telomeric DNA cooperatively and with exceptionally high sequence specificity and a preference for repeats terminating in a 3′ end (Lei et al. 2002, 2003). However, full-length Pot1 displays diminished cooperativity and broader sequence-specificity, allowing it to bind multiple variants of the degenerate *S. pombe* telomere repeat (Trujillo et al. 2005). Pot1 variants containing point mutations that disrupt single-stranded DNA binding are unable to complement a *pot1Δ* phenotype (discussed below), suggesting that single-stranded DNA binding occurs in vivo and is critical for function, although SpPot1 may also resemble its human ortholog in making contacts with protein components of the double-strand telomere-binding complex (Loayza et al. 2004). The crystal structure of the Pot1^{DBD}/DNA complex is remarkable for the extensive base-pairing and -stacking interactions that Pot1 induces among residues within the nucleic acid single strand, forming a highly specialized single-stranded DNA conformation that confers exquisite specificity both for sequence and for DNA versus RNA (Lei et al. 2003).

The Ku heterodimer, composed of *S. pombe* Ku70 (SpKu70) and SpKu80, is found at telomeres as well as double-strand breaks (DSBs) in diverse eukaryotes including fission yeast (Baumann and Cech 2000;

Nakamura et al. 2002). Biochemical and crystallographic studies of Ku reveal that its two subunits form a ring that binds and encircles the end of a DNA molecule, spooling it through the ring (Walker et al. 2001). On the basis of these studies, it seems likely that Ku binds telomeric DNA by virtue of its inclusion of a DNA "end." Ku may also bind to telomeric proteins, as suggested by interactions between budding yeast Ku and Sir4 (Tsukamoto et al. 1997; Martin et al. 1999; Mishra and Shore 1999; Bertuch and Lundblad 2003). However, the absence of Ku at internal stretches of telomeric repeats (Miyoshi et al. 2003) suggests that any protein-mediated recruitment of fission yeast telomeric Ku must involve an end-restricted telomere protein, rather than a double-strand telomere-binding protein.

Telomerase-mediated Telomere Maintenance and Life without Telomeres

The catalytic protein subunit of telomerase, telomerase reverse transcriptase or Trt1, is a basic 116-kD protein with 30% identity (47% similarity) to human TERT within the seven reverse transcriptase domains and the so-called T domain (Nakamura et al. 1997). This gene was first isolated on the basis of its homology with the gene encoding the catalytic subunit of *Euplotes aediculatis* telomerase. $trt1^+$ contains 15 introns, the record-largest number detected in any *S. pombe* gene. Despite the observation that fission yeast telomerase activity is completely RNase-sensitive, the telomerase RNA subunit has yet to be identified.

Deletion of $trt1^+$ leads to a senescence phenotype reminiscent of those found in diverse eukaryotes, with survival of telomerase loss occurring via an unusual strategy made possible by the small chromosome number of *S. pombe* (Nakamura et al. 1998). For the first ≥50 generations following $trt1^+$ deletion, cells are healthy and grow as well as wild type; nonetheless, telomere length has started to decline progressively. Eventually (after ~100 generations), telomeres erode to a point where telomere sequences are undetectable by Southern hybridization. At this time, cell viability is severely reduced and many cells appear elongated, indicating that a checkpoint-mediated cell cycle arrest precedes cell death. However, a subpopulation of cells survives the loss of telomerase. Strikingly, these survivors typically contain circular chromosomes in which all telomere repeats and 5–7 kb of subtelomeric sequences have been lost (Nakamura et al. 1997, 1998). As telomeres disappear, so do the telomere-specific binding proteins that prevent further telomere shortening and chromosome end-joining. Since fission yeast contain only three chromosomes, a reasonable chance exists that each chromosome's opposing ends fuse to each other, rather than to an end of a different chromosome.

This chromosome circularization phenotype is seen following any genetic manipulation that results in a complete loss of telomeric DNA (discussed below). Following $trt1^+$ deletion, circularization is the primary survival mode when cells are cultured on solid media. However, when $trt1\Delta$ cells are cultured in liquid media, survivors are recovered in which telomeres of heterogeneous length persist, presumably via a recombinational mode of telomere maintenance similar to the ALT (alternative lengthening of telomeres) telomere maintenance pathway seen in some human telomerase-minus cells or the "type II" pathway used in budding yeast survivors. A likely explanation for the emergence of these linear chromosome-containing survivors in liquid is that although they are initially rare, they grow faster than cells harboring circular chromosomes and eventually overtake the culture. Deletion of $taz1^+$ greatly increases the prevalence of homologous recombination (HR)-based $trt1\Delta$ survival regardless of culture conditions (Nakamura et al. 1998; K. Miller et al., in prep.), reflecting the role of Taz1 in preventing telomeric HR (see below).

In addition to their slow growth rate, circular chromosome-containing cells display several other defects. They are hypersensitive to DNA-damaging agents, markedly cold-sensitive, and virtually unable to undergo successful meiosis (Naito et al. 1998; Miller and Cooper 2003; A. Hebden et al., in prep.). Current studies aim to decipher whether these problems stem from topological problems inherent to chromosome circularity or from a lack of telomere repeats.

Fission yeast Est1 was identified by its limited homology with budding yeast Est1, a protein that binds single-strand telomeric DNA and is absolutely required for telomerase activity in vivo (Chapter 12). Like its budding yeast counterpart, SpEst1 is essential for telomerase to function in vivo, with the $est1\Delta$ phenotype being virtually identical to a $trt1\Delta$ phenotype. Nonetheless, Est1 is dispensable for in vitro telomerase activity. Unlike budding yeast Est1, SpEst1 has so far not revealed the ability to bind DNA directly in vitro (Beernink et al. 2003).

Regulation of Telomere Length and 3' Overhang Formation

Telomere-length homeostasis is regulated by a large assemblage of proteins, as numerous mutations are known to result in telomere shortening and several in telomere lengthening. The general properties of telomere-length regulation in fission yeast suggest the existence of a *cis*-acting feedback network similar to that inferred from studies in budding yeast (Marcand et al. 1997; Teixeira et al. 2004) and humans (van Steensel and de Lange 1997), although equivalents of the elegant genetic studies of budding yeast telomere-length control have yet to be reported. Deletion of *taz1*[+] leads to a dramatic increase in telomere length and length heterogeneity (from ~250+/−50 bp in wild-type cells to 1–5 kbp in *taz1Δ* cells)(Cooper et al. 1997). As this telomere elongation occurs in a *trt1*[+] background but not a *trt1Δ* background (Nakamura et al. 1998), Taz1 regulates telomere length at least in part by controlling telomerase action and is likely to be a key component of the machinery that counts telomere repeats and exerts negative feedback control of synthesis of its own binding sites, analogous to budding yeast Rap1 or human TRF1.

Genetic studies of the Taz1-interacting proteins Rap1 and Rif1 show that they also participate in telomere-length control and that Taz1 has both negative and positive roles in telomere elongation (Kanoh and Ishikawa 2001; Miller et al. 2005). *rap1Δ* cells show dramatically increased telomere lengths that slightly exceed those of *taz1Δ* cells. However, double-mutant *rap1Δtaz1Δ* cells show telomere lengths equivalent to those of *taz1Δ* single mutants, suggesting that the excessive lengthening of *rap1Δ* telomeres requires Taz1. *rif1Δ* cells show a modest (about twofold) increase in telomere length, whereas *rif1Δtaz1Δ* telomeres are indistinguishable from *taz1Δ* telomeres (Kanoh and Ishikawa 2001). Interestingly, the longest fission yeast telomeres to date are seen in *rap1Δrif1Δ* double mutants, but the additional loss of Taz1 (in *rap1Δrif1Δtaz1Δ* triple mutants) restores a *taz1Δ* telomere length. Thus, the inhibition of telomere lengthening by Taz1 is exerted via multiple pathways, including a Rap1-dependent pathway and a separate Rif1-dependent pathway. A positive role for Taz1 is revealed by the restraint of telomere elongation seen upon loss of Taz1 from *rap1Δ* or *rap1Δrif1Δ* mutants. The positive effect of Taz1 on telomere elongation may stem from its role in promoting replication fork passage through telomeres (see below).

Unusually short but stable telomere lengths are seen upon deletion of numerous genes associated with DNA repair and checkpoint pathways. Loss of pKu70 or pKu80 leads to telomere shortening, whereas *lig4*[+] deletion has no effect on telomere length, indicating that Ku affects telomere length in a manner independent of its role in nonhomologous end-joining (NHEJ) (Baumann and Cech 2000; Miyoshi et al. 2003). It remains unknown whether fission yeast Ku makes specific contacts with the telomerase RNA subunit, which may confer a stimulatory effect on telomerase activity, as it does in budding yeast (Peterson et al. 2001; Bertuch and Lundblad 2003; Stellwagen et al. 2003) and possibly in humans (Ting et al. 2005).

As has been observed in budding yeast (Carson and Hartwell 1985; Adams and Holm 1996; Adams Martin et al. 2000), mutations in genes encoding primase subunits and DNA polymerases α and δ lead to varying degrees of telomerase-dependent telomere elongation (Dahlen et al. 2003). These mutations may confer defective coordination between telomerase and the lagging-strand synthesis machinery, leading to G-strand synthesis without sufficient C-strand synthesis to provide binding sites for the regulatory double-strand telomere-binding complex, as suggested for budding yeast (Diede and Gottschling 1999). Further analysis, including examination of the 3′ overhang in DNA replication mutants, will provide information on whether these ideas are applicable to fission yeast as well.

Telomeres shorten in cells lacking components of the so-called 9-1-1 complex (Rad1, Rad9, and Hus1) that is thought to form a proliferating cell nuclear antigen (PCNA)-like clamp around DNA at sites of damage, as well as in cells lacking the replication factor C (RFC)-like protein Rad17, which is thought to be involved in recruiting 9-1-1 to sites of DNA damage (Dahlen et al. 1998; Nakamura et al. 2002). The involvement of these DNA-repair factors in telomere-length regulation supports the idea that the pathway leading to telomerase engagement at chromosome ends shares features with the pathways engaged at damage-induced DNA ends. However, downstream players in the telomerase engagement pathway clearly differ from those in checkpoint activation, as downstream effector molecules such as Chk1 and Crb2 are dispensable for telomere maintenance (Dahlen et al. 1998; Matsuura et al. 1999; Nakamura et al. 2002), whereas telomerase is dispensable for checkpoint function (A. Hebden et al., unpubl.).

The theme of participation of the DNA repair/checkpoint machinery in telomere maintenance is dramatically illustrated by the effects of the ATM (ataxia telangiectasia-mutated) homologs on telomeres. Cells lacking the fission yeast ATR (ataxia telangiectasia-related) homolog, Rad3, exhibit short stable telomeres, whereas cells lacking the ATM homolog, Tel1, display normal telomere lengths (Nakamura et al. 2002). Likewise, Rad3 is required for full lengthening of *taz1Δ* telomeres and Tel1 is not. Strikingly, however, loss of both Rad3 and Tel1 leads to an immediate and complete loss of telomeric DNA with survival occurring only via chromosome circularization (Naito et al. 1998). Telomere loss and chromosome end-fusion also accompany simultaneous loss of budding yeast Mec1p (ATR) and Tel1p (ATM), leading to lethality (Craven et al. 2002; Mieczkowski et al. 2003). The fission yeast MRN complex, composed of Rad32 (the Mre11 homolog), Rad50, and Nbs1, acts in the same pathway as Tel1 for telomere-length maintenance, as deletion of any one of these genes has no effect on telomere length, whereas deletion of any in combination with Rad3 leads to abrupt telomere loss and chromosome circularization (Nakamura et al. 2002). Notably, the relative severity of telomeric defects in the Rad3 pathway versus the Tel1/MRN pathway is opposite to that observed in budding yeast, in which Tel1p/MRX has a more prominent role in telomere-length regulation than Mec1p. As viability declines much more rapidly upon simultaneous loss of Rad3 and Tel1 than upon $trt1^+$ deletion, $rad3\Delta tel1\Delta$ cells suffer an end-protection defect not found in $trt1\Delta$ cells. It remains unclear whether the need for either Rad3 or Tel1/MRN for telomere maintenance reflects not only a role in preventing rampant telomere attrition, but also a role in telomerase activation.

The length of the telomeric 3′ single-stranded overhang is still unknown, but several factors that govern the strength of the total 3′-overhang signal detected by native gel analysis have been identified. Although an overhang signal is detected only during S phase in wild-type cells, a persistent, slightly elevated signal is detected in $pku70\Delta$ cells. In $taz1\Delta$ cells, the 3′-overhang signal is even more intense and persists in G_1-arrested cells as well as through G_2 (Tomita et al. 2003; Ferreira and Cooper 2004).

Three nonmutually exclusive processes may contribute to 3′-overhang generation: (1) removal of the RNA oligonucleotide that primes synthesis of the terminal Okazaki fragment produced by lagging-strand replication, a process that, by definition, applies to only one end of each chromosome; (2) telomerase-mediated telomere repeat synthesis coupled with incomplete fill-in synthesis; and/or (3) nucleolytic processing of the chromosome end, akin to the 5′ resection seen at damage-induced DSBs. In human and budding yeast, nucleases generate overhangs at both chromosome ends (Wellinger et al. 1993, 1996; Makarov et al. 1997; Chai et al. 2005).

Overhang formation at wild-type fission yeast telomeres has not been analyzed in detail, but several studies of overhang formation in a $taz1\Delta$ background have shed light on the pathways that converge on dysfunctional telomeres. The copious $taz1\Delta$ overhang signal depends on both $trt1^+$ and DNA-processing activities (Tomita et al. 2003), indicating that at least processes 2 and 3 above contribute to $taz1\Delta$ overhang generation.

In this setting, interesting parallels and contrasts can be drawn with those processes that generate 3′ single-stranded overhangs at DSBs. Like DSB resection, the telomeric 3′ signal elevation in $taz1\Delta$ cells requires

MRN, but not the residues that confer Rad32 nuclease activity in vitro (Tomita et al. 2003). In the absence of MRN, the overhang signal diminishes, but it is restored upon deletion of *pku70*, suggesting that Ku prohibits the activity of a 5′ nuclease that is independent of MRN. In triple mutants missing Rad32, Ku, and Taz1, resection of genomic DSBs requires the 5′ to 3′ exonuclease ExoI, whereas resection of telomeric 5′ ends does not. Conversely, a mutation in the flap endonuclease, Dna2, confers defective telomeric 3′-overhang generation without affecting resection of genomic DSBs (Tomita et al. 2004). Thus, telomere 3′-overhang generation is distinct from the 5′ resection of DSBs in being controlled by Taz1, but telomeres made dysfunctional by *taz1*$^{+}$ deletion are also processed differently from DSBs. The foregoing data would be consistent with a model in which telomeric resection comprises unwinding by the MRN complex, followed by endonuclease cleavage by Dna2, with the timing and extent of the process being controlled by both Ku and Taz1. Interestingly, the role of MRN in preventing catastrophic telomere loss in *rad3Δ* cells may be independent of its role in 3′-overhang generation, as Tel1 is dispensable for overhang generation in *taz1Δ* cells, and Dna2 is required for overhang formation but not for telomere maintenance in a *rad3Δ* background (Tomita et al. 2004).

Human Pot1 appears to have a key role in telomere-length regulation, as it binds both the double-stranded telomere-binding complex and the single-stranded overhang, and expression of full-length or truncated forms of Pot1 results in dramatic alterations in telomere length (Colgin et al. 2003; Loayza and de Lange 2003). Budding yeast single-stranded telomeric overhangs are bound by Cdc13, which regulates not only telomerase access, but also the extent of 5′ resection (Garvik et al. 1995; Chandra et al. 2001; Pennock et al. 2001). SpPot1 is likely to serve at least some of these roles in fission yeast, based on its single-stranded DNA-binding activities, its OB-fold homology with Cdc13 and more extensive homology with hPot1, and the drastic telomere-loss phenotype of *pot1Δ* mutants (see below). The identification of *pot1* alleles that retain partial function will allow assessment of its roles in telomerase control and/or overhang maintenance.

Telomere Protection and Dysfunction

A central role of the telomere complex is to prevent chromosome ends from experiencing the same fates as damage-induced DSBs, a task made particularly complex by the telomeric localization and activation of numerous factors that process DSBs. For instance, elements of the DSB resection machinery are engaged at telomeres (see above), and the single-stranded DNA that they generate presumably provides a critical substrate for telomerase and a binding site for Pot1. However, excess resection would engender the loss of binding sites for double-stranded telomeric DNA-binding proteins, the consequent loss of their protective functions, and checkpoint activation. A breakdown in control of resection activities has been elegantly demonstrated for budding yeast *cdc13* mutants (Garvik et al. 1995) and may well be recapitulated in fission yeast *pot1Δ* mutants. These cells display abrupt and complete telomere attrition, with survival occurring only via chromosome circularization (Baumann and Cech 2001). Abrupt telomere attrition may occur via rampant nucleolytic degradation, chromosome breakage, or HR events that excise the telomere, coupled with an inability to engage compensatory telomerase activity. As is the case in humans, we are still in the early stages of understanding the factors that bind the telomeric overhang and the molecular events that follow their loss.

Nonetheless, fission yeast has been a particularly amenable model for studying other aspects of telomeric protection, a prominent example being protection from chromosome end-fusion, a function that is accomplished with remarkable similarity in fission yeast and humans. It is logical that end-fusions form in cells that have lost all telomeric DNA and, by extension, the entire protective telomere-binding complex. On the other hand, the identification of end-fusions in which telomeric DNA is retained flanking the fusion point provides information about the specific members of the telomere complex that prevent fusion. The identification of end-fusions is particularly straightforward in fission yeast as the small number of chromosome ends yields a finite set of discrete fusion bands on pulsed-field gels.

End-fusions that retain telomeric DNA are observed in G_1-arrested *taz1Δ* cells (Fig. 4) (Ferreira and Cooper 2001). These fusions are strictly dependent on the NHEJ machinery, as they are absent in *taz1Δ* cells lacking Ku or Lig4. Thus, although Ku has important roles in normal telomere metabolism (see above), its NHEJ activities need to be tightly restrained by Taz1. NHEJ-mediated telomere fusions are also seen in human cells lacking the function of the Taz1 ortholog, TRF2, and although budding yeast appears to lack a Taz1/TRF ortholog, inhibition of ScRap1 leads to telomere-telomere fusions, detectable using a polymerase chain reaction (PCR) assay (Pardo and Marcand 2005).

In contrast to *taz1Δ* telomere fusions, the intrachromosomal end-fusions that follow complete telomere attrition in *trt1Δ* and *pot1Δ* cells can form independently of Ku-dependent NHEJ and Rad22-dependent

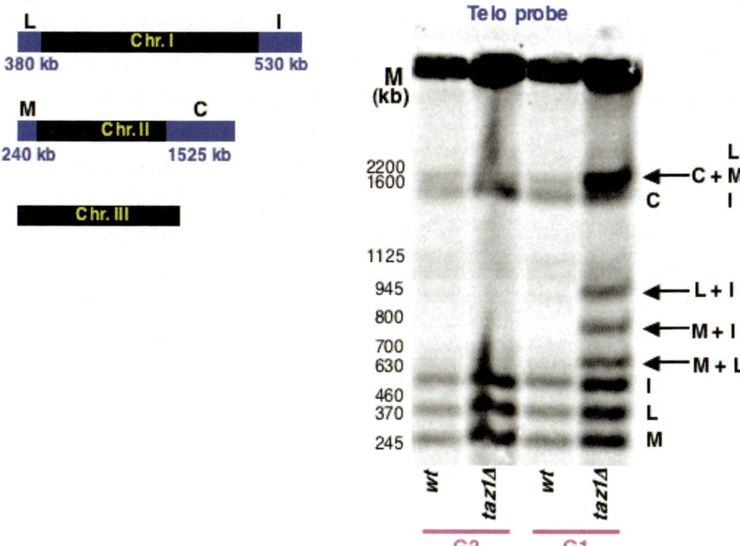

Figure 4. Telomere fusions occur during G_1 arrest in *taz1Δ* cells. (*Left*) Diagram of telomeric *Not*I restriction fragments on chromosomes I and II; chromosome III lacks *Not*I restriction sites. *Not*I digestion releases the four telomere fragments, L, M, I, and C, that can be resolved by pulsed-field gel electrophoresis (PFGE). (*Right*) Southern blot following PFGE of *Not*I-digested *S. pombe* DNA, probed with a telomeric oligonucleotide. Fusion bands are indicated by arrows to the right of the blot. (Reprinted, with permission, from Ferreira and Cooper 2001 [©Elsevier].)

HR (Baumann and Cech 2000; P. Baumann, pers. comm.), although the efficiency of end-joining may diminish in the absence of these pathways. This observation highlights the differences between chromosome ends lacking the double-stranded telomere-binding complex and those lacking all telomeric repeats. Pot1 and/or Ku may remain bound to *taz1Δ* telomeres and may block Ku-independent end-joining even though they cannot prevent Ku-dependent NHEJ.

A crucial and illuminating feature of *taz1Δ* telomere fusions is the restriction of their occurrence to G_1-arrested or HR-compromised cells (Ferreira and Cooper 2001). This restriction reflects cell cycle regulation of DNA repair and the primarily G_2 cell cycle of vegetatively growing fission yeast (Ferreira and Cooper 2004). *S. pombe* grow during G_2, and by the time they complete mitosis and cell division, they have reached the cell-size threshold that permits S-phase initiation. Thus, although fission yeast prefer haploidy, they lack a true G_1 phase, and FACS (fluorescence-activated cell sorting) analysis of log-phase fission yeast cultures confirms that virtually 100% of cells contain a 2C DNA content. The preponderance of G_2 character in fission yeast dictates the dominant DNA DSB repair mode. Instead of using NHEJ, cycling fission yeast favor HR (Ferreira and Cooper 2004). This makes teleological sense, since its utilization of genetic information from an intact sister makes HR highly accurate, whereas NHEJ is potentially mutagenic. However, in G_1-arrested haploid cells, HR is not a generally feasible option and indeed cells up-regulate NHEJ by approximately tenfold in G_1 versus G_2; correspondingly, HR levels are tenfold lower during G_1 than during G_2.

At telomeres, HR is not necessarily problematic, as HR between telomeres leads to exchanges of telomere repeats between different chromosome ends, not to end-fusions. Intramolecular HR between a chromosome end and more proximal telomere repeats can lead to telomere excision (Li and Lustig 1996; Wang et al. 2004), but any telomeres that undergo such excision reactions can presumably be readily resynthesized by deregulated telomerase activity in the absence of Taz1. Indeed, subtelomeric sequences are continuously rearranged in a Rhp51-dependent manner in *taz1Δ* but not in *taz1*[+] cells, indicating that *taz1Δ* telomeres undergo continuous HR (M.G. Ferreira and J.P. Cooper, in prep.). Thus, telomeres lacking Taz1 are vulnerable to DSB repair reactions and are subjected to the cell-cycle-dictated prevailing repair mode, HR in cycling cells and NHEJ in G_1-arrested cells. *taz1Δ* telomere fusions confer lethality upon resumption of growth following G_1 arrest, with viability being restored by deletion of *pku70*[+], *pku80*[+], or *lig4*. An interesting conundrum is how NHEJ joins *taz1Δ* telomeres given that they contain extensive 3′ overhangs during G_1 as well as during G_2 (Ferreira and Cooper 2004). Presumably, the NHEJ process includes either overhang removal or a fill-in step.

Cells lacking pKu70 or pKu80 also display elevated subtelomeric HR that depends on Rad22 (the fission yeast Rad52 ortholog) and Rhp51 but not Rad50 (Kibe et al. 2003). The persistent telomeric 3′ overhangs seen in both *pkuΔ* and *taz1Δ* cells may contribute to their elevated HR. Whether HR at dysfunctional telomeres is governed by a single or multiple mechanisms and what molecular events are involved are fascinating outstanding questions.

As is the case in human cells, the idea that dysfunctional telomeres are treated as DSBs is supported not only by their engagement of inappropriate DNA repair reactions, but also by their recruitment of DNA damage foci (Takai et al. 2003; M.G. Ferreira and J.P. Cooper, in prep.). Foci containing Rad22 and Rad11 (the RPA ortholog) are present at *taz1Δ* telomeres throughout the cell cycle. Nonetheless, *taz1Δ* cells do not activate the DNA-damage checkpoint when grown under optimal conditions. Thus, a telomere-specific factor that remains present at *taz1Δ* telomeres may block checkpoint activation while allowing activation of DNA-repair activities, or alternatively, the perceived level of DNA damage in *taz1Δ* cells may be sufficient to activate repair pathways but not cell-cycle-arrest pathways.

The mainly G_2 cell cycle and resulting predominance of HR allows *taz1Δ* cells to thrive despite having profoundly dysfunctional telomeres, permitting a study of the outcomes of telomere dysfunction without the inconvenience of lethality due to end-fusions. Against this backdrop, examining these cells through the cell cycle reveals challenges at each stage. Intriguingly, during S phase, an excess of stalled replication forks is seen at *taz1Δ* telomeres, suggesting that unprotected telomeres hinder fork progression (K. Miller et al., in prep.). This could stem from the telomere sequence itself; perhaps the G-rich repeats are difficult to unwind or prone to spurious intra- or interstrand base-pairing interactions once unwound. Such difficulties have been invoked to explain defective telomere replication in mammalian cells lacking the RTEL or WRN helicases (Crabbe et al. 2004; Ding et al. 2004b). Stalled replication forks at *taz1Δ* telomeres may contribute to their hyperrecombination as well as triggering strand breakage. Indeed, loss of Trt1 from *taz1Δ* cells leads to an abrupt decline in telomere length (Nakamura et al. 1998; K. Miller et al., in prep.), suggesting that telomere breakage events are common in *taz1Δ* cells but are normally masked by robust telomerase activity.

Telomeric replication fork stalling may be the basis for the cold sensitivity of *taz1Δ* cells. At temperatures of ≤20°C, they display abnormal mitoses with anaphase bridges and fragmented DNA (Miller and Cooper 2003). The appearance of these mitotic defects requires that the preceding S phase occurred at 20°C; if *taz1Δ* cells are transferred to the cold at G_2/M, the first mitosis is normal and aberrant patterns of chromosome segregation appear only in the second mitosis, whereas transfer to 20°C at G_1/S confers segregation defects at the first mitosis. The DNA in these cells forms entangled masses and DSBs, and both the DNA damage and spindle assembly checkpoints are activated. A speculative scheme to explain these phenotypes is as follows: As replication forks approach *taz1Δ* telomeres, they stall. At the stalled forks, the unwound and unreplicated G-rich DNA strands invade sister or nonsister duplexes, generating the entangled masses of DNA. The entanglement can be resolved at normal growing temperatures, but not at 20°C.

Of the Taz1-interacting proteins, Rap1 mediates protection from telomere fusions, whereas Rif1 is dispensable for this function (Miller et al. 2005). However, *rap1Δ* cells display neither cold sensitivity, pronounced telomeric fork pausing, nor elevated levels of HR-based survivors of Trt1 loss. Thus, some but not all of the functions of Taz1 are mediated by recruitment of Rap1. To address this issue directly, a chimeric protein was engineered in which Rap1 was fused to the carboxy-terminal 167 amino acids of Taz1 (which includes the 63-amino-acid DNA-binding domain)(Chikashige and Hiraoka 2001). This fusion protein localizes to telomeres and largely restores meiotic telomere function (see below) in *taz1Δ rap1Δ* strains, suggesting that a key function of Taz1 during meiosis is to recruit Rap1. However, the same cells have elongated telomeres, extensive 3′-overhang signals, and fusions during G_1 arrest (Miller et al. 2005). Hence, ectopic recruitment of Rap1 (along with the amino terminus of Taz1) cannot confer correct telomere length regulation or prevention of end-fusion.

Intriguingly, Rap1 and Rif1 can also function independently of Taz1. Rap1 shows a residual localization to telomeres in *taz1Δ* cells, binding to at least one of the 12 telomeres in one-third of all zygotes (Chikashige and Hiraoka 2001; Kanoh and Ishikawa 2001). Furthermore, deletion of $rap1^+$ markedly enhances the cold sensitivity of *taz1Δ* cells, indicating that in the absence of Taz1, Rap1 protects cells from telomeric entanglement and its consequences. Strikingly, however, $rif1^+$ deletion suppresses the cold sensitivity of *taz1Δ* cells. As Rif1 shows only a moderate enrichment at telomeres (see above) and is absent from *taz1Δ* telomeres, it may be a *trans*-acting factor that affects telomere function indirectly, perhaps by regulating DNA damage checkpoint or repair responses in a manner akin to the role of human RIF1 (Silverman et al. 2004).

Higher-Order Telomere Structure and Telomeric Silencing

The telomere is one of the most prominent regions of heterochromatin in *S. pombe*, along with the centromere and the silent mating-type locus. Genes placed adjacent to telomeres are transcriptionally repressed (Nimmo et al. 1994), as are telomere-proximal genes in budding yeast and humans. This repression requires Taz1 (Cooper et al. 1997), Rap1 (Chikashige and Hiraoka 2001; Kanoh and Ishikawa 2001), and a group of heterochromatin proteins including Swi6, Rik1, Clr1, Clr2, Clr3, Clr4, and Csp4 (Allshire et al. 1995). Several of these factors have been detected at the subtelomeric repeat region by chromatin immunoprecipitation. Although Taz1 can be detected up to 10 kb from the telomere, the HP1 (heterochromatin protein 1) homolog, Swi6, appears to bind the terminal 50 kb of the chromosome (J. Kanoh and F. Ishikawa, pers. comm.). The association of Taz1 with subtelomeric repeats is presumably through protein-protein interactions, since telomere repeat tracts are absent from subtelomeric and internal regions (although 13-bp stretches of telomeric sequence are present in the rDNA)(Sugawara 1988). Swi6 is associated with transcriptionally silent regions, and its subtelomeric localization is consistent with gene expression levels determined by DNA microarray analysis (Y. Chikashige and Y. Hiraoka, unpubl.). Swi6 binding requires two other heterochromatin proteins, Clr4 and Rik1 (Ekwall et al. 1996; Nakayama et al. 2000). Clr4 contains a SET domain and is a methyltransferase that modifies histone H3 (Nakayama et al. 2001). Rik1 recruits the Clr4 protein to chromatin, and modification of histone H3 by Clr4 triggers binding of Swi6, which polymerizes to spread heterochromatin.

At centromeres and the mating-type locus, the RNAi machinery (Ago1, Dcr1, and Rpd1) drives the formation of heterochromatin (Hall et al. 2002; Schramke and Allshire 2003; Volpe et al. 2003). The RNAi machinery is dispensable for telomeric silencing, but it may nonetheless be present at telomeres, as its inhibition results in some telomeric alterations. During interphase, the RNAi mutants show dispersal of telomeric signals; although telomeres localize to two to four spots at the nuclear periphery in wild-type cells, five to six peripheral telomere spots are seen in $ago1\Delta$, $dcr1\Delta$, or $rpd1\Delta$ mutants. Thus, an RNA component may contribute to the "glue" that holds telomeres together (Hall et al. 2003).

Ku is dispensable for telomeric silencing in fission yeast (Baumann and Cech 2000; Manolis et al. 2001). This contrasts with Ku's key role in silencing genes at a subset of budding yeast telomeres (see Chapters 10 and 12). These distinctions are consistent with the observation that unlike budding yeast Ku, fission yeast Ku does not appear to have a role in the peripheral localization of telomeres, nor does it localize to internal telomeric repeats, despite the fact that these internal telomere sequences confer Taz1-dependent silencing of nearby genes (Miyoshi et al. 2003).

Micrococcal nuclease (MNase) footprinting studies indicate that fission yeast telomeres are packaged into an unusual chromatin structure, pronouncedly distinct from that of canonical nucleosomal arrays (Chikashige et al. 1989; C. Tuzon et al., in prep.). Although the telomere repeats themselves comprise only the terminal approximately 300 bp of chromosomal DNA, the terminal kilobase of DNA is protected from MNase digestion. This protected structure depends on the presence of Taz1, as it disappears in cells lacking $taz1^{+}$ (C. Tuzon et al., in prep.).

In contrast, it remains intact upon deletion of $rik1^+$, indicating that histone H3 methylation is not required for the telomere chromatin footprint. Hence, telomeric chromatin appears to be subject to at least two levels of organization: First, histones at or adjacent to telomeres are subject to heterochromatin-associated modification. Second, Taz1-bound telomeric chromatin appears to fold into an MNase-resistant structure, regardless of the local histone modification state. The molecular nature of this structure remains unknown. Conceivably, it may represent a specific nucleosome-packing pattern in which linker DNA is occluded by Taz1 binding or, alternatively, a t-loop of discrete size in which the linker DNA forming the loop is hidden from MNase.

By analogy with the t-loop structure of telomeres isolated from human cells (Griffith et al. 1999; Nikitina and Woodcock 2004), a t-loop model for fission yeast telomeres has been explored in a recent in vitro study (Tomaska et al. 2004). t-loops were observed by electron microscopy upon addition of recombinant Taz1 to a telomeric repeat array composed of GGTTAC repeats. Interestingly, Taz1 itself appeared as a hexameric ring wrapping around the DNA. It also bound preferentially to the single-stranded overhang, despite its clear preference for double-stranded DNA in previous binding assays (Vassetzky et al. 1999; A. Deshpande and J.P. Cooper, unpubl.). It will be informative to extend this approach to study the interaction between Taz1 and a native telomere sequence that it binds with higher affinity and to apply electron microscopy analysis to telomeres isolated from live fission yeast.

Meiotic Telomere Function

Clustering of telomeres in meiosis, or bouquet formation, is observed in a wide variety of eukaryotes. *S. pombe* shows a striking example of telomere clustering during meiotic prophase. The nucleus oscillates between the cell poles throughout meiotic prophase, and telomeres remain clustered and associated with the spindle pole body (SPB, the fission yeast centrosome) at the leading edge of the moving nucleus (Fig. 5). Oscillatory movement of the nucleus is mediated by microtubules (Ding et al. 1998) with cytoplasmic dynein as a motor protein (Yamamoto et al. 1999). Protein components involved in the meiotic telomere-SPB cluster and nuclear movements are summarized in Figure 6.

Studies of mutants have demonstrated that telomere clustering and the subsequent nuclear movements promote the pairing of homologous chromosomes (Shimanuki et al. 1997; Cooper et al. 1998; Nimmo et al. 1998; Yamamoto et al. 1999; Chikashige and Hiraoka 2001; Kanoh and

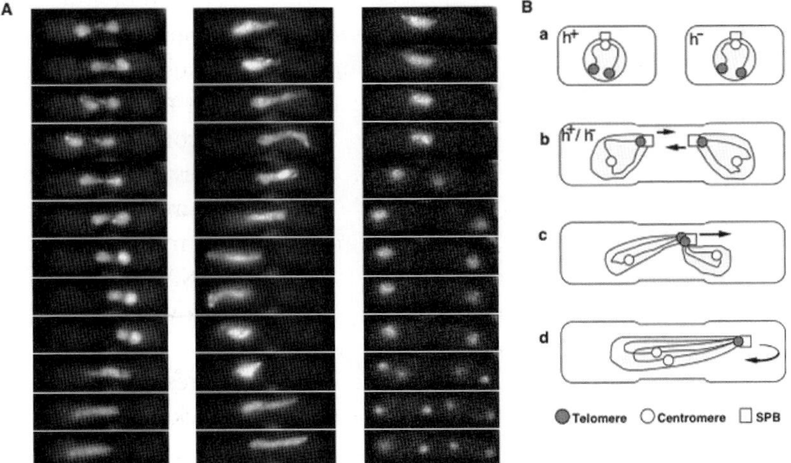

Figure 5. Telomere clustering and nuclear movements in meiosis. (*A*) Nuclear movements in living meiotic cells during karyogamy (*left column*), the horse-tail period (*middle column*), and meiotic divisions (*right column*). DNA was stained with Hoechst 33342. (*B*) Upon nitrogen starvation, haploid cells of the opposite mating type, h^+ and h^-, conjugate with each other and enter meiosis. (*a*) At the time of meiotic induction, centromeres form a single cluster near the SPB and telomeres are separated from the SPB. (*b*) In the haploid nucleus of the conjugated zygote, telomeres form a cluster close to the SPB and centromeres become separated from the SPB. (*c*) During nuclear fusion, the telomere is the site of initial contact between homologous chromosomes. After nuclear fusion, the SPB moves back and forth between the cell ends, pulling the entire nucleus. (*d*) In the oscillating nucleus, homologous sets of chromosomes attain corresponding orientations of polarized telomere-centromere configuration. (*A*, Reprinted, with permission, from Chikashige et al. 1994 [©AAAS].)

Ishikawa 2001; Ding et al. 2004a). In particular, loss of Taz1 or Rap1 results in defective telomere clustering and meiosis. It is important to note that these mutants suffer not only a telomere clustering defect, but also telomere fusions during the G_1-arrest period that necessarily precedes meiosis (Ferreira and Cooper 2001; Tuzon et al. 2004). Nonetheless, molecular fusion of Rap1 with the carboxy-terminal 167 amino acids of Taz1 containing its DNA-binding domain largely complements both the defective telomere clustering and the low spore viability of *taz1Δrap1Δ* cells (Chikashige and Hiraoka 2001). Thus, binding of Rap1 to the telomere is a key requirement for telomere clustering at the SPB and progression of meiosis.

The importance of telomeres for successful meiosis is also supported by another line of evidence. Loss of telomerase (Trt1), Rad3/Tel1,

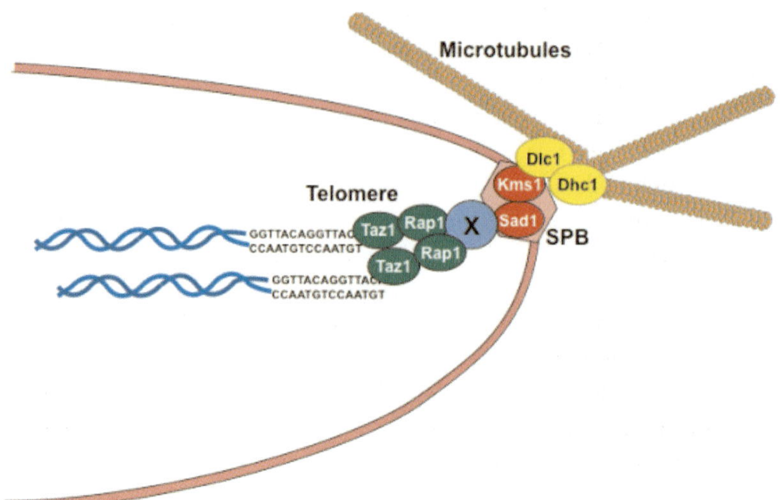

Figure 6. The meiotic telomere-SPB complex. The meiotic SPB acts as a nucleation center for astral microtubules, which mediate oscillatory nuclear movement during meiotic prophase (Chikashige et al. 1994; Ding et al. 1998). The Taz1 and Rap1 proteins are constitutive telomere components (see text). The Sad1 and Kms1 proteins are both constitutive SPB components (Hagan and Yanagida 1995). Kms1 is specifically required for meiosis; loss of Kms1 results in dispersal of SPB and telomere signals, reduced levels of homologous recombination and low spore viability (Shimanuki et al. 1997). Dhc1 (dynein heavy chain) and Dlc1 (dynein light chain) constitute the dynein motor protein complex, which drives the nuclear oscillation during meiotic prophase (Yamamoto et al. 1999; Miki et al. 2002). Although Dlc1 is constitutively expressed, Dhc1 is expressed only in meiosis. Loss of Dhc1 or Dlc1 results in reduced frequency of homologous recombination and low spore viability. The proteins that connect Rap1-bound telomeres to the SPB remain unknown.

or Pot1 leads to loss of the telomeric repeats and chromosome circularization as described above. Although those cells with circularized chromosomes are viable during the mitotic cell cycle, they show severe defects in meiosis, being devoid of telomere clustering and viable spore formation in most cases (Naito et al. 1998). In some cases, however, the circular chromosomes cluster to the SPB through Taz1-dependent heterochromatin that is maintained on subtelomeric repeats even after the loss of telomeric repeats (Sadaie et al. 2003). In this setting, the efficiency of homologous pairing of circularized chromosomes correlates with the efficiency of association between Taz1-bound subtelomeric repeats and the SPB, supporting the idea that telomere-SPB clustering is critical for homolog pairing.

The foregoing results show that the telomeric repeat DNA is not absolutely necessary for telomere clustering once heterochromatin structures are formed. Recently, it has been demonstrated that the Rik1 and Clr4 heterochromatin proteins, but not Swi6, are required for meiotic telomere clustering (Tuzon et al. 2004). Thus, a Swi6-independent function of heterochromatin operates at telomeres during meiosis. Since heterochromatin function at centromeres and the silent mating-type locus requires Swi6 as well as Rik1 and Clr4, this function of heterochromatin at telomeres is distinct from that at centromeres and the mating-type loci. Mutants in the RNAi pathway also show a small defect in meiotic, as well as mitotic, telomere clustering (see above).

Candidate protein components that tether telomeres to the SPB are only beginning to emerge. Although Rap1 is a key component for telomere clustering, it does not interact directly with Sad1, a constitutive SPB component, or with Kms1, a meiosis-specific SPB component (Fig. 6). To find proteins connecting telomeres to the SPB, proteins that interact with Sad1 and Kms1 have been sought by a yeast two-hybrid assay; 26 and 7 gene products were identified to interact with Sad1 and Kms1, respectively. However, loss of these proteins did not affect telomere-SPB clustering (Miki et al. 2004). Another potential participant in this process, Ccq1, was identified in complexes containing the SPB component Pcp1 (Flory et al. 2004). Ccq1 localizes to telomeres, and zygotes containing only one copy of $ccq1^+$ show dispersion of both Ccq1 and Taz1 signals, raising the possibility that Ccq1 may be important for formation of a stable meiotic telomere-SPB complex.

As telomere clustering is induced by the mating pheromone response (Chikashige et al. 1997; Yamamoto et al. 2004), telomere-SPB connectors might be expected among the proteins that are expressed under mating pheromone signaling. A genome-wide search for such proteins found two novel proteins, Bqt1 and Bqt2, that connect Rap1 to Sad1 when mating pheromone signaling is induced (Y. Chikashige and Y. Hiraoka, in prep.). Neither Bqt1 nor Bqt2 alone functions as a connector, but the two proteins together form a bridge between Rap1 and Sad1. Thus, Bqt1 and Bqt2 may regulate the transition between mitotic and meiotic telomere complexes.

SUMMARY

S. pombe provides a nearly complete suite of homologs of human telomere proteins combined with the formidable power of yeast genetics. The virtual lack of a G_1 stage in the fission yeast cell cycle and the resulting dearth of NHEJ permit viability in the presence of mutations that lead to inappropriate DNA-repair reactions at telomeres. These observations have highlighted the key role that cell cycle stage has in dictating the outcome of telomere deprotection. The ability of fission yeast to survive telomerase loss by circularizing its chromosomes provides an unusual opportunity to study eukaryotic life without telomeres. The presence of histone-modifying factors and the RNAi machinery at chromosome ends allows dissection of the heterochromatic nature of telomeres. The prominent role of telomeres during meiotic prophase in *S. pombe* makes it an ideal organism in which to probe the contribution of telomeres to the dynamics of homolog pairing and chromosome segregation during sexual development and meiosis. At the same time, huge gaps remain in our knowledge of *S. pombe* telomeres, from the identity of the telomerase RNA to the mechanisms of telomerase activation to the three-dimensional structure of fission yeast telomeres. As these gaps become filled in, the principles uncovered in fission yeast will likely continue to provide hints about telomere function in humans.

ACKNOWLEDGMENTS

We thank all members of the Cooper and Hiraoka laboratories for discussions over the years; T. Nakamura and C. Pitt for help with Figure 3B; and P. Baumann, F. Ishikawa, and J. Kanoh for personal communications. Current research in the Cooper laboratory is supported by Cancer Research UK and in the Hiraoka laboratory by National Institute of Information and Communications Technology and the Japan Science and Technology Agency.

REFERENCES

Adams A.K. and Holm C. 1996. Specific DNA replication mutations affect telomere length in *Saccharomyces cerevisiae*. *Mol. Cell. Biol.* **16:** 4614–4620.

Adams Martin A., Dionne I., Wellinger R.J., and Holm C. 2000. The function of DNA polymerase alpha at telomeric G tails is important for telomere homeostasis. *Mol. Cell. Biol.* **20:** 786–796.

Allshire R.C., Nimmo E.R., Ekwall K., Javerzat J.P., and Cranston G. 1995. Mutations derepressing silent centromeric domains in fission yeast disrupt chromosome segregation. *Genes Dev.* **9:** 218–233.

Baumann P. and Cech T.R. 2000. Protection of telomeres by the Ku protein in fission yeast. *Mol. Biol. Cell* **11:** 3265–3275.

———. 2001. Pot1, the putative telomere end-binding protein in fission yeast and humans. *Science* **292:** 1171–1175.

Beernink H.T., Miller K., Deshpande A., Bucher P., and Cooper J.P. 2003. Telomere maintenance in fission yeast requires an Est1 ortholog. *Curr. Biol.* **13:** 575–580.

Bertuch A.A. and Lundblad V. 2003. The Ku heterodimer performs separable activities at double-strand breaks and chromosome termini. *Mol. Cell. Biol.* **23:** 8202–8215.

Carson M.J. and Hartwell L. 1985. CDC17: An essential gene that prevents telomere elongation in yeast. *Cell* **42:** 249–257.

Chai W., Shay J.W., and Wright W.E. 2005. Human telomeres maintain their overhang length at senescence. *Mol. Cell. Biol.* **25:** 2158–2168.

Chandra A., Hughes T.R., Nugent C.I., and Lundblad V. 2001. Cdc13 both positively and negatively regulates telomere replication. *Genes Dev.* **15:** 404–414.

Chikashige Y. and Hiraoka Y. 2001. Telomere binding of the Rap1 protein is required for meiosis in fission yeast. *Curr. Biol.* **11:** 1618–1623.

Chikashige Y., Ding D.Q., Imai Y., Yamamoto M., Haraguchi T., and Hiraoka Y. 1997. Meiotic nuclear reorganization: Switching the position of centromeres and telomeres in the fission yeast *Schizosaccharomyces pombe*. *EMBO J.* **16:** 193–202.

Chikashige Y., Ding D.Q., Funabiki H., Haraguchi T., Mashiko S., Yanagida M., and Hiraoka Y. 1994. Telomere-led premeiotic chromosome movement in fission yeast. *Science* **264:** 270–273.

Chikashige Y., Kinoshita N., Nakaseko Y., Matsumoto T., Murakami S., Niwa O., and Yanagida M. 1989. Composite motifs and repeat symmetry in *S. pombe* centromeres: Direct analysis by integration of *Not*I restriction sites. *Cell* **57:** 739–751.

Colgin L.M., Baran K., Baumann P., Cech T.R., and Reddel R.R. 2003. Human POT1 facilitates telomere elongation by telomerase. *Curr. Biol.* **13:** 942–946.

Cooper J.P., Watanabe Y., and Nurse P. 1998. Fission yeast Taz1 protein is required for meiotic telomere clustering and recombination. *Nature* **392:** 828–831.

Cooper J.P., Nimmo E.R., Allshire R.C., and Cech T.R. 1997. Regulation of telomere length and function by a Myb-domain protein in fission yeast. *Nature* **385:** 744–747.

Court R., Chapman L., Fairall L., and Rhodes D. 2005. How the human telomeric proteins TRF1 and TRF2 recognize telomeric DNA: A view from high-resolution crystal structures. *EMBO Rep.* **6:** 39–45.

Crabbe L., Verdun R.E., Haggblom C.I., and Karlseder J. 2004. Defective telomere lagging strand synthesis in cells lacking WRN helicase activity. *Science* **306:** 1951–1953.

Craven R.J., Greenwell P.W., Dominska M., and Petes T.D. 2002. Regulation of genome stability by TEL1 and MEC1, yeast homologs of the mammalian ATM and ATR genes. *Genetics* **161:** 493–507.

Dahlen M., Sunnerhagen P., and Wang T.S. 2003. Replication proteins influence the maintenance of telomere length and telomerase protein stability. *Mol. Cell. Biol.* **23:** 3031–3042.

Dahlen M., Olsson T., Kanter-Smoler G., Ramne A., and Sunnerhagen P. 1998. Regulation of telomere length by checkpoint genes in *Schizosaccharomyces pombe*. *Mol. Biol. Cell* **9:** 611–621.

Diede S.J. and Gottschling D.E. 1999. Telomerase-mediated telomere addition in vivo requires DNA primase and DNA polymerases alpha and delta. *Cell* **99:** 723–733.

Ding D.Q., Chikashige Y., Haraguchi T., and Hiraoka Y. 1998. Oscillatory nuclear movement in fission yeast meiotic prophase is driven by astral microtubules, as revealed by continuous observation of chromosomes and microtubules in living cells. *J. Cell Sci.* **111:** 701–712.

Ding D.Q., Yamamoto A., Haraguchi T., and Hiraoka Y. 2004a. Dynamics of homologous chromosome pairing during meiotic prophase in fission yeast. *Dev. Cell* **6:** 329–341.

Ding H., Schertzer M., Wu X., Gertsenstein M., Selig S., Kammori M., Pourvali R., Poon S., Vulto I., Chavez E., Tam P.P., Nagy A., and Lansdorp P.M. 2004b. Regulation of murine telomere length by Rtel: An essential gene encoding a helicase-like protein. *Cell* **117:** 873–886.

Ekwall K., Nimmo E.R., Javerzat J.P., Borgstrom B., Egel R., Cranston G., and Allshire R. 1996. Mutations in the fission yeast silencing factors clr4+ and rik1+ disrupt the localisation of the chromo domain protein Swi6p and impair centromere function. *J. Cell Sci.* **109:** 2637–2648.

Fairall L., Chapman L., Moss H., de Lange T., and Rhodes D. 2001. Structure of the TRFH dimerization domain of the human telomeric proteins TRF1 and TRF2. *Mol. Cell* **8:** 351–361.

Ferreira M.G. and Cooper J.P. 2001. The fission yeast Taz1 protein protects chromosomes from Ku-dependent end-to-end fusions. *Mol. Cell* **7:** 55–63.

———. 2004. Two modes of DNA double-strand break repair are reciprocally regulated through the fission yeast cell cycle. *Genes Dev.* **18:** 2249–2254.

Flory M.R., Carson A.R., Muller E.G., and Aebersold R. 2004. An SMC-domain protein in fission yeast links telomeres to the meiotic centrosome. *Mol. Cell* **16:** 619–630.

Funabiki H., Hagan I., Uzawa S., and Yanagida M. 1993. Cell cycle-dependent specific positioning and clustering of centromeres and telomeres in fission yeast. *J. Cell Biol.* **121:** 961–976.

Garvik B., Carson M., and Hartwell L. 1995. Single-stranded DNA arising at telomeres in *cdc13* mutants may constitute a specific signal for the RAD9 checkpoint. *Mol. Cell. Biol.* **15:** 6128–6138.

Griffith J.D., Comeau L., Rosenfield S., Stansel R.M., Bianchi A., Moss H., and de Lange T. 1999. Mammalian telomeres end in a large duplex loop. *Cell* **97:** 503–514.

Hagan I. and Yanagida M. 1995. The product of the spindle formation gene *sad1+* associates with the fission yeast spindle pole body and is essential for viability. *J. Cell Biol.* **129:** 1033–1047.

Hall I.M., Noma K., and Grewal S.I. 2003. RNA interference machinery regulates chromosome dynamics during mitosis and meiosis in fission yeast. *Proc. Natl. Acad. Sci.* **100:** 193–198.

Hall I.M., Shankaranarayana G.D., Noma K., Ayoub N., Cohen A., and Grewal S.I. 2002. Establishment and maintenance of a heterochromatin domain. *Science* **297:** 2232–2237.

Kanoh J. and Ishikawa F. 2001. spRap1 and spRif1, recruited to telomeres by Taz1, are essential for telomere function in fission yeast. *Curr. Biol.* **11:** 1624–1630.

Kibe T., Tomita K., Matsuura A., Izawa D., Kodaira T., Ushimaru T., Uritani M., and Ueno M. 2003. Fission yeast Rhp51 is required for the maintenance of telomere structure in the absence of the Ku heterodimer. *Nucleic Acids Res.* **31:** 5054–5063.

Konig P., Fairall L., and Rhodes D. 1998. Sequence-specific DNA recognition by the myb-like domain of the human telomere binding protein TRF1: A model for the protein-DNA complex. *Nucleic Acids Res.* **26:** 1731–1740.

Larrivee M., LeBel C., and Wellinger R.J. 2004. The generation of proper constitutive G-tails on yeast telomeres is dependent on the MRX complex. *Genes Dev.* **18:** 1391–1396.

Lei M., Baumann P., and Cech T.R. 2002. Cooperative binding of single-stranded telomeric DNA by the Pot1 protein of *Schizosaccharomyces pombe*. *Biochemistry* **41:** 14560–14568.

Lei M., Podell E.R., Baumann P., and Cech T.R. 2003. DNA self-recognition in the structure of Pot1 bound to telomeric single-stranded DNA. *Nature* **426:** 198–203.

Li B. and de Lange T. 2003. Rap1 affects the length and heterogeneity of human telomeres. *Mol. Biol. Cell* **14:** 5060–5068.

Li B. and Lustig A.J. 1996. A novel mechanism for telomere size control in *Saccharomyces cerevisiae*. *Genes Dev.* **10:** 1310–1326.

Li B., Oestreich S., and de Lange T. 2000. Identification of human Rap1: Implications for telomere evolution. *Cell* **101:** 471–483.

Loayza D. and de Lange T. 2003. POT1 as a terminal transducer of TRF1 telomere length control. *Nature* **423:** 1013–1018.

Loayza D., Parsons H., Donigian J., Hoke K., and de Lange T. 2004. DNA binding features of human POT1: A nonamer 5′-TAGGGTTAG-3′ minimal binding site, sequence specificity, and internal binding to multimeric sites. *J. Biol. Chem.* **279:** 13241–13248.

Makarov V.L., Hirose Y., and Langmore J.P. 1997. Long G-tails at both ends of human chromosomes suggest a C-strand degradation mechanism for telomere shortening. *Cell* **88:** 657–666.

Manolis K.G., Nimmo E.R., Hartsuiker E., Carr A.M., Jeggo P.A., and Allshire R.C. 2001. Novel functional requirements for non-homologous DNA end joining in *Schizosaccharomyces pombe*. *EMBO J.* **20:** 210–221.

Marcand S., Gilson E., and Shore D. 1997. A protein-counting mechanism for telomere length regulation in yeast. *Science* **275:** 986–990.

Martin S.G., Laroche T., Suka N., Grunstein M., and Gasser S.M. 1999. Relocalization of telomeric Ku and SIR proteins in response to DNA strand breaks in yeast. *Cell* **97:** 621–633.

Matsuura A., Naito T., and Ishikawa F. 1999. Genetic control of telomere integrity in *Schizosaccharomyces pombe*: rad3(+) and tel1(+) are parts of two regulatory networks independent of the downstream protein kinases chk1(+) and cds1(+). *Genetics* **152:** 1501–1512.

Mieczkowski P.A., Mieczkowska J.O., Dominska M., and Petes T.D. 2003. Genetic regulation of telomere-telomere fusions in the yeast *Saccharomyces cerevisiae*. *Proc. Natl. Acad. Sci.* **100:** 10854–10859.

Miki F., Kurabayashi A., Tange Y., Okazaki K., Shimanuki M., and Niwa O. 2004. Two-hybrid search for proteins that interact with Sad1 and Kms1, two membrane-bound components of the spindle pole body in fission yeast. *Mol. Genet. Genomics* **270:** 449–461.

Miki F., Okazaki K., Shimanuki M., Yamamoto A., Hiraoka Y., and Niwa O. 2002. The 14-kDa dynein light chain-family protein Dlc1 is required for regular oscillatory nuclear movement and efficient recombination during meiotic prophase in fission yeast. *Mol. Biol. Cell* **13:** 930–946.

Miller K.M. and Cooper J.P. 2003. The telomere protein Taz1 is required to prevent and repair genomic DNA breaks. *Mol. Cell* **11:** 303–313.

Miller K.M., Ferreira M.G., and Cooper J.P. 2005. Taz1, Rap1 and Rif1 act both interdependently and independently to maintain telomeres. *EMBO J.* **24:** 3128–3135.

Mishra K. and Shore D. 1999. Yeast Ku protein plays a direct role in telomeric silencing and counteracts inhibition by rif proteins. *Curr. Biol.* **9:** 1123–1126.

Mitton-Fry R.M., Anderson E.M., Theobald D.L., Glustrom L.W., and Wuttke D.S. 2004. Structural basis for telomeric single-stranded DNA recognition by yeast Cdc13. *J. Mol. Biol.* **338:** 241–255.

Miyoshi T., Sadaie M., Kanoh J., and Ishikawa F. 2003. Telomeric DNA ends are essential for the localization of Ku at telomeres in fission yeast. *J. Biol. Chem.* **278:** 1924–1931.

Naito T., Matsuura A., and Ishikawa F. 1998. Circular chromosome formation in a fission yeast mutant defective in two ATM homologues. *Nat. Genet.* **20:** 203–206.

Nakamura T.M., Cooper J.P., and Cech T.R. 1998. Two modes of survival of fission yeast without telomerase. *Science* **282:** 493–496.

Nakamura T.M., Moser B.A., and Russell P. 2002. Telomere binding of checkpoint sensor and DNA repair proteins contributes to maintenance of functional fission yeast telomeres. *Genetics* **161:** 1437–1452.

Nakamura T.M., Morin G.B., Chapman K.B., Weinrich S.L., Andrews W.H., Lingner J., Harley C.B., and Cech T.R. 1997. Telomerase catalytic subunit homologs from fission yeast and human. *Science* **277:** 955–959.

Nakayama J., Klar A.J., and Grewal S.I. 2000. A chromodomain protein, Swi6, performs imprinting functions in fission yeast during mitosis and meiosis. *Cell* **101:** 307–317.

Nakayama J., Rice J.C., Strahl B.D., Allis C.D., and Grewal S.I. 2001. Role of histone H3 lysine 9 methylation in epigenetic control of heterochromatin assembly. *Science* **292:** 110–113.

Nikitina T. and Woodcock C.L. 2004. Closed chromatin loops at the ends of chromosomes. *J. Cell Biol.* **166:** 161–165.

Nimmo E.R., Cranston G., and Allshire R.C. 1994. Telomere-associated chromosome breakage in fission yeast results in variegated expression of adjacent genes. *EMBO J.* **13:** 3801–3811.

Nimmo E.R., Pidoux A.L., Perry P.E., and Allshire R.C. 1998. Defective meiosis in telomere-silencing mutants of *Schizosaccharomyces pombe*. *Nature* **392:** 825–828.

Pardo B. and Marcand S. 2005. Rap1 prevents telomere fusions by nonhomologous end joining. *EMBO J.* **24:** 3117–3127.

Pennock E., Buckley K., and Lundblad V. 2001. Cdc13 delivers separate complexes to the telomere for end protection and replication. *Cell* **104:** 387–396.

Peterson S.E., Stellwagen A.E., Diede S.J., Singer M.S., Haimberger Z.W., Johnson C.O., Tzoneva M., and Gottschling D.E. 2001. The function of a stem-loop in telomerase RNA is linked to the DNA repair protein Ku. *Nat. Genet.* **27:** 64–67.

Sadaie M., Naito T., and Ishikawa F. 2003. Stable inheritance of telomere chromatin structure and function in the absence of telomeric repeats. *Genes Dev.* **17:** 2271–2282.

Schramke V. and Allshire R. 2003. Hairpin RNAs and retrotransposon LTRs effect RNAi and chromatin-based gene silencing. *Science* **301:** 1069–1074.

Shimanuki M., Miki F., Ding D.Q., Chikashige Y., Hiraoka Y., Horio T., and Niwa O. 1997. A novel fission yeast gene, *kms1+*, is required for the formation of meiotic prophase-specific nuclear architecture. *Mol. Gen. Genet.* **254:** 238–249.

Silverman J., Takai H., Buonomo S.B., Eisenhaber F., and de Lange T. 2004. Human Rif1, ortholog of a yeast telomeric protein, is regulated by ATM and 53BP1 and functions in the S-phase checkpoint. *Genes Dev.* **18:** 2108–2119.

Spink K.G., Evans R.J., and Chambers A. 2000. Sequence-specific binding of Taz1p dimers to fission yeast telomeric DNA. *Nucleic Acids Res.* **28:** 527–533.

Stellwagen A.E., Haimberger Z.W., Veatch J.R., and Gottschling D.E. 2003. Ku interacts with telomerase RNA to promote telomere addition at native and broken chromosome ends. *Genes Dev.* **17:** 2384–2395.

Sugawara N. 1988. "DNA sequences at the telomeres of the fission yeast *S. pombe*." Ph.D. thesis, Harvard University, Cambridge, Massachusetts.

Takai H., Smogorzewska A., and de Lange T. 2003. DNA damage foci at dysfunctional telomeres. *Curr. Biol.* **13:** 1549–1556.

Teixeira M.T., Arneric M., Sperisen P., and Lingner J. 2004. Telomere length homeostasis is achieved via a switch between telomerase-extendible and -nonextendible states. *Cell* **117:** 323–335.

Ting N.S., Yu Y., Pohorelic B., Lees-Miller S.P., and Beattie T.L. 2005. Human Ku70/80 interacts directly with hTR, the RNA component of human telomerase. *Nucleic Acids Res.* **33:** 2090–2098.

Tomaska L., Willcox S., Slezakova J., Nosek J., and Griffith J.D. 2004. Taz1 binding to a fission yeast model telomere: Formation of t-loops and higher order structures. *J. Biol. Chem.* **279:** 50764–50772.

Tomita K., Kibe T., Kang H.Y., Seo Y.S., Uritani M., Ushimaru T., and Ueno M. 2004. Fission yeast Dna2 is required for generation of the telomeric single-strand overhang. *Mol. Cell. Biol.* **24:** 9557–9567.

Tomita K., Matsuura A., Caspari T., Carr A.M., Akamatsu Y., Iwasaki H., Mizuno K., Ohta K., Uritani M., Ushimaru T., Yoshinaga K., and Ueno M. 2003. Competition between the Rad50 complex and the Ku heterodimer reveals a role for Exo1 in processing double-strand breaks but not telomeres. *Mol. Cell. Biol.* **23:** 5186–5197.

Trujillo K.M., Bunch J.T., and Baumann P. 2005. Extended DNA binding site in Pot1 broadens sequence specificity to allow recognition of heterogeneous fission yeast telomeres. *J. Biol. Chem.* **280:** 9119–9128.

Tsukamoto Y., Kato J., and Ikeda H. 1997. Silencing factors participate in DNA repair and recombination in *Saccharomyces cerevisiae*. *Nature* **388:** 900–903.

Tuzon C.T., Borgstrom B., Weilguny D., Egel R., Cooper J.P., and Nielsen O. 2004. The fission yeast heterochromatin protein Rik1 is required for telomere clustering during meiosis. *J. Cell Biol.* **165:** 759–765.

van Steensel B. and de Lange T. 1997. Control of telomere length by the human telomeric protein TRF1. *Nature* **385:** 740–743.

Vassetzky N.S., Gaden F., Brun C., Gasser S.M., and Gilson E. 1999. Taz1p and Teb1p, two telobox proteins in *Schizosaccharomyces pombe*, recognize different telomere-related DNA sequences. *Nucleic Acids Res.* **27:** 4687–4694.

Volpe T., Schramke V., Hamilton G.L., White S.A., Teng G., Martienssen R.A., and Allshire R.C. 2003. RNA interference is required for normal centromere function in fission yeast. *Chromosome Res.* **11:** 137–146.

Walker J.R., Corpina R.A., and Goldberg J. 2001. Structure of the Ku heterodimer bound to DNA and its implications for double-strand break repair. *Nature* **412:** 607–614.

Wang R.C., Smogorzewska A., and de Lange T. 2004. Homologous recombination generates T-loop-sized deletions at human telomeres. *Cell* **119:** 355–368.

Wellinger R.J., Wolf A.J., and Zakian V.A. 1993. *Saccharomyces* telomeres acquire single-strand TG1-3 tails late in S phase. *Cell* **72:** 51–60.

Wellinger R.J., Ethier K., Labrecque P., and Zakian V.A. 1996. Evidence for a new step in telomere maintenance. *Cell* **85:** 423–433.

Wood V., Gwilliam R., Rajandream M.A., Lyne M., Lyne R., Stewart A., Sgouros J., Peat N., Hayles J., Baker S., et al. 2002. The genome sequence of *Schizosaccharomyces pombe*. *Nature* **415:** 871–880.

Wright W.E., Tesmer V.M., Huffman K.E., Levene S.D., and Shay J.W. 1997. Normal human chromosomes have long G-rich telomeric overhangs at one end. *Genes Dev.* **11:** 2801–2809.

Yamamoto A., West R.R., McIntosh J.R., and Hiraoka Y. 1999. A cytoplasmic dynein heavy chain is required for oscillatory nuclear movement of meiotic prophase and efficient meiotic recombination in fission yeast. *J. Cell Biol.* **145:** 1233–1249.

Yamamoto T.G., Chikashige Y., Ozoe F., Kawamukai M., and Hiraoka Y. 2004. Activation of the pheromone-responsive MAP kinase drives haploid cells to undergo ectopic meiosis with normal telomere clustering and sister chromatid segregation in fission yeast. *J. Cell Sci.* **117:** 3875–3886.

17

Plant Telomeres

Dorothy E. Shippen

Department of Biochemistry and Biophysics
Texas A&M University
College Station, Texas 77843-2128

THE FIELD OF TELOMERE BIOLOGY WAS PIONEERED, in large part, through the seminal studies of Barbara McClintock in maize more than 60 years ago. Although many of the key insights in telomere research in the ensuing years have derived from other organisms, there is renewed interest in plant telomere biology as our understanding of the intimate relationship between cellular proliferation and telomere maintenance evolves. This interest is fueled by the striking contrast between the plasticity of plant development and genome architecture, and the much more deterministic nature of mammals. In contrast to mammals, plants produce new organs throughout their lives from meristematic proliferation zones, and many cells in the plant body are totipotent. The plant genome also displays an exceptional plasticity and tolerance to genome stresses, including changes in ploidy, DNA methylation, chromosomal rearrangements, and transposition. Taken together, these features raise fundamental questions about the contribution of telomeres and telomerase in facilitating cell proliferation and genome stability in plants.

In the last few years, plant telomere biology has begun to blossom, owing in large part to research with the small flowering plant of the mustard weed family, *Arabidopsis thaliana*. The completion of the *Arabidopsis* genome sequencing project and the availability of T-DNA insertion lines that allow for gene knockouts have enabled researchers to make rapid progress in deciphering the functions of several telomere-related genes and in uncovering plant responses to telomere dysfunction. In this review, I focus on recent studies of telomere architecture and function and illustrate how plants are providing surprising insights into telomere biology.

TELOMERIC DNA

The first telomere sequence cloned from a higher eukaryote, TTTAGGG, was obtained from *Arabidopsis* (Richards and Ausubel 1988). Although most plant species, including green algae and most dicots (plants with embryos containing two true leaves or cotyledons), harbor this sequence (Cox et al. 1993; Fuchs et al. 1995), there are some notable exceptions (Table 1). The majority of taxa within the monocot order (plants with embryos containing a single cotyledon), Asparagales, including Alliaceae (Fuchs et al. 1995; Pich et al. 1996; Pich and Schubert 1998), Aloe (Adams et al. 2000), and Hyacinthaceae (Weiss and Scherthan 2002; Puizina et al. 2003), lack TTTAGGG repeats at their chromosome termini. It was initially proposed that telomeres in these organisms might contain satellite repeats, ribosomal DNA, or mobile DNA (Fuchs et al. 1995; Pich et al. 1996; Pich and Schubert 1998). However, recent studies have uncovered canonical telomere repeats, but of a different nucleotide sequence. In several cases, the vertebrate hexameric sequence, TTAGGG, was found at chromosome ends (Weiss and Scherthan 2002; Weiss-Schneeweiss et al. 2004). Consistent with this observation, sequence analysis of telomerase elongation products generated with *Othocallis siberica* extracts showed synthesis of TTAGGG repeats (Weiss-Schneeweiss et al. 2004). Hybridization

Table 1. Telomere length and repeats in various plant species

Organism	Telomere repeat	Telomere length (kb)	References
Chlorella vulgaris	T_3AG_3	0.5	Higashiyama et al. 1995
Arabidopsis thaliana	T_3AG_3	2–9	Richards and Ausubel 1988; Shakirov and Shippen 2004
Tobacco	T_3AG_3	40–160	Fajkus et al. 1995
Silene latifolia	T_3AG_3	4.5	Riha et al. 1998
Barley	T_3AG_3	20–80	Kilian and Kleinhofs 1992; Kilian et al. 1995
Maize	T_3AG_3	2–40	Richards and Ausubel 1988; Burr et al. 1992
Pisum sativum	T_3AG_3	100	Cesare et al. 2003
Tomato	$T_2(T/A)AG_3$	20–50	Ganal et al. 1991
Othocallis siberica	T_2AG_3	10	Weiss-Schneeweiss et al. 2004
Aloe, *Hyacinthella dalmatica*	T_2AG_3	ND	Weiss and Scherthan 2002; Puizina et al. 2003
Chlamydomonas reinhardtii	T_4AG_3	ND	Petracek et al. 1990

ND, not determined.

studies for species within *Allium* detected variant telomeric repeats, including TTGGGG and TTAGGG (Sykorova et al. 2003). Sequence analysis of telomerase products revealed synthesis of TTAGGG repeats as well as an unexpectedly high frequency of degenerate telomere repeats, suggesting that the *Allium* telomerases may be error-prone, as reported for the *Paramecium* telomerase (McCormick-Graham et al. 1997).

In most eukaryotes, telomere-associated sequences consist of highly repetitive satellite sequences that are often abutted by degenerate telomere repeats. Although the telomeres of several angiosperms (flowering plants) display this organization, *Arabidopsis* is unusual in that its telomere tracts are flanked by unique subtelomeric sequences on at least seven of the ten chromosome arms (Arabidopsis Genome Initiative 2000; Heacock et al. 2004). Ribosomal RNA genes lie immediately adjacent to the telomeres on the short arms of chromosomes 2 and 4 (Copenhaver and Pikaard 1996). The telomere association of nucleolar organizer regions may provide an explanation for the surprising observation that telomeres in *Arabidopsis* are associated with the nucleolus for most of mitotic and meiotic cell cycles (Armstrong et al. 2001; Fransz et al. 2002). In contrast, telomeres in other eukaryotes, including other plant species, are located at the nuclear periphery. It is notable that chromosome organization varies widely among plant species. For example, a subset of plants like *Arabidopsis* and others with small genomes do not display the classic Rabl configuration of telomeres during interphase (Dong and Jiang 1998).

Telomere length varies widely among different plant species (Table 1). In *Arabidopsis,* telomeres span 2–9 kb (Richards and Ausubel 1988; Shakirov and Shippen 2004), whereas in pea they are considerably longer at 20–40 kb on average (Cesare et al. 2003), and in tobacco even more extended, reaching 160 kb (Fajkus et al. 1995). Telomere length differs not only between evolutionarily distant organisms, but also within the same species. In *A. thaliana* plants of Wassilewskija (WS) ecotype, telomeres range from 2.5 to 8 kb (Gallego and White 2001; Bundock and Hooykaas 2002; Riha et al. 2002; Gallego et al. 2003a; Shakirov and Shippen 2004), but in the Columbia ecotype, telomeres are confined to 2–5 kb (Richards and Ausubel 1988; Shakirov and Shippen 2004). A survey of 11 different *A. thaliana* accessions indicates a twofold difference in telomere length (Shakirov and Shippen 2004). Maize telomeres are much more variable in length, ranging from an average of 1.8 to 40 kb in inbred maize lines (Burr et al. 1992). Genetic mapping of a recombinant inbred maize population revealed that three genetic loci account for 50% of telomere length variability. In all cases, the length of the telomere tract is strictly maintained at a specific set point, and hence length homeostasis is achieved. Recent studies indicate that a staggeringly large number of genes regulate telomere length in yeast (Askree et al. 2004). Thus, telomere length in plants is also likely to be controlled by many genes.

Single-strand G-overhangs of 12–30 nucleotides (nt) have been detected on telomeres in *Silene* and in *Arabidopsis* (Riha et al. 2000). Consistent with studies in yeast and mammals (Dionne and Wellinger 1996; Hemann and Greider 1999), these structures are present in both telomerase-positive and -negative tissues (Riha et al. 2000). Unexpectedly, however, G-overhang signals could be detected on only half of the chromosome ends in *Arabidopsis* (Riha et al. 2000). Whether this observation reflects the limit of current technology or an inherent asymmetry in the structure of *Arabidopsis* telomeres remains to be determined. Electron microscopy (EM) studies reveal the presence of large t-loops on pea telomeres (Cesare et al. 2003). Some loops encompass more than 80 kb, approximately ten times the size of t-loops in mammalian cells (Griffith et al. 1999), which may reflect the much larger size of telomere tracts in peas.

TELOMERE DYNAMICS

The dynamic nature of telomere length regulation in *Arabidopsis* was recently illustrated by following the fate of individual telomere tracts in parents and progeny through successive plant generations using probes directed at unique subtelomeric regions (Shakirov and Shippen 2004). Siblings derived from different parents display significant variability in individual telomere tracts, although all telomeres are maintained within the ecotype-specific range. For plants derived from the same parent, the overall telomere length is more consistent. The parent-progeny analysis also revealed that very short telomeres are almost always lengthened in the subsequent generations until they reach a length optimum, whereas telomeres above this optimum typically shortened in the next generation by 200–500 base pairs (bp). This rate of loss is consistent with the complete absence of telomerase action (Fig. 1) (Fitzgerald et al. 1999; Riha et al. 2001). These data in *Arabidopsis* agree well with recent studies in yeast (Teixeira et al. 2004) and argue that telomere length homeostasis is achieved by the intermittent, but preferential, action of telomerase on the shortest telomeres in the population (Fig. 1) (Shakirov and Shippen 2004).

Telomere length in barley is altered during cellular differentiation. In embryos, telomeres reach 80 kb, but in leaves are shortened to 23 kb (Kilian et al. 1995). This dramatic decline is unlikely to result from the end-replication problem and instead may reflect a process akin to telomere rapid deletion (TRD) described for budding yeast (Li and Lustig

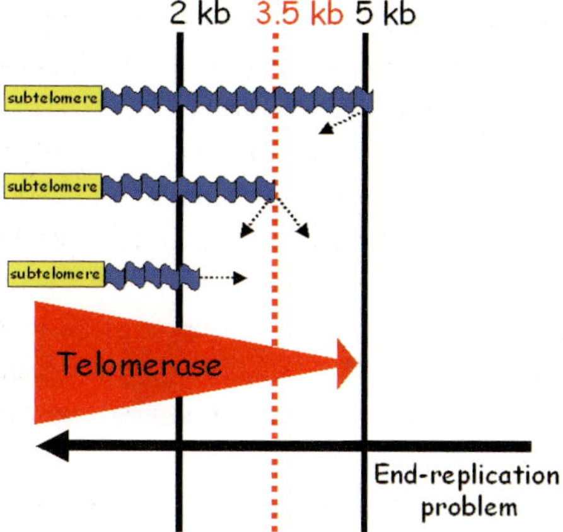

Figure 1. Model for telomere length homeostasis in *Arabidopsis*. Telomeres achieve an optimal, ecotype-specific length by the opposing forces of lengthening by telomerase and shortening by the end-replication problem. When telomere length reaches the lower threshold (2 kb in the Columbia ecotype), telomerase is more likely to act (*bottom* telomere). As telomeres approach the optimum size (3.5 kb), there is an equal probability of telomerase action or inaction, resulting in telomere splitting (*middle* telomere). At the maximum size (5 kb; *top* telomere), telomerase is less likely to act, and telomeres shorten from the end-replication problem. (Reprinted, with permission, from Shakirov and Shippen 2004 [©American Society of Plant Biologists].)

1996; Bucholc et al. 2001). TRD is a homologous recombination mechanism that resets very long telomeres to the wild-type length in a single step, producing extrachromosomal telomeric DNA as a by-product. A similar homologous recombination event has recently been shown to delete t-loop-sized segments from mammalian telomeres (Wang et al. 2004). Telomere dynamics in wheat are also consistent with TRD, as discrete extrachromosomal fragments carrying telomeric repeats are present in dormant wheat embryos, but are lost during germination (Bucholc and Buchowicz 1995). In addition to programmed telomere shortening, barley telomeres are also subject to dramatic elongation. In callus culture, telomeres expand from ~10 kb to >150 kb (Kilian et al. 1995).

Such remarkable changes in bulk telomere length are not characteristic of *Arabidopsis*. Individual telomere tracts display almost no size variation throughout the life span. Typically, terminal restriction fragments are represented by a single discrete band for a given telomere tract, despite the different proliferation history of cells in an *Arabidopsis* plant. This observation means that telomere tracts on the ends of homologous chromosomes must be processed in a uniform fashion throughout development (Shakirov and Shippen 2004). In wild-type plants, individual telomeres span a 1 kb range (Riha et al. 2001), but in telomerase-deficient mutants they become even more compressed (Fig. 2), arguing that

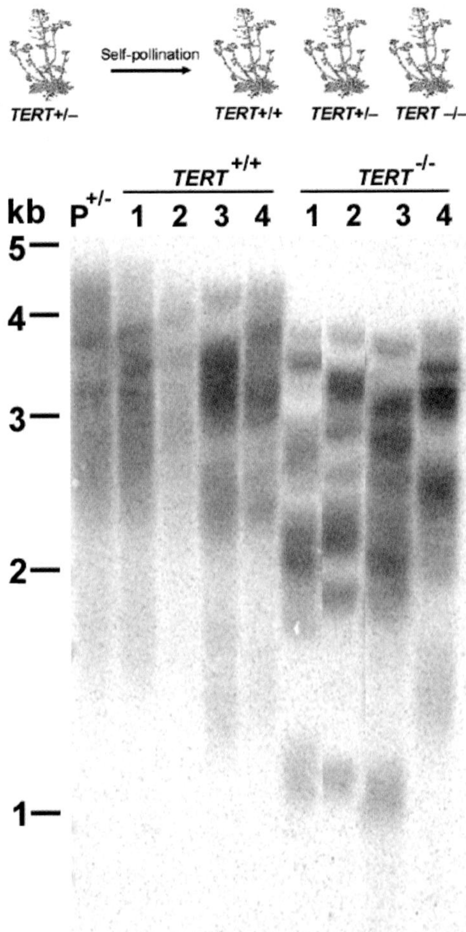

Figure 2. Telomere shortening and decreased heterogeneity of individual telomere tracts in telomerase-deficient *Arabidopsis*. Self-pollination of plants heterozygous for a T-DNA insertion in *TERT* creates homozygous wild-type, heterozygous, and homozygous mutant progeny (*top* panel). For the first five generations *tert* mutants display no morphological defects (Riha et al. 2001). TRF analysis of the heterozygous *tert* parent (P$^{+/-}$), and its four wild-type or homozygous *tert*$^{-/-}$ progeny (*bottom* panel). Telomeres in the homozygous mutants are ~200–500 bp shorter than the parent and display much less heterogeneity.

telomerase is primarily responsible for the length heterogeneity of *Arabidopsis* telomeres (Fitzgerald et al. 1999; Riha et al. 2001). The basis for the striking difference in telomere length regulation in barley and *Arabidopsis* is unknown, but it may be that the smaller size of *Arabidopsis* telomere tracts requires a tighter length regulation.

PLANT TELOMERASES

After the highly sensitive polymerase chain reaction (PCR)-based telomere repeat amplification protocol (TRAP) was introduced to detect telomerase activity in mammalian cells (Kim et al. 1994), it became feasible to assay for plant telomerases by simply modifying the reverse primer to match the plant telomere repeat sequence. Soon telomerase activity was detected in a wide variety of monocots and dicots (Fajkus et al. 1996; Fitzgerald et al. 1996; Heller et al. 1996; Kilian et al. 1998; Riha et al. 1998). TRAP enabled researchers to address a fundamental question: Is the activity of plant telomerases restricted to actively dividing cells as in humans or is there a broader profile of expression that reflects the increased flexibility of plant development? Several studies revealed that telomerase activity is highly regulated in plants, with enzyme activity limited to cells with high proliferation capacity (e.g., reproductive organs, embryos, and immortalized, dedifferentiated cells grown in culture) (Fitzgerald et al. 1996; Heller et al. 1996; Kilian et al. 1998; Riha et al. 1998). Kilian and colleagues found that telomerase is highly active in maize embryos, but is only transiently expressed in endosperm (Kilian et al. 1998). Hence, McClintock's early observation that the "healing" of broken chromosomes occurs in the embryo, but not in the endosperm (McClintock 1939, 1941), correlates well with the profile of telomerase expression in these settings.

In situ hybridization has uncovered several examples of chromosome-healing events in plants (Murata et al. 1992; Werner et al. 1992; Tsujimoto 1993). Broken wheat chromosomes are healed early in development during the first mitotic divisions of the sporophyte, when telomerase is likely to be highly active. Nevertheless, several rounds of the breakage-fusion-bridge (BFB) cycle occur before the acquisition of a fully capped telomere (Friebe et al. 2001), implying that de novo telomere formation is inefficient. Sequence analysis of the healed breakpoints in wheat chromosomes is consistent with telomere addition by telomerase (Tsujimoto et al. 1997, 1999; Friebe et al. 2001), and biochemical studies show that many plant telomerases can efficiently incorporate telomeric repeats onto nontelomeric DNA in vitro (Fitzgerald et al. 2001).

Thus, the formation of telomeres on double-strand breaks (DSBs) in plants appears to be controlled primarily by the presence or absence of telomerase.

THE TELOMERASE RNP AND MECHANISMS OF TELOMERASE REGULATION

Relatively little biochemical information is available concerning the telomerase ribonucleoprotein particle (RNP) complex from plants. Partial purification of the enzyme from cauliflower suggests a molecular mass of ~670 kD (M. Fitzgerald and D. Shippen, unpubl.). The tobacco telomerase appears to be a similar size, although a minor fraction of the activity associates with a smaller, 250-kD complex (Yang et al. 2004). Genome database searches have revealed several promising candidates for telomerase-associated factors (Table 2). The most well-characterized of these is the telomerase catalytic subunit, telomerase reverse transcriptase (TERT). TERT has been cloned from *Arabidopsis* (Fitzgerald et al. 1999; Oguchi et al. 1999) and from rice (Heller-Uszynska et al. 2002), and in both cases encodes a 131-kD protein bearing canonical reverse transcriptase motifs and the TERT-specific T-motif. *Arabidopsis* mutants bearing a T-DNA insertion in TERT display progressive telomere shortening, consistent with an essential role for this gene in telomere maintenance (Fitzgerald et al. 1996; Riha et al. 2001).

Additional proteins shown to associate with the yeast and human telomerase RNPs include Est1 (Lundblad 2003), dyskerin (Mason 2003), and PinX1 (Lin and Blackburn 2004). Putative homologs for all of these genes are present in *Arabidopsis* (Table 2). Although the telomere-related functions of dyskerin and PinX1 have not yet been explored in plants, neither of the two *Arabidopsis* sequence homologs of the *EST1* gene (*EST1a* and *EST1b*) appears to play a significant role in telomere biology. In budding yeast, Est1 is required for telomerase recruitment (Lundblad 2003) and in mammals for telomere length homeostasis and end protection (Reichenbach et al. 2003; Snow et al. 2003). In contrast, *Arabidopsis* mutants null for *EST1a* or *EST1b*, or both genes, have fully capped telomeres of wild-type length (R. Idol and D. Shippen, unpubl.), implying that the *EST1* genes do not play the same role in plants as in other organisms.

The telomerase RNA subunit has not yet been identified in any plant species. Bioinformatics strategies to identify the telomerase RNA gene are confounded by the exceptionally rapid divergence of telomerase RNA sequences (Chen et al. 2000) and, in *Arabidopsis*, by the presence of

Table 2. List of human and *Arabidopsis* proteins implicated in telomere biology

Human protein	*Arabidopsis* protein	Gene number	References
Artemis	Artemis1a[a]	At1g19025	J. Watson and D. Shippen,
	Artemis1b[a]	At1g27410	unpubl.
	Artemis1c[a]	At1g66730	
	Artemis1d[a]	At3g26680	
	Artemis1e[a]	At2g45700	
ATR	ATR[a]	At5g40820	Culligan et al. 2004
ATM	ATM[a]	At3g48190	Garcia et al. 2000
Dyskerin	Dyskerin[a]	At3g57150	Maceluch et al. 2001
Est1	Est1a[a], Est1b[a]	At1g28260	R. Idol and D. Shippen,
		At5g19400	unpubl.
ERCC1/XPF1	ERCC1/XPF1[a]	At3g05210	Hefner et al. 2002
Ku70	Ku70	At1g16970	Bundock et al. 2002; Riha
Ku80	Ku80	At1g48050	et al. 2002; Tamura et al. 2002; Gallego et al. 2003a
Mre11	Mre11	At5g54260	Hartung and Puchta 1999; Bundock and Hooykaas 2002
Nbs1[b]			
PARP2	PARP2[a]	At4g02390	Lepiniec et al. 1995
PinX1	PinX1[a]	At5g26610	E. Shakirov and D. Shippen, unpubl.
Pip1/PTOP[b]			
Pot1	Pot1[a]	At2g05210	Baumann et al. 2002;
	Pot2[a]	At5g06310	Shakirov et al. 2005
Rad50	Rad50	At2g24420	Hartung and Puchta 1999; Gallego and White 2001; Gallego et al. 2001
Rap1[b]			
Tankyrase[b]			
TERT	TERT	At5g16850	Fitzgerald et al. 1999; Oguchi et al. 1999
TIN2[b]			
TRF1/TRF2	TRFL2[a], 4[a], 1[a], 9[a], TBP1[a], TRP1[a]	At1g07540	Chen et al. 2001; Hwang
		At3g53790	et al. 2001; Karamysheva
		At3g46590	et al. 2004
		At3g12560	
		At5g13820	
		At5g59430	

[a]These proteins have not been shown to function at telomeres in vivo.
[b]Sequence homologs are absent in *Arabidopsis*.

abundant interstitial telomeric DNA sequences (Regad et al. 1994; Tremousaygue et al. 1999).

Several aspects of telomerase regulation are similar in plants and animals. Interestingly, however, the modes by which *TERT* expression influences enzyme activity differ in rice and *Arabidopsis*. In *Arabidopsis*, *AtTERT* mRNA abundance parallels telomerase activity (Fitzgerald et al. 1999; Oguchi et al. 1999), but in rice the steady-state level of *OsTERT* mRNA is the same in telomerase-positive and telomerase-negative tissues (Heller-Uszynska et al. 2002). On the other hand, rice harbors several differentially spliced *OsTERT* mRNAs (Heller-Uszynska et al. 2002), suggesting that alternative splicing could contribute to telomerase regulation. In contrast, there is no evidence for alternative splicing of *AtTERT* mRNA (Fitzgerald et al. 1999). These differences in *TERT* regulation may not be so dramatic considering that monocots (rice) and dicots (*Arabidopsis*) diverged more than 100 million years ago (Chaw et al. 2004).

Studies in tobacco show that telomerase activity is cell-cycle-regulated. In synchronized cell suspension cultures, telomerase activity is low or undetectable outside of S phase (Tamura et al. 1999). The induction of telomerase activity correlates with entry into S phase, but appears to be unlinked to DNA replication. Phytohormones strongly influence telomerase expression (Tamura et al. 1999; Yang et al. 2002). Exogenous application of auxin increases telomerase activity during S phase, whereas abscisic acid, a phytohormone that induces a cyclin-dependent protein kinase inhibitor, blocks both auxin and S-phase-mediated telomerase activation (Yang et al. 2002).

Further evidence that auxin plays a key role in telomerase regulation was obtained through characterization of the *TELOMERASE ACTI-VATOR1* (*TAC1*) gene in *Arabidopsis* (Ren et al. 2004). McKnight and colleagues used TRAP to screen a large collection of activation-tagged mutants for plants that inappropriately express telomerase in their leaves. Among the mutants they recovered, one overexpresses *TAC1*, which encodes a small zinc-finger protein. The *TAC1* expression profile mirrors that of *TERT*, with *TAC1* mRNA present only in actively dividing organs. In the mutant, overexpression of *TAC1* correlates with increased expression of *TERT* mRNA, and although telomerase activity is now more widespread, telomeres do not increase in size. For the most part, *TAC1* overexpressing mutants display a wild-type appearance, and DNA content of the cells is unaltered. However, responses to auxin are slightly perturbed in these plants, indicating that *TAC1* impinges on the auxin-signaling pathway that regulates telomerase expression. Ectopic expression of *TAC1* in rice also induces telomerase activity in leaves

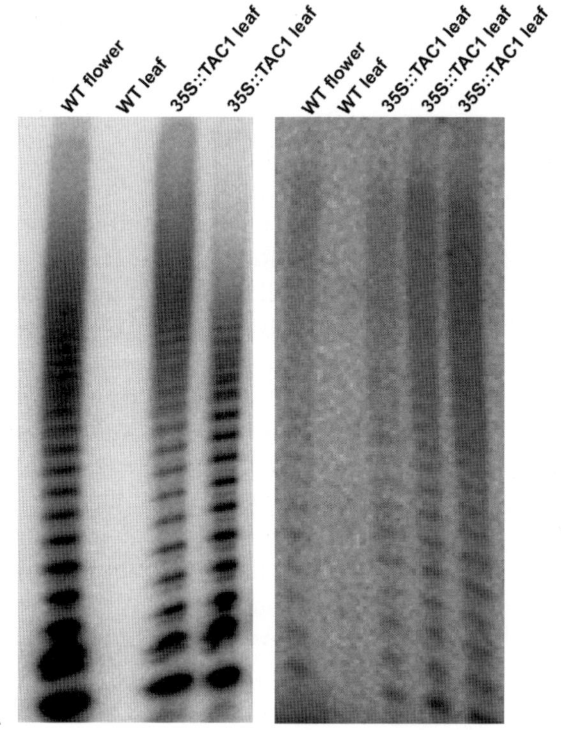

Figure 3. Telomerase activation in the leaves of plants overexpressing the TAC1 protein. TRAP assay results for *35S::TAC1* transgenic individuals of *Arabidopsis* (*A*) and rice (*B*) are shown. The figure was contributed by Shuxin Ren and Tom McKnight.

(Fig. 3), arguing that the function of this gene is broadly conserved among flowering plants.

The last common ancestor of plants and animals was likely to have been a unicellular eukaryote that lived more than 1.5 billion years ago (Meyerowitz 2002). If so, then plants and animals developed into multicellular organisms along separate evolutionary trajectories and have independently derived strategies for regulating telomerase. Because constitutive expression of telomerase has no significant impact on plant growth and development, the striking pattern of telomerase repression and activation is curious. In contrast to humans, plants do not develop cancer, and hence there is no obvious selective pressure to suppress telomerase activity in nondividing cells. One possibility is that telomerase inactivation in leaves occurred stochastically and, in the absence of a detrimental outcome, was allowed to persist (Ren et al. 2004). Alternatively, the consequences of constitutive telomerase expression may be too subtle to detect in the short term, but could be detrimental over evolutionary time.

CONSEQUENCES OF A TELOMERASE DEFICIENCY IN *ARABIDOPSIS*

Arabidopsis mutants that harbor a T-DNA insertion in the *TERT* gene display telomere shortening at a rate of ~200–500 bp per plant generation (Fig. 4) (Fitzgerald et al. 1999; Riha et al. 2001). Plants exhibit no defects in growth or development until the fifth or sixth generation, when they begin to show morphological abnormalities in leaves and reduced fertility. At this stage, the shortest telomeres reach a size of ~1.2 kb (M. Heacock and D. Shippen, unpubl.), suggesting that below this threshold telomeres fail to protect the chromosome terminus. In support of this interpretation, bridged chromosomes are first observed in anaphases of G_6 telomerase mutants (Riha et al. 2001).

Despite the onset of genome instability, plants can survive for up to five additional generations while their telomeres continue to shorten. The mutants endure increasing levels of developmental anomalies in both vegetative organs and reproductive systems and, by the tenth generation, they arrest at a terminal vegetative state harboring shoot meristems that are grossly enlarged, disorganized, and, in some cases, transformed into a callusoid mass (Riha et al. 2001). The shortest telomeres detected in terminal plants are ~300 nt in length (Fig. 4) (Heacock et al. 2004), only slightly longer than the telomeres of budding yeast.

In terminal mutants, up to 44% of anaphases contain bridged chromosomes (Riha et al. 2001). Evidence that these structures represent covalent fusion of telomeric DNA was obtained using a PCR strategy that exploits the unique subtelomeric DNA sequences to amplify chromosome fusion junctions (Heacock et al. 2004). Although examples of telomere–telomere fusion were found, the most frequent events involved fusion of telomere tracts directly to subtelomeric DNA. This finding indicates that chromosome ends are subject to extensive exonucleolytic digestion prior to fusion. In yeast and mammals, the nonhomologous end-joining (NHEJ) pathway is used to fuse dysfunctional telomeres (Williams and Lustig 2003). The data from *Arabidopsis* strongly support a role for NHEJ components in telomere fusion, and they also reveal a striking degree of redundancy in DNA repair pathways (Heacock et al. 2004). Chromosome fusion junctions could be amplified in DNA samples from *tert* and *ku70 tert* mutants, but in *ku70 tert* mutants there was a greater incidence of microhomology at the fusion junction than in *tert* mutants, arguing that plants, like yeast (Baumann and Cech 2000; Mieczkowski et al. 2003) and

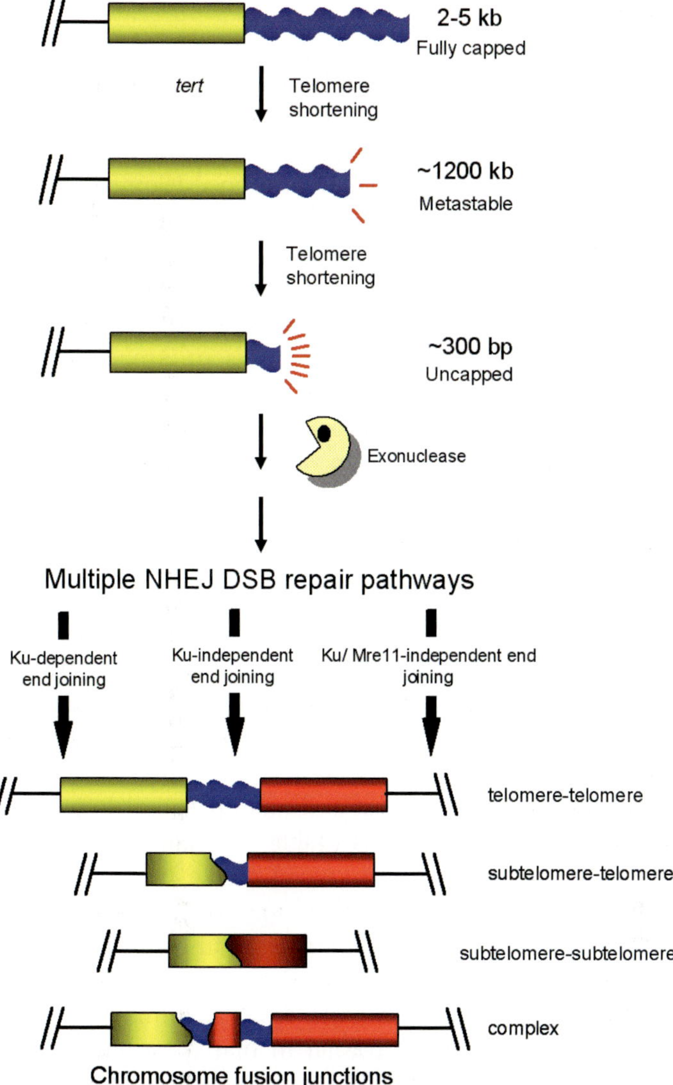

Figure 4. Model of the molecular and cytogenetic consequences of telomere shortening in *Arabidopsis*. In telomerase mutants, telomeres progressively shorten and chromosome fusion events are first detected by PCR when the shortest telomeres are ~1200 bp (M. Heacock and D. Shippen, unpubl.), a "metastable" length where they fluctuate between a capped and uncapped state. Plants survive for up to five additional generations. In the terminal generation, the shortest telomeres reach their minimal functional length, ~300 bp (Heacock et al. 2004). Sequence analysis of the chromosome fusion junctions shows evidence for extensive exonucleolytic digestion prior to fusion. Multiple, yet distinct, NHEJ (nonhomologous end-joining) pathways contribute to the fusion of dysfunctional telomeres in *Arabidopsis* (Heacock et al. 2004). Junctions involving fusion of telomeric DNA of one chromosome to subtelomeric DNA of another chromosome are prevalent in all genetic backgrounds examined; complex fusion junctions that apparently represent secondary fusion events are rare. (*Blue wavy line*) Telomeric DNA.

mammals (Espejel et al. 2002), possess both Ku-dependent and -independent pathways for joining dysfunctional telomeres. In yeast, the Ku-independent pathway for NHEJ is a more homology-driven process involving the Mre11 complex (Ma et al. 2003). Remarkably, uncapped telomeres in *Arabidopsis* that arise in *ku70 tert mre11* triple mutants still fuse, but microhomology at the junction is no longer favored (Heacock et al. 2004). Thus, *Arabidopsis* harbors at least three different pathways for NHEJ (Ku-dependent, Ku-independent, and Ku/Mre11-independent) (Fig. 4). The robust and redundant nature of end-joining pathways in *Arabidopsis* is further illustrated by the capacity of *Arabidopsis ku70, ku80,* and *lig4* mutants to integrate T-DNA, a pathway known to be mediated by NHEJ (Freisner and Britt 2003; Gallego et al. 2003b; van Attikum et al. 2003).

TELOMERE-BINDING PROTEINS

Although a variety of different telomeric DNA-binding activities have been identified in plants (Zentgraf 1995; Kim et al. 1998; Lee et al. 2000; Yu et al. 2000; Zentgraf et al. 2000; Chen et al. 2001), bioinformatics approaches have offered the most fruitful avenues to investigate telomere proteins. The sequenced genomes from *Arabidopsis* and rice have been mined to uncover genes with similarity to single- and double-strand telomere-binding proteins in yeast and mammals based on conservation within the DNA-binding domains (Table 2). Figure 5 shows a diagram of the *Arabidopsis* telomere with several of the proteins implicated in telomere biology.

The oligonucleotide/oligosaccharide-binding (OB) fold is a structurally conserved feature associated with the single-strand telomere-binding protein Cdc13 from yeast and Pot1 from mammals (Theobald and Wuttke 2004). *Arabidopsis* is unusual in that it harbors two Pot1-like genes dubbed *AtPOT1* and *AtPOT2* (Baumann et al. 2002; Shakirov et al. 2005). *AtPOT1* and *AtPOT2* encode proteins bearing two OB folds near their amino terminus and a carboxy-terminal extension (Fig. 6). The OB-fold regions of these genes display 31% identity to each other and 17% identity to human Pot1. Outside this domain the similarity with human Pot1 drops considerably. Both *AtPOT1* and *AtPOT2* are constitutively expressed at low levels (Shakirov et al. 2005), and both are alternatively spliced (Tani and Murata 2005). Overexpression of dominant negative alleles in transgenic plants indicates that AtPot1 and AtPot2 are functionally nonredundant. AtPot1 influences telomere length regulation, whereas AtPot2 contributes to chromosome end protection (Shakirov et al. 2005). Interestingly, the rice genome contains only a single *POT* gene,

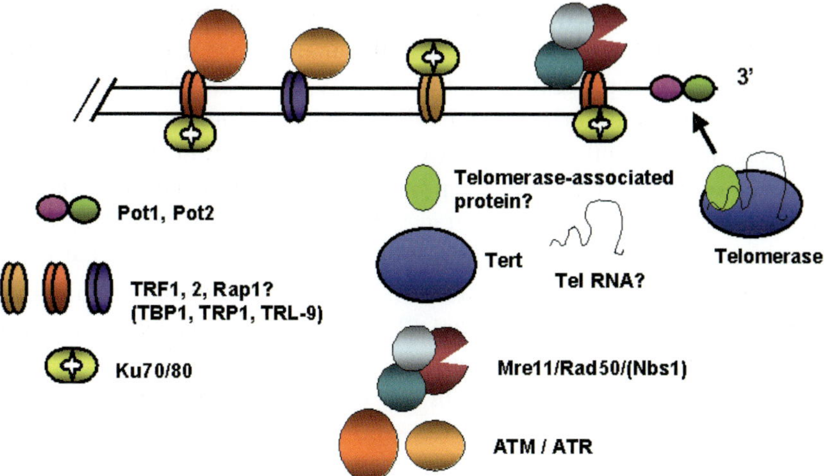

Figure 5. Diagram of the *Arabidopsis* telomere with factors implicated in telomere biology. Depicted are proteins encoded by the *Arabidopsis* genome that have been shown to affect plant telomeres or are implicated in telomere capping and maintenance in other systems. See text for details. Most of the information concerning telomere-associated factors in plants derives from genetic analysis; the physical interactions depicted here have not been verified in vivo. Recent studies indicate that ATM and ATR homologs physically associate with telomeres in yeast (Takata et al. 2005), but the role of these proteins in *Arabidopsis* telomere biology has not yet been explored.

which displays slightly more sequence similarity to *AtPOT2*. Further analysis of Pot proteins in plants is likely to provide new insight into the evolution of single-strand telomere-binding proteins.

Genomics strategies to identify double-strand telomere-binding proteins have focused on genes encoding myb-like DNA-binding domains, a hallmark of this class of proteins, which includes Rap1 and Taz1 from budding and fission yeast, respectively, and the vertebrate TRF1 and TRF2

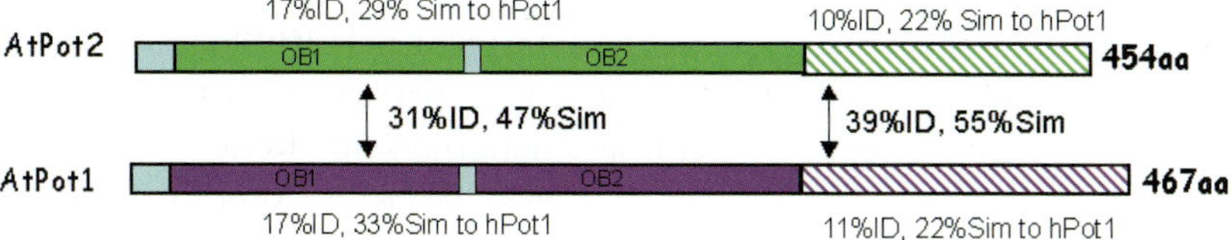

Figure 6. Two POT genes in *Arabidopsis*. Schematic diagram of AtPot1 and AtPot2 proteins compared to each other and to human Pot1. AtPot1 and AtPot2 are predicted to contain two oligonucleotide/oligosccharide-binding (OB) folds near their amino terminus. The carboxy-terminal portion of the proteins is less well conserved. Both AtPo1 and AtPot2 more closely resemble mammalian orthologs than Pot1 proteins from fungi.

(for review, see Rhodes et al. 2002). Although *Arabidopsis* contains more than 100 myb-related genes, biochemical analysis has identified a small subset that encode proteins with the capacity to specifically bind double-stranded plant telomeric DNA in vitro. Such proteins include several relatively small (~30 kD) proteins with a single amino-terminal myb domain (Marian et al. 2003; Schrumpfova et al. 2004), as well as somewhat larger proteins, like TRP1 (Chen et al. 2001) and TBP1 (Hwang et al. 2001) from *Arabidopsis* and NgTRF1 from tobacco (Yang et al. 2003), that more closely resemble TRF1 and TRF2 in size and in architecture (Table 2; Fig. 5). Among this latter group are 12 *Arabidopsis* genes bearing carboxy-terminal myb-like domains, designated TRF-like (TRFL) genes (Karamysheva et al. 2004). TRFL proteins fall into two families based on amino-acid sequence and biochemical properties, and only TRFL family 1 members, which include the previously characterized TRP1 and TBP1, display specificity for double-strand plant telomeric DNA in vitro. Interestingly, the minimal DNA-binding domain for this class of proteins extends beyond the conserved myb-like motif (Karamysheva et al. 2004). Three of the six TRFL family 1 genes appear to have arisen from gene duplication events. Therefore, functional redundancy may account for the inability to observe a telomere-related phenotype with mutants bearing T-DNA insertions in TRFL family 1 genes (L. Vespa and D. Shippen, unpubl.).

The most promising candidate for a double-strand telomere protein from plants is NgTRF1 from tobacco (Yang et al. 2003, 2004). NgTRF1 was identified from a collection of cDNAs encoding myb-containing proteins with similarity to human TRF1 and TRF2. NgTRF1 displays an intriguing cell-cycle-regulated profile of expression that is opposite that of telomerase (Yang et al. 2003), with increased expression in nonproliferating cells. Recombinant NgTRF1, and specifically the myb domain, inhibits telomerase activity on partially double-stranded oligonucleotide substrates in vitro (Yang et al. 2004), although it is unclear whether this is through a direct interaction with the enzyme or via steric hinderance. Overexpression of NgTRF1 in cultured tobacco cells results in telomere shortening, whereas expression of an antisense NgTRF1 construct leads to telomere lengthening. Remarkably, cells overexpressing either wild-type or antisense NgTRF1 undergo apoptosis. In both settings, cultures show diminished viability and TUNEL (terminal deoxynucleotidyl transferase biotin-UTP nick end labeling)-positive cells, as well as classic DNA fragmentation ladders and cytochrome C release from mitochondria (Yang et al. 2004). Together, these findings argue that NgTRF1 not only contributes to telomere length regulation, but also may be important for genome stability and cell cycle progression.

In yeast and mammals, double-strand and single-strand telomere proteins interact with a suite of factors that promote chromosome end protection and telomere replication. Among the interaction partners for TRF1 are Tin2 (Kim et al. 1999), tankyrase (Smith et al. 1998), Pot1 (Loayza and de Lange 2003), and Pip1/PTOP (Liu et al. 2004; Ye et al. 2004). TRF2 binds Rap1 (Li et al. 2000), and both TRF1 and TRF2 associate with a number of proteins involved in the DNA damage response (Hsu et al. 2000; Song et al 2000; Zhu et al. 2000; Kishi et al. 2001). Remarkably, several of the genes encoding key components of mammalian telomeres are absent in *Arabidopsis*; no obvious homologs are evident for tankyrase, Rap1, Tin2, or PTOP/Pip1 (Table 2). Thus, a different ensemble of proteins must have evolved to protect and maintain plant telomeres.

TELOMERES AND THE DNA DAMAGE RESPONSE IN *ARABIDOPSIS*

One of the fascinating conundrums in telomere biology is that proteins required for the DNA damage response, and specifically for NHEJ, associate with telomeres and are necessary to block end-to-end chromosome fusion in wild-type cells. In *Arabidopsis*, three of these factors—Ku70/80 complex, Mre11, and Rad50—have been examined (Table 2).

Ku is an essential component of NHEJ that binds DSBs and facilitates end joining by ligase 4. Ku is also required for telomere protection and maintenance in yeast and mammals, although somewhat different phenotypes are associated with a Ku deficiency in these organisms (Cervantes and Lundblad 2002; Espejel et al. 2002). *Arabidopsis* mutants with a T-DNA insertion in either *Ku70* or *Ku80* are viable and fertile, but they exhibit increased sensitivity to gamma irradiation and methyl methane sulfonate (Bundock et al. 2002; Riha et al. 2002). Unexpectedly, TRF analysis reveals dramatic telomerase-dependent elongation of telomere tracts in both Ku mutants with telomeres reaching a size of more than 20 kb (Bundock et al. 2002; Riha et al. 2002; Gallego et al. 2003a). In the presence or absence of telomerase, Ku70-deficient *Arabidopsis* have extended G-overhangs (Riha and Shippen 2003), suggesting that Ku is required for maintenance of the telomeric C-strand.

Double mutants lacking both Ku70 and Tert display accelerated telomere shortening at a rate that is two to three times faster than in *tert* single mutants (Riha and Shippen 2003). By the third generation, most *ku70 tert* mutants have reached the terminal phenotype and exhibit massive genome instability (Riha and Shippen 2003). Thus, Ku proteins in *Arabidopsis* act synergistically with telomerase in telomere maintenance, but the details of the interaction are unknown.

Another protein complex required for DNA repair and telomere maintenance is MRX (Mre11, Rad50, and Xrs2/Nbs1). The MRX complex, which functions in DSB repair through NHEJ and homologous recombination, localizes to human telomeres and may be involved in telomere replication or t-loop formation (Lydall 2003; Callen and Surralles 2004). Mutations in MRX components are extremely deleterious in mammals, but in *Arabidopsis rad50* and *mre11* mutants are viable (Gallego et al. 2001; Bundock and Hooykaas 2002), providing a unique opportunity to investigate the role of this complex in higher eukaryotes. Plants with a T-DNA disruption in the carboxyl terminus of *RAD50* are sterile, but otherwise show no defects in growth under standard cultivation conditions (Gallego et al. 2001). Interestingly, although telomere length is unperturbed in *rad50* mutants, callus derived from these plants fails to proliferate and exhibits a sudden loss of telomeric tracts after 10 weeks of cultivation. Long-term survivors possess extended telomeres (Gallego and White 2001), indicating that Rad50 contributes in some manner to telomere maintenance in *Arabidopsis*.

Three *mre11* mutants have been described in *Arabidopsis* (Bundock and Hooykaas 2002; Heacock et al. 2004; Puizina et al. 2004). In *mre11-1* and *mre11-3* mutants, a T-DNA insertion disrupts a conserved phosphoesterase motif in the amino terminus, leading to sterility and severe developmental defects. In a third mutant, *mre11-2*, the T-DNA insertion lies near the carboxyl terminus, and plants are fertile, suggesting that the carboxy-terminal domain of Mre11 is dispensable for meiotic functions (Puizina et al. 2004). The contribution of Mre11 to telomere maintenance in *Arabidopsis* is unclear. One study reported that *mre11-1* mutants possessed longer telomeres (Bundock and Hooykaas 2002), whereas analysis of *mre11-3* mutants found no difference in telomere length relative to wild type (Heacock et al. 2004). Nevertheless, because *Arabidopsis mre11* mutants are viable, unlike their counterparts in vertebrate cells (Yamaguchi-Iwai et al. 1999), it should be possible to thoroughly dissect the contribution of the Mre11 complex in telomere biology.

The extraordinary tolerance to genome instability shown by plants raises questions about the interaction of DNA damage response proteins and dysfunctional telomeres. In late-generation telomerase mutants that exhibit multiple anaphase bridges, the mitotic index is similar to wild type, suggesting that prolonged BFB cycles can proceed virtually unchecked in *Arabidopsis* (Riha et al. 2001). In situ hybridization experiments using rDNA probes confirm that chromosomes in late-generation telomerase mutants undergo repeated rounds of BFB cycles (Siroky et al. 2003). In contrast, telomerase-deficient mice harboring chromosome fusions survive only in the absence of an intact checkpoint response (Artandi and DePinho 2000; Gisselsson et al. 2000). A major difference in how plants and mammals respond to dysfunctional telomeres is that mammals react through p53-mediated pathways, frequently initiating programmed cell death. Plants lack a p53 homolog, which may account for the extended survival of telomerase-deficient *Arabidopsis*. In addition, partial aneuploidy may be tolerated, because a significant portion of the *Arabidopsis* genome has been duplicated. The totipotency of plant cells could also enhance survival as cells with severely compromised genomes can be replaced.

SUMMARY

Our understanding of the composition and maintenance of plant telomeres has greatly increased over the past few years. The picture that emerges is one of intriguing divergence in the function of plant telomere components relative to their counterparts in yeast and mammals. Furthermore, whereas plant genomes harbor many sequence homologs of telomere-related genes, several key players in the mammalian telomere complex are conspicuously absent. Further analysis of plant telomeres is likely to uncover a host of novel factors and to reinforce the need for comparative telomere biology to distinguish broadly conserved features from lineage-specific adaptations.

ACKNOWLEDGMENTS

I am grateful to Tom McKnight and members of the Shippen lab for many illuminating discussions and for critically reading the manuscript. Special thanks go to Eugene Shakirov, Michelle Heacock, and Shuxin Ren for help in the preparation of figures. Work in my lab is supported by grants from the National Institutes of Health and the National Science Foundation.

REFERENCES

Adams S.P., Leitch I.J., Bennett M.D., and Leitch A.R. 2000. *Aloe* L.—a second plant family without (TTTAGGG)$_n$ telomeres. *Chromosoma* **109:** 201–205.

Arabidopsis Genome Initiative. 2000. Analysis of the genome sequence of the flowering plant *Arabidopsis thaliana*. *Nature* **408:** 796–815.

Armstrong S.J., Franklin F.C., and Jones G.H. 2001. Nucleolus-associated telomere clustering and pairing precede meiotic chromosome synapsis in *Arabidopsis thaliana*. *J. Cell Sci.* **114:** 4207–4217.

Artandi S.E. and DePinho R.A. 2000. A critical role for telomeres in suppressing and facilitating carcinogenesis. *Curr. Opin. Genet. Dev.* **10:** 39–46.

Askree S.H., Yehuda T., Smolikov S., Gurevich R., Hawk J., Coker C., Krauskopf A., Kupiec M., and McEachern M.J. 2004. A genome-wide screen for *Saccharomyces cerevisiae* deletion mutants that affect telomere length. *Proc. Natl. Acad. Sci.* **101:** 8658–8663.

Baumann P. and Cech T.R. 2000. Protection of telomeres by the Ku protein in fission yeast. *Mol. Biol. Cell* **11:** 3265–3275.

Baumann P., Podell E., and Cech T.R. 2002. Human Pot1 (protection of telomeres) protein: Cytolocalization, gene structure, and alternative splicing. *Mol. Cell. Biol.* **22:** 8079–8087.

Bucholc M. and Buchowicz J. 1995. An extrachromosomal fragment of telomeric DNA in wheat. *Plant Mol. Biol.* **27:** 435–439.

Bucholc M., Park Y., and Lustig A.J. 2001. Intrachromatid excision of telomeric DNA as a mechanism for telomere size control in *Saccharomyces cerevisiae. Mol. Cell. Biol.* **21:** 6559–6573.

Bundock P. and Hooykaas P. 2002. Severe developmental defects, hypersensitivity to DNA-damaging agents, and lengthened telomeres in *Arabidopsis* MRE11 mutants. *Plant Cell* **14:** 2451–2462.

Bundock P., van Attikum H., and Hooykaas P. 2002. Increased telomere length and hypersensitivity to DNA damaging agents in an *Arabidopsis* KU70 mutant. *Nucleic Acids Res.* **30:** 3395–3400.

Burr B., Burr F.A., Matz E.C., and Romero-Severson J. 1992. Pinning down loose ends: Mapping telomeres and factors affecting their length. *Plant Cell* **4:** 953–960.

Callen E. and Surralles J. 2004. Telomere dysfunction in genome instability syndromes. *Mutat. Res.* **567:** 85–104.

Cervantes R.B. and Lundblad V. 2002. Mechanisms of chromosome end-protection. *Curr. Opin. Cell Biol.* **14:** 351–356.

Cesare A.J., Quinney N., Willcox S., Subramanian D., and Griffith J.D. 2003. Telomere looping in *P. sativum* (common garden pea). *Plant J.* **36:** 271–279.

Chaw S.-M., Chang C.-C., Chen H.-L., and Li W.-H. 2004. Dating the monocot-dicot divergence and the origin of core eudicots using whole chloroplast genomes. *J. Mol. Evol.* **58:** 424–441.

Chen C.M., Wang C.T., and Ho C.H. 2001. A plant gene encoding a Myb-like protein that binds telomeric GGTTAG repeats *in vitro. J. Biol. Chem.* **276:** 16511–16519.

Chen J.L., Blasco M.A., and Greider C. 2000. Secondary structure of vertebrate telomerase RNA. *Cell* **100:** 503–514.

Copenhaver G.P. and Pikaard C.S. 1996. RFLP and physical mapping with an rDNA-specific endonuclease reveals that nucleolus organizer regions of *Arabidopsis thaliana* adjoin the telomeres on chromosomes 2 and 4. *Plant J.* **9:** 259–272.

Cox A.V., Bennett S.T., Parokonny A.S., Kenton A., Callimassia M.A., and Bennett M.D. 1993. Comparison of plant telomere locations using a PCR-generated synthetic probe. *Ann. Bot.* **72:** 239–247.

Culligan K. Tissier A., and Britt A. 2004. ATR regulates a G2-phase cell-cycle checkpoint in *Arabidopsis. Plant Cell* **16:** 1091–1104.

Dionne I. and Wellinger R.J. 1996. Cell-cycle regulated generation of single-stranded G-rich DNA in the absence of telomerase. *Proc. Natl. Acad. Sci.* **93:** 13902–13907.

Dong F. and Jiang J. 1998. Non-Rabl patterns of centromere and telomere distribution in the interphase nuclei of plant cells. *Chromosome Res.* **6:** 551–558.

Espejel S., Franco S., Rodriguez-Perales S., Bouffler S.D., Cigudosa J.C., and Blasco M.A. 2002. Mammalian Ku86 mediates chromosomal fusions and apoptosis caused by critically short telomeres. *EMBO J.* **21:** 2207–2219.

Fajkus J., Kovarik A., and Kralovics R. 1996. Telomerase activity in plant cells. *FEBS Lett.* **391:** 307–309.

Fajkus J., Kovarik A., Kralovics R., and Bezdek M. 1995. Organization of telomeric and subtelomeric chromatin in the higher plant *Nicotiana tabacum. Mol. Gen. Genet.* **247:** 633–638.

Fitzgerald M.S., McKnight T.D., and Shippen D.E. 1996. Characterization and developmental patterns of telomerase expression in plants. *Proc. Natl. Acad. Sci.* **93:** 14422–14427.

Fitzgerald M.S., Shakirov E., Hood E., McKnight T.D., and Shippen D.E. 2001. Different modes of *de novo* telomere formation by plant telomerases. *Plant J.* **26:** 1–12.

Fitzgerald M.S., Riha K., Gao F., Ren S., McKnight T.D., and Shippen D.E. 1999. Disruption of the telomerase catalytic subunit gene from *Arabidopsis* inactivates telomerase and leads to a slow loss of telomeric DNA. *Proc. Natl. Acad. Sci.* **96:** 14813–14818.

Fransz P., de Jong J., Lysak M., Castiglione M.R., and Schubert I. 2002. Interphase chromosomes in *Arabidopsis* are organized as well-defined chromocenters from which euchromatin loops emanate. *Proc. Natl. Acad. Sci.* **99:** 14584–14589.

Friebe B., Kynast R.G., Zhang P., Qi L., Dhar M., and Gill B.S. 2001. Chromosome healing by addition of telomeric repeats in wheat occurs during the first mitotic divisions of the sporophyte and is a gradual process. *Chromosome Res.* **9:** 137–146.

Friesner J. and Britt A. 2003. Ku80- and DNA ligase IV-deficient plants are sensitive to ionizing radiation and defective in T-DNA integration. *Plant J.* **34:** 427–440.

Fuchs J., Brandes A., and Schubert I. 1995. Telomere sequence localization and karyotype evolution in higher plants. *Plant Syst. Evol.* **196:** 227–241.

Gallego M.E. and White C.I. 2001. RAD50 function is essential for telomere maintenance in *Arabidopsis. Proc. Natl. Acad. Sci.* **98:** 1711–1716.

Gallego M.E., Jalut N., and White C.I. 2003a. Telomerase dependence of telomere lengthening in Ku80 mutant *Arabidopsis. Plant Cell* **15:** 782–789.

Gallego M.E., Bleuyard J.Y., Daoudal-Cotterell S., Jallut N., and White C.I. 2003b. Ku80 plays a role in non-homologous recombination but is not required for T-DNA integration in *Arabidopsis. Plant J.* **35:** 557–565.

Gallego M.E., Jeanneau M., Granier F., Bouchez D., Bechtold N., and White C.I. 2001. Disruption of the *Arabidopsis* RAD50 gene leads to plant sterility and MMS sensitivity. *Plant J.* **25:** 31–41.

Ganal M.W., Lapitan N.L., and Tanksley S.D. 1991. Macrostructure of the tomato telomeres. *Plant Cell* **3:** 87–94.

Garcia V., Salanoubat M., Choisne N., and Tissier A. 2000. An ATM homologue from *Arabidopsis thaliana*: Complete genomic organisation and expression analysis. *Nucleic Acids Res.* **28:** 1692–1699.

Gisselsson D., Pettersson L., Hoglund M., Heidenblad M., Gorunova L., Wiegant J., Mertens F., Dal Cin P., Mitelman F., and Mandahl N. 2000. Chromosomal breakage-fusion-bridge events cause genetic intratumor heterogeneity. *Proc. Natl. Acad. Sci.* **97:** 5357–5362.

Griffith J.D., Comeau L., Rosenfield S., Stansel R.M., Bianchi A., Moss H., and de Lange T. 1999. Mammalian telomeres end in a large duplex loop. *Cell* **97:** 503–514.

Hartung F. and Puchta H. 1999. Isolation of the complete cDNA of the Mre11 homologue of *Arabidopsis* (accession No. AJ243822) indicates conservation of DNA recombination mechanisms between plants and other eukaryotes. *Plant Physiol.* **121**: 312.

Heacock M., Spangler E., Riha K., Puizina J., and Shippen D.E. 2004. Molecular analysis of telomere fusions in *Arabidopsis*: Multiple pathways for chromosome end-joining. *EMBO J.* **23**: 2304–2313.

Hefner E., Pruess S.B., and Britt A.B. 2002. *Arabidopsis* mutants sensitive to gamma radiation include the homologue of the human repair gene ERCC1. *J. Exp. Bot.* **54**: 669–680.

Heller K., Kilian A., Piatyszek M.A., and Kleinhofs A. 1996. Telomerase activity in plant extracts. *Mol. Gen. Genet.* **252**: 342–345.

Heller-Uszynska K., Schnippenkoetter W., and Kilian A. 2002. Cloning and characterization of rice (*Oryza sativa* L) telomerase reverse transcriptase, which reveals complex splicing patterns. *Plant J.* **31**: 75–86.

Hemann M.T. and Greider C.W. 1999. G-strand overhangs on telomeres in telomerase-deficient mouse cells. *Nucleic Acids Res.* **27**: 3964–3969.

Higashiyama T., Maki S., and Yamada T. 1995. Molecular organization of *Chlorella vulgaris* chromosome 1: Presence of telomeric repeats that are conserved in higher plants. *Mol. Gen. Genet.* **246**: 29–36.

Hsu H.-L., Gilley D., Galande S.A., Hande M.P., Allen B., Kim S.-H., Li G.C., Campisi J., Kohwi-Shigematsu T., and Chen D.J. 2000. Ku acts in a unique way at the mammalian telomere to prevent end joining. *Genes Dev.* **14**: 2807–2818.

Hwang M.G., Chung I.K., Kang B.G., and Myeon H.C. 2001. Sequence-specific binding property of *Arabidopsis thaliana* telomeric DNA binding protein 1 (AtTBP1). *FEBS Lett.* **503**: 35–40.

Karamysheva Z., Surovtseva Y., Vespa L., Shakirov E.V., and Shippen D.E. 2004. A C-terminal Myb-extension domain defines a novel family of double-strand telomeric DNA binding proteins in *Arabidopsis*. *J. Biol. Chem.* **279**: 47799–47807.

Kilian A. and Kleinhofs A. 1992. Cloning and mapping of telomere-associated sequences from *Hordeum vulgare* L. *Mol. Gen. Genet.* **235**: 153–156.

Kilian A., Heller K., and Kleinhofs A. 1998. Development patterns of telomerase activity in barley and maize. *Plant Mol. Biol.* **37**: 621–628.

Kilian A., Stiff C., and Kleinhofs A. 1995. Barley telomeres shorten during differentiation but grow in callus culture. *Proc. Natl. Acad. Sci.* **92**: 9555–9559.

Kim J.H., Kim W.T., and Chung I.K. 1998. Rice proteins that bind single-stranded G-rich telomere DNA. *Plant Mol. Biol.* **36**: 661–672.

Kim N.W., Piatyszek M.A., Prowse K.R., Harley C.B., West M.D., Ho P.L.C., Coviello G.M., Wright W.E., Weinrich S.L., and Shay J.W. 1994. Specific association of human telomerase activity with immortal cells and cancer. *Science* **266**: 2011–2015.

Kim S.H., Kaminker P., and Campisi J. 1999. TIN2, a new regulator of telomere length in human cells. *Nat. Genet.* **23**: 405–412.

Kishi S., Zhou X.Z., Ziv Y., Khoo C., Hill D.E., Shiloh Y., and Lu K.P. 2001. Telomeric protein Pin2/TRF1 as an important ATM target in response to double strand DNA breaks. *J. Biol. Chem.* **276**: 29282–29291.

Lee J.H., Kim J.H., Kim W.T., Kang B.G., and Chung I.K. 2000. Characterization and developmental expression of single-stranded telomeric DNA-binding proteins from mung bean (*Vigna radiata*). *Plant Mol. Biol.* **42**: 547–557.

Lepiniec L., Babiychuk E., Kushnir S., Van Montagu M., and Inzé D. 1995. Characterization of an *Arabidopsis thaliana* cDNA homologue to animal poly(ADP-ribose) polymerase. *FEBS Lett.* **364:** 103–108.

Li B. and Lustig A.J. 1996. A novel mechanism for telomere size control in *Saccharomyces cerevisiae*. *Genes Dev.* **10:** 1310–1326.

Li B., Oestreich S., and de Lange T. 2000. Identification of human Rap1: Implications for telomere evolution. *Cell* **101:** 471–483.

Lin J. and Blackburn E.H. 2004. Nucleolar protein PinX1p regulates telomerase by sequestering its protein catalytic subunit in an inactive complex lacking telomerase RNA. *Genes Dev.* **18:** 387–396.

Liu D., Safari A., O'Connor M.S., Chan D.W., Laegeler A., Qin J., and Songyang Z. 2004. PTOP interacts with POT1 and regulates its localization to telomeres. *Nat. Cell Biol.* **6:** 673–680.

Loayza D. and de Lange T. 2003. POT1 as a terminal transducer of TRF1 telomere length control. *Nature* **424:** 1013–1018.

Lundblad V. 2003. Telomere replication: An Est fest. *Curr. Biol.* **13:** R439–R441.

Lydall D. 2003. Hiding at the ends of yeast chromosomes: Telomeres, nucleases and checkpoint pathways. *J. Cell Sci.* **116:** 4057–4065.

Ma J.-L., Kim E.M., Haber J.E., and Lee S.E. 2003. Yeast Mre11 and Rad1 proteins define a Ku-independent mechanism to repair double-strand breaks lacking overlapping end sequences. *Mol. Cell. Biol.* **23:** 8820–8828.

Maceluch J., Kmieciak M, Szweykowska-Kulinska Z., and Jarmolowski A. 2001. Cloning and characterization of *Arabidopsis thaliana* AtNAP57—a homologue of yeast pseudouridine synthase Cbf5p. *Acta Biochim. Pol.* **48:** 699–709.

Marian C.O., Bordoli S.J., Goltz M., Santarella R.A., Jackson L.P., Danilevskaya O., Beckstette M., Meeley R., and Bass H.W. 2003. The maize single myb histone 1 gene, Smh1, belongs to a novel gene family and encodes a protein that binds telomere DNA repeats *in vitro*. *Plant Physiol.* **133:** 1336–1350.

Mason P.J. 2003. Stem cells, telomerase and dyskeratosis congenita. *BioEssays* **25:** 126–133.

McClintock B. 1939. The behavior in successive nuclear divisions of a chromosome broken at meiosis. *Proc. Natl. Acad. Sci.* **25:** 405–416.

———. 1941. The stability of broken ends of chromosomes of *Zea mays*. *Genetics* **26:** 234–282.

McCormick–Graham M., Haynes W.J., and Romero D.P. 1997. Variable telomeric repeat synthesis in *Paramecium tetraurelia* is consistent with misincorporation by telomerase. *EMBO J.* **16:** 3233–3242.

Meyerowitz E.M. 2002. Plants compared to animals: The broadest comparative study of development. *Science* **295:** 1482–1485.

Mieczkowski P.A., Mieczkowska J.O., Dominaska M., and Petes T.D. 2003. Genetic regulation of telomere-telomere fusions in the yeast *Saccharomyces cerevisiae*. *Proc. Natl. Acad. Sci.* **100:** 10854–10859.

Murata M., Nakata N., and Yasumoro Y. 1992. Origin and molecular structure of a midget chromosome in a common wheat carrying rye cytoplasm. *Chromosoma* **102:** 27–31.

Oguchi K., Hongtu L., Tamura K., and Takahashi H. 1999. Molecular cloning and characterization of *AtTERT*, a telomerase reverse transcriptase homolog in *Arabidopsis thaliana*. *FEBS Lett.* **457:** 465–469.

Petracek M.E., Lefebvre P.A., Silflow C.D., and Berman J. 1990. Chlamydomonas telomere sequences are A+T rich but contain three consecutive G-C base pairs. *Proc. Natl. Acad. Sci.* **87:** 8222–8226.

Pich U. and Schubert I. 1998. Terminal heterochromatin and alternative telomeric sequences in *Allium cepa*. *Chromosome Res.* **6:** 315–321.

Pich U., Fuchs J., and Schubert I. 1996. How do Alliaceae stabilize their chromosome ends in the absence of TTTAGGG sequences? *Chromosome Res.* **4:** 207–213.

Puizina J., Siroky J., Mokros P., Schweizer D., and Riha K. 2004. Mre11 deficiency in *Arabidopsis* is associated with chromosomal instability in somatic cells and Spo11-dependent genome fragmentation during meiosis. *Plant Cell* **16:** 1968–1978.

Puizina J., Weiss-Schneeweiss H., Pedrosa-Harand A., Kamenjarin J., Trinajstic I., Riha K., and Schweizer D. 2003. Karyotype analysis in *Hyacinthella dalmatica* (Hyacinthaceae) reveals vertebrate-type telomere repeats at the chromosome ends. *Genome* **46:** 1070–1076.

Regad F., Lebas M., and Lescure B. 1994. Interstitial telomeric repeats within the *Arabidopsis thaliana* genome. *J. Mol. Biol.* **239:** 163–169.

Reichenbach P., Hoss M., Azzalin C.M., Nabholz M., Bucher P., and Lingner J. 2003. A human homolog of yeast Est1 associates with telomerase and uncaps chromosome ends when overexpressed. *Curr. Biol.* **13:** 568–574.

Ren S., Johnston J.S., Shippen D.E., and McKnight T.D. 2004. Telomerase activator 1 induces telomerase activity and potentiates responses to auxin in *Arabidopsis*. *Plant Cell* **16:** 2910–2922.

Rhodes D., Fairall L., Simonsson T., Court R., and Chapman L. 2002. Telomere architecture. *EMBO Rep.* **3:** 1139–1145.

Richards E.J. and Ausubel F.M. 1988. Isolation of a higher eukaryotic telomere from *Arabidopsis thaliana*. *Cell* **53:** 127–136.

Riha K. and Shippen D.E. 2003. Ku is required for telomeric C-rich strand maintenance, but not for end-to-end fusions in *Arabidopsis*. *Proc. Natl. Acad. Sci.* **100:** 611–615.

Riha K., Fajkus J., Siroky J., and Vyskot B. 1998. Developmental control of telomere length and telomerase activity in plants. *Plant Cell* **10:** 1691–1698.

Riha K., McKnight T.D., Griffing L., and Shippen D.E. 2001. Living with genome instability: Plant responses to telomere dysfunction. *Science* **291:** 1797–1800.

Riha K., Watson J.M., Parkey J., and Shippen D.E. 2002. Telomere length deregulation and enhanced sensitivity to genotoxic stress in *Arabidopsis* mutants deficient in Ku70. *EMBO J.* **21:** 2819–2826.

Riha K., Fajkus J., McKnight T.D., Vyskot B., and Shippen D.E. 2000. Analysis of the G-overhang structures on plant telomeres: Evidence for two distinct telomere architectures. *Plant J.* **23:** 1–11.

Schrumpfova P., Kuchar M., Mikova G., Skrisovska L., Kubicarova T., and Fajkus J. 2004. Characterization of two *Arabidopsis thaliana* myb-like proteins showing affinity to telomeric DNA sequence. *Genome* **47:** 316–324.

Shakirov E.V. and Shippen D.E. 2004. Length regulation and dynamics of individual telomere tracts in wild-type *Arabidopsis*. *Plant Cell* **16:** 1959–1967.

Shakirov E.V., Surovtseva Y.V., Osbun N., and Shippen D.E. 2005. The *Arabidopsis* pot1 and pot2 proteins function in telomere length homeostasis and chromosome end protection. *Mol. Cell. Biol.* **25:** 7725–7733.

Siroky J., Zluvova J., Riha K., Shippen D.E., and Vyskot B. 2003. Rearrangements of ribosomal DNA clusters in late generation telomerase-deficient *Arabidopsis*. *Chromosoma* **112:** 116–123.

Smith S., Giriat I., Schmitt A., and de Lange T. 1998. Tankyrase, a poly(ADP-ribose) polymerase at human telomeres. *Science* **282:** 1484–1487.

Snow B.E., Erdmann N., Cruickshank J., Goldman H., Gill R.M., Robinson M.O., and Harrington L. 2003. Functional conservation of the telomerase protein Est1p in humans. *Curr. Biol.* **13:** 698–704.

Song K., Jung D., Jung Y., Lee S.G., and Lee I. 2000. Interaction of human Ku70 with TRF2. *FEBS Lett.* **8:** 81–85.

Sykorova E., Lim K.Y., Chase M.W., Knapp S., Leitch I.J., Leitch A.R., and Fajkus J. 2003. The absence of *Arabidopsis*-type telomeres in *Cestrum* and closely related genera *Vestia* and *Sessea* (Solanaceae): First evidence from eudicots. *Plant J.* **34:** 283–291.

Takata H., Tanaka Y., and Matsuura A. 2005. Late S phase-specific recruitment of Mre11 complex triggers hierarchical assembly of telomere replication proteins in *Saccharomyces cerevisiae*. *Mol. Cell* **17:** 573–583.

Tamura K., Liu H., and Takahashi H. 1999. Auxin induction of cell cycle regulated activity of tobacco telomerase. *J. Biol. Chem.* **274:** 20997–21002.

Tamura K., Adachi Y., Chiba K., Oguchi K., and Takahashi H. 2002. Identification of Ku70 and Ku80 homologues in *Arabidopsis thaliana*: Evidence for a role in the repair of DNA double-strand breaks. *Plant J.* **29:** 771–781.

Tani A. and Murata M. 2005. Alternative splicing of Pot1 (Protection of telomere)-like genes in *Arabidopsis thaliana*. *Genes Genet. Syst.* **80:** 41–48.

Teixeira M.T., Arneric M., Sperisen P., and Lingner J. 2004. Telomere length homeostasis is achieved via a switch between telomerase-extendible and -nonextendible states. *Cell* **117:** 323–335.

Theobald D.L. and Wuttke D.S. 2004. Prediction of multiple tandem OB-fold domains in telomere end-binding proteins Pot1 and Cdc13. *Structure* **12:** 1877–1879.

Tremousaygue D., Manevski A., Bardet C., Lescure N., and Lescure B. 1999. Plant interstitial telomere motifs participate in the control of gene expression in root meristems. *Plant J.* **20:** 553–561.

Tsujimoto H. 1993. Molecular cytological evidence for gradual telomere synthesis at the broken chromosome ends in wheat. *J. Plant Res.* **103:** 239–244.

Tsujimoto H., Yamada T., and Sasakuma T. 1997. Molecular structure of a wheat chromosome end healed after gametocidal gene-induced breakage. *Proc. Natl. Acad. Sci.* **94:** 3140–3144.

Tsujimoto H., Usami N., Hasegawa K., Yamada T., Nagaki K., and Sasakuma T. 1999. *De novo* synthesis of telomere sequences at the healed breakpoints of wheat deletion chromosomes. *Mol. Gen. Genet.* **262:** 851–879.

van Attikum H., Bundock P., Overmeer R.M., Lee L.Y., Gelvin S.B., and Hooykaas P.J. 2003. The *Arabidopsis AtLIG4* gene is required for the repair of DNA damage, but not for the integration of *Agrobacterium* T-DNA. *Nucleic Acids Res.* **31:** 4247–4255.

Wang R.C., Smogorzewska A., and de Lange T. 2004. Homologous recombination generates T-loop-sized deletions at human telomeres. *Cell* **119:** 355–368.

Weiss H. and Scherthan H. 2002. *Aloe* spp.—plants with vertebrate-like telomeric sequences. *Chromosome Res.* **10:** 155–164.

Weiss-Schneeweiss H., Riha K., Jang C.G., Puizina J., Scherthan H., and Schweizer D. 2004. Chromosome termini of the monocot plant *Othocallis siberica* are maintained by telomerase, which specifically synthesizes vertebrate-type telomere sequences. *Plant J.* **37:** 484–493.

Werner J.E., Kota R.S., and Gill B.S. 1992. Distribution of telomeric repeats and their role in the healing of broken chromosome ends in wheat. *Genome* **35:** 844–848.

Williams B. and Lustig A.J. 2003. The paradoxical relationship between NHEJ and telomeric fusion. *Mol. Cell* **11:** 1125–1126.

Yamaguchi-Iwai Y., Sonoda E., Sasaki M.S., Morrison C., Haraguchi T., Hiraoka Y., Yamashita Y., Yagi T., Takata M., Price C.M., Kakazu N., and Takeda S. 1999. Mre11 is essential for the maintenance of chromosomal DNA in vertebrate cells. *EMBO J.* **18:** 6619–6629.

Yang S.W., Sung K.K., and Kim W.T. 2004. Perturbation of NgTRF1 expression induces apoptosis-like cell death in tobacco BY-2 cells and implicates NgTRF1 in the control of telomere length and stability. *Plant Cell* **16:** 3370–3385.

Yang S.W., Jin E., Chung I.K., and Kim W.T. 2002. Cell cycle-dependent regulation of telomerase activity by auxin, abscisic acid and protein phosphorylation in tobacco BY-2 suspension culture cells. *Plant J.* **29:** 617–626.

Yang S.W., Kim D.H., Lee J.J., Chun Y.J., Lee J.-H., Kim Y.J., Chung I.K., and Kim W.T. 2003. Expression of the telomere repeat binding factor gene NgTRF1 is closely coordinated with the cell division program in tobacco BY-2 suspension culture cells. *J. Biol. Chem.* **278:** 21395–21407.

Ye J.Z., Hockemeyer D., Krutchinsky A.N., Loayza D., Hooper S.M., Chait B.T., and de Lange T. 2004. POT1-interacting protein PIP1: A telomere length regulator that recruits POT1 to the TIN2/TRF1 complex. *Genes Dev.* **18:** 1649–1654.

Yu E.Y., Kim S.E., Kim J.H., Ko J.H., Cho M.H., and Chung I.K. 2000. Sequence-specific DNA recognition by the Myb-like domain of plant telomeric protein RTBP1. *J. Biol. Chem.* **275:** 24208–24214.

Zentgraf U. 1995. Telomere-binding proteins of *Arabidopsis thaliana*. *Plant Mol. Biol.* **27:** 467–475.

Zentgraf U., Hinderhofer K., and Kolb D. 2000. Specific association of a small protein with the telomeric DNA complex during the onset of leaf senescence in *Arabidopsis thaliana*. *Plant Mol. Biol.* **42:** 429–438.

Zhu X.D., Kuster B., Mann M., Petrini J.H.J., and de Lange T. 2000. Cell-cycle-regulated association of Rad50/Mre11/Nbs1 with TRF2 and human telomeres. *Nat. Genet.* **25:** 347–352.

Appendix: A Personal Account of the Discovery of Telomerase

Elizabeth H. Blackburn

Department of Biochemistry and Biophysics
University of California, San Francisco
San Francisco, California 94143–2200

DID A NEW ENZYME EXIST IN CELLS THAT COULD SYNTHESIZE TELOMERIC DNA?

The molecular features of telomeric DNA were not consistent with any of the properties expected from the models for telomere replication that had become current in the 1970s. In the early 1980s, four lines of evidence led me to think that a new enzyme activity might work on telomeres to elongate them. Each consisted of molecular observations on telomeric DNA that could not be readily explained otherwise.

First, telomeric GGGGTT repeat tracts on minichromosomes in the ciliates *Tetrahymena thermophila* and *Glaucoma chattoni* were heterogeneous in length (Blackburn and Gall 1978; Yao et al. 1979; Katzen et al. 1981). Second, telomeric GGGGTT repeat tract DNA was found added to various sequences in ciliate minichromosomes as a result of the process by which new telomeres formed on chromosomes during development of the somatic nucleus. This stage in the ciliate life cycle had an intriguing parallel in the postfertilization phase when maize chromosomes could heal and when *Ascarid* chromosomes fragmented to form new chromosomes with stable ends. Meng-Chao Yao, working in Martin Gorovsky's lab and then Joe Gall's lab, had found such telomeric DNA acquisition for *Tetrahymena* rDNA minichromosomes, and in my lab at the University of California at Berkeley, we had made similar observations for other, non-rDNA telomeres of the somatic nucleus (Yao et al. 1979; King and Yao 1982; Blackburn et al. 1983).

Figure 1. (*Left* to *right*) Peter Challoner, Janis Shampay, and Carol Greider, at the time of the research described in the text.

Simultaneously, David Prescott's group in Colorado had made the corresponding observation for the telomeres of somatic minichromosomes in a ciliate of a different class, the hypotrichous ciliate *Oxytricha* (Boswell et al. 1982). Third, Piet Borst and his collaborators in Holland had found that a gene of trypanosomes encoding one of its variant surface antigens—a group of antigens that has important roles in the parasite's ability to evade the host's immune response—was located on a telomeric restriction fragment. Following this gene's restriction fragment over time, they observed that the telomeric DNA apparently gradually grew longer as trypanosome cells multiplied (Bernards et al. 1983). Fourth, sequencing by graduate student Janis Shampay in my lab (Fig. 1) had demonstrated that yeast telomeric TG_{1-3} repeat DNA was added directly to the ends of *Tetrahymena* T_2G_2 repeat telomeres maintained in yeast (Szostak and Blackburn 1982; Shampay et al. 1984), as described in Chapter 1.

Finally, a letter to me from Barbara McClintock in 1983 spurred my thinking that a new activity might exist and act on telomeres. In it, she told that long ago she had identified a maize mutant that, by cytogenetic criteria, had lost the normal capacity to heal broken chromosome ends early in plant development (B. McClintock, pers. comm.; McClintock's carbon copy of this letter is reproduced in http://profiles.nlm.nih.gov/LL/B/B/D/W/). I was struck by the implication that a fully functional telomere ("healed end") was generated from a broken chromosome end not just by chance, but by an active, developmentally controlled process. This suggested that there is a gene (which by definition could be mutated) associated with the ability to heal. Together, I believed that these lines of evidence were pointing to the possibility that a new enzyme activity might exist that could add telomeric DNA to DNA ends.

My conviction—that an activity that could make telomeric DNA should be sought—was further galvanized by a conversation on this topic I had with Barbara McClintock on a visit to Cold Spring Harbor Laboratory, at around the time when the other lines of evidence were accumulating. Thus, I became determined to look for such an enzyme. I was also encouraged by a more biochemically concrete suggestion of an enzyme activity in *Tetrahymena* that could make telomeric DNA longer. This had come from a serendipitous result obtained in 1982 by one of my first graduate students, Peter Challoner (Fig. 1). Peter was doing experiments in which he incubated DNA molecules in extracts of mating *Tetrahymena* cells, prepared at the stage when telomeres might be added to newly forming somatic chromosomal ends.

The most obvious time for telomere addition, I reasoned, would be at a particular stage after mating, when *Tetrahymena* was in the process of generating its hundreds of new telomeres in the newly forming somatic nucleus, and also replicating its rDNA minichromosomes into thousands of copies. If there were an enzyme that synthesized telomeric DNA, this would be a time when great demands would be placed on it, and it presumably would be ramped up to high levels. In addition, through the work of David Nanney, Peter Bruns, Ed Orias, Sally Allen, and their colleagues in the ciliate genetics field, *Tetrahymena* had been made into a powerful experimental system in which the whole cell population could be synchronized for studying developmental processes following mating. We already knew when the rDNA was first amplified and when the DNA fragmentation that leads to new telomere synthesis was occurring. In contrast to the need to wait for many generations before one could observe yeast transformants, these aspects of ciliate cell reproduction took place over approximately 1 hour. That was the time to catch *Tetrahymena* cells and make extracts to look for any of this hypothetical activity.

In such extracts, Peter Challoner had incubated 1.6-kb "mini-rDNA" molecules he had constructed in vitro, which had tracts of CCCCAA/ GGGGTT repeats at both ends. He adapted a reaction mix that Tom Cech and colleagues (Zaug and Cech 1980) had used to study rRNA transcription and processing in vitro (and with which they discovered self-splicing RNA). Peter had wanted to see if he could persuade these mini-rDNAs to turn into palindromic molecules. Drawing from the known biochemical needs of bacterial or bacteriophage DNA replication

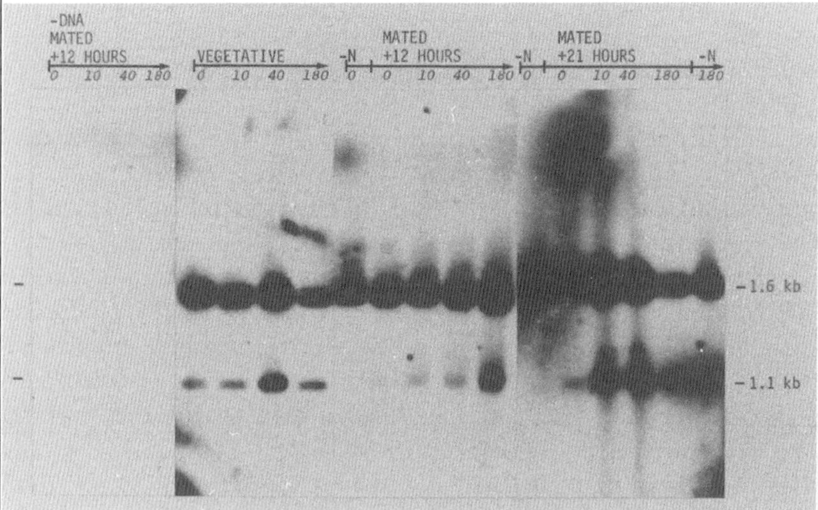

Figure 2. Figure and text reprinted, with permission, from Peter B. Challoner's thesis (Challoner 1984).

A second activity observed in the nuclear extracts can be seen in the difference between the 1.1 kb cleavage products in the 180 minute time points of vegetative versus 12 hour mated cells. While the 1.1 kb molecule remained homogeneous throughout the vegetative incubation, it was converted into a heterogeneous size class from ≈ 1.25 to 1.1 kb during incubation in the nuclear extract from 12 hour mated cells. Thus, this second activity is specific to nuclei from mating cells, and causes an apparent increase in the molecular weight of the linear 1.1 kb cleavage product. Although not proven by the data in Figure 35, such a result is consistent with the addition of telomeric CCCCAA repeats to free DNA termini, as is required during macronuclear development. Therefore, while these preliminary _in vitro_ incubation experiments do not provide conclusive information concerning the mechanism of rDNA palindromization, they do suggest that nuclear extracts may be exploited to study other macronuclear development processes.

Figure 2. (_Continued_)

in vitro and the knowledge that DNA replication involves discontinuous synthesis by a primase-initiated process, in which short RNA primers are used to start each growing DNA polynucleotide chain, he included all four dNTPs as well as rNTPs. Although he did not find palindromic molecules, intriguingly he did note that after 180 minutes of incubation, the terminal restriction fragment bands (which ended in a tract of CCCCAA/GGGGTT repeats) got fuzzier-looking when fractionated by gel electrophoresis; thus, they had become heterogeneous during the course of the incubation with mated cell extract (Fig. 2). They also migrated slower, as though they were an average of about 100 bp longer than the original input telomeric fragments (Challoner 1984). Although not unambiguous evidence of addition of telomeric DNA, a lead had nonetheless been inadvertently provided by these seemingly unrelated experiments.

I decided first to look for more direct evidence that telomeric repeats were added to DNA fragments in such extracts. It was logical to presume that a still-hypothetical telomere-addition activity was acting in a setting in which DNA synthesis was occurring, because both DNA strands of the telomeres would ultimately have to be made. Also, DNA replication needs energy, so I included an ATP-regenerating enzyme system as well as all four dNTPs and rNTPs. Finally, my idea was that telomeric DNA would be best added to preexisting telomeres, but perhaps also to the nontelomeric ends, because they are the substrates for telomere addition during somatic nuclear development. I decided to cover all bases by starting with a mixture of DNA restriction fragments that had both a telomeric end and various nontelomeric ends as substrates.

I incubated extracts made from freshly mated *Tetrahymena* cells with the mixture of DNA restriction fragment substrates, and other dNTPs, rNTPs, Mg^{++} ion and other salts, buffers, and the ATP-regenerating enzyme and its substrate. Then, after different times of incubation, the DNA products were purified, fractionated by gel electrophoresis, and analyzed by Southern blotting with probes for the input restriction fragments and, most critically, with a probe for telomeric CCCCAA repeats. Two things emerged. Several of the fragments acquired CCCCAA hybridization. Even better, at least one new band that hybridized with the CCCCAA probe appeared. It was fuzzy, with an appearance reminiscent of telomeric DNA fragments, just as one would expect if newly added telomeric DNA were being added to a cleavage or rearrangement product of a restriction fragment that I had put into the mix. Its intensity increased with time of incubation with the extract, consistent with enzyme action in a biochemical experiment. The next month I showed the resulting autoradiograms in a slide at the April 1984 UCLA Keystone Symposium on Genome Rearrangements in Steamboat Springs, Colorado (**Fig.** 3).

In April 1984, Carol Greider joined the lab as a graduate student, and to my delight, she was willing and eager to start work on this project (see Fig. 1). As recounted more fully elsewhere (Greider and Blackburn 2004), the assay for a telomere synthesis enzyme had to be greatly streamlined by laborious trial and error. It was necessary to identify the optimal DNA substrate for a potential telomere synthesis enzyme and to test other *Tetrahymena* extract and reaction conditions. The innovation of using DNA sequencing-type gels to resolve the reaction products greatly helped, as well as including a radiolabeled dNTP, ultimately $[^{32}P]dGTP$, as the labeled nucleoside triphosphate substrate.

December of 1984 brought the clearest sign to date of a potential telomerase activity (although it was not yet named telomerase, a name coined by Claire Wyman, who was another graduate student in the lab). A synthetic DNA oligonucleotide of the telomere G-strand TTGGGG was tested. Such a DNA primer substrate could now be added at micromolar

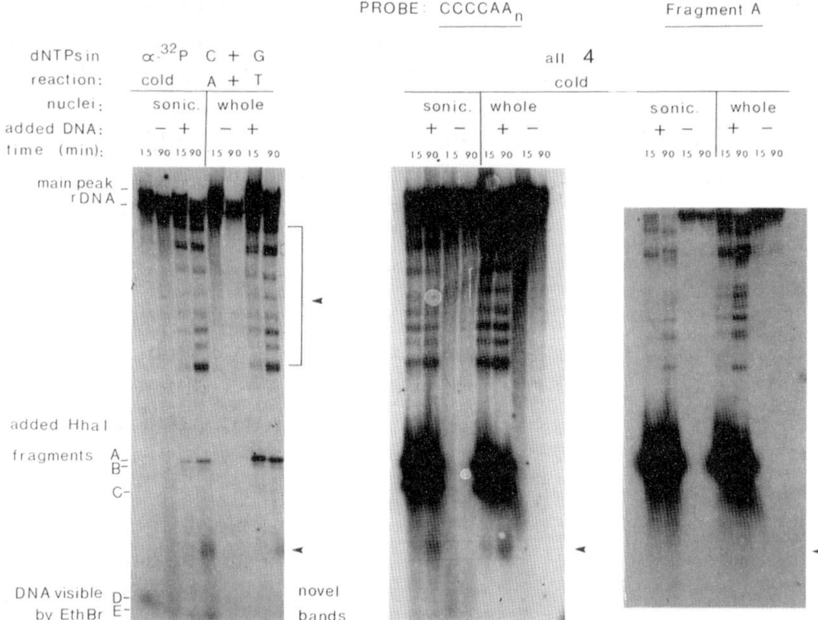

Figure 3. Early experiments by E.H.B. incubating restriction fragments with *Tetrahymena* extracts, showing acquisition of CCCCAA repeat hybridization. Fragment A was a DNA fragment from a bacterial plasmid.

concentrations, rather than the nanomolar concentrations which had been the highest concentrations that could be achieved with restriction fragments. This (TTGGGG)$_4$ oligomer primed synthesis of DNA products that formed a repeating pattern, with an apparent six-base periodicity visible, extending up to the top of the gel—precisely what was expected if a telomere synthesis enzyme had been acting in the assay (shown schematically in Fig. 4A). Furthermore, the activity was stronger when the extracts were made from mated cells during the period when telomeric repeats were added to the newly fragmenting chromosomes, although repeat synthesis was also clearly evident in vegetatively growing cells.

Several months of work ruled out all the potential artifacts we could think of that might have been potential confounding explanations for these novel findings and established that the added repeats were indeed of the sequence (GGGGTT)$_n$. In addition, a yeast telomeric sequence oligonucleotide worked well as a substrate for a *Tetrahymena* extract activity in vitro (Fig. 4B). As described in Chapter 1, we knew that in vivo, yeast added yeast repeats to linear minichromosomes carrying heterologous (ciliate) telomeres. We did not have yeast telomerase activity

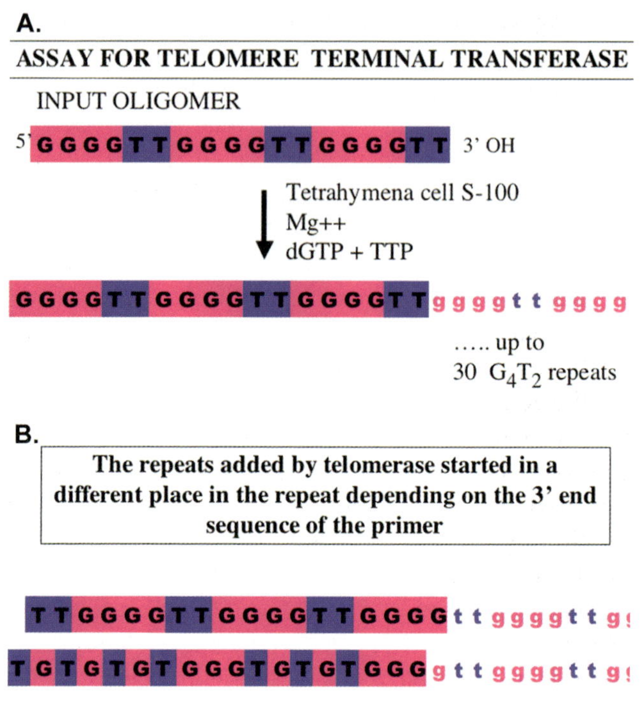

Figure 4. The *Tetrahymena* telomerase reaction, shown schematically. (*A*) Assay using a *Tetrahymena* telomeric sequence primer. (*B*) Assay using a yeast telomeric sequence primer (Greider and Blackburn 1985, 1987).

in vitro then (this had to await my postdoctoral fellow Marita Cohn's subsequent development of an activity assay for this species' telomerase [Cohn and Blackburn 1995]). Therefore, we devised the converse in vitro reaction: A yeast sequence DNA oligomer primer was added to the *Tetrahymena* extract. Strikingly, this TG-rich yeast sequence oligonucleotide was elongated, and tandem TTGGGG hexanucleotide repeats were added, just as they were to the TTGGGG repeat primers. We liked the symmetry of this test of the idea that the basis for the functional conservation of telomeric DNA in vivo was that yeast telomerase had elongated the foreign telomere, extending it with yeast repeats.

The yeast primer experiment had another bonus. Whereas a *Tetrahymena* oligonucleotide primer ending in four Gs at its 3′ end was elongated starting with the apparent addition of two Ts, the yeast primer, which happened to end in three Gs, but had the same length as the all-*Tetrahymena* sequence primer, primed a product that started with an added G, then two Ts, making a product pattern that was offset by one base (Fig. 4). This observation was critical in the recognition that the sequence at the 3′ end of the oligonucleotide determined which bases in the TTGGGG repeat would be added and pointed to what we eventually found was the RNA template in the RNA moiety of telomerase.

In the meantime, the discovery of this new telomeric DNA-synthesizing activity was published in December 1985 (Greider and Blackburn 1985). Excitingly, RNase inactivated the activity. During the next 2 years, the activity was partially purified and its reaction properties were extensively characterized (Greider and Blackburn 1987). However, the RNA was a low-abundance species in the cell and difficult to obtain, and although partial sequence information was obtained by direct RNA sequencing, cloning the gene that encoded the RNA was not accomplished until Carol had left Berkeley and become a Cold Spring Harbor Laboratory Fellow. By 1989, the presence of a 9-nucleotide sequence, 5′-CAACCCCAA-3′, within the RNA had suggested a template mechanism, which was confirmed by biochemical experiments (Fig. 5) (Greider and Blackburn 1989).

Tetrahymena telomerase, we found, added tandem repeats of the *Tetrahymena* telomeric DNA sequence, TTGGGG, onto the 3′ end of any G-rich strand telomeric oligonucleotide primer, independently of an exogenously added nucleic acid template, from nearly all eukaryotes tested (Greider and Blackburn 1985, 1987; Blackburn et al. 1989). As telomerase synthesized only the G-rich strand of telomeres, we proposed that synthesis of the complementary C-rich strand is carried out by primase-polymerase-mediated discontinuous synthesis, typical of semiconservative DNA replication mechanisms, using the extended G-rich strand as a template (Shampay et al. 1984; Greider and Blackburn 1985).

Similar findings were subsequently made for telomerase activities from the ciliates *Oxytricha* and *Euplotes* by Dorothy Shippen-Lentz in my lab and by Alan Zahler in David Prescott's lab, respectively (Zahler and Prescott 1988; Shippen-Lentz and Blackburn 1989), and from human cells by Gregg Morin in Joan Steitz's lab (Morin 1989). Each telomerase synthesized its own species-specific sequence—GGGGTTTT repeats for the hypotrichous ciliates and AGGGTT for human—and had primer-recognition properties similar to those we had found for the *Tetrahymena* telomerase. Like the ciliate telomerase activities, the human cell activity was also ribonuclease-sensitive (Morin 1989), suggesting that it too was

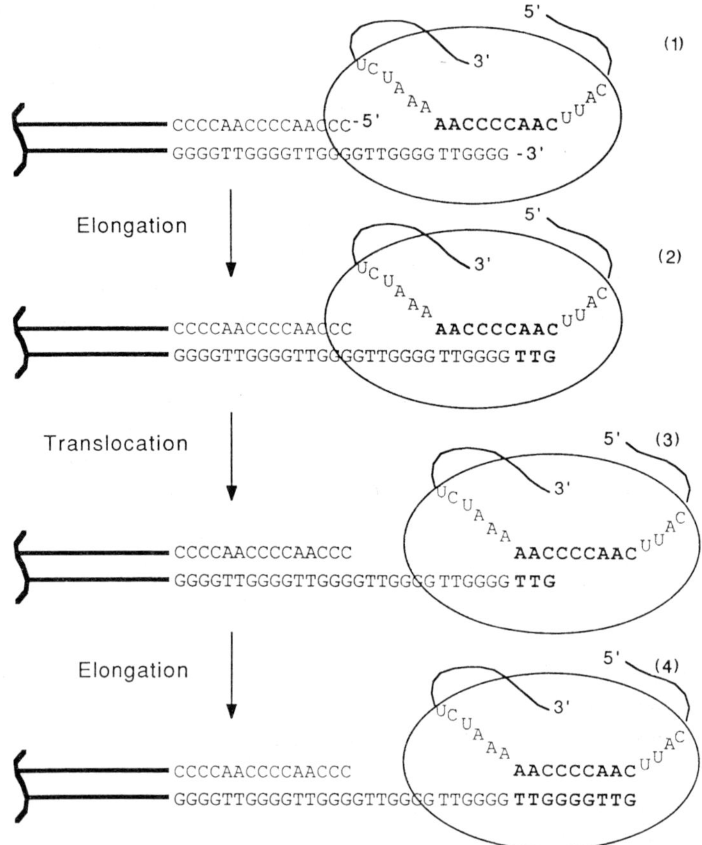

Figure 5. A model for the action of the *Tetrahymena* telomerase. (Reprinted, with permission, from Greider and Blackburn 1989 [©Nature Publishing Group].)

a ribonucleoprotein. Identification of telomerase activity in human cells, in particular, highlighted the generality of this enzyme in eukaryotes.

The RNA moiety of the telomerase of the ciliate *Euplotes* was quickly identified (Shippen-Lentz and Blackburn 1990). It too had a counterpart of the CAACCCCAA sequence of *Tetrahymena* telomerase RNA (Greider and Blackburn 1989), 5'-CAAAACCCCAAAA-3', which Dorothy Shippen showed by functional analyses in vitro was the templating domain for synthesis of GGGGTTTT repeats (Shippen-Lentz and Blackburn 1990), the telomeric sequence of *Euplotes*.

These findings established telomerase as a widespread cellular reverse transcriptase, unusual in synthesizing short repeats and in carrying its own internal RNA template for DNA synthesis. The presence of such an RNA component prompted the speculation that telomerase may be a relic from the time of the evolutionary transition from RNA to DNA genomes (Blackburn 1990). The analogy was drawn to the ability of the self-splicing ribozyme of *Tetrahymena* to polymerize RNA, using an internal RNA template (Zaug and Cech 1986). This suggested the possibility that telomerase is a relic of an RNA replicase ribozyme that became able to synthesize DNA. In this model, as protein components took over this catalytic role, they gave rise to the purely protein reverse transcriptases and DNA-dependent DNA polymerases found in modern eukaryotes and prokaryotes, but in the case of telomerase, the template RNA was retained and is still used to maintain the ends of eukaryotic chromosomes.

DEMONSTRATION OF THE ROLE OF TELOMERASE IN VIVO

We now had good evidence that the enzyme telomerase could synthesize the G-rich strand of telomeric DNA. Proof that the 5′-CAACCCCAA-3′ sequence in the *Tetrahymena* telomerase RNA gene is the template for telomere synthesis came from site-directed mutagenesis of this sequence and overexpression of the mutated gene in *Tetrahymena* cells. These experiments, which established the in vivo role of telomerase, were done by Guo-Liang Yu in my lab, with help from his fellow graduate students Laura Attardi and John Bradley. Guo-Liang injected gene constructs containing template mutant versions of the telomerase RNA gene (TER), borne on a special vector system that he had devised for *Tetrahymena* transformation (Yu et al. 1988). Altered telomere repeats specified by the mutant gene appeared in the telomeres that he cloned out of the transformant cells. This proved that telomerase is indeed the cellular reverse transcriptase enzyme that synthesizes telomeres in cells by copying its own internal RNA template within the TER moiety of the enzyme complex. Importantly, one template mutation caused failure to add any of the predicted sequence DNA onto telomeric ends. Instead, the telomeres progressively shortened as the cells grew for about 20 to 25 more cell divisions, and then the cells senesced. This showed that interference with the telomerase function in the otherwise effectively immortal *Tetrahymena* cells could limit their life span. It established the first role for telomerase in cellular life span and the first indication that its presence was necessary for the replicative immortality of cells (Yu et al. 1990).

Telomerase action, then, could explain how replication of the chromosomal DNA could continue without the progressive loss of terminal sequences predicted to result from normal semiconservative DNA replication; addition of telomeric DNA to the chromosomal ends by telomerase could counterbalance this terminal DNA attrition. Evidence for such attrition had been previously seen for broken *Drosophila* chromosomes that lost terminal DNA sequences progressively (Biessmann and Mason 1988; Levis 1989), and in the dynamic behavior of individual *Saccharomyces cerevisiae* telomeres in vivo (Shampay and Blackburn 1988). The senescence phenotypes of the *Tetrahymena* telomerase RNA mutants (Yu et al. 1990) provided an experimental basis for the suggestion that cell senescence in other organisms could be related to similar attrition of telomeres.

ACKNOWLEDGMENTS

I am grateful for support over the years from many members of my laboratory who have contributed to the work cited here. I appreciatively acknowledge funding support from the National Institutes of Health since 1978 and more recently from the Steven and Michele Kirsch Foundation and the Osher Foundation.

REFERENCES

Bernards A., Michels P.A., Lincke C.R., and Borst P. 1983. Growth of chromosome ends in multiplying trypanosomes. *Nature* **303:** 592–597.

Biessmann H. and Mason J.M. 1988. Progressive loss of DNA sequences from terminal chromosome deficiencies in *Drosophila melanogaster*. *EMBO J.* **7:** 1081–1086.

Blackburn E.H. 1990. Telomeres and their synthesis. *Science* **249:** 489–490.

Blackburn E.H. and Gall J.G. 1978. A tandemly repeated sequence at the termini of the extrachromosomal ribosomal RNA genes in *Tetrahymena*. *J. Mol. Biol.* **120:** 33–53.

Blackburn E.H., Greider C.W., Henderson E., Lee M.S., Shampay J., and Shippen-Lentz D. 1989. Recognition and elongation of telomeres by telomerase. *Genome* **31:** 553–560.

Blackburn E.H., Budarf M.L., Challoner P.B., Cherry J.M., Howard E.A., Katzen A.L., Pan W.C., and Ryan T. 1983. DNA termini in ciliate macronuclei. *Cold Spring Harbor Symp. Quant. Biol.* **47:** 1195–1207.

Boswell R.E., Klobutcher L.A., and Prescott D.M. 1982. Inverted terminal repeats are added to genes during macronuclear development in *Oxytricha nova*. *Proc. Natl. Acad. Sci.* **79:** 3255–3259.

Challoner P.B. 1984. "The organization of the ribosomal RNA genes of Tetrahymenids." Ph. D. thesis, University of California, Berkeley.

Cohn M. and Blackburn E.H. 1995. Telomerase in yeast. *Science* **269:** 396–400.

Greider C.W. and Blackburn E.H. 1985. Identification of a specific telomere terminal transferase activity in *Tetrahymena* extracts. *Cell* **43:** 405–413.

———. 1987. The telomere terminal transferase of *Tetrahymena* is a ribonucleoprotein enzyme with two kinds of primer specificity. *Cell* **51:** 887–898.

———. 1989. A telomeric sequence in the RNA of *Tetrahymena* telomerase required for telomere repeat synthesis. *Nature* **337:** 331–337.

———. 2004. Tracking telomerase. *Cell* **116:** S83–S86, S87.

Katzen A.L., Cann G.M., and Blackburn E.H. 1981. Sequence-specific fragmentation of macronuclear DNA in a holotrichous ciliate. *Cell* **24:** 313–320.

King B.O. and Yao M.C. 1982. Tandemly repeated hexanucleotide at *Tetrahymena* rDNA free end is generated from a single copy during development. *Cell* **31:** 177–182.

Levis R.W. 1989. Viable deletions of a telomere from a *Drosophila* chromosome. *Cell* **58:** 791–801.

Morin G.B. 1989. The human telomere terminal transferase enzyme is a ribonucleoprotein that synthesizes TTAGGG repeats. *Cell* **59:** 521–529.

Shampay J. and Blackburn E.H. 1988. Generation of telomere-length heterogeneity in *Saccharomyces cerevisiae*. *Proc. Natl. Acad. Sci.* **85:** 534–538.

Shampay J., Szostak J.W., and Blackburn E.H. 1984. DNA sequences of telomeres maintained in yeast. *Nature* **310:** 154–157.

Shippen-Lentz D. and Blackburn E.H. 1989. Telomere terminal transferase activity from *Euplotes crassus* adds large numbers of TTTTGGGG repeats onto telomeric primers. *Mol. Cell. Biol.* **9:** 2761–2764.

———. 1990. Functional evidence for an RNA template in telomerase. *Science* **247:** 546–552.

Szostak J.W. and Blackburn E.H. 1982. Cloning yeast telomeres on linear plasmid vectors. *Cell* **29:** 245–255.

Yao M.C., Blackburn E., and Gall J.G. 1979. Amplification of the rRNA genes in *Tetrahymena*. *Cold Spring Harbor Symp. Quant. Biol.* **43:** 1293–1296.

Yu G.L., Hasson M., and Blackburn E.H. 1988. Circular ribosomal DNA plasmids transform *Tetrahymena thermophila* by homologous recombination with endogenous macronuclear ribosomal DNA. *Proc. Natl. Acad. Sci.* **85:** 5151–5155.

Yu G.L., Bradley J.D., Attardi L.D., and Blackburn E.H. 1990. In vivo alteration of telomere sequences and senescence caused by mutated *Tetrahymena* telomerase RNAs. *Nature* **344:** 126–132.

Zahler A.M. and Prescott D.M. 1988. Telomere terminal transferase activity in the hypotrichous ciliate *Oxytricha nova* and a model for replication of the ends of linear DNA molecules. *Nucleic Acids Res.* **16:** 6953–6972.

Zaug A.J. and Cech T.R. 1980. In vitro splicing of the ribosomal RNA precursor in nuclei of *Tetrahymena*. *Cell* **19:** 331–338.

———. 1986. The intervening sequence RNA of *Tetrahymena* is an enzyme. *Science* **231:** 470–475.

Index

A

Aging
dyskeratosis congenita as model,
114–115
mice versus humans, 89
prospects for telomere studies, 131–132
telomere position effect, 299–300
telomere shortening, 82, 85–89
ALT. *See* Alternative lengthening of
telomeres
Alternative lengthening of telomeres
(ALT)
cancer
prognostic value, 185
telomerase inhibitor development
considerations, 97, 185,
187–188
tumor distribution, 184–186
gene identification, 182–184
history of study, 163–164
promyelocytic leukemia body
association, 166–168
repressor genes
evidence, 178
identification, 180–182
telomerase activity effects,
178–180
telomere evolution implications,
219–220
telomere length phenotype, 164–166
telomeric exchange rates, 168–171
telomeric recombination
evidence, 171–172
linear extrachromosomal telomeric
repeat DNA, 176
rolling circle, 175

t-loops, 174–175
templates, 173–174
yeast telomerase deletion mutant
studies, 176–178
Anaphase bridging, cancer,
128–129
Aplastic anemia
clinical features, 151
gene mutations, 151–152
telomere maintenance deficiency,
152
ATM
telomerase double knockout mouse,
120–121
telomere capping function in
Drosophila, 448–450
telomere damage response,
409–411
yeast homolog. *See* Tel1

B

BIR. *See* Break-induced replication
Blackburn, Elizabeth, telomerase discovery
reflections, 551–562
BLM
alternative lengthening of telomeres,
183
mutation. *See* Bloom's syndrome
telomere function, 404
Bloom's syndrome, clinical features,
117
Boveri, Theodor, telomere biology
contributions, 1
Breakage–fusion–bridge cycle
cancer changes, 129
elucidation, 3–6, 111–112

Break-induced replication (BIR)
    overview, 201–202
    recombination-mediated telomere
        maintenance relationship,
        206–207
Budding yeast. *See Kluyveromyces lactis*
    telomeres; *Saccharomyces*
    cerevisiae telomeres

C

Cancer
    alternative lengthening of telomeres
        prognostic value, 185
        telomerase inhibitor development
            considerations, 97, 185,
            187–188
        tumor distribution, 184–186
    comparative genomic hybridization
        studies, 128
    episodic instability model of epithelial
        carcinogenesis, 126–128
    lifetime risk, 89
    mice versus humans, 89
    mouse models of age-associated cancer,
        125–129
    prospects for telomere studies,
        131–132
    stem cells, 89–90
    telomerase
        diagnosis and prognosis applica-
            tions, 90–94
        expression, 82–83
        inhibitor targeting, 94
        knockout mouse susceptibility
            studies, 122–123
        regulation, 84–85
    telomere shortening as tumor
        suppression mechanism,
        85–88
Cdc13
    end structure regulation, 378
    localization on telomeres, 356–357
    recombination-mediated telomere
        maintenance, 215–216
    telomerase interactions, 32, 35
    telomere function, 354–355, 371–372
    three-dimensional structure studies of
        DNA interactions, 321–323,
        356–357

Challoner, Peter, telomere biology
        contributions, 552–553, 555
Ciliate telomeres. *See also Tetrahymena*
    telomeres
    abundance, 466
    associated proteins and features of
        complex, 468–470
    Euplotes, Oxytricha, and *Stylonychia*
        chromosome fragmentation,
        474–476
        genomic organization and
            macronuclear development,
            472–474
        internal eliminated sequences,
            473–474
        synthesis, 476–478
        telomerase biochemistry, 478–480
    G-strand overhang, 468
    genome sequencing, 466
    maintenance, 470–472
    nuclei types, 465–466
    processing, 471–472
    prospects for study, 486–487
    repeat sequences, 466–468
    telomerase
        genes, 471
        processivity, 471
        RNA biogenesis, 68
        RNA structure, 57, 62
Cirrhosis, hepatocellular carcinoma
    association, 130
Cohn, Marita, telomere biology
    contributions, 558

D

DC. *See* Dyskeratosis congenita
DCIS. *See* Ductal carcinoma in situ
DNA sequencing, history of telomere
    studies, 10–12
Dot1p, telomere position effect role,
    283–284
Drosophila telomeres
    capping proteins
        ATM, 448–450
        HOAP, 445–447
        HP1, 440–441, 444–447
        MRN complex, 447–449
        overview, 438–440
        UbcD1, 450–451

comparison with other insects,
437–438
length regulation, 436–437
modeling of capping and elongation,
451–455
retrotransposon elements *HeT-A* and
TART
developmental expression,
435–436
evolution, 437
protein products, 433
species distribution, 437
transcription, 433–435
telomere position effect
modeling, 455
mutant studies, 454–455
nuclear localization of telomeres,
290–291
position effect variegation, 262, 264,
453–454
subtelomeric structure and
phenotype, 278–279
telomere length effects, 274–275
Ductal carcinoma in situ (DCIS), telomere
erosion and anaphase
bridging, 128
Dyskeratosis congenita (DC)
autosomal dominant disease,
146–149
autosomal recessive disease, 150
clinical features, 139–141
course, 141–142
diagnostic testing, 150
heredity, 140–141, 143
management, 142–143, 150
mortality, 142
prospects for study, 153–154
telomerase deficiency, 139
telomerase knockout mouse
comparison, 114–115, 149
X-linked disease and dyskerin
mutations, 144–146
Dyskerin
dyskeratosis congenita mutations,
144–146
protein–protein interactions, 145
structure, 144–145
telomerase interactions, 33, 38, 72, 146
yeast homolog, 145

E

End-replication problem, history of study,
8–10
ERCC1/XPF
telomere function, 406–407
telomere damage response, 409, 414,
417–418
Esc1, telomere position effect role,
287–289
Est1
mutant studies, 358
discovery, 15, 35
cell cycle changes in levels, 374
mammalian orthologs, 407, 420
Schizosaccharomyces pombe, 503
telomerase regulation, 33, 36–37,
359–363
Est2, chromatin association, 373–374
Est3, telomerase regulation, 32, 35,
358–359, 362–363
Euplotes. See Ciliate telomeres

F

Facioscapulohumeral muscular dystrophy
(FSHD), subtelomeric D4Z4
repeat, 280, 291–292
Fission yeast. *See Schizosaccharomyces
pombe* telomeres
14-3-3, telomerase reverse transcriptase
trafficking, 70
FSHD. *See* Facioscapulohumeral muscular
dystrophy

G

GAR1
dyskerin interactions, 145
telomerase interactions, 33, 38
G-overhang strand
budding yeast, 349–351
ciliate telomeres, 468
interacting proteins in budding yeast,
354–367
mammalian structure and POT1
binding, 389–390
plants, 528
POT1 binding, 321–323, 325–329, 390
recognition, 321–329
Saccharomyces cerevisiae

G-overhang strand (*continued*)
    binding proteins, 354–357
    structure, 349–351
    telomere end-binding protein binding,
        320–321, 469–470
    three-dimensional structure,
        318–319
    t-loop formation, 350–351
G-quadruplex
    single-stranded G-overhang. *See*
        G-overhang strand
    telomerase inhibition, 319–320
    telomere structure, 317–321
Greider, Carol, telomere biology
        contributions, 552, 556, 559

H

HeT-A. See Drosophila telomeres
Heterochromatic transcriptional silencing
    overview, 261–262
    telomere position effect. *See* Telomere
        position effect
Histones, modification and telomere
        position effect, 281–284
HML, gene silencing, 281
HMR, gene silencing, 281, 285
HOAP
    chromosome distribution, 441
    HP1 interactions, 446–447
    telomere capping function,
        445–446
Homologous recombination
    Schizosaccharomyces pombe telomeres,
        510–511
    sister chromatid exchange, 414–415
    t-loop deletion, 415–417
Hoyeraal-Hreidarsson syndrome
    clinical features, 150–151
    pathophysiology, 151
HP1
    chromosome distribution, 441
    discovery, 440
    HOAP interactions, 446–447
    lamin B receptor interactions, 451
    structure, 440–441
    telomere capping function, 444–445
    telomere fusion types in deficiency,
        444
    telomere position effect role,
        271, 290

Hsp90, telomerase interactions, 34, 38
Human telomeres. *See also* Mammalian
        telomeres
    cell cycle and replication, 391–392
    telomerase
        associated proteins, 33–34, 36–37
        multimerization, 39
        RNA biogenesis, 69
    telomere position effect
        nuclear localization of telomeres,
            291–292
        overview, 265–266
        subtelomeric structure and
            phenotype, 279–281
        telomere length role, 273–274

I

Ink4a/Arf, telomerase double knockout
        mouse, 120, 122–123

K

Kluyveromyces lactis telomeres,
        recombination-mediated
        telomere maintenance,
        208–212, 214
Knockout mouse, telomerase studies
    ATM double knockout, 120–121
    cancer susceptibility, 122–123
    dyskeratosis congenita comparison,
        114–115, 149
    Ink4a/Arf double knockout, 120,
        122–123
    p53 double knockout and telomerase
        checkpoint, 119–120,
        123–124
    phenotype, 111–113
    premature aging, 115–117
    Wrn helicase double knockout, 117
Ku
    Arabidopsis thaliana, 541
    essentiality, 405
    recombination-mediated telomere
        maintenance, 215–216
    Schizosaccharomyces pombe,
        501–502
    semiconservative DNA replication
        function, 366
    subunits, 363
    telomere interactions, 32, 35–36,
        363–366, 405

telomere position effect role, 272–273,
    285–289, 293, 368–369
telomeric silencing role, 513
TLC1 interactions, 364–366

L

Lamin B receptor (LBR), HP1
    interactions, 451
LBR. *See* Lamin B receptor

M

Mad1, telomerase expression regulation,
    85
Mammalian telomeres. *See also* Human
    telomeres
  associated proteins
    BLM, 404
    ERCC1/XPF, 406–407
    MRN complex, 406
    nonhomologous end-joining
        pathway factors, 405
    poly(ADP-ribose) polymerase,
        401–403
    RAD51D, 404
    shelterin. *See* POT1; Rap1; TIN2;
        TPP1; TRF1; TRF2
    WRN, 403–404
    yeast telomeric protein orthologs,
        407–408
  attrition response, 412
  damage response
    overview, 408–411
    repression by shelterin, 411
  nucleosome association, 391
  overhang structure and POT1 binding,
    389–390
  recombination
    homologous recombination
      sister chromatid exchange,
          414–415
      t-loop deletion, 415–417
      interstitial telomeric DNA,
          417–418
      nonhomologous end-joining at
          dysfunctional telomeres,
          412–414
  silencing, 391
  telomerase RNA component structure,
      57–58, 63
  t-loop structure, 390

TTAGGG repeat array, 387–389
McClintock, Barbara, telomere biology
    contributions, 3–7, 83, 111,
    525, 552–553
Mec1
  recombination-mediated telomere
      maintenance, 205, 215
  telomere length control, 375–377
Meiosis
  axial element proteins, 229
  chromosome and telomere dynamics,
      227–229
  initiation, 227
  mitosis comparison, 227
  nuclear envelope attachment of
      telomeres
    imaging, 231–233
    intact telomere requirement,
        230–231
    telomere-associated proteins,
        233–240
  postmeiotic cells and telomere
      distribution, 245
  Schizosaccharomyces pombe telomere
      function, 514–517
  stages, 227, 229
  telomere clustering
    homolog pairing, 241–243, 246
    nuclear movements, 240–241
    recombination independence,
        243
    resolution coordination with
        recombination progress,
        243–245
Menin, telomerase expression regulation,
    85
Microtubule-organizing center (MTOC),
    telomere clustering,
    225
Mismatch repair proteins, alternative
    lengthening of telomeres
    repression, 181–182
Mlp1
  nuclear envelope attachment of
      telomeres, 235
  telomere position effect role, 287
Mlp2
  nuclear envelope attachment of
      telomeres, 235
  telomere position effect role, 287

MRE11/RAD50/NBS1 (MRN) complex
  alternative lengthening of telomeres,
      182
  telomere capping function in
      Drosophila, 447–449
  telomere function, 406
MRE11/RAD50/Xrs2 (MRX) complex
  end structure regulation, 377
  plants, 542
  recombination-mediated telomere
      maintenance, 205,
      366–367
  telomere rapid deletion role, 219
MRN complex. *See* MRE11/RAD50/NBS1
      complex
MRX complex. *See* MRE11/RAD50/Xrs2
      complex
Msp3, nuclear envelope attachment of
      telomeres, 234
MTOC. *See* Microtubule-organizing
      center
Muller, Hermann, telomere biology
      contributions, 2–3, 83

N

Nbs1, telomere length regulation,
      420
Ndj1, nuclear envelope attachment of
      telomeres, 234
NHEG. *See* Nonhomologous end-joining
NHP2
  dyskerin interactions, 145
  telomerase interactions, 33, 38
9-1-1 complex, *Schizosaccharomyces pombe*
      telomere maintenance,
      505
Njd1, telomere rapid deletion role,
      219
Nonhomologous end-joining (NHEJ)
  dysfunctional telomeres, 412–414
  pathway factors associated with
      telomeres, 405
  Schizosaccharomyces pombe telomeres,
      508–510
NOP10
  dyskerin interactions, 145
  telomerase interactions, 33, 38
Nucleosome, t-loops and dynamics,
      336–337, 391

O

OB fold, G-overhang binding proteins,
      321–323, 325–329
Oxytricha. See Ciliate telomeres

P

p53
  activation by dysfunctional telomeres,
      121
  alternative lengthening of telomeres
      repression, 180–181
  cancer studies, 123–124, 131
  telomerase double knockout mouse
      and telomerase checkpoint,
      119–120, 123–124
  telomere damage response, 410
PARP. *See* Poly(ADP-ribose) polymerase
PEV. *See* Position effect variegation
PinX, telomerase regulation, 70
Plant telomeres
  binding proteins, 538–541
  DNA damage response,
      541–543
  dynamics and homeostasis,
      528–530
  G-overhang strand, 528
  length variability, 527–528
  model systems, 525
  prospects for study, 543
  repeat sequences, 526–527
  telomerase
    associated proteins, 532–533
    components, 532
    deletion mutant studies, 536, 538
    functional overview, 531–532
    regulation, 534–536
Plasmodium falciparum, telomere
      localization and gene
      expression, 298
PML body. *See* Promyelocytic leukemia
      body
Poly(ADP-ribose) polymerase (PARP),
      substrates and telomere
      function, 401–403
Position effect variegation (PEV), history
      of study, 262, 264, 454–454
POT1
  Arabidopsis thaliana, 538

discovery, 399
DNA sequence specificity, 399
knockout *Tetrahymena*, 470
protein–protein interactions, 399
Schizosaccharomyces pombe, 501, 504,
    507
shelterin complex and functional
    overview, 392–393
single-stranded G-overhang binding,
    321–323, 325–329, 390
splice variants, 399–400
telomere damage response, 410
telomere length regulation, 390, 400,
    419–420
POT2, *Arabidopsis thaliana*, 538–539
Prescott, David, telomere biology
    contributions, 552, 559
Processivity, telomerase, 54–56
Promyelocytic leukemia (PML) body
    alternative lengthening of telomeres
        association, 166–168
    associated proteins, 168
Rabl configuration, 231, 527
RAD proteins
    alternative lengthening of telomeres,
        182–183
    promyelocytic leukemia body
        association, 168
    RAD51D function, 404
    recombination-mediated telomere
        maintenance in yeast
        break-induced replication
            relationship, 206–207
        overview, 200–201
        roll-and-spread mechanism,
            207–208
        type I pathway, 202–204
        type II pathway, 204–206
    telomere evolution implications,
        219–220

Rap1
    discovery, 15
    DNA-binding sequence, 346
    double-stranded telomeric DNA
        recognition, 329, 331–333,
        335
    mammalian protein structure and
        function, 397–398, 401
    nuclear envelope attachment of
        telomeres, 234
    Schizosaccharomyces pombe, 499–501,
        504, 512
    shelterin complex and functional
        overview, 392–393
    Taz1 interactions, 499–501
    telomere position effect role, 266, 272,
        368
    three-dimensional structure studies of
        DNA interactions, 317,
        331–333
Rb. *See* Retinoblastoma protein
Recombination-mediated telomere
    maintenance
    alternative lengthening of telomeres
        evidence, 171–172
        linear extrachromosomal
            telomeric repeat DNA,
            176
        rolling circle, 175
        templates, 173–174
        t-loops, 174–175
    atypical maintenance, 214–217
    Kluyveromyces lactis, 208–212,
        214
    Saccharomyces cerevisiae
        break-induced replication
            relationship, 206–207
        overview, 200–201
        roll-and-spread mechanism,
            207–208
        survival by terminal palindrome
            formation, 217–218
        type I pathway, 202–204
        type II pathway, 204–206
Retinoblastoma protein (Rb), alternative
    lengthening of telomeres
    repression, 180–181
Reverse transcriptase. *See* Telomerase
    reverse transcriptase
Rif1
    deletion mutants, 504
    mammalian orthologs, 407–408
    Schizosaccharomyces pombe, 501, 504,
        512
Rik1, nuclear envelope attachment of
    telomeres, 235
RNA, telomerase. *See* Telomerase RNA
    component

Roll-and-spread model,
        recombination-mediated
        telomere maintenance in
        yeast, 207–208, 210–211
Rpd3
    aging studies, 300
    telomere position effect role, 289

S

Saccharomyces cerevisiae telomeres
    end structure regulation, 377–378
    G-overhang interacting proteins,
        354–357
    history of study, 12–14
    length regulation
        cell cycle dependence, 372–373
        lagging-strand DNA synthesis
            coupling, 370–372
        nonextendible and extendible state
            switching, 369–370
        Tel1 role, 375–377
        telomerase component association
            with chromatin, 373–375
        telomeric recombination, 375
    recombination-mediated telomere
        maintenance
        break-induced replication
            relationship, 206–207
        overview, 200–201
        roll-and-spread mechanism,
            207–208
        survival by terminal palindrome
            formation, 217–218
        type I pathway, 202–204
        type II pathway, 204–206
    structure
        consensus repeat, 346–348
        sequence degeneracy in budding
            yeast, 346–347
        single-stranded G-overhang,
            349–351
        subtelomeric repeat region,
            348–350
    telomerase
        associated proteins
            gene identification, 351–352, 354
            overview, 32, 35–36
        regulatory subunits, 359–363
        structure, 357–359

telomerase RNA
    biogenesis, 68–69
    structure, 58–59, 62–63
telomere position effect
    biological functions, 294–297,
        299–300
    nuclear localization of telomeres,
        284–290
    overview, 264–265
    subtelomeric structure and
        phenotype, 275–278
    telomere length role, 272–273
    trans-acting factors, 266–272,
        367–368
Schizosaccharomyces pombe telomeres
    chromatin structure, 513–514
    chromosome arrangement, 495
    double-strand breaks, 507–508
    G-strand overhang, 506–507
    length regulation, 503–507
    meiotic telomere function, 514–517
    prospects for study, 517–518
    recombination, 508–510
    repeat sequences, 496–497
    silencing, 512–514
    telomerase
        associated proteins, 33, 497,
            499–502
        deletion mutant studies, 502–503
Senescence, telomere shortening, aging,
        and cancer prevention, 82,
        85–88
Sgs1, recombination-mediated telomere
        maintenance, 205
Shampay, Janis, telomere biology
        contributions, 552
Shelterin. *See* POT1; Rap1; TIN2; TPP1;
        TRF1; TRF2
SIP1, telomerase expression regulation, 85
Sir2
    aging studies, 299–300
    telomere-chromatin interactions, 337
    telomere position effect role, 266, 271,
        282, 285, 295, 297
Sir3
    telomere-chromatin interactions, 337
    telomere position effect role, 266, 271,
        282–283, 285–286, 288,
        295–296

Sir4
    telomere-chromatin interactions, 337
    telomere position effect role, 266, 271,
        282, 285, 288, 295
Sister chromatid exchange
    alternative lengthening of telomeres
        analysis, 169–171
    mammalian telomeres, 414–415
SMN protein. *See* Survival of motor
        neuron protein
Sm proteins, telomerase interactions, 32,
        36
SP100, alternative lengthening of telom-
        eres, 182
Stn1
    recombination-mediated telomere
        maintenance, 216–217
    telomere capping activity, 355–356,
        371–372
Stylonychia. See Ciliate telomeres
Survival of motor neuron (SMN) protein,
        telomerase assembly role,
        70–71

T

TAC1, telomerase regulation in plants,
        534–535
Tankyrases, substrates and telomere
        function, 401–403
TART. See Drosophila telomeres
Taz1
    deletion mutants, 504, 506–511
    double-stranded telomeric DNA
        recognition, 330, 335
    nuclear envelope attachment of
        telomeres, 234
    Rap1 interactions, 499–501
    structure, 497, 499
    telomere length regulation, 504
    telomere position effect role, 272
    t-loop induction, 350
    TRF protein homology, 499
TEBP. *See* Telomere end-binding protein
tefu. See ATM
Tel1
    recombination-mediated telomere
        maintenance, 205, 215
    Schizosaccharomyces pombe telomere
        maintenance, 505–506
    telomere length control, 375–377
Tel2, mammalian orthologs, 407–408
Telomerase
    assay of activity, 51, 53–54, 557
    assembly, 70–71
    associated proteins
        ciliate telomerases, 30, 35
        human telomerase, 36–38
        table, 31–34
        yeast telomerases, 35–36
    cancer expression. *See* Cancer
    catalytic activity
        cleavage, 57
        nucleotide incorporation, 50, 52
    components. *See* Telomerase reverse
        transcriptase; Telomerase
        RNA component
    deficiency and disease. *See* Aplastic
        anemia; Dyskeratosis
        congenita; Hoyeraal-
        Hreidarsson syndrome
    discovery, 15, 83, 551–562
    function, 81
    G-quadruplex inhibition, 319–320
    inhibition
        alternative mechanisms for
            telomere maintenance
            considerations, 97, 185,
            187–188
        approaches, 94–95, 98–100
        clinical trials, 99, 101
        complications in cancer
            management, 96–97
        prospects, 100–101
        screening, 98
        small molecule inhibitors, 94, 96
    knockout mouse
        ATM double knockout, 120–121
        cancer susceptibility, 122–123
        dyskeratosis congenita comparison,
            114–115, 149
        Ink4a/Arf double knockout, 120,
            122–123
        p53 double knockout and
            telomerase checkpoint,
            119–120, 123–124
        phenotype, 111–113
        premature aging, 115–117
        Wrn helicase double knockout, 117

multimerization, 38–39
plants, 531–536
prospects for study, 39–40, 72
reconstitution, 52–53
Telomerase reverse transcriptase (TERT)
   biogenesis and trafficking, 69–70
   domains and structure
      amino-terminal domain, 65
      carboxy-terminal domain, 28,
         66–67
      carboxy-terminal extension, 28–29
      catalytic domain, 66
      dimerization domain, 67
      overview, 27–28, 64–65
      RNA-binding domains, 29, 65–66
   processivity and translocation, 54–56
   reconstitution of telomerase, 52–53
   regulation of expression, 84–85
   sequence homology between species,
      26, 49, 63–64
Telomerase RNA component (TERC)
   biogenesis
      ciliates, 68
      human, 69
      yeast, 68–69
   5′-boundary elements, 25–26
   dyskeratosis congenita mutations,
      146–148
   homology between species, 22, 50, 57,
      59
   knockout mouse. *See* Telomerase
   protein-binding sites, 26
   reconsitution of telomerase, 52–53
   size variability, 57
   structure
      ciliates, 57, 62
      domains, 24
      overview, 22–24, 50
      pseudoknots, 24–25
      vertebrates, 57–58, 63
      yeast, 58–59, 62–63
   template boundary definition, 59–62
   template region repeat sequence, 50
Telomere end-binding protein (TEBP)
   function studies, 470
   POT1 homology, 321, 469
   single-stranded G-overhang binding,
      320–321, 469–470
   subunits, 321, 469

three-dimensional structure studies of
      DNA interactions, 317,
      321–323, 327–329, 469
Telomere length
   alternative lengthening of telomeres
      cell phenotype, 164–166
   cancer, 123, 130–131
   Cdc13 regulation, 355–356
   Drosophila regulation, 436–437
   heritability, 93
   humans versus mice, 110
   mammalian telomere regulation,
      418–420
   regulation in yeast
      cell cycle dependence, 372–373
      lagging-strand DNA synthesis
         coupling, 370–372
      nonextendible and extendible
         state switching,
         369–370
      Tel1 role, 375–377
      telomerase component association
         with chromatin, 373–375
      telomeric recombination,
         375
   Schizosaccharomyces pombe regulation,
      503–507
Telomere position effect (TPE)
   biological functions
      aging, 299–300
      parasites, 297–299
      yeast, 294–297, 299–300
   cell cycle
      requirements for silencing,
         293–294
      timing, 292–293
   definition, 262
   discovery, 262, 264–266
   Drosophila model, 455
   histone modifications
      chromatin structure and telomere
         position effect status,
         281–283
      nonacetylation modifications,
         283–284
   nuclear localization of telomeres
      Drosophila, 290–291
      humans, 291–292
      yeast, 284–290

species distribution and comparison,
262–263
subtelomeric structure and phenotype
Drosophila, 278–279
human diseases, 279–281
yeast, 275–278
telomere length role
Drosophila, 274–275
humans, 273–274
yeast, 272–273
trans-acting factors, 266–272
Telomere rapid deletion (TRD)
discovery, 218–219
mechanism, 219
telomere length control, 375
Telomere shortening
aging and cancer prevention, 82,
85–88
cancer risks, 93–94
Telomeric repeat amplification protocol
(TRAP), principles,
53–54
Ten1, telomere capping activity, 355–356,
371–372
TERC. *See* Telomerase RNA component
TERT. *See* Telomerase reverse transcriptase
Tetrahymena telomeres. *See also* Ciliate
telomeres
abundance, 466
chromosome fragmentation and
telomere synthesis,
482–483
genomic organization and
macronuclear development,
480–481
history of study, 10–13, 21, 465
internal eliminated sequences, 480
molecular genetics studies,
481–482
prospects for study, 486–487
telomerase
associated proteins, 30–31, 35
biochemistry, 484–485
TIN2
function, 397
protein–protein interactions, 396–397,
401
shelterin complex and functional
overview, 392–393

telomere damage response, 410
TLC1, Ku interactions, 364–365
t-loop
homologous recombination and
deletion, 415–417
nucleosomes, 336–337
single-stranded G-overhang in
formation,
350–351
structure, 335–336, 390
telomeric recombination in alternative
lengthening of telomeres,
174–175
TPE. *See* Telomere position effect
TPP1
discovery, 398
function, 398
mutation, 398
shelterin complex and functional
overview, 392–393
Translocation, telomerase processivity,
54–56
TRAP. *See* Telomeric repeat amplification
protocol
TRD. *See* Telomere rapid deletion
TRF1
alternative lengthening of telomeres,
183
dimerization, 394
discovery, 394
DNA activity remodeling in vitro,
400–401
DNA-binding sequences, 394–395
double-stranded telomeric DNA
recognition, 329–335
nuclear envelope attachment of
telomeres, 233–234
plant homologs, 540–541
poly-ADP ribosylation, 402–403
promyelocytic leukemia body
association, 167–168
protein–protein interactions,
395
shelterin complex and functional
overview,
392–393
structure, 394
telomere length regulation, 395,
418–419

TRF2
   alternative lengthening of telomeres,
      183
   DNA activity remodeling in vitro,
      400–401
   DNA-binding sequences, 396
   double-stranded telomeric DNA
      recognition,
      329–335
   essentiality, 396
   nuclear envelope attachment of
      telomeres, 234
   plant homologs, 540–541
   poly-ADP ribosylation, 403
   promyelocytic leukemia body
      association, 167–168
   protein–protein interactions,
      396
   shelterin complex and functional
      overview, 392–393
   structure, 396
   telomere damage response, 408–409,
      411
Trypanosoma brucei, telomere localization
      and gene expression,
      297–298

U

UbcD1, telomere capping function,
      450–451

Ulcerative colitis, cancer association, 130

W

Werner's syndrome
   clinical features, 116, 118
   genetics, 115
   mouse models, 115–117
   telomere dysfunction, 118
WRN
   mutation. *See* Werner's syndrome
   telomere function,
      403–404
Wrn helicase
   alternative lengthening of telomeres,
      183–184
   telomerase double knockout mouse,
      117
Wyman, Claire, telomere biology
      contributions, 556

X

XPF. *See* ERCC1/XPF
Xrs2, recombination-mediated telomere
      maintenance, 205

Y

Yu, Guo-Liang, telomere biology
      contributions, 561–562